# RECENT TRENDS AND CONCEPTS IN GEOGRAPHY

VOLUME ONE

## About the Editors

**Dr. R.B. Mandal** (*b.* 1948) teaches geography at R.D. and D.J. College Monghyr and at the Post-graduate Department of Geography Bhagalpur University. He had his education from graduation to post-doctoral work at the Patna University. He has also been associated with research projects sponsored by the University Grants Commission and the Rockefeller Foundation.

**V.N.P. Sinha** is currently on the faculty of the Department of Geography, University of Cape Coast, Ghana. He holds a Ph.D. degree from Patna University.

# Recent Trends and Concepts In Geography

## Volume One

*Edited by*
Ram Bahadur Mandal
Vishwa Nath Prasad Sinha

**CONCEPT PUBLISHING COMPANY PVT. LTD.**
**NEW DELHI-110059**

**ISBN-13: 978-81-7022-853-0 (Set)**
**ISBN-13: 978-81-7022-854-7 (Vol. I)**
**ISBN-13: 978-81-7022-855-4 (Vol. II)**
**ISBN-13: 978-81-7022-856-1 (Vol. III)**

First Published 2016

*Published and Printed by*

Concept Since 1974

**Concept Publishing Company Pvt. Ltd.**
A/15-16, Commercial Block, Mohan Garden
New Delhi-110059 (India)
*T*: 25351460, 25351794, *F*: 091-11-25357109
E: publishing@conceptpub.com, W: www.conceptpub.com
**Editorial Office:** H-13, Bali Nagar, New Delhi-110 015, India.

**Cataloging in Publication Data**--Courtesy: D.K. Agencies (P) Ltd. <docinfo@dkagencies.com>

**Recent trends and concepts in geography** / edited by Ram Bahadur Mandal, Vishwa Nath Prasad Sinha.
3 volumes cm
Reprint. Originally published: 1980.
Includes bibliographical references and indexes.
ISBN 9788170228530 (set)
ISBN 9788170228547 (vol. I)
ISBN 9788170228554 (vol. II)
ISBN 9788170228561 (vol. III)

1. Geography. I. Mandal, R. B., 1948- editor. II. Sinha, V. N. P. (Vishwa Nath Prasad), editor.

G62.R43 2016 DDC 910 23

# Preface

GEOGRAPHY is a unique science in the sense that it integrates various disciplines, encompassing as it does the physical, economic, social, political, historical and medical sciences in relation to man, space and environment.

With the widening scope and increasing dimensions of geography, both in respect of its conceptual and systematic branches, today, there is available a huge amount of literature on this subject. Considerable changes have been made in the methods of study adopted by modern geographers who have spared no effort to make theirs as exact a science as possible.

In trying to make choices between old fashioned and modern geography, descriptive and quantitative geography, spatial analysis and interdisciplinary approaches and so on, geographers today reveal an eagerness to enlarge their horizons and to keep abreast with the latest developments in the subject.

There is also visible a good deal of enthusiasm for deeper research into topics already touched upon with the application of newer techniques to geographical findings such as statistical methods, model building, remote sensing, the use of data matrix, and a host of other sophisticated tools. All this adds up to what can be termed as a conceptual revolution in the field of geography.

What is important today is the dissemination of such knowledge so that both concepts and techniques can be studied and utilized by an ever widening group of geographers and not limited to a coterie of professionals and specialists. It is with this purpose in mind that the essays in these three volumes have been collected together and published. The readings suggested at the end of chapters as well as the references, provide a comprehensive bibliography of research materials.

The three volumes in this series cover the conceptual, physical, resources, agricultural landuse and human geography aspects in their broad outlines. Woven into the main fabric are the social, political, historical and allied problems which interact and provide complexity to the main themes.

All the three volumes contain brief introductions to each part which gives the reader glimpses into the topics dealt with in each of them.

It is hoped that this unique collection of essays will serve the purpose of enlightening the reader regarding the kind of research and the variety of topics which are being taken up by modern geographers today.

The extensive use of maps, figures and diagrams will help illustrate the problems that have exercised the minds of scholars and researchers in this wide-ranging subject.

These volumes have been made possible because of the unstinting cooperation of a host of people too numerous to acknowledge individually. Gratitude must be expressed, however, to Prof. P. Dayal, Head, Department of Geography and Dr L.N. Ram, from the same department at Patna University for their valuable suggestions and the latter for sponsoring this project on behalf of the Geographical Research Centre, Patna. The credit goes to the contributors who contributed scholarly papers on a variety of topics with great enthusiasm in spite of other urgent engagements. To friends and critics like M. Jha, G.N. Singh, P.N. Mandal and Suresh Kumar go special thanks for constructive suggestions and generous help at all times. Without the cooperation of all of them these volumes would not have seen the light of day.

MONGHYR
*18 May 1979*

RAM BAHADUR MANDAL
VISHWA NATH PRASAD SINHA

# Contents

## SECTION TWO

## PHYSICAL GEOGRAPHY

# List of Contributors

1. *Davgun, Satish K.*,
   Department of Geography, Kent State University, Kent, U.S.A.
2. *Gopal Krishan*
   Department of Geography, Punjab University, Chandigarh, India.
3. *Kumar, Anil*,
   Reader in Geomorphology, Bhagalpur University, Bhagalpur, India.
4. *Mandal, R.B.*,
   Lecturer in Geography, R.D. & D.J. College, Monghyr, India.
5. *Mukhopadhyay, S.C.*,
   Reader in Geography, Calcutta University, Baligunj Circular Road, Calcutta, India.
6. *Pandey, R.N.*,
   Research Department of Geography, Bhagalpur University, Bhagalpur, India.
7. *Peters, Gary L.*,
   Associate Professor in Geography, California State University, Long Beach, California, U.S.A.
8. *Prasad, Nageshwar*,
   Postgraduate, Department of Geography, Bhagalpur University, Bhagalpur, India.
9. *Robinson, Brian*,
   Department of Geography, Saint Mary's University, Halifax, Canada.

10. *Sahu, Shukdeo,*
    Research Scholar, Bhagalpur University, Bhagalpur, India.
11. *Rasheed, K.B. Sajjadur,*
    Department of Geography, University of Maiduguri, Nigeria.
12. *Satpathy, Debidutta P.P.,*
    Reader in Geography, Ranchi University, Ranchi, India.
13. *Singh, G.N.,*
    Lecturer in Geography R.D. & D.J. College, Monghyr, India.
14. *Singh, O.P.,*
    Lecturer in Geography, H.R. Degree College, Khalilabad, Basti, India.
15. *Singh, R.L.,*
    U.G.C. Scientist, Department of Geography, B.H.U., Varanasi, India.
16. *Singh, Rana P.B.,*
    Lecturer in Geography, B.H.U., Varanasi, India.
17. *Sinha, V.N.P.,*
    Lecturer in Geography, University of Cape Coast, Cape Coast, Ghana.
18. *Shrivastava, V.K.,*
    Reader in Geography, Gorakhpur University, Gorakhpur, India.
19. *Stadel, C.,*
    Chairman, Department of Geography, Brandon University, Manitoba, Canada.
20. *Thakur, Rameshwar,*
    Research Fellow, University of Delhi, Delhi, India.

# List of Figures

# SECTION ONE
## CONCEPTUAL DEVELOPMENT

# Introduction

THE ORIGINS of the study of geography may be traced back to medieval Greece. The medieval period is said to be an age of search and research for the advancement and systematic development of geography through inland and overseas mercantile activities. But modern geography has developed into a subject with a more complex content, sophisticated methodology and definite direction. In recent years the overwhelming interest in interdisciplinary approaches to geographical problems has led to the emergence of locational *versus* cause and effect relationship ; ideographic as opposed to a nomothetic approach ; and as an integrating science *versus* an independent geographic discipline. Modern geography, to be very precise, may be conceived as the scientific study of spatial distribution of physical and cultural features of landscape and biosphere in relation to the races of mankind.

The collection of articles and studies brought together in this Reader represents an admittedly self-conscious and in many respects venturesome enterprise.

The main purpose, however is to help the student keep abreast of recent trends and developments in the realm of geography. It is hoped that the book will open up new vistas of systematic thinking thus extending the frontiers of knowledge.

As regards the theme of the book, it starts with a brief note on the nature, scope and concepts in modern geography. Apart from this, contributions made by some of the leading pioneers of geography like Humboldt, Ritter, Ratzel and Semple are brought into focus.

Satpathy has made an attempt to trace the history of geography in the ancient scriptures of India. The author argues in the paper that in almost all the works of Kalidasa, there are sufficient descriptions about the physical, economic, political and human geography of India to justify the belief that geography was studied in all its aspects at that time.

The micro-geographic framework, the systematic arrangement of facts and figures, and the analysis of real world situations, according to Sinha, is contended to have changed the dimensions of geography in relation to its sister disciplines. The use of spatial analysis and systems approach in comparisons with hypothetical thinking has dragged this subject nearer to the interdisciplinary field, rather than to an independent and isolated branch of knowledge.

The author of the paper on "Sense of Place" pleads that a study of the kind attempted in his article needs to be viewed in a broader perspective within a comparative framework. The concept of Sense of Place is also to be discussed in relation to time as well as preservation and planning of landscape features.

Advertising, tourism, pop songs, women's fashions, "westerns", popular travelogues (such as the *National Geographic*) may not normally be considered to be sources of *bona fide* geography, and no such claim is made here. However, the desires and redundancies of modernity are relevant to cultural geography because they are part of the restlessness of a culture whose constancy is only to replace itself (justification through erasure). A cultural geography which grounds itself in roots in the soil and direct unfoldings from nature to culture (cultural landscapes) is uneasy when it attempts to grapple with such popular ephemera. A structuralist rather than a phenomenological approach has the advantage of being aware of its own nostalgia and of ironically accepting symbolic impoverishment as part and parcel of a fragmented age. These alternative approaches are contrasted. However, since the overall concern of the paper is to point out a cultural crisis, the content is a mosaic of possibilities and examples rather than a strict methodological explication. The mosaic metaphor is a paradigm of the crisis. Like a kaleidoscope, it rests just this side of its own decentered universe. It becomes myth when it creates a fresh surface on the reality it proclaims to have revealed. This dialectic of contradictions is one that cannot replace itself but instead tends to confirm the inauthentic as the driving force behind modernity's stasis. For example, the tourist experiences neither being at home nor the exotic. Yet he manages to live that experience, in fact, to have his cake and eat it. By its very nature spurious experience must be voracious. By definition it is a phenomenon both loaded with and empty of meaning.

Morphological analysis is recognized as the concept integrating causal relationship among the three groups of knowledge—biosciences, physical sciences, and geonomy. Both the visions of reality and the tangibility of proxemics are expressed with this concept in a systematic manner.

The systems approach is a new development in the field of geography, which is a method advanced by Bertalanffy as "General System Theory". The present day world can be viewed as interconnected with a system hierarchy, i.e., subsystems, systems, super systems, etc. It can be grouped into isolated, closed and open systems. On the basis of internal complexity, systems can also be classified as morphological, cascading, process-responses, control, etc. It has been felt that a system approach overcomes the descriptive compilation in geography.

Behavioral geography is a new concept in geography and this has been brought over from psychology and sociology. It is an integral approach to perceive things for a particular behaviour. Behaviour is the dictates of environment but a man perceives it according to his ability in relation to education, economic status, race, culture, etc. in a territorial space. The behaviour of man is determined by nature but possibilists compromise with nature after controlling the gift of nature, i.e., solar energy. At the end, environment has been classified into geographical, operational, perceptual and behavioral which influences geographic space in different fashions in relation to man and social development.

Development is a change which is indigenous to the organic and inorganic world and is an irreversible process. Thus development may lead towards increase or decrease of any social factor in space and time. Social development in space comes under inorganic development but, like the law of the allometric growth of zoology, social development on a particular territorial space has dimensions both towards increment when the expansion in space is possible or decrease in size. For geographers, regional disparities are of special interest ; these are measured with the help of various techniques and interpreted in the light of the physical resources base, technological advancement, institutional framework and external exposure of different areas in temporal dimensions.

The use of the computer in geographical data analysis has become closely linked since the 1950s. The computer as an electronic device, is capable of storing numerical data and of performing certain operations on that data. The instructions are generally given in the form of a programme in a coded language so as to receive the required result.

In a paper on sociometric analysis, an attempt has been made to explain interaction between cultures *via* social psychological techniques. This has been made for Bangladesh where the sociometric technique has been fruitfully used to identify and measure the levels of attraction and repulsion between culture groups occupying different territories. This study has been contemplated under the following sub-heads : introduction of the problem, sociometric measurement, results, interpretation, determinants of choice, and epilogue. The Sociometric test is a potential field of investigation which is generally used by the social psychologist in dealing with the perception, attitude, behaviour and social distance etc.

Maps are the tools in the hands of a geographer through which he expresses his ideas about the size, shape, movement and oscillation of the earth. The logic of cartography includes the designing, construction and production of maps through data processing, field work and final printing. The elements of cartography cover a number of specialised technical phases like the scale of the map, the method of projection, symbolization of data in preparing a map, the preparation of the map design and the presentation of the map for wider application. Nowadays, with the development of science and technology, interest in and the significance of cartographic methods is changing towards a more quantitative analysis, thematic mapping with high precision instruments, computer programming for quick results, and mathematical model building.

1

# Origin and Growth of the Discipline of Geography

*—Ram Bahadur Mandal*

THE WORD *Geography* is derived from two Greek words *geo* and *graphos*. *Geo* means earth and *graphos* means description. Thus, geography is a science of description of the earth's surface. In order to have a definite subject matter to be studied in geography, it is defined as "a study of the earth's surface in its areal differentiation as the home of man".[1]

Sometimes, it is asked whether geography is a science or an art. A few decades ago there was a notion that the listing of facts about places and peoples could be included in geography. This has sometimes been dubbed as gazetteer geography. In this sense, geography is said to be a 'descriptive art', while the distribution of any item and its causal explanation as 'the cause and effect relationship' has placed geography in the category of science. Hence, from the descriptive and analytical point of view, geography is an art, a science and a philosophy, all at the same time. The aim of this paper is to trace the growth and development of the discipline of geography.

## THE FIELD OF GEOGRAPHY

In delineating the sphere of the geographers' interest, Monkhouse has rightly shown the field of geography after drawing inferences from earth science, social science and geometrical science. Here, in between the transition zone of earth science and social science lies human ecology which shows the relationship between man and his environment, while economic geography is entirely the product of the economic activities of any society. Similarly, cartography is dependent upon geometrical science, and physical geography on earth science.

The field of geographical study is really wide and the subjects with which geographers are concerned are as follows :

The earth's size, shape and movement.

The location and distribution of continents and water bodies.

The structure and relief of the earth's surface in relation to rocks.

The characteristics and movements of the waters of oceans, lakes and rivers.

The atmospheric conditions in relation to climate and weather.

The distribution of vegetation, animals, population and characteristics of social, cultural and political organizations on the earth's surface.

The activities and livelihood of men including the different kinds of settlements in which they live.

All these show that geography has many angles and facets and, therefore, it should not be made out that it is a confused subject. In short, geography may be said to study the "where", "why", "how", and "when" of things over the earth's surface.

## CONCEPTS OF GEOGRAPHY

The geographer's aim is to explore, interpret and understand the features of the earth's surface and in doing so, the following concepts must be kept in mind :

### Spatial Relationships

These involve the time and distance between two items distributed on the earth's surface. Thus any item having no spatial locational dimensions cannot be treated as the subject matter of geography.

### Man-Land Relationship

Different categories of land offer varying conditions of adjustment for man with nature. In this sense each and every study of geography should be related with man and his habitat, because without man there cannot be a geography of a region.

### Environmental Unity

This shows that the integrated view of a place including terrain, soil, climate, vegetation, animal life, people, settlements, etc. should be treated as geography, because the forces and processes of geography link the various units into an integrated form. Because of these characteristics of geography, the different phenomena of space have been differentiated and this helps in the growth of regional geography.

## GROWTH OF THE DISCIPLINE OF GEOGRAPHY

Geography is as old as creation itself but the growth of geography as a scientific discipline and as we know it today, has developed only after the 1950s.

### The Early Period

The Greeks first studied geography scientifically. Thales of Miletus was a pioneer in determining the earth's spherical form and Aristotle demonstrated its sphericity along with delimiting the habitable parts. Strabo's interest centred around the adaptation of human beings in relation to physical environment and this was primarily concerned with mathematical relationships. The work of Ptolemy was the culmination of Greek geographic learning.

The people who took up the torch of knowledge next were the Romans during the dark ages, there was geography without geographers in Europe. Only written accounts of travellers and gazetteers were available.

### The Medieval Period

In the medieval period, the growth of geographic interest beyond the Muslim world followed the travels of Marco Polo, the increase of trade with the East and voyages during the discovery of America. Further, the sixteenth and seventeenth centuries gave rise to a sound theoretical base in geography with the publication of text-books like *Varenius Geographia Generalis* and maps (Mercator).

It can, therefore, be said that A.D. 1400 to 1650 was the age of great maritime discovery and this was the period when the study of geography really began. At that time the trend was : what was not European was not familiar. The seventeenth century in Europe was a period of consolidation of gains in the field of geography. This century created a spirit of enquiry about the earth itself by traders, missionaries and expeditions. It was in A.D. 1660 that the Pythagoras' Society was established by the Royal Geographical Society. Bernard Orinius, the father of geography was also born in the seventeenth century. He classified geography as general, special and comparative. General geography deals with the globe, special geography deals with the country and in comparative geography latitudinal and longitudinal comparisons are made. Bernard Orinius tried to classify the mountains also on the basis of size and height.

1650-1850 was the age of fact-finding for the advancement and systematic growth of geography. Ethnography developed at this period. This was the age of mercantile activities and commerce. Quantitative methods in geography also emerged in this period at the time of Humboldt and Ritter.

In the medieval period explorers and travellers were also great naturalists who added much to the distribution of physical and human phenomena. Among them the following are important :

(a) Alexander Von Humboldt, a German, who travelled widely in South America and Siberia.
(b) Sir Joseph Hooker, who travelled in northern India and visited Antarctica.
(c) A.R. Wallace, who carried out important work in biology in South East Asia.
(d) Charles Darwin, who voyaged around the world in *The Beagle*.

### The Modern Period

The modern period really begins at the end of the eighteenth century with the work of Alexander Von Humboldt and Karl Ritter. Since then the two principal methods of approach to geography can be distinguished : the systematic and the regional.

There are two main branches of geography—physical and human. Among these the first studies the world's landforms, while human geography places man in his physical setting and studies his relations with the environment. Economic geography, political geography and social geography are offshoots of human geography.

The regional concept was brought into geography by Ritter. He has considered minerals, vegetation and animals in association to formulate geographical regions.

After A.D. 1800, geography became a subject with content, methodology and definite purpose. The year 1859 was a landmark with the death of Humboldt. In the same year *The Origin of Species* by Charles Darwin was published. According to him, the world has come to be what it is by natural selection. The book thus set people rethinking, questioning and experimenting. The mountain building theory and the origin of the species brought geography and history together. It was also felt that environment should be properly assessed to know about an area.

Ratzel focused the role of man in geography. He was the first author to have written a book on *Anthropogeographie*. Barrows used the term 'human ecology' in 1925. In 1925, C.O. Saur emphasized that attitudes, aims, objectives and technology are more important than deterministic approaches in geography. During this period culture was more important than land. With Paul Vidal de la Blache, regional concepts of differentiation came to be studied in geography. The town is on the apex of a region. Thus the important concepts in geography were developed by Baron Von Richthofen, Albrecht Penck, Friedrich Ratzel in Germany, P.M.J. Vidal de la Blache and Jean Brunhes in France, William Morris Davis and Ellen Churchill Semple in the United States.

In the last 25 years great changes and a virtual academic explosion of geographical knowledge have taken place. To understand this, we may look at the activities and themes of geography of this century. Firstly, great specialization tends to know more and more about less and less. Secondly, the inter-disciplinary approach of study has increased such as, geopolitics, zoo-geography, life science and rural development, etc. It is felt generally that view-wise and period-wise there is no uniformity in the discipline of geography. We should, therefore, include the materials of each and every branch of literature and science in geography in order to enrich our discipline.

SOME IMPORTANT CONTRIBUTORS OF GEOGRAPHY

**Alexander Von Humboldt (1769-1859)**

Humboldt was a great German scholar who investigated into the fields of geology, physics, chemistry, botany, biology, geography and history. He wrote almost 40 books in science and travelled widely. First of all, Humboldt travelled in Holland, Belgium, Britain and France in 1790. In 1799, he travelled extensively in the Venezuela, Amazon Valley, the Andes and Peru in South America besides Mexico, the United States of America and France. In 1829, on the invitation of the U.S.S.R. Government, he travelled through the mineral belts of the Ural and Altai mountains.

Humboldt had written a scholarly book entitled *Cosmos* in five volumes which incorporates the geographical account of the entire world. In each book he has discussed the geographical elements of latitude, longitude, height from the sea level, magnetic difference, types of rocks, temperature, pressure, direction of wind, and vegetation on the earth's surface in relation to man.

The geographical ideas of Humboldt were very much like those of Ritter. Humboldt clarified the difference between natural history and regional distribution. In *Cosmos*, the analyses of the difference of physical history of the world and physical geography were made in adherence to the views of Kant. He has studied the inter-relation between inorganic and organic materials of the world. This concept of human geography is a great contribution of Humboldt as he said : "the geography of organic being—plants and animals—is connected with delineation of the organic phenomena of our terrestrial globe".[2]

Humboldt held that the differences in human traits were mainly due to the influence of the physical environment on them and because the inhabitants tried to modify the physical forces according to their need. In relation to plants and wild animals, man becomes less akin to his natural environment.

Due to mental superiority, the man tries to lessen the control of physical forces. In spite of this, man is highly related with the terrestrial life.

Humboldt has adopted a comparative method in geographical studies. As for example in his paper on grasslands and deserts, he compared pampas, prairies, steppes, ilanos, etc. and in each region he discussed the effects of climate, soil, height from the sea level, etc. The areas covered by Humboldt in his travels were divided in vegetation regions and the maps were included in the *Atlas Geographique* published in 1814-19. It was the belief of Humboldt that from the North Pole to the South Pole there is one life and hence there is unity in nature.

He emphasized the systematic study of geography in comparison with regional geography but he had published the regional geographical accounts of Mexico, Cuba, the Plateau of Brazil, etc. Humboldt first of all piloted the cause of systematic geography such as geomorphology, climatology and plant geography while Ritter emphasized regional geography.

**Karl Ritter (1779-1859)**

Ritter was the founder of modern geography. He belonged to Germany. His works represent the first phase of systematic geography in the classical period. In 1804, Ritter wrote his book *Europe : A Geographical, Historical and Statistical Painting.* After two years he published six maps of Europe.

The celebrated book of Ritter entitled *Erdkunde* was published in 19 volumes from 1817 onwards. The first volume contains geographical accounts of Africa, while the second volume relates to the continent of Asia. These publications brought great fame for Ritter. In 1820, he joined as Professor of Geography in Berlin University. There he served the department till his death in 1859. In 1827, he came in contact with Humboldt which helped a great deal in the progress and prosperity of geography.

Ritter has made significant contributions to the development of human geography. According to him, the earth and man are correlated and it is difficult to analyse them separately. Ritter approached the problem of geography with an interest in history. He believed in the comparative study of geography in order to find some inherent laws. In his early days, the interest of Ritter centered around systematic geography and for this fact he classified the relief of the land into mountains, plateaux, plains and good land. He also discussed this classification in his regional geography. *Erdkunde* is related with the regional approach of geography. *Landerkunde* deals with the physical division of a region instead of political divisions. He defined geography as "the department of science that deals with the globe in all its features, phenomena and relations, as an independent unit, and shows the connections of this unified whole with the man and man's creator".

**Friedrich Ratzel (1844-1904)**

As a human geographer, Ratzel founded the base of political geography in the nineteenth century. He served as Professor of Geography at the Polytechnic Institute of Munich and later at the University of Leipzig in Germany. His first systematic analysis of the subject *Politische Geographie* published in 1897 delineated "the state as an organism fixed in the soil". He considered that the state, a union of the people and the land, was a spatial organism ; and for this fact in 1896, he published a paper entitled "The Laws of the Territorial Growth of States". According to Ratzel, space was a political force of significance and he asserted that the states decay as a result of the declining conception of space. He considered frontier as the changing zone of assimilation, which is reflected in the expansive force of state boundaries through war. The idea of living space is associated with the theory of the organic state and the dynamic boundary.

It is incorrect to say that Ratzel was determinist : he was more of a possibilist. In this connection his other contributions to geography were the opposition of the rationalistic concept of man, the recognition of the role of culture and culture diffusion, and the analysis of population and its movement for understanding the functional areas. His thought was influenced by the humanities and social sciences and he attempted to formulate the laws of human behaviour. The first volume of Ratzel's *Anthropogeographie* was published in 1882. He attempted to show that the distribution of man on the earth is more or less controlled by the natural forces. This volume was a reworking of the theme developed by Ritter in his *Erdkunde*. The second volume published in 1891, described the distribution of population for the period under study.

The last publication of Ratzel was *Die Erde und Leben Eine Verglei Chende Erdkunde* (1901-2) in which he showed the earth as an integrated whole and an interacting unity. Ratzel's work differs from Ritter's in two respects. Firstly, Ritter treated human geography as systematic and not regional; and secondly, he held that man was the end-product of evolution of species as controlled by the surrounding environment. In this way a deterministic tint can always be seen in Ratzel's writings.

Ratzel considered it necessary to study the relations between the state and the earth's surface. Under the influence of the evolutionary theory of biology, Ratzel favoured the organic theory of state and society.

Ratzel was the teacher of Ellen Churchill Semple, who carried his philosophy of geography to the United States of America, and the same was incorporated in her monumental work of *Anthropogeographie*. Semple revived Ratzel's concept as a basic principle in geopolitics. Hence the organic theory of state forms part of political geography. Further, he investigated the growth and character of political units, their position, size, boundaries and relation to the sea on the earth's surface.

Ratzel's approach to geographical analyses was through direct investigation of landscape as did Humboldt. In his own words : "I travelled, I sketched, I described". Due to his wider interest, he did research in physical geography specially on Fiords, the snow cover of German mountains. He also edited the book *Bibliothek Geographischer Handbucher*.

In this way, Ratzel investigated geography from the modern scientific points of view. He constructed his system according to the principles of evolution and his conclusions were based on worldwide inductions. Due to the vastness of the subject which he chose, his conclusions were never final.

**Paul Vidal de la Blache (1845-1918)**

Blache was an influential student of human geography, who provided great encouragement to geography in France through approaches of regional, historical and archaeological studies. He opposed Ratzel's view of man-space determinism in favour of man's choice of action. The followers of Blache have continued this thought in their publications like *Geographie Universale* published by Armand Coline in Paris.

According to Blache, it is essential for the geographer to have a thorough grounding in geology and that should be the basis for regional studies. He inspired the publication of a number of regional monographs of France prepared by his disciples. His own *Tableau de la Geographie de la France* (1903) is the classic exposition of regional description. A knowledge of history is essential for a geographer and on this account he has published an *Atlas of Historical Geography*, (1894). His lectures were published entitled *Principles de Geographie Humaine* (1921) which described man as the central point in regional environment in geography, who chooses certain opportunities according to his inclination as offered by the environment. In this way, a group of people may react in a number of different ways, depending upon their cultural heritage and group organisation. Vidal's approach to the study of human geography was through the comparative study of different parts of the earth. For example, according to him, "the field of study par excellence of geography is the surface, that is, the ensemble of phenomena which lies in the zone of contact between the solid, liquid and gaseous bodies which form the earth".[3] In his book *Principles af Human Geography*, Blache emphasized the role of physical environment but cautioned against explaining all human facts in terms of environmental control. Further, he also considered that "present forms are unintelligible unless their past evolution is known".

REFERENCES

1. R.E. Dickinson, *The Makers of Modern Geography*, Lyall Book Depot, Delhi, 1973.
2. A.V. Humboldt, *Cosmos.*, Vol. I.
3. Paul Vidal de la Blache, *An Atlas of Historical Geography*, 1894.

# 2

# Geography in Kalidasa's works

*—D.D.P.P. Satpathy*

KALIDASA was a poet of unrivalled and many splendoured genius. In his descriptions of nature we get a glimpse of the various aspects of geography. Besides the phenomena of physical geography we get sufficient references to economic, human and biogeography. In many of his works, particularly in the *Raghuvamsham* and *Meghadutam*, we get a detailed description of the various *janapadas* or kingdoms of contemporary India. The contents of his works act as a mirror of life reflecting the geography of the country and customs and manners prevailing in his times ; "that is why we come to the conclusion that he must have travelled widely throughout the length and breadth of the country".[1]

Although Kalidasa is credited with writing a large number of epics and dramas, experts believe that he wrote four epics—*Raghuvamsham*, *Kumarasambhavam*, *Meghadutam* and *Ritu Samhara*; and the three dramas—*Abhijnana Shakuntalam*, *Malabikagnimitram* and *Vikramorvashiyam*. Our study will be restricted to these seven works. Although there are different opinions about the place of origin and time of the poet, we will stick to the popular belief that Kalidasa was the most priceless gem in Emperor Vikramaditya's (Chandra Gupta II) court.

Geography has been poeticised in Kalidasa's epics. In *Meghadutam*, the advancement of the monsoon in the garb of a cursed *yaksha* sending his heart's message to his beloved residing in the Himalayan town, Alakapuri, and the various types of landscapes and regions it meets on its way are described. In *Reghuvamsham*, he has given a vivid description of the important kingdoms of contemporary India. In *Kumarasambhavam*, he has given a clear picture of the Himalayan region, its lofty snow-capped peaks, the frozen ground, the formation of the rivulets by the melting of the snow, etc. In *Ritu Samhara*, he has attempted to describe the six Indian seasons in great detail. In comparison to the epics, the descriptions of geographical environment and the occurrences of place names are not so prominent in the dramas.

## METEOROLOGY AND INDIAN CLIMATE

Kalidasa had a profound knowledge of meteorology. Meteorological phenomena like the velocity and direction of winds, origin, types and chemistry of clouds, the pleasant and oppressive effects[2] of tropical sunshine from season to season, absorbing of moisture by the clouds from water bodies like the oceans,[3] the seas, the rivers and lakes, variation of the lengths of days and nights[4] during different seasons of the year, etc.

*Ritu Samhara* (in six cantos) is the poetical description of six Indian seasons into which the Hindu year is divided. In this mini-epic (*Khanda Kavya*), we find a systematic description and characteristics of the six seasons namely summer (*grishma*), rainy (*prabrut*), autumn (*sarata*), late autumn (*hemanta*), winter (*sisira*) and spring (*vasanta*).

Summer is characterized by intense heat,[5] bright sunshine profuse sweating, dearth of water due to drying up of water bodies and laziness and fatigue common to men and animals. During this season forest fires are common features in the forested hill slopes. Being fanned by the wind, these fires consume grasses, bamboos and wild animals.

The rainy season is characterized by a sky overcast with black clouds, intense humidity, the occurrence of rainbows, flashes of lightning, thunders, the flowering of various plants, erosion of the banks by the rivers and suspension of a great amount of dust and clay particles in the river water.

Autumn is free from clouds, the sky is clear, the waters of the lakes and ponds are clear, the impurities settle down, the river water moves slowly[6] due to the heavy load of sediments, ducks swim in groups, a variety of seasonal flowers spread their fragrance in the air, and paddy starts ripening in the field.

In the *hemanta* season glamour is added to the rural landscape by the existence of golden-hued paddy fields where the paddy has ripened. In this season, the young wheat saplings sprout forth, the morning due falls on the grasses, the cranes flock near the water bodies and the sunshine becomes pleasant.

During winter, the nights are longer[7] and days are shorter. Fire and sunshine are pleasant. Due to oppressive cold, people live behind closed doors. They wear heavy clothes. This is the season of paddy harvesting. The sugarcane plants adorn the fields at places.

Spring is the king of the seasons. It is the most pleasant and enjoyable of all seasons. It is characterized by mangoes, the cuckoo's songs, and various seasonal flowers. The days are warm but the nights are cool. The morning and evening are pleasant. Kalidasa was fond of describing this season. He has described it on four occasions—(a) *Kumarasambhavam*, 3rd canto; (b) *Malabikagnimitram*, 3rd Act; (*c*) *Ritu Samhara*, 6th canto; and (d) *Raghuvamsham*, 9th canto.

From the descriptions of the seasonal characteristics in *Ritu Samhara*, Hara Prasad Shastri[8] who was confident about the fact that Kalidasa was born in western Malwa, concluded that "the fact that all the natural objects mentioned in the *Ritu Samhara* are to be found together only in one district of India that is western Malwa". He further says that "in no other part of India are the traditional six seasons so well defined". A perusal of the Mandasore inscriptions shows that describing the seasons was a fashion among the poets of that region.

Kalidasa has mentioned the causes of rainfall. According to him the clouds devoid of water vapour go near the ocean[9] and they draw so much water vapour from the latter that they can distribute water to different parts of the world. He has also discussed the types of clouds. According to modern meteorology, there are ten types of clouds (cumulonimbus, stratus, altostratus, cirrus, etc.) but according to the Hindus, there are four types of clouds namely—*Pushkara*, *Abarta*,[10] *Sambarta* and *Drona*.[11] He has referred to the meteorological truth that when the cloud becomes saturated with moisture it causes rain when a hill or mountain comes athwart and the rain-bearing clouds come down and remain nearer to the surface of the earth. According to him, the cloud is propelled by wind and the monsoon moves slowly to the north from the Vindhyas.

Kalidasa has also discussed the chemistry of clouds. According to his view cloud is a mixture of smoke, light, water and wind (*Dhrumajyotih Salila marutam sannipat kvam madhah* : P. *Megha*, 12). This is nothing but the vapour imperfectly condensed, having flashes of lightning and being driven by the wind acting upon it.

The monsoon is the harbinger of rain in India. The appearance of black clouds during the summer season heralds the coming of the monsoon. Kalidasa was aware of these facts. He was also aware of the date and time of the onset of the monsoon. According to him mid-June[12] was the approximate date of the onset of monsoon. To some extent the path determined by him for the cloud whom he wants to play the role of a messenger, roughly corresponds with the direction of movement of the Arabian Sea branch of the South-west monsoon (*Fig. 2.1*). The cloud in *Meghadutam* starts its journey from the Vindhya range and after making rain on the mountains, plateaux and plains and absorbing moisture from different rivers on the way it reaches the Himalayan region *via* Kurukshetra and Kanakhal.

*Science* : Kalidasa had some scientific concepts too. Besides the reference to the chemistry of clouds, he has also referred to an established physical law that a magnet attracts iron.[13] He has also mentioned about the process of evaporation. When rain falls according to him, on the parched stones or lands, it is converted into vapour.[14] He also referred to the popular law of science that the water takes a lower course.

*Cosmography and Oceanography* : Although Kalidasa's works do not contain many ideas about cosmography, the poet has, on one occasion, stated that the moon was formed by that part of the earth's surface which has now become an ocean.[15] References to the tides[16] are found at two places in *Raghuvamsam*. There are also references in his works to shorelines and coastlines which bring the moisture of the ocean waters.

*Astronomy and Astrology* : The astronomical or celestial features like winter and summer solstices have found a place in his works. The mention of certain familiar planets like the Venus, Mercury and Mars is frequently seen. Besides the phenomena like the occurrence of eclipses, comets, meteors etc. have also been referred to. Mention of various *rashis, nakshatras* and their effects on human activities mentioned in his writings clearly indicate that he had a profound knowledge of astrology and astronomy. It is also conjectured that Kalidasa wrote a book on astrology named *Jyotirvidyavarna.*

## PHYSICAL GEOGRAPHY AND GEOMORPHOLOGY

The features of physical geography like mountain peaks, mountain slopes, plains, plateaux, caves, springs, waterfalls, passes, rivers, lakes, oceans and seas have been used several times in his writings. Kalidasa was aware of the fact that all big rivers met with the oceans or seas.[17] He has mentioned about the vastness[18] of the water bodies—the oceans. He has also referred to the existence of hills in the oceans.[19]

Kalidasa had also observed the fluvial processes of erosion and deposition. According to him, during the rainy season the river water erodes sediment from both the banks. During this period the water level in the rivers is sufficiently high and the rivers exhibit bankful discharge. The river water, particularly during the month of *Sravana*[20] (mid July to mid August), remains saturated with the sediment. He has also mentioned about the formation of deltas at the mouths of the rivers

Besides the features produced by fluvial erosion and deposition, Kalidasa also has referred to the landscapes produced by corals.[21] Glacial topography and the formation of rivers due to the melting of glaciers, etc. have been described by him. By referring to the mirage (Mrigatrishna), he has shown his acquaintance with desert ecology.

*Landscape Painting :* Kalidasa was a pastsmaster in landscape painting. In *Meghadutam* (*Purbamegha*), we find a detailed geographical description of the Vindhyan and Malwa Plateau regions. In *Kumarasambhava,* he has described the Himalayan region in detail highlighting its various aspects of physical, economic and human geography. In *Raghuvamsham,* while describing King Raghu's *Digvijaya* and *Indumati's swayambara,* he has tried to present a detailed spatial location of the different kingdoms or *janapadas* of contemporary India with their geographical peculiarities.

In *Purbamegha* (first part of *Meghadutam*) we come across about 32 names of places, almost half of them being the names of rivers, and the rest being the names of mountain ranges, hills, plateaux, capital and other towns of some regions, etc. Let us attempt to identify the places mentioned in *Meghadutam* in the present geographical set-up of India.

*Ramagiri* : This is the place according to the poet where the cursed and banished *Yaksha* was staying. According to Mallinatha, the famous commentator on Kalidasa's works, it is the present Chitrakuta mountain which is situated on the bank of the Ganga south of Allahabad in the Banda district of U.P. But Wilson[22] identified it with the present Ramtek mountain situated to the north of Nagpur and which is simply a Maratha equivalent of the Sanskrit Ramagiri.

*Malakesetra* : This is identified with the present Malwa Plateau. According to Chaturvedi,[23] it stretched from the origin of the river Narmada to the neighbourhood of the Vindhyas. It was a part of the former Rewa state. Mallinatha takes "Mala" to mean "a plateau or an elevated hilly spot". However, Wilson observes : "It is not easy after the lapse of ages to ascertain precisely the site of several places enumerated in the poem before us. The easterly progress of the cloud and the subsequent direction by which it is to reach the mountain of Amarakuta prove that the place mentioned here must be somewhere in the vicinity of Ratanpur, the northern half of the province of Chhattisgarh. . . . The only modern traces that can be found of it are in a place called Malwa, a little to the north of Ratanpur". Perhaps, he meant the Malwa Plateau region by "Malakesetra" with which he was thoroughly acquainted.

*Amarakutaka* : This is identified with the present Amarakutaka mountain which lies to the south-east of the Vindhyas. Perhaps, Amarkantaka is the corrupt version of Amarakutaka. It is also known as the Somaparvata and Surathadri.[24] This hill is a part of the Mekhala hills in Gondwana in the territory of Nagpur in which the rivers Narmada and Son take their rise.[25] This mountain range is situated in the former princely state of Rewa in the Bundelkhand region and in the present

State of Madhya Pradesh. It forms the easternmost extremity of the Maikal range. This mountain range which attains an elevation of 1057 metres is a sacred place for the Hindus as it contains the sacred Chandikatirtha.[26]

*Narmada* : This is the Namados of Ptolemy and is identified with the Peninsular river of the same name which originates from the Amarkantaka mountain and empties into the Arabian Sea near Broach after flowing for a distance of 1312 km westward between the Vindhya range of hllls to the north and the Satpuras to the south. Its other names are Somodvaba, Rewa and Mekhalasuta. Kalidasa has mentioned the rock-strewn bed of this river. According to him it is narrow at the foot of the Vindhyas.[27] He might be pointing at the narrow deep gorge at Harinphal.

*Dasarna*[28] : This Kingdom where the Dasarna river (modern Dhasorn river) was flowing existed to the south-east of the Vindhya region or the Madhyadesha. According to Wilson, it is the Dosarene of Ptolemy Periplus and may be identified with the modern Chhattisgarh region. This region forms the eastern part of Malwa. The Dasarna river (little river with ten tributaries) used to flow in this country. Its name is mentioned in *Anguttara Nikaya* and in the *Mahabharata* (II, 5-10).

*Vidisha* : This was the capital of the Dasarna country and is identified with modern Bhilsa which is located 26 miles north-east of Bhopal on the bank of the Betwa river.

*Vetravati* : This is identified with the modern Betwa river. It rises on the north of the Vindhya chain and pursuing a north-easterly course of 340 miles traverses the Malwa plateau and passing through the south-west corner of Allahabad falls into the Yamuna below Kalpee. We find the mention of the river in the *Markandeya Purana*. According to the *Puranas*, this river, receiving numerous tributaries before joining with the Yamuna, originated from the Paripatra of Pariyatra mountains representing a western branch of the Vindhya range, which formed the western boundary of the ancient kingdom of Avanti and may be taken to extend north-east through Rajputana under the known name of Aravalli.[29]

*The South-west Monsoon in Meghadutam*

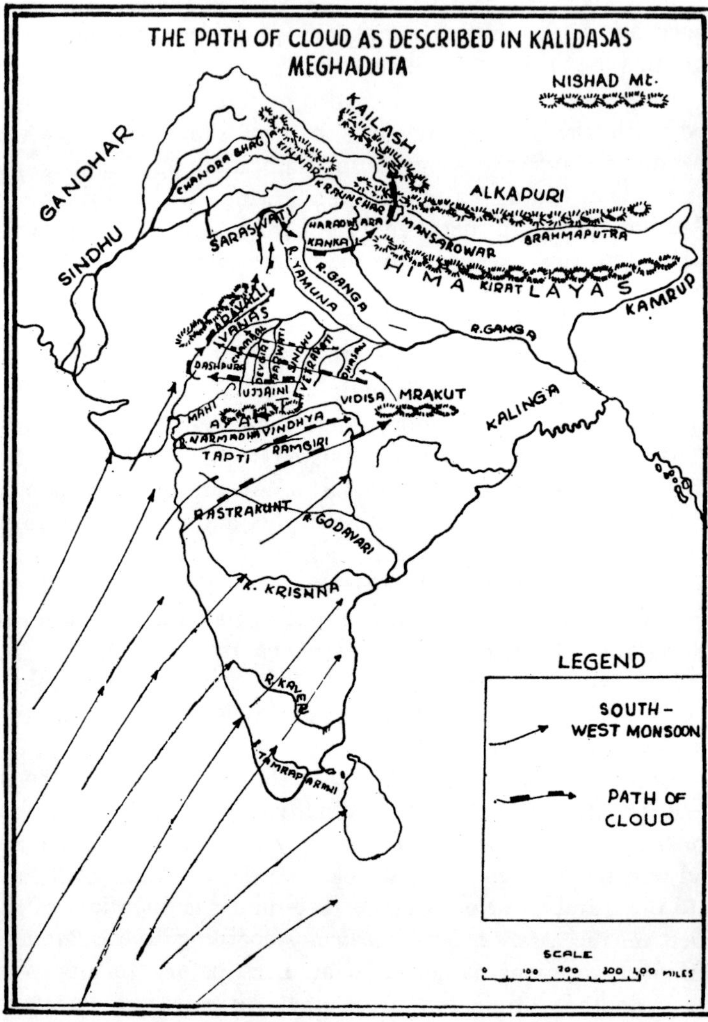


FIG. 2.1

*The Nich Hills* : This is a mountain in the vicinity of Vidisa. This is one of the branches of the Vindhyan range. It is so called because of its low elevation.

*Ujjayini* : This was the capital of Avanti and is identified with modern Ujjain. Its other ancient names were Avantika and Pushpakarandini.[30] It is situated to the south of the river Sipra. Its name is mentioned in the inscriptions of Nasik cave. According to Ptolemy, it was the capital of king Chastana (A.D. 150). "Huien Tsang describes the capital of U-She-Yen-na or Ujjaini, as 30 li or 5 miles in circuit, which is only a little less than its size at the present day".[31] Modern Ujjain is about a mile south of the ancient city. It is one of the seven sacred cities of India.[32] About this town Wilson remarked that "few cities perhaps can boast of a more continuous reputation, as it has been a place of great note from the earliest periods of Hindu tradition down to the present day". There is a vivid description of this ancient seat of Indian culture in the *Avanti Khanda* of the *Skanda Purana*. From the intimate description of the geographical environment of this town by Kalidasa, it seems that he must have lived here for a pretty long time. The Mahakal temple of this ancient town is still there as a witness of its glorious past. Due to its antiquity, this temple has become a geographical entity.

*Bana Nadi* : Although Mallinatha takes this in the sense of 'forest rivers' in general, some other commentators have thought it to be a proper name and have tried to identify it with the Nabanadi or Naga Nadi. The latter name is preferred by Wilson who thinks it may be the name of small river west of the Betwa and is also named as Parvati which originating from the Vindhya chain runs north-west till it joins another called the Sipra and the two together fall into the Chambal. According to Wilson, Parvati and Naga Nadi are synonymous.

*Nirvindhya* : This is a small stream which rises in the Vindhyas and flows northward between Ujjaini and Vetravati. Although in the *Matsya*, *Vayu* and *Brahmanda Puranas* it has got the same name but in some other scriptures it is known as Newaj. Its present name is Newaj.

*Sindhu* : There is a controversy about the identification of Sindhu river mentioned in *Purbamegha*. Some people think that there is only one Sindhu river in India which is known to the world as Indus. Others, however, believe that by Sindhu, Kalidasa referred to the river Kali Sindh in Malwa which springs from Bangi and falls into the Chambal river. According to Nandargikar "the little Sindhu which springing from

Devas and passing along with many other minor streams, by Ujjain, falls into Chambal".

*Avanti* : It is the name of a country which existed to the north of the Narmada river and formed the eastern part of Malwa Plateau. This country was bounded to the west by the Chambal river, to the north by the kingdoms of Mathura and Jaihoti, to the east by Maheswarpura and to the south by the Satpura mountains running between the Narmada and the Tapti. This country was about 900 miles in circuit. Avanti has been called Malava since the seventh or eighth century A.D. Panini mentioned it in one of his *Sutras* (IV, I. 176). It was also mentioned by Patanjali (IV, I. 1.9.36) and the authors of *Skanda Purana* (ch. I. 1.923) and the *Bhagavata Purana* (X. 45.31).

*Sipra* : Ujjain is situated on the banks of this river. Its name in the *Vayu Purana* is Madra; in the *Matsya Purana* it is Ksipra; in the *Brahamanda Purana,* it is Madra; and in other *Puranas,* Sighroda or Sighra. From this very name, it seems that it was a swift flowing stream. This river rises in the Paripatra range. It flows into the Chambal a little below Sitaman.

*Vatsa :* This country was situated to the north of the Vindhyas, Kausambi was its capital which is identified with modern Kosam, situated on the Yamuna and is about 38 miles from Allahabad above the Yamuna river. Anguttara Nikkaya mentions the kingdom of Vatsa or Vamsa as one of the sixteen *mahajanapadas* of India. It is also mentioned in the *Mahabharata* (II, 30, 10-11) and the *Harivamsa* (29.73).

*Gandhavati* : There was a small stream of this name near Ujjaini. In course of time its size and importance decreased considerably. Hara Prasad Shastri observes thus : "What was a beautiful small stream in Kalidasa's time is now a drain running through the heart of the modern city".

*Gambhira* : Like the Gandhavati this is also a small stream in Malwa. Its name has been mentioned by Jinasena in his *Adipurana* (ch. 29). It is a tributary of the Chambal.

*Devagiri* : Wilson identifies it with Devgad situated to the south of Chambal at the centre of the Malwa Plateau. Dr. Fleet thought it to be situated about 60 miles to the south-west of Jhansi. The temple of Skanda (Kartikeya) is situated here. H.P. Shastri observes that "the hill is still there and the deity is still there worshipped by the *ahirs* and *khanderao*".

*Charmanvati* : It is identified with the modern Chambal river. It was also known as the *Charawala* or *Shivnad*. The ancient Dasapura town was situated on this river. This is a tributary of the Yamuna. It rises in the Janapaw mountain of Indore region and falls in the Yamuna. According to a Puranic legend, this river was formed by the blood of numerous cows which King Rantideva used to sacrifice everyday to feed the guests. According to the *Mahabharata* and the *Puranas*, this king was "enormously rich, very religious, charitable and profuse in sacrifices." (Dawson).

*Dasapura* : It is identified with the modern Dasor of Mandasor on the left or north bank of the river Sivana ; and is the chief town of the Mandasor District in Scindia's dominations in the western Malwa Division of Central India. It must have been an important city in ancient times as it is mentioned in the *Mahabharata*, the Nasik inscriptions and the Gupta inscriptions.[33] Wilson identified it with the modern Rantipore or Rantampore as it was the capital of King Rantideva in ancient times. This site lying to the north of the Chambal is in the line from Ujjain to Thaneshwar and this lies in the course of the cloud's progress (*Meghadutam*). This town is supposed to be the native place of poet Kalidasa. H.P. Shastri, from Kalidasa's detailed description of the geographical environment in and around Dasapura remarks that "more intimate acquaintance with this part of the country than with the rest of Malwa. . . his native city was either Dasapura or some place near it."

*Brahmavarta* : This ancient holy land existed between the divine rivers Saraswati and Drisadvati.[34] This region lay to the north-west of Hastinapura and contiguous to Kurukshetra.

*Kurukshetra :* This was the site where the Mahabharata war was fought between the Pandava and the Kaurava. This is one of the sacred places of India and lies a little to the southeast of Thaneshwar and is drained by the river Saraswati. To the south-east of *Kuluta* beyond *Trigarta*, lay the district of Kurukshetra. In the *Manusmriti*, it is described with other districts.[35] In its limited sense, it is the tract near the holy lake still known by the same name to the south of Thanesnwar. In the larger sense it extended from the south of Saraswati to the north of Drisadvati[36] (Anand Mohan Barua). This flat and extensive plain tract now lies to the north of Delhi in the State of Haryana.

*Saraswati* : This was one of the sacred river of *Uttarapatha*. This historical river is particularly holy in Kurukshetra[37], the north-west part of which is drained by this river. The Sarasvati is described in the *Milindapanha* as a Himalayan river and its source is traceable to the Himalayan range above the Simla Hills.[38] This river is lost amidst the sands of the northern part of the Rajputana desert at some distance from Sirsa. Now it is a sub-terranean river whose characteristic is aptly described in *Siddhanta Siromani*[39] (*Goladhyaya-Bhubanakosa*). Although it was rather out of the path of the cloud's advancement (*Meghadutam*), yet Kalidasa perhaps could not resist the temptation to mention this river which occupied a unique place among all other Indian rivers.

*Kanakhal* : It is the name of the sacred place near Haridwar where the Ganges descends on the plain lands after traversing a great distance in the Himalayas taking its origin from the Gangotri glacier. Mallinatha takes it to refer to the mountain there through the gorge of which the Ganga descends. It is mentioned in the *Mahabharata* as a mountain.[40] It is called the gate of the Ganga.[41] Wilson observes, "It is rather extraordinary that Kalidasa should have omitted the name of Haridwar and preferred Kanakhal, especially as the former occurs in the *Puranas*. . ." It is also conjectured that gold was found[42] in that place and the word Kanakhal is the corrupt form of "Kanakakhal".

*Ganga* : This sacred river is the life-line of northern India and is the heroine of numerous legends of the *Puranas*. Kalidasa has referred to this river as *Janhu's daughter*. Wilson remarks, "Janhu's daughter is the Ganga or the Ganges river which after traversing its way through an extensive tract of mountainous country, here first enters the plain". By here he means at Kanakhal.

*Himalaya* : This is the invincible natural wall to the north of the country stretching from north-west to south-east, for a distance of about 1600 miles. The Gangotri glacier, the place of origin of the Ganga river, is located here. It contains several snow-clad, lofty peaks and is the birth-place of almost all important rivers of northern India.

*Sangama* : This is the confluence of the three sacred rivers of India—the Ganga, the Yamuna and the subterraneous Saraswati at Allahabad. This is a sacred place of the Hindus and here according to *Puranasamvasva*[43] the Ganga is not easily accessible. It has been nicely described in *Raghuvamsham* (13.54-58).

*Charananyasa* : The Harki-Pairi or the feet of Hara which is a small hillock at Haridwar may be identified as *Charananyasa* because in *Sambhurahasya* it has been referred to as *Sricharananyasa*.

*Krauncharandhra* : This is a mountain pass and is identified with the Mananiti pass in the Himalayas. It is one of the six important passes of the Greater Himalayas. In the *Taittiriya, Aranyaka* (1.31.2) the name of the Krauncha mountain is mentioned. According to Pargiter, this mountain appears to be a part of the Mainaka mountain range in the Greater Himalayan mountain system. According to the *Ramayana* (*Kiskindhyakanda*, ch. 44) it was a part of the Kailasha mountain where the lake Manasarovar exists. It has also been mentioned in *Harivamsa* (viii, 941-42). However, the creation of the Krauncharandhra as related to the heroic deeds of the two famous heroes of the *Puranas*—Lord Kartikeya and Lord Parsurama. This pass lying in Krauncha Parvata provides passage to Tibet from India. It is also known as the *Hansadwar*, the gate of the flamingos, because it provides a passage to the flamingos from central and northern India to the lake Manasarovar during the rainy season.

*The Kailasha Mountain* : The Kailasha mountain is one of the three mountains of the trans-Himalayan zone described in the ancient literature of India. It is one of the highest peaks (6,714 m) of the Greater Himalayan Zone. It is supposed to be the abode of Lord Shiva. The whiteness of its snow-capped peaks has been compared to the accumulation of loud laughters of Lord Shiva.[44] This mountain is located about 40 kilometres to the north of the Manasarovar. It is the

Kangrinpoche of the Tibetans and may be identified with the Vaidurya Parvata. It is also referred to as the Hemakuta.[45] Literally, Kailasa Parbat means the pleasure-mountain. A detailed description of this mountain is found in *Kumara-sambhava.*

*Manasarovar* : This sacred lake is situated between the Himalayas and the Kailasha mountain. It is also called the Brahma Sara.[46] According to Valmiki's *Ramayana*, the ancient river Saraju or the present river Gogra originated from it. But in reality no such river originates from it. The river Sutlej flows from another lake of larger dimension called the Ravana Hrada which lies close to the west of Manasa. The work entitled *Jambudiva-Pannatti* traces the source of the four great rivers called the Ganga, the Rohita (i.e. the Lohitya or the Brahmaputra), the Sindhu (Indus) and the Harikanta to the twin lotus lakes (Padma Hrada) one on the side of the lesser and the other on that of the Greater Himalayan range.[47] The Jaina lotus lake is no other than what is called the Bindusara in the great epic. *Anatotta* in Pali literature and Manasarovar in popular parlance, it is the favourite haunt of the flamingos which migrate to it at the commencement of the rainy season.

*Alakapuri* : It is very difficult to identify this town now. According to the Puranic accounts, it was the capital of Kubera, the god of wealth and the regent of the north. According to Kalidasa, it was situated on the slope of the Kailasha mountain. It is so called because of its grandeur. The *Brahmanda Purana* and the *Bhagavata*[48] have described its splendour, grandeur and wealth. Some commentators thinks it to be a fictitious city and Kalidasa has practically described the grandeur of Ujjayini in the name of Alaka.

Among the 32 places mentioned above, 16 are in or in the neighbourhood of the Central Provinces of India. It seems that the poet was well acquainted with the Vindhya region which compelled Hara Prasad Shastri to believe that the poet had shown great partiality to the region lying to the north of the Vindhyas which is at present known as the Malwa Plateau and Kalidasa has described the same as the Malakshetra.

In the *Raghuvamsam*, Kalidasa has given a vivid geographical description of India—its important kingdoms or *janapadas* mountains, rivers and lakes. We find references of the geographical place names on three occasions in this epic : King Raghu's *Digvijay* (canto 4), *Indumati's swayambara* (canto 6), and the aerial journey of Lord Rama and Goddess Sita over land and sea (canto 13). Kalidasa inherited some geographical names from Valmiki whom he has faithfully followed in the *Raghuvamsham* barring a few deviations in detail. In this great epic we get the descriptions of the snow-clad Himalayan ranges, the monstrous beasts of the watery wastes, the Gangetic alluvium and many other aspects of physical geography. Here we also find a glimpse of the varied people of different regions of India and their cultures and customs. Besides Valmiki's *Ramayana*, he must have taken some place names from the Puranas as well. From these descriptions we get an idea of the India of Kalidasa's time. King Raghu commences his campaign from Ayodhya which was his capital of Uttara Kosala. From there, he proceeds to the east where he subdues the *Sumhas* and the *Vangas*. Crossing the Kapisha river, he moves along with his army towards the south. He wins over the kings of *Utkala*, *Kalinga* and *Pandva*. Then he crosses the *Sahyadri* and subjugates the *Aparanta* districts along the western coast of India. After that he makes a northward journey along the coastal route. After crossing the *Trikuta* mountain, the deserts of Rajputana and Baluchistan regions, he reaches *Parasya* and defeats the "Parasikas". He crosses the *Sindhu* river and reaches the lands of *Kambojas* and *Huns* and vanquishes them in a fight. Then he reaches the Himalayan region and overpowers the *Utsawa Samketas*. Then leaving aside the Kailasha Mountain, he moves east and reaches *Pragjyotisha* and *Kamarupa* kingdoms which offered no resistance and presented Raghu with various forest products. Thereafter he came back to his capital and thus completed his cherished *Digvijaya*. Let us try to identify the places mentioned in this epic with the modern names.

*Sumha* : This may be identified with the present Rarh region of West Bengal occupying the plains to the north of the Damodar river. Kale thought that the Sumhas lived in a country abounding in cane-plants which grow well along the bank of the Ganges.

*Vanga* : This country is identified with modern Bangladesh or erstwhile East Pakistan and former East Bengal.

*Gangasagar :* This is the place where the Ganga meets with the Bay of Bengal. This is one of the sacred places of the Hindus. Every year during the *Makar Sankranti* a big fair is held here. To mark his victory King Raghu erected pillars[49] on the delta of the Ganga and on the islands near the Gangasagara.

*Kapisha River* : This is identified with the modern *Kangsabati* or Kasai river. Originating in the Chota Nagpur Plateau and flowing through the southern part of the Midnapore district of Bengal it empties into the Bay of Bengal.

*Utkala* : This is identified with the northern part of Orissa. It was also known as the *Odradesa* and stretched between the lower course of the Suvarnarekha river to the north and the Mahanadi to the south. It was bounded on the west of Gondwana, on the north by Jashpur and Singhbhum districts, on the east by the sea and on the south by the Ganjam district. This *janapada* was approximately 1167 miles in circuit.[50]

*Kalinga* : This is the *Kie-ling-kia* of Hiuen Tsang. It roughly extended between the Mahanadi in the north and Godavari in the south. Its circuit was approximately 800 miles. It was famous for beautiful elephants, coconut and betel groves.

*Mahendragiri* : This is identified with the modern Eastern Ghat ranges. Perhaps along with the ocean it was the most prominent geographical feature of Kalinga and for that reason Kalidasa styles the king of Kalinga as the lord of the Mahendra[51] and the Mahodadhi (Bay of Bengal). The Mahendragiri is referred to as separating Ganjam from the valley of the Mahanadi. It has been mentioned in the *Yogini-tanta* (2.4.128ff). In *Vishnu Purana,* it has been referred to as the source of the Rushikulya River. Pargiter thinks that this mountain range should be limited to the hills between the Mahanadi, the Godavari and the Wainganga and "may perhaps comprise the portion of Eastern Ghats north of the Godavari."[52] According to Bana[53] it joined with the Malava Paravata. Besides its economic importance, the most characteristic feature of this mountain is that : "it has preserved its name unchanged from the time of the composition of the *Mahabharata* to the present day."[54]

*Kaveri* : This is the sacred river of south India. It originates from the western Ghats (Sahyadri), and flowing south-east through Karnataka, falls into the Bay of Bengal.

*Malayachala* : It is the *Bettigo* of Ptolemy and is called *Podigei* or *Padigai* in Tamil. Pargiter identifies it with the southern part of the Western Ghats from the Nilgiris to Cape Comorin. It is noted for its pleasant cool breeze. It abounds in black pepper[55] creepers.

*Pandya* : This was the southernmost kingdom of India which was noted for its pearls.[56]

*Tamraparni* : This river "must have flowed below the southern boundary of the kingdom of Pandya and may be identified with the modern *Tambravari* or with the combined stream of this river and the chittar."[57] According to Ptolemy the port of *Korkai* (modern Kilakarai) was situated at the mouth of this river. "This may justify us in identifying it with the Gundur, the name under which the combined water of three streams flow in two streams into the sea."[58] This river originating in the Malaya range flows mainly in the Tirunelveli district of Tamilnadu and empties into the Bay of Bengal.

*Dardura Hill* : This was a part of the Western Ghat range and formed the eastern boundary of the former Travancore estate. It is situated to the south of the Kaveri river.

*Sahyadri :* This is identified with the Western Ghat range stretching from the Tapti river in the north to Kanyakumari in the south.

*Kerala* : The strip of landmass lying between Gokarna in the north, Kanyakumari in the south, Western Ghats to the east and the Arabian Sea to the west is known as the Kerala country.

*Murala River :* This river might have been flowing in the Surala or Sirala or Murala region which in all probability coincided with the lower Tapti basin centred round Surata and Nabasari. It might be another name of the Tapt· river.

*Trikuta Parvata* : From its very name it is evident that this mountain had three prominent peaks. According to S.M. Ali[59] this hill was very high, full of precious metals and it contained numerous caves. This hill was located in the *Malasaa Dwipa* and was inhabited by *yakshasas.* Sitaram Chaturvedi[60] identifies it with the Girnar hill of Gujarat.

*Aparanta* : This is identified with the western territories comprising Konkan, Kanara and Bombay districts. This is one of the seven *janapadas* of the Puranic period. According to the Puranic legends the coastal belt existing between the Western Ghats and the Arabian Sea was given as a gift to Parsurama by the ocean. It was therefore known as *Parsurama Kshetra.*

*Parasya* : This country lay to the west of the Indus. It is identified with modern Iran but in the past the Persian empire extended much closer to the Indian boundary.

*Sindhu Nadi* : This is the well known Indus. It is a Trans-Himalayan river. It is the greatest known river of Uttarapatha after which the Indus group is named. It was rivalled only by the Ganges. According to the *Matsya Purana,* it is one of the seven celestial rivers. Originating from the Himalayas and passing through Kashmir, the Punjab and Sind, it falls into the Arabian Sea.

*Kamboja* : This includes that part of Afghanistan which lies nearer to Kandhar. According to *Saktisangam Tantra*[61], it extended from the south-east of the Arab countries to the Punjab. It was famous for very good horses.

*The Territory of the Huns* : One of the two groups of the ferocious central Asiatic tribes—the Huns—settled in the north-eastern part of Kamboja. They occupied the areas near Peshawar. Sialkot town of west Punjab was their capital for some time. This tribe created great trouble for the Gupta kings.

*Lohitya* : It is identified with the Brahmaputra river. It is so called because of the red tint of its water.

*Pragjyotisha* : This ancient *janapada* is identified with the provinces of eastern Assam. It corresponds roughly with the Middle Brahmaputra valley. "This was a famous *janapadas* on the fringe of the 'Eastern Country.'"[62] According to Kalidasa, it lay to the north of the Brahmaputra river and "therefore probably included the whole belt of alluvial land north of the Brahmaputra from the Tista eastwards".

*Kamarupa* : This ancient kingdom, famous for elephants

is identified with western Bhutan and parts of Assam.

In other cantos of *Raghuvamsham* the following place-names are met with :

*Malyavaya Parvata :* Like the Gandhamadana range, the location of the Malayayana Parvata is highly controversial. According to the *Vishnu Purana*, the Meru is confined between the Nila and the Nisadha mountains on the north and south and Malyavana and Gandhamadana on the west and east.

*Pampa Lake* : It existed near the Rushyamuka hill and the Pampa river. The Rushyamuka has been identified by Partiger with the range of hills stretching from Ahmednagar beyond Naladrug and Kalyani dividing the Manjira and the Bhima rivers. The river Pampa rises in this mountain and falls into the Tungabhadra.

*Godavari River* : It is identified with the river of the same name. It was also known as the Gautami river.

*Chitrakuta Hill* : It is the modern Chitrakut hill lying 100 km. W-S-W of Allahabad.

*Mandakini River* : This river used to flow to the north of the Chitrakut hill. Nearer to that hill, the river Mohini[63] was also flowing.

*Govardhana Hill* : This hill which is connected with the Krishna legend is 30 km. away from Brindavana in the Mathura district of U.P.

*Ayodhya* : Situated on the banks of the river Sarayu (modern Gogra), it was the capital of Kosala. It is still there but lacks ancient grandeur and importance.

*Uttara Kosala* : The territory to the north of Varanasi was known as Uttara Kosala. It was the kingdom of Lord Rama and his ancestors. The river Sarayu nourished this kingdom.

*Nandigrama*: This is a small village near about 8 miles away from Ayodhya.

*Panchavati* : This was a place in Dandakaranya.

*Naimisharanya* : This is modern Nimikharu which is situated in the Sitapur district of U.P. It is a sacred place for Hindus.

*Tamasa* : This river is identified with the Tons or Little Saryug river. This river after flowing through Azamgarh and Balia districts falls into the Ganga.

*Gopratara* : This is a sacred place on the banks of the

Sarayu (modern Gogra).

*Mithila Puri* : This was the capital of King Janaka. It may be identified with modern Darbhanga district.

*Janasthana* : This was a region near Dandakaranya where *rakshasas* used to live.

*Gandhamadana Hill* : This hill is a part of the Rudra Himalayas and a part of the Kailasha range.[61] It lies to the east of the Meru. Besides *Raghuvamsham* Kalidasa has mentioned it in *Kumarasambhava* and *Vikramorvashiyam.* It is one of the most important hills of the Puranic age. It has also been mentioned in *Yoginitantra* (1.15), *Jataka* (VI, 519), *Harivamsam* (ch. 26, 5-7), *Padmapurana* (ch. 133), Banabhatta's *Kadambari* and *Divyavadana* (p 157).

Besides the above place names, the following have also found place in his works :

*Gokarna* : This is a sacred place for Hindus and is situated about 16 km north of the Kunta town in North Kanara district of Maharashtra.

*Pratisthanapura* : This was situated near Allahabad and was the capital of King Pururaba. It is identified with modern Jhansi.

*Mahakosi* is a Himalayan river.

*Barada* is a Himalayan river on the banks of which their is the image of a goddess with eighteen hands.

*Somatirtha* is a sacred place near Pindapuri in Karnataka.

*Meru* : B.C. Law identifies it with the Rudra Himalaya in Garhwal. It is probably the Mount Meros of Arrian. According to the *Bhagavatapurana,* this mountain is situated at the middle of *Ilavarta.* According to Hindu tradition, it was the rendezvous of the gods. It was visited by Vaishampayana and sage Salankayana meditated here. S M. Ali[65] has identified it with the Great Pamir knot of Asia.

*Amaravati* : This was the legendary capital of Heaven. Some thinkers have tried to identify it with the place Indralaya near Bokhara in Turkistan.

*Ausadhi Prastha* was a Himalayan township. The Ganga used to flow[66] near it and this township was surrounded by medicinal herbs.

*Krathakaushika* was another name for Vidarbha, so named because it was divided between two princes Kartha and Kaushika.

*Badarikashrama* is a sacred place for Vaishnavite Hindus and is situated in the Himalayan region. According to old scriptures this place is situated between Kanvashrama and the Nanda Parvata.

*Malini* : This river drained the terai region of the Himalayas. It is identified with the river of the same name which flows through the Bijnor district of U.P. The hermitage of sage Kanva was situated on the banks of the river Besides these, mention has also been made of the Gourishikhara peak which may be identified with the Gourishankar peak of the Himalayas. Besides, he has referred to the Kanvashrama, the Sachitirtha, Agastyoshrama etc., which were the famous places of pilgrimage in ancient India.

It is evident from his writings that Kalidasa was familiar with King Vikramaditya. But it is not certain whether he was the king of the first century A.D. (*Shakari Vikarama*) or the Vikramaditya or Chandragupta II (A.D. 385-415) of the Gupta dynasty. The last hypothesis has been accepted by Macdonnel, Keith, Bhandarkar, Mirasi etc. Fergusson and Max Muller thought that Kalidasa belonged to the sixth century A.D. But this is not accepted by the people now-a-days. In his writing, direct and indirect references to the Gupta dynasty[67]—its kings like Samudragupta[68], Chandragupta, Vikramaditya[69], Kumaragupta[70] etc. are to be found. It is supposed that the epic *Kumarasambhava* was presented to Vikramaditya by the poet on the auspicious occasion of Kumaragupta's birthday. In the drama *Vikramorvashiyam*, there is no character like Vikrama. It is apparent that the poet wrote the drama to please emperor Vikramaditya. *Kumarasambhava* glorifies the heroism of Kumara i.e. Skanda—the general of the gods. During the Gupta period Skanda worship was popular in India.

King Raghu's *Digvijaya* in the fourth canto of *Raghuvamsham* is not an historical fact. But the various countries or *janapadas* mentioned in that context were actually the kingdoms or *janapadas* into which India was divided during Kalidasa's period. In King Raghu's *Digvijaya* the remarkable omission was the kingdom of Magadha. After starting from Uttar Kosala King Raghu's army should have first met with the army of Magadha. But this was not so. It may be conjectured that the poet was obliged to the king of Magadha. Therefore, he did not want to show the defeat of that king. Besides, Patliputra was the capital of Samudragupta. It was Vikramaditya who shifted the capital from Patliputra to Ujjayini. Kalidasa's weakness for Magadha is evident from the fact that he has shown queen Sudaksina and queen Sumitra to be the princesses of Magadha although in the latter case Valmiki is silent. Although in the *Digvijaya* neither at the time of proceeding nor at the coming back of Raghu's army was Magadha mentioned ; though in the sixth canto while describing Indumati's *swayambara*, the poet has mentioned the name and powers of *Magadheshwara*[71] first of all.

Another remarkable omission in the fourth and sixth cantos of *Raghuvamsam* was the Punjab. This country at that time was either occupied by some outsiders—most likely *yavanas* or the Huns even if it was a Hindu kingdom, its kings must have been insignificant in power and wealth.

In Indumati's *swayambara* (VI canto) in *Raghuvamsham* mention of the kings of Magadha, Anga, Ujjayini, Surasena, Anup, Kalinga and Pandya has been made. It seems that during Kalidasa's time the Gupta kings either had friendly relations with those countries or they were quite powerful and prosperous. In that canto it is also indicated that Vidarbha and Kamrupa countries had very friendly relations with the Gupta kings.

It is evident from both the fourth and sixth cantos of *Raghuvamsham* that Pandya was the most important kingdom in south India. This country was famous for pearls from the ocean, cardamom, cloves and black pepper. At that time the Chedi kings of Kalinga had become very strong and so the lord of Mahendra as he has been designated by Kalidasa, tried to resist the forces of Raghu with full might. During that period the Nagas of central India were gradually becoming powerful. In this sixteenth canto of *Raghuvamsam*, Kalidasa has narrated the marriage of Kusa, son of Lord Ramchandra with Kumudvati, the Naga princess.

During that period the barbaric Huns tribe of Central Asia

could not enter the heart of the country but were staying in its fringe regions. Raghus forces fought with the Huns outside Indian territory. In Indumati's *swayamvara*, no king from the Indus valley and the Punjab participated. These areas might have been occupied by the Huns. In *Vikramorvashiyam*, Kalidasa has hinted at the dominance of a barbaric tribe to the north-east of the country. These people like the Huns used to keep their hair long and also grew beards. From the same drama we get an indication of the attack on the country by some alien power. The coming back of emperor Kusa from Kusavati to Ayodhya also points towards an historical fact of great importance, perhaps it indicated that the sovereignty of the country was in jeopardy.

There are three theories regarding the accurate description of various kingdoms of India with their distinct geographical features by Kalidasa in such a period of Indian history when the means of transport were insufficient, inefficient and at places non-existent.

In describing king Raghu's *Digvijaya* :—

1. Kalidasa must have got inspiration from Samudragupta's victory over Deccan and Vikramaditya's victories over various countries.

2. Secondly, the *Ramayana* and the *Mahabharata* are the storehouses of ancient place-names. Kalidasa possibly took the materials of his epics and dramas from these two epics. In *Raghuvamsham* he has faithfully followed Valmiki in 10-15 cantos of *Raghuvamsham*. The stories of *Abhigyana Shakuntalam*, *Vikramorvashiyam*, *Malabikagnimitram* have been taken from the *Mahabharata*. It may be possible that Kalidasa collected different place-names and their geographical characteristics from these two epics and some *Puranas*.

3. Thirdly, Kalidasa must travelled widely in India with a keen observer's eye. His descriptions of Ujjayini or of the Vindhya region are eloquent about the fact that the poet had a very detailed knowledge of these regions.

In the sixth canto, Kalidasa has described the geographical characteristics of seven countries namely Magadha, Kalinga, Pandya, Anga, Surasena, Anupa and Vidarbha. While the first three have already been identified, we shall try to identify the last four. The Angadesha existed between

Magadha in the west and Banga in the east. It may be identified with North Bengal. The Surasena Kingdom is identified with Mathura and its surrounding regions in U.P. The Anupadesha existed to the north of the Gulf of Cambay and the Narmada river. The Vidarbha kingdom is identified with the north-eastern part of the modern Maharashtra State.

While describing Rama's aerial way back from Ceylon to Ayodhya (*Raghu* canto 13) on the *Puspaka* plane Kalidasa has given a vivid description of the Indian landscape—as if he had taken aerial photographs of certain chosen spots. Rama and Sita start from Lanka (Ceylon), cross Setubandha (Adam's bridge) and then in succession cross the Malayagiri, the Malabar Hill, the Pampa lake, the Godavari river, Panchavati, the ashrama of sage Agastya, the Panchapsara lake of sage Satakarni, the garden of sage Sarabhanga, the Chitrakuta hill, the Mandakini river, sage Atri's hermitage, the Sangama and finally reach Ayodhya—the capital of Uttara Kosala—situated on the banks of the river Sarayu.

In the aerial journey Kalidasa has described the ocean—the watery waste in a splendid manner. According to the poet, in the oceans untold treasures are ever on the increase.[72] The sea contains the submarine fire *Baraba.* The pleasant luminary (moon) has emerged from the mighty deep. Its vastness on all sides is beyond determination. The frightful ocean receives numerous tributaries, its bright water contains the gigantic whales and various kinds of reptiles, its swirling waves roll over the coasts which are dotted with the forests of areca and palm trees.[73]

The background of the story of the epic *Kumarasambhava* is the Himalayas. Here the Himalayas are described as the god of the distant north[74] spanning the eastern and western oceans and acting as a measuring rod of the earth and being repository of medicinal herbs and precious stones. The Himalayas are rich in mineral resources[75] and are the source of thousands of rivers. The poet has described in detail the snowy, lofty peaks, the dark, long caves, various tribes living inside the caves and on the mountain slopes, and various types of Himalayan flora and fauna. Although king Himalaya, the queen Mena and their son Mainaka have been personified in this epic, we can establish their geographical reality. The

Mainakagiri has been identified with the Siwalik range of the outer Himalayas.[76] Its name is also mentioned in *Yoginitantra* (2.4, p 128-29) and in Bana's *Kadamabari*.

## Zoogeography

Kalidasa has referred to various kinds of animals, reptiles, birds and insects. Among the land animals lions, elephants, horses and different species of deer have been described frequently. He has also referred to yaks, rhinoceros and *sarabhas*. The first two are found today but the last one which had a peculiar physique is extinct now. *Sarabha* was a kind of deer. It had eight legs—four downwards like other quadrupeds and four upwards growing on the back. It was a very strong animal and could even kill a lion. Therefore, it was also known as the lion-killer. In those times encounters between lions and elephants was a common feature of the jungles. From Kalidasa's descriptions, it follows that the Asiatic lion which is now restricted to the Girnar forest of Gujarat had a wider distribution in India and its number then was probably much higher in comparison to its present strength.

It is very strange that among the four varieties of big cats only the lion could find a place in his works—the other three varieties namely the tiger, the leopard and the cheetah do not. It may be possible that the number of tigers, during Kalidasa's time was insignificant.

The number and variety of animals were much greater in the Himalayan region than in other parts of the country. Deer were almost omnipresent. Apart from the forests they were present in sizable number in the *ashramas* of the sages. There is reference to the *kasturi mriga*, from the navel of which a scented substance called *kasturi* is obtained. He has also described a special variety of deer—*chamari mriga*, whose tail they used as *chamara*[77] or hairy fan. Among the domestic animals cows were the most common. Cow worship was widely prevalent at that time throughout the country. In the first canto of *Raghuvamsham* we find that king Dilip worships and takes all possible steps to please the celestial cow *Nandini*—the daughter of *Kamadhenu*—to get an heir to the throne.

Among the animals of water we find mention of whales

(*Raghu*, 13.10), crocodiles, alligators (*Raghu*, 16.55.79) sea-snakes (*Raghu*, 16.76) etc. in Kalidasa's works. We also find mention of two kinds of fishes—*Rohita* or *Rohu* (*Abhigyana* VI Act) which is supposed to be the best among all the fishes in taste and *safari* or *saphara* (*Megh*, 143) which is a small white glistening fish that darts rapidly through the water and is almost omnipresent in Indian water bodies. Kalidasa has also mentioned the corals and conch shells that are found (*Raghu*, 13.13).

Among the birds peacocks, pigeons, cranes, doves, cuckoos, *chataks*, *chakravakas*, waterfowl, hawks, etc., and among the insects *bhramara*, *birabahuti*, and *dangsa* (*Raghu*, I) find a place in his works. Although hunting was a sport for the kings and nobles in those days, in the Yavnas animals were sacrificed and *brahmins* used to take meat but Kalidasa advocated the conservations of animal wealth. He denounced the killing of animals in *Raghuvamsham* and *Abhigyana Shakuntalam*. This is probably because of the fact that Kalidasa was a Shaivite or due to the influence of Buddhism and Jainism.

**Plant Geography**

Kalidasa has also referred to various types of plants in his works. During his time many more areas of India were covered with forests in comparison with the present day and being a widely travelled man, Kalidasa must have come across numerous trees, plants and flowers. During Kalidasa's time, the most important tree was the *sal* (*shorea robusta*). It is also called *vanaspati* (the king of the jungle) bamboo (*dendrocalamus strictus*), was found in almost all the regions of India in the Indo-Gangetic plain, Central Indian Plateau or in the mountainous terrain of the Himalayas. Kalidasa was fond of describing the musical sound produced by the wind on entering the holes of the bamboos or canes. Besides these, trees like *deodar* (*pinus longifolia*) of the Himalayan region, the *aguru*, trees of Assam forests, *arjuna*, *ashoka*, *asara* (*sal*) *khadira* (*acacia catechu*), *saptaparna* of the Himalayan region, palm and areca trees of the eastern and southern sea coasts, *nameru* (*multani champa*) have also found place in Kalidasa's works. The trees like *palasa* (*butea monosperma*), *kadamba*, *tamarind*, *priyal*, *parijat*, *mandara*, *lodha*, *tamala*, *sallaki*, *sirisha*,

***shalmali, shami***, etc., the flowering plants like ***shefalika, bakula, mallika, kunda, kumuda, kurubaka, kusuma*** etc., grasses like ***mustaka, munja, narakata, nagarmusta*** etc. legumes like ***privangu*** and the creepers like *malati, madhavi* etc., found a place in his writings. Besides these he has mentioned on several occasions about the mushrooms (*kandali kusuma*) which sprout during the rainy season. In addition to these, he has referred to the herbs which have medicinal value and those which produced *light*. The Himalayan town Aushadiprastha described by him in *Kumarasambhava* was surrounded by various medicinal and phosphorescent herbs. References to the canes are numerous in his work. *Bhojapatra* which is actually the bark of a tree, and which was used in ancient India as paper, has been mentioned in *Raghuvamsham* (canto I).

Like *Kumarasambhava*, the background of the stories of *Shakuntalam, Meghadutam* and *Vikramorvashiyam* are the mighty Himalayan snowy peaks, lashed with green valleys, snow-fed rivers, beautiful waterfalls, thickly forested hill-slopes and its calm scenes and romantic environment.

If we analyse all his works then we see that the poet had an intimate knowledge of the two regions—the Vindhya and the Himalayas. From the descriptions in *Meghadutam*, it is evident, that the poet knew every inch of the land of the Vindhyan region. From his descriptions of the Himalayan region in *Kumarasambhava* and other works, his knowledge of that region seems to have been quite perfect. Besides, he was also aware of the fact that the Himalayas are some of the highest mountains in the world. The expressions like *Nagadhiraja, Shailadhiraja, Gouri-guru, Giriraja*, etc., are quite eloquent about this. The poet seemed to have had a special fascination for that part of the Himalayas where the Ganga was flowing.

## Economic Geography

Although hunting was a sport for kings and the royal family and the means of procuring food for the hilly tribes, farming, mining and catching of elephants from the forests were the main economic activities. He has mentioned paddy and wheat cultivation and sugarcane growing. Besides, these he has referred to the production of betelnut on the hilly slopes of the *Mahendra Giri* and cardamom and black pepper on the hills of the south. The betel, coconut and palm groves of Kalinga and the other kingdoms in the south, the sandal-wood of the forest of Karnataka, the *aguru* and *kalaguru* of Assam forests, the pearls from the oysters in the Tamraparni coastal region also attracted the poet's attention. During his aerial journey over the southernmost part of the country, Lord Rama says to Goddess Sita about the *Pushpak* : "It now passes over the outskirts of the southern coast where pearls and gems lie scattered and where are seen the countless rows of fruit-laden areca trees". According to Kalidasa, the Himalayas and the oceans are the storehouses of vast resources.

During Kalidasa's time clothes of different qualities were manufactured. References to rough and fine cotton clothes and silken clothes are found. The weavers of that period were so expert in weaving that they could weave extremely fine cloth.

During Kalidasa's time, elephants, horses, and yaks were the principal pack animals and they also used to carry human beings. The nobles, chiefs and important persons used to move in chariots. Although those were the chief means of transport of men and materials, references to water vessels and even aeroplanes are prominently found in his works. Land routes were the chief routes of movement of goods and human beings but water routes and air routes were also in use.

People were aware of engineering and the importance of the construction of bridges although Kalidasa has not made mention about the materials of the bridge. Rowing and boating were done for recreation. Mention of water vessels are also found in the fourth canto of *Raghuvamsham*. The soldiers of the Banga King came in water vessels to fight with Raghu's forces. The people must have established trade relations with the neighbouring islands of India through water vessels.

In Kalidasa's *Raghuvamsham* (canto 13), there is mention of the aeroplane (*pushpak vimana*). But Kalidasa has borrowed its idea from Valmiki, the author of the *Ramayana*. However, there is no doubt that Kalidasa lived in a period of Indian history when Indian civilization had reached great heights.

## Human Geography

Kalidasa has referred to several tribes of contemporary India living in different parts of the country. The Utsavasanketas, the hill tribe of Laddakh and some parts of nothern Kashmir ; the Kiratas, who lived in the Himalayan region from Kashmir to Assam and also in Nepal and the Vindhya region (*Shakti Sangam Tantra*) have been mentioned on several occasions. According to Pargiter, those tribes lived in Uttarapatha. Ptolemy identifies Kirshadia to be the 'land of the Kiratas'. They also lived in Eastern India and in the Brahmaputra valley. This tribe has been mentioned in the *Mahabharata* (XII, 207, 43). According to *Srimad Bhagavad Gita* (II, 2-18) they lived outside the Aryan fold. Besides these, Kalidasa also mentions the Nishadas, the Nagas, the Vidyadharas, the Huns, the Kambojas, the Mlechhas, and the Yavanas.

From the descriptions of Ujjayini and Alakapuri, it seems that during Kalidasa's time India had reached an advanced stage of urbanisation. The lofty buildings, the built roads and lanes, the gardens and temples, etc. of these two cities described, speak eloquently about the well-developed social and urban systems. In the absence of an air-conditioning system, people made mats of (*khas khas*) the roots of a kind of fragrant grass to cover the doors and windows during the summer season to protect themselves from the scorching heat of the tropical sun and to keep their rooms cool. In the bathroom, there was provision for showerbaths. Therefore, the poet used the words *yantradharagriha* for bathrooms.

Apart from the Aryan and non-Aryan races of India, Kalidasa has also mentioned some external martial races like Yavanas, Pallavas, Huns, etc. The society of that time was quite liberal according to accounts. As a consequence, inter-caste marriages were allowed. Due to this several alien customs, manners, social taboos were assimilated into the Indian culture.

## Regionalism

During Kalidasa's time, India was broadly divided into two macro regions—*Aryavarta* and *Dakshinapatha* which were separated by the Vindhyan range extending east to west for a long distance in the central part of India, and acting as a great dividing line between the two distinct cultural and population groups. Besides these two broad divisions, Northern India was divided into many sub-divisions like Brahmavarta, Uttarpatha etc. Among the physiographic regions, the Himalayan region, the Indo-Gangetic plain region, the Vindhyan plateau region, the Western Ghats and south Indian mountain region, and the coastal regions, were quite prominent. These macro-regions and physiographic regions were subdivided into various countries or *janapadas*. The centre of political activity was northern India but south India had great economic importance which has been highlighted by poet Kalidasa by referring to the different economic resources like pearls of the Tamraparni coast, the cardamom, the black pepper, cloves and other spices, the coconuts, the betelnuts, etc., of the mountain slopes and coastal regions of the south.

Kalidasa, therefore, while writing some of the best pieces of Sanskrit literature has touched almost all aspects of geography in a splendid manner. He may thus be styled as the "founder of Indian geography". It is a pity that in spite of the great geographical sense of our ancient poets and thinkers as well as the availability of a great deal of literature full of geographical concepts and ideas we could not take full advantage of them and there is a long gap between those ancient writings and recent literature on Indian geography.

## REFERENCES


1. P. Ranganathan, *Kalidasa's Raghuvamsa*, The Gondals Press, New Delhi, 1964, p 5.
2. *Raghu*, 16-44.
3. *Raghu*, 17-72.
4. *Raghu*, 16.45.
5. *Ritu*, I.
6. *Ritu*, 3-3.
7. *Ritu*, 5-7.
8. H.P. Shastri, *Kalidasa His Home*, J. Bors, 1915, pp 197-212.
9. *Raghu* 17 72.
10. *Purba Megha*, 6.
11. *Jyotish Tatwa*.
12. *Meghadutam* : *Purba Megha* I.
13. *Raghu*, 17.63.

14. *Megha*, 1 : 127.
15. *Raghu*, 13-4.
16. *Raghu*, 16.27.
17. *Raghu*, 13.9.
18. *Raghu*, 13.7.
19. *Raghu*, 17.41.
20. *Raghu*, 6.52.
21. *Raghu*, 13.13.
22. H.H. Wilson, *Vishnu Purana.*
23. S.R. Chaturvedi, *Kalidasa Granthawali*, notes at the end.
24. Pargiter, *Markandaya Purana*, ch. 57.
25. B.C. Law, *Mountains and Rivers of India*, National Committee for Geographers, Calcutta, 1968 ; p 31.
26. *Padma Purana*, ch. 133, v. 21.
27. *Purba Megha*, 51.
28. Lit. the country having ten strongholds or citadels.
29. B.C. Law, "*Rivers of India as in Ancient Literature*", in Mountains and Rivers of India, Calcutta, 1968, p 203.
30. Hem Chandra.
31. A. Cunningham, *The Ancient Geography of India.*
32. Ibid.
33. M.R. Kale, *The Meghaduta of Kalidasa*, Motilal Banarsidas, 1969, p 86.
34. *Manusmriti*, II-17.
35. *Manu.*, S. II, 19.
36. *Mahabharata*, "Vana Parba", 83.4.
37. *Mahabharata.*
38. B.C. Law, "Rivers of India as in Ancient Literature" *Mountains and Rivers of India*, Calcutta, 1968, p 185.
39. *Siddhanta Siromani.*
40. *Mahabharata*, "Vana Parba", CXXXV 5.
41. *Harivamsa.*
42. *Katha Sarita Sagara*, III 4.5.
43. *Puranasamvasa.*
44. *Purba Megha*, p 61.
45. *Mahabharata*, Bhismaparva, ch. 6 of Kalidasa's *Abhigyan Shakuntalam*, p 237.
46. *Raghuvamsham*, XIII, 60.
47. *Jambudiva Pannatti.*
48. *Bhagavata*, II. 9.
49. *Raghu*, IV.
50. A. Cunningham, *The Ancient Geography of India*, Bharatiya Publishing House, 1975, p 430.
51. *Raghu*, 6.54.
52. *Markandaya Purana*, p 305 Notes.
53. *Harsh Charita*, ch. VII.
54. *The Ancient Geography of India*, p 435.

55. *Raghu*, IV-46.
56. *Raghu*, IV-50.
57. B.C. Law, "Rivers of India as in Ancient Literature," *Mountains and Rivers of India*, p 210.
58. Ibid.
59. S.M. Ali, *The Geography of the Puranas*, p 180.
60. S.R. Chaturvedi, *Kalidasa Granthavali.*
61. *Saktisangam Tantra.*
62. S.M. Ali, *The Geography of the Puranas*, p 151.
63. *Ramayana*, "Ayodhya Kanda", LIV, 39, LVI 7-8.
64. B.C. Law, *Mountains and Rivers of India*, 1968, p 11.
65. S.M. Ali, *The Geography of the Puranas*, p 101.
66. *Kalika Puranania*, 41 canto.
67. (a) *Raghu*, 3/8.
    (b) *Raghu*, 4/26.
68. *Raghu*, 1/5.
69. *Raghu*, 4/12.
70. (a) *Raghu*, 2.
    (b) *Raghu*, 5/36.
    (c) *Raghu*, 7/16.
71. *Raghu*, 6/21.
72. *Raghu*.
73. *Raghu*, 13.
74. *Kumara*, I.I.
75. *Kumara*, I.
76. *Mountains and Rivers of India*, p 6.
77. The *chamara* or *chauri* is made of the bushy tail of the *chamari mriga* or *chamari gabhi* (a kind of cow) or of a bundle of peacock's feathers, generally set in a handle of silver and waved before kings or the images of gods to keep off flies and other insects or to serve as a fan.

# 3

# Changing Methods and Techniques in Geography

—*V.N.P. Sinha*

GEOGRAPHY as an academic subject has experienced revolutionary changes since its initiation and geographers have become more or less accustomed to the idea of progress. More particularly, methodology and techniques have been the subject of profound change and have greatly altered the subject matter of geography but the changes that have taken place so far concerning teaching and research methodology have not yet reached the senile stage. The subject is gaining more and more new ideas and traditional methods of geographical study have correspondingly been subordinated.

Earlies, geography was considered like a traditional gezetteer with descriptions of geology, landforms, climate, soil, natural vegetation, agriculture and associated human activities. More precisely, it was considered to be the writing about the earth and the subject was quite helpful in making an attempt to measure things and to place them on maps especially during the great age of exploration. Now geographers are more concerned with the understanding provided by the systematic sciences and the emphasis in geographic study is not only on the theoretical operation of a process in isolation but also on the areal variation development on the face of the earth as a result of the operation of the processes of change. Consequently, profound changes have occurred in the means of multiplying knowledge and in the methods of diffusing it. Almost all branches of geography have developed to distinguish themselves and new associations have developed to help in making attempts to measure things and to place them on maps especially during the great age of exploration. Now geographers are more concerned with the understanding provided by the systematic sciences and the emphasis in geographic study is not only on the theoretic operation of a process in isolation but also on the areal variation developed on the face of the earth as a result of the operation of the process of change. Consequently, profound changes have occurred in the means of multiplying knowledge and in the methods of diffusing it. Almost all branches of geography have developed to distinguish themselves, and new associations have developed with sister sciences like statistics, sociology, hydrology, architecture, planning, geology, etc. The subject as such has started accumulating its literature from many sources. Modern geographers have also deviated from the traditional gazetteer descriptions and have introduced more statistical methods in geographical analyses like regression, correlation, variance and covariance, graphic space models, simulation, probability and chance (Amoah, 1975, p 56).[1]

Recent developments at research level in geography and to some extent in the structure of university courses have enabled geographers to go into more effective use of statistical materials and methods. The introduction of statistical methods has made the analyses more scientific and logical. As such the application of quantitative methods in geographical analyses has become more intense during the last two decades and has overthrown the historically imposed traditional methods from earlier geographers. Now geographers are taking much interest in the formalisation of ideas into theories and models. The radical transformation of the spirit and language of geography due to mathematisation has enabled geographers to become programmers in their own right in computer mapping. Although the computer is an enormous aid to the development of geography, computerisation of material for its own sake emerges as an occupational hazard. Advanced level mathematics or statistics are a highly desirable qualification for a potential geographer.

The main purpose of the paper is to describe precisely the changes that have taken place in geographical teaching and research methods. Geography has widened its scope and spread its branches and has linked its relationship with its sister sciences but this paper analyses only a few important aspects. It has been aimed to clarify the changes in methodological problems although methodology does not require any special tool, skill or technique of field of study. Sometimes some people confuse methodology with different techniques of data collection and map-making, audio-visual aids and methods of teaching ; methodology is used in the sense of the logic of justification in the context of validation (Harvey, 1969, p 6).[2]

The literature on the methodology of geography in the English language is relatively small compared with that on substantive work. Wooldridge and East (1952),[3] James and Jones (eds, 1954),[4] Taylor (ed, 1957),[5] Freeman (1962),[6] and Brock (1965),[7] have provided the historical background of the developments in the subject and its various branches. For more than thirty years, however the best known methodological treatment of the subject has been Richard Hartshrone's *Nature of Geography* (1939).[8]

More recently up-to-date material have been obtained from Harvey's recent treatise on *Explanation in Geography* (1969),[9] and from Davies' *The Conceptual Revolution in Geography* (1971).[10] Hartshorne has also revised some of his conceptions of geography as evidenced by his *Perspective on the Nature of Geography* (1959) ; his methodology is seen by many among the rising younger generation of geographers as being partly responsible for perpetuating historicism and anti-scientific attitudes in modern geographical learning.

Almost all geographers agree that regional geography is an essential part of the discipline ; some geographers utilize the word 'region' as a device for selecting and studying areal groupings of the complex phenomena found on the earth (Whittlesey, 1954, p 21)[11] because no two places on the earth are exactly alike. The establishment of regions provides a method by which generalisations regarding the characteristics and relationships of areas can be formulated. Regions are considered as areas of any size that are homogeneous with respect to specific criteria and the size of the regions depends upon the amount of generalisation that is desired.

Until a few year ago, the great majority of the regional studies of the world were based upon climatic or vegetation region and it was only later that geographers started taking an interest in cultural regions. Russell and Kniffen (1951)[12], recognised seven large culture worlds having distinctive traits and similar ways of living in their environment like the European, the Oriental, the African, the American, the Pacific, the Dry and the Polar.

The focus of interest in historical geography has shifted a great deal from time to time across a broad range of inquiries. Historical geography includes gazetteers of the ancient world, reconstructions of past geographies, studying in sequent occupance, chronicles of geographical change through time, retrogresssive narratives, investigations of relict features, critical appreciations of the perceptions of the past and essays in theory (Darby, 1953).[13] Now most historians have led the way in describing geographies of the past in order to set historical actions in appropriate geographical context.

The old dictum of Herodotus that history must be presented geographically, and geography must be presented historically is being given consideration in modern geographical studies. Now geographers interested in historical geography are paying more attention to re-creation of past geographies and the observation of geographical changes through time (James, 1959, p 7).[14] Consequently, some people even think that history is controlled by geography and man lives in a strait-jacket of environment. Certainly a historical approach to geography enables us to see the human value of the subject better and helps us to see the development of human responses and the changing significance of the resources that are available to man.

During recent years the study of historical geography has been focusing on the speculative side as future changes in the patterns of population and settlement, the arrangement of cities and towns, of industrial and commercial establishments and of areas of specialised economic production because of the present landscape provide many keys to the past. The reconstruction of the past from the present, the retrogressive method of Maitland (1897)[15] proceeds from the better known to the less well-known, from the visible real world to the obscurity of the Dark Ages. All features in the present landscape are relict features, survivals from some past period.

Constant reference to past events is necessary to understand how they come to occupy their present positions, but not all past events are equally important. The depth to which the past is fathomed depends on the level of causation being sought.

Most geographers also agree that geography is by tradition a man-centered science. In Germany, Friedrich ; Ratzel in the United States, Ellen C. Semple ; in France, Paul Vidal de la Blache and Jean Brunhes each in his or her own way re-emphasised the geography of man. Most of the research of American geographers is in various specialised branches of human geography, such as economic, social, cultural and political geography or even more restricted subdivisions concerning, for instance, recreation, manufacturing and settlements (Brock, 1959, p 34).[16] Economic geography is another branch of human geography which monopolises production, trade and consumption and manufacturing. There are two valid approaches to the study of economic geography—the macroscopic and the microscopic—one is macro-economic geography which has developed mainly to establish and coordinate relationships among observed phenomena, while micro-economic geography has developed relating to site characteristics and manufacturing. Cultural geography concentrates almost exclusively on the material man-made features which give character to an area, custom, religion, etc. Political geography includes the analysis of traditions and social institutions which markedly divide the world into nation states.

Among the recent trends in human geography is the application of suitable quantitative methods and models to make the analysis scientific. Such trends are indicative of the growing use of scientific methods of inquiry in the investigations of geographical problems. The new approach in reasoning and inductive inquiry, allows the truth of our ideas about certain aspects of the world to be tested rigorously in the light of factual evidence (Campbell and Wood, 1969, p 81).[17]

Until quite recently the place of population studies in geography was uncertain. By the year 1950 there was indication of a growing branch of geography distinct enough to have its own body of methodological and substantive literature (Trewartha, 1953).[18] Since then the ground has solidified with a general primer of programmatic statement (Clarke, 1966).[19] The objective of population geography has become to define and to bring forth the significance of differences from place to place in the number and kind of human inhabitants.

The more important ideas and methods used by American geographers in studies of population geography were imported from Europe. For the past three decades the methods used in the mapping of population in America have been strongly influenced by European scholars like Stendegeer whose discussions of the procedures and categories he used in mapping the population of Sweden (1919) and appeared in the *Geographical Review* in 1922.[20] He made use of dots for rural people and globes of proportional volumes for urban concentrations. Aurousseau[21] published several papers on population geography, emphasised the need for more attention to this field and suggested some of the basic concepts and methods.

The geographical study of rural settlements also began with Ritter's work in the early nineteenth century. Since then both the content and methodology of this study have been developed principally within a west European, particularly German and French context (Baker, 1969, p 123).[22] Geographers have studied rural settlements at a number of levels. At the lowest level, they have been interested in the characteristics and qualities of fences and fields. At another level they have been concerned with the functional inter-relations of these structures that give rise to contrasting and changing patterns of rural settlements (Gleave, 1962).[23]

Most of the earliest works in settlement geography was carried out by Germans on two main subjects—house types and urban centres. Then Jordan (1966)[24] claimed that the concern of settlement geography should be the study of the form of the cultural landscape, involving its orderly description and attempted explanation. Afterwards a number of geographers stressed the need of the purely locational relationships of rural settlements and spatial distribution of human activity which reflect an ordered adjustment to the factors of distance, that locational decisions are generally taken so as to minimise movements. At the same time locations are endowed with a greater degree of accessibility than others (Garner, 1967).[25] Distance factors in rural settlement patterns have been discussed by

Chisholm (1962, p 126)[26] who suggested that this was a very fruitful method of classifying settlement patterns between farmstead and field, which would provide material invaluable for planning and development and be significant in describing the relationship between man and his physical environment probably be more useful than is vouchsafed by the currently popular method of measuring the degree of dispersion and nucleation.

Urban geography is not only a very rapidly growing field of study, it is also a remarkably diverse one. The interests of urban geographers vary from time to time almost with the fashion. In spite of its popularity, urban geography does not possess a definite methodology. There has been a long-standing preoccupation among geographers with the morphology of cities or the townscape.

The analysis and interpretation of cities by both European and American geographers are largely a product of the twentieth century although methods of studying the natural pattern or layout of cities, and the treatment of cities as part of the landscape were developed in 1899 by Schluter. Urban geographers are more concerned with the examination of the forms and patterns of settlement as of today; they trace back the evolution of the phenomena of settlement to their origin, and forecast the changes to come. Furthermore, they study cities for the purpose of formulating basic concepts of city growth, city location or city character or they may study cities in order to contribute to the solution of practical problems of urban planning.

From the analysis of the location of individual cities and of city patterns in larger areas have emerged some generalisations, empirical formulae, and deductive theories on the distribution and size of cities in general. Notable among these is the work of Christaller's hexagonal arrangement of central places.[27] Several other theories relating to the spacing and size of settlements have been formulated.

Modern studies of urban landuse have become more focused on specific problems, particularly those found around the fringes of urban development. The delimitation of central business districts has frequently used landuse information as the most important criterion. In recent years geographers have become more concerned with the processes that lie behind the formation and characteristics of urban landuse regions than with the detailed delimitation of their boundaries.

Several generalised concepts of city structure have been developed in recent decades. The three principal generalisations are (i) the concentric circle or zonal theory ; (ii) the wedge of sector theory, and (iii) the multiple nuclei theory. The concentric circle theory of urban growth and structure was developed largely by Burgess (1925).[28]

During recent years the study of transport geography has received much attention especially in the field of transport network. The application of the graph theory by Garrison (1960)[29] to study the transport network in order to measure geometrically the manner in which places in a network are connected by transport links. It has long been recognised that economic development is intimately related to the growth of efficient transport systems.

Until recently the study of industrial geography has been more concerned with description than with explanation. People have tried to concentrate on the relationship between industrial activities and the physical environment rather than on economic and social phenomena. The need for a theoretical basis for the study of industrial location has only recently been recognised. Location theories were developed to balance the supply of production factors and the demand for products with an established areal pattern of production.

The pioneer work of Weber (1909)[30] was concerned with the principles by which costs could be minimised while locating industry. The optimum location is defined as the point at which transport costs per unit of production are least, or where the benefits of agglomeration and of labour availability are sufficient to effect the disadvantages caused by net locating where the transport costs are least. Lösch (1940)[31] attempted to incorporate demand into location theory by considering the size of the market area. In his opinion optimum location is that which commands the largest market area and this brings the highest possible revenue. Another pioneer attempt at testing theory was that of Harris (1954).[32] He aimed to incorporate both population potential, as measure of market size, and transport cost minimisation into the analysis of location patterns of industries.

Much methodological research has been directed towards the problem of measurement, especially of locational shifts of industrial growth and its secondary impacts. The localisation of industrial index, first developed by Sargent Florence (1948)[33] is a notable study of the relationships between size of plant, investment and location has been a favourite measure. He developed a "locational quotient"—a device for comparing a region's percentage share of a particular activity with its percentage shares of some basic aggregate such as employment or income—as a measure of regional industrial concentration.

Changing methodology and techniques have greatly altered the study of geomorphology. Some geomorphologists have been moving from cyclic to non-cyclic interpretation of land-forms. Changes include the introduction of precision, attention to detail, experimentation, the use of quantitative methods and the pruning of some inaccessible old material and ideas. American geomorphologists have developed advanced concepts regarding the nature and origin of landforms. William Morris Davis[34] suggested the cyclic theory of landforms. New concepts regarding the cycle of erosion have been developed by the South African geomorphologists namely L.C. King[35] and T.J.D. Fair.[36] Both these scholars have published a number of articles concerning the development of slopes which suggest a sequence of stages quite different from the familiar youth, maturity, and old age described by Davis.

Familiarization with the latest developments in climatology has added a wide variety of new methodology but much of the research has been concerned with various aspects of cloud physics, wind circulation, etc. Fronts are now regarded as secondary consequences rather than cause of circulation patterns not necessarily created by juxtaposition of unlike air masses. Air masses themselves have been presented in a new light allowing a greater flexibility in description which discards the rigid source region criterion of definition and places more stress on synoptic circulation patterns (Sutcliffe, 1963, 113-18).[37]

A great deal of attention has been paid in recent years to the nature of the mid-latitude tropospheric wave patterns and their relationship to the surface synoptic system (Hare, 1960,

345-67).[38] Tremendous strides in the understanding of the earth's atmospheric circulation have been taken in the last few decades.

During recent years again hydrology has become an important branch of geography and has received universal recognition. Greater advances have been made in the last four decades than during the whole of the subject's previous history. The study of surface water and ground water specially have received the greatest attention. A more conventional approach is to regard hydrology as that discipline which deals with the potable water resources of the earth, their occurrence, circulation, and distribution, their physical and chemical properties, and their interactions with the physical and biological environment, including their responses to human activity (Jones, 1969, p 58).[39] Early biogeographers were mainly dealing with purely descriptive inventories of plant communities. Clements (1916)[40] introduced the concept of plant succession in which plant communities are recognised as developing in complexity over a period of time until an ultimate or climax vegetation is achieved. The climax vegetation is viewed as a community in equilibrium with its environment, and is hence self perpetuating. Tansley (1935) suggested a polyclimax theory in which climax communities were defined as those in equilibrium with the entirety of their particular environmental conditions.

Applied geography has also attracted the attention of geographers and occupies its fair share in teaching and research. Geography until recently was still a largely descriptive subject with a very modest technical basis and the possibility of applying the type of material which it produced was limited. In 1930, Stamp[41] began the land utilization survey in co-operation with many professional and amateur geographers. The second world was presented problems of many kinds which geographers were able to tackle especially naval intelligence, terrain study, air photograph analysis and cartographic work.

Geographers today are taking much interest in the study of town and country planning. Most branches of geography provide a lot of ideas useful in physical planning. Economic geography particularly in its resources agricultural and industrial branches has been one of the major suppliers of concepts. Urban geography, transport geography, social geography, population geography are no less significant. Central place studies deal with the location, size, nature and spacing of production and market nodes. Geographers are well reputed for technical competence in the handling of data and mapping and ground survey. Landuse surveys provide a lot of information for physical planning. Most geographical works have now become problem oriented. Since human geographers have a deep interest in the problems of cities, towns, and villages and are much concerned with population distribution, location and functioning of economic phenomena, these surveys are extremely useful for the physical planner.

## SUMMARY

Geography has experienced revolutionary changes since its initiation into the development of geographical knowledge particularly in methodology and research techniques. It has deviated far from the traditional gazetteer type description of particular region to a scientific analysis of its various aspects. The subject has become wide spread and has become enriched with literatures from its sister sciences. Now almost all branches of geography have formulated a great many theories and models and have become problem oriented. The maximum use of statistical methods in almost all branches of geography has made the subject more useful and enhanced its status as applied science.

## REFERENCES

1. F.E.K. Amoah, "Recent Trends in the Methodology of Geography", *The Oguaa Educator*, vol. 5, No. 2, 1975, pp 56-67 Ref. p 56.
2. D. Harvey, *Explanation in Geography*, Butler and Janner Ltd. London, 1969, p 6.
3. S.W. Wooldridge, and W.G. East, *The Spirit and Purpose of Geography*, Hutchinson University Press, London, 1952.
4. P.E. James, and C.F. Jones, (Eds.), *American Geography : Inventory and Prospect*, Syracuse University Press, Syracuse, 1954.
5. G. Taylor, (ed)., *Geography in the Twentieth Century*, Methuen, London, 1957.
6. T.W. Freeman, *A Hundred Years of Geography*, Methuen, London, 1962.

7. J.O.M. Brock. *Geography—Its Scope and Spirit*, C. Merrill Co , 1965.
8. R. Hartshorne, *The Nature of Geography*, Lancaster, P, 1939.
9. D. Harvey, *op. cit.*, p 6.
10. W.K.D. Davies, *The Conceptual Revolution in Geography*, Haizell Watson and Viney, 1971.
11. D. Whittlesey, "The Regional Concept and the Regional Method" in *American Geography : Inventory and Prospects* edited by P.E. Janes and C.F. Jones, Syracuse University Press, New York, p 10.
12. R.J. Russell, and F.B. Kniffen, *Culture Worlds*, The Macmillan Company, New York, 1951, p 620.
13. H.C. Darby, "On the Relations of Geography and History", *Transactions of the Institute of British Geographers*, 1953, pp 1-11.
14. P.E. James, "The Hard Core of Geography" in *New Viewpoints in Geography*, National Council for the Social Studies, Washington, pp 1-9, Ref. p 7.
15. F.W. Maitland, *Domesday Book and Beyond*, Cambridge University Press, Cambridge, 1897.
16. J.O.M. Brock, "Progress in Human Geography", in *New Viewpoints in Geography* by P.E. James (ed.) National Council for the Social Studies, Washington, 1959, pp 34-53, Ref. p 34.
17. W.J. Campbell and P.A. Wood, "Quantification and the development of theory in Human Geography", *Trends in Geography* by (eds.) R.U. Cooke and J.H. Johnson, Pergamon Press, London, 1969, pp 81-89, Ref. p 81.
18. G.T. Trewartha, *A Case for Population Geography*," A.A.A.G., 1953, vol. 43, pp 71-97.
19. J.I. Clarke, *Population Geography*, Pergamon, 1966, Oxford.
20. D.G. Sten, "A Map of the Distribution of Population in Sweden : Method of Preparation and General Results", *Geographical Review* 1922, vol. 12, pp 72-83.
21. M. Aurousseau, "The Distribution of Population : A Constructive Problem" *Geographical Review*, 1921, vol. II, pp 563-592.
22. A.R.H. Baker, "The Geography of Rural Settlements", in *Trends in Geography : An Introductory Survey* by R.U. Corke and J. H. Johnson, Pergamon Press, London, pp 123-132, Ref. p 123.
23. M.B. Gleave, "Dispersed and Nucleated Settlements in the Yorkshire Welds 1770-1850", *Transactions of the Institute of British Geographers*, 1962, 30, pp 105-118.
24. T.G. Jordon, "On the Nature of Settlements Geography" *Profcssional Geographer*, 1966, vol. 18, pp 26-28, Ref. p 27.
25. B. Garner, "Models in Urban Geography Settlement Location" in *Models in Geography* (eds) R.J. Chorley, and P. Haggett, Methuen, London, 1967, pp 303-360.
26. M. Chisholm, *Rural Settlement and Landuse*, Hutchinson, 1962, London, p. 126.
27. W. Christaller, *Die Zentralen Orte, in Suddeutschlan da*, Jena, 1935.

28. E.W. Burgess, "The Growth of the City" in *The City*, R.E. Park, E.W. Burgess, and R. D. McKenzie, Chicago, 1925, pp 47-62.
29. W.L. Garrison, "Connectivity of the Inter-state Highway Systems", *Papers and Proceedings*, Regional Science Association, 1960, No. 6, pp 121-37.
30. A. Weber, Theory of the Location of Industries, University of Chicago, Press, Chicago, 1909.
31. A. Lösch, *The Economics of Location*, Yale University Press, 1940, New Haven.
32. C.D. Harris, "The Market as a Factor in the Location of Industry in the U.S., *AAAG*, 44, 1954, pp 315-48.
33. F.P. Sargent, *Investment, Location and Size of Plant*, Cambridge University Press. Cambridge, 1948.
34. W.M. Davis. *Geographic Essays*, Ginn & Co., Boston, 1909.
35. L.C. King, "Canons of Landscape Evolution", *Bulletin of the Geological Society of America-64*, 1953, pp. 721-52.
36. T.J.D. Fair, "Slope from and Development in the Coastal Hinterland of Natal", *Transaction of the Geological Society of South Africa*, 51; 3347, 1948.
37. R.C. Sutcliffe, *Weather and Climate*, Weidenfeld and Nicolson, 1966, London, pp 113-18.
38. F.K. Hare, "The Westerlies", *Geographical Review*, 1960, vol. 50, pp 345-367.
39. G.P. Jones, "Progress in Hydrology" in *Trends in Geography : An Introductory Survey*, by R.U. Cooke and J.H. Johnson (eds), Pergamon Press, London, pp 58-69.
40. F.E. Clements, *Plant Succession : An Analysis of the Development of Vegetation*, Publication of the Carnegie Institute of Washington, No. 242.
41. L.D. Stamp, *The Land of Britain—Its Use and Misuse*, Longmans, 1948, London.

# 4

# Geography and the Sense of Place

—*Gary L. Peters*

THE PURPOSE of this essay is twofold : to review selected examples of the study of "sense of place" and to suggest possible directions for future research. According to Tuan, "Space and place together define the nature of geography."[1] Though spatial analysis has received most of the attention during the last two decades, the study of place continues to occupy the time of a significant coterie of researchers.

The essence of cultural geography was suggested by Wagner in the following statements :

> Every tribe or nation creates its own geography. Wherever human beings live there are different evaluations, embellishments, enhancements and exploitations of environments. Every people knows a proper kind of site for its houses, size for its settlements, shape for its fields, surface for its roads, sanctuary for rest and reverence. The form thus given to environment bespeaks not only economic ends but the ethical traditions, aesthetic values and ecological concepts of its people.[2]

And, we might add, the geographic study of these elements occurs within the context of place. On the one hand, place is simply a geographic location. On the other hand, however, place connotes much more than just location. As Prince noted :

> We cannot know a place until we discover its literature, its arts and its sciences; nor, conversely, can we understand literature or art or science without some knowledge of geography. A knowledge of places is an indispensable link in the chain of knowledge.[3]

Furthermore, Prince suggested, "Both regions and writers, place and person, are unique, and it is in their distinctive qualities that we may find their essential character."[4] It is this "essential character" of places that is the central theme of "sense of place" studies.

According to Crary, "Regions, like people, have personalities."[5] The personalities of places are likely to be complex and multi-faceted and vary according to the observer. Briggs noted that, "The same environment is not necessarily experienced in the same way by different individuals or, extending the argument, by different groups."[6] Thus, for example, there is not one San Francisco, but many. The San Francisco of the Beatniks was not the city experienced by tourists on the Grayline. San Francisco today is not the same city to both those in the Tenderloin and those on Twin Peaks. As Hurst commented :

> Every place has a special character for its residents, its visitors, and those who 'know' it only second-hand. Each place has a personality of its own, buried perhaps beneath prejudice and emotion, derived not just from buildings, but also from contours, street patterns, drama, colour, surprises, smells, noises, and so on.[7]

The differentiation of place in the purely locational sense from place as personality was clearly described by Langer as follows :

> As *scene* is the basic abstraction of pictorial art, and *kinetic volume* of sculpture, that of architecture is an *ethnic domain*. Actually, of course, a domain is not a "thing" among other "things"; it is the sphere of influence of a function, or functions; it may have physical effects on some geographical locality or it may not. Nomadic cultures, or cultural phenomena like the seafaring life, do not inscribe themselves on any fixed place on earth. Yet a ship, constantly changing its location, is nonetheless a self-contained place, and so is a Gypsy camp, an Indian camp, or a circus camp, however often it shifts its geodetic bearings. Literally, we say the camp is in a place, culturally, it is a place. A Gypsy camp is a different place from an Indian camp, though it may be geographically where

> the Indian camp used to be. A place, in this non-geographical sense, is a created thing, an ethnic domain made visible, tangible, sensible. As such it is, of course, an illusion. Like any other plastic symbol, it is primarily an illusion of self-contained, self-sufficient, perceptual space. But the principle of organization is its own : for it is organized as a functional realm made visible—the centre of a virtual world, the "ethnic domain," and itself a geographical semblance.[8]

The sense of place, thus, derives not only from the characteristics of place, but also from the ways that places are perceived by various individuals and groups.

The importance of perception in the social sciences has emerged in recent years.[9] In geography the study of perception has ranged from mental maps to environmental hazards, from landscapes to place.[10] In an interesting discussion of various ways to view landscapes Meinig reminded us that "even though we gather together and look in the same direction at the same instant we will not—we cannot—see the same landscape".[11]

Perception is important in studying sense of place. In turn perception is influenced by experience. As Tuan recently noted, "Place is a centre of meaning constructed by experience".[12] Our experience of places ranges, according to Tuan, from the highly theoretical to the visceral, though for most people places are perceived to be somewhere in between.[13]

The study of place and the sense of place require a different approach from that used in the study of space, or spatial analysis. As Relph said, "Place and sense of place do not lend themselves to scientific analysis for they are inextricably bound up with all the hopes, frustrations, and confusions of life, and possibly because of this social sciences have avoided these topics".[14] Relph's book *Place and Placelessness*, is the most complete study of the sense of place available. In the book the following four themes are pursued : (1) relationships between space and place, (2) components and intensities of place experience, (3) the nature of the identity of places, and of people within places, and (4) the ways in which attachments to place, and the sense of place, are involved in the making of places. His approach to places is phenomenological rather than scientific. According to Relph :

> Whether place is understood and experienced as landscape in the direct and obvious sense that visual features provide tangible evidence of some concentration of human activities, or in a more subtle sense as reflecting human values and intentions, appearance is an important feature of all places. But it is hardly possible to understand all place experiences as landscape experiences.[15]

Relph identified and discussed several types of spaces including the following : pragmatic space, perceptual space, existential space, architectural space and cognitive space. Place is discussed with respect of location, landscape, time, community and various aspects of its relationship to humans. He suggested that "To be attached to places and have profound ties with them is an important human need".[16] The need for an attachment to place is important to most people, perhaps more important than we sometimes realize. As Brower noted, "The places that we have roots in, and the flavour of their light and sound and feel when things are right in those places, are the wellsprings of our serenity".[17] The overwhelming success of Alex Haley's *Roots* reflected, at least in part, people's needs for ties to other times and places.[18] Rootlessness has been suggested as a problem of importance and concern. Packard stated that :

> While the footlooseness of Americans as pioneers was a source of vitality and charm, several of the new forms that the accelerating rootlessness of Americans is taking should be a cause for alarm. Great numbers of inhabitants feel unconnected to either people and places and throughout much of the nation there is a breakdown in community living. In fact there is a general shattering of small-group life. A number of forces are promoting social fragmentation.[19]

It is difficult, of course, to measure the contribution that rootlessness makes to social ills because we know so little about its importance.

Relph went on to a detailed discussion of the identity of

places, mainly through a dichotomy of insideness *versus* outsideness. According to him, "Identity of place is as much a function of inter-subjective intentions and experiences as of the appearances of buildings and scenery, and it refers not only to the distinctiveness of individual places but also to the sameness between different places."[20]

Relph also focused attention on placelessness, the diminishing differences between places, and the effect of other-directed architecture. He was critical of both modern business and the modern consumer with respect to their contributions to placelessness and described processes such as "disneyfication" and "museumization". There is no doubt that industrialization has carried with it a tendency to homogenize societies. However, there are counter-tendencies as well, as suggested in Nairn's statement as follows :

> But surely all America is the same ? Well, yes and no. On one level of course it is—the same paperbacks in the drugstores, the same kind of food in the Deep-Freeze, the same cars on the road. Yet in fact, this is largely a surface gloss of sameness, under which people and places are as different as ever. . . . . .[21]

Nairn's comment seems more hopeful for geographers somehow than Relph's statement that "The trend is towards an environment of few significant places—towards a placeless geography, a flatscape, a meaningless pattern of buildings".[22] At the very least there is a considerable amount of work left for geographers to do if we are to understand the patterns and processes involved. The intrusion of industry into the landscape, and its creation of placelessness, is discussed in a subsequent section. Modern landscapes, regardless of their appearance, must be viewed as desirable to someone, whether for profit or convenience. As Clay pointed out :

> The strip is trying to tell us something about ourselves : namely, that most Americans prefer convenience; are determined to simplify as much of the mechanical, service, and distribution side of life as possible; and are willing to patronize and subsidize any informal, geographic behaviour setting that helps.[23]

Too often ugliness persists in industrial and commercial landscapes despite the best efforts of critics to point to the detrimental effects of ugliness on a human existence. Unfortunately, as Lewis commented, "We are so accustomed to environmental ugliness that we are surprised and often indignant when the fact is called to our attention".[24] Both beauty and ugliness are part of, and contribute to, the sense of place, of course, though geographers have too often remained aloof from the need for aesthetic judgments concerning landscapes.[25]

## TIME AND THE SENSE OF PLACE

We are all aware that some places persist for very long periods of time while others are ephemeral.[26] Monuments, for example, are sometimes maintained well after their original significance has been lost or forgotten. On the other hand, especially in American cities, buildings and other landmarks quickly give way to the bulldozer artists and are replaced by new buildings—some of which turn out now to be kitschy replicas of old buildings. Too often today people return to the places of their childhood only to find that the places are gone. The loss of places in our past becomes a loss of a part of our personal past. Such experiences suggest an admixture of romantic fantasy and reality; nostalgia surrounds us. As Lowenthal commented, "Nowadays the terrible future gains on us; we look back to old familiar landscapes in the fear that the comforts of the past may be vanishing before our eyes".[27] And so they are, especially in urban areas, though we would deceive ourselves if we sought to replace all that is with all that was before. However, we need some constants in an otherwise everchanging landscape, as noted in the following comment by Lowenthal :

> We need the past, in any case, to cope with present landscapes. We selectively perceive what we are accustomed to seeing; features and patterns in the landscape make sense to us because we share a history with them.[28]

We previously mentioned the importance of experience in determining the sense of place. Experience in turn is a function of time. An understanding of a place requires more than just a casual visit. According to Briggs, "A basic distinction must be drawn, of course, between impressions of the identity of places as set out in travellers' tales and sense of identity felt by those who live in an environment and experience it directly and continuously".[29] The strong attachment to place that many people develop results from an intimate knowledge of, and identification with, place that can only come about as a result of long-time residence.

Shorter visits to places do provide some sense of what they are, however. Especially valuable for geographers are the accounts of some literary figures. Examples abound, but particularly impressive for their descriptions of places are Durrell,[30] Miller,[31] and Kazantzakis.[32] Good writers, who are also generally accurate and sensitive observers, offer us considerable insight into the nature of places and their inhabitants. In a provocative essay on literature and landscape Salter and Lloyd argued that :

> The search for meaning and order in the landscape—that is, the desire to see landscape more clearly and completely—is a primary concern of geography. This search leads to landscape description that looks beyond the more obvious forms and functions into the deeper human implications of the world around us. When we apply creative writing to support our geographic vision, we have gained a powerful ally toward our goal of communicating a sensitive, articulate image of the phenomenon we call landscape.[33]

On the one hand, time is necessary for the experience of sense of place, but on the other hand time alters places as we experience them. Ritual and tradition often have the effect of deepening the ties that people have to places. According to Relph, "When the rituals and myths lose their significance and the people cease to participate fully in them the places themselves become changeable and ephemeral".[34]

### INDUSTRIAL INTRUSION AND THE SENSE OF PLACE

Geographers are well aware of the importance of the Industrial Revolution and its impact on the socio-economic fabric of societies, as well as the evolution of that impact in both time and space. However, the impact of industrialization on the landscape has been inadequately explored. Modern manufacturing and big business are major contributors to a modern sense of placelessness.

In discussing the impact of modern industry on the creation of placelessness, Relph noted that :

> Steel mills, oil refineries, light engineering works, quarries, waste disposal sites, all have an appearance that is quite independent of location. Furthermore the sheer scale of modern mining, manufacturing, and business enterprises tends to obliterate places, whether through flooding by dam construction, digging them up for minerals, burying them beneath slag heaps, or simply building over them.[35]

The impact of industrialization on the landscape has been noted by people in many different fields of study. For example, Giedion stated that :

> The Industrial Revolution, the abrupt increase in production brought about during the eighteenth century by the introduction of the factory system and the machine, changed the whole appearance of the world. . . .[36]

Other examples include Hawkes,[37] Sussman,[38] and Peters and Anderson.[39] The image of the machine as an intruder in the landscape has been discussed at length by Marx.[40]

Aside from the role of industrialization as a homogenizing agent, an inevitable outcome of much industrial development is ugliness. Blighted landscapes are not only aesthetically abominable but also, as Lewis noted, "The highest costs of ugliness, however, show up in the palpable but unmeasured demoralization of people who live in ugly places and have given up hope of improving them".[41] From inner cities to small towns in mined-out rural areas ugliness, hopelessness and despair seem to be silent partners.

It could probably have been different. It still could be different if we were willing to expand the necessary time and energy and to make the essential commitment to human landscapes. As Abbey commented :

> Not that technology and industrialism are evil in themselves. The problem is to get them down to human scale, to keep them under human control, to prevent them from ever again becoming the self-perpetuating, ever-expanding monsters we have allowed them to become. What we need is an optimum industrialism, neither too much nor too little, a truly sophisticated, unobtrusive, below-ground technology.[42]

We cannot do away with industry and its landscapes, but we can certainly humanize them. We can better use information about them in making locational decisions, especially where there are likely to be incompatabilities between industrial landscapes and adjacent land uses. In doing so, however, we need to reconsider our use of zoning laws. Too often these laws lead to the destruction of places and the sense of place. For example, changes in zoning, from residential to commercial, often leave a landscape of intermixed residences and commercial structures and a neighbourhood whose identity has been needlessly sacrificed. As Eckardt warned us :

> Within the residential areas zoning establishes a caste system more rigid than any that ever prevailed in India. There are high-income neighbourhoods and middle-income neighbourhoods and the ghettos for the poor. It's all devised to protect property values, not human values.[43]

Industrialization has changed the essence of many places and our perception of them. Urbanization has isolated increasing numbers of people from the land and all that an intimate knowledge of the land provides. Writers have tried many times to tell us what it is that we have lost, but few listen. For example, Steinbeck noted that :

> There is a warmth of life in the barn, and the heat and smell of life. But when the motor of a tractor stops, it is as dead as the ore it came from. The heat goes out of it like the living heat that leaves a corpse. Then the corrugated iron doors are closed and the tractor man drives home to town, perhaps twenty miles away, and he need not come back for weeks or months, for the tractor is dead. And

> this is easy and efficient. So easy that the wonder goes out of work, so efficient that the wonder goes out of land and the working of it, and with the wonder the deep understanding and the relation.[44]

Places, and the identity that they provide, are likely to remain important, whether or not we can accurately quantify their contribution to our well-being. As Rees recently commented, "A sense of place is a human imperative without which there can be no integration of the self with the environment".[45] Artists, poets and novelists continue to offer us images of ourselves and our landscapes, but somehow in the social sciences we have been led astray by an unswerving dedication to the scientific method and objectivity. However, as Roszak suggested, "Not only do novelists make better political weather vanes than our social scientists, they calculate the human costs with more precision".[46] Any consideration of the sense of place in such areas as urban and regional planning requires an acceptance of subjective feelings about places, not as a replacement for other planning considerations but as an additional element.

## PLANNING, PRESERVATION AND THE SENSE OF PLACE

Place and its influence on the way people think and act should be a consideration in the planning of cities. The failure of planning and the deterioration of urban evironments and the associated increase in placelessness have been subjects of many recent studies including those by Nairn[47] and Blake.[48]

Among geographers, Ford was one who related preservation and the sense of place.[49] He focused attention on the evolving nature of places and the need for management of change, not just the preservation of places for the sake of preservation. Ford focused on downtown and its evacuation as people and economic activities head for the suburbs. He noted that "Geographers have spent a great deal of time studying the functions of the centre city and their locations but very little time studying the character of the settings and the attitudes toward those settings which affect location decisions".[50] Ford pointed to changes taking place in several cities, including San Francisco and New Orleans ; especially warehouse renovation, such as has occurred in Ghirardelli Square, where old buildings have been refurbished and converted into shops and office spaces. After answering some of the criticisms of renovation and restoration Ford concluded that :

> The negative image of the zone of discard seems to be changing as it is realized that in that zone are some of the best reminders of a city's many eras and unique historical evolution. With the renovation of old warehouses, factories, and markets, many office occupants have begun to perceive the zone of discard as an ideal location to enjoy a human scale setting and a sense of temporal continuity and place.[51]

In a more recent article, Ford studies the changes which occurred in three cities : Bath, Chester and Norwich. He was especially concerned with variations in the management of change and the preservation of place and concluded that :

> Our cities should provide visible clues to where we have been and where we are going. Just as we should not erase our past and destroy our roots, we cannot live only in a frozen, romanticized setting of the past. We must learn to design cities that can maintain a sense of place while accepting gradual change.[52]

Not everyone agrees on the role of preservation, however. All too often inauthenticity dominates and places take on an air of unreality. We might wonder, for example, what Steinbeck would think if he could visit Cannery Row today. Relph commented that :

> The possibilities for maintaining and reviving man's sense of place do not lie in the preservation of old places—that would be museumization ; nor can they lie in a self-conscious return to the traditional ways of place-making—that would require the regaining of a lost state of innocence. Instead, placelessness must be transcended.[53]

Careful planning practices are necessary if change is to be managed in a beneficial way and the preservation of features which contribute to the genuine essence of place should be part of the mosaic.

Nearly two decades ago Lynch argued that "The urban landscape, among its many roles, is also something to be seen, to be remembered, and to delight in".[54] Fortunately Lynch has continued to maintain a holistic view of regions and has helped guide the way toward humane planning. In one recent volume Lynch pointed out that :

> Access and territory are aspects of the mental image of space, considered as potential movement and action. But the identification of places, as well as their organization into mental structures, not only allows people to function effectively but is also a source of emotional security, pleasure and understanding. Orientation in space (and time) is the framework of cognition. We have powerful abilities for recognizing places and for integrating them into mental images, but the sensory form of those places can make that effort at understanding more or less difficult . . So we take delight in physically distinctive, recognizable locales and attach our feelings and meanings to them. They make us feel at home, grounded. Place character is often recalled with affection ; its lack is a frequent subject of popular complaint. People are pleased to "know" a great city, or to understand its history. Indeed, a strong sense of place supports our sense of personal identity.[55]

Lynch's suggestion about the mental structures in which people organize places is an area where geographers have been active.

Jordan's recent study of perceptual regions in Texas is an excellent example of what geographer's have to contribute.[56] Jordan collected 3,860 questionnaires from college students in 30 different schools. Students were asked about vernacular regions relative to their home counties. These data were used to produce a map of perceptual regions in Texas. Regions of differing size and type were found. Two-thirds of the regions were identified by environmental names, though Jordan found that environmental names were becoming less popular and political names were becoming more common designators of regional identity. He suggested that :

> As geographers, we ought to know more about perceptual regions than we do, at various scales and in different

parts of the world. Spatial patterns, behaviour, organization, and flows are among our traditional concerns, and we should be able to understand these better if we know how populations perceive regions.[57]

One of the problems facing planners who wish to incorporate perceived qualities of places into the planning process is the difficulty of handling data which is not easily quantified or broken down into costs and benefits in dollar terms. As Berry pointed out :

> The ultimate decision on protection of a landscape is usually political in that there are non-aesthetic costs and benefits to consider and a number of interested parties that must be taken into account. Aesthetic values then constitute one kind of claim in a larger decision-making process. . . . To incorporate aesthetic images as completely as possible into the decision-making process, it is necessary to address these images on their own terms, i.e., in an analytical way which can encompass the variety of cognitive and valuational relationships within the aesthetic structure of the landscape.[58]

The difficulties that planners face with respect to assessing the values of different projects are undoubtedly complex. It is easy to agree with the following statement by Fairbrother :

> The problem is formidable, but unless cost-benefit analysis can progress beyond finance it can never develop its full potential as an instrument of planning (and a method ready-made for computers). Certainly in landscape the most comprehensive and enlightened analysis is needed, since both costs and benefits exist in many contexts and in terms of the whole community.[59]

More often than not the central concern of planners is with the efficient arrangement of space and places are mass-produced with little or no consideration of the ways in which people will experience them. As Relph suggested, "There is little scope in this for treating places as centres of existential significance, or even in terms of their qualities of appearance —rather such approaches require the use of quantitative techniques of analysis and manipulation based on the averages of economic man and a dispassionate and impersonal approach to both place and people. . . "[60]

## CONCLUSIONS

The central purpose of this brief essay was to provide an introduction to some small portion of the growing body of literature, especially in geography, related to the theme of the sense of place. Only the surface has been scratched, and that only lightly. However, hopefully, this will stir some to further explore.

The sense of place is largely experiential. As Aiken phrased it, "Ultimately landscape sensibility and a 'sense of place' are grounded in personal, subjective, emotional and idiosyncratic experiences".[61] As geographers, our approaches to landscape and place have been many and varied. In recent years phenomenological and humanistic approaches have predominated. But if our findings are to move from the esoteric realm of academia to the world of planners and managers we have a long way to go.

Perhaps it is still too soon to think of some larger framework within which to consider the sense of place and perceptual regions, but sooner or later it needs to be done. In spatial analysis one of the most successful frameworks for the study of places is central place theory, with its neatly formed hierarchies of nested places. Its success lies in a simple universal, the essence of which is suggested by Koestler as follows :

> If we look at any form of stable social organization, from the insect state to the Pentagon, we shall find that it is hierarchically structured ; the same applies to the individual organism, and, less obviously, to its innate and acquired skills.[62]

Koestler introduced the idea of the holon and stated that :

> We have seen that biological holons, from organisms down to organelles, are self-regulating entities which manifest both the independent properties of wholes and the dependent properties of parts. This is the first of the

> general characteristics of all types of holarchies to be retained; we may call it the *Janus principle*.[63]

Holarchic structuring is common to all societies and without it chaos would result. Human use of the landscape is structured and perhaps we could view places as holons. In other words, we could begin to integrate different findings regarding the sense of places in some hierarchic fashion. To some extent this has already been done, though it perhaps should be made an explicit guiding framework.

Each place, at least in a geographic sense, may be viewed at once as a whole and as a part of a larger whole. Thus, like Janus, place has two faces. According to Koestler, "Infinity yawns both at the top and bottom of the stratified hierarchies of existence, and the dichotomy of self-assertive wholeness and self-transcending partness is present on every level, from the trivial to the cosmic."[64] Places, as we have viewed them, are somewhere in between.

Because the sense of place is an experiential phenomenon we find that the strongest attachment to place comes at the smallest practical level of a place hierarchy, and that as the size of a region expands our knowledge of it diminishes. A logical place to begin, then, is with the home. As Relph stated, "Home in its most profound form is an attachment to a particular setting, a particular environment, in comparison with which all other associations with places have only a limited significance".[65] In one recent study Porteous commented that :

> Home provides both the individual and the small primary group known as the family with all three territorial satisfactions. These satisfactions derive from the control of physical space, and this control is secured by two major means. The personalization of space is an assertion of identity and a means of ensuring stimulation. The defense of space is the means by which stimulation is achieved and security assured. If we exclude personal space because of its mobile nature, the average citizen appears to expend more effort personalizing and defending the home than any other level of fixed physical space.[66]

Home is the focus for most of our activities and provides us with identity. It is often a refuge from the madness of the world outside. It is the centre of our patterns of circulation, from the journey to work or shop or school to longer journeys for business or pleasure. However, not everything about home is pleasing for everyone. As holon, home is at once a self-contained and independent place, in which individual action and identity are paramount, and part of a neighbourhood. With respect to neighbourhoods, homes help provide identity, but they are not completely independent. Home owners are expected, at least in typical American suburban neighbourhoods, to perform certain rituals to maintain a part of the neighbourhood's image. For example, lawns are to be mowed, broken windows replaced, and untidy places tidied up. The ideal, of course, does not always correspond with reality.

In turn neighbourhoods are collected into cities, cities and their hinterlands into counties or other regions, which in turn become parts of a nation. Nations are, of course, holons in that they at once are both independent wholes and parts of a world system of nations, connected *via* patterns of communications, trade and travel.

An unfinished task, it seems, is to begin to collect the many and varied studies of places and the ways in which they are perceived into a larger whole. If we are to have an influence on the way places develop in the future, we need a broader awareness of the ways that places affect people. It is time for social scientists to admit the evidence of poets, novelists and artists as legitimate ways to further an understanding of the human condition. Despite the material gains that industrialization has given us it is time to pause and reflect on its human costs, especially as they are reflected in the accelerated creation of placelessness. The increasing mobility of Americans, coupled with their increasing affluence, threaten to rapidly alter the geography of North America, an alteration which more and more reflects the personal preferences of those involved. As Zelinsky discovered :

> Our evidence indicates, however, clumsily and coarsely, that a hitherto uncharted, multidimensional domain of geographic phenomena can be detected hovering over the surface of the United States : the world of virtually un-

> constrained personal impulse. Although there are still many things to be learned, especially at different spatial scales, in various time periods, and using additional source materials, perhaps as glimpsed *via* other perspectives, a new land has been sighted. Expanding our zone of scientific attentiveness in matters human and geographic beyond the familiar chambers of the economic, demographic, social, and cultural into the deep corridors of personal perception and the appetite for self-creation is one way to enrich our understanding of the former.[67]

If we are to continue to live in desirable places, and to derive identity and satisfaction from them, then we must do our best to understand the essential sense of places and communicate this understanding to decision-makers at all levels. If we do not, places may well lose their significance.

## REFERENCES

1. Yi-Fu. Tuan, "Space and Place : Humanistic Perspective," *Progress in Geography*, Vol. 6, 1975, p 213.
2. Philip L. Wagner, "America Emerging," *Landscape*, Vol. 13, 1963, p 22.
3. Hugh C. Prince, "The Geographical Imagination," *Landscape*, Vol. 11, 1961, p. 22.
4. *Ibid.*, p 24.
5. Douglas Crary, "A Geographer Looks at the Landscape," *Landscape*, Vol. 9, 1959, p 23.
6. Asa Briggs, "The Sense of Place," in Smithsonian Institution; *The Fitness of Man's Environment*, Harper Colophon Books, Harper and Row, Publishers, New York, 1968, p 79.
7. M.E. Eliot Hurst, "A Sense of Place," in M.E. Eliot Hurst (ed.) ; *I Came to the City : Essays and Comments on the Urban Scene*, Houghton Mifflin Company, Boston, 1975, p 41.
8. Susanne K. Langer, *Feeling and Form*, Charles Scribner's Sons, New York, 1953, pp 94-95.
9. Kenneth Boulding, *The Image*, University of Michigan, Ann Arbor, 1961.
10. Thomas F. Saarinen, *Environmental Planning : Perception and Behavior*, Houghton Mifflin Company, Boston, 1976.
11. D.W. Meinig, "The Beholding Eye : Ten Versions of the Same Scene," *Landscape Architecture*, 1976, p 47.
12. Yi-Fu Tuan, "Place : An Experiential Perspective," *The Geographical Review*, Vol. 65, 1975, p 152.
13. *Ibid.*

14. E. Relph, *Place and Placelessness*, Pion Limited, London, 1976, Preface.
15. *Ibid.*, p 31.
16. *Ibid.*, p 38.
17. David Brower, Foreword, in Allan Gussow, *A Sense of Place*, Friends of the Earth, San Francisco, 1971, p 12.
18. Alex Haley, *Roots*, Doubleday, Garden City, New York, 1976.
19. Vance Packard, *A Nation of Strangers*, David McKay Company, Inc., New York, 1972, pp. 1-2.
20. E. Relph, *Place and Placelessness*, Pion Limited, London, 1976, p. 44.
21. Ian Nairn, *The American Landscape*, Random House, New York, 1965, p. 4.
22. E. Relph, *Place and Placelessness*, Pion Limited, London, 1976, p. 117.
23. Grady Clay, *Close-up : How to Read the American City*, Praeger Publishers, 1973, p 108.
24. Peirce F. Lewis, "The Geographer as Landscape Critic," in P.F. Lewis, D. Lowenthal and Y-F Tuan, eds. ; *Visual Blight in America*, Commission on College Geography Resource Paper No. 23, *Association of American Geographers*, Washington, D.C., 1973, p. 2.
25. David Lowenthal "Not Every Prospect Pleases," *Landscape*, Vol. 12, 1962-63, pp. 19-23.
26. Robert M. Newcomb, "The Persistence of Place," *Landscape*, Vol. 17, 1967, pp 24-26.
27. David Lowenthal, "Past Time, Present Place : Landscape and Memory," *The Geographical Review*, Vol. 65, 1975, p 1.
28. *Ibid.*, p. 5.
29. Asa Briggs, "The Sense of Place," in Smithsonian Institution; *The Fitness of Man's Environment*, Harper Colophon Books, Harper and Row, Publishers, New York, 1968, p 88.
30. Lawrence Durrell, *Spirit of Place : Letters and Essays on Travel*, Thomas Alan G. (ed.), E.P. Dutton & Co., Inc., 1971.
31. Henry Millar, *The Colossus of Maroussi*, A New Directions Paperbook, New Directions Publishing Corporation, New York, 1941.
32. Nikos Kazantzakis, *Report to Greco*, Translated from the Greek by P.A. Bien, Bantam Books, New York, 1966.
33. C.L. Salter, and W.J. Lloyd, *Landscape in Literature*, Resource Papers for College Geography, No. 76-3, Association of American Geographers, Washington, D.C., 1977, p. 2.
34. E. Relph, *Place and Placelessness*, Pion Limited, London, 1976, p. 33.
35. *Ibid.*, p 109.
36. Siegfried Giedion, *Space, Time and Architecture*, Fourth Edition, Harvard University Press, Cambridge, Mass., 1963, 1976, p 163.
37. Jacquetta Hawkes, *A Land*, Random House, New York, 1951.
38. Herbert L. Sussman, *Victorians and the Machine : The Literary Response to Technology*, Harvard University Press, Cambridge, Mass., 1968.

39. Gary L. Peters and Burton L. Anderson, "Industrial Landscapes : Past Views and Stages of Recognition", *The Professional Geographer*, Vol. 28, 1976, pp 341-348.
40. Leo Marx, *The Machine in the Garden : Technology and the Pastoral Ideal in America*, Oxford University Press, New York, 1964.
41. Peirce F. Lewis, "Aesthetic Pollution : When Cleanliness is not Enough," *Public Management*, 1970, p 10.
42. Edward Abbey, *Desert Solitaire : A Season in the Wilderness*, Ballantine Books, New York, 1971, pp. 46-47.
43. Wolf Von. Eckardt, *A Place to Live*, A Seymour Lawrence Book, Delacorte Press, New York, 1967, p 15.
44. John Steinbeck, *The Grapes of Wrath*, The Viking Press, New York, 1939, p. 157.
45. Ronald Rees, "Images of the Prairie," *The Canadian Geographer*, Vol. 20, 1976, pp. 276-277.
46. Theodore Roszak, *The Making of a Counter Culture*, Doubleday and Company, Inc., Garden City, New York, 1968, p. 182.
47. Ian Nairn, *The American Landscape*, Random House, New York, 1956.
48. Peter Blake, *God's Own Junkyard : The Planned Deterioration of America's Landscape*, Holt, Rinehart and Winston, New York, 1964.
49. Larry R. Ford, "Historic Preservation and the Sense of Place," *Growth and Change*, Vol. 5, 1974, pp 33-37.
50. *Ibid.*, p 34.
51. *Ibid.*, p 36.
52. Larry R. Ford, "Continuity and Change in Historic Cities: Bath, Chester, and Norwich," *The Geographical Review*, Vol. 68, 1978, p 273.
53. E. Relph, *Place and Placelessness*, Pion Limited, London, 1976, p 145.
54. Kevin Lynch, *The Image of the City*, The Technology Press and Harvard University Press, Cambridge, Mass., 1960, p v.
55. Kevin Lynch, *Managing the Sense of a Region*, The MIT Press, Cambridge, Mass., 1976, p 23.
56. Terry G. Jordan, "Perceptual Regions in Texas," *The Geographical Review*, Vol. 68, 1978, pp 293-307.
57. *Ibid.*, p 307.
58. David Berry, "Landscape Aesthetics and Environmental Planning : A Critique of Underlying Premises", *Regional Science Research Institute*, Discussion Paper No. 85, Philadelphia, 1975, p 27.
59. Nan Fairbrother, *New Lives, New Landscapes : Planning for the 21st Century*, Alfred A. Knopf, New York, 1970, p 168.
60. E. Relph, *Place and Placelessness*, Pion Limited, London, 1976, p. 88.
61. S. Robert Aiken, "Towards Landscape Sensibility," *Landscape*, Vol. 20, 1976, p 27.

62. Arthur Koestler, *Janus : A Summing Up*, Random House, Inc., New York, 1978, p 33.
63. *Ibid.*, p 34.
64. *Ibid.* p 67.
65. E. Relph, *Place and Placelessness*, Pion Limited, London, 1976, p 40.
66. J. Douglas Porteous, "Home : The Territorial Core," *Geographical Review*, Vol. 66, 1976, p. 383.
67. Wilbur Zelinsky, "Selfward Bound? Personal Preferenee Patterns and the Changing Map of American Society," *Economic Geography*, Vol. 50, 1974, p 175.

# 5

# Spurious Geography

—*Brian Robinson*

THIS ESSAY is a cultural geographer's assessment of the spurious as a feature of contemporary culture. The opening section considers the sources and validity of the uses of the term spurious. It then goes on to suggest that a structuralist rather than a phenomenological methodology is the one best suited to deal with what is by definition a phenomenon both loaded with and emptied by meaning.

What is meant by "spurious" may be exemplified in a quotation from Wilbur Zelinsky on the subject of types of cultural region : "A . . . trivial type of region might be noted then summarily dismissed : the spurious or synthetic region".[1] Included within this rubric are most of the state regions of the United States, and "the concoctions of the advertising agency . . . 'Chicagoland', Delmarva, TVA country. 'Pittsburg Pirate Land' and the like".[2] Zelinsky's strictured use of the word spurious notwithstanding it is suggested here that these admittedly bastard types are worthy of more attention than summary dismissal. Their significance is advocated, however, not out of a sense of specific criticism aimed at Zelinsky but instead from a general admission that our culture has, as a root problem, an inability to authenticate public events and places. For example, the philosopher Stanley Cavell, has summarized the crisis of the conventional in what amounts to an entire dismissal :[3]

> In a time of slogans, sponsored messages, ideologies, psychological warfare, mass projects, where words have lost touch with their sources or objects and in a phonographic culture where music is for dreams, or for kissing,

> or for taking a shower, or for having your tooth drilled, our chances seem to be those of silence, or nihilism (the denial of the value of shared meaning altogether), or statements so personal as to form the possibility of communication without the support of convention . . .

Understandably the philosopher of language questions a culture which has learnt only to misuse its linguistic resources. The question for the cultural geographer is, from what touchstone does a sense of the geographically spurious and inauthentic arise ? In Zelinsky's particular case it was derived from a not flattering comparison with two other forms of culture area which he termed "the traditional" and "the voluntary region".[4] In both cases Zelinsky's main presuppositions were that "emotional and social roots were sunk into the land" and that "man and habitat interacted in endlessly complex fashion . . . ".[5] The difference between the two regional types was that "the . . . American case (of the voluntary region) simulates, but cannot match, the premodern culture area in purity and longevity".[6]

This presupposes a continuum from the traditional to the spurious. There is nostalgia for a time when there were roots in places and rituals in time and, following on from Cavell, the spurious is an alienation of the self-maintained by disconnected worlds which leave the individual with but private incommunicable experiences. There are no agreed logical principles and emotional symbols strong enough to bear additional public references. Instead symbols present themselves as a mosaic of adventitious attachments whose combined effect is too commercial, bureaucratic, or plain functional and obvious to be meaningful (the spatial metaphor of the mosaic used above does however remined us that the root meaning of the word symbol is *to throw together*).[7]

### CRITICAL TERMS

All of which is rather negative. Indeed it has become a conventional aspect of augst and alienation in modern society. In addition, as MacCannell has suggested, a critical language which has resorted to the epithets "pseudo" and "pretentious" and, by implication, "spurious" has merely demythologized its subject without addressing itself to the real problem which is that of a myth too easily seen through.[8] Our very terms seem inappropriate. A simple example may help clarify this impasse.

The crisis of modernity has long been that of "the machine in the garden".[9] The strains and tensions endured have been those of the pastoral unhinged by the machine and they are therefore the product of an inventory of contradictions which may be listed as urban *versus* rural, class against class, frontier *versus* metropolis, North against South, and Black against White. But many of these contradictions may be shelved thanks to the mediating influence of myth. The cowboy is therefore not seen as a workman or hierling let alone as a black, but as a living legend whose individualism shores up the myth of freedom. The frontier thereby maintains its hold despite the fact that metropolitan dominance might be a more truthful representation. The cowboy is therefore also a buffer against urbanism. At the end of the day there are the concoctions of western movies and the advertising industry whose pandering intents are too obvious. But there remains, despite the naiveté of the example and our capacity to see through the myth, the question, what is a concoction ? It is one thing to suggest that it is derived from nostalgia, another to say that it may be maintained by myth, but in neither case is it possible to employ these two words without implying that the phenomena involved are somehow illegitimate. Many of the terms used to this point have, if not a moralistic, certainly a critical ring to them. Indeed a neutral approach hardly seems conceivable when meaning itself is at risk.

**Phenomenology, Structuralism and Cultural Geography**

Two approaches may be considered. One would be to interpret the symbolic content of myth. The other (and this is more a tactic than anything else) would be to suspend meaning altogether, accept symbolic impoverishment, and treat the spurious as part and parcel of a fragmented age. In the field of the analysis of myths Eliade would be an example of the first and Levi-Strauss of the second.[10] Within geography phenomenological approaches are close to Eliade's.

hermeneutics of meaning, but there doesn't seem to be an alternative cultural geography equipped to deal with the debris of our age at its own level.[11] On the contrary one of the practical successes of "topophilia" is that it has become a kind of pilgrimage in search of authentic place. In the phenomenology of Gaston Bachelard we find a moving example of the difficulty of authenticating the city. For him city houses are "incomplete" because "home has become mere horizontality".[12] The city has no dimension except in a purely exterior sense. The number of the street and the floor (the fundamental abstract sign system which makes the city legible) merely defines location.[13] The houses "have no roots and, what is quite unthinkable for a dreamer of houses, skyscrapers have no cellars".[14] For Bachelard the city is literally the unimaginable space of rootlessness. Our only chance is to naturalize it.

Will love of place be equal to the task? For example, Philip Wagner in a somewhat apocalyptic essay has tentatively suggested that "the creative or expressive urge (in other words, love) may . . . emerge in sublimited, or rather perhaps in restored and more authentic forms, before catastrophe occurs".[15] The connection made between "restored" and "authentic" tells us much about the link between nostalgia and authenticity but what is more important is that Wagner goes on to advocate an approach to symbol and meaning which comes close to entertaining an alienated dichotomy between the "objectivity" of technology and the "subjectivity" of creative artists. As he has put it :[16]

> Artifacts and acts function first of all as symbols and the landscape is a kind of theatre possessing many stages. A man learns his cues memorizes his lines, and interprets his roles. This image of drama comes much closer to the "model" (or myth) than do cold technologies of code and channel.

Cultural geographers currently prefer the existential, phenomenological, sensory and humanistic approach to environments to a more abstract structuralist one, though Y.F. Tuan has described the difference as "a paired" rather than a polar "opposition".[17]

However, the worst of all possible worlds would be one in which so called subjective and objective approaches would end up vying with one another for centre stage. Phillip Rieff has summarized this as the guru *versus* the data-bank and has set a plague on both houses.[18] Standing back from false dichotomies, his own approach is inevitably laced with considerable irony. For him the unacceptable alternative would be to drift into a "cultureless" age. His response to Yeats' famous warning "the centre cannot hold mere anarchy is loosed upon the world", was "Yeats was mistaken; all centres hold equally well. Therefore, there is no centre".[19] What Rieff has suggested, especially in his work on Freud, is that the machine *versus* the garden, its myth and its morality, has been transformed not simply into the same-only-different opposition between data-bank and guru, but also into an irony that can stand back and examine the rootless centreless worlds of plastic and instant transcendence.[20] From this perspective the spurious is not a wasteland. It is neither over-valued fashion nor "trivial". On the contrary the spurious is curiously both the heart of things and marginal. Meaning can be brought to bear on it only when it is realized just how vulnerable meaning may have to be. In a disinherited time it is the improviser who, like Levi-Strauss' bricoleur, uses ends as means whom we turn to as a type figure.[21] Structuralism may be defined as the latent methodology of improvisation.

**Structuralist Possibilities**

Structuralism as a tactic has two specific and relevant characteristics. First of all the structuralist is intensely self-conscious concerning nostalgia. And second, structuralists accept the fragmented nature of the human structure of the world. Moreover in view of what has been said concerning subject *versus* object dichotomies, structuralism is not something we apply to the phenomena at hand. What Barthes maintains, for example, is "the structuralist activity", an act which is as creative as its source.[22] At least, creation becomes reconstitution and the subject-object dichotomy is no longer relevant. In suggesting this the present paper goes further along the road of anti-humanism than Y.F. Tuan was prepared to in his papers on structuralism.[23] Perhaps the most important result of this expansion of the field is that Tuan's phenomenological and eclectic approach to such ideas as nostalgia and authenticity is radically recentered. To coin a phrase Tuan's "activity" was to authenticate. Like Cavell he questioned a society which has lost its roots thanks to "the unreflective acceptance of society's" conventions, including reflex response to words".[24] No such authenticity is maintained by structuralists. The rootlessness of both words and contemporary experience is accepted. Since, as the phenomenologist Merleau-Ponty suggested, language does not "model itself on the natural aspect of the world," this means that, as Tuan also recognized, language can turn its back on the world in the act of signifying.[25] The tendency and the risk is towards the inauthentic. Structuralist thought on the other hand, even as it accepts this turning away, looks back to a time when language was in the world, when there was the possibility of a direct unfolding from object to word. For example, both Derrida and Foucault turn to the sixteenth century as a time when it was possible to establish connections between man and the world in terms of Being, reference, source and origin. Derrida sees the geographical discoveries as not simply a geographical reorientation but more radically, as an unhinging of thought which destroyed all centered (ethnocentric) systems.[26]

> One can . . . assume that ethnology could have been born as a science only at the moment when a decentering had come about : at the moment when European culture . . . had been *dislocated*, driven from its locus, and forced to stop considering itself as the culture of reference.

It matters little whether one agrees or disagrees with Derrida on this historical point, the nostalgia for a locus and the references a centre makes possible, is the crux. Similarly Foucault saw the period as :[27]

> A culture in which the signification of signs did not exist, because it was reabsorbed into the sovereignty of the like; but in which their enigmatic, monotonous, stubborn, and primitive being shone in an endless dispersion.

Because that "being" was "the being of language" the sixteenth century "geographer" was as much an etymologist as a chorographer. The search was for the unique name in a landscape of constant references and many meanings. Despite "endless dispersion" authority authenticated. As I have put it elsewhere (without realizing the implications) : "sources and authorities were closer to the original state of affairs and could therefore be assumed to have had prior knowledge of the actual nature of the place when the first name was given".[28] Authority was the direction in which thought went, and there was a primordial being towards which thought could be directed. The ultimate language of geography was the corrupted "first name".

For Derrida and Foucault, the relevance of the ensuing decentering of thought is that thought proceeded without an Origin. That is with Derrida :[29]

> There will be no unique name, not even the name of Being. It must be conceived without nostalgia; that is, it must be conceived outside the myth of the purely maternal or paternal language belonging to the lost fatherland of thought.

Nostalgia is nevertheless present because what we are presented with is a monster—the unnameable. As Terdiman has put it in a different but not inappropriate context :[30]

> Beckett's haunting "Where now ? Who now ? When now ?" at the opening of *The Unnameable* asserts this fiction's radical denial of the sure grasp of external reality, referable at all times to objective geography, identity, (and) chronology . . .

The nostalgia is for among other things geography, or, at least, the geography of identity, of the traditional unadulterated undifferentiated culture region replete with names in the land.

Is it possible to talk about a nameless fragmented space ? Nostalgia informs us that our geographies are mere locations rather than named identities and that we must be content with "context" how things function and connect. "There may be a (space) where the words pass, and it may even be geographically located, but it is only the context of the narrative and not its content".[31] Location is never an immediate experience, it makes sense only, in relation to some other location. One city implies repetitively other cities and so on *ad infinitum*. As a consequence the structuralist is interested in "the alphabetical order of sense as a mediating function rather than as a direct meaning."[32] E. E. Said therefore termed the structuralist's culture(s), "Abecedarium Culturae".[33] Delmarva and Boswash make alphabetical and functional sense but it could be inappropriate to expect them to mean much else. Further, that these new names are attempting to connect places which at present are not part of rooted meaning, is not a new phenomenon. Neologisms have been part of our functional experience ever since cities lost their capacity to act as references in their own right. Reference is now adventitious and supplemental to the city. Names and numbers in the city landscape have become more pervasive as sign systems have had to substitute for the disorders of the city. Now it is to the signs that we refer, not the city itself. In effect the city has disappeared beneath a welter of signs. It has become more and more anonymous as signs have driven urbanism towards repetitive standardized technology-oriented architecture (Bachelard's skyscrapers without cellars), which in turn has made the city less and less legible.[34] It is a vicious circle which no amount of nostalgia for what the semiologist Choay terms "culturalist" solutions, can hope to retrieve.[35] By "culturalist" Choay means a message bearing capability revealed simply through contiguity, though, in the case of urban systems, reference to other buildings, streets, squares, gateways, etc. The structuralist approach, on the other hand, accepts the fact that urban spaces become meaningless the more signs seek to make them readable. The structuralist asserts the meaningless of the city without nostalgia for reference and syntagm. On the contrary, it is possible to welcome the new freedom which the greater flexibility of sign systems makes possible. In effect signs have opened the city up to infinity, pointing one, interminably in continual deference, to another, and, after Derrida, this is something the reader of signs affirms "with a certain laughter and with a certain dance".[36] If unity is to return it is in the full knowledge that it is a game, a mine and "a pretended . . . unity" recovered without nostalgia, in awareness of nostalgia.[37]

The work of Derrida and Foucault has established the breadth of the problem. They have shown how we got here and they have suggested where we might go in the future (if anywhere). Levi-Strauss and Barthes on the other hand have given us works which are less philosophical in their scope. By comparison they are more positivistic. "Positivistic" because along-side nostalgia is a world reduced to fragments by the acts of the demythologist. When meaning disappears what we are left with is the rules connecting the pieces. This may be referred to, after Heller, as a "positivistic impoverishment" in which order is presumed to rest among the things themselves bereft of transcendent references.[38] Order, as conceived by the structuralist, is not the opposite of science's disorder, but of nothingness. Order is expressed as a trace just this side of silence. If structuralists are to write without nostalgia then they should, in Derrida's words, surrender themselves to "indetermination, to the *seminal* adventure of the trace."[39] Levi-Strauss does not go that far, indeed Derrida has accused him of a kind of duplicity based on nostalgia (of in fact using a provisional tactic) nevertheless Levi-Strauss' attitude towards disorder is critical if we are to understand the structuralist view of the relation between order and fragmentation.

By coincidence Levi-Strauss chose landscape to demonstrate the relationship. Geography is for Levi-Strauss "an immense disorder which may be sorted out however we please".[40] No self-respecting geographer would agree with this, nonetheless the grain of truth contained in it is the assertion that only the most elegant explanation will suffice and that accordingly to sketch out the history of a landscape's evolution, and to ponder the ups and downs of its history and prehistory though valid are unsatisfactory as compared to the structure revealed in "the search, on a limestone plateau in Languedoc, for the line of contact between two geological strata".[41] Once again there is no reason for the geographer to agree with this. What is of interest is to follow the line of thinking. The jumble is what we perceive "at first glance", the structure is to be reconstructed not from "the accidents of geography" but from the immense sedimentation that underlies geography.[42] Of course this is, as they say, "only an analogy" but what we must trust is the thought that makes analogy possible. This is the reason why Levi-Strauss admires that other great analogist Sigmund Freud.[43] Significantly Freud began *Civilization and its Discontents* with a geographical analogy between the mind and the sedimentation induced by the evolution of a landscape, in this case, the city of Rome.[44]

The analogy most closely associated with the positivist's approach is the work of the bricoleur, the improviser of order, mentioned earlier. Put crudely the bricoleur is an odd-job-man for whom every piece of discarded junk is a likely part in a future project. Or, to reverse the analogy, the bricoleur manufactures order in a world where the plans are missing because what he has to mend is quite often still in use but no longer serviced by the manufacturer. The spare parts are deriveable but solely as redundancies. And since the final project is unpredictable "every litter bin counts".[45] Translating this into structuralist terms, as summarized in Said's review, "for the structuralist all details have the status of information . . . and every . . . item . . . has the dignity of message-bearing capability".[46] Dignity may even be restored to such diverse and seemingly spurious activities as those of the tourist and the women's fashion industry.

**Spurious Experiences**

MacCannell's view of the link between fragmentation and order, seen through the eyes of the tourist, is relevant here.[47] Like Levi-Strauss he began his task with a reference to "first glance" chaos. As MacCannell put it "modernity first appears . . . as disorganized fragments, alienating, wasteful, violent, superficial, unplanned, unstable and inauthentic".[48] But, he suggested, it is precisely this kind of world with which the tourist is best equipped to deal. Ignoring the prejudice which dismisses the tourist's image as an inauthentic one (in Samuel Beckett's summary dismissal "Baedeker is the end rather than the means")[49] MacCannell says :[50]

> The concern of moderns for "naturalness" their nostalgia and their search for authenticity are not merely casual

> and somewhat decadent, though harmless, attachment to the souvenirs of destroyed cultures and dead epochs. They are also components of the conquering spirit of modernity —the grounds of its unifying consciousness.

The implications are radical, but the chief value of Mac Cannell's acceptance of the inauthentic as the conventional is that the various negatives attached to our experience (the inauthentic) becomes irrelevant. The spurious becomes a root experience of modernity.

Take for example the Beatles song "Two of Us".[51] In it there is the extraordinary line : "you and I have memories longer than the road that stretches out ahead". The search is for memories; a kind of instant souvenir which is there without reflection. The connection is often made between the "instant" and the inauthentic. It is, of course, used in a condemnatory sense. But the Beatles' use of memory as prescience destroy any possibility of a criticism based on rooted experience ("emotional . . . roots . . . sunk into the land"). Nostalgia pervades as we act, it is not something recollected in tranquillity.

Also nostalgia is a source of action which destroys the work ethic. For example, the line, "spending someone's hard earned pay", uses the language of work ethic criticism in order to disarm that criticism not just by being cheeky and adolescent but by charmingly converting it into nostalgia. Expanding the argument further, the link between road, journey, and trip is easily made but it was Richard Poirier who pointed out that the Beatles' *Sgt. Pepper* albumn intertwines the "trip" metaphor with a mosaic metaphor of modernity :[52]

> The Beatles' songs . . . work for a kaleidoscope effect, for fragmented patterns and sound that can bring historic masses into juxtaposition only to let them be fractured by other emerging and equally evocative fragments.

Nostalgia is one of the means by which the Beatles juxtapose and thereby open out the frame of reference. Nostalgia becomes, if not irony, certainly, as Poirier has also suggested, parody, and of course parody has its own set of references which, in turn, establishes a new myth.[53]

None of the above on travel as inauthentic experience was taken from MacCannell, however his approach may be appreciated if we marry it to that of another analyst of modern mythology, Roland Barthes. The example taken is from Barthes' book *Systeme de la Mode*[54] (rather than his more readable and epigrammatic *Mythologies* which contains an interesting discussion of the *Guide Bleu*[55]). In Barthes discussion of the rhetoric of women's fashion we find a passage that enables us to see the breadth of MacCannell's understanding of the part played by travel in a context which is wider than the one either author originally intended. Thus, according to Barthes, for fashion *place is travel.* It is something to be *reached,* across a disrupted landscape, elsewhere. Place is a value whereas space is melted away or placed in limbo rather than something to be overcome. Place can only stand out and its value can never be less than absolute. Here I'm going somewhat beyond Barthes, nevertheless in his words, "la Mode is a rapid succession of absolute places.[56] Significantly the means of getting there becomes a place, a "Capri" or a "Monte-Carlo" perhaps. The automobile has been *there* all the time. In the meantime place exists by being taken out of its environment. It is put back into *context* as a relation across meaningless space. It is as an elsewhere that place is *related.* Notice the emphasis on the activity that makes structuring possible. The movement is from place in its own geography to essence *in relation* to whatever, depending on context, that is, depending on current fashion. The concept of place for fashion is one that automatically decenters itself. It is first loaded with, but later it is emptied by, meaning.

**Myths from National Geographic**

There are topics which are "naturals" for myths. The example of the cowboy has already been mentioned. *National Geographic* resurrects him as a "living legend" whose "partnership" with a horse still "endures" despite the attractions of the "less lonely" life of the city.[57] It is not suggested that there is any misrepresentation here. On the contrary by definition a myth is made from facts combined to produce a set of connections with greater validity than the facts themselves. It is how the relationships are structured that counts. The sentence can be reworked to show the implicit assumptions of the myth's structure (Fig. 5.1)

*The Structure of a Myth*

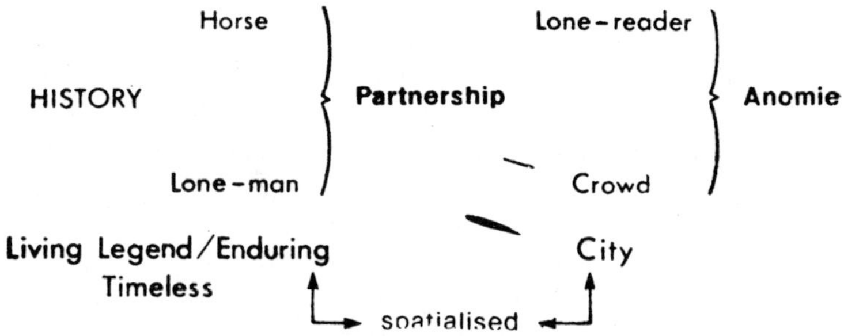


FIG 5.1

The timeless nature of the legend creates its reader's space by spatialising the relationship between the ever changing anomie of the city and the changeless individualism of historic roots. Partnership between man and horse reinforces the natural root of that individualism. History is thereby naturalised as well.

Civilizations are most easily handled when they are dead ones. For instance the Inca civilization is kept alive by Andean peasants. The photograph attributes them with "living Inca ways—caught in the amber of tradition."[58] The other side of this living legend is a child who had apparently been frozen five hundred years ago but who is now, labelled, "seemingly about to awaken."[59] The combination of the two may be summarized diagrammatically (Fig 5.2).

*Myth and Metaphor*

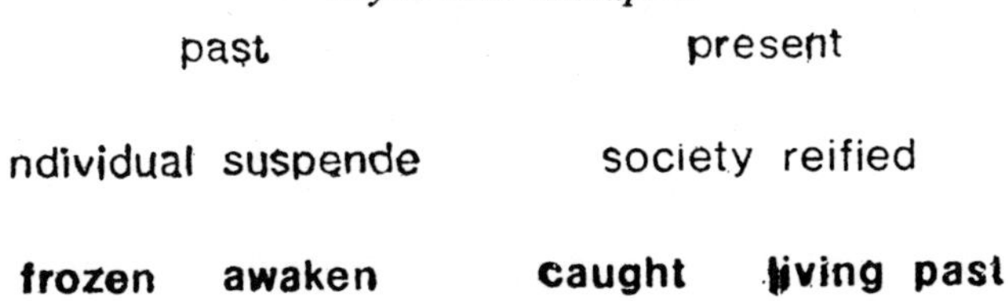


FIG. 5.2

The metaphor, (attributed to the photography because the photograph cannot speak for itself) mythologises the present by suspending it and the past by awakening it. Active history and tradition has died in order that it should be appropriated in the silence of the museum. History is turned into property. The reader is a vicarious curator.

By way of contrast primitives are naturally changeless. The less they are touched by civilization the better. But if they are recently "discovered" better still. If they are not pristine when discovered then change is subsumed by the category the "inevitable". What makes these sullied members of the race still available for myth is that primitives, more than any other type of man, are men reduced. They are either literally vanishing or else they can be reduced to simple elements like the family. Such captions as "happiness is a father to comfort you" and "I'm a good wife, hard-working and virtuous" introduce us to a universal attribute, the family.[60] Happiness and smiling also proclaim the universality of the same institution. The extraordinary caption to a picture in which a "Namba girl affectionately clutches her father's hand" suggests ghoulishly that she is an "heir" to this "dying culture".[61] The category "tradition" is thereby retained despite the obvious fact of a cultural world maintained by the last few in a best familial, at worst biological, conditions.

Exotic cities cannot be reduced. Instead they are clothed in people whose work is termed "bustle" and whose movements become "perpetual".[62] The Exotic exudes business in the form of vitality. The presentable poor are lined up—peddlars, hawkers and rickshaw wallahs.[63] Commerce and transportation are *humanized*, as are the streets. Calcutta's "streets by day" are turned into "bedrooms by night", with the notable advantage that they are "rent free"[64]. The myth exports balance sheet home economics to Calcutta and finds the figures infallible. After Barthes this may be termed doing "the accounts of the ineffable".[65] Paradoxically this provision of rent-free-homes appears in a demythologising article subtitled "the world's most maligned metropolis.'.[66] The world has been made that little bit safer for common-sense-economics, not to mention home-economics.

To sum up these four examples, the forms mythological activity take are attribution and appropriation, the impositions of a vicarious coloniser. The continents have been provisioned not inhabited. Unlike real colonization however myth does not favour exploration of the truly barbaric or the exotic. If the alternatives to the exotic are surrender at one extreme and domestication at the other, then *National Geographic* is in the latter category. It is by no means committed to the fabulous, compulsive travel, or to homelessness. It is the comforting myth of the homeless finding a home (even in Calcutta). The redundant become living legends if they represent a lost value (individualism). If they are primitives we give them families, rather than individualism, in place of tradition. If they have tradition we reify their "customs" forever. *National Geographic* is an extreme in culturalism. It culturalises.

**Culturalist Possibilities**

Reference has been made to the sixteenth century's concern for language as a theory of knowledge. As far as geography is concerned this produced a culturalist geography which married language to nationalism. As I've suggested elsewhere :[67]

> Nations unlike titled families or family firms have difficulty in establishing their precise origins, so we should be neither dismayed nor surprised when naming techniques refer to pseudo-origins in etymology and mythology. On the contrary, this kind of naming is the raison d'etre of corographical geography.

Or, modern geography has as one of its main traditions a source in the rise of vernacular languages and nationalism. Unfortunately that tradition has lost most of its momentum within academic geography. The title of the popular magazine *National Geographic* is symptomatic of where that venerable tradition still maintains itself. The question, what is the link between language, nation and geography is apparently beyond the academic pale. But surely what this apparently pretentious connection would be about is the homology between language and landscape as repositories of national experience. For example, Inglis has suggested that the morphology of the English landscape is a product of what he calls "privacy" and "politeness".[68] Inglis is writing within a particular English studies tradition of culture as literacy, but he also uses the phenomenology of observation and description to give his landscape concrete form. That is to suggest his culturalism is a combination of language as literacy and landscape as language (the structuralist would concentrate on the latter alone). For Inglis the message written into the street scene is *there* in the contiguity of forms built into the landscape. Even terrain plays its part, so that there is a direct unfolding from nature to culture :[69]

> Tenterden is a street. But the steep hill halfway along the High Street means that from either end the heart of the town is its terminus, either because from up above, the heart is concealed in the fold, or from below the heart is tucked under the curve of the hill.

Admirable as this may be it should be pointed out that nearly all of Inglis' examples are taken from the country-townscape, and if a criticism could be levelled at this type of literary-cum-phenomenological description it is that it does not come to grips with the industrial city. The eye of description cannot follow when contiguity had been fragmented by explosive, if repetitive, growth. Examples of this inadequacy may be found in many sources. It is, as E.A. Wrigley has pointed out, a general failing of French regional description.[70] Outside geography Hoskins, in *The Making of the English Landscape*, is much more comfortable when he has culturalized the industrial scene, when for example he has discerned roots in such previous forms as field and road systems.[71] But perhaps the best example of the problems of culturalism is that of Lewis Mumford's *The Culture of Cities*, in which the approach is entirely nostalgic and teleological.[72] This is seen not only in his well-known admiration of the medieval city but also in the text appended to the photographs contained within the book (this Mumford himself describes as a separate text[73]). First, it is noteworthy that the photographs change from initial street scenes of medieval towns (the equivalent of Inglis' integrated description of the English townscape) to, in industrial cities, either individual buildings or oblique air photographs of complete urban scenes. This is more than symptomatic of the scale problem implicit in Inglis' description, it is also the problem itself. On one level Mumford is right. As the city disintegrates there is an hiatus between the designed house and the concrete jungle. The street scene is somewhere between. But surely it is in photographs of the street that the activity of the modern city is best seen, hence *National Geographic's* photographic myth of the city as bustle and perpetual motion. Mumford (and this is the second contrast brought out by his photographs) when he portrays the street illustrates it in the form of illuminated street signs, which he finds wanting as compared to the silent symbols of the medieval city.[74] The structure of this culturalist nostalgia may be illustrated diagrammatically (Fig. 5.3).

*Myth and Nostalgia*

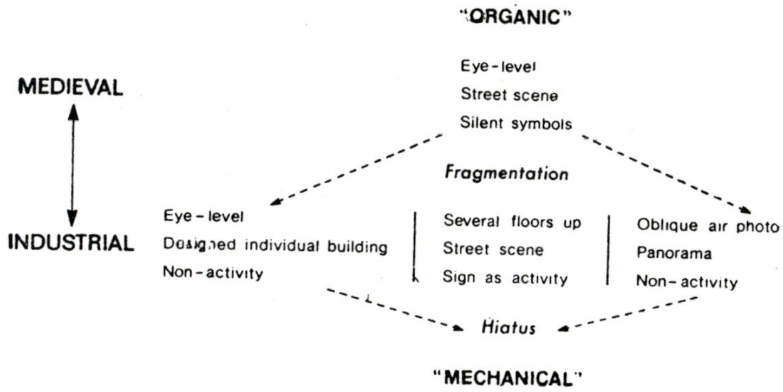


FIG. 5.3

The crucial element is Mumford's misplacement of the sign as activity. The very garishness (the cliched epithet) is the link holding the fragments together. As every graffiti artist knows it is no longer the message borne in unadulterated contiguity between buildings but the power of literacy that holds the city together.[75] For example literacy makes market competition possible simply because thanks to the hand distributed flier it widens the sphere of retailing beyond local communities.[76] At the level of window shopping marked prices turn verbal haggling into market competition and create the city as display.[77] Display in turn creates an emphasis on marginal selling characteristics (exoticism, eroticism etc.) as opposed to "true" worth and content. Daniel Defoe pointed this out in the London of over two-hundred years ago :[78]

> . . .a hundred pound rent for a house to sell jellies and apple pies, two hundred pound to set up a brandy shop and afterwards not a hundred pound stock to put into it . . . a pastry cook's shop, which 20 pounds would effectively furnish at a time, with all needful things for sale, yet that fitting up one of these shops should cost upwards of £300, Anno Domini, 1720, let the year be recorded.

Of course this spurious aspect of display can be dismissed but it would be at the expense of understanding the dynamic of modernity. The sign is no longer a silent representation of activity, it has become activity itself. Signs are the communication channel between individual (building) and totalities (megalopolititan neologisms). They make privacy within anomie plausible if not always tolerable (which is why on television as opposed to less visual media they are accused of invading privacy).

For Mumford the city must contain culture or else it is nothing. The assertion, "with language itself, it remains man's greatest work of art", is a culturalist teleology which inevitably leaves the modern city with little to say for itself.[79] Further it ignores the nation as the true container for modern culture. The nation rather than the industrial city has grown at the expense of Mumford's medieval city which would suggest that an attempt to combine nation and city together would pose considerable problems. Yet this is what Jecquetta Hawkes has set out to do in her book *A Land*.[80] Like Levi-Strauss and Freud she sees the link between geology and individual consciousness as a deep sedimentation. Unlike the former two however she relies on a revealed unfolding from geology to human artifact and consciousness. Like another archaeologist, Tielhard de Chardin, her culturalism is evolutionary.[81] Thus the city is "a deposit" which for the nation becomes "the symbol for the unity of a people and its land—and therefore the symbolic centre of my own theme".[82] Authenticity is the unity that brings together land, nation and city, and since this is a kind of legibility or at least a mark and a deposit, it is natural that it should be culturalised with the term "symbolic". Culturalism may be defined as the Orphic revelation of such deep seated deposits.

Not all deposits on the landscape are as open to the probes of the culturalist, Mumford's "insensate industrial town", Coketown, would hardly be said to be an unfolding—more likely a rupture or, perhaps most commonly, a cancer.[83] Still Inglis has managed to incorporate industrial housing within his theme. Here is his description of an English "back-to-back" slum :[84]

> *The closeness* : the dark corners for lovers, cats and bikes; the common passageways. The stubborn inwardness and grim good humour in the shoulders humped against the drizzle off the Atlantic. The regular punctuation of bright, net-curtained windows as the light fades on a winter afternoon. The brown, beaten lots and scruffy kids kicking about on their lumpy surfaces.

The culturalism is worth analyzing. There is an emphasis on contiguities which are physical, "closeness", and cultural, "stubborn inwardness". There is the Bachelardian use of the idea of the corner which brings together a set of domesticates including a homely machine.[85] Notice the use of an inventory of non-sequiturs, "lovers, cats and bikes", whose very inconsequence makes them evocative. Darkness and the elements are employed in the same fashion. There are, "the dark corner", the light that fades, the sadness of drizzle and brown beaten lots. Bright net curtained windows penetrate the elemental gloom with their domesticity and the need for privacy or more correctly for a display of private cleanliness which is semi-public. This is paired with the dark corner of semi-private love making. Domestic animals must copulate there as well. Privacy has its edges. The square patch of semi-public domestic light has its companion, a dark corner for semi-private behaviour.

Topography has its place and its margins as well. "Shoulders humped against the drizzle" is homologous to the "lumpy surfaces" kids play on. Culture is transformed and naturalized as terrain. People are as resistant to the forces of erosion as the hills. But kids whose play is marginal to dour resistance must play in the left overs. Stubbornness is to play what humped shoulders are to waste ground. Richard Hoggart, who is the original inspiration for Inglis, is clearer on the marginal characteristic of spaces as non-places :[86]

> The nearest park or green open space is some distance away, but the terraces are gap-toothed with sour and brick-bespattered bits of waste ground and there is a piece of free ground half a mile away called 't' Moor. Evocative name : it is a clinkered six-acre stretch surrounded by works and grimy pubs, with a large red-brick urinal at its edge.

The irony of the attempt to naturalize the waste cannot obscure the fact that this non-place is for waste. Evocative power also resides in marginal behaviour and the elimination of urine. The "dark corners" Inglis rejoices is served the same function as well.

From the point of view of analyzing "the spurious" the key word is surely "evocative". Because at the base of contemporary culture there is no longer historical roots in nature but that root in self that reverts to childhood.[87] The invocation, "We may come closer by looking through the eyes of the child" is itself suggestive.[88] The emphasis is on a lost world of smells, noises and play :[89]

> Most attractive of all, a scene with noise, light and smell—between eleven and twelve on a sunny Sunday morning, when all doors are open and most steps occupied; the roast beef gives out its flavour from almost every house, the wirelesses mix their noise with each other, you can hear families talking or laughing or quarrelling. But there is little quarrelling just now; over almost all is a sense of ease, recreation and good food to come.

We domesticate Mumford's insensate streets. On Sunday the family ritually inhabit the streets, but it is the child who is the repository of this experience. Childhood is the world we must loose hence its air of authenticity. Evocation, therefore, becomes to the culturalist what fragments are to the bricoleur. The culturalist cultivates the street, the structuralist gleans the margins.

## CONCLUSION

The problems examined in this paper can be summarised in the different approaches of structuralists and phenomenologists to the phenomena of space and place. For the phenomenologist :[90]

> Space is abstract. It lacks content; it is broad, open, and empty, inviting the imagination to fill it with substance and illusion; it is possibility and beckoning future. Place by contrast is the past and the present, stability and achievement.

The structuralist notes the increasing abstraction of space. For him this is acceptable because reference is not contingent upon history and/or objects. Space, while it remains empty, is not compared to place as if it were an absence waiting to be filled by the plenitude of place. Instead it is nothingness over which signs hover, reflecting not the world, but simply themselves. The structuralist, therefore, tends to see attempts to fill space with place as a kind of nostalgia. On the other hand, the phenomenologist sees space without place as the inauthentic. Structuralists would respond that sources of rooted symbolism are not varied enough in the over-simplified world of commercialized communication for new "symbols of possibility" to arise.[91]

The structuralist's approach to history is a direct challenge to much of traditional cultural and historical geography. The phenomenological approach sees time as an unfolding from rock to clay and from clay to human artifact, whereas the structuralist concentrates on the difference between clay and pot and wonders how the connection could have been made at all. It is the difference between history as erasure and history as palimpsest. The former creates fragments and ruptures, the latter roots and hearths. One improvises his world as best he can, the other "authenticates", hence the charges of rootlessness and homelessness the "culturalist" levels at modern experience.

Tourism, women's fashion, pop songs and popular geography may be considered to be sources of the inauthentic. In this paper they have been treated as message systems rather than as spurious attempts to be something that they cannot.

It may well be that the contradictions built into these systems are so great that, as Barthes put it, for the observer sarcasm is a condition for truth (irony and parody are useful safety valves as well), nevertheless there is more to the tourist's view than reductions to image and perception would allow.[92] Much more is connected than the photograph taken and memory stored away. Tourism, women's clothes, a pop song and automobiles vibrate with significance, even as they are merely juxtaposed.

The inclusion of lost tribes, hidden empires, living legends and exotic cities widen the picture. The commercialized geography from which the armchair tourist's symbols are derived was seen to be the incurious activity of appropriation and provisioning. The symbols are one-dimensional but the message is far from spurious. Or, a fragmented age achieves its mimed sense of unity by means of kaleidoscopic associations of views, slides and posters. The metaphor of a mosaic also applies. At any rate the system is spatial. It rests just this side of its own fragility creating a fresh opaqueness of the surface it proclaims to reveal.

It is truism that rootlessness is the search for authenticity among the redundancies speeded up sign-systems have tantalisingly made available. As a quest the ultra-conventionalism of commercialized geographies is no less true to the duplicity of concealing the expected than any other loaded yet doubtful union. Commercialized religion also comes to mind. It is the peculiar relationship of the conventional and the doubtful that makes them interesting. As myths they are perpetually in need of shoring up but as "vested interests" provisioning new worlds they may be more relevance to an understanding of modern consciousness than a nostalgia which perpetuates little else but its own alienation.[93]

Joseph Frank summarized the situation in a much wider context :[94]

> Ever since the Renaissance, modern man has cultivated both the objective visual imagination (the ability to portray space) and the objective historical imagination (the ability to apprehend chronological time); both have now been abandoned.

The result is a fragmentation of experience. But across this decentered landscape we find a congruence of examples. Both T.S. Eliot and the Beatles present their materials in terms of juxtaposition.[95] "The meaning-relationship is completed only by the simultaneous perception in space of word-groups that have no comprehensible relation to each other when read consecutively in time".[96] The effect is to mythologize, to destroy time and to spatialize. Meaning in such a vision is not built in. It restlessly remakes itself in a continual state of difference (fragmentation) and deference (sign-systems). Like a kaleidescope it erases itself in its own reorientation.

Fragmented as this vision might appear it does give space a privileged, virtually a monopoly, location in our thinking. The journey alone is real and one is always nearer by not keeping still. Hassan writes; "In our mind Silence and Eternity belong to Space".[97] Space is perhaps the most spurious of all the new geographies. Yet it is also the one that comes closest to an understanding of the infinite silence that makes the meaningless possible and paradoxically, conceivable.

Whether or not phenomenology can fill or naturalize that space is an open question. In Sartre's view : "A good part of our life is passed in plugging up holes, in filling empty places, in realizing and symbolically establishing a plenitude".[98] If that is the case then nostalgia for plenitude and rooted place is at the heart of a basic contradiction within our society: one that leaves us open to the risk of spurious geographies.

## POST-SCRIPT

Culturalism and structuralism have been opposed in this essay within a particular context, the western nation and city and the flow of ephemera that connect them. But there are contexts where such oppositions are unthinkable because they are luxurious irrelevances. A structuralist analysis of how the one-dimensional thought of Chairman Mao reduced China to a cultural desert has little to recommend it, unless it is a perverse pleasure in the degree to which new signs, poster campaigns and the alphabet, have erased the past. Simon Leys reports that Peking's walls, which were "not so much a medieval defense apparatus as a depiction of a cosmic geometry, a graphic of the universal order", have been destroyed.[99]

But it is not nostalgia that has prompted Leys' culturalist response :[100]

> For those who knew it in the past Peking now appears to be a murdered town. The body is still there, the soul has gone. The *life* of Peking, which created never-ending theatre in its streets and squares, the noisy and enjoyable life of the city has gone, leaving only the physical presence of a mute and monochromatic crowd, oppressed by a silence broken only by the tinkle of bicycle bells.

With a vengeance this is history as erasure. The problem of the spurious is a western luxury. In Susan Sontag's view the west's luxurious capacity to feed so rapaciously from spurious images has been bred by photography. In contrast China has need only for a few images and this in turn has made the manufacture of any other images suspect. Chinese being photographed in Tien An Men Square are one with the monologue. Their precisely posed image cannot be prostituted to sell a hairstyle or cigarettes or shoes—not even sunshine. In Sontag's own words :[101]

> The limits placed on photography in China only reflect the character of their society, a society unified by an ideology of stark, unremitting conflict. Our unlimited use of photographic images not only reflects but gives shape to this society, one unified by the denial of conflict. Our very notion of the world—the capitalist twentieth century's "one world"—is like a photographic overview. The world is "one" not because it is united but because a tour of its diverse contents does not reveal conflict but only an even more astounding diversity. This spurious unity of the world is affected by translating its contents into images. Images are always compatible, even when the realities they depict are not.

If then "the concoctions of the advertising agency" are too easily see through they nevertheless have the resilience of replication, adaptability and redundancy. In their hit and miss way, and despite visions of an apocalypse choked by redundancies, discards, replacements, and pollutants, they form the ground of modernism, its conquering spirit and its unifying consciousness.

## REFERENCES

1. W. Zelinsky, *The Cultural Geography of the United States*, Prentice Hall Englewood-Cliffs, 1973, p 112.
2. Zelinsky, op. cit. (footnote 1) p 112. Rather than these regional examples what would be more difficult to authenticate than America? Having lost the innocence of Whitman's "roughs and beards and space and ruggedness and nonchalance that the soul loves" the United States has become a mask. The artist Saul Steinberg's contemporary America consists of "individuals unmasking themselves only to reveal other masks, verbal cliches masquerading as things, a countryside that is an amalgam of all imported styles, an outlook that is at once conventional and futuristic. . . ." Harold Rosenberg, *Saul Steinberg*, Knopf, New York, 1978, p 14.
3. S. Cavell, *Must We Mean What We Say*, Cambridge University Press, New York, 1976, pp. 201-202.
4. Zelinsky, op. cit. (footnote 1) pp 110-111.
5. Zelinsky, op. cit. (footnote 1) p 111.
6. Zelinsky, op. cit. (footnote 1) p 111,
7. Greek *Ballo* throw, and *sun* with, *The Pocket Oxford Dictionary.*
8. D. MacCannell : *The Tourist*, Schocken Books, New York, 1976, p 105.
9. L. Marx, *The Machine in the Garden*, Oxford University Press, New York, 1967. The present essay begins where Marx leaves off with the problem of what he has termed "symbols of possibility". Marx was optimistic as was Robinson in "Some Fragmented Forms of Space", *Annals*, Association of American Geographers, Vol. 67, 1977, pp 449-463. Neither author however addressed themselves to the problem of how symbols manage to take hold.
10. There is little point in referencing all the works of these two prolific authors. The most relevant sources are M. Eliade, *The Sacred and the Profane*, Harcourt, Brace and World, New York, 1959, pp 20-65; *Cosmos and History*, Harper and Row, New York, 1959, pp 3-48. C. Levi-Strauss, "The Structural Study of Myth", *Structural Anthropology*, Basic Books, New York, 1963, pp 206-231, and "Overture", *The Raw and the Cooked*, Cape, London, 1970, pp. 1-32.
11. For recent critical appraisals of phenomenology in geography see J. N. Entrikin, "Contemporary Humanism in Geography", *Annals*, Association of American Geographers, Vol. 66, 1976, pp 615-632; M. Billinge, "In Search of Negativism", *Journal of Historical Geography*, Vol. 3, 1977, pp 55-67. Both of these are somewhat pessimistic. On the other hand Buttimer is more enthusiastic : "Grasping the Dynamism of Lifeworld", *Annals*, Association of American Geographers, Vol. 66, 1976, pp 277-292. Gregory has compared structuralism and phenomenology. His conclusion is that "a genuinely structural historical science must be grounded in an explicit problematization of discourse, one which *refuses* to take constructs

and typifications of either past or present life-world for granted, but which *relates* them to social practices whose textures and rhythms reveal and re-form their constitutive social structures". This makes him especially critical of idealist positions, including Levi-Strauss' formalism : D. Gregory, "The Discourse of the Past", *Journal of Historical Geography*, Vol. 4, 1978, pp 161-173, reference on p 173.

12. G. Bachelard, *The Poetics of Space*, Orion Press, New York, 1964, p 27.
13. Bachelard, op. cit. (footnote 12) p 27. But if this is *mere* location it should also be realized people are also located by names and numbers. Of course, to another age it would not have been conceived that individuals could have several names let alone numbers. The use of age as an identifying characteristic is a recent attribute. "In the Middle Ages the Christian name had been considered too imprecise a description, and it had been found necessary to complete it with a surname, a place name in many cases. And now it has become advisable to add a further detail, the numerical character, the age. The Christian name belongs to the world of fancy, the surname to that of tradition. The age, a quantity legally measurable to within a few hours comes from another world, that of precise figures." P. Aries, *Centuries of Childhood*, Vintage Books, New York, not dated, p 15. Since Age calculation depends on Place calculation, presumably both grew together. Peter Laslett suggests that naming customs may have varied in their rate of change from one European culture to another. "In Holland for example the common folk did not acquire second names until the time of Napoleon" ! P. Laslett, *The World We Have Lost*, Scribner's, New York, 1965, p 44.
14. Bachelard, op. cit., (footnote 12) p 27.
15. P.L. Wagner, *Environments and People*, Prentice Hall, Englewood-Cliffs, 1972, p XII.
16. Wagner, op. cit., (footnote 15) p 5.
17. Y-F. Tuan, "Structuralism existentialism and environmental perception", *Environment and Behaviour*, Vol. 4, 1972, p 331.
18. P. Rieff, *Fellow Teachers*, Harper and Row, New York, 1973; p 13. "It is almost as if a pendulum were endlessly moving from technology to mysticism and back, with little hope of breaking the vicious circle," B. Marchand. "A Dialectical Approach in Geography", *Geographical Analysis*, Vol. X (No. 2), 1978, p 105.
19. P. Rieff, op. cit., (footnote 18) p 39.
20. P. Rieff, op. cit., (footnote 18) p 14.
21. C. Levi-Strauss, *The Savage Mind*, University of Chicago Press, Chicago, 1966, p 21.
22. R. Barthes, ' L'activité structuraliste", *Essais Critiques*, Editions du Seuil' Paris, 1964, pp 213-220.
23. Y.F. Tuan, op. cit., (footnote 17) ; "Geography, Phenomenology, and the Study of Human Nature", *Canadian Geographer*, Vol. XV,

1971, pp 181-192 and "Humanistic Geography", *Annals*, Association of American Geographers, Vol. 66, 1976, pp 266-276.

24. Y.F. Tuan, op. cit. (footnote 23) 1971, p 191.
25. M. Merleau-Ponty, "An Unpublished Text" *The Primacy of Perception*, Northwestern University Press, Evanston, 1966, p 7.
26. J. Derrida, "Structure, Sign and Play", in R. Macksey and D. Eugenio, eds, *The Languages of Criticism and the Sciences of Man*, The John Hopkins University Press, Baltimore, 1970, p 251.
27. M. Foucault, *The Order of Things*, Vintage Books, New York, 1973, p 43.
28. B. Robinson, "A Note on Geographical Description in the Age of Discoveries", *Professional Geographer*, Vol. XXIII, 1971, p 209.
29. J. Derrida, *Speech of Phenomena*, Northwestern University Press, Evanston, 1973, p 159.
30. R. Terdiman, *The Dialectics of Isolation*, Yale University Press, New Haven, 1976, p 241.
31. P. Ackroyd, *Notes for a New Culture*, Vision, London, 1976, p 59.
32. E.W. Said, "Abecedarium Culturae, in J.K. Simon, ed., *Modern French Criticism*, University of Chicago Press, Chicago, 1972, p 363.
33. Ibid.
34. F. Choay, "Urbanism and Semiology" in C. Jencks and G. Baird, eds., *Meaning in Architecture*, Barrie and Rockliff London, 1969, pp 26-37. Another semiology of the cityscape is Roland Barthes" "Semiologie et Urbanisme", *L'Architecture d'Audjourd' hui*, Vol. 42, 1970-71, pp 11-13. For recent collection of semiotic essays on space see "Semiotique de l'espace", *Communications*, No. 27, 1977, *passim.* Amos Rapoport reports that he finds semiotic analysis "abstract". This is understandable because his own approach is that of a culturalist. A. Rapoport, *Human Aspects of Urban Form*, Pergamon, Oxford, 1977, p 329.
35. Ibid. p. 36.
36. Derrida, op. cit., p 159.
37. Choay, op. cit., p 36.
38. E. Heller, *Kafka*, Fontana, London 1974, pp 117-118.
39. Derrida, op. cit., p 265.
40. C. Lévi-Strauss, *Tristes Tropiques*, Criterion Books, New York, 1961, pp 59-60.
41. Lévi-Strauss, op. cit., p 59.
42. Lévi-Strauss, op. cit., p 60.
43. Lévi-Strauss, op. cit., p 60-61.
44. S. Freud, *Civilization and its Discontents*, *The Complete Works*, Vol. 21. Hogarth Press, London, 1951, pp 69-71. For a list of Freud's analogies see A. Wilden. *System and Structure*, Tavistock Publications, London, 1972, pp 125-126. "Metaphor as such is Freud's profoundest paradigm" Ibid., p 126.
45. Said, op. cit., p 361.
46. Said, op. cit., p 361.

47. MacCannell, op. cit.
48. MacCannell, op. cit., p 2.
49. S. Beckett, Proust, Grove Press, New York, not dated, original 1931, p 11
50. MacCannell, op. cit., p. 3.
51. The Beatles, Let It Be, released May 1970.
52. R. Poirier, *The Performing Self*, Chatto and Windus, London, 1971, p 139.
53. Poirier, op. cit., p 125.
54. R. Barthes, *System de la Mode*, Editions du Seuil, Paris, 1967.
55. R. Barthes, *Mythologies*, Paladin, St. Albans, 1973, pp 74-77.
56. Barthes, op. cit., p 254.
57. *National Geographic*, Vol. 144, No. 4 ; October 1973 ; pp 498 and 495.
58. *National Geographic*, Vol. 144, No. 6 ; December 1973 ; p 755.
59. *National Geographic*, *op. cit.*, p. 733.
60. *National Geographic*, Vol. 141, No. 1 ; January 1972 ; p 64.
61. *National Geographic*, *op. cit.*, p. 83.
62. *National Geographic*, *op. cit.*, p. 508.
63. *National Geographic*, Vol. 143, No. 4 ; April, 1973 ; p 543
64. *National Geographic*, op. cit., p 540.
65. Barthes, op. cit., (footnote 55) p 76.
66. *National Geographic*, op. cit., p 534.
67. B. Robinson, "Elizabethan Society and its Named Places", *Geographical Review*, Vol. 63, 1973, p. 332.
68. F. Inglis, "Nation and Community", *New University Quarterly*, Autumn, 1976, p. 446. I am grateful to K. Snyder, Dept. of English, St. Mary's University, Halifax, for drawing my attention to this article.
69. Inglis, op. cit., p. 449.
70. E. Wrigley, "Changes in the Philosophy of Geography", in R.J. Chorley and P. Haggett, eds., *Frontiers in Geographical Teaching*, Methuen, London, 1965, pp 3-20.
71. W. Hoskins, *The Making of the English Landscape*, Penguin Books; Harmondsworth, 1970, pp 270-271.
72. L. Mumford, *The Culture of Cities*, Harcourt Brace Jovanovich, New York, 1970.
73. Mumford, op. cit., p XVII.
74. Mumford, op. cit., p. 20. Tom Wolfe reports. "Las Vegas is the only town in the world whose skyline is made up neither of buildings, like New York, nor of trees, like Wilbraham, Massachusetts, but signs." T. Wolfe, *The Kandy-Kolored Tangerine-flake Steamline Baby*, Simon and Schuster, New York, 1965, p. 6. This is not to deny "the contemporary problem we face of bland buildings that lack the necessary elements for interaction". E. N. Bacon, *The Design of Cities*, Penguin, New York, 1976, p 76. What cannot be advocated is a nostalgia which equates "the demands of the fast movement on

the expressway with that which was needed to create harmony along the Panathenaic Way." Bacon *loc. cit.*

75. D. Ley and R. Cybriwsky, "Urban Graffiti as Territorial Markers", *Annals*, Association of American Geographers, Vol. 64, 1974, pp 491-505.
76. D. Alexander, *Retailing in England During the Industrial Revolution*, Athlone Press, London, 1970, pp 231-238.
77. Alexander, op. cit.
78. Cited in M.D. George, *England in Transition*, Penguin, Harmondsworth, 1953, references on pp 33 and 34.
79. Mumford, op. cit., p 5.
80. J. Hawkes, *A Land* Cresset Press, London, 1951.
81. P. Teilhard de Chardin, "The Antiquity and World Expansion of Human Culture", in W.L. Thomas, Jr. ed., *Man's Role in Changing the Face of the Earth*, University of Chicago Press, Chicago, 1970, pp 103-112. Mumford in the same volume asserted that he felt as if he were speaking with "the voice of another man. . . the voice of Father Teilhard de Chardin". *Ibid.*, p 1152.
82. Hawkes, op. cit., p 221.
83. Mumford, op. cit. (footnote 72), pp 143-222. Mumford's evaluation of Bloom in James Joyce's *Ulysees* must be the most mistaken view of both a city and Joyce's Everyman in print. To describe him as "a dissociated mind in a disintegrated city" is a travesty. *Ibid.*, p 271. There is little doubt that dissociation plays its part in the novel however, as Raymond Williams has put it, "the genius of *Ulysees* is that it dramatises three forms of consciousness. . . Bloom, Stephen and Molly. Their interaction but also their lack of connection is the tension of composition of the city itself". R. Williams, *The Country and the City*, Chatto and Windus, London 1973, p. 245. Vladamir Nabokov felt enough confidence in Joyce to demand a map of Dublin in examinations from his students. V. Nabokov, *Strong Opinions*, McGraw-Hill, New York, 1973, p 55. It is hardly surprising therefore that the White's place Mumford's antipathetic views of disgust in the wider context of "almost every variety of anti-urban feeling since the Civil War." Morton and Lucia White, *The Intellectual Versus the City*, New American Library, Toronto, 1964, p 206. The phenomenologist John McDermott puts the whole thing in perspective : "Many critics of the city have no sense of the pulse of city life except that prescribed by those cultural canons continuous with the lives of the critics. Given such an interpretive context, as we move through the alleged travail of everyday experience, we seem reduced to the impropiety and therefore the social inability to generate non-organic aesthetic metaphors able to provide the security and affection nostalgically traceable to nature metaphors". J.J. McDermott, *The Culture of Experience*, New York University Press, New York, 1976, p 194.
84. Inglis, op. cit., p 543.
85. Bachelard, op. cit., pp 136-147.

86. R. Hoggart, *The Uses of Literacy*, Chatto and Windus, London, 1957, p 52.
87. A collection of readings much influenced by the importance of childhood and by Jacquetta Hawkes is P. Shepard and D. McKinley, eds., *The Subversive Science*, Houghton Mifflin, Boston, 1969.
88. Hoggart, op. cit., p 56.
89. Hoggart, op. cit., p 56.
90. Y.F. Tuan, "Place and Space," *Geographical Review*, Vol. 65, 1975, pp 164-165.
91. Marx, op. cit., p. 365. The reference is to Marx's sanguine use of the word symbol.
92. Barthes, op. cit., p 12.
93. Zelinsky, op. cit., p 112.
94. J. Frank, "Spatial Form in Modern Literature", *The Widening Gyre*, Indiana University Press, Bloomington, 1968, pp 59-60.
95. Poirier, op. cit., pp 138-139.
96. Frank, op. cit., p 13.
97. I. Hassan, *The Dismemberment of Orpheus*, Oxford University Press, 1971, p 21.
98. J.P. Sartre, *Being and Nothingness*, Washington Square Press, New York, 1966, p 781.
99. S. Leys, *Chinese Shadows*, The Viking Press, New York, 1977, p 55.
100. S. Leys, op. cit., p 60.
101. S. Sontag, *On Photography*, Farrar, Strauss and Giroux, New York, 1978, p 174.

6

# Concept of Morphological Analysis—Some Theoretical Postulations

—*R.L. Singh and Rana P.B. Singh*

> The waves of the sea, the little ripples on the shore, sweeping curve of the sandy bay between the headlands, the outline of the hills, the shape of the clouds, all these are so many riddles of form, so many problems of morphology . . . .
>
> —D'Arcy Thompson on *Growth and Form*

THE REALITY of this world is the stability of mankind, as such man is considered as the core concern of any study. This stability may be disturbed with three characteristics of human life : reproduction—ability to produce an organism similar to oneself ; heredity—ability to transmit parental characters to the offspring; and variability—ability to exhibit variation.[1] These characteristics form the way of existence, continuity and variety which jointly process the development of life. This system is extremely complex—its levels, orderings and identical features are the only visualization through man's proxemics. Lowenthal has rightly remarked : "The lineaments of the world we live in are both seen and shaped in accordance, or by contrast, with images we hold of other worlds—better worlds, past worlds, future worlds We constantly compare the reality with fancy. Indeed, without the one we could neither visualize nor conceptualize the other".[2] Actually what a man perceives is his world. According to Thom "We perceive beings, objects, things to which we give names. These beings or things are forms or structures endowed with a degree of stability; they take up some part of space and last for some period of time".[3] This manifestation of relations and its awareness can be better understood by the way of space-time-continuum (as postulated by Albert Einstein)[4], which can be visualized in a system of very complex and sophisticated organization in a multidimensional space. To analyse this complex system, the concept of morphological analysis has been postulated.

The concept of morphology has dual expression—as generic quality it views the non-existent processes, and as visual imprints it shows man's achievement along with the organization system. The generic quality is expressed through the time-series while the latter phenomenon is associated with the structural stability in spatial context.

Three basic approaches—retrogressive, retrospective and prospective—are complementary views to each other in the time-series. The former method projects backwards into the past; the second approach throws light on the present form; while the last is concerned with the future.[5] Since these three methods jointly analyse the regular action or succession of actions taking place, it could always be conceived as stochastic.[6]

The concept of structure is the second characteristic which is the visible aspect of anything; it often implies a value judgment like orderly arrangement or regularity.[7] It provides the base to explore the inter-relationship between positivism, materialism and phenomenology, finally resulting in the interpretation of the social reality. It is more than the Euclidean space ($E^3$) as its attributes—length, width and depth are perceived within the fourth dimension, i.e., time. In fact, space constitutes the natural environment in which man lives, acts and rests. With this conception Euclidean dimensions of the typology of space are analysed in corroboration to the concept of territory which is perceived and visualized through different lenses of thinking and proxemics; it has been rightly noted that "the spatial and social isolation of a territorially based group of people is reflected in the way that they view the outside world".[8] It is this notion that each settlement reflects its own spatial behaviour with three characteristics—form, visibility and use—which are the reflection of sequent occupance, value system and selectivity.

To integrate the dual frame of space and time—existence and prediction, form and function, layers and levels, structure and system, stability and growth, a morphological analysis is a necessary consideration as it represents the totality at any level.

## APPROACHES TO MORPHOLOGICAL ANALYSIS

Morphological analysis is recognized as the concept integrating causal relationship among the four groups of knowledge—biosciences, physical sciences, humanities and geonomy. The visions of reality and tangibility of proxemics are both expressed with this concept

The fundamental consideration of bioscientists is to analyse the structure-cells which form a total system when they join together. One may note that "the liver, spleen, kidneys, stomach, intestines and sexual glands are attached by the folds of the peritoneum to the inner surface of a large cavity whose bottom is formed by the pelvis, the sides by the abdominal muscles, and the roof by the diaphragm". Here two disciplines—genetics and anatomy study the 'why' and 'how' respectively. On the micro-level the morphological analysis is concerned with the structural relationship among the various cells making a form. Although various components of the form are analysed separately, basically cells and medium, structure and function cannot be separated at any level—the isolation of cells from their natural environment is altogether unwarranted.[9] In fact, each cell is a very complex organism. Cells belong to many different races or types as defined by both their structural and their functional characteristics. Taking this distinctiveness the cell type is roughly divided into two classes—the fixed or basic cells whose associations form the tissues and the organs, and the mobile cells which travel throughout the entire organism. It is well-known that cells unite in dense masses, the tissues and organs, whose architectonic depends on the structural and functional needs of the organism in its totality.[10]

In developmental biology the above totality is analysed within a three-phase growth (i) histogenesis, referring to the changes occurring with time in a cell or group of cells in the biological realm, (ii) pattern formation, referring to differences in space rather than in time, and finally (iii) morphogenesis, referring to the processes by which the embryo changes its shape. With this view it is evident that morphogenesis is the process by which the final form is produced after the formation of the skeleton.[11]

In the physical sciences, morphological analysis is applied very widely in corroboration to structural analysis. Basically, it deals with the form, reactions and relationship among various compounds and components. The results finally lead to the preparation of a taxonomy showing distinction and similarities.

In atom-analysis, Mendeleef's modified law that "the physical and chemical properties of elements are periodic formation of their atomic number (i.e., number of protons)" is very significant. Depending upon the number of electrons, protons and neutrons, the three parts of an atom, the atom character will vary. Even protons and neutrons may be further subdivided into particles known as *quarks* which have to be 1/3 of fraction charge to that of an electronic charge (this is yet to be experimentally proved). Due to some similar arrangements of electrons, protons and neutrons, there exist similar properties among the atoms; on the basis of atomic properties the total elements (105) are arranged into seven groups.

Morphological analysis is more profoundly used in identifying the characteristics of compounds and crystals. X-rays of a compound are taken in the context of the difference between lattice points in order of wavelength and thus the structures are fixed. Depending upon the parameters of unit cells (it is the smallest repeating unit in space the lattice which when repeated over and over again results in a crystal of the given substance), i.e., the intercepts (a, b, c) and the interfacial angles ($\alpha$, $\beta$, $\gamma$) the crystals are divided into seven systems—cubic, tetragonal, orthorhombic, rhombohedral, monoclinic, triclinic and hexagonal.* Similarly, with the help of X-rays, physicists are unveiling the architectonic of the molecules of the simpler substances of the tissues, i.e., the spatial relations of the atoms constituting those molecules. With the aid of most complicated and varied techniques, studies are now going on concerning the larger structures resulting from the aggregation and organization of molecules, the cells of the tissues and of the blood. Within such contexts morphological analysis are

* It is noted that the distances a, b and c give the length of the edges of the unit cell and their angles $\alpha$, $\beta$ and $\gamma$ give the angles between the three imaginary axes.

applied to study structural-functional inter-relationship among the organisms forming its totality.

In the humanities the concept of morphology is applied to study those social institutions phenomenologically which are inter-related with socio-economic norms and bonds and their functional as well as interactional system supporting decisions, transformation and communications among the complex structures of various groups or cultures within the spatial matrix. No social group or unit is as simple as it appears. With the morphogenetic analysis one may find that the roots of the processes and norms responsible for the growth of that system are attached to the ancient past with the maintenance of inter-linkage from one generation to another adding certain new traditions and beliefs. This relationship is the causal factor for developing mutual cohesiveness among the persons within or beyond a social unit in spite of stratifications found at many levels. Actually stratification is considered to be the main distributive arrangement of society accounting for "who gets what, how and why."[12]

In geonomy the basic consideration lies on the morphological analysis in terms of processes and materials within a time frame. In a very simple way, thus, morphology (M) may be defined[13] as the function of processes (P) and materials (A) over time (t), i.e.,

$$M=f\,(PA)\,dt$$

Projecting the growth trend with a consideration that time is the controlling factor the equation may be changed as :[14]

$$\frac{dM}{dt}=\frac{d}{dt}(PA)$$

With this notion landscape is viewed as "the function of process, stage and structure"—the aspects leading to the crystallization of distinctive structure and patterns. Such distinctions are perceived by geographers in their concept of 'areal differentiation'.

## The Morphology of Landscape

Since its existence as an independent discipline, geography has been recognised as the science of the earth's surface similar to the German view of *Erdkunde*, the science of earth in general; but it does not mean that general geography is general earth science; it expresses rather the areal differentiation found over the space—that is locational characteristics. This could be analysed through the systematic study of various geo-elements within a regional context, which are visualized in the frame of morphologic systems. In this system the relationship of elements to each other is searched in a manner that shows the trend of either growth or decline. Any disturbance in the basic scale of consideration will cause disturbance in the pattern of trends, as such it is very important to make the basic scale. The system of the growth and existence of any landscape may be divided into three types—morphologic, process-response and control. These systems are responsible for the existence, development and maintenance of the balance (equilibrium) either in the physical or the cultural landscape, e.g., river basin or human settlement. Such an approach provides a base for the realisation of the reality of the union of physical and cultural elements. Sauer has noted that "The content of landscape is found, therefore, in the physical qualities of area that are significant to man and in the forms of his use of the area, in facts of physical background and facts of human culture."[15] It will be worthwhile to examine the concept of morphology in the case of settlements in somewhat greater detail as a specific example.

## MORPHOLOGY OF SETTLEMENTS : A GEOGRAPHER'S APPRAISAL

Settlements are perceived as one of the basic elements of the system between Man and Society. According to Doxiadis, "the space, natural or man made, used by man, is a part of the settlement . . . (and) the space of a settlement accommodates all functions of man and can be recognised by his imprints on nature".[16] Phenomenologically it refers to the manner of identification, classification, regionalization, structural analysis, system analysis as well as the analysis of other associated components. Thus, it provides a meaningful answer to what, which and how of settlement system.[17]

In the settlement context the built up area is the most intensively used part where dense expression of various functions of physical and cultural attributes can be seen. Thus the built up area becomes the nodal point for a functioning community which covers the surrounding space. Morphology of settlement is the expression of these two norms of the system, which can be divided into four basic parts—(a) homogeneous part (fields/houses), (b) the central part (built up area/shop), (c) the circulatory part (road/lanes) and (d) the special part (any unit of distinction).[18] These elements are closely associated with three dimensions, *height*, *length* and *width*, and since settlement cannot exist without their function and without people, that cannot be separated from the fourth dimension, *time*.

Applying the biologic sense of organic analogy, society may be conceived as a complex organization within the dynamic cycle of mechanism and process of change; this cycle is further operated at different levels by two sub-systems : (i) organizational and (ii) transactional. This operation is carried out "through the supporting processes of *decisions*, *transformations* and communications, all of which are embedded in a spatial matrix".[19] The other face of the cyclic developmental processes is the understanding of all the attributes through various dimensions, expressed as *Total Quantum* by Stafford Beer, which contains three facets of the morphology in total—plurality, unit and energy; plurality denotes the identity of various elements, unit shows the level of organization and its structure, and energy bounds the elements and organizational level into one system.[20]

The organizational pattern is interrelated with distance and space-articulation which are the functions of the dimensions of the process of interaction; it includes at least three ingradients of human group territoriality.[21]

(*i*) *Sense of spatial identity* : Representing an extension of personal space into larger socio-spatial sphere and usually manifested in the development of territorial symbolism or iconography. This notion is related with physical distance and its correlates, e.g., the time distance, cost distance, perceived distance, etc.

(*ii*) *Sense of exclusiveness* : Showing the socio-cultural distance, because human territoriality is usually associated with an areal homogenization of various distinctive features : race, ethnic group, economic class, political ideology, etc.

(*iii*) *Compartmentalization or channelizing of human interaction in space* : This concept is associated with activities and communication within localized areas inserting boundaries as it were into normal distance—decay relationship. This concept may also be termed as *functional distance*.

These ingredients are the subject of time, which further affects the patterning and crystallization of territoriality and spatial, functional systems, vigorously. These ingredients also tend to form a joint structure to be called as "spatial pattern" with a supplement of "spatial processes." It has rightly been advocated that "people generate spatial processes in order to satisfy their needs and desires and these processes create spatial structures which in turn influence and modify spatial processes".[22] Winston Churchill has rightly remarked that "We shape buildings, buildings shape us".

Taking functions and activities, settlements are broadly grouped into two types—rural and urban; the former is characterised by the dominancy of primary activities while the latter is represented by a concentration of tertiary activities.

The most visible imprint over the landscape is the buildings "whose dimensions are defined by the dimensions of man as related to his functions, and not by man alone".[23] The basic scale in this concern starts from the primary activities, but it is overthrown by the environmental vision in comprehension, as such settlement may be defined as a multi-dimensional unit. For example, an Indian village may be considered as,

> . . . . one of the basic units of Indian culture in itself, represents the enduring works of humanity passing through the corridors of time and testifies to the success of Indian civilization in managing the problems of survival. Thus, the Indian village further represents the impact of Indian civilization on particular places and explains the pattern of life in rural India contributing the energy for its maintenance. Through their fundamental identity, the concept of Indian village-ecology may be further elaborated which symbolizes the Hindu sense of unity with nature and provides a recognition of the integration of life.[24]

The concept of social and functional space in an Indian village constitutes together the culture of the area—the intervening variable between "human organism" and environment". This ideology can also be used in acculturation studies. This postulation (may also be termed behavioral-spatial structure) in the cultural personality of the Indian village is the result of two norms, e.g., religio-ritual norm as a model of distance maximization, secular norm as a model of distance minimization; these jointly together form equilibrium in the socio-spatial organization.[25]

The concept of urban morphology mainly refers to the form (including shape) or structure in spatial context.[26] In the context of time it also refers to pattern formation along with growth. The physical and biological concepts of form evaluation have been applied in urban areas quite elaborately elsewhere.[27]

## MORPHOLOGY : COMPONENTS AND THEORY

Applying the system of ekistic units, the basic components of morphology can be divided into five categories : nature, man, society, shells and networks; these joined together to form the optimization of the quality of man's relationship with his environment. The scale of settlement and its morphological structure varies from a very simple and small structure, a room, to a very complex and sophisticated structure, a universal city that Doxiadis called *Eucumenopolis*.[28]

With a view to elaborating the Sauerian concept, Schaefer considered geography essentially as morphological with no reference to time or change.[29] Similar observations have also been postulated by Bunge, who reiterates this outlook with the remark that "geography is the science of locations".[30] Such conceptions have not considered the concept of biological evolution which forms the organism, while Sauer has emphasised the morphological concept in corroboration to the growth and development of biological organism. According to Sauer the notion of morphology rests on the following postulates : "(1) that there is a unit of organic and quasi-organic quality, that is structure to which certain components are necessary, these component elements being called 'forms' . . . .; (2) that similarity of form in different structures is recognised because of functional equivalence, the forms then being 'homologous';

and (3) that the structural elements may be placed in series, especially into developmental sequence, ranging from incipient to final or completed stage".[31] Although Schaefer has accepted his analogy of development and time in context with processes, he basically stressed upon the analysis of forms only. Bunge has also supported this idea. But the emerging trends in geography have been more conscious towards the'process model' approach, which certainly "involve inclusion of a time element, either explicitly or implicitly. . . .; time, like space, may also be conceived as relative as well as an absolute quality".[32] This further innovates the concept of "process-form-function" which is purely a comprehensive morphological consideration.[33] It is similar to that of the Davisian phase that "landscape is the function of process, 'stage and structure'. The Darwinian theory of evolution, Mainx's recognition of biology as a synthesis of morphology, physiology and genetics, are some of the similar contemporary ideas.

In order to comprehend the notion of morphology in terms of its properties, its distinctive features and the causes behind its evolution, it is analysed through two approaches—*atomistic*, and *holistic*. In the former approach "components" of structure and their properties are identified in order to understand how they come together to form the whole; it means that one must find the components first before knowing what the structure is? Atkin has rightly remarked that, "it is almost as if we expected to invent the land before the man".[34] The *holistic* approach proceeds to obtain a more precise view of structure—how the structure might be changing with time; here the same ordering and hierarchies may be derived to know the trend of growth and formation of the whole. In the spatial context it can be shown that the atomistic approach considers the locational characteristics in three-dimensional space ($E^3$), while the holistic view rests upon multi-dimensional space. This latter approach is a sign of the creative emergence of scientific investigation as well as the introduction of the integration of essence of various disciplines.[35]

It is to be noted that each structure, however, simple it may be, certainly forms un iquenessin itself in spite of differences in the degree of individualization of its components (e.g., in the case of the Indian village as discussed in the sequel). This uniqueness has a double origin : it comes simultaneously from the constitution of the basic structure in spatial frame, and from its development and history. Taking a component at micro level, e.g., socio-spatial structure, it is to be found that it depends on the affinity between norm and values of the residents and the environment surrounding them. Even at a further micro level taking human behaviour, it may be expressed that the perceptive view (proxemics of the residents) seems to be transmitted like electromagnetic waves, from one spatial unit to another.[36] The equilibrium of the society within the Morphological context consists of a well integrated whole of man including his moral and aesthetic values, and his organic activities. The intensity of equilibrium determines the social level of an individual in the spatial unit of occupance. Reciprocally, the behavioral norms are mainly influenced by the social environment as profoundly as by the fluids of the body.

The application of time factor into morphological analysis introduces the concept of morphogenesis which refers to the analysis of the form and system or network of socio-cultural attributes of a settlement unit in a spatio-temporal frame.[37] This hypothesis further assumes that in each structural system, there is a socio-cultural unity in which all components contribute in a harmonious way to its existence and continuity. Morphology as a biological organ may be compared to man who is constituted in his fourth dimension (time) by a series of forms following, and blending into each other. Similarly, the morphological growth of a village may be identified with phases or stages like egg, embryo, infant, adolescent, adult, mature and old; all these stages are the expression of physical and cultural elements.

To develop the language of morphological structure, mathematics can provide the most potential base, especially topology which can easily express the interactional pattern and the degree of connectivity. Finally it would be helpful in identifying the structural as well as functional characteristics. With the application of topologic techniques, even the smallest particle of occupance can be filmed and magnified to such an extent that each layer could be identified very properly. This technique will play "an important healing role"[38] in analysing morphological character. The application of catastrophic theoretic frame is another useful base in morphological analysis, especially where genetic factors are involved. Very frequently it is found that some form of uncertainty lies in pattern formation and pattern regulation, which causes sudden disturbance in the equilibrium. Of course, such events are very familiar but their explanation could be scientifically presented through the catastrophe theory.[39] The internal and external variability making equilibrium in the morphological system, tends to play the role of cause and effect, the former representing controlling attributes influencing the latter, which represents the resulting behaviour pattern. In such a system the model of catastrophic changes may be useful to analyse the level of uncertainty. These applications will provide methodological insights into newer spaces characterizing human society and environment.

## CONCLUSION

To consider geography as a dynamic science, it becomes necessary to view morphology from a new attitude based on harmonious relationship of space-time-continuum rather than conquest. Similar to ecological systems diversity and adaptations are necessary in morphological considerations which are, in turn, helpful for the survival as well as growth of a discipline. Since the concept of morphology shows changing trends, growth character in totality through different angles whatsoever its size and level, it is useful for prediction. The morphologic generalization will be further applicable to new structures to some extent.

In conclusion, it may be synthesised that morphology may be conceived as the areal reality, structural stability and equilibrium among the components and the transactional system analogous to that of Buckley's concept of "homeostatic and deviation-amplifying processes"[40] and Maruyama's morphostasis and morphogenesis.[41] Similarly, Ashby (1951) constructed a dynamic device that operates the process of adaptation to its environment, known as *homeostat*, (from *homeostasis*) the maintaining of the properties of a system without certain desired limits.[42] This is also similar to Eichenbaum and Gale's "process-form-function"[43] conception as well as the Singhs' concept of morphogenesis of landscape (physical and cultural)."[44]

## REFERENCES

1. L. Rastrigin, *This Chancy, Chancy, Chancy World*, Moscow, Mir Publishers, 1973, p 232.
2. David Lowenthal, "Introduction" in *Geographies of the Mind, Essays in Historical Geography*, eds. David Lowenthal and Martyn J. Bowden, New York: Oxford Univ. Press, 1976, p 3.
3. Rene Thom, *Structural Stability and Morphogenesis*, Reading : Benjamin Inc., 1975.
4. See R.L. Singh and Rana P.B. Singh, *Spatial Planning in Indian Perspective*, Varanasi, N.G.S.I, Res. Bulletin No. 28, 1978, p 7-9.
5. R.L. Singh, K.N. Singh, Rana P.B. Singh, eds., *Geographic Dimensions of Rural Settlements*, Varanasi, N.G.S I. Res. Pub. No. 16, 1976, XV.
6. J. Eichenbaum and S. Gale, "Form, Function and Process : A Methodological Inquiry, "*Economic Geography*, 47-4, 1971, p 526.
7. *Ibid.*
8. R.M. Downs and D. Stea, *Maps in Minds, Reflections on Cognitive Mapping*, New York, Harper & Row, 1977, p 78.
9. Alexis Carrel, *Man—the Unknown*, London, Hamish Hamilton, 1959, p 77.
10. Carrel, p 81.
11. L. Wolpert, et. al. "Cell Movement and Cell Contact in Sea Urchin Morphogenesis," *Endeavour*, 26, 1967, p 85-90.
12. D.G. Mandelbaum, *Society in India*, Berkeley : University of California Press, 1970, p 661.
13. K.J. Gregory, "A Physical Geography Equation" *National Geographer*, 7-2, 1977, 137-141.
14. See J.B. Thornes and D. Brunsden, *Geomorphology and Time*, London, 1977.
15. Carl O. Sauer, *Land and Life*, ed. John Leighly, Berkeley : University of California Press, 1965, p 325.
16. C.A. Doxiadis, *Ekistics : An Introduction to the Science of Human Settlements*, New York, Oxford Univ. Press, 1968, p 24.
17. R.G. Golledge *Process Approaches to the Analysis of Human Spatial Behaviour*, Ohio, Ohio State Univ. Discussion Paper in Geography, No. 16, 1970, p 2.
18. C.A. Doxiadis, pp 27-28.
19. R.J. Johnson, *Spatial Structures*, London, Methuen, 1973, p 4.
20. Stafford Beer, *Platform for Change*, London, John Wiley, 1975, p 27.
21. E. Soja, *The Political Organization of Space*, Washington D.C., Asso. Am. Geog., Res. Paper No. 8, 1971, p 34; also Rana P.B. Singh *Clan Settlements in the Saran Plain*, Varanasi, N.G.S.I. Res. Pub. No. 18 1977, p 118.
22. R. Abler, J.S. Adam, P. Gould, *Spatial Organization*, Englewood Cliffs, Prentice Hall, 1971, XIII.
23. Doxiadis, p 23.

24. Rana P.B. Singh, p 121.
25. Idem, pp 140-145.
26. See T.G. Jordan and L. Rowtree, *The Human Mosaic, A Thematic Introduction to Cultural Geography*, San Francisco, Canfield Press, 1976, p 103.
27. See for details, R.L. Singh, "Social Factors in the Morphogenesis of Varanasi : A Suggestion in Technique for Form Evaluation" in *Urban Geography in Developing Countries*, ed. R.L. Singh, Varanasi, N.G.S.I. Res. Pub. No. 13, 1974, 3-27; also K.L. Taneja, *Morphology of Indian Cities*, Varanasi : N.G.S.I. Res. Pub. No. 8, 1971.
28. Doxiadis, pp 34-35.
29. F.K. Schaefer, "Exceptionalism in Geography : A Methodological Examination", *Annals, Asso. Am. Geogr.*, 43, 1953, 243.
30. W. Bunge, *Theoretical Geography*, Lund Studies in Geography, Series C No. 1 1966, p 199.
31. C.O. Sauer, ref. 16, p 326.
32. B.N. Boots and A. Getis, "Probability Model Approach to Map Pattern Analysis", *Progress in Human Geography*, Vol. 1-2, 1977, 267; see also A Getis and B.N. Boots, *Models of Spatial Processes : An Approach to the Study of Point, Line and Area Patterns*, Cambridge Univ. Press, 1977.
33. J. Eichenbaum and S. Gale, ref. 6, p 537.
34. R.H. Atkin, "An Approach to Structure in Architectural and Urban Design", *Environmental and Planning*, B, 1, 1974, p 52.
35. See R.A. Atkin, *Mathematical Structure in Human Affairs*, London, Heinemann Educ, Books, 1974; also his *Combinatorial Connectivities in Social Systems*, Basel : Birkhauser Verlag, ISR No. 34, 1977.
36. It is comparable to A. Carrel's remark, see ref. 9, p 245.
37. R.L. Singh and Rana P. B. Singh, "Some Approaches to the Morphogenesis of Indian Village" in *Readings in Rural Settlement Geography*, eds, R.L. Singh, K.N. Singh, Rana P.B. Singh, Varanasi, N.G S.I. Res. Pub. No. 14, 1975, p 367.
38. Peter Gould, "Mathematics in Geography : Conceptual Revolution or New Tool ?" *International Soc. Sc. Jl.*, 26-2, 1975, p 322.
39. See Rene Thom, also E.C. Zeeman, "Catastrophe Theory" *Scientific American*, 234-4, 1976, p 65-83.
40. W. Buckley; *Sociology and Modern Systems Theory*, Englewood Cliff : Prentice Hall, 1967.
41. M. Maruyama; "Morphogenesis and Morphostasis", *Methods*, 12-48, 1960, p 251-96.
42. L. Rastrigin; ref. 1, p 239.
43. J. Eichenbaum and S. Gale; ref. 6, pp 525-544.
44. R.L. Singh and Rana P.B. Singh; pp 367-375, also R.L. Singh.

7

# Systems Approach in Geography

*—Satish K. Davgun*

APPLICATION of "System Analysis" to solve many problems, which the scholars in different branches of science face, is comparatively a new development. It was mostly after the Second World War that such notions as "General System Theory", "System" and "System Analysis" became quite popular and common in scientific literature. General System Theory was first developed in the United States and has been introduced to other parts of the world in recent years. "In the early 1950s the availability of large digital computers with enormous data-processing capabilities gave great impetus to the data-oriented systems approach."[1] Bertalanffy may be the pioneer in popularizing the "General System Theory".

The notion of "System" is of central significance in present day science, life and society. "In many fields of endeavour, the necessity of a 'system approach' or 'systems thinking' is emphasized, new professions called 'systems engineering', 'systems analysis' and the like have come into being, and there can be little doubt that this concept marks a genuine, necessary and consequential development in science and world view."[2] This new development of science is becoming quite popular in many branches of knowledge.

Many scientists used to investigate part of the "whole" of the complexity of the real world. In this way a group of scientists specializes in one branch of knowledge. The investigations in different branches, if compiled together, used to know about the "whole". Instead of inspecting the "reality" from different points, one alternative is to study it in its totality. "All (such) studies have, as their central theme, the analysis of the manner in which the components of each section fo the real world are internally structured (i.e. organized, linked, or related) and, further, how each section links to other structures."[3] These structures are "systems". Our world can be viewed as made up of interconnected systems, at different scales, forming a systems hierarchy i.e. sub-systems, systems, super-systems, etc. While there are many scientists who are in favour of this new philosophy of science, some others are not quite ready to accept this philosophy.

### WHY GENERAL SYSTEM THEORY ?

It has been realized that one cannot find answers to questions about "reality" as long as one remains within the confines of a single discipline. The knowledge of the various branches of science (knowledge) is to be pooled together to understand this reality. During the early period of development of the General System Theory, there have been many objections against its usability, but gradually it was realized that such objections missed the point of what the systems theory stands for, namely, attempting scientific interpretation and theory where previously there was none, and higher generality than that in the special sciences."[4]

## Systems, their Attributes and Classification

A system is a set of inter-related elements. It is a set of components interacting with each other and operating together as a complex whole, according to some observed pattern. A system has a boundary, so have its components, beyond which they do not or cannot function. The components can be termed sub-systems. The social structure of the Hindu society, is a good example of a system. A particular Hindu caste itself is a system, with sub-castes as components. These sub-castes interact with each other and operate within the caste complex. The periodical health centres in rural areas of India, are the components of the rural health care system. The components of a system mutually take inputs and produce output. For example, new medicines are provided to rural health care system (organization) which is a sort of input; now the distribution of that medicine to rural people is output, and in this process all components (periodical health centres) take part. A "system" is, in fact, a conceptual entity rather than a clearly defined empirical one.[5]

Chorley and Kennedy[6] have classified systems in two ways—one depending upon function, another depending upon the structure. Based on functional division, systems can be labelled as isolated, closed and open systems. An isolated system is that which does not receive any input and thus has no output. This type of system can only be had in the laboratories, instead of in the real world. Hence, in social sciences we do not have isolated systems.

Closed systems are those systems with such boundaries which prevent the import and export of mass, but not of energy. In a closed system the only change of state that can occur is a result of the work that is done, as any initial differences in energy in various parts of the system are equalized that is; as the system tends toward entropy.[7] Chisholm concludes that there is no system which can be classified as closed system. Open systems are characterized by an exchange of both mass and energy with their surroundings. The majority of the systems are open systems. "An open system is defined as a system in exchange of matter with its environment, presenting import and export, building up and breaking down of its material components."[8] A developing country is a good example of an open system. The technical know-how or technicians which she imports are the inputs, and the exports of some finished products are the output, while the raw material, labour, factories and other elements act as components. Similarly, viewing it from another level, an industry itself is an open system; raw materials and energy are the inputs, manufactured products are the output, and different sections of production (machines etc.) are the components. Another example is of the canal system (for irrigation): release of water into canals from headworks is input and release of water from canals to fields (farms) is the output. Open systems are the commonest type of systems, classified on the basis of functions. There can be another type of classification of systems.

On the basis of their internal complexity, systems can be grouped as morphological systems, cascading systems, process-responses systems, control systems and others. A morphological system has the physical properties such as geometry, composi tion, strength and others. One can quote the example of a city. As more and more people move away from the central part of the city to the outskirts, it results in new housing units at the periphery and deterioration of living conditions in the central part of the city, giving place to non-residential land uses. And at the outskirts, there will be more construction work for houses, shopping centres and others. "The relationships between the (se) parameters (of a morphological system) can be expressed by a web of correlations, . . ."[9]

In the case of cascading systems, the output from one system becomes input to another system. It is like the product of one industry being processed in another industry to produce something. Step-wise migration is an example of cascading systems. It is like the development of new agricultural technique in a research system and then that innovation becomes input in the extension system. The commodities which a country imports, become the output of the international trade system, and when these commodities are handed over to the internal distributional system it becomes input in that system.

The process response systems are formed by the intersection of morphological and cascading systems. In this type of system, the external input in any of the components transmits changes in the system. Let us take for example, the case of the transportation system in an area with rail, road and air transportation. If, due to a strike, the airlines stop their flights, it will result in changes in road and rail transportation. In the same way if the number, timing and direction of rail traffic is altered, it will affect the road and/or river transportation. Another example of process-responses system lies in the changes in the agricultural techniques during the Green Revolution in N.W. India which is resulting in changes in the whole agricultural system there.

Control systems, on the other hand, are designed to create equilibrium in the input and output of that system. The example can be given of the cotton textile industry. If due to some reasons, the production of raw cotton is less in one year, so the input in the cotton textile mills will be less, which will automatically result in less production. There can be more classes of systems depending upon internal complexity.

Based upon the scale, there can be a hierarchy of systems. Laszlo[10] has pointed out different levels of systems. Systems belonging to a level below that of any chosen level are called sub-systems in relation to the system of the chosen level, and the system belonging to the next higher level is the supra-system in relation to it. In the case of migration, our chosen level is migration from small city (town) to big cities; so the migration from villages to towns is sub-system, and from city to a cosmopolitan area is supra-system. In nature, there are different levels of systems from atoms to system of metagalaxy. But geographers are concerned with the system in the middle of this hierarchical system.

## APPLICABILITY OF G.S.T. IN SOCIAL SCIENCES

Now the question arises whether General System Theory is applicable in social sciences or not. There are some scholars who do not accept that the System Theory can be applied in social sciences, while a number of social scientists believe that General System Theory and System Analysis have application in social sciences.

Chisholm[11] is one of those geographers who do not accept the General System Theory in geography. His first objection is about the delimitation of boundaries of the systems. He says it is arbitrary and objective. But to study the complex nature of the whole, one needs to break it down into segments. Another criticism which he raises is the problem of defferentiating between open and closed systems, and he remarks that in geographical studies all systems are open; so it is rather difficult to distinguish one open system from another system. Chisholm concludes that as all systems are open systems, which are, according to him, arbitrarily defined and the main concentration is on processes and relationships between different systems, then it is just plain common sense instead of the General System Theory.

LaPatra[12] points out some objections towards the applicability of a systems approach in social sciences. Social processes are not amenable to the methodology of a systems approach. Social phenomena are too complex to be analyzed. If one has to analyze the employment system, one will have to include other systems such as educational system and system of individual behaviour. System analysis has intruded into many traditional social science areas, but the results have not been appreciable. Studies of crime, waste management, mass transportation, malnutrition and others have been done, without giving due consideration to political and economic factors and so these previous studies have caused a sort of disillusionment among people.

There are some scholars who advocate the use of system theory and system analysis in geography. It has been pointed out that there is a great potential for applying "system" in geography. Very few research work has been done in geography with the orientations of system theory. Foote and Greer-Wootten[13] have concluded in their study, that a systems approach overcomes the descriptive-compilation in geography. Geography, according to them is concerned with the analysis of the vast system of the earth's surface comprising man and the natural environment.[14] Though there are many difficulties involved in applying systems concepts to geographic research, yet for the fullest development of geography as a science the systems theory and system analysis are quite necessary in geographic research.

## REFERENCES

1. Jack W., LaPatra, *Applying the Systems Approach to Urban Development* (Dowden, Hutchinson & Ross Inc., 1973, p 10.
2. Ervin Laszlo, *Introduction to Systems Philosophy : Towards a New Paradigm of Contemporary Thought* (New York : Harper Torchbooks), p xvii.
3. Richard J. Chorley and Kennedy, A. Barbara, *Physical Geography : A System Approach* (London : Prentice Hall International Inc., 1971, p 1.
4. Ludwig von, Bertalanffy, *General System Theory; Foundations, Development, Applications* (New York : George Braziller, 1968), p 14.
5. T.J. Wilbanks and R. Symanski "What is System Analysis," *Professional Geographer*, Vol. 20, 1968, p 82.
6. Chorley and Kennedy, op. cit., pp 2-10.
7. M. Chisholm "General Systems Theory and Geography *Transactions of Institute of British Geographers*, No. 42 (1967), p 46.
8. Bertalanffy, op. cit., p. 141.
9. Chorley and Kennedy, op. cit., p 5.

10. Laszlo, op. cit., p 29.
11. Chisholm, op. cit., pp 45-52.
12. LaPatra, op. cit., pp 11-14.
13. Don C. Foote and Bryn Greer Wootten, "An Approach to Systems Analysis in Cultural Geography," *Professional Geographer*, Vol. 20, 1968, p 89.
14. Foote and Greer-Wootten, op. cit., p 86.

# 8

# Behavioral Geography—Defined and Defended

—*Rana P.B. Singh*

> From the neglected womb of spatial analysis thus sprung the unwanted child of Behavioral Geography
>
> —Gunnar Olsson, 1974

THE VAST content of geographic researches may be grouped, in terms of approaches, into two: *Spatial analyst* and *structuralist*, in which the former searches for causal explanation, i.e. cause-effect relationship, and the latter concerned with the human action and activities; grouped in fact, for practical understanding. Within these two schools there has developed a large number of systematic fields, whose orientation may be grouped into three types : (i) the treatment of geography as a behavioral and social science, in spite of a potential platform for physical components, (ii) consideration of geography as a research and educational discipline, and (iii) geography as the byproduct or selectived activization.[1] Among these typologies the first group has received much attention with a view to promoting geography as an integrating science.

## DEFINING APPROACH

It is reasonable to conceive behavioral geography as the core concern of social geography. In fact, as recently defined, social geography seems to concern itself "with human spatial behaviour and the derived geographical patterns from the point of view of society;"[2] whatever may be the conception and compass of the human geographic approach, it certainly rests upon man and his products in context with the locational behaviour where policies are formulated and finally decisions have to be made. Proceeding in this direction the acquaintance with the human behaviour pertaining to the *decidability* and consequently their *proxemics* with the surrounding environment is an essential task. The development of such a thinking is the base of behavioral geography. Even the earlier approaches of environmental determinism treating nature as the controlling factor, or environmental possibilism treating man as the modifier, are both interrelated in a different form with the human's behavioral relationship to the environment.[3]

Emerging approaches as grouped into two classes are further related with the dialectics of analysis like deductive-inductive, descriptive-prescriptive, static-operative provokings. During the 1960s these ideas were treated in context with the simple dress of physical distance which in due course of *spatio-temporal continuity and existence* were shed for the more elaborate costumes having threads of functional distance, perceived distance and such patterns of non-Euclidean geometry. It would be quite appealing to expose the explicit behavioral assumption in geographical investigation on the grounds of both logical structure and axiology. Adaptation of techniques of sister disciplines is not an absolute task, albeit critical evaluation in search for more adequate behavioral models with greater explanatory ability is necessary. Particular attention should be given to theory development; in fact "the development of theory was ultimately perceived to rest on an explicit development of behavioral models."[4] Harvey has rightly expressed, "An explicit development of scientific theory in geography automatically entails, among other things an explicit analysis of the behavioral postulates on which that theory can be based."[5]

Behavioral geography particularly concerned itself with the aspects, attributes and components of perception with a view to providing a base for understanding how men structure in their own minds the world around them.[6] The methodological philosophy in this context is related to three problems—models of mind, procedures for recovering and measuring perceptions and preferences.[7] The relations between mind and behaviour in behavioral geography are represented through a schematic model (Fig. 8.1).

*Mind and Behaviour*

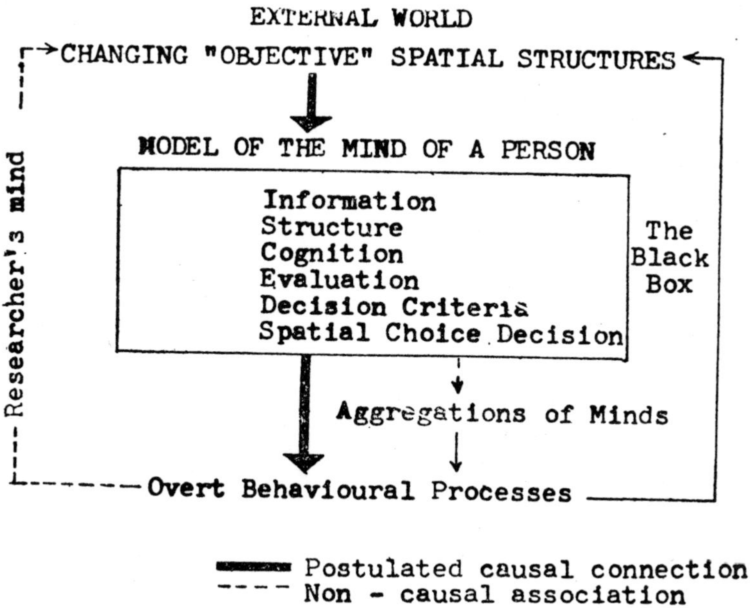


FIG. 8.1

Perception, structure and behaviour are the triological frame in behavioral geography, which are the outcome of the interaction of two basic variables—individual thinking, and environmental constraints confronting those individuals. It may be expressed that within the environmental envelope man structures his world in accordance to proxemics which in turn are reflected in behaviour. This idea is comparable to the four-fold division of environment suggested by Sonnenfeld who suggested behavioral environment as the core while the rest converge around that—perceptual, operational and geographic in ascending order (Fig. 8.2). As one tends towards the analysis from behavioral to geographical the level of sophistication and complexity increases.[8] Such a broad spectrum of interrelationship may be observed in daily human consequences like space preference, movement alternatives, etc.[9]

The foregoing concepts and visualization are further expressed in a spatial context. It is noteworthy that "human

*A Four-fold Division of Environment*

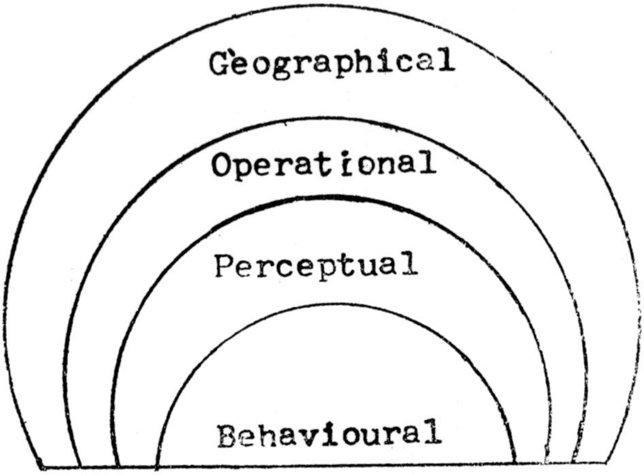


Nested set of Environments

FIG. 8.2

spatial awareness involves constant reorientation as places and self images change". In this context varying ways of spatial knowledge and its organization has to be highlighted. A five-fold categorization of spatial knowledge may be expanded as:[10]

(1) *The Pragmatic Space* of physical action in which sensory motiric responses are made.
(2) *The Perceptual Space* of immediate orientation based on the experience of sensory imagery. Here the individual views him/herself as separate from the environment but still part of it.
(3) *The Existential Space* which forms the individual imagery of the environment. This is a mental image or imitation of the perceptual world involving social constructs of reality within a cultural context.
(4) *The Cognitive Space* of the physical world involving imagery specifically related to perceived relationships. Within this mode the individual is able to think about spatial relations.
(5) *The Abstract Space* of pure relationships which has no necessary tie to perceptual or sensory motoric

> skills. Imagery and symbolism have meaning in and of themselves : within this context the individual is able to conceptualise other concepts of space and objects in space.[11]

While recognizing the identical distinction, the main concern of behavioral geography is with the context, character and organization of spatial information in the mind of the individual rather than the *skill* ; skill and tools are to be taken only as explanatory means.

The comprehension of treating organization in spatial imagery is critically expanded and reviewed by Golledge who theorized the working hypothesis in three ways :[12]

(i) We can accept a developmental theory of environmental learning as a normative theory in the initial stages of our investigation into the process of comprehending the environment.

(ii) We suggest, however, that socio-cultural and cross-cultural differences are sufficient to ensure that many individuals will not pass through all the stages suggested by Piaget, and that these stages will be meaningful only in culture-specific contexts.

(iii) We also adopt the working hypothesis that converging towards the understanding of space using Euclidean geometrical relations is again culture-specific and need not dominate the final stages of environmental comprehension.

All these theorizing presumptions can be represented as a set of progressions involving a continuing synthesis and re-synthesis of cognized relationship among points (P), line (L) and areas (A).[13]

## Some Recent Implications

With the initiation of the concept of "Mental Map" advocated by Peter Gould, it has gained a wide popularity. This concept deals with the "ways in which distance between the person and the place affects the manner of construction and the final product of the image-building process."[14] The level of perception basically reflected within factors like chance and uncertainty. The solution in this context is provided to a certain extent by using multidimensional scaling, which is the result of at least trio-reasonings : (i) with the adoption of a cognitive philosophy,[15] the mind is now seen as a crucial variable in the explanation of spatial behaviour in geography,[16] (ii) scaling is a primary technique employed in geographic research into mental structure and process, notably in respect to the cognition and evaluation of environmental stimuli such as shops, residence and cities,[17] (iii) geographers have made no serious attempt to examine the interface between scaling procedures and the mind.[18]

The context and themes of spatial choice and spatial behaviour may be clearly grouped into three parts[19] : (i) philosophical problems related to the work in general, complexity of the human mind and method of extracting and representing cognitive phenomena; it also includes such themes like relationship between individuals and their environmental and technical aspects like the use of cartography and mathematics; (ii) comprehension of the perspectives on analysis preferences and choices ranging from experimental design for recovering cognitive configurations of places, decomposing space preference functions, variations in the quantity of cognitive information and reaching to the hypothesized relations with spatial behaviour; and (iii) practical implications in skill and relevance. These parts are interrelated with some common themes and objects, e.g. (a) a deep concern for rigorous objectivity, and a scientific approach to the analysis of behavioral problems, (b) a concern for broadening the scope of existing geographical theory with the inclusion of process variables, (c) value-judgement of quantitatively analysed behavioral data, (d) an expression of full awareness of relative state of mind—either philosophical or behavioral, and (e) expression of cautions optimum to expand the relevance of geographical investigation in behavioral problems. Thinking on these lines was first provoked in the late 1970s,[20] but now they have received serious consideration.

One of the best uses of behavioral models has been introduced to corroborate the Central Place Theory. The studies of shopping trips reveal that contrary to the assertions of Central Place Theory, people often do not go to the nearest central place to purchase goods but prefer to travel further to alternative places.[21]

The spatial characteristics of environment have been comprehended with a view to promoting the decision-making orientations, which are the subject of information availability, communication, learning, space perception, space preferences, and so on. With such an analogy there appear to be four visualizations :[22]

(1) more controlled and detailed studies of individual behaviour will need to be undertaken before any substantive theoretical developments appear;
(2) more concern will be given to the stimuli which lead to decisions, actions and the sequence of events leading to specific geometric outcomes;
(3) more research will be directed toward the difficult questions of man-environmental and non-machine interfaces; and finally
(4) more work will emerge on the relationship between "objective" and "perceived" criteria of behaviour and decision-making.

With the proper evaluation, synthesis and rationalization of such predictive trends, behavioral geography can be strengthened as a distinct and applied branch.

## DEFENDING APPROACH

In spite of rapid inclination in this recent field of research, there exist limitations both in form of *unit of analysis* as well as in its *forms;* to quote Gunnar Olsson, it can be observed seriously:

> Although this intellectual offspring never really developed, its conception and birth had important epistemological consequences. This is because we quickly came to realize that in the extension of spatial analysis into behavioral geography lies nothing less than the ghost of geographic inquiry. This elusive creature has been called by many names but it is usually referred to either as the ecological fallacy or as the geographic inference problem of form and process. Its message is that even though we

> sometimes can deduce statements about spatial form from knowledge of generating processes, we cannot invert the reasoning procedure to infer process from form. Put differently, we have not found a way of preserving the truth of premises derived from large scale data in conclusions about small-scale behaviour.[23]

The limitation of the subject is primarily based on individual's attitudes, norms and conditions which surrounds him, while in a particular territory like household, government bodies and such allied organizations, all make decisions creating the environment in which we all mutually interact and economically transact. As such, the basic consideration of an individual's proxemics at any level in planning purpose is considered as a controversial matter.

Recent literature in behavioral geography shows much stress on viewing environmental change almost entirely as a function of a private economy; in fact, there appears to be a good deal of concern with public economy which also affects the environmental change.

To solve the above problems, the comprehension of approach is the basic need. In fact "our real need is for a new philosophy of man. In that philosophy, the human condition will be conceived as an evolving relationship between the contradictory forces of certainty and ambiguity, man and men, men and society, freedom and necessity."[24] Within such an approach Cox has rightly proposed that it concerned : "(i) not only with adaptation to environmental constraints but also with change in those constraints; and (ii) not only with individuals as decision-makers responsible for spatial patterns but also organizations such as governments and bureaucracies."[25] With such thinking behavioral geography could be more helpful in planning postulations. It can easily make a strong case as for example Downs and Stea have stated that "Cognitive mapping is a necessary requirement for human survival,"[26] whatever its level or territoriality may be like, political reforms or a particular nation or region. In any locational decisions, a consideration of the mental map would be a basic need for a rational policy, because mental maps reflect the biases and values of their holds, as Gould and White have remarked that "planning in a humane economy, that places people first, must surely consider the mental images of places as a crucial input to policy decisions which affect in such deep ways the patterns and satisfactions of individual lives."[27]

The implication of the ecological approach in behavioral geography provides more potentiality for its relevance. With the consideration that each component of man, especially behavioral, as a part of the system, it may be expected that "a change introduced at any point would have reverberations through the system."[28] Even some intriguing examples of "system effects have also to be seen in the smaller segments of the built and social environment."[29] Such implications indicate the applicability of behavioral postulates even at micro level.

As "a person is a biological organism, a social being and a unique individual, perception, attitude and value reflect all three levels of being,"[30] which may be further expressed in landscapic visualization—"what we see, what we study, and the way we shape and build in the landscape, is selected and structured for each of us by custom, culture, desire and faith."[31]

Let us end the paper with an observation by Lowenthal : "The lineaments of the world we live in are both seen and shaped in accordance, or by contrast, with images we hold of other worlds—better worlds, past worlds, future worlds."[32]

Further, "the geography of the world is unified only by human logic and optics, by the light and colour of artifice by decorative arrangement and by ideas of the truth, the good and the beautiful" which is borrowed from the Indian philosophic thought—'Satyam, Sivam, Sundaram'.

## REFERENCES


1. E.J. Taaffe, *Geography*, N.J. Englewood Cliff, Prentice Hall 1970, p 1.
2. J.A. Jackle, et al, *Human Spatial Behaviour, A Social Geography*, North Scituate, Mass : Duxbury Press, 1976, p 7.
3. K.R. Cox, "The Behavioral Revolution in Geography : Definition and Evaluation" *Geographical Perspectives*, No. 33, 1974, p 41.
4. Ibid, p 42.
5. David Harvey, "Behavioral Postulates and the Construction of Theory in Human Geography" *Geographica Polonica*, No. 18, 1970, p 27.
6. Taaffe, p 31.

7. P. Burnett, "Behavioral Geography and the Philosophy of Mind" in *Spatial Choice and Spatial Behaviour,* eds. R.G. Golledge, G. Rushton, Columbus, Ohio State Univ. Press, 1976, p 24.
8. See, T.F. Saarinen, *Environmental Planning, Perception and Behaviour,* Boston, Houghton Miff in Com., 1976, pp 7-8.
9. G. Rushton, "Analysis of Spatial Behaviour by Revealed Space Preference", *Annals Asso. Am. Geogr.*, 59, No. 2, 1969, pp 391-400.
10. R. G. Golledge, A.N. Spector, The Masterly Approach, *Comprehending the Urban Environment : Theory and Practice,* Mimeo. Univ. of California Santa Barbara, 1978, pp 7-8.
11. C. Norberg-Schulz, *Existence, Space and Architecture,* New York, Praeger (1971), 11; See also J. Piaget, *Structuralism,* New York, Basic Books, 1970.
12. Golledge and Spector, p 10.
13. For the most comprehensive survey on this aspect especially statistical concern ; see A. Getis ; B. Boots ; *Models of Spatial Processes ; An Approach to the Study of Point, Line and Area Patterns* ; Cambridge, The Univ. Press, 1978.
14. P. Gould, R. White, *Mental Maps,* Harmondsworth, Penguin Books, 1974, p 17.
15. R. M. Downs, D. Stea, *Maps in Minds* ; New York, Harper & Row, 1977 ; especially Chapter 1.
16. R.M. Don's, "Geographic Space Perception : Past Approaches and Future Prospects" *Progress in Geography,* 2, 1969, p 65—108.
17. See R. G. Golledge, G. Rushton, *Multidimensional Scaling,* Washington, Assoc. Am. Geogr., C. C. G. Techn. Paper 10, 1972.
18. E.J. Harman, and J.F. Betak, "Behavioral Geography, Multidimensional Scaling and the Mind" in Golledge and Rushton, p 3.
19. See, Golledge and Rushton, Introduction, pp xi-xii.
20. See, K.R. Cox, R. G. Golledge, eds *Behavioral Problems in Geography : A Symposium,* Evanston, North Western Univ. Series in Geogr. No. 17, 1969.
21. L.J. King and R.G. Golledge, *Cities, Space and Behaviour,* Englewood Cliffs, N.J., Prentice Hall, 1978, p 161-162.
22. L.S. Bourne, "Through the Looking Glass: Comments on Behavioral Geography" in *Behavioral Perspective in Geography, A Symposium,* eds. J. Rees, P. Newby, Middlesex, The Polytechnic, 1974, p 103-104.
23. Gunnar Olsson, "The Dialectics of Spatial Analysis" *Antipode,* Vol. 6, No. 3, 1974, p 52.
24. *Ibid.*, p 59.
25. Cox, p 49.
26. Downs and Stea, p xiii.
27. Gould and White, p 178.
28. Saarinen, p 241.
29. *Ibid.*, p 242.

30. Y.F. Tuan; *Topophilia*, Englewood Cliff, N.J., Prentice Hall, 1974, p 245.
31. D. Lowenthal, ed. *Environmental Perception and Behaviour*, Chicago, Univ. of Chicago, Research in Geog, No. 109 (1967), 1; also see L. Van Ryzin, ed. *BERM Behaviour Environment Research Methods*, Madison, Univ. of Wisconsin, 1977; G. Moore, R.G. Golledge, eds. *Environmental Knowing*, Stroudsberg, Penn, Dowden, Hutchinson and Ross, 1976.
32. David Lowenthal's Introduction in *Geographies of the Mind*, ed., D. Lowenthal, M.J. Bowden, New York, Oxford Univ. Press, 1976, p 3.

9

# Development, Social Development and Geography

—*Gopal Krishan*

THE OBJECTIVES of this paper are to (i) examine the differences in perspectives of various social sciences on development, (ii) place social development in the ambit of development, (iii) state the relationship between geography and development, and (iv) establish a case for a Geography of Development.

One of the major concerns of social sciences recently has been "a quest for a multidisciplinary theory which explains the process of development."[1] Such an interest was the outcome of three main factors : (i) the striking, conspicuous disparity in the standard of living of people in the developed and developing countries, (ii) conscious efforts of the developing countries at ameliorating the lot of their poverty stricken masses through planned development, and (iii) keenness on the part of the social sciences to contribute to this grand task of reconstruction. The first factor aroused the intellectual curiosity for explanation, the second stimulated a search for models of development that could be adopted by developing countries for their uplift, and the third in the emergence of a variety of perspectives in which the development process could be viewed.

## DEVELOPMENT AND THE SOCIAL SCIENCES

Among all the sciences concerned with development, economics has achieved the most in generating theory and policy prescriptions.[2] The most pressing problems of the developing world revolve round poverty. Their overriding priority is an increase in production. So it is the economic analysis, economic policy, economic planning and economic aid that appear as paramount.[3]

By comparison, sociologists have focused primarily on conceptualization of the development process. Their three-fold grouping of societies into the folk, the agrarian and the modern is well-known. They specify integration, goal gratification, adaptation and maintenance of cultural patterns as the four functional properties of any social system.[4] In the folk society, the integrative sub-system is dominant; in the agrarian society, the goal attainment sub-system; and in modern society, the adaptive sub-system.[5] The value maintenance sub-system is common to all societies. Sociologists display a distinct theoretical thrust in their works and often do not make pretensions of presenting their research efforts as applied in nature. They have been concerned with describing and explaining the law of the social events without making any great claims that they could change the course of events.[6]

Political scientists, in their analysis of development, highlight the role of political economy. The level of development of a society and the prevailing disparities within it are viewed as the function of the production and distribution system in terms of the who controls the means of production, who produces and for whom ? Taking into account the relative capacity of different political systems to absorb change and generate further innovation, three development types were distinguished by Apter : mobilization system as in the Soviet Union, reconciliation system as in India, and the modernizing autocracy as in Iran.[7]

Scholars like McClelland[8] and Hagen[9] have emphasized the role of psychological factors in the determination of development and underdevelopment. Two important personality traits of the societies are said to be basic to the process of development : achievement motive in order to seek solution to problems and other-directedness or sensitivity to opportunity. Industrialization, group activity and proper child-rearing are suggested as instruments for breaking down traditional thinking and encouraging other-directedness.

It is primarily through the development of sub-histories, particularly economic history and social history, that historians inject theory into the various facets of their discipline, including matters relating to development.[10] Karl Marx's ideas were distilled from his comprehensive survey of the economic history of mankind. Social history, by comparison, remained confined to the study of social conditions at particular times in various areas. The interweaving of threads of social history was lacking; hence the historians of this branch missed the opportunity of coming out with any outstanding theory on development.

Worthwhile contributions by geographers in development studies are still awaited. This is amazing in the context of the objective of geography to comprehend the totality of character of different areas. So far they have contented themselves with description of regional disparities in development, without any serious effort at theorization.[11] The theoretical work on central place hierarchy[12] and spatial diffusion of innovations[13] stands out as an exception.

In spite of a considerable amount of multidisciplinary research on the subject, development defies any precise definition. There is a great variety of terminology—social change, modernization and development—to make the task of definition more difficult. An attempt at making some distinction between these terms will be in order. Social change stands for the alterations occurring in the structure and functioning of a society. It is a value-free term. Modernization is a sequential process of change generated by the interaction of economic and cultural innovations impinging on the traditional economy, polity and society, with feedback effects on the innovating activities.[14] The crux of modernization is innovating—break in tradition—innovation. Modernization and aspirations to modernity are probably the most overwhelming and the most permeating features of the contemporary scene.[15] Development, in geographical parlance, denotes the quality of functioning of a regional system in terms of economic progress, social advancement, political maturity and environmental conservation. Economic progress may be measured in terms of the standard of living of people, full and rational utilization of manpower, and degree of diversification of occupational structure. Social advancement may be referred to matters of social justice, social cohesion and quality of population. Political maturity may be judged from the degree of decentralisation of power, administrative efficiency and mass participation in welfare activities.[16] Environmental conservation stands for the state of ecological balance with effective check on erosion of soil, contamination of water and pollution of air, apart from maintenance of the quality of the habitat.[17]

The geographic concept of development is a synthesis of view points of various social sciences.

### SOCIAL DEVELOPMENT

This is to say that social development is one of the elements of development. In sociological literature, social development has been defined as the development of men in their mutual relations.[18] A social system advances as it absorbs a larger amount of energy into its being (scale), as this energy is more adequately directed to the maintenance and further growth of the system (efficiency), as each part is so formed as to respond most exactly to the needs of others with the least loss of energy through mutual constraint (freedom), and thereby secures its own support (mutuality).[19] In other words, energy, organisation and vital harmony are the hallmarks of social development.

Economists have long viewed social development as a byproduct of economic development in contrast to the sociologists' view point of holding social development as supreme and basic to economic development. Rather traditionally, social development, has been seen as competing with economic development. This is manifest in many arguments raised about allocation of scarce resources, e.g. between building of factories or building of schools. Of late, economists have shown greater awareness of social reality. Nevertheless, their analysis of social development remains loaded with input-output considerations. This is evident from their studies in the form of economics of education.

May we then define social development, from the point of view of political scientists, as the degree of socialism a society has achieved? Socialism itself may be hard to define. In simplistic terms, it may be construed as reduction in disparities, equality of opportunities and decentralization of the means of production. Among the various objectives of any development, a high premium is placed on minimization of inequalities.

Psychologists take a dynamic view of social development. It is meant to create developmentally oriented people by the engineering of new attitudes and behavioral patterns. These have to be induced even in infants and children. This makes social development a long range endeavour that takes future human resources and the future social structure as its goal.

In actual practice, social development has been measured in terms of a variety of quantifiable criteria. These include nutrition, housing, health, education, transport and population mobility.[20] The contention is that these indicators, among others, reflect the quality of functioning of a social system, efficiency of its economy, welfare bias of its polity, and orientation of the attitudinal and behavioral patterns of its people. And this is the way the different social scientists would like to view the process of social development.

## DEVELOPMENT AND GEOGRAPHY

It remains to examine the relevance of geography to development. Geography is essentially an area-characterizing and area-differentiating science. One of its major functions is to point out differences between various parts of the world, including those in regard of their level and rate of development. A primary task for the geographer is, therefore, to find ways and means of providing a more realistic description of the current and changing situation in development over the world in spatial perspective.[21] Such studies can serve as a corrective to the popular tendency of treating the under-developed world as one monolithic block with much the same problems throughout.[22] Each country with its various regions has a particular combination of problems; the keys to development are different and regionally unique.[23]

Secondly, geographical studies of spatial organisation and of spatial diffusion of innovations can be of a great help in understanding the development processes. Johnson rightly argues that the way terrestrial space has been organised reflects the *degree* of development and underdevelopment in an area.[24] The manner in which the innovations diffuse over space is an index to the *rate* of development.

Lastly, geography can continue contributing to development studies by way of mapping and resource inventories. Geographers should evolve new cartographic and surveying techniques relevant to such studies. By virtue of their integrative approach, geographers are in a better position to throw light on key processes in developed and developing societies.

**Geography of Development**

There is evidently a strong case for developing the Geography of Development as a special branch of our discipline. What could be the broad outline of such a course? One could perhaps begin with a search for a definition of development in geographical terms. The models of development particularly those which are spatial in nature as those of Hirschman, Myrdal, Friedmann and Perroux, could be the next item of discussion.[25] A critical examination of the role of agriculture, industry and infra-structural facilities in development could follow.[26] A study of the relationship between development and population, including demography, settlement patterns and cultural lineaments, would also be necessary.[27] Equally essential is the comprehensive understanding of the role of political economy since the map of "power" is fundamental to the nature, level and rate of development in any area.[28] A consistent spatial perspective is to be maintained while assessing the role of the various factors in development. Regional disparities in development are of special interest to a geographer. These disparities are not only to be measured through suitable techniques but are also to be interpreted in the light of the physical resource base, technological advancement, institutional framework and external exposure, all in temporal dimension, of different areas. Formulation of spatial theories of development would be a laudable contribution of geographers to development studies. As an applied exercise, geographers could prepare regional development plans of the areas of their interest.[29]

A course in the Geography of Development is going to be intellectually stimulating and operationally profitable. It would expose geographers to varied aspects of relative development and underdevelopment in different parts of the world. An inquiry into the processes involved would enable them to make a worthwhile contribution to development strategies.

REFERENCES

1. Khodadad Farmanfarmaian (ed.), *The Social Sciences and the Problems of Development*, Princeton University Programme of North Eastern Studies, 1976.
2. George Dalton, "Economic anthropology : Traditional and Modernizing Communities" in Manfred Stanley (ed.) : *Social Development*,

Basic Books, New York, 1972, p 223.

3. *Ibid.*, p 224.
4. Talcott Parsons, Robert F., Bales and Edward A. Shils, *Working Papers in the Theory of Action*, The Free Press Glencoe, III, 1949, pp 79 and 183.
5. Bert F. Hoselitz, "Development and the Theory of Social Systems" in Manfred Stanley (ed.) : *Social Development*, Basic Books, New York, 1972, p 54.
6. T.B. Bottomore, "Sociology in Norman MacKenzie (ed.), *A Guide to Social Sciences*, Winsley Street, London, 1966, p. 93.
7. David E. Apter, *Some Conceptualization Approaches to the Study of Modernization*, Prentice Hall, Englewood Cliffs, New Jersey, 1968, pp 276-283.
8. David C. McClelland, *The Achieving Socicty*, Van Nostrand, Princeton, 1961.
9. Everett E. Hagen, *On Theory of Social Change : How Economic Growth Begins*, Dorsey Press, Homewool, 1962.
10. Asa Briggs, "History and Society", in Norman MacKenzie (ed.), *A Guide to Social Sciences*, op. cit., p 38.
11. W.F. Heinemeyer and H. Van Der Wusten, "Regional detail in development", *Tijdschrift voor Economische en Sociale Geografie*, 59, 1968, pp 282-288.
12. Walter Christaller, *Central Places in Southern Germany* (translated by C.W. Baskin), Englewood Cliffs, New Jersey, 1966.
13. T. Hagerstrand, *Innovation Diffusion as a Spatial Process*, Chicago University Press, Chicago, 1967.
14. George Dalton, op. cit., p 234.
15. S.N. Eisenstadt, *Modernization : Protest and Change*, Prentice Hall, International, London, 1966, p 1.
16. For a detailed discussion on economic, social and political indicators of development see I. Adelman and C.T. Morris, *Society, Politics and Economic Development*, Johns Hopkins Press, Baltimore, 1967, pp 9-129.
17. Harold A. Wood, "Toward a geographic concept of development", *Geographical Review*, 67, 1977, p 465.
18. L.T. Hobhouse, *Social Development*, George Allen & Unwin Ltd., London, 1966, p 76.
19. *Ibid.*, p 84.
20. United Nations, *Population Aspects of Social Development, Report of a Regional Seminar and Selected Papers*, Bangkok, 1972.
21. Richard Hartshorne, "Geography and Economic Growth" in N. Ginsburg (ed.), *Essays on Geography and Economic Development*, Chicago Press, Chicago, 1960, p 21.
22. David Grigg, "Geographical Studies of Economic Development with Special Reference to Agriculture" in Michael Chisholm and Brian Rogers (eds.) : *Studies in Human Geography*, Heinemann Educational Books, London, 1973, p 59.

23. *Ibid.*, p 59.
24. E.A.J. Johnson, *The Organisation of Space in Developing Countries*, Harvard University Press, Cambridge, Harvard, 1970.
25. D.E. Keeble, "Models of Economic Development" in Richard J. Chorley and Peter Haggett, *Socio-economic Models in Geography* Methuen, London, 1970, pp 243-302.
26. Guy Hunter, *Modernizing Peasant Societies*, Oxford University Press, London, 1969.
27. Warren C. Robinson (ed.), *Population and Development Planning*, The Population Council, New York, 1975.
28. Charles K. Wilbur (ed.), *The Political Economy of Development and Underdevelopment*, Random House, New York, 1973.
29. R.P. Misra, K.V. Sundaram and V.L.S. Prakasa Rao, *Regional Development Planning in India*, Vikas Publishing House, 1974, pp 239-369.

# 10

# Use of the Computer in Data Analysis

*—V.K. Shrivastava*

THIS PAPER intends to make the reader conversant with the preliminaries of computer usage for data analysis particularly in the field of social science research. The computer as an electronic device is capable of strong numerical data and of performing certain operations on that data. Computers simply follow instructions commuted to them in a coded language. The language should be readable to the machine. The instructions are given in the form of a programme. The 'think computers think' label cautions the user to prepare a logical and pertinent programme as per the technical requirements of the computer machine.

The use of the computer in data analysis hardly needs any emphasis. The social sciences, particularly geography, have become closely linked since the 1950s. The use of quantitative methods became so prevalent and widespread that even a misnomer in the title of Quantitative Geography emerged.[1] Not only is the use of statistical methods prevalent but the study of networks[2] modelling[3] and simulation techniques have become popular too.[4] Such fast developments in geographical methodology have made it obligatory on the part of the researcher to carry out complex and lengthy calculations. On the other hand, carrying out such tedious calculations means parting with much valuable time of research in arithmetical operations and getting very little time to concentrate on the meaning and interpretation of results. The use of digital computers may thus become unavoidable.

## A BRIEF HISTORY OF COMPUTER USAGE

The computer as an aid to calculation was used by the ancient Greeks. The mechanical calculators of Pascal and Leibniz may be said to be the fore-runners of the modern computer. Till the nineteenth century calculating devices depended on the human brain, particularly in storing data and instructions. It was an Englishman known as Charles Babbage who conceived the idea of building a machine that is now known as a computer.[5] This was around the 1920s, in the very beginning of the Industrial Revolution. He first constructed a mechnical device known as the Difference Engine. Later on a more modified version of the same machine was produced and it was known as the Analytical Engine. It had five components i.e. a memory, an arithmetic unit, an input section, an output section and a control unit. The first commercial computer was completed in the United Kingdom in 1951 by Ferranti Ltd. in association with Manchester University. Nowadays several computer machines of IBM, UNIVACI, etc. are in use. Modern 'mini-computers' have greater capabilities of speedy handling and execution of instructions. It is important to note, however, that all modern computers still incorporate the basic five components of the Analytical Engine.

## COMPUTING AND THE GEOGRAPHER

A major drawback in the computer usage by the geographer emanates from the geographer's ignorance about the capabilities (and limitations) of the computer in solving his peculiar needs. A computer can tabulate, cross-tabulate, analyse and store all sorts of data. It can also generate graphs and maps showing the spatial dimensions of the results of data analysis. On the other hand, computer experts are rarely well informed about the nature of geographical problems. Specific programmes which are generally available in computer centre libraries may not serve the need of a social scientist or a geographer.

This problem has been solved by the Statistical Package for the Social Sciences (SPSS)[6]. This is an integrated system of computer programmes for the analysis of social science data. This comprehensive programme explains the procedures and features of the SPSS and exemplifies the organization and coding of data, preparation of control cards and running programmes. This paper, however, deals with the purpose, procedures and features of the SPSS Programmes.

## THE STATISTICAL PACKAGE FOR SOCIAL SCIENCES

Normally, in computer usage, two types of data application are met with, i.e., business type applications, which require a large amount of input and output data and a small amount of calculation, and secondly, scientific type applications, which typically involve relatively small amounts of input and output data and large amounts of calculation. The analysis of data of social sciences may be said to lie somewhere between these two extremes. This is because the quantity of input data in the social sciences is large and the calculations needed may be large or small in amount depending on the nature of the analysis required. Besides, data analysis in these sciences involves the repeated and routine application of a number of procedures. Over and above this, the data in social sciences is mostly related to the diverse economic and cultural forms of human behaviour. Such data is not acquired by conducting tests in laboratories as in the pure sciences. The social scientists have neither the training nor the background to gain the technical grasp of conventional computer usage to perform the data analysis task required by their research. It is, therefore, necessary that the data analysis process should be simplified so that the social science user himself could gain control over the manipulation and analysis of his data. The SPSS system of programmes is the result of such attempts of simplification of data analysis process.

### THE PURPOSES

Computers are widely used for data processing and data analysis. The former involves such routine tasks as classification, sorting, storing and retrieval of data which have been presented to the computer in a suitably coded form. Data analysis, on the other hand, combines data processing with mathematical or statistical manipulation. The computer is capable of carrying out lengthy arithmetical operations at a high speed. Data processing and data analysis form an important part of activity in the empirical sciences. These are particularly useful in the social sciences where the amount of data is large and the main problem is to reduce the data to a manageable size.

A computer does not operate on its own. Though it performs a number of procedures, at each stage of the procedure it must be fed with the exact sequence of steps to be followed. Such a sequence of steps is called a programme. A programme is applicable to many different sets of data. Minor adjustments in the programme are also possible without disturbing the input data with programme control cards. Computer centres maintain a library of programmes which may be used to carry out one or another standard procedure.

Data analysis may require repeated recourse to a variety of procedures. The output data of one programme may become the input data of another programme. For example, after cross-tabulation of a set of data, rank-order correlation or regression may be needed. In such cases a long chain of tasks is required. It then becomes important that the output data of one programme should be compatible with the input data for another. If the programmes required differ from one another, one has to master the details of all of them leading to possible conclusion.

Under these circumstances a system of related programmes, which perform a related set of procedures and which share a common set of conventions in which the set manipulates data, becomes necessary. The SPSS system is such a set of related programmes for the manipulation and statistical analysis of many types of data with a particular emphasis on the needs of the social sciences. It permits one to execute a sequence of tasks with a minimum of manual intervention. It is not necessary to re-enter the data at any time. The system stores and retrieves the appropriate data when required. The retrieval of data enables the user to make good use of some other programme, if needed. Additional procedures which have not been already provided can also be included.

The SPSS system involves the use of simplified language which closely corresponds to the natural language a social scientist might use.

## STATISTICAL PROCEDURES IN SPSS

As has already been stated, the purpose of any process of data analysis is to condense information contained in a body of data into a form which can be easily comprehended and interpreted. This process may be used in two ways : (i) to describe a body of empirical data or (ii) to search for the meaningful patterns of relationship among sets of variables leading to the building and testing of social theory. The numbers computed from the data during the process of analysis are called statistics. A wide variety of statistical procedures are available in the social sciences enabling the interpretation of data, testing and building of theories, etc. The SPSS contains many of the most common statistical procedures employed by social scientists. These procedures may be distinguished as parametric and non-parametric procedures. The non-parametric statistical procedures do not make a series of assumptions about the distribution or level of measurement of the variables. These are generally applied to nominal and ordinal data as they do not have well established matrices. The parametric procedures require more stringent assumptions concerning the distribution of data and they are generally applied for data with an interval matrix. However, the statistical procedures available in the SPSS are best catalogued according to the function they usually perform in the process of data analysis.

**One-way Frequency Distributions, Measures of Central Tendency and Dispersion**

In order to examine the characteristics of the distribution of each of the independent and dependent variables, SPSS contains three statistical procedures :

(a) **Condescriptive** is used with interval scale variables which assume a large number of values having a continuum, e.g. per capita income, production values, etc.

(b) and (c) **Codebook and Marginals** are designed for use with variables which assume only a limited number of values, e.g. grouped statistics. They can also be applied to nominal variables and produce descriptive frequency such as religions, affiliations castes, race, etc.

All these sub-programmes can produce statistics such as mean, median, mode, standard deviation, etc. Codebook can also produce elaborate tables with appropriate labelling, frequency and can also produce a histogram of frequency distributions. The information so generated provides a reference which the investigator will consult frequently. The output becomes too voluminous.

Marginals produce similar information in a condensed but unlabelled format.

### Table Displays of Relationships between Two or More Variables

The information about relationship may be obtained through correlation or through some form of table display. Three sub-programmes are available in SPSS.

(a) **Crosstabs** permits the user to compile two-way to n-way cross tabulations of variables. It also computes a variety of non-parametric statistics based on these tables. A sequence of two-way tables shows the values of two variables along two axes. The frequency counts are expressed as a percentage of row total, column total, table total or in any combination. In addition, the statistics measuring the degree of association of the two variables based on the distribution of frequency counts are also available.

(b) **Fastabs** produces similar output but operates faster and on numeric data only.

Both the sub-programmes have complete variable and value label capabilities. They also permit the user to process missing data by a number of different options.

(c) **Breakdown** is analogous to Crosstabs but the means, standard deviations and variances are presented as a logical breakdown or tree—first for groups, then for sub-groups within groups, next for sub-groups within sub-groups, etc. up to five independent variables.

### Bivariate Correlation Analysis

The SPSS system furnishes two programmes for performing correlation analysis PEARSON CORR and NONPAR CORR. The former produces product moment correlation coefficients which are best suited for normally distributed data with an interval scale, while the latter computes either Spearman or Kendall's rank-order correlation coefficient.

Both the programmes provide thc user with significance tests and have the capability of producing matrices for input into other programmes.

**Multivariate Correlation and Regression**

Explaining and predicting relationship among variables is a crucial requirement of social science research. SPSS provides two such programmes which enable data analysis helpful in explaining such relationship. PARTIAL CORR provides a single measure of association (the partial correlation coefficient) describing the linear relationship between two variables. It is analogous to n-dimensional cross tabulation for continuous and interval variables. This programme can operate on raw data or from matrices generated by PEARSON CORR or NONPAR CORR.

Multiple regression allows the researcher to study the linear relationship between a set of independent variables and a dependent variable while taking into account the interrelationships among the independent variables. The procedure REGRESSION enables a variety of multiple-regression calculations. Output from the programme includes a listing of the variables included in the regression at each stage, the coefficients in the regression equation, their standard error, and the significance level of the coefficients.

**Guttman Scaling and Factor Analysis**

Guttman Scale Analysis is a means of analysing the underlying operating characteristics of three or more items.

In the course of data analysis, the social science researcher may construct many types of indexes. An index is any measure which combines the values of several variables into a composite manner. Indexes are used to predict or gauge some underlying continuum.

Social scientists who are particularly interested in psychological and attitudinal research, have developed many techniques for combining individual items into scales and indexes. Guttman Scaling is one such widely used technique. It must be undimensional, that is the component items must all measure movement toward or away from the same underlying object. Secondly, Guttman scales must be cumulative, i.e., the component items can be ordered by degree of difficulty. The respondents who reply positively to a difficult item will always reply positively to a less difficult one and vice-versa.

The programme GUTTMAN SCALE[7] provides the user with the capability of producing up to 50 separate scales on a single task. The scales are evaluated by the Goodenough Technique.[8]

**Factor Analysis**

The distinctive character of this programme is its data reduction capability. The technique enables one to see whether an underlying pattern of relationship exists so that the data may be reduced to a smaller set of factors or components that may be taken as source variables accounting for the observed interrelation in the data.

The programme FACTOR performs a variety of factor-analytic techniques. Five factoring methods are available :

(i) Principal factoring without iteration,
(ii) Principal factoring with iteration,
(iii) Rao's cononical factoring,
(iv) Alpha factoring, and
(v) Image factoring.

FEATURES OF SPSS

(1) **Sequencing Calculations** : The SPSS is driven through its various functions by a sequence of control cards which the user must prepare. The control programme in SPSS reads control cards, decodes them, and causes the appropriate function called for by the control card to be executed. Arranging of the control cards in the appropriate sequence is important.

(2) **Entering and Processing Data** : Data is organised within the SPSS system in units called files. A file consists of user's data along with associated information. The file may be permanently stored as an SPSS system file on tape or disc.

(3) **Subfiles** : Data entered into the SPSS system may be sub-structured into groups called subfiles. They can be sampling points.

(4) **Missing Data** : In social science research it is generally not possible to obtain a complete set of data for every case in the file. SPSS processes such variables as missing

data. Each variable may have up to three values, i.e. '0' for not determined, '1' refused to answer, '2' do not know, etc.

(5) **Recording Data** : The researcher suitably codes the variables selected by him. After entering the coded data he may find it necessary to modify it. This can be done by RELODE process. The change can be temporary as well as permanent.

(6) **Variable Transformation** : Two types of variable transformation are possible. The unconditional transformation are defined by COMPUTE card and cause a new variable to be constructed from the values of the other variables—

| 1 | 16 | |
|---|---|---|
| compute | A = B+C | 'A' needs to be defined from existing B, C. |

The 'IF' card enables the user to test if a certain condition is true, if true, a transformation is performed.

The Statistical Package for Social Scientists Programme also enables sampling of data, selecting specific variables for processing and weighting. The criteria and conditions of sampling may also be specified.

Lastly, the system enables retrieval of data as well. It requires that the user should communicate to the computer to save file after analysis has been performed on a set of data.

The SPSS generates printouts automatically so that the user knows how the system has operated on the data.

## REFERENCES

1. J.P. Cole and C.A.M. King, *Quantitative Geography,* Wiley, Chichester, 1968.
2. (a) P. Haggett, *Locational Analysis in Human Geography,* Arnold, London, 1965.

   (b) P. Haggett and R.J. Chorley, *Network Analysis in Geography,* Arnold, London, 1969.
3. D. Harvey, *Explanation in Geography,* E. Arnold, London, 1969.
4. J.P. Cole, *Situations in Human Geography, A Practical Approach,* Blackwell, Oxford, 1975.
5. (a) M.P. Mather, *Computers in Geography : A Practical Approach,* Basil Blackwell, Oxford, 1976.

   (b) P.E. Morrison, *Charles Babbage and his Calculating Engine,* New York, 1961.

6. Norman Nie, D.H. Bent and C.H. Hull, *Statistical Package for the Social Sciences*, McGraw Hill Book Company, New York, 1970.
7. R.E. Anderson, "A Computer Programme for Guttman Scaling with the Goodenough Technique", *Behavioral Sciences* Vol. II, No. 3, 1966, p 235.
8. W.H. Goodenough, 'A Technique for Scale Analysis', *Edu. Psycho. Measut.*, 1944.

# 11

# The Use of Sociometric Measurement in Cultural Geography

—*K.B. Sajjadur Rasheed*

## THE PROBLEM

IT IS generally agreed by all practitioners of cultural geography that this branch of geography involves the application of the concept of culture to geographic problems. Cultural geography is thus intimately involved with culture—the sum total of human experience. Since man is the only animal to possess culture, he has been able to leave his imprint on the natural environment. And a good number of cultural geographers have so far sought a spatial understanding of those features on the earth's surface which are produced or moulded by human action. Mankind is viewed as an active agent on the surface of the earth and cultural geography is deeply concerned with the human response to and the use of particular environment. This emphasis is understandable since culture is justifiably viewed as a powerful component in the man-land equation. One of the basic themes before cultural geographers is the manner in which particular culture groups perceive and evaluate the physical attributes of their environs; and it is common knowledge with us that dissimilar cultures react differently to similar physical-biotic environments. This line of thinking has introduced cultural geography to a behavioral approach in that man's perception of his environment is the function of his attitudes, values, norms, traditions, etc. And differing perceptual characteristics produce varied response patterns to the environment.

It is argued here that an exciting frontier for research in cultural geography exists in understanding how culture groups perceive and interact with, other spatially separated culture groups. Every culture has its own distinctive personality comprising various kinds of traditions, customs, norms, values and the like. Consequently, different human groups with varying cultures perceive and interact with each other differently. Interaction between culture groups cane be xplained in terms of their attitude, behaviour, goals, preferences and prejudice. Intergroup relations between cultures occupying different territories can be summed up in terms of attraction and repulsion between them. Every culture lives in contact with its neighbouring cultures, and cultural geography can seek to probe as to how a culture group perceives another culture group during the process of interaction between them. The mode of reasoning in this kind of study would be similar to that in the study of man's perception of the environment, though the methodology might differ. Interaction between culture groups is indeed a geography problem since different culture groups might occupy different regions and the perception of one culture group by another group, in its broadest sense, portrays the perception of one region by another region.

Intergroup relations, viz., attraction and repulsion, have a chance to develop from contact and association. However, such opportunities are limited for the great majority of people when they are separated by international boundaries. Vague thoughts of attraction and repulsion between culture groups may be observed in such instances, which are, in fact, based on incomplete and indirect information rather than on contact and association. On the other hand, subgroups within a national culture are often known to exist. These groups may occupy different regions within a state, yet they have a far greater opportunity for contact and association among themselves than culture groups living across state boundaries. Population groups living in different parts of the same country often exhibit preferences for association with certain groups and feelings of repulsion for other groups. The use of the concept of culture has already drawn cultural geography close to several other social sciences, and the study of intergroup relations of regionally separated cultures can be fruitfully accomplished through cultural geography's relation to social psychology. A social psychologist studies the behaviour of individuals in social contexts. One major focus of his study is group structure and process, stressing features of interaction between groups. And when such culture groups have different regional origins, an investigation of their attitudes, perceptions, thoughts and feelings, i.e., their total behavioral disposition toward each other is indeed a cultural-geographical problem. In such problems, the cultural geographer can employ the sociometric technique—used by the social psychologists and sociologists to measure interpersonal attraction—to unfold as well as measure the pattern of regional attraction, repulsion and indifferences among different culture groups.

## SOCIOMETRIC MEASUREMENT

Sociometry is described as a method of presenting the total structure of relations existing at a given time among the members of a given group (Jennings, 1948). The social psychologist is interested in the total picture of interpersonal relations, attractions and repulsions among individuals in a group, and he uses sociometric techniques to measure them. It was Moreno (1953) who for the first time introduced the technique of sociometry in social sciences. Moreno (1953) founded sociometry as an intellectual movement within the behavioral sciences, and he developed the sociometric test as a technique to measure the attractions and repulsions within a given group. Any specific group of persons would normally exhibit distinctive patterns of association among its members. Given an opportunity for free choice, some members of the group are chosen more frequently than others, while some are always ignored. Such choices are, in effect, the function of liking or *positive affect* toward the person chosen. The patterns of attraction characterising regular associations among group members is termed the *affect structure* of a group and such regularities of association have been conceptualized in terms of *sociometric structure* (Secord and Backman, 1964).

The sociometric test is a technique of obtaining quantitative data on the preferences of group members for associating with other members. In this test, members of a group are asked to indicate their choices and rejections of other members of the group in certain specified context. It is assumed that members of the group have been in association for a sufficient time to permit the development of affective ties and repulsions within the group. Data obtained from the sociometric questionnaire may be summarised in the form of a graph known as a *sociogram*. A sociogram consists of a set of points (each point representing a person) connected by two types of lines, one denoting choice and the other rejection. This graphic presentation brings a number of features of the sociometric structure into light. For instance, certain persons receiving many 'choice' lines are referred to as sociometric 'stars' while others who receive and make few choices are 'isolates'. Since the construction of the sociogram is the result of trial-and-error process and it becomes unmanageable when the group is large, the sociometric data can be best presented in the form of a sociometric matrix. (Proctor and Loomis, 1951). The construction of the matrix is a simple operation with the choosers arranged on one axis and the recipients of choices on the other.

Sociometric technique has been employed by social psychologists in varied situations as informal groups, school classes, army, prisons and industry. The considerable latitude that exists in the application of sociometry has evoked an interdisciplinary popularity and scope for usage of these measures. These measures enable the investigator to evaluate the perception of the social environment by an individual; and cultural geography can gainfully utilize these techniques to understand the perceptions of individuals in an interacting social environment.

## APPLICATION OF THE SOCIOMETRIC MEASUREMENT : AN EXAMPLE

Sociometric measure, as described above, does indeed possess wide potentials in assessing interaction between culture groups. Each point in the sociogram or each cell in the sociometric matrix may be conceived as representing a cluster of individuals from a specified culture group and the entire sociogram or matrix representing the larger group; and an interculture interaction may thus be measured. It has been stated earlier that such interaction studies are more feasible and meaningful for culture groups within a national boundary because cultures within a state are more likely to be in association with each other than cultures from different states. The sociometric test was applied in this paper as a means to measure interaction of culture groups in Bangladesh. Although the national culture of Bangladesh is compact and homogeneous, yet persons from different districts within the country exhibit differences in attitude, behaviour, general life style or dialect. Besides, persons from different districts often show discernible amount of regional consciousness. Hence, groups of persons belonging to different districts can be identified in Bangladesh as regionally separated culture sub-groups. In the present study, the sociometric technique was employed to identify and measure the attractions and repulsions between the people of various districts in Bangladesh.

Bangladesh has a total of 19 districts, and all but one were included in this study to measure preferences and dislikes of the people of one district for those of another district. The district of the Chittagong Hill Tracts was excluded from the study for two reasons. First, this district is largely populated by tribal people who are ethnically different from the rest of the population of Bangladesh. Secondly, these tribal groups are less mobile than other Bangladeshis and hence, they have far less opportunity of associating with peoples of other districts.

The attractions and repulsions between persons of different districts were recognised by means of a sociometric questionnaire. It sought to elicit two primary pieces of information : (a) the most-preferred district, i.e., the district with whose residents the respondent is most inclined to associate, and (b) the most-rejected district, i.e., the district in which residents are least inclined to associate. In order to avoid bias, the respondents were asked to exclude their own district when deciding upon their most-preferred or most-rejected districts. The respondents were also offered six structured statements each for choice and rejection, and were asked to cite which statements best explain their choices and rejections of particular district. Besides, the questionnaire asked the respondents to name the districts of their fathers and mothers as well as the districts within Bangladesh where they lived for less than a year and more than a year.

The sociometric questionnaire was administered to 780 undergraduate and postgraduate students of the University of Dacca belonging to all the 18 districts of the study area. The University of Dacca was an ideal place to carry out this test since this institution receives students from all parts of Bangladesh and provides good opportunities for their association. Understandably, some districts like Comilla and Dacca were more represented than other districts in the original sample of 780. Subsequent to the filling of the questionnaires, 20 questionnaires of respondents belonging to each of the 18 districts were chosen by the use of a random table to yield a final sample of 360 respondents for the sociometric analysis of choice and rejection.

**Results**

The information obtained from each of the 360 questionnaires (20 from each district) on the chosen districts and the rejected districts was summarized in matrix form (Table 11.1). The difference from a traditional sociometric matrix was that, instead of individual persons arranged on the two axes as choosers and recipients of choices, 18 districts were arranged in the same order on the vertical axis and the horizontal axis. In this modified form of sociometric matrix, rows indicate the numbers and directions of outgoing choices and rejections by respondents of each district, while columns indicate the numbers and sources of choices and rejections received by respondents of each district. (Figures in parentheses in Table 11.1 designate the number of rejections). The matrix, therefore, demonstrates how the sample of 20 respondents of each district identified the districts with whose residents they would like to associate and they would like to avoid. For instance, the entries in the first row indicate that, of the 20 respondents from Dinajpur, 6 had chosen persons of Rangpur, 3 each of Bogra, Pabna and Rajshahi, 2 each of Kushtia and Khulna, and 1 of Jessore, while the same group of respondents from Dinajpur had 10 of them rejecting persons of Noakhali, 3 of Chittagong, 2 each of Dacca and Faridpur, and 1 each of Khulna, Tangail and Sylhet. On the other hand, the entries in the first column indicate that the district of Dinajpur was chosen by 5 respondents of Rangpur, 2 each of

Bogra and Rajshahi, and 1 each of Khulna and Kushtia, while Dinajpur was rejected by 3 respondents of Tangail, 2 of Sylhet, and 1 each of Mymensingh and Comilla. The column totals show the number of choices and rejections received by each district and this result is shown in Figure 11.1. Although

*Bangladesh : Choice and Rejection of Districts*

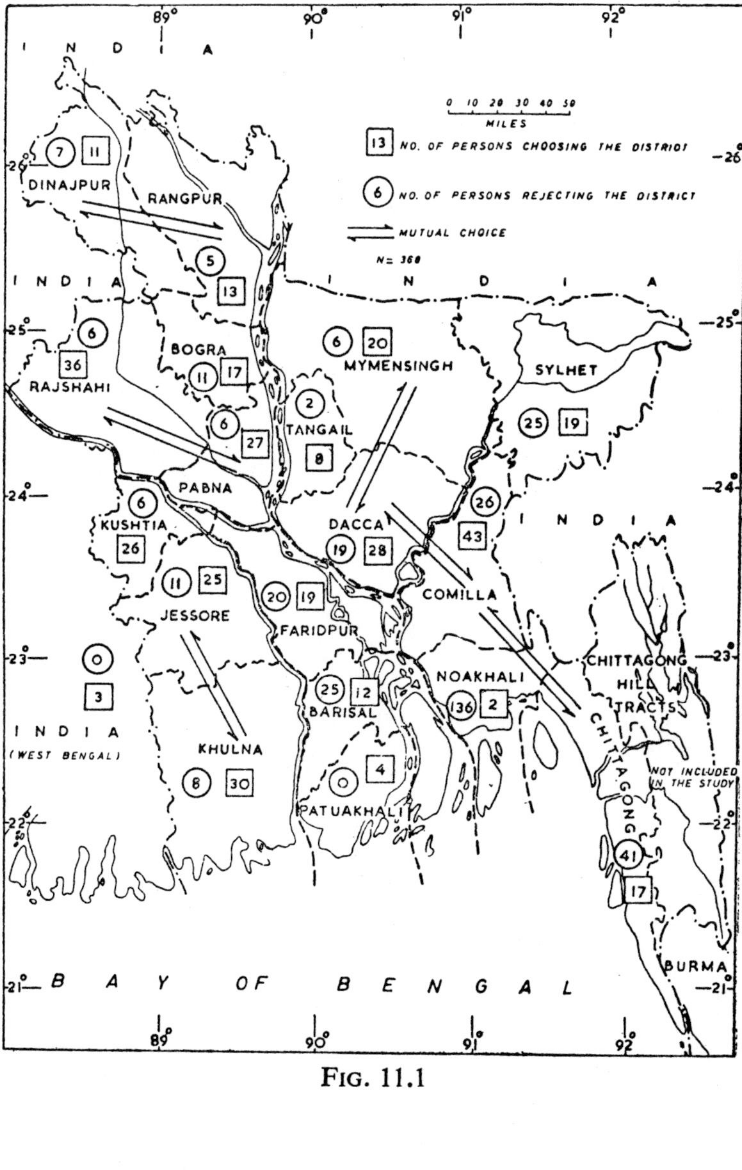


FIG. 11.1

TABLE

*Sociometric*

| Districts | Dinajpur | Rangpur | Bogra | Pabna | Rajshahi | Jessore | Kushtia | Khulna | Barisal |
|---|---|---|---|---|---|---|---|---|---|
| 1 | 2 | 3 | 4 | 5 | 6 | 7 | 8 | 9 | 10 |
| Dinajpur | — | 6 | 3 | 3 | 3 | 1 | 2 | 2 (1) | — |
| Rangpur | 5 | — | 3 | 2 | 2 | 1 | (1) | 3 | — |
| Bogra | 2 | 3 | — | 3 | 5 | 2 (2) | 1 | 1 | (2) |
| Pabna | — | — | 2 | — | 6 | 3 | 6 | 2 | — |
| Rajshahi | 2 | 3 | 3 | 5 | — | 2 | 2 | 2 | (2) |
| Jessore | — | — | 2 | 2 | 4 | — | 5 | 5 | (2) |
| Kushtia | 1 | 1 | — | 4 | 6 | 4 | — | 4 | (2) |
| Khulna | 1 | — | — | — | 2 | 6 | 4 | — | 1 (2) |
| Barisal | — | — | — | — | 2 | 1 | — | 2 (2) | — |
| Patuakhali | — | — | (1) | (1) | — | — | 2 (1) | 3 | 9 |
| Faridpur | — | — | (1) | 3 | 1 | 5 | (1) | — | (2) |
| Tangail | (3) | (2) | 2 | 4 | — | (2) | — | — | (2) |
| Mymensingh | (1) | — | — | 1 | 2 | (2) | 1 | (2) | (2) |
| Dacca | — | — | (3) | (2) | 1 | — | — | — | — |
| Sylhet | (2) | (2) | (2) | (1) | (4) | — | 2 | 2 | — |
| Comilla | (1) | (1) | (2) | — | — | (3) | (1) | (3) | (4) |
| Noakhali | — | — | (2) | (2) | (2) | (2) | (2) | — | 2 |
| Chittagong | — | — | 2 | — | 2 | — | 1 | 4 | (5) |
| Total choices and rejections received | 11 (7) | 13 (5) | 17 (11) | 27 (6) | 36 (6) | 25 (11) | 26 (6) | 30 (8) | 12 (25) |

(Figures in parentheses indicate rejection)

11.1
*Matrix*

| Patuakhali | Faridpur | Tangail | Mymensingh | Dacca | Sylhet | Comilla | Noakhali | Chittagong | (West Bengal) India |
|---|---|---|---|---|---|---|---|---|---|
| 11 | 12 | 13 | 14 | 15 | 16 | 17 | 18 | 19 | 20 |
| — | (2) | (1) | — | (2) | (1) | — | (10) | (3) | — |
| — | (2) | — | 3 (1) | (1) | (3) | 1 | (8) | (4) | — |
| — | (2) | — | 2 | — | (2) | 1 (2) | (9) | (1) | — |
| — | (1) | 1 | (1) | (2) | (2) | (2) | (8) | (4) | — |
| — | (3) | — | — | (1) | (1) | (2) | (10) | 1 (1) | — |
| — | 2 | — | — | (1) | (2) | (4) | (9) | (2) | — |
| — | (4) | — | — | (1) | — | (5) | (6) | (2) | — |
| — | — | — | (2) | (3) | 1 | 2 (2) | (11) | 3 | — |
| 4 | 4 (1) | — | — | 5 | (5) | (2) | 1 (4) | 1 (6) | — |
| — | 4 (1) | — | -- | 2 | (3) | (2) | (6) | (5) | — |
| — | — | 1 | — | 6 | (2) | 4 | (11) | (3) | — |
| — | — | — | 5 | 2 (3) | 3 | 4 | (8) | — | — |
| — | — | 4 (1) | — | 6 | (2) | 6 | (8) | (2) | — |
| — | 4 | 2 | 5 | — | (2) | 6 | (12) | 2 (1) | — |
| — | — | — | 1 | (3) | — | 8 (2) | (4) | 4 | 3 |
| — | 2 | — | 4 | 5 | 4 | — | (3) | 5 (2) | — |
| — | 3 | — | (2) | 2 | 7 | 5 (3) | — | 1 (5) | — |
| — | (4) | — | — | (2) | 4 | 6 | 1 (9) | — | — |
| 4<br>(0) | 19<br>(20) | 8<br>(2) | 20<br>(6) | 28<br>(19) | 19<br>(25) | 43<br>(26) | 2<br>(136) | 17<br>(41) | 3 |

the choices and rejections were supposed to be within Bangladesh, 3 respondents from Sylhet had chosen districts from West Bengal (India) and the last column in Table 11.1 was thus necessary to indicate this fact. The sociometric matrix revealed that Comilla received the highest number of choices, while its neighbour—Noakhali—received the highest number of rejections and the lowest number of choices. The 18 districts of the study area are arranged in Table 11.2 in order of the number of choices received.

TABLE 11.2

| Districts | Number of choices received |
|---|---|
| Comilla | 43 |
| Rajshahi | 36 |
| Khulna | 30 |
| Dacca | 28 |
| Pabna | 27 |
| Kushtia | 26 |
| Jessore | 25 |
| Mymensingh | 20 |
| Faridpur | 19 |
| Sylhet | 19 |
| Bogra | 17 |
| Chittagong | 17 |
| Rangpur | 13 |
| Barisal | 12 |
| Dinajpur | 11 |
| Tangail | 8 |
| Patuakhali | 4 |
| Noakhali | 2 |

In an analysis of interpersonal attraction, it is often observed that two persons might choose each other and the pair is said to be a pair of mutual choice. However, in an intergroup investigation, as is done in this study (each group representing a particular district), it is extremely unlikely that all respondents of District X would choose District Y and *vice versa*. Nevertheless, it was found that, in some cases, a large number of respondents of a district had chosen a particular district and the former district was also chosen by the latter in a similar fashion. In order to locate these mutual choices, the pairs of districts where attraction is mutual for at least 25 per cent of the respondents were identified, and these pairs are shown in Figure 11.1 as districts of mutual choice. Six such pairs of districts of mutual choice were found in Bangladesh, viz., Dinajpur-Rangpur, Pabna-Rajshahi, Jessore-Khulna, Mymensingh-Dacca, Dacca-Comilla and Comilla-Chittagong. It means, to cite one example, that at least 5 respondents of Dinajpur had chosen Rangpur and *vice versa*.

The sociometric matrix does offer possibilities for several types of operations to identify its structural aspects. Since the present study was of an exploratory nature, it was attempted to discern from the matrix if any specific trend in choice and rejection patterns existed. A careful examination of the cell entries in Table 11.1 would reveal that some districts had a tendency of making choices from particular district (usually in proximity to each other), while clearly avoiding associations with other districts. This is an indication of regional cohesiveness where some districts within particular regions of Bangladesh are more associated with each other than with districts of other regions. An inspection of the sociometric matrix yielded two regions in Bangladesh with at least 75 per cent of the respondents of the component districts of each region choosing from among the districts of the region. These two regions, comprised districts of affinity, are called Affinitive Regions, and they are shown as Affinitive Region I and Affinitive Region II on Figure 11.2. For instance, each of the eight districts in Affinitive Region I had 75 per cent of the respondents making a choice of a district from within the same region. The same is true for each of the five districts of Affinitive Region II. The remaining five districts of Barisal, Patuakhali, Noakhali, Chittagong and Sylhet failed to meet the criterion for inclusion in or formation of any Affinitive Region, and hence, were termed Non-Affinitive Districts.

**Interpretation**

The sociometric matrix summarizing the results of the data on inter-district attraction and repulsion revealed a number of interesting facts about the cultural relationship of regional population clusters within Bangladesh. It was expected that all persons of all districts in the sample would not have equal amount of attractions for all districts, i.e., some districts would be chosen more than others. In the study of interpersonal attraction within a group, the social psychologist terms the highly chosen persons as sociometric *stars*, while persons who receive few choices and rejections are called *isolates*. In similar vein, the district of Comilla—being the recipient of highest number of choices—assumed the rank of sociometric star in this study. On the other hand, both Patuakhali and Noakhali emerged as isolates with very few choices received by them. However, there is one interesting difference between the characteristics of Patuakhali and Noakhali. Patuakhali received only 4 choices and no rejections; while Noakhali received 2 choices, but received 136 rejections. Patuakhali has been presumably ignored because of its relative remoteness and lack of transport facilities out of the district and consequently, few people in the sample had any familiarity or experience with persons of that district. The high rejections for Noakhali are perhaps the result of negative reactions from contact and experience with that district. Noakhali had received rejections from all districts in the study area. At the other end of the scale, Comilla was not chosen by persons of all districts; it received larger number of choices from the districts of eastern half of Bangladesh. In general, a pattern of regional grouping is clearly discernible in the matrix with the districts of north-western and western Bangladesh choosing from within that region and rejecting the districts of eastern Bangladesh, and *vice versa*. This cohesiveness aided the delineation of two affinitive regions—one on the western part and the other along the central part of the country (Figure 11.2). The districts within each affinitive region share common borders and a general agreement of outlook and values have helped in a desire to associate among themselves. However, the districts of Barisal, Patuakhali, Noakhali, Chittagong and Sylhet failed to meet this criterion, and it is possible that their individualistic and isolated orientation did not allow them to be included in or constitute any affinitive region.

As regards the pairs of districts of mutual choice, it was observed that five of the six pairs of such districts (Figure 11.1) are physically juxtaposed, which leads one to the inference that opportunities for greater contact between them have con-

*Bangladesh : Regional Choice Pattern*

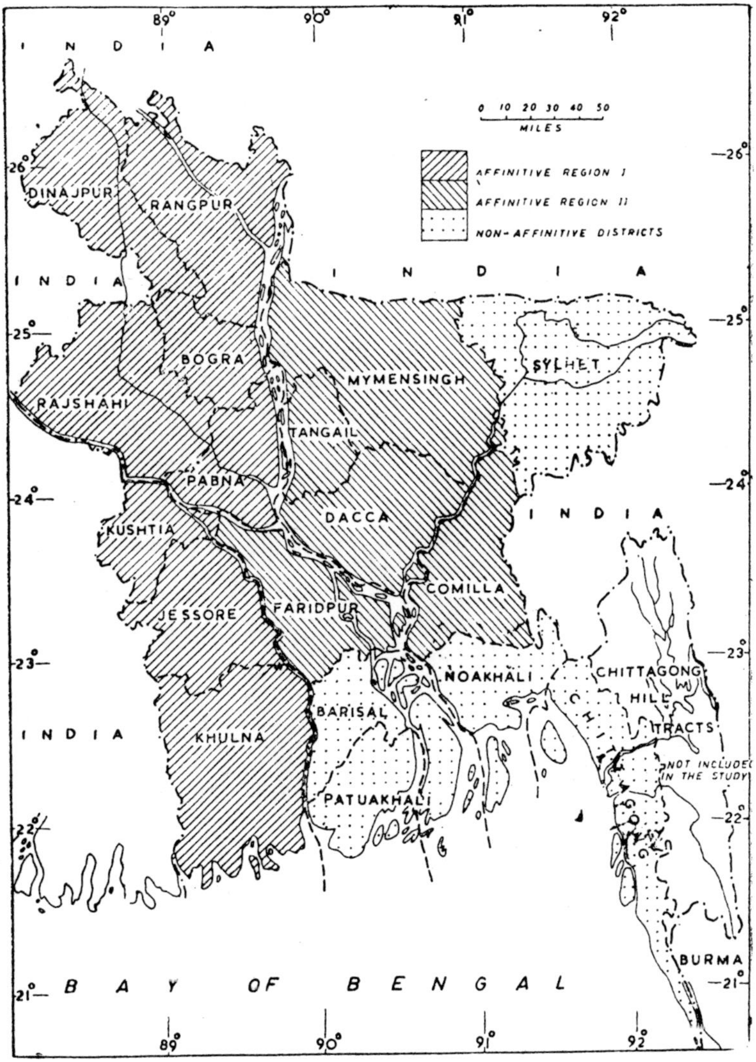


FIG. 11.2

tributed to the development of bonds in the pairs. The only exception was the Comilla-Chittagong pair, separated by the district of Noakhali. The latter district, being a recipient of many rejections, might have indirectly aided Comilla and Chittagong to develop a pair of mutual choice.

The respondents were also presented in the questionnaire with six structured statements as explanations for the choice and rejection of a particular district. Each respondent was asked to put a cross mark against the statement or statements which best explain the reason for his/her choice of a particular district; as well as against the statement or statements that best explain the reason for his/her rejection of a district. These statements and the frequency of their citation by the respondents for their choices and rejections are summarized in Table 11.3. It is clear from the table that most of the respondents (85.6 per cent) admired the quality of amiable disposition of the people of their chosen districts. This was not unexpected since the aim of

*Table 11.3*

| | Percentage of respondents citing this reason |
|---|---|
| *Reasons for Choice* | |
| Natural environment is fascinating | 39.2 |
| Climate is attractive and pleasant | 20.3 |
| People are generally well educated | 35.3 |
| People are amiable and their outlook is acceptable | 85.6 |
| Social customs are agreeable | 43.1 |
| Regional dialect is intelligible and delightful | 45.7 |
| *Reasons for Rejection* | |
| Natural environment is not fascinating | 10.5 |
| Climate is unattractive and unpleasant | 7.2 |
| People are generally less educated | 15.0 |
| Behaviour and outlook of the people are unacceptable | 87.2 |
| Social customs are not agreeable | 47.7 |
| Regional dialect is unintelligible and not delightful | 64.7 |

the questionnaire was to identify the district with whom association and contact are desired, and hence, the respondent would naturally affects value to the behaviour, outlook and temperament of the chosen persons. Next on the list of reasons for choice are general intelligibility of the regional dialect, and agreeable social customs of the chosen district. Among the reasons which explain rejections, 87.2 per cent of the respondents cited disagreeable behaviour and outlook of the persons of the rejected district. Hence, it can be concluded that amiable disposition and temperament are the qualities that influence persons of one region in choosing or rejecting persons of another region. Table 11.3 further shows that a large number of respondents (64.7 per cent) also mentioned unpleasant nature and unintelligibility of the regional dialect of the rejected district as a major reason for rejection. In fact, the positive and negative aspects of (a) behaviour and outlook, (b) regional dialect, and (c) social customs had been cited, in that order of frequency, by the respondents as the reasons for choice and rejection respectively.

It is common knowledge that choice and rejection of a person is generally influenced by association with the chosen or rejected person. In the context of the present study of inter-regional attraction and repulsion, such choice and rejection might be influenced by contact when persons from one region travel to another region. This contact through travel was presumed to have resulted from two types of events. Firstly, if the districts of two parents are different, the children might have visited at times the maternal grandparents' home and acquired some knowledge of that district. Secondly some persons, not necessarily of the preceding category, might have travelled to or lived in a district other than their own and thus, gained knowledge of that district. The sample of this study revealed that 21.2 per cent of the respondents had their parents coming from different districts, but only one-third of this category of respondents had chosen the districts of their mothers. Hence, on the basis of the present sample, the influence of the mothers' districts on the choice of a district does not seem to be very great. On the other hand, 54.2 per cent of the persons interviewed had travelled to or lived in their chosen districts. In a further breakdown of the duration of contact with the chosen district, it was found that 22.2 per cent of the respondents had actually lived in their chosen districts for a period of more than one year. It is, therefore, argued here that inter-regional contact through travel and living experience might have reasonable amount of influence in choosing a particular district.

**Determinants of Choice : Summary**

Several determinants of choice and rejection of persons from different regions can be identified from the sociometric test in this study. Firstly, nearness is a powerful determinant of choice. Persons of contiguous regions have the greatest opportunity to interact and hence, persons of any particular region would normally prefer to associate with persons of neighbouring regions. Similarly, lack of knowledge or misinformation about persons of distant regions often generates repulsion for them. Secondly, those regions are more likely to be chosen whose people posesses the most valued characteristics for association and company, viz., pleasant behaviour and outlook. The opposite of this would be true for rejection. Thirdly, there is a trend for persons to choose those regions which are similar to themselves in terms of dialect, speech and social customs. The absence of this similarity would result into rejection. Lastly, the regions where the choosers have travelled or lived in, would have a greater chance of being chosen because first-hand experience has given them an opportunity for interaction.

## EPILOGUE

In this paper, only an attempt had been made to introduce the scope of analysing interaction between cultures via social psychological techniques. It was empirically demonstrated that the sociometric technique can be fruitfully used to identify and measure the levels of attraction and repulsion between culture groups occupying different territories. Social psychologists have developed a number of theories to explain interpersonal attraction. No attempt was made here to employ any of them in order to account for the choices and rejections. Nevertheless, it remains a potential field for future investigation. The sociometric test is but one of the many tests done by the social psychologist. The cultural geographer has the scope of using many other tests and techniques of the social psychologist in dealing with parallel problems involving perception, attitude and behaviour, social distance, and the like. A great deal of empirical investigation is, however, needed. Before, these tests can be fully sharpened to the requirements of the cultural geographer.

## SUGGESTED READINGS

Helen Hall, Jennings, *Sociometry in Group Relations*, Washington, D.C., American Council on Education, 1948.

J.L. Moreno, *Who Shall Servive*? Beacon, N.Y., Beacon House, Inc., (2nd edition), 1953.

C.H. Proctor and C.P. Loomis, "Analysis of Sociometric Data", In Marie Jahoda, M. Deutsch and S.W. Cook (eds), *Research Methods in Social Relations*, Vol. 2, New York, Holt, Rinehart and Winston, Inc., 1951, pp 561-585.

P.F. Secord and C.W. Backman, *Social Psychology*, New York, McGrow Hill Book Company, Inc., 1964.

12

# Nature and Trends of Cartography

*—Rameshwar Thakur*

GEOGRAPHY is concerned with different branches of knowledge, thought and logic. Cartography is one of them. It is concerned with the designing, constructing and producing of maps. This includes the whole process from original field work to final printing. But Mishra[1] suggests that it is also a science of human communication. Cartography now includes scientific, technical and artistic efforts in producing a map. Naturally it demands skills which are more fundamentally cartographic than the skills of making original drawings. Thus, the important aspect of cartography is the co-ordination of the entire map-making process, scientific as well as artistic.[2]

It is generally stated that the role of scientific skill is greater than the artistic skills in the map making process. This is due to the fact that scientific training for planning, designing and executing a map is more difficult to acquire than the artistic skills, that is, the manual skills needed to convert a well thought out map into a well drawn map. Thus, one can be designated as a cartographer even if one lacks manual skills for drawing maps while a good draftsman without the intellectual skills needed to plan and design a map cannot be called a cartographer. It does mean, however, that the scientific aspect of cartography, i.e., the aesthetic sensibility and the principle of constructing maps is given more importance than the manual skills for making fair drawings.[3] From this it should not be inferred that the artistic aspect of cartography has been devalued or deemphasized.

Art in cartography has much value and importance because the purpose of cartography is to improve the graphic representation of the earth, which is partly artistic in character. In other words, it will not be wrong to state that the study of cartography is partly a study of map graphics. And since art is the highest form of graphics, it is natural that a good map must be artistic in nature. But this is a controversial and debated question; doubts are often raised about the legitimacy of art and artistic talent in the process of map making.

In the past, cartography was definitely an art. It becomes quite clear when one views the products of earlier cartographers. They embellished their maps with all sorts of imaginative things along with fancy scroll work, ornate lettering and intricate compass roses.[4] By means of these ingredients and special colouring instruments they tried to decorate the maps so that it could be hung on a wall. The major reason behind this was that the cartographer refrained from making the map ugly. But by adorning and beautifying maps, their basic purpose was lost. Since most maps are functional and reveal some concept or relationship, it is possible that the viewer could be stimulated first by its beauty and fail to realise the conceptual relationship. Here lies the demerit of giving too much emphasis on artistic learning in cartography. Even so, the creative kind of endeavour by producing something that has never been done before satisfies the reader.[5] In this respect, cartography is a creative and imaginative art. It is then vitally important that the cartographer should have a good command of the principles of representations to construct a map.[6] Therefore, it is essential to have new combinations of techniques and skills in the creative representation of maps.

From the preceding discussion it has become clear that cartography is an art. But it can also be called a science that has its own individuality. The important point to remember is that it is neither an experimental science like physics or chemistry nor a social science like economics or sociology. Nevertheless, it employs the scientific method in devising ways and means of bringing order and system, generality and simplicity, refinement and legibility to an almost incomprehensible range of complex details.[7] The scientific method generally falls into two categories. In the first place, cartography is a geographic science; and in the second, it is an auxiliary science. In the first category the main purpose of cartography is to represent realistically the surface of the earth as well as other heavenly bodies. The realistic representation of the surface demands a good knowledge of the characteristic features of that area.

In other words, the cartographer must be familiar with the location and spatial distribution of objects according to a given scale. Cartographic representation also includes the selection of the important and elimination of the non-essential factors. This is an essential aspect of generalisation of data. However, generalizing the data is one of the difficult tasks for the cartographer since only a master of the subject can generalize well.[8] It also involves logic, synthesis and training in reasoning. In that way cartography is a kind of practical science very much like engineering. It is an auxiliary science because it acts as a bridge between techniques, art and earth sciences.[9] The cartographer has to collect complex details of the earth surface in the generalization of map. This means that he must be familiar with topographic mapping, geodetic foundation, census data, air photographs, documentary materials and a host of other kinds of sources.[10] These different sources help differently in the presentation of maps. The geodetist and the surveyor provide an idea of the shape and size of the earth, geographer about the location of surface features, while others provide the subject matter. Ultimately the cartographer classifies, generalises and simplifies the data and converts them into mappable form. Thus cartography is based on sound principles and seeks to accomplish its aims by means of intellectual and visual logic and in this way it is scientific in nature.[11] Comparatively speaking, cartography is more a science than an art and it employs scientific methods and logic.

## CARTOGRAPHY AS A MEANS OF COMMUNICATION

Maps are vehicles in the process of communication in cartography. In the widest sense the function of a map is communication. The maker of the map, the cartographer, supplies information and it is his task to present this information adequately. The information is supplied clearly and carefully through a combination of drawings, words and symbols. Viewed in this way a map becomes a medium of communication and cartography a communication science. In other words, the main purpose of cartography is the communication of facts and ideas about the earth. However, there is a dispute on this point. Some geographers and cartographers stress the point that the representation of the earth is more significant than the aspect of communication. Such disputes and objections do not seem to be valid. Cartography was most effective in the representation of the parts of the earth in the past when man did not know much about the shape, size and contents of the earth. Also this aspect was important since the major portion of the earth was left unsurveyed and the geographer was keen to let people know the discovered part of the earth through maps. Now this phase of discovery has passed and the shape, size and contents of the earth surface are known to us. Therefore, the character of cartography has changed. The emphasis is now on the effective portrayal of the facts and ideas about the earth. In this sense the aspect of communication is getting dominant place in the function of maps.[12]

In order to achieve this the geographer should be aware of the rules of communication. Koeman[13] states the doctrine of communication in general as "how do I say what to whom". This is an important item in modern cartography and one which a cartographer should know. In general saying is effectuated by using graphic symbols and a great series of modes of expression are at the disposal of a cartographer. Communication by means of graphic symbols is the most important form of transferring information and when this form is applied we speak of thematic maps.[14] Maps result from the then "how do I say what to whom". In other words the map maker should show what has been surveyed on the terrain or in the areal photograph or in his opinion has been represented by the past category of land surveyors, photogrammetrists and photographers.[15] The word to whom applied to the map-user. The language of graphic symbols should not only be understood by the map-maker but also by the map-user (Fig 12.1). The map should be used by different people representing different levels of mental development. For example, it should be used by school children, college students, research authorities, academicians and by any expert. Therefore, cartographers and map editors should tune the map accordingly. This cartographic process has been compared and linked with the theory of communication systems.

*The Language of Graphic Symbols*

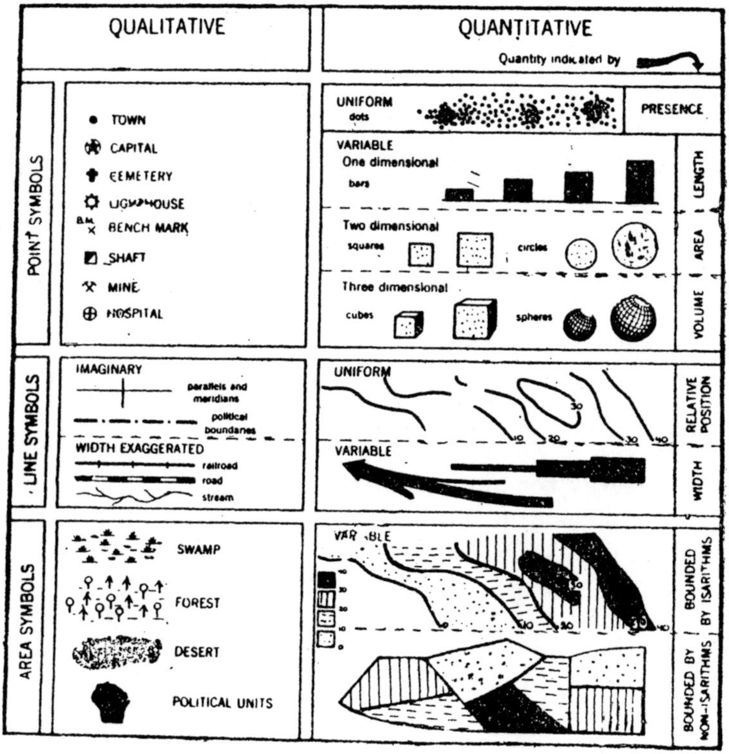


FIG. 12.1

A communication system contains five elements. These are : (a) source, (b) transmitter, (c) channel, (d) receiver, and (e) distribution. The message comes from the source and goes to its destination passing through various phases. The transmitter gets a message from the source and encodes this message into a language that can be fed into the channel selected for the transmission of the message, and finally the receiver reads this message and decodes it into a common language.[16] This arrangement of communication system is linked with the cartographic process and a similarity is found between them. The ideas and facts of the earth's surface constitute the source of message. The cartographer converts these ideas and facts into words, drawings and symbols and can be very well compared to a transmitter. Maps and other cartographic products are the channels, while the symbols etc. perceived by the map user are the received symbols. Finally map users all over the world can be compared to the destination of the communication system. It is thus clear from the above comparison that perception plays an important role because it is this through which almost all learning and understanding is accomplished. By sensory mechanism we receive the objects and events depicted on the map and then we acquire a perfect knowledge of the earth and its surface features.[17] In this way cartography, the art and science of making maps, is also a science of communication.

## ELEMENTS OF CARTOGRAPHY

As soon as the data are assembled by different sources : by surveying, by compilation from various sources or by census taking. The map data must be graphically portrayed. In other words, the techniques of representation must be considered. It covers a number of specialized technical phases.[18]

(i) the scale of the mapping,
(ii) the method of projection,
(iii) the symbolization of the map data being mapped,
(iv) the map design, and
(v) the presentation of the map for application.

1. **The Scale of the Mapping** : All maps must be made to scale. Scale is the means through which the part or whole of the earth is reduced to a convenient map size. After reduction the map becomes scientific and handy. The second important use is particularly in transference of the spherical earth drawn on a flat surface. Therefore, there are many aspects of scale that complicate their use. Generally, scale is defined as the ratio between the same dimension on the map and the corresponding dimension on the earth stated as a proportion, such as 1 : 2500, 1 : 100000, or 1 : 250,000,000. The first part relates to the map while the second part to the earth. It must be, however, pointed out that the amount of departure varies greatly with the system of map projection.

2. **The Method of Projection** : The work of a cartographer would have been easier had our earth been flat. Because of the spherical shape of the earth and the plane surface on which this shape has to be represented a cartographer has to advise complex graphical, geometrical and mathematical methods of transferring the earth.[19] This process of transformation is termed as map projection. A large number of systems of projection are possible, but relatively few are regularly employed. Some of them are azimuthal (angles), conformal (area), and equivalent (shapes) which are popularly used. Cartographers from the very beginning have been developing a projection to represent the above qualities. Still, to date, it has not been possible to develop a projection satisfying all the properties.

3. **The Symbolization of the Data** : Every map contains some data to show. It is, therefore, imperative to assume that mapped information must be symbolized in a perfect manner so that their locational relations could be generalized. Symbols constitute a kind of code by which the cartographer can present the most effective story in the limited space available.[20] This generalization should go hand in hand with the scale of the map and must involve a selection from among the elements of each category and a simplification of their representation.[21] Map symbols can be classified into two parts. The first is qualitative and the second quantitative. Each of these are again classified into three parts. Practically three symbols are used to convey the information to the map used. They are generally categorised as point, line and area symbols Fig. (12.1). This map shows the possible variations in each of these subcategories. Wright[22] has clearly stated that the geographical data can be reduced to dimensional terms, i.e., a point refers to a location, a line one-dimensional, an area or surface two-dimensional, or a volume three dimensional. It has also been stated that the first-three geographical quantities, i.e., point, line and area symbols can be shown easily and the last one, geographical volume puts problems before the cartographer.

The difference between point, line and area symbols is clear and straight. Point symbols, such as dots, circles and triangles may show qualitative as well as quantitative aspects. Qualitative symbols show the existence of an object but they do not represent any quantitative data. On the other hand, the quantitative symbols show the length, size and volume of an object. For example, bars indicate the length, circles the size and spheres of the volume. Line symbols range from the simple line representing a linear quality, such as a road, an air route, streams, coast lines or a boundary to a class having the generic name in arithmetic. The former indicates qualitative data while the later quantitative. The isarithm is commonly used to portray distributions that vary in amount from place to place, such as elevation of the land surface above sea level, but it may be employed for more abstract concepts.[23] Lastly, area symbols also represent qualitative or quantitative data and use both point and line symbols to give a combined effect of the area spread of the objects.[24] Quantitative area symbols are those used for the distribution of swamps, forests, deserts, political units or soil types. Quantitative area symbols are used also to give the relative density of the occurrence of a phenomenon whether by administrative units or by isarithmic lines.[25] It is clear from the above discussion that qualitative area symbolization relates to the variation in pattern while quantitative area symbolization indicates variation in value changes.

4. **The Map Design** : Among the many aspects of the map making process, map design is an important one. This is because of the fact that it is crucial and complex. For purposes of scientific communication, the map must depict information with clarity, simplicity and accuracy. Naturally then, a map maker has to follow certain principles. He must follow proper line, weights, lettering styles, projection, scale, distribution, colours and patterns. If these are not employed the map reader may be easily misled. Therefore, the map maker must be cautious against the possibility of subjective impressions.

5. **Reproduction of Map** : Maps are generally constructed for reproduction. Before the invention of the printing machine, duplicate copies were prepared by the same map maker and this used to take a lot of time. Now many duplication machines have been invented to get this done easily. This is based largely on physical, chemical and technical processes. This forms a separate branch called polygraphy, a branch of techniques that makes various printed products.[26] Recently, it has become an important aspect of the changing nature of

cartography. Where, however, maps have to be produced in a reduced size as in a book, it is imperative to take this factor into account while drawing the original, because many details are apt to be lost while reducing them, generally by photographic processes.

## CHANGING ASPECT OF CARTOGRAPHY

These days with the scientific-technical revolution in general, the character of most of our sciences is changing. Cartography is no exception to this rule. The subject and method of cartography are changing fast. Therefore, this requires special interest and significance.

In the past, cartography was more concerned with the measurement and the representation of the shape and size of the earth. It was, therefore, considered as a subdivision of geodesy or its scientific interests were limited to cartographic projections. Cartographers were busy with the construction of projections revealing a narrow mathematical problem. The facts presented on maps were mostly the salient visible features of the earth's surface. Those features were water bodies, mountains, towns, roads, forests and settlements. They were mostly represented by pictorial embellishments. Naturally they added charm to the maps and were very pleasing to the eye. This pictorial decoration was retained till the first quarter of this century. Today these broad details are known to us and we have at our disposal minute details of the earth surface features. Therefore, modern cartography is attempting to analyze complex data and is trying to present them in a more comprehensible form to map users.[27]

By the middle of this century the pictorial decoration was almost gone. The artistry that the early cartographers lavished on decoration has been replaced by ingenuity.[28] Further, they have been replaced by new symbols with which to map them.[29] Now, every map presents a definite amount of information. A large number of diverse thematic maps—geologic, pedologic, climatic, economic and settlement maps are coming up. It is in this context that modern cartography is a graphic science and presents the details of the earth surface.[30] The intent and purpose of the modern cartographer are now not merely to study the astronomical aspects and shape and size of the earth, rather to make a better representation of the details of the earth. This has been made possible with the development of science and technology particularly with the application of photogrammetry, remote sensing and the use of electronic devices.[31] Cartography has come to be understood as a discipline that studies and devises the scientific-technical methods and processes of map compilation and reproduction.[32] This theme is quite opposite to the earlier geometric interpretation of cartography.

Another development of modern changing cartography is the map that shows by means of isolines or other symbols standing for specified quantities. These quantities could be elevation, depth, temperature, population densities, settlements and distribution, etc. Such quantitative maps were totally unknown before 1900. Now they have become immensely important tools of research and exposition in the various earth sciences.[33] Many other social scientists also, such as historians, anthropologists, economists, demographers and geologists depend upon census and similar statistical inventories.

As Berry has remarked, modern cartography has a well developed base of mathematics, just like econometrics and psychometrics, unfortunately we are slow to coin and apply a term like geometrics to the science of mathematical geography, or we may say, statistical geography.

Recently, research in cartography is being carried out in many directions. The most important is related to the map as a research tool and to its employment as a medium of communication. Based on a thorough review of the research done in cartography, Robinson[34] has indicated the following directions : (a) investigation over the methods of deriving the error factor in quantitative mapping, (b) the design of graduated quantitative point symbols, e.g. circles, (c) the employment of statistical words in the spatial context, (d) the role and practice of cartographic generalization ranging from linear elements as coastlines, drainage and boundaries to the analysis of isarithmic and chroplethic intervals, (e) design and employment of colours, patterns and tones as area symbols, (f) investigation over cartographic, graphic methods, (g) devising methods of map reproduction and duplication. All these aspects require a thorough investigation at the present moment to bring up this branch as more fruitful and important.

### THE DANGER AHEAD

"Computer Mapping has brought and will bring many more changes and advances to cartography and map applications." Because of the complexity in package programming, and the mathematical and cartographic knowledge required for map programming, it is probably unrealistic to expect all geographers and many other map makers to master the art in the seventies. Many of us will have to reconcile ourselves to the adoption of programmes written by abler persons, which we will use simply by following the "write-up". The inherent danger of this approach is serious. One may devalue all decision-making in the mapping process to the programmer, and finally trust the impressively lengthy computer programme to grind out magic maps which are often described as efficient and accurate. The write-up usually contains a series of instructions on programme use. Although many conceptual assumptions and judgments on cartographic processes are employed in the programming, information on these is seldom included in the write-up. But if a map maker considers himself a researcher, and intends to use the output maps for analysis and interpretation, then he must know how the maps are derived and what conceptual models are used in the mapping process. The model of interpolation is significant for isarithmic mapping, but Taylor's discussion of Symap reveals little about its interpolation model.

Taylor has claimed that "the computer will make the map a much more meaningful tool than it has ever been before". Computer application can advance mapping techniques under the following conditions : (1) the programmer is competent in thematic cartography; (2) the user understands the basic algorithm and cartographic assumptions employed in the programmes; and (3) the user exercises good judgment in choosing the programmes and their options.

Geographers will inevitably come more and more to rely on packaged programmes for their mapping needs. Indeed, this is the great challenge and opportunity for geographic cartographers in the years ahead. We need not fear the monster but we do need to fear the prospect that many geographers may unthinkingly accept mapping programmes without understanding their underlying logic or appreciating fully the options available. What has probably already occurred with regard to map projections, could now occur to the generalization process. Geographers have enough problems without being their own victims. "Quick, Igor, Stop the machine for a while so we can think."[35]

## REFERENCES

1. R.P. Mishra, "Fundamentals of Cartography" (*Prasaranga* : Univ. of Mysore, 1969), p 12.
2. Ibid., p 12.
3. Ibid., p 13.
4. A. H. Robinson, *Elements of Cartography* (New York, John Wiley & Sons, Inc., 1962) p 15.
5. Ibid., p 15.
6. Ibid., p 15.
7. R.P. Mishra, op. cit., p 14.
8. A.H. Robinson, op. cit., p 13.
9. R.P. Mishra, op. cit., p 14.
10. A.H. Robinson, op. cit., p 14.
11. R.P. Mishra, op. cit., p 15.
12. C. Koeman, "The Principles of Communication in Cartography" in E. Aruberger and F. Aurada, (Eds.), *International Year Book of Cartography*, (London, George Philip & Son Ltd., 1971), p 170.
13. Ibid., p 171.
14. Ibid., p 172.
15. Ibid., p 173.
16. R.P. Mishra, op. cit., p 16.
17. Ibid., p 17.
18. A.H. Robinson, "Cartography" in D.L. Sills (Ed.) *International Encyclopaedia of the Social Sciences*, vol. 2 (New York, The Mac Millan Company & the Free Press, 1968), pp 325-29.
19. R.P. Mishra, op. cit. p 80.
20. A.H. Robinson, op. cit. p 136.
21. Ibid., p 327.
22. J.K. Wright, "Crossbreeding Geographical Quantities, *The Geographical Review*, vol. 45, 1955, pp 52-65".
23. A.H. Robinson, op. cit., p 328.
24. R.P. Mishra, op. cit., p 261.
25. Ibid., p 261.

26. K.A. Salichtcher, "The Subject—Method of Cartography : Contemporary Views", *The Canadian Cartographer*, vol. 7, 1970, p 86.
27. R.P. Mishra, op. cit., p 20.
28. J.K. Watson, "Map", *Encyclopaedia Britanica,* vol. 14 (Chicago, William Benton, 1967), p 836.
29. Ibid., p 836.
30. R.P. Mishra, op. cit., pp 20-21.
31. J.K. Watson, op. cit., pp 836-37.
32. K.A. Salichtcher, op. cit., pp 77-78.
33. J.K. Watson, op. cit., pp 336-37.
34. A.H. Robinson, op. cit., p 328.
35. Mei-Ling Hsu, and P.W. Porter, "Computer Mapping and Geographic Cartography". *Annals of the Association of American Geographers,* vol. 61, No. 4, 1971, pp 796-799.

# SECTION TWO

# PHYSICAL GEOGRAPHY

# Introduction

THE CONSTANTLY changing dimensions of physical geography lead to changes in the conceptual development with the study of hazards of nature, lithospheric conditions and the expanding needs of mankind. Although in the western world quantitative methods have been adopted by geographers in the name of scientific precision, these do not show the spatial relationship, the man-land relationship and the environmental integrity, which are the primary concepts of geography.

The papers presented here are helpful in the understanding of the mountain regions, geomorphology of the drainage begins neotectonic effect in moulding the landscape features, continental drift and palaeomagnetisors, evolution of drainage lines, movement of jet streams and weather forecasting, Paleo-climatic development of landscape, climate-genetic geomorphology and the contributions of L.C. King in geomorphic thought.

Mountain geography emphasises the discrepancy between the importance of mountain regions on a world scale and the apparent inferior role of mountain geography as an academic study. An attempt is made here to describe the nature of mountains and to establish criteria related to the physical and human environment that can be used for a definition of mountains. It is further shown that the delicately balanced mountain environment is easily altered and disturbed by physical forces and human interference. The paper concludes with a survey of recent problem oriented mountain research and international cooperation in the field of Applied Mountain Geography.

"Geomorphology of Drainage Basins in Palamau Upland" is devoted to finding out the statistical significance of the variations in drainage behaviour for ten river basins selected from Palamau Upland, Bihar. The quantitative approach throws light on various aspects of drainage analyses such as stream order, drainage frequency, bifurcation ratio, drainage length, drainage density, sinuosity index, basin shape and size, stream gradient and stage of geomorphic development. An attempt has also been made to correlate hypometric integral and some morphometric attributes.

In analysing the recent trends of geomorphology, systematic stratigraphic, neotectonic, quantitative along with various morphogenetic analyses on the fluvial landforms in general, and terraces of flood plains in particular of the Tista and Subarnarekha basin areas have been emphasized here. This reveals the positive role of crustal tilting or warping along with the effects of pleistocene climatic changes (Gunz, Mindel, Riss and Wurm) on the regime of the river systems concerned. Tertiary warping and quarternary tectonic movements have far reaching effects both on the Tista (the Brahmaputra Basin) and the Subarnarekha drainage basin. These earth movements have also brought remarkable changes in different mountains and ocean basins of the world.

The continents started drifting some 200 million years ago at the end of the Carboniferous period. Regarding this, several theories and modifications along with evidences have been presented by Bacon, Wegener, Ahmad, S.W. Carey, Holmes, Schiller, etc. In this study recent advances have been analysed with the determination of the alignment of magnetic properties of the earth surface during different geological periods and the concept of global plate tectonics.

The study of Palaeomagnetism is a significant aspect of continental drift which is intimately related with the alignment of rocks of different geological periods. Holmes has investigated the magnetic poles and fields of the earth surface in different parts of the world. According to his view rocks of different periods are differently aligned from the stage of solidification of magma on the earth surface. This has helped in the drifting of continents in almost all directions. Evidences are given from North and South Atlantic Oceans, Australia, the Deccan Plateau of India, British Trias, African Karoo Highlands and the Europe Alps.

The recent theory of 'Plate Tectonics' enunciated by a term of American and Russian scientists has been offered a more penetrating analysis than the earlier ones.

The resources of the Indian Ocean in relation to plate tectonics has been analysed with special reference to the evolution of geological structure and topography of the ocean floor. The study of Plate tectonics provides a clue in locating the hydrothermal minerals along its three types of Plate boundaries.

The rivers of the North Indian plains originated from the Himalayas in Eocene, Miocene and Pliocene periods successively according to the formation of three successive mountain systems. Thus the problem arises as to whether the Himalayan rivers flow into the Bay of Bengal or Arabian Sea respectively or only towards the Bay of Bengal as the master stream of Ganges does today. On this controversial issue western geologists are in favour of the single river theory, while Indian scientists feel that the rocks of the Siwalik Himalayas have been deposited by the joint efforts of Himalayan and extra-Peninsular drainage, and the same discharge water into the Bay of Bengal after the formation of the Rajmahal gap.

Weather forecasting means predicting the probable weather conditions for short and long terms with the help of past and present weather maps and upper air observations. Accurate weather forecasting has been a problem for centuries for aviators, sailors, businessmen and agriculturists. The paper aims at analysing the principles and methods of weather forecasting for students of geography and climatology.

The study on jet stream deals with the high speed wind blowing at an height of 2000 feet to 60,000 feet below the Tropopause in the zone of the Troposphere. There are several jet streams in the world e.g. arctic, polar, sub-tropical, etc. This wind blows in a serpentines way with discontinuous concentration of high velocity wind and the same is responsible for air disaster in the mid sky in case the plane is trapped in the zone of lesser velocity wind. It has also good effects, as it modifies the weather conditions of several countries of the world.

The Paleo-climatic development of landscape in Hazaribagh Plateau is a model study which shows the need of landscape identification, developed under some post climatic condition. In this plateau region various topographic forms e.g. weathered profiles, residual hills, gullied surfaces, rock benches, terraces, flood plain deposits and deep soil cover confirm the changes of climatic condition and these help in the evolution of landscape in different global periods.

The study on "Contributions of L.C. King in Geomorphic Thought" attempts a critical appraisal of geomorphic ideas. King occupies an important place in the galaxy of modern geomorphologists as a propounder of the concept of pediplanation cycle. Being an adherent of climatic geomorphology he believes in the significance of climatic variations and their effect on land slope evolution. His major contribution is the recognition of four-fold slope elements i.e. the waxing slope, free face, constant slope and the waning slope. He regards two essential processes in pediplanation cycle—the parallel retreat of slope and the pedimentation. His study of geomorphology encompasses the evolution of orogenic belts, shields, plainlands and ocean basins. Thus he is an advocate of world geomorphology of which regional geomorphology is a part. Furthermore, King has interpreted the erosion surfaces of Africa, Asia, North America, Europe, South America and Australia as ancient pediplains dating back to the Cretaceous period.

'Climato-genetic geomorphology' focuses upon the effect of past-climatic conditions on the geomorphology of the Barakar Basin. The paper traces the tectonic history in close connection with the erosional surfaces, antecedent drainage, effects of running waster and climatic control in geomorphic processes of the region. In conclusion, it is observed that the cycle of landscape erosion is an outcome of both the post-climatic conditions and the tectonic character of rocks.

# 13

# Nature of Mountain Regions—Concepts of Mountain Geography

—*Christoph Stadel*

HIGHLAND REGIONS with an elevation of over 1000 m occupy approximately one quarter of the earth's surface. They are found in every continent in different climatic zones and in a wide array of cultural environments. Some mountains have been foci for highly developed civilizations, mainly in subtropical and tropical regions. It is estimated that about 5 per cent of the world's population lives in mountainous regions; while an even larger proportion of mankind lives in adjacent lowlands affected by processes and developments in mountains. Some mountain states are densely populated, in particular the tropical and subtropical mountains of Latin America, Africa and the Indian subcontinent.

The importance of mountain regions becomes even more evident when one takes into account that mountains and adjacent lowlands are interdependent regional systems with a multitude of linkages. The geomorphic and climatic conditions of mountains are directly linked with features and events in adjacent lowlands. Moreover, in a cultural, social and economic sense, active interaction can frequently be observed between mountains and lowlands. Although remote mountainous areas may experience cultural isolation, mountain valleys and passes have, in many instances, acted as gateways for cultural interaction and diffusion. A prominent element in the inter-relationship between mountains and lowlands are the movements of population—both temporary and permanent—between the two environments. In an economic sense, mountains often assume an important role as resource regions for lowland cores. The attractiveness of mountain resources; for example stone, gravel, wood, water, minerals, agricultural products, and more recently, the recreational potential, have often been exploited or controlled from the lowlands. This interference has at times been a threat to the distinctiveness of the mountain environment and its people.

In spite of the world-wide distribution of mountains and their direct and indirect importance for a large proportion of mankind the systematic study of mountains has attracted comparatively few scholars, and mountain geography is hardly recognized as a genuine subfield of geography. Yet, research in this field can be highly rewarding; many questions concerning geomorphology and microclimatology of mountains are still to be answered. In addition, although there has been increasingly intensive use of mountain space and resources, little work has been done to study the human impact on mountain regions. The complex aspects and ramifications of resource and infra-structure developments, as well as landscape capability thresholds and social compatibility factors have received little attention.

## THE NATURE OF MOUNTAINS AND CRITERIA FOR A DEFINITION

A consensus may be found for a rather general definition of mountains but the formulation of precise criteria that are globally applicable for mountains is difficult. It has been said that "mountain areas are as different from one another as from lowland areas" (Hewitt, 1966, p 18). It is generally accepted that mountains are regions of steep and irregular topography, at considerable absolute (above sea level) or relative (above the surrounding areas) elevation. These situations lead to the development of distinct climatic and biotic conditions and to specific forms of human activity and utilization of space and resources.

When more specific definitions are attempted, a distinction is frequently made between low, medium and high mountains. Also, a distinction between arctic and subarctic mountains; temperate mountains of the oceanic or continental variety; and subtropical and tropical mountains may be appropriate. In terms of human life and space utilization, mountains in the "developed industrialized" realm should be distinguished from those in the "lesser developed" largely agricultural realm.

The description of high mountains has been of particular interest to geographers. Among the scholars who have pioneered in this field Norbert Krebs, Carl Troll, Germaine and Paul Veyret, G. Furrer and Ken Hewitt rank prominently. Most attempts to define the nature of high mountains have been focused on their natural environment, their geomorphology, their climate, and their ecology. In a paper given in 1940, Carl Troll attempted for the first time to define the nature and the lower limit of high mountains in different climatic zones of the earth (Troll, 1941). In this paper, he postulated a three-dimensional analysis of climates, vegetation forms, and landscapes, and argued that each lowland climate has an associated range of mountain climates—a concept which was missing in Koppen's climatic classification. Troll discredited the common assumption that the latitudinal range of climates and vegetation forms from the equator to the poles is equivalent to a vertical zonation of climate and vegetation from the base to the summit of a high mountain. In later works (Troll, 1943, 1951), he established the following specific geographic factors (*Geofaktoren*) which he thought to be of particular interest for an analyses of high mountains : diurnal and seasonal march of temperature; the freeze-thaw pattern; diurnal wind systems; the snowline; the form of the snow cover due to ablation processes; humid and arid seasons; solifluction and patterned soils; ecological aspects. As a result of his studies, Troll defined high mountains as areas that rise above the timberline, above the snowline of the last Pleistocene glaciation period and above the lower limit of periglacial soil processes. He attached little importance to absolute and relative elevation, criteria which have been used by most other scholars and which are most significant factors in the general perception most people have of mountains. Hewitt (1966) characterizes mountain environments as having a combination of high relief, steep slopes and altitudinally varying local climates.

Mountains are highly active geomorphic regions. Generally in the mountains, extreme temperature, moisture, and wind conditions prevail : large diurnal temperature amplitudes with a pronounced freeze-thaw action at higher altitudes favour gelifraction. Mountains also receive large amounts of orographic precipitation, high rates of insolation and strong winds.

The "high energy condition" in mountains leads to high rates of gravitational stress in the form of denudation erosion and debris transport. The gravitational stress is frequently further accentuated by extreme climatic conditions prevailing at high altitudes, as well as by distinctive lithology and vegetation conditions. Research on geomorphic processes in mountains has emphasized the role of high-intensity transfer processes. Fast transfer processes may reach catastrophic proportions in the form of rockfalls, mudflows, flashfloods and avalanches. Slower transfer processes include the various forms of solifluction, soil creep and other denudation forms. Caine (1974) proposed several models for the geomorphic activity in mountains. He considered an alpine geomorphic system consisting of two dynamic subsystems, i.e., slopes and drainage channels, with both these subsystems supplemented by three sediment subsystems, i.e., a geochemical subsystem, a fine sediment subsystem and a coarse detritus subsystem. Caine admits, however, that "it is difficult to define a general model for alpine geomorphic activity" (p 741). He also stresses the importance of biotic elements as geomorphically significant factors. Pronounced vegetation changes may have an influence on varying forms and rates of geomorphic activity; for example, the timberline is often a significant boundary line of geomorphic action. He observes finally that "it is through a variety of controls and responses (and the effects of time in introducing a 'historical hangover') that the diversity of alpine scenery is produced rather than through the operation of geomorphic processes peculiar to the alpine" (p 743).

It has been said that "the high mountain of morphologists, glaciologists and climatologists begins where man disappears" (G. Veyret, 1975, p. 180). Germaine Veyret has argued, however, that a definition of the mountain environment should include reference to human activities and landuse. But inclusion of human factors in formulating general criteria that are applicable to the various mountain environments of the world is problematic.

Relief and climate have been major factors in determining and shaping the human landscapes of mountains although societal changes and new technologies may have lessened these constraints.

Taking into account the fact that mountains of lower latitudes have traditionally been the preferred and most densely settled highland regions, it would appear that climate, and temperature in particular, has been a determining factor for mountain living. In mid-latitudes, cold and snow severely restrict human activities at higher altitudes. Settlements in mid-latitude mountains are rarely found above 2000 m (e.g. in the European Alps), whereas in the drier tropical mountains, human residences may be found at elevations above 4000 m (e.g. semi-arid Andes). At extreme elevations, the reduction in atmospheric pressure produces the well-known physiological effects of *hypoxia.* This 'rarefication of air' sets an altitudinal limit for permanent settlement even in the climatically favoured tropical mountains, although long-time residents have over generations developed a special degree of acclimatization, and their bodies have adapted to high mountain living.

Although it is difficult to identify human criteria for a global definition of mountains, some aspects of the human environment are characteristic for mountains. Because of their specific physical conditions, mountains have always played a significant role as source regions of a variety of resources. As this wealth was frequently exported, the mountain population frequently remained in a state of relative underdevelopment and developed attitudes of self-reliance, individualism and conservatism.

The most important basis for mountain living in most areas consists of agricultural activities. Because of the considerable relief variations, these activities are usually characterized by a vertical zonation of different use systems with an intensive field cultivation in valley regions and lower altitudes and extensive livestock grazing and forestry at higher elevations. Field cultivation, because of the broken relief configuration and social traditions, is frequently carried out in small, often fragmented holdings, and machinery is used at best on a small-scale. The steep slopes and their associated problems of denudation and erosion, remain major constraints for mountain agriculture. The altitudinal spread of agricultural uses and the different climatic conditions at each level, necessitate considerable vertical mobility of the population on a daily or seasonal basis.

In mid-latitude mountains, the upper levels of the mountain ecumene can be used only for a short period of time during the summer months with the winter interrupting most agricultural activity even at low altitudes. Most mountain regions have traditionally experienced seasonal migration flows between mountains and the adjacent lowlands. In the Cordilleras of Latin America, mountain dwellers work in lowland plantations on a seasonal basis; in the European Alps, an outmigration into the industrial centres of the lowlands during the winter months was characteristic, before the development of winter tourism changed the traditional migration pattern.

The spatial or seasonal limitation of the human utilization of mountains and the difficult or costly installation and maintenance of infra-structures created in many areas a considerable lag in levels of development between lowlands and neighbouring mountain regions. Mountain people have been labelled as the "end link in the chain of under-privileged people" (Lampe, 1975, p 11). Furthermore, in many mountain regions, population growth rates have remained high. This has led to additional problems relating to relative over population, although population densities may be considerably lower than in the plains. Today, mountain regions are often threatened by ecological or economic overutilization of their resources—in the 'Third World' largely because of population pressure, in the 'Industrial World' because of frequently reckless exploitation of space and resources.

Relief and climate place serious constraints on the development of a transportation system in mountains. Although technology has to a certain extent overcome the barrier effect of mountains, a too uninhibited 'transportation conquest' of mountains has disturbed the unstable or fragile environment; erosion has often damaged or destroyed transportation lines that were considered safe. In mid-latitudes, snow still blocks many mountain roads during the winter season, whereas in tropical regions, seasonal torrential rains may endanger transportation lines.

A new development that profoundly changed the environment, human activities and landuse systems, has been the rapid rise of mass tourism in some mountain regions of the world. Mountain tourism has created much environmental damage or destruction. On the other hand, properly planned and managed tourism has brought alternative sources of employment and income to mountain peoples without necessarily being detrimental to the environment. In subtropical and tropical mountains, summer "sightseeing tourism" prevails, although the "season" may last throughout the year. In temperate mountains, a clear distinction exists between summer and winter tourism; the European Alps may be the only major mountain region where winter tourism has become the dominant form of tourism and the principal source of income. Here, the two traditional constraints of relief and snow have become the major assets although they may continue to be hazards for transportation lines, and avalanches still endanger human life and settlement. Winter tourism has stimulated the development of a "new generation" of high altitude settlements and infra-structures, often above the treeline and above the upper limit of the agricultural ecumene Winter tourism has also weakened the role of mountain agriculture, particularly of high altitude pastoral activity. As winter has been turned into a "booming season", the traditional seasonal migration flows have been substantially altered if not completely reversed. In the seasonal cycle of vertical migration, temporary high altitude settlements and grazing grounds used to be abandoned during the winter months with the migrant population moving to lower altitudes. These high altitude areas today have excellent potential for winter tourism: an abundance of snow; open spaces with a variety of slope conditions; and high insolation rates. The development of winter tourism also created massive new horizontal population flows between lowlands and mountains, with well-pronounced peaks during weekends and holidays. In addition to its economic and ecological consequences, the new mass tourism tended to "erode" the basis of a traditional and largely conservative society by the introduction of a new value system and of lifestyles which led to "mental urbanization" of the population.

### PROBLEMS IN MOUNTAIN ENVIRONMENTS AND PROBLEM-ORIENTED MOUNTAIN RESEARCH

As shown above, the specific physical factors of mountains created a delicately balanced environment that is easily altered, disturbed and threatened by physical influences, human developments, or by a combination of both. The specific problems in mountain regions are varied as they are influenced by different locational, cultural and economic aspects. The catalogue of problems relating to mountains within the "Developed Realm" includes inadequate use of land and water resources, and the danger of ecological damage brought about by a too massive introduction of industries, tourism, new settlements and transportation lines. There appears to be a need for the development of ecologically and socially acceptable models of human use adapted to the fragile mountain environment. Overutilization of mountains does not only create a danger for the environment but also leads to a deterioration of the "quality of life" for the indigenous population. Among the major problems of mountains in the "Developing Realm", the population pressure combined with a limited resource base has created a degradation of the environment, a growing pauperization of the population, and a continued need for emigration. Particularly detrimental has been the large-scale destruction of forests, as a source of local fuel and timber and as a means of creating additional space for agriculture. The vicious cycle of over population, scarcity of land, destruction of forests, erosion, pauperization can only be broken if a solution to the population problem and an improvement of the subsistence economy can be found, and if alternative sources of fuel can be made available to the mountain population.

Although problems in mountain regions have been described in many case studies, there have been few attempts to undertake comparative studies of human use systems in major mountain realms of the world and to identify development problems of mountains on a global basis. In contrast, the natural environment of mountain regions has received considerable attention by scholars. Carl Troll, in particular, has pioneered in systematic studies of the "geo-ecology" of mountain regions, especially in the tropical realm. He has also been instrumental in creating a Commission of the International Geographical Union on "High-Altitude Geo-Ecology."

In 1973, within the UNESCO International Programme on 'Man and the Biosphere' (MAB), Project-6 was initiated dealing with the 'Impact of Human Activities on Mountain and

Tundra Eco-systems'. A study group met in February 1973 for the first time in Salzburg, Austria, and later in the year in Lillehammer, Norway defined the following problem areas for a study in different mountain regions of the world :

(1) Problems of resource development and human settlement in tropical mountain regions.
(2) Problems of tourism, technology and landuse alternatives in temperate mountains.
(3) Problems of landuse in high latitude mountain and Tundra eco-systems, with special reference to grazing, industrial development and recreation.

Subsequently a number of regional meeting were held in various mountain regions of the world, and an attempt was made to establish specific pilot study projects in representative mountain areas.

In December 1974, the Ford Foundation in cooperation with the Deutsche Stiftung für Internationale Entwicklung (German Foundation for International Development) organized an international workshop on "Development Problems in Mountain Regions" in Munich. This interdisciplinary workshop attempted for the first time to examine on a global scale the complex aspects of the development and utilization of mountain resources and to formulate strategies emphasizing integration of environmental and social concerns in the planning process. The "Club of Munich" formulated the following resolutions (Ives, 1975, pp 101-102) :

(1) Future mountain development is environmentally and socially acceptable will require a combination of research, application of research, and dissemination of information at various levels.
(2) The UNESCO Man and the Biosphere Programme (MAB) Project-6, was adopted unanimously as a potentially effective structure for fully integrated interdiscipli nary research . . . It was agreed that acceptance of and support for the proposed MAB-6 pilot project in the Andes and Himalayas could be augmented by the addition of applied, developmental projects such that

operational models of local and regional significance could attract support for progressively more ambitious undertakings.

(3) International and regional dissemination of effective information should be planned using a variety of methods and approaches.

(4) Identification of mountain research resources, in terms of manpower and laboratory and training facilities, would be needed to form a consultative resource pool and workforce.

(5) An organizational structure, or structures should be developed for furthering the Munich Workshop objectives.

Among the more specific resolutions was a proposal for creating a small Task Force of scientists and planners to develop detailed short and long-term objectives and budgets; as well as the establishment of an international association for mountain research and development.

Mountains are complex and diversified landscapes, and an identification of globally applicable criteria for defining mountain environments is problematic. Mountains are an important component of the human ecumene disposing of a variety of significant resources. Mountain regions are today affected by many problems that tend to threaten the natural and human environment. In spite of some evidence for growing recognition of the field of mountain geography, it is still surprising to see that systematic and integrated studies on mountain environments remain a "step-child" in the large home of geography.

## *SUGGESTED READINGS*


Académie Polonaise des Sciences, *L'Aménagement de la Montagne*, compte rendu du IIIe Colloque franco-polonais de Géographie, 1969, Panstwowe Wydawnictwo Naukowe, Warsaw, 1971, pp 227.

A.B. Bagdassarian, "The Evaluation of Mountain Landscapes for Recreational Purposes," in : ADAMS, P.W., HELLEINER, F.F. Eds., *International Geography*, University of Toronto Press, Toronto, 1972, Vol. 1, pp 659-662.

P.T. Baker, M.A. Little, Eds., *Man in the Andes, A Multidisciplinary*

*Study of High-altitude Quechua*, Dowden, Hutchinson and Ross, Stroudburg, Pennsylvania, 1976, pp 482.

H. Berger, "Beitrag zur Hochgebirgsforschung," *Mitteilungen der Osterreichischen Geographischen Gesellschaft*, Vol. 102, 1960, pp 87-92.

H.H. Boesch, "Die Formen des Hochgebirges," *Die Alpen*, Vol. 22, 1946, pp 293-299.

H. Böhm, "Die geländeklimatische Bedeutung des Bergschattens und der Exposition für das Gefüge der Natur-und Kulturlandschaft," *Erdkunde*, Vol. 20, 1960, pp 81-93.

N. Caine, "The Geomorphic Processes of the Alpine Environment," in : J.D. Ives, R.G. Barry, *Arctic and Alpine environments*, Methuen, London, 1974, pp 721-745.

C.J. Cool, Probleme der Bergregionen, in : W. Danz, Ed., Entwicklungsprobleme in Bergregionen, *Schriftenreihe des Alpeninstituts*, Vol. 3, 1975, p 12.

J. Corbel, "Vitesse de l'érosion," *Zeitschrift für Geomorphologie*, Vol. 3, 1959, pp 1-28.

Council of Europe, Economic and Social Problems of Mountain Regions, *Study Series Local and Regional Authorities in Europe*, Vol. 5, 1973, pp 1-199.

W. Danz. Ed., Entwicklungsprobleme in Bergregionen, *Schriftenreihe des Alpeninstituts*, Vol. 3, 1975, pp 1-88.

H. Flückiger, *Gesamtwirtschaftliches Entwicklungskonzept für das Berggebiet*, Bern, 1970, pp 420.

R. Fouet, C. Pomerol, *Les montagnes*. Presses Universitaires de France, Paris, Que sais-je? No. 682, 1965, pp 126.

G. Furrer, P. Fitze, Die Hochgebirgsstufe-ihre Abgrenzung mit Hilfe der Solifluktionsgrenze, *Geographia Helvetica*, Vol. 25, 1970, pp 156-161.

G. Furrer, G. Dorigo, Abgrezung, und Gliederung der Alpen mit Hilfe von Solifluktionsformen, *Erdkunde*, Vol. 36, No. 2, 1972, pp 98-107.

I.P. Gerasimov, et al, The French-Soviet Symposium, "The Alps—The Caucasus," *Arctic and Alpine Research*, Vol. 8, No. 2, 1976, pp 219-226.

J. Gardner, "Mountain Temperatures," *Canadian Alpine Journal*, Vol. 51, 1968, pp 224-228.

Giunta Regionale della Lombardia, *Le Alpi e l'Europa*, 5 Vols., Laterza, Milano, 1974-75.

G. Glaser, International Research on the Problem of Mountain Environment, in particular the programme on 'Man and Biosphere' (Project Area 6), in: W. Danz, ed. "Entwicklungsprobleme in Bergregionen, "*Schriftenreihe des Alpeninstituts*, Vol. 3, 1975, pp 33-37.

K. Hermes, "Die Lage der oberen Waldgrenze in den Gebirgen der Erde und ihr Abstand zur Schneegrenze, *Kölner Geographische Schriften*, Vol. 5, 1955, pp 1-277.

K. Hewitt, The freeze-thaw environment of the Karakoram Himalaya,

*Canadian Geographer*, Vol. 12, 1968, pp 85-98.

K. Hewitt, "The Mountain Environment and Geomorphic Processes," in : H.O. Slaymaker, H.J. McPherson, Ed., "Mountain Geomorphology: Geomorphological Processes in the Canadian Cordillera," *B C. Geographical Series*, Vol. 14, 1966, pp 17-34.

P. Hollermann "Some Reflections on the Nature of High Mountains, with Special Reference to the Western United States," *Arctic and Alpine Research*, Vol. 5, No. 3, Part 2, 1973, pp A149-A160.

A. Hurtado, "The Influence of Altitude on Man," *WHO Chronicle*, August, 1972, pp 354-359.

International Geographical Union, Commission High-Altitude Geoecology, Geoecology of the High-Mountain Regions of Eurasia, Proceedings, Symposium IGU, Commission High-Altitude Geoecology 1969, *Erdwissenschaftliche Forschung*, Vol. 4, 1972, pp 1-300.

J.D. Ives, "The Development of Mountain Environments," Munich International Workshop, *Arctic and Alpine Research*, Vol. 7, No. 1. 1975, pp 101-102.

J.D. Ives, Ed., "Proceedings of the Symposium of the International Geographical Union, Commission on High-Altitude Geoecology, Calgary, Alberta," 1972, *Arctic and Alpine Research*, Vol. 5, No. 3, Part 2, 1973, pp A1-A199.

J.D. Ives, R.G. Barry, *Arctic and Alpine Environments*, Methuen, London, 1974, pp 1024.

J.D. Ives, P.V. Krebs, "Natural Hazards Research and Landuse Planning Responses in Mountainous Terrain: the Town of Vail, Colorado, Rocky Mountains, U.S.A.," *Arctic and Alpine Research*, Vol. 10, No. 2, 1978, pp 213-222.

J.D. Ives, et al, "Natural Hazards in Mountain Colorado, *Annals Association of American Geographers*, Vol. 66, No. 1, 1976, pp 129-144.

J.D. Ives, R.P. Zimina, Eds., "Mountain Geoecology and Landuse Implications," Proceedings of the Symposium of the International Geographical Union, Commission on High-Altitude Geoecology, Caucasus Mountains, USSR, 1976, *Arctic and Alpine Research*, Vol. 10, No. 2, 1978, pp 159-548.

N. Krebs, "Die Karte der Reliefenergie Süddeutschlands," *Petermann's Geographische Mitteilungen*, Vol. 68, 1922, pp 49-53.

K. Lampe, "Know-how Bank for Mountain Environment," in Danz, W., Ed., Entwicklungsprobleme in Bergregionen, *Schriftenreihe des Alpeninstituts*, Vol. 3, 1975, pp 10-12.

B. Messerli, "Problems of Vertical and Horizontal Arrangement in the High Mountains of the Extreme Arid Zone (Central Sahara)", *Arctic and Alpine Research*, Vol. 5, No. 3, 1973, pp 139-147.

W.T.W. Morgan, "Altitude and Crop Zones. A Transact in the Kenya Highlands," *Die Erde*, Vol. 108, 1977, pp 139-147.

W. Mörgeli, *Beitröge zur Abgrenzung und Differenzierung des Berggebietes mit quantitativen Methoden*, Juris Druck, Zurich, 1973, pp 71.

F. Mühlemann, "Die Zukunft der Berggebiete im Lichte der Leitbilder," *Schriftenreihe zur Orts., Regional-und Landesplanung*, Vol. 16, 1973, pp 36-48.

R. Peattie, *Mountain Geography : A Critique and Field Study*, Greenwood Press, New York, 1936, pp 257.

C. Rathjens, "Neuere Entwicklung und Aufgaben einer vergleichenden Geographie der Hochgebirge, *Geographisches Taschenbuch*, 1966/69, 1966, pp 199-210.

S.A. Schumm, "The Disparity between Present Rates of Denudation and Orogeny," *U.S. Geol. Survey, Prof. Paper*, 1963, 454-H.

H.O. Slaymaker, H.J. McPherson, Eds., "Mountain Geomorphology: Geomorphological Processes in the Canadian Cordillera," *B.C. Geographical Series*, Vol. 14, 1966, pp 1-274.

J.A. Stein, "Internationales Program für 'Man and Mountains'—eine Einführung,'' in DANZ, W., Ed., Entwicklungsprobleme in Bergregionen, *Schriftenreihe des Alpeninstituts*, Vol. 3, 1975, pp 14-23.

J.A. Stein, "Tourism and Mountain Environment," in W. Danz, Ed., Entwicklungsprobleme in Bergregionen, *Schriftenreihe des Alpeninstituts*, Vol. 3, 1975, pp 63-65.

A.N. Strahler, "Quantitative Slope Analysis, *Bull. Geol. Soc. Amer.*, Vol. 67, 1956, pp 571-596.

P.H. Temple, "Monuments which do not decay," *Geographical Magazine*, Vol. 44, No. 8, 1972, pp 767-771.

W.F. Thompson,, "Cascade Alp Slopes and Gipfelfluren as Climageomorphic Phenomena," *Erdkunde*, Vol. 26, 1962, pp 81-94.

C. Troll, "Die Stellung der Indianer-Hochkulturen im Landschaftsaufbau der tropischen Anden, *Zeitschrift der Gesellschaft für Erdkunde Berlin*, Vol. 3, 1943, pp 93-128.

C. Troll, ed., "Die tropischen Gebirge, ihre dreidimensionale klimatische und pflanzengeographische Zonierung," *Bonner Geographische Abhandlungen*, Vol. 25, 1959, pp 1-93.

C. Troll, Ed., "Geo-ecology of the Mountainous Regions of the Tropical Americas," *Colloquium Geographicum*, Vol. 9, 1968, pp 1-223.

C. Troll, "Okologische Landschaftsforschung und Vergleichende Hochgebirgsforschung." *Erdkundliches Wissen*, Vol. 11, 1966, pp 1-366.

C. Troll, "Strukturböden, Solifluktion und Frostklimate der Erde," *Geologische Rundschau*, Vol. 34, 1943/44, pp 545-694.

C. Troll, "Studien zur vergleichenden Geographie der Hochgebirge der Erde," *Bonner Mitteilungen*, Vol. 21, 1941, pp 1-50.

C. Troll, "Über das Wesen der Hochgebirgsnatur, "*Jahrbuch Deutscher Alpenverein* (*AV-Zeitschrift*, Vol. 80), 1955, pp 142-157.

C. Troll, "Vergleichends Geographie der Hochgebirge der Erde in landschafts-ökologischer Sicht," *Geographische Rundschau*, Vol. 5, 1975, pp 185-198.

L.A. Valessyan, "An Economic-Geographical Study of Mountain Relief as a factor for the territorial organization of productive forces," in

P.W. Adams, F.M. Helleiner, Eds., *International Geography 1972*, University of Toronto Press, Toronto, Vol. 1, pp 617-618.

G. Veyret, "La haute montagne. Essai d'une définition, son évolution et ses problémes," in G. Veyret, *Les hommes et les Alpes, recherches geographiques*, Allier, Grenoble, 1975, pp 181-189.

G. Veyret, "Le probléme de l'équilibre démographique en montagne," in G. Veyret, *Les hommes et les Alpes, recherches géographiques*, Allier, Grenoble, 1975, pp 51-62.

P. Veyret, "La haute montagne; réflexions sur son relief et son évolution," *Acta Geographica*, Vol. 51, 1964, pp 21-25.

P. Veyret, G. Veyret, "Essai de définition de la montagne," *Revue de Geographie Alpine*, Vol. 50, 1967, pp 5-35.

14

# Geomorphology of Drainage Basins in Palamau Upland

*—O.P. Singh*

QUANTITATIVE EVALUATION of geomorphic characteristics of the drainage basins and their correlation to basin hydrology have now-a-days become the principal objective of research in fluvial geomorphology. A great impetus was given to fluvial morphometry by Horton's researches (1945). He has suggested the method[1] of quantitative analysis of morphometric characteristics of drainage basins. Following Horton's considerable work on the quantitative analysis of the morphometric characteristics of drainage basins several works have been done by Strahler[2] (1952 and 1957), Schumm[3] (1956), Chorley[4] (1959), Morisawa[5] (1962) and in India by Singh[6] (1967), Singh[7] (1969), Pal[8] (1973), Dixit[9] (1976), and others. The result of these studies indicates that morphometric attributes like bifurcation ratio, stream length, drainage density, etc. will substantially contribute to evaluate the hydrological characteristics of a basin. Keeping these in view, an attempt has been made to analyse the morphometric attributes and their correlation with hypsometric integrals of ten drainage basins, ranging in area from 38.78 sq. km. (Sakheri) to 102.91 sq. km. (Satbahini), selected in Palamau Upland (Fig. 14.1) a part of Chotanagpur Plateau. This Upland, "a broken and divided country" of Palamau District, Bihar is characterised by rugged terrain, poor soil, poor agriculture and scrub forest experiencing tropical monsoon climate. After feeding the data of mean annual rainfall and temperature in Peltier's scale of morphogenetic region and processes, the Palamau Upland exhibits moderate morphogenetic region characterised by moderate chemical weathering, minimum wind action, but maximum pluvial erosion.[10]

*Palamau Upland*

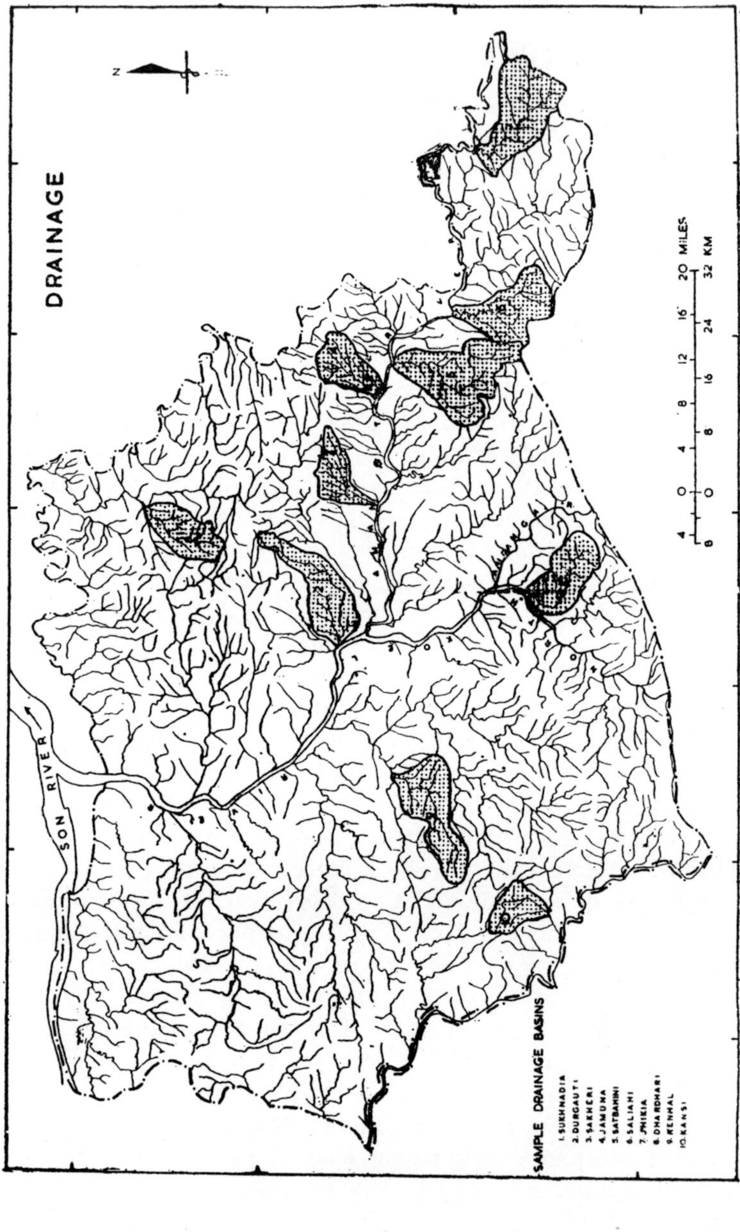


FIG. 14.1

## MATERIALS AND METHODS

For a quantitative study of the geomorphic characteristics of streams and their inter-relationships, the topographical sheets of 1 : 63360 scale and aerial photographs of 1 : 40,000 scale were used as basic material for the present study. In this paper, the streams have been ordered after Strahler's (1952) adaptation of Horton's (1954) scheme of classification. The order number of the basins, stream length, basin area, relief ratio, elongation ratio, stream frequency, drainage density, sinuosity index and hypsometric integral have been calculated and were checked in the fields at selected points. All these attributes have been classified, tabulated and reproduced in the form of maps and diagrams by suitable cartographic devices. An attempt has also been made to correlate all morphometric properties with hypsometric integrals of the drainage basins.

## RESULTS AND DISCUSSIONS

### Stream Order and Number

The first step in drainage basin analysis is designation of stream order. The first order streams are those which have no tributaries. At the junction of any two first order segments, a channel of second order is produced and extends to the point, where it joins another second order segment whereupon a segment of third order results and so forth."[11] Out of ten selected drainage basins, Saliahi and Satbahini rivers stand at the highest order stream having been tipped at 5th order and others contain channels upto the 4th order only. The graphs (Fig. 14.2) well represent that the number of channel segments decreases as the order value increases.

### Bifurcation Ratio

The bifurcation ratio ($R_b$) has been worked out as a ratio between the number of a given order (u) and the number ($N_u$) of the next higher orders of the following formula

$$R_b = \frac{N_u}{N_u + 1}$$

The average bifurcation ratios of the drainage basins stand at 3.92, 3.01, 3.34, 4.30, 2.71, 3.65, 2.69, 3.31, 3.08, 4.27 respectively (Table 14.2). These values of the basins are more or less

Stream Order in Drainage Basins

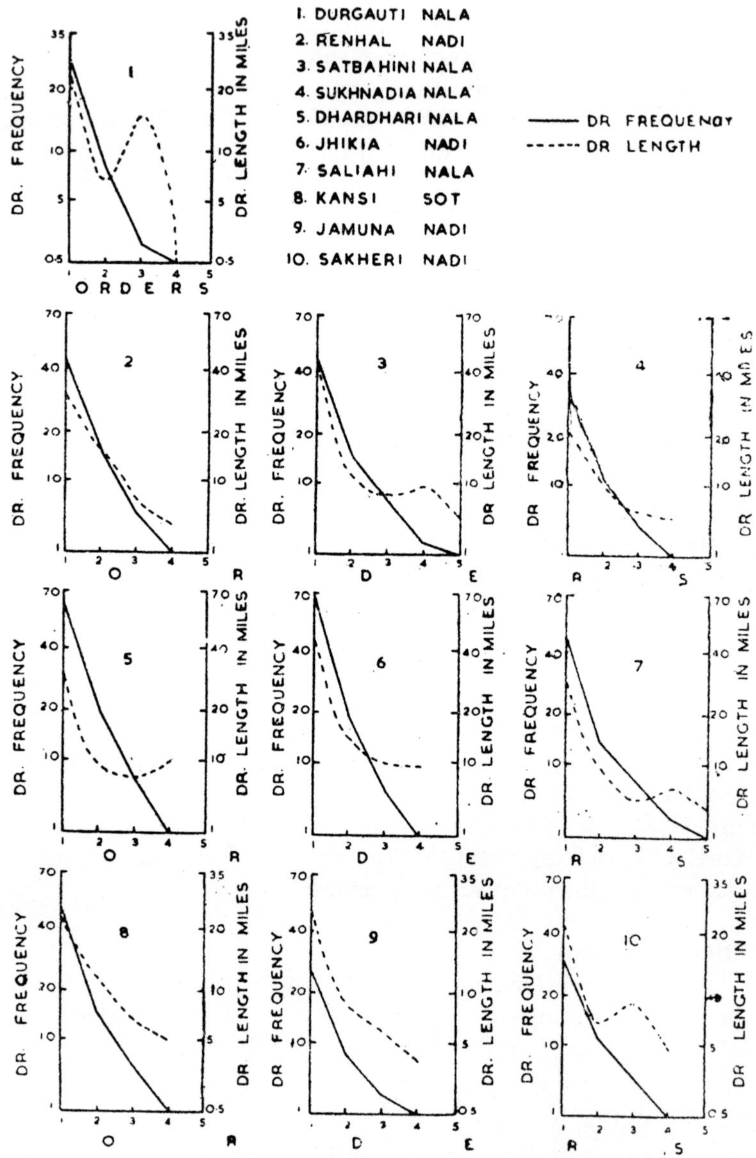


FIG. 14.2

TABLE 14.1

*Orderwise Length and Frequency of Sample Drainage Basins*

| Rivers | Orderwise Length in Kms | | | | | Orderwise Frequency | | | | |
|---|---|---|---|---|---|---|---|---|---|---|
| | I | II | III | IV | V | I | II | III | IV | V |
| Kansi | 35.20 | 18.40 | 11.2 | 8.2 | — | 49 | 15 | 6 | 1 | — |
| Jamuna | 40.00 | 13.60 | 9.6 | 5.6 | — | 27 | 8 | 3 | 1 | — |
| Sakheri | 35.20 | 11.20 | 14.4 | 7.2 | — | 31 | 11 | 5 | 1 | — |
| Jhikia | 71.20 | 23.20 | 16.0 | 15.2 | — | 69 | 19 | 6 | 1 | — |
| Saliahi | 48.00 | 15.20 | 8.0 | 10.4 | 7.2 | 49 | 14 | 7 | 3 | — |
| Renhal | 49.60 | 25.60 | 11.2 | 6.4 | — | 44 | 16 | 5 | 1 | — |
| Satbahini | 72.00 | 18.00 | 12.8 | 15.2 | 8.0 | 47 | 15 | 7 | 2 | 1 |
| Sukhanadia | 34.40 | 16.00 | 9.6 | 8.0 | — | 35 | 11 | 4 | 1 | — |
| Durgauti | 35.20 | 11.20 | 24.0 | 1.6 | — | 26 | 8 | 2 | 1 | — |
| Dhardhari | 48.80 | 15.20 | 12.0 | 16.0 | — | 61 | 20 | 7 | 1 | — |

normal according to Horton's law, as they range between 3 and 5. The little variations in the values are due to the differences in their stage of geomorphic development and differences in topographic characters.

TABLE 14.2

*Bifurcation Ratios of Sample Drainage Basins*

| Drainage Basins | Bifurcation Ratio | | | | Average Bifurcation Ratio |
|---|---|---|---|---|---|
| | I/II | II/III | III/IV | IV/V | |
| Kansi | 3.27 | 2.5 | 6.00 | — | 3.92 |
| Jamuna | 3.37 | 3.67 | 3.00 | — | 3.01 |
| Sakheri | 2.82 | 2.20 | 5.00 | — | 3.34 |
| Dhardhari | 3.05 | 2.86 | 7 00 | — | 4.30 |
| Saliahi | 3.50 | 2.00 | 2.35 | 3 | 2.71 |
| Renhal | 2.70 | 3.20 | 5.00 | — | 3.65 |
| Satbahini | 3.13 | 2.14 | 3.50 | 2 | 2.69 |
| Sukhanadia | 3.18 | 2.75 | 4.00 | — | 3.31 |
| Durgauti | 3.25 | 4.00 | 2.00 | — | 3.08 |
| Jhikia | 3.63 | 3.17 | 6.00 | — | 4.27 |

**Stream Length and its Relation to Stream Order**

Table 14.3 reveals that the mean channel length for first order ranges from 0.72 km (Kansi Basin) to 1.53 km (Satbahini

TABLE 14.3

*Morphometric Properties of Drainage Basins*

| Drainage Basins | Area in sq. km. | Total stream Length in km. | Total number of stream | Mean stream Length | | | | |
|---|---|---|---|---|---|---|---|---|
| | | | | I | II | III | IV | V |
| Kansi | 44.03 | 72.8 | 61 | 0.72 | 1.21 | 1.87 | 8.00 | — |
| Jamuna | 41.42 | 68.8 | 39 | 1.47 | 1.70 | 3.20 | 5.60 | — |
| Sakheri | 38.78 | 68.0 | 48 | 1.13 | 1.02 | 2.88 | 7.20 | — |
| Dharadhari | 47.79 | 92.0 | 89 | 0.80 | 0.75 | 1.71 | 16.00 | — |
| Jhikia | 94.72 | 125.6 | 95 | 1.02 | 1.21 | 2.67 | 15.20 | — |
| Saliahi | 65.33 | 88.8 | 74 | 0.97 | 1.09 | 1.13 | 3.47 | 7.2 |
| Renhal | 65.66 | 92.8 | 66 | 1.12 | 1.60 | 2.24 | 6.40 | — |
| Satbahini | 102.91 | 46.4 | 72 | 1.53 | 1.21 | 1.82 | 7.60 | 8.0 |
| Sukhanadia | 45.80 | 68.0 | 51 | 0.97 | 1.45 | 2.40 | 7.20 | — |
| Durgauti | 61.20 | 73.3 | 37 | 1.36 | 1.25 | 12.00 | 1.60 | — |

Basin). The mean channel length for the 2nd order ranges between 0.75 km (Dhardhari) to 1.70 km (Jamuna) and for 3rd order between 1.13 km (Saliahi) to 12.00 km (Durgauti). An increase in mean channel length for the 3rd order of Durgauti Basin is due to the presence of a longer stream and not because of an increase in the number of streams of that order.[12] For the 4th order, the mean channel length exhibits more variation with the minimum value of 1.6 km (Durgauti) and the maximum value of 16 km (Dhardhari). The maximum value is due to the meandering nature of the stream although the number of this stream-order is only one. The 5th order lies in two basins only (Saliahi and Satbahini Basins). The overall average stream length for the 1st order segment stands at 1.10 km; for the 2nd, 3rd, 4th and 5th orders, the same stands at 1.26, 3.18, 7.9 and 7.6 km respectively.

The proportion of increase of mean length of segments of two successive orders becomes length ratio ($R_L$) and is calculated as given below :

$$R_L = \frac{\bar{L}_u}{\bar{L}-u^2} \quad \text{when } \bar{L}_u = \frac{L_u}{N_u}$$

When $\bar{L}_u$ is the mean length of all streams, segments of order (u), '$L_u$' is the sum of lengths of all stream segments and '$N_u$' is the number of the stream segment of a given order. The length ratio ranges between 1.45 and 3.71. This variation of $R_L$ from one basin to the other and even in a single basin (from one order to the other) are because of lithological controls which have restricted the branching of the rivers.

The relationship between stream order and mean channel length (cumulative) shows a contrast to the general law of Horton (1945, p 279). It is noted (Fig. 14.2) that these curves do not show linear relationships. It shows that the mean channel length of higher order increases but not in geometric progression. On an average the first order stream segments are of the shortest length and as the order increases the average length also increases but Durgauti river registers a departure from this general rule as the average length decreases with increasing order. The irregular fashions of the curves (Fig. 14.2) contrary to Horton's law are, probably, due to decreasing stream length of the higher orders because the hard and tenacious granites and gneisses have compelled the rivers to have a sudden turn and meet the trunk rivers at sharp bends thus reducing the length in the higher order.

**Drainage Density**

Drainage density is simply a length per unit area which is expressed as :

$$D_d = \frac{\Sigma L}{A_d}$$

where $D_d$=Drainage Density,

$\Sigma L$=Length of stream,

$A_d$=Area of drainage basin.

The drainage density values of all basins (Table 14.4) range between 1.88 (Durgauti) and 3.08 (Dhardhari), indicating low density. The low density is due to structural control as the predominance of granites and gneisses of Archaean Formations and poor supply of rain water have restricted the branching of the rivers and their tributaries. Relative case of infiltration of precipitation into the ground surface is another important controlling factor. The poor supply of rain from erratic and unreliable monsoon has tended to give lower drainage density.

TABLE 14.4

*Morphometric Characteristics of Sample Drainage Basins*

| Basins | Drainage Density km/sq. km. | Relief Ratio | Elongation Ratio | Stream Gradient | Hypsometric Integral (%) |
|---|---|---|---|---|---|
| Kansi | 2.64 | 0.030 | 0.52 | 1/33 | 41.20 |
| Jamuna | 2.66 | 0.016 | 0.51 | 1/63 | 44.20 |
| Sakheri | 2.80 | 0.015 | 0.47 | 1/65 | 38.50 |
| Dhardhari | 3.08 | 0.028 | 0.58 | 1/35 | 28.20 |
| Jhikia | 2.12 | 0.011 | 0.54 | 1/92 | 46.20 |
| Saliahi | 2.17 | 0.019 | 0.41 | 1/54 | 68.00 |
| Renhal | 2.26 | 0.021 | 0.46 | 1/80 | 50.50 |
| Satbahini | 1.96 | 0.022 | 0.44 | 1/46 | 30.00 |
| Sukhanadia | 2.37 | 0.018 | 0.53 | 1/57 | 41.50 |
| Durgauti | 1.88 | 0.010 | 0.32 | 1/94 | 28.00 |

**Relief Ratio and its Relation to Drainage Density**

The height-length relationship of a basin parameters is expressed as "relief ratio" (Morisawa, 1955, p 1028) which may be obtained by the following formula :—

$$R_y = \frac{R_u}{L_b}$$

where $R_y$ = Relief Ratio

$R_u$ = Value of relief of a basin.

$L_b$ = Basin Length measured in a straight line approximately parallel to the major drainage channel.

The value of relief ratio ranges between 0.01 (Durgauti) and 0.03 (Kansi) indicating late-mature stage of geomorphic development.

The relationship of relief ratio with drainage density (Fig. 14.3) is not harmonious in most of the basins. Kansi Basin, having higher relief ratio has 2.64 km/sq km drainage density whereas Jamuna Basin with low relief ratio (0.016) has a comparatively higher drainage density (2.66 km/sq km) and Durgauti Basin with low relief ratio (0.01) and poor drainage density (1.88 km/sq km) show comparatively a positive relationship.

**Elongation Ratio and its Relation to Relief Ratio**

The shape of a drainage basin is expressed by its "elongation ratio" (Schumm 1956, p 622) which is obtained by the following formula :

$$R_E = \frac{C_d}{L_b}$$

where, $R_E$ = Elongation ratio,

$C_d$ = Diameter of a circle with the same area as the basin.

$L_b$ = Length of the basin as measured for the relief ratio.

The values, ranging between 0.32 and 0.58 show that Durgauti, Saliahi, Satbahini, Renhal and Sakheri Basins are comparatively triangular in shape while others are more or less square. While comparing the elongation ratio with the relief ratio, the

*Relief Ratio and Drainage Density*

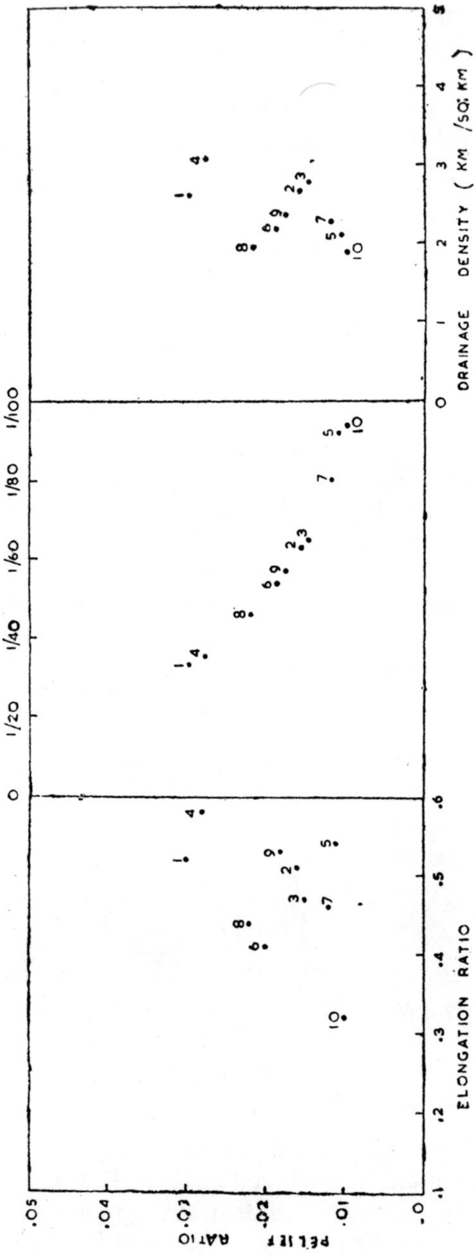


FIG. 14.3

map (Fig. 14.3) shows the positive and negative both relationships. The relationship found in the basins of Kansi (comparatively higher relief ratio with moderate elongation ratio) and Jhikia basin (comparatively low relief ratio with higher elongation ratio) is just reverse to Schumm's (1956, p 608) relationship, i.e., "as the relief ratio increases, the drainage basins become more elongated." But Durgauti Basin, having comparatively low relief ratio and low elongation ratio exhibits the positive relationship that as the relief ratio decreases, the value of elongation ratio also decreases.

**Stream Gradient and its Relation to Relief Ratio**

Stream gradient is obtained by dividing the stream length measured from mouth to source by the elevation difference. The gradient ranges from 1/94 to 1/33 (Table 14.4). The map (Fig. 14.3) shows a positive relationship that as the stream gradient increases, the relief ratio also increases.

**Sinuosity Index**

The sinuosity index has been very much in use in understanding the geomorphological character of a region. The meaning of the sinuosity of a stream is associated with bending, wedding, curving and deviating of the same. This ratio is obtained by dividing the channel length (CL) by valley length (VL). "No straight river occurs in the state of nature. Even young consequent rivers, newly established on a land surface not previously subjected to river action, must inevitably have an irregular initial course due to the unavoidable inequalities of the land surface.... All such surfaces have sufficient inequality to give the consequent rivers which are established upon them some significant departures from the most unnatural of all courses a straight line."[13] Thus it is evident that all streams possess some amount of sinuosity, although the degree of their sinuosity varies with their stage, age and topographic and geological backgrounds. This variation in sinuosity has been analysed by several researchers like Mueller,[14] Schumm,[15] Leopold[16] and Singh and Sharma.[17] Following Mueller's lines, sinuosity indices of all ten rivers have been calculated (Table 14.5) which vary between 1.09 and 1.23. These values indicate the sinuous pattern of rivers because the value of 1.5 or above is taken as a criterion for meandering and below 1.5 as sinuous (Leopold *et. al.*, 1964, p 281).

All the streams have both hydraulic and topographic sinuosity (HSI and TSI) because "the streams are prisoners of the valleys which they cut and occupy."[18] Therefore, in the present study, HSI and TSI have been tested for all streams on the lines of Mueller (1968, pp 371-385) who states that TSI is outstanding during youth when HSI is negligible, conversely, HSI is outstanding during old age after the most of TSI has been removed. The values of HSI and TSI (Table 14.5) indicates

TABLE 14.5

*Sinuosity Indices**

| Drainage Basins | CL (km.) | VL (km.) | Air (km.) | CI | VI | HSI % | TSI % | SSI |
|---|---|---|---|---|---|---|---|---|
| Kansi | 12.8 | 11.6 | 10.0 | 1.27 | 1.16 | 41 | 59 | 1.09 |
| Jamuna | 13.6 | 11.4 | 10.0 | 1.35 | 1.13 | 63 | 37 | 1.19 |
| Sakheri | 13.6 | 12.0 | 10.5 | 1.29 | 1.14 | 52 | 48 | 1.13 |
| Dhardhari | 20.0 | 12.1 | 9.4 | 2.12 | 1.81 | 28 | 72 | 1.17 |
| Jhikia | 24.5 | 20.3 | 14.2 | 1.72 | 1.43 | 41 | 59 | 1.20 |
| Saliahi | 18.4 | 16.8 | 13.7 | 1.34 | 1.22 | 35 | 65 | 1.09 |
| Renhal | 18.7 | 15.5 | 13.9 | 1.34 | 1.10 | 70 | 30 | 1.22 |
| Satbahini | 30.8 | 12.3 | 11.6 | 1.30 | 1.05 | 83 | 17 | 1.23 |
| Durgauti | 25.1 | 21.3 | 19.0 | 1.32 | 1.11 | 66 | 34 | 1.18 |
| Sukhnadia | 15.2 | 12.3 | 11.7 | 1.30 | 1.05 | 83 | 17 | 1.23 |

* CL=Channel Length; VL=Valley Length; Air=The Shortest Air Distance Between the Source and Mouth of the Streams; CI=Channel Index (CL/Air); VI=Valley Index (VL/Air); HSI=Hydraulic Sinuosity Index × [(% equivalent of (CI—VI)/CI—1)]; TSI=Topographic Sinuosity Index [% equivalent of (VI—1)/CI—1)]; SSI=Standard Sinuosity Index (CL/VI).

that most of the streams except a few are passing through their young stage. The Dhardhari Basin has the highest TSI (72 per cent) and lowest HSI (28 per cent) which indicate that about one-fourth of the hydraulic sinuosity has been removed. The lowest HSI value of this river may be attributed to the course of the river over resistant granite and gneiss beds on one hand and the rejuvenation due to successive uplifts of the landscape on the other.[19] The TSI value is lowest (17%) cent in the basin of

Sukhnadia river which indicates that this river flows in such an area where the morphometric attributes are insignificant. This is mainly because the river has started developing a flood plain primarily due to the geological structure of the area where the river has cut through the sandstones and established its course on the shale beds which has greatly accentuated hydraulic action.

**Hypsometric Analysis**

Hypsometric analysis, in general use for calculation of Hydrologic information, takes a complete drainage basin above a selected point on a main stream as the area of study. The percentage hypsometric method used in this investigation relates the area enclosed between a given contour and the upper (headward) segment of the basin perimeter to the height of that contour above the basal plane. The method has been used by Langbein[20] (1947) and Strahler[21] (1952) for hydrologic investigation. The ratios are involved which are as follows :

(1) a/A ('a' is the cumulative area enclosed between consecutive contours and 'A' is the total basin area) represented on the abscissa, and

(2) h/H ('h' is the cumulative height between successive contours above the base and 'H' is the total basin height) plotted on the ordinate.

Actual measurement and calculation of hypsometric data have been done in the following manner :

With a planimeter, the area of the entire basins has been measured first, then the area enclosed between each contour and the upper perimeter has been calculated. Ratios are computed, ranging from 1 to 0. Where relief is strong and contours closely crowded, every second or third contour is used except near the summit where all available contours are used.

Height ratios are obtained first by determining the total range between basin mouth and summit point. The height of each measured contour above the mouth elevation is then determined and ratios to total basin height computed. These values range from 0 to 1 in inverse series to the area ratios. The ratios are plotted on convenient cross-section. paper and the curves have been drawn (Fig. 14.4) smoothly with the aid of a draftsman's curves.

For the computation of the volume of earth material contained between the ground surface and the bottom, the lines made by Strahler (1952, p 1120) have been adopted which are as follows :

$$\text{(I)} \qquad V = \int_{\text{Base } el}^{\text{Summit } el} a\, dh$$

where, V=Volume

a=area

h=height

If we now divide both sides of this equation by H and A, which are constant terms,

$$\frac{V}{HA} = \frac{1}{HA} \int_{\text{Base } el}^{\text{Summit } el} a\, dh$$

$$\text{or} \qquad \frac{V}{HA} = \int_{\text{Base } el}^{\text{Summit } el} \frac{a}{A}\, d\frac{h}{H}$$

$$\text{If} \qquad \frac{a}{A} = x \text{ and } \frac{h}{H} = y, \text{ then}$$

$$\text{(II)} \qquad \frac{V}{HA} = \int_0^1 x\, dy$$

Thus, if the hypsometric function, x=f(y), is integrated between the limits of x=0 and x=1.0, a measure of landmass volume remaining with respect to volume of the entire reference solid is obtained. This integral is here designated as "hypsometric integral" and is equivalent to the ratio of area under the hypsometric curve to the area of the entire square. It is expressed in percentage units and is obtained from percentage hypsometric curve by measuring the area under the curve with a planimeter (Table 14.4).

The integral value ranges between 28% and 68%, indicating old, mature and young stages of the rivers. The upper portion of the curve for Saliahi Basin is convex but the lower portion has concavity (Fig. 14.4) which shows that the river in the lower part is more active for erosion than the upper part. The curves for all drainage basins except Dhardhari and Sukhnadia

*Hypsometric Analysis of Drainage Basins*

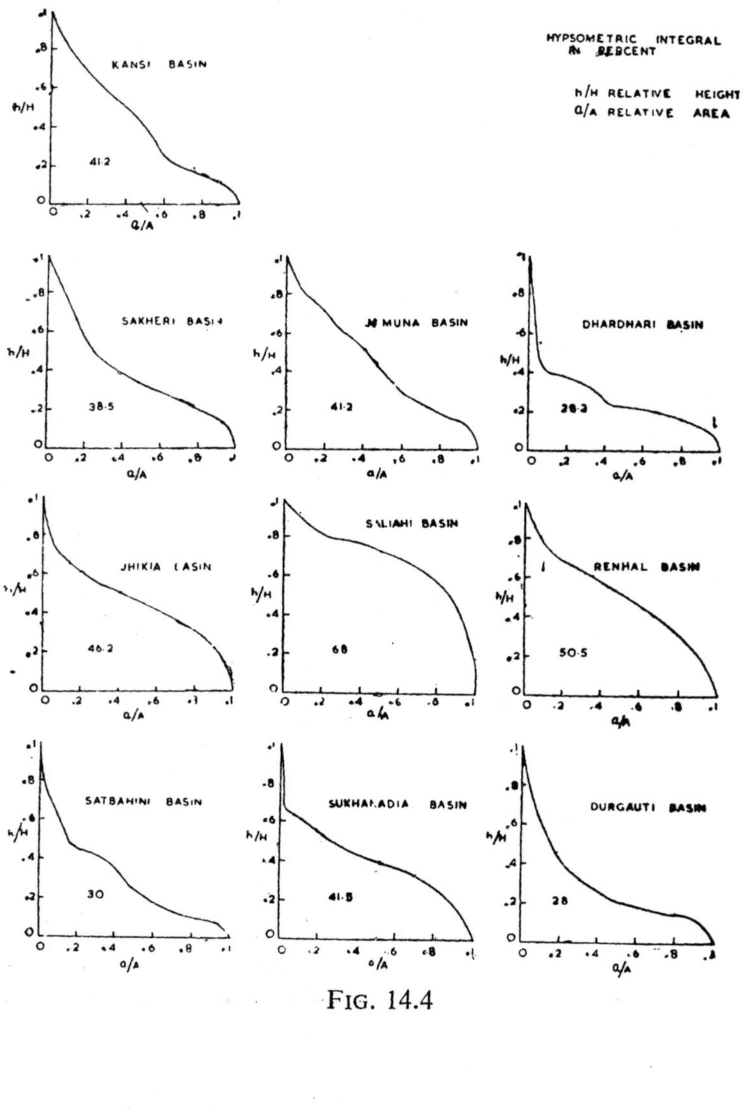


FIG. 14.4

have no major irregularities. The upper portions of the curves of Dhardhari and Sukhnadia Basins are steep due to the presence of isolated hillocks and steep—sided spurs at higher altitudes covering a small percentage of the basin areas. The curves for Sakheri, Kansi, Sukhnadia, Jamuna, Jhikia and Renhal Basins pass approximately across the centre of the diagram with the hypsometric integral values of 38.5%, 41.2%, 41.5%, 44.2%, 46.2% and 50.50% respectively indicating mature stage. The low integral values in the basins of Satbahini, Dhardhari and Durgauti indicate the late-mature stage where about 70%, 71.8% and 72% of their landmass respectively have been eroded away.

**Relationship Between Hypsometric Integral and Other Morphometric Variables**

It is rather difficult to establish definite relationship between hypsometric integral and other morphometric variables. However, some facts observed may be mentioned here. Durgauti Basin has the lowest hypsometric integral with low relief ratio, low stream gradient, low elongation ratio and low drainage density, indicating old stage of geomorphic development. Only three drainage basins (Dhardhari, Kansi and Saliahi) show the negative correlation with the hypsometric integral. On the basis of the value of the hypsometric integral (28.2%), Dhardhari Basin is in the old stage, although its relief ratio, stream gradient and elongation ratio are higher. These are, probably, because that areal coverage in the upper reaches is restricted and proportionally lesser confined due to hillocks on one side of the stream. Kansi Basin represents a mature stage with the integral value of 41.4 per cent. There are erosional scarps of the more or less cresentic divide enclosing the upper reaches of this stream whereas in the lower reaches, several hillocks are observed. Here, the higher relief ratio and higher stream gradient are probably due to erosional scarps in the upper reaches. Saliahi basin has moderate relief ratio, moderate stream gradient and moderate elongation ratio but its integral value is higher (68 per cent), indicating the young stage of the river. It has a steep sided valley like a canyon in the middle reaches which is natural in the region under study because Peninsular India in general and Chotanagpur in particular have recorded three uplifts concurrent with the Himalayan uplifts during Oligocene, Mid-miocene and Pleistocene with the result that some rivers like Saliahi are rejuvenated.[22]

## CONCLUSIONS

The conclusions drawn from the analysis are given below:—

(1) The number of channel segments decreases as the order value increases.

(2) The bifurcation ratios of the drainage basins are more or less normal according to Horton's law, as they range between 3 and 5. The little variations in the values of two basins (Saliahi and Satbahini) are due to the difference in their stage and topographic characters.

(3) The mean channel length of a higher order increases but not in geometric progression.

(4) The low drainage density observed in the basins is mainly due to the structural control of Archaean Formation, poor supply of rain water along with infiltration capacity of surface rock and old stage of geomorphic development.

(5) The sinuosity indices of the rivers range between 1.09 and 1.23, indicating a sinuous pattern.

(6) The basins generally exhibit broken and undulating terrain.

(7) The relationship of relief ratio with drainage density is not harmonious in most of the cases barring a few.

(8) There is no definite relationship between elongation ratio and relief ratio.

(9) The channel gradient increases with an increase in elongation ratio.

(10) Sinuosity index of stream is a key to the geomorphological character of a region.

(11) The hypsometric integrals of the basins range between 28 per cent and 68 per cent, indicating old, mature and young stages of the rivers. The old stage of the rivers is but natural in accordance with the age of Palamau Upland, a part of Peninsular India, belonging to the Pre-Cambrain Era. The younger stage of the rivers do speak of the rejuvenation of this Pre-Cambrain Land Surface due to Tertiary Uplifts during

Olioceene, Mid-miocene and Pleistocene times.

(12) The form of the hypsometric curve and the integral value are important elements in topographic form and show marked variations in regions differing in the stage of geomorphic development and geological structure.

(13) There is no significant relationship between hypsometric integral and other morphometric variables.

(14) Quantitative determination of different morphometric characteristics of the drainage basins and their inter-relationship have a practical application in basin hydrology.

(15) The drainage basin is an appropriate areal unit for the organisation of human activities and for regional planning due to its structural as well as physiographical and hydrological unity.

## REFERENCES

1. R.E. Horton, "Erosional Development of Streams and their Drainage Basins; Hydrological Approach to Quantitative Morphology," *Bull. Geol. Soc. Ame.* Vol. 56, pp 275-370.
2. A.N. Strahler, "Quantitative Geomorphology of Erosional Landscapes," *19th International Geological Congress*, Algiers, Sec. 13, 1952, pp 341-354.; "Quantitative Analysis of Watershed Geomorphology," *Trans. Ame. Geophys. Uni.*, Vol. 38, 1957, pp 913-920.
3. S.A. Schumm, "The Evolution of Drainage Systems and Slopes in Badlands at Perth Amboy, New Jersey," *Bull. Geol. Soc. Ame.*, Vol. 67,1956, pp 597-646.
4. R.J. Chorley, "Drainage Basin as the Fundamental Geomorphic Unit" in *Water, Earth and Man,* Edited by R.J. Chorley 1959, pp 85-89.
5. M.E. Morisawa, "Quantitative Geomorphology of Some Watersheds in the Applachian Plateau," *Bull. Geol. Soc. Ame.*, Vol. 73, 1962, pp 1025-1046.
6. K.N. Singh, "Southern Upland of Eastern U.P. : Landforms and Settlements," Unpublished Ph.D. Thesis, Deptt. of Geography, B.H.U., 1967, pp 1-415.
7. S. Singh and B. Ghose, "Quantitative Geomorphology of Drainage Basins in Semi-arid Environment." *Ann. Arid Zone*, Vol. 8, 1969, pp 37-44.
8. S.K. Pal, "Quantitative Geomorphology of Drainage Basins in the Himalaya," *Geographical Review of India*, Vol. 35, 1973, pp 81-101.
9. K.R. Dixit, "Drainage Basins of Konkan : Forms and Characteristics," *National Geographical Journal of India*, Vol. 22, 1976, pp 79-105.
10. Savindra Singh, and O.P. Singh, "Morphometric Controls on Rural Settlements of Palamau Upland : A Quantitative Approach," *The*

*National Geographer*, Vol. XI, 1976, p 124.
11. A.N. Strahler, *Physical Geography*, John Wiley & Sons, New York, 1969, p 413.
12. O.P. Singh, "Palamau Upland : A Study in Landforms and Settlements," unpublished Ph.D. Thesis, Deptt. of Geography, B.H.U., 1977, p 77.
13. W.M. Davis, "Meandering Valleys and Underfit Rivers," *Annals, A.A.G.*, Vol. 3, 1913, pp 4-5.
14. J.E. Mueller, "An Introduction to the Hydraulic and Topographic Sinuosity Indices," *Annals, A.A.G.*, Vol. 58, 1968, pp 371-385.
15. S.A. Schumm, "Sinuosity of Alluvial Rivers of the Great Plains *Bull. Geol. Soc. America*, Vol. 74, 1963, pp 1089-1099.
16. L.B. Leopold, M.G. Wolman and J.P. Miller, *Fluvial Processes in Geomorphology*, W.A. Freeman & Co., 1964, pp 281-298.
17. D.K. Singh and H.N. Sharma, "Sinuosity Index of Streams as a Key to the Geomorphological Character of a Region," paper read before *21st International Geographical Congress*, held in the Deptt. of Geography, Gauhati University, Assam and published in the proceedings, 1968, pp 50-57.
18. J.B. Leighly, "Meandering Arroys of the Dry South-West," *Geographical Review*, Vol. 26, 1936, p 272.
19. O.P. Singh, "Sinuosity Index of Rivers as a Key to the Geomorphological Character of a Region : A Case Study of Palamau Upland," *The Geographical View Point*, Vol. 5, 1974, p 53.
20. W.B. Langbein, "Topographic Characteristics of Drainage Basins, *U.S. Geol. Surv.* W.S. Paper 968-C, 1947, pp 125-157.
21. A.N. Strahler, "Hypsometric (Area-altitude) Analysis of Erosional Topography", *Bull. Geol. Soc. America*, Vol. 63, 1952, pp 1117-1142.
22. O.P. Singh, "Evolution of Physical Landscape in Palamau Upland, *The Deccan Geographer*, Vol. 15, 1977, p 262.

# 15

# Some Techniques and Recent Trends in Geomorphological Studies

*—Subhash Chandra Mukhopadhyay*

## INTRODUCTION

THE TISTA and the Subarnarekha basins with their variegated landscape-elements have become conspicuous geomorphic units in the Eastern Himalayan zone, and in the Chotanagpur Plateau respectively. Here, only some relevant points on various morphological facets of the Subarnarekha as analogy or secondary examples with a detailed reference to the Tista, and supporting evidences for neotectonic movements are referred. There are similarities in some areas of evolutionary processes between these two critical drainage basins set in polycyclic landscapes of contrasting fields in Eastern India. This study on selected drainage basins have been made on the basis of advanced methods and techniques in order to illustrate the recent trends in geomorphology.

The Tista river is included in the system of the eastern Himalayan drainage comprising the Kosi, the Arun, the Mahananda, the Jaldhaka and the Tista itself [Fig 15.1 (a)]. The dynamism is the changes in the course of the Tista in the piedmont and the adjoining plain areas of North Bengal and poses a spatial type of problem. This problem is due to the variation of the denudational or depositional processes[1] relating to the local climatic and topographic conditions available in the Eastern Himalayas and adjoining plains. It is of particular geomorphic interest specially associated with the development of the essential elements of the picturesque landscape being aided by tectonic actions.

The Tista flows in the upper reaches through a widespread sheet of gravel (Post-Glacial) which has constituted mainly the high-level river terraces. It is typified by the incised meanders along with two sets of alluvial and gravelly terraces in the upper reaches while the lower part to the south of the piedmont areas the channel is transformed into a braided pattern with terraces of alternate scourage and valley filling process [Fig. 15.1 (a)].

*Eastern India : Location of the Tista-Subarnarekha Drainage Basin*

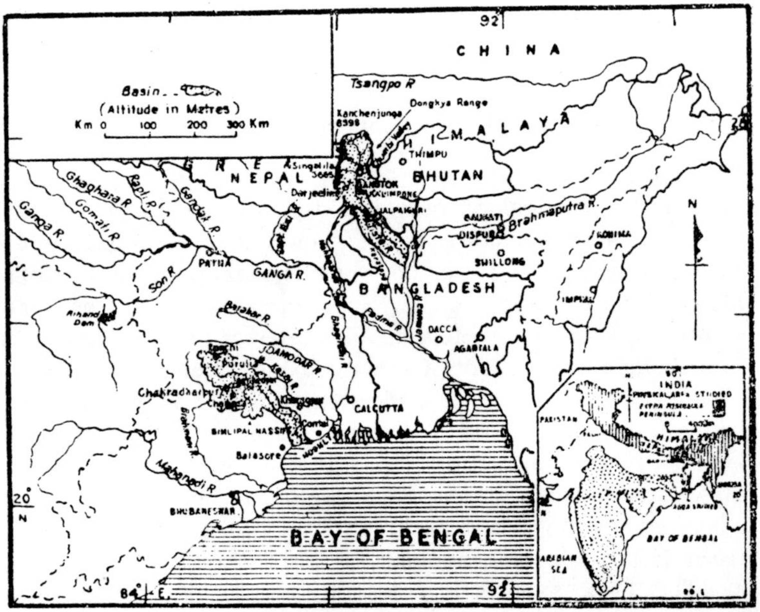


FIG. 15.1 (a)

This paper embodies observations on the studies of meanderings, river terraces, flood plains, geomorphic surfaces with a thick mantle of debris, deeply incised valleys, gorges, dissected undulating terrain, antecedent drainage of the Tista basin and superposition in the Subarnarekha; and the wide level surfaces at different altitudes relating to the progress of cycle of erosion and landform developments. It also includes results of quantitative analyses along with statistical tests of the relevant data obtained from published works, toposheets ERTS LANDSAT imagery (Remote sensing), and field work in the Tista and Subarnarekha basins and surrounding areas.

## OBJECTIVE

The present study comprises mainly the aspects of erosional facets by which rocks of different geological formations are truncated in relation to the Subarnarekha (Plateau) and Tista (Himalaya) drainage systems being influenced by tectonic movements. These two basins exhibit the typical topographic forms under fluvial cycles of erosion which illustrate the precise relationship between tectonic adjustments and landform developments. Morphological along with morphogenetic analyses are made of the causes and characteristic of the diversity of the terrain within the Tista and Subarnarekha basin to illustrate the modern trends in geomorphology.

## METHODS OF ENQUIRY

In this piece of geomorphological investigation, the author depended mainly on geomorphic mapping based on ERTS LANDSAT imagery,[2] Survey of India toposheet analyses (73 and 78 series) and intensive field work. The study is thus based on available data and empirical observations. Most of the observations in this study, however, have been based upon intensive field work in the areas under considerations. Stratigraphical, sedimentological or chemical analysis of the field evidences and samples on terraces and flood plain deposits have also been made. Mention should be made about the morphometric and statistical analysis of the fundamental geomorphic units within the two drainage basins. During the period of field study the mapping and other related survey works have been accomplished. Various calculations were made of average height, volume, average slope and construction of hypsographic clinographic curves, profiles both actual and computed (exponential curve) and different morphological parameters like (a) relative relief, (b) drainage density, (c) slope (Wentworth, 1930)[3], (d) roughness index (Hook, 1955)[4] and (e) dissection index in connection with statistical tests i.e. Koloogorov-Smirnov statistics (Miller and Kahn, 1962)[5] Chayes-Kruskul test (1966)[6] along with a series of 't' tests amongst the different geomorphic sectors for each parameter (a, b, c and d) and analysis of variance, etc. with a view to ascertaining the geomorphological characteristics of the Tista basin.

Regarding the variation of sculpturing processes of the landscape, different climographs and technical diagrams along with rating curves and harmonic analysis, etc. on the basis of climatic and hydrological data (climato-genetic geomorphology, Budel, 1970)[7] of the area have been prepared.

The outcome was preparation of a number of maps and diagrams depicting morphographic, morphometric and morphogenetic features of the two drainage basin areas.

## GEOLITHOLOGY STRUCTURE AND LANDFORMS

It is necessary to present a brief account of the areal distribution of the various rock-groups or structure, and to indicate the role of these rock formations relating to the study of geomorphological problems both of the Tista and Subarnarekha basins. The study area is a part of the Indian subcontinent, made up of rocks of ages ranging from precambrian to quanernary in different sections of the (Iron Ore Series, Dharwar, Chaibasa stage, etc.) Chotanagpur Plateau (the Subarnarekha basin and its fringe areas), the lofty Himalaya mountains and the Gangetic-Brahmaputra Plains in eastern India (Mukhopadhyay 1970).[8] Regarding the Extra Peninsula except the alluvial plains of the lower Tista sub-basin in north Bengal and Bangladesh to the south the area is made of three distinct geological formations : (i) the Siwaliks, (ii) the Gondwanas or Damudas, and (iii) the Dalings, etc.

The unequal uplifts or tilts on the different parts of these two studied basins have also caused the developments of striking difference in the topographic expressions like a series of hills of various origins, gorges, redge and vales, escarpments, terraces and flood plain levels, meanders, composite and complex type of basins and undulating surfaces etc. within and the surroundings areas of the Tista basin.

Except for the lower reaches, the greater part of the basin consists of plateau and mountainous tracts, and both the basin areas have been warped or disturbed by tectonic movements since the early Tertiary period.

The areas covered by these two drainage basins are typified by folded structure and varied lithology with older rocks occupying the upper structural levels (superposition) specially in the Tista valley section. It bears the evidences of several persistent thrust or high angle reversed faults, imbricate thrusting, etc. delimiting different units e.g. recumbent folding of Dalings and Darjeeling gneisses and over thrust on the foothill of Damudas, the main boundary fault limiting the boundary of the Siwaliks to the north, etc. There are occasional tectonic intermixing of underlying and overlying formations. The northern thrust of the foothill Damudas is a true nappe and essentially autochthonous or para-autochthonous (Acharya 1970).[9] In the Darjeeling hills, the overriding Daling slates are more deformed, and the dynamic metamorphism of the foot-hills Damudas is essentially due to this thrusting. The resulting folds are generally open and broadly plunge upstream at moderate angles. One of these large scale and spectacular folds has formed the present Tista valley antiform where foliation and bedding of the Dalings, on both sides of the river bank, are found to dip into the hills (Darjeeling). Development of various types of lineation, parallel to the directions of tectonic transport, are present. Such features are well developed on the overlying Dalings and also in the Damudas close to the thrust. Along the plane of the thrusting, pebbles of the upper pebbly sandstones are sliced out within the Tista basin.

The Subarnarekha basin is made of distinct tectonic units e.g (i) Southern granitic platforms, (ii) a northern mobile belt, (iii) an intervening Singhbhum shear zone, etc., particularly in the middle segment—the Sangai-Kharkai-Garra sections of the valley. As a whole, there is a broad control of the structure of the overlying rocks particularly on the plane of the upthrust which in turn is reflected in the development of morphological features both in the Tista and the Subarnarekha basin areas [Fig. 15.1 (b).]

## ANALYSIS

This study reveals many interesting features of drainage characteristics specially in terms of length, basin shapes, ruggedness, etc. It indicates that the Tista basin (excluding the uppermost part) consists of five major and two minor hyper-

*Geological Map of North Bengal* (*The Tista Basin*)

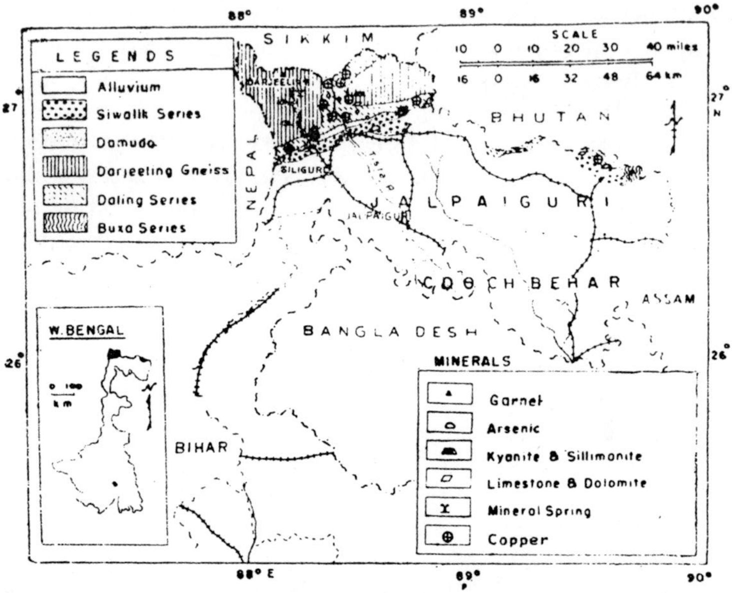


FIG. 15.1 (b)

bolic segments which intersect at four and one different elevations respectively in between 4000 to 300 metres a.s.l. (Fig. 15.2). Like the profiles of the other Himalayan rivers, the Tista and its tributaries are not simple but are disturbed by a number of breaks or knick-points.

The profiles both of the Tista and the Subarnarekha have also been studied (Fig. 15.2) in terms of mathematical curve with satisfactory fit (exponential curve $Ye = ab^x + k$) to have an idea of the implication of tectonic activities in the upper and fall in sea levels in the lower reaches. It has been found that the "breaks" in the profiles seem to occur in groups where the average gradients are relatively steep and each group of partial profile, as has been presumed, belongs to a single former base level. In the lower reaches, the rivers flowing on their flood plains have been intermittently affected by the fall in the level of the Bay of Bengal. Consequently, it has cut through the floodplain related to each level of the sea and the remains of each floodplain survive as "terraces" along the present river

*The Long Profile of the River Subarnarekha*

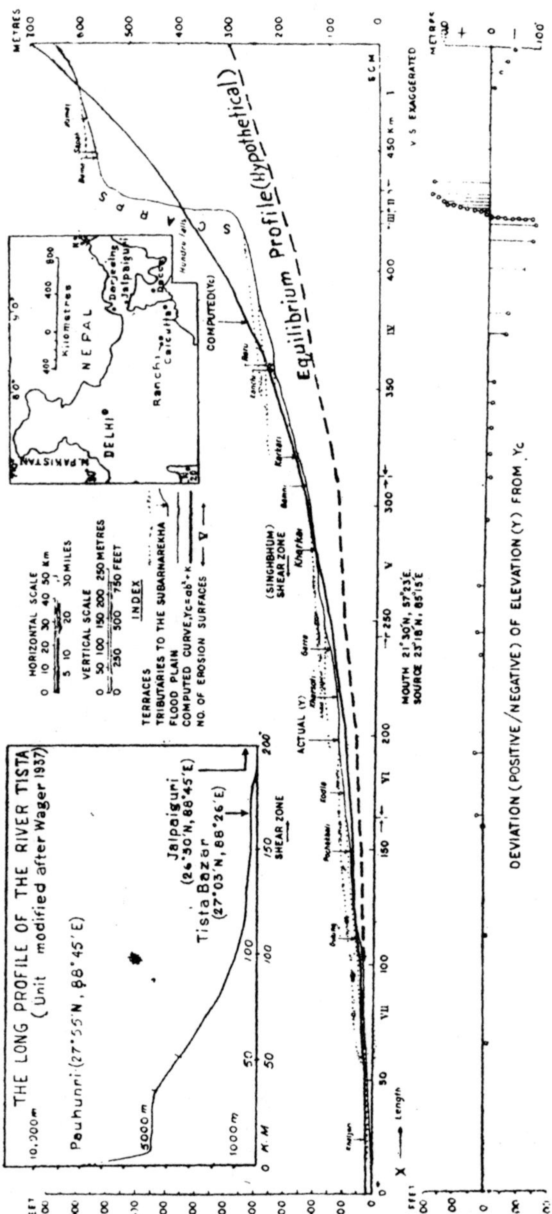


FIG. 15.2

valleys as discussed in the next part. The statistical analysis of the various morphological parameters and profiles at chosen confidence levels (Miller and Kahn, 1962) have served to demonstrate the genuineness of the observed anomalies in the drainage characteristics of the Tista and the Subarnarekha basins. The most plausible explanation of these sharply segmented profiles is that successive uplifts have occurred in the basin areas. As a first approximation the long profile of the uppermost segment represents the relict of the earliest river profile at grade. It enables a rough estimate being made of the amount of relative uplift at successive stages—the uplift in this case is estimated as more than 200 m and 300 m for the Tista and the Subarnarekha respectively.

**Altimetric Analysis**

The Subarnarekha basin set in the multicyclic Chotanagpur Plateau, Bihar, Orissa and West Bengal, is characterised by higher degree i.e. 7th of complexity in its landscape evolution. On the contrary, the present part of the Tista basin has attained the fifth cycle of development, and so it is at present a landscape of that complexity. The number of cycles would be increased if the whole drainage area (i.e. the uppermost part) were to be considered. Five distinct cyclic surfaces could be detected possibly because other surfaces have been excluded or effaced due to its shifting nature and one of these surfaces being developed under the current cycle or epi-cycle of erosion. The Tista and the Subarnarekha basins, like the other Himalayan and Plateau drainage basins, also conform with complex peculiarities in their evolution and have experienced several cycles of erosion onward from Eocene to late Tertiary periods. Studies on the terrace sequences and stratiography indicate that for both the drainage basins, the phases of planation and deposition have alternated with recurrent down cutting associated with interruptions in the progress of cycles.

Distinct poly cycle-forms in terms of breaks or knick-points in the profiles of the Subarnarekha and Tista, their tributaries and hills are common within these basins. It has been deciphered in the field that variation of rock resistance is not always the prime cause of 'break' and other anomalies found in the profiles. The long-profile of the Subarnarekha and Tista rivers exhibit such knicks even in the lower reaches irrespective of the variations of channels or drainage patterns and width of the valleys throughout the basin areas. The occurrences of these breaks at some critical points, the various channel and drainage patterns along with other drainage anomalies are significant particularly in the study of implications of neotectonics.

## STATISTICAL TESTS

The calculations of 'percentage of total sum of squares' (Yule and Kandal, 1950)[10] accounted for by the computed curve ($Y_c$) relating to the statistical analysis of profiles (the Subarna-rekha, Tista and their tributaries) and distinct geomorphic surfaces both of erosion and accumulation within these two drainage basins, are indicative of the 'significant' and greater degree of disturbance in part of the middle section living at 100 m, and more than 900 m above sea level for the Sabarna-rekha and the Tista respectively.

A comparison among the (S1-S7) seven of the Subarnarekha and five geomorphic surfaces (S1-S5) of the Tista has been made in respect four (a and c, b and d) of the five parameters a, b, c, d and e viz. relative relief, drainage density, slope roughness index and dissection index respectively by applying Kolmogorov-Smirnov Statistic Test (Miller and Kahn 1962).[11] It would be noted that (i) the frequency distribution for both (a) relative relief and (c) surface slope (average) are almost similar for both the parameters in each of the surfaces, and for (b) drainage density and (d) roughness index dissimilarities in most of the surfaces are common (Fig. 15.3)

Regarding the Tista basin in particular the differences (a and c) are highly significant among the upper surfaces, the confidence level in such cases being greater than 97 per cent confidence level. (iv) On the other hand, the mutual differences between the lower three surfaces are not statistically significant even at 90 per cent confidence level. These three surfaces actually represent successively lower reaches of the Tista sub-basin, hence the lack of significant differences between these lower surfaces, is of geomorphic interest so far as the stages of development are concerned. It is suggestive of the contention that these

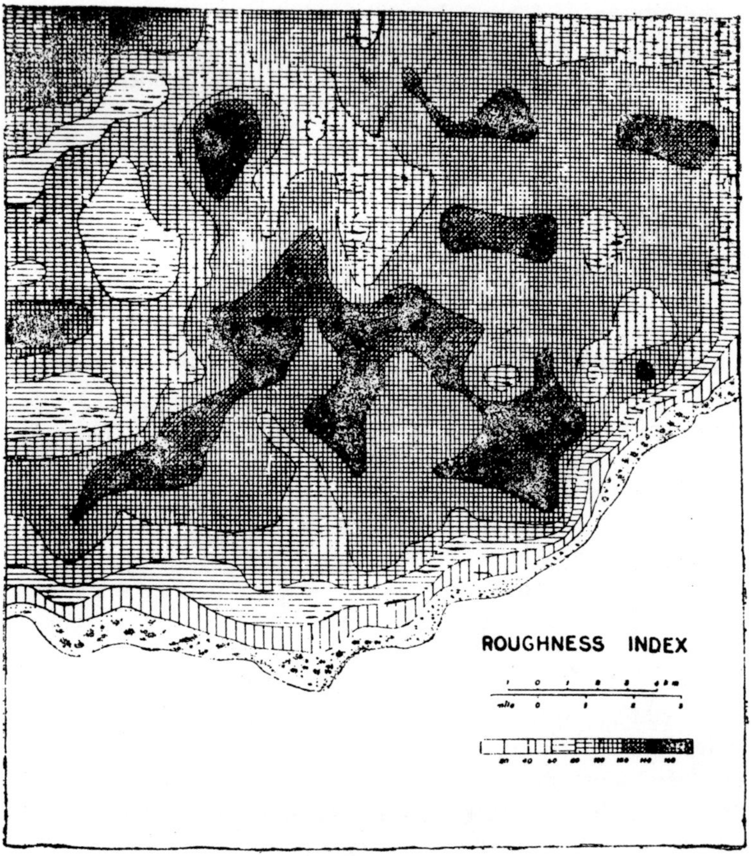


FIG. 15.3

surfaces are associated with the effects of differential upheavals. (v) The fourth surface is marked by comparatively smaller ratio of gentle slope and moderately low roughness index less than (5°) five degree and 20 respectively than the upper one. These anomalies support the idea that these surfaces at different altitudes have developed within a river basin which experiences 'disturbances' several times since its inception. The multiple surfaces, set in a tectonically active area, represent parts of a continuous geomorphic surface within a drainage basin being subjected to differential upliftments, large scale dissection and distractions, etc.

Mention should be made of the 'thrust zones' in an almost east-west direction both in the Tista and the Subarnarekha basins as evidences of 'structural instability' in the different geological formations which have actually constituted parts of the eastern Himalaya and the Chotanagpur Plateau. An important feature of the drainage basins or sub-basins under consideration is the way in which the geological structure and tectonics, as mentioned above, are expressed in the landforms. The distinct undulating downs, meandering channels, ridge and vales, gorges, deeply incised valleys, terraces etc. of the upper part with steeper gradients differ strikingly from the landforms of the lower surfaces. These surfaces or sectors represent a part of the middle section with critical structural or lithologic units of variable resistance for the anomalies in the parameters like relatively high mean gradient or slope and roughness index in the lower sectors than the upper ones. Regarding a series of 't' tests amongst the surfaces for each parameter, it is obtained that each surface is significantly different from the next higher surface relating to single or more parameters. It has been found that the maximum difference lies in between the first two surfaces and the least difference between the last two surfaces, it is somewhat different for drainage density and roughness index which lie between second and third, and third and fourth respectively. From the "analysis of variance" it has been observed that there is an intrinsic difference between the four geomorphological parameters—relative relief, drainage density, slope and roughness index for the five distinct surfaces. Considering single variable at a time, it is suggestive of highly significant difference between the surfaces. It is true specially for the variable 'roughness index' and also 'drainage density.' Besides these, correlation coefficients of modal variables from different sections of terraces and floodplains constituents have been considered and tested for significance adopting the Chayes method (1960).[12]

It has been found that the number of occurrences of different geomorphic surfaces which are obtained from the graphical analysis, do conform to the results of statistical and other quantitative analysis. The number of divisions i.e. five distinct geomorphic surfaces of the Tista, and the seven surfaces of the Subarnarekha basin areas, gains support also from the penta-modal and hepta-modal altimetric frequency-curves respectively. These are located approximately at 100-200, 400-500, 700-800, 1000-1100 and 1500-1700 m; other minor peaks are found mainly in between 2000 m to 3000 m, and 50 m—100 m in the upper and lower reaches of the Tista valley respectively (Fig. 15.4). The cumulative frequency curves (ogives) are drawn on the basis of maximum, minimum and spot-heights. Three separate curves show the fundamental surfaces (altitudes) of the geomorphic surfaces in terms of the breaks which conform with the above mentioned altitudes both in the Tista and the Subarnarekha drainage basins.

Bivariate analysis and other methods of correlation have been made to find out the quantitative nature and degree of relationship between the variables like the relative relief (X) and drainage density (Y), slope (X) and roughness index (Y), relative relief (X) and slope (Y), etc. It has been established that these variables are closely related with a high degree of positive correlation and with the help of regression or estimating line it is possible to discern average probable change in the Y series given a certain amount of change in X i.e. the regression equation of y on x. The equation is expressed as $Yc=a+bx$, where a and b are two unknown constants or parameters of the line. This tool of regression has also been extended to three variables. The nature of relationship or the cause and effect relationship between the morphometric attributes, is more clearly indicated through regression analysis than by correlation which gives only the degree of relationship as stated above. The diagrams and the analysis depict the high degree of positive correlation for all the variables, except the variables like relative relief (X) and drainage density (Y) which show a relatively low degree of positive correlation (Fig. 15.5). The morphometric analysis (Singh, 1967)[13] of the landforms has been adopted in terms of (a) relative relief, (b) drainage density, (c) slope (average), (d) roughness index and (e) dissection index, etc. attributes of a critical parts of middle and lower reaches of the Tista sub-basin and its surroundings in between 20 m to 3000 m a.s.l. Besides these the elevation-relief index, drainage

*Cumulative Frequency Graph : The Kolmogorov Smirnov Test for Erosion Surfaces*

*(The Karkot Basin)*

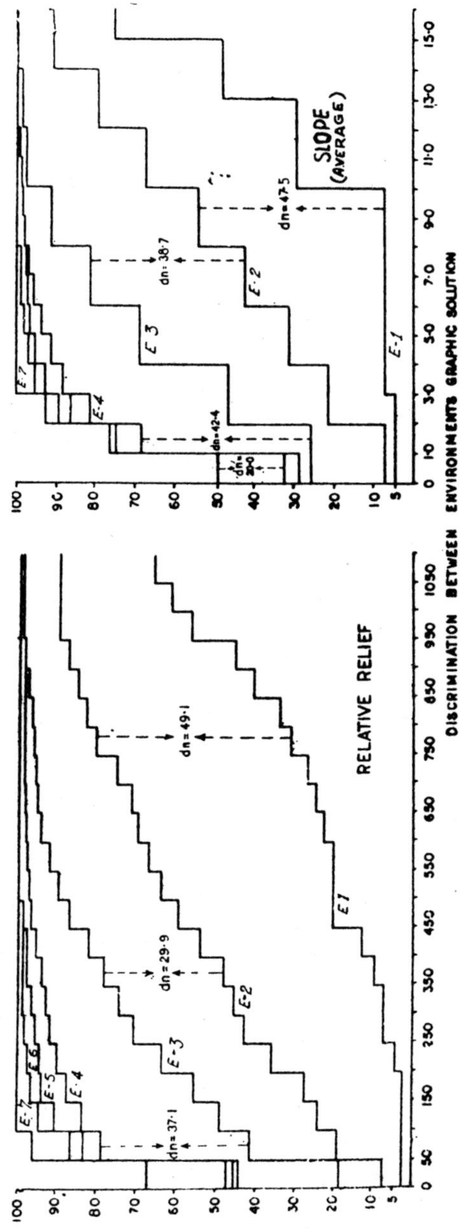


FIG. 15.4

*Scatter diagram (The Tista Basin)*

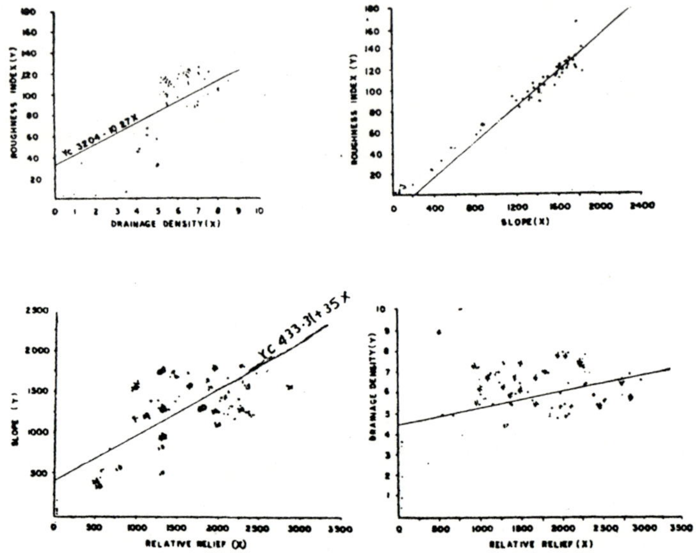


FIG. 15.5

texture, ruggedness numbers, constant of channel maintenance or inverse of drainage density, etc. have also been used. The area thus selected has been divided into 2.54 m. (one inch) grids (1100 such) included in 78 A and B series of (Survey of India) toposheets at a scale of 1 : 63,360 and 1 : 50,000 within Sikkim, Darjeeling and Jalpaiguri districts. Relative relief measures the varying amplitude of relief of the part of the Eastern Himalayas and adjoining plains in terms of absolute relief. This morphometric attribute actually exhibits the aspects of uniformities, discontinuities or graduality in the gradient of change in the spread of landform units. Different methods of slope analysis as a part of clinographic curves have been used. But in the present study the average slope (grid-wise) in degrees has been adopted. An attempt has been made to show the nature and degree of correspondence among the relative relief, slope (average), drainage density, roughness index, dissection index, etc. attributes relating to the analysis of landforms in different morphological facets and magnitudes. But, here, only the attributes like slope and roughness index of the Tista basin in particular are discussed for the purpose of brevity.

**Average Slope**

Field studies reveal much more degree of slope ranging between 40° to 65° specially in the scarp zones or gorges or deeply incised river bank divides and knick point areas etc. being situated in Sikkim and Darjeeling divisions of parts of the Eastern Himalayas.

Higher altitude, folded structure, local disturbances along with the sculpturing effects under glacio-fluvial processes are the dominant factors in the development of higher degree slopes of 35° and above. For example, the elevated sections of the rugged terrain lying above 2000 metre above mean sea level in Senchal mountain (2615 metre above sea level), and the western divides, etc. area typify the attainment of average slope of more than 40 degree. Contrasts in slope development, steep and moderate grades are found in the nearby areas of Sevok Bazar (26°55′N; 88°29′E; 260 m a.s.l.), a critical zone relating to the "grade condition' of river profile. Here, slope (average) variation is observed to lie between 20° to 30° towards upstream while the average slope is reduced to 1° successively in the opposite direction. In the north-central sections of the upper and middle reaches of the Tista valley areas, a slope-range of 30° to 40° has been found. This section covers major parts of the tributary valleys of Rangit, Great Rangit, Rangitkhola, Rangbangkhola, Ratho-khola, Gitkhola, Jhepijhora, Kalijhora, etc., and is made of green quartzitic-schists, slates, quartzites (Daling series) felds-pathic or shaley micaceous brownish sandstone, carbonaceous shale and coal seams (Damudas), Darjeeling gneiss etc. rocks. Some pockets of steeper or very steep slopes are located in the nearby areas of sharp-edged peaks and the dividing ranges in Darjeeling Himalaya e.g. Phalut (27°11′ N; 88°01′ E, 359 m a.s.l.) eastern Singalila (27°15′N; 88°02′E,3679 m), Sandakphu (3630 m) etc. the bordering scarp and spur zone, the dissected flanks and the narrow or gorge-like valleys of the Tista. The right bank tributary valleys like Rathakhola (2994 m), Jhepijhora or Rangbang, etc. being situated in the upper levels (2800 m and above) are very steep (above 40°) in the slope development. Singtam (27°14′ IN; 88°25′E, 2640 m) hill displays a moderate slope. On the other hand, the sections away from the piedmont areas are marked by a lower range of slope-variation not more than 5°. The gentle undulating surfaces and riverine flats constituting the lower part of the basin are characterized by the lowest slope group i.e. below 2°.

**Roughness Index**

The values of roughness index portray the differential developments of weathering status even in the intramontane sections of the Sikkim and Darjeeling Himalayas and parts of their surrounding watersheds. It provides a fair idea about the nature of variation of erosional features, the depth of denudational strength, the role of weathering and erosional agents engaged in the work of incision or sculpturing the land mass mainly in terms of the varying amplitude of valleyside-slopes and resulting configuration. This index is being widely used by the earth scientists in connection with the morphological studies of terrain, and it leads to a better understanding of the surface configuration evolved under a complex process. Actually, it is more than the development of 'slope' or 'dissection index' as it incorporates a number of determining factors in the evolution of the landforms. This index reflects the combined effects of evolutionary rhythmic processes in the development of relief.

Roughness index follows almost the same trend and indicates a close relationship with the other morphometric attributes like slope, dissection index, relative relief and drainage texture, etc. Patches of highest category roughness index are developed in some rugged mountainous tracts like Menam (27°20′N; 88°25′ 3234 m.a.s.l.), Singtam, Senchal, Mahaldiram, Sitong, Sorang (27°10′N; 88°12′E), etc. and their surrounding edges being marked by deeper dissection, higher altitudes, steeper gradients or slopes and higher frequency or fine drainage texture. The range of roughness index variation is from 100 to 170 in the middle and parts of the adjoining rugged hills or dissected valleys lying at levels of 1000 m and above towards upstream. The denuded hills drained by the Rangbang, Rathokhola, Rangitkhola, etc. tributaries display the index from high to very high. It has been observed that high index is associated with high rainfall, high frequency or fine texture of drainage, mature relief and very steep slopes of the hills or valley sides located primarily in and around the thrust or boundary zones of the several rock formations e.g. the Siwaliks, the Dalings, the Damudas, etc. [Fig. 15.1(b)]. This elevated critical area of the Tista basin has undergone neotectonic movements and a much longer phase of gradational processes by elongated steeper gradient of streams.

Surfaces of the highest group of roughness index, 140-160 and above, are found in the dissected flanks of southern hilly tracts around Sinchal and Mahaldiram area (26°56′N; 88°20′E), hillocks and scarps, gorges, elevated divides or watershed flanks, southern spurs, etc. in the south western segment of the Middle Tista basin. Riverine flats, confluence zone and wide valley areas towards the lower parts of the basin show essentially the lower values 20 and below, while the adjoining piedmont and foothill sections are typified by contrasting higher to moderate index as 120-80. The figure of roughness index actually manifests the intra-valley variation in the denudational activities relating to the rhythmic processes responsible for sculpturing surfaces of this part of the Eastern Himalayas. It also helps to study the resulting configuration, as portrayed in the basins' area along with the differential erosional activities being engaged in different micro-units. Occasional patches of higher index 120-160 have brought about interruptions in the general trend—surface of roughness index. It is suggestive of the possible occurrences of crustal warping or tilting in some critical areas, thus, registering the effects of the interactions of so many factors in the evolution of landscape. Higher trend of roughness index strikingly corresponds with the greater amplitude of relief, drainage frequency and ruggedness number in various sections of the Tista basin. The index reflects the combined effects of structural and erosional processes which are normally expressed in terms of relative relief, slope-angles and drainage-texture within the basin.

## FIELD SCIENCE VIS-A-VIS QUANTITATIVE SCIENCE

Field methods have been employed to study particularly the process as behaving dynamically to model the landscape of the Tista basin areas. Field observations and identification have revealed many peculiarities in landform development. The constituent elements of the multicyclic landscape of these two selected parts of Eastern India are—(i) ridge and vales, (ii) antecedent drainage for the Tista and superposed for the Subarnarekha, (iii) gorges or deeply incised valleys, (iv) terraces, continuous or discontinuous and floodplains, (v) along the rejuvenated valleys, (vi) level surfaces at different altitudes, and (vii) well-marked scrap and spur, etc. Regarding the *morphogenetic analysis* of the terraces and floodplains of the Tista basin—as a case study, three (morphogenetic) types of terraces characterise the basin area e.g. (i) VM, (ii) FA and (iii) REA (Table 15.1). Valley and meander terraces (VM)—a largely erosional flood plain are evolved in the intramontane Section of the Tista in the Sikkim-Darjeeling Himalayas. Floodplain of accumulation (FA)—includes the alluvial fan of the Tista in the piedmont segment of Darjeeling. Rhythmical erosional accumulational floodplain (REA) is associated with the Tista Section in part of the great North Bengal Plain. Here, the alluvial fans and floodplains with meander terraces are found as interwoven in space and time (Pecsi 1968).[14] Floods of the Tista as high as 6 to 9 m are frequent during the rainy season when deposition of suspended load becomes heavy all over the floodplain with a fairly smooth relief. The floodplain deposits in REA group of the Tista has again been sub-divided into two broad categories of grain size, these are (i) the coarse gravel and sand of the deep bed and of the depressions, kettle-shaped 2 to 3 m, and (ii) the finer sand and silt of the shoal parts of the bed and floodplains proper (Fig. 15.6). It has been found in stratigraphic succession that mostly coarse grained units are followed by fine grained and silty or clayey sediments with gradational contact. The main bedding types are large scale crossbedding in coarse sand and pebbles, small-scale cross-bedding in medium to fine sand and well developed parallel lamination, convolute laminations, wavy lamination, lenticular lamination and climbing ripple lamination in silty and clayey sediments. Stratigraphic and sedimentological analysis of the gravelly-alluvial terraces and flood plains suggest that (a) they are depth controlled, and (b) are associated with both (i) normal fluvial cycle and (ii) catastrophic flood cycles. Three

*Longitudinal Profile across the Tista Valley (Right Bank Side R.E.A.F. Type)*

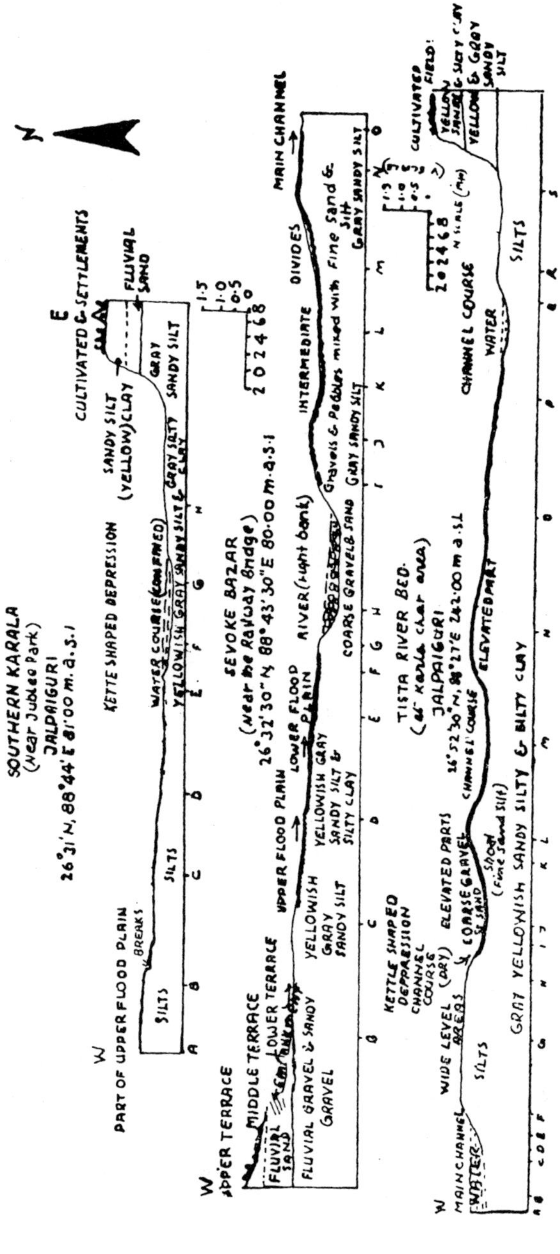


FIG. 15.6

fluvial cycles are catastrophic flood cycles of recent to sub-recent in age, are deciphered in the analysis of terrace sediments. To assess the importance of structural or tectonic activities in the development of river profiles set in different geomorphic surfaces at different altitudes, statistical techniques, mathematical (exponential) curves, correlations between variables and modal variables in morphometric attributes (a, b, c, d and e) terrace-elements and drainage basins tests like Kolmogorov-Smirnov Statistics, Chayes-Kruskal, a series of 't' tests, analysis of variances and percentage of sum of squares, harmonic analysis of rainfall, etc. have been used. For the purpose of assessing the systematic relationships amongst the modal variables, an attempt to compute their correlation with the elements or attributes which are included in the dissimilar closed table, has been made. In a table of mean closed and open variances (vector) of modal variables from the terrace elements of different levels, occurrences of negative element is common. The apparent negative correlation may be due to the closure effect rather than to any true correlation between the pairs of variables. In order to test the observed correlation coefficient (r) between original proportion against the null value (P), and to calculate the approximate null value (P), an equation given by Chayes and Kruskal 1966 along with remaining-space transformation method are adopted.

Disastrous floods or inundations are common both in the Tista and the Subarnarekha basins and it poses problems particularly in the lower Tista with much more intensity and frequency. The occurrences of seasonal floods are associated with heavy amount of annual average rainfall between 3000-3400 mm being accompanied by tropical disturbances under the easterly Jet stream particularly in summer as a part of the monsoon climate regime. Harmonic analysis (Fourier Series) on rainfall reveals that in this monsoon climatic regime (i) annual precipitation variance of the southern parts of West Bengal of sub-Himalayan zone, is typified by dominantly annual tendency, (ii) while only a fraction of the variance is accounted for by other harmonics.

The landforms of the Subarnarekha basin covering south-eastern segments of the Chotanagpur Plateau, are of polycyclic and polygenetic characters. As mentioned above, the wide level surfaces or the step-like features of the surfaces at various altitudes along with terraces are related to a complex geomorphological history involving several cycles of erosions, with these surfaces representing the remnants of former peneplains. On the other hand, the different floodplain levels of the Tista basin, specially of REA group, are in consequence, syngenetic forms with no substantial difference between them. Incidentally in alluvial-fan type valley (F.A.) sections traces of late Pleistocene and early Holocene deposits are found to occur side by side with younger sediments.

## MAIN FINDINGS AND OBSERVATIONS

Present analysis of landforms of the Subarnarekha and the Tista basins as a subject primarily of field and quantitative science portray morphology of macro-meso and micro units. Apart from field observations in some critical areas various quantitative parameters are measured and computed in an attempt to pinpoint a precise relationship between tectonic adjustments and landform developments.

Like the seven geomorphic surfaces of the Subarnarekha, the Tista basin has been demarcated into five distinct geomorphic surfaces conforming five hyperbolic segments relating to the attempt of a satisfactory fit for the long profile of exponential type ($Y_c = ab^x + k$) and also clinographic-altimetric frequency curves (Fig. 15.7).

A range of data has been aggregated also for morphometric analysis, and samples collected mainly from terraces and floodplains of these to basins under study for stratigraphic and sedimentological analysis.

These studies have revealed a few other points which are not at once apparent from a qualitative study : (i) the geomorphic surfaces at various altitudes which also conform with the sharply segmented long profile is suggestive of the successive uplifts which have occurred in the two basin areas. As a first approximation, the long-profile of the upper segment represents the relict expression of the early river profile at grade. It also enables a rough estimate being made of the amount of relative uplift at successive stages e.g. about 300 m for the Subarnarekha and above 200 m. for the Tista.

*Altimetric Frequency Diagram (The Tista Basin)*

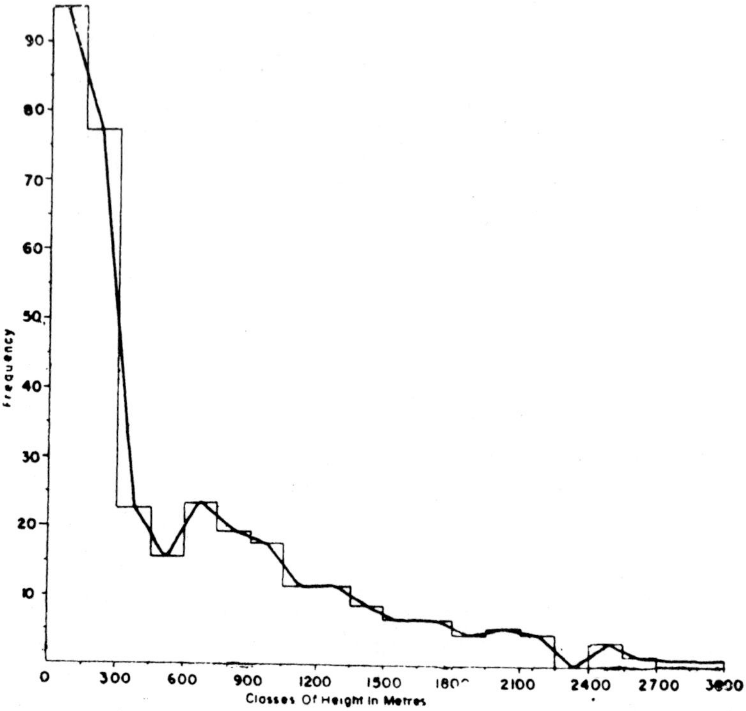


FIG. 15.7

These 'surfaces' are related to a complex geomorphological history involving several cycles of erosions. Actually they represent the remnants of former planation surfaces or peneplains.

The mean gradient as well as mean slope of the successive segments within the Tista basin should normally decrease continuously with distance from the source. But these parameters are higher in the sectors down stream than over sectors up stream. The former represents boundary zones of geological formations marked by structural disturbances. These are evidences of neotectonic uplifts which presumably account for the relatively higher mean gradient and mean slope, etc.

The 'meander index' of the Tista as measured from the topo-sheet (78 A and B Series) for each segment does not vary largely (1.35—1.55) except in the higher segment where it is much lower. This again shows an anomalous feature in the drainage basin.

The influence exerted by the tectonic activities practically dominates over the bed rock on the geomorphic surfaces although the Tista flows over several lithologic units of variable resistance to weathering and erosion. However, the mean gradient, meandering index, meandering length etc. tend to be relatively lower over the phase of Darjeeling gneisses, mica-ceous fine grained sand stones, conglomerate layers etc., and over the shaley-micaceous-sandstone, carbonaceous shale, schist, Slate, quartzitic and coal bearing sandstone, shale sequence of Damuda lie over the other rocks (the Dalings, etc.). Incidentally, in the Tista Valley good exposure of the Siwaliks are found, and these are overthrust by the Damuda (lower Gondwana) along the foothills of Darjeeling district. The Darjeeling Damuda (Gondwana) are highly sheared and tectonically active areas exhibiting a greater deformed retic with a more reduced section within the Eastern Himalayas.

## CONCLUSIONS

Quantitative analysis, correlation and linear regression, analysis of frequency distributions to several variants of the general linear modal and amongst the modal variables, and analysis of variance followed by computation of drainage density, ruggedness number, channel width, meander lengths, stream length ratios, stream number orderwise, bifurcation ratios, average slope, roughness index, relative relief, dissection index, etc. all taken together has served to demonstrate at high confidence level, the justification of the observed drainage anomaly in the Tista basin area. Statistical tests indicate that for all the parameters (a, b, c, d and e) the individual geomor-phic surfaces are all mutually distinct at a high confidence level.

There are bed-rock or gravelly-alluvial variety of terraces along the Tista even along Lachenchu, the uppermost reaches of the valley in north Sikkim and Darjeeling Himalayas. One may attribute the development of these terraces to changes of base levels of erosion by tectonic movements.

Besides the quantitative analysis, the stratigraphic and terrace evidences of the Tista basin seem to recognize the tectonic control of base level consistent with rejuvenation of the drainage by repeated earth movements in recent times, and the effects of pleistocene climatic changes.

There are instances of late Pleistocene terrace isles—elongated isles flanked by living and dammed up river branches in REA type specially in the nearby areas of Jalpaiguri (26°30′N, 88°45′E; 87 m a.s.l.) and Kamarjani (25°25′N, 89°48′E 22 m a.s.l., Bangladesh). Again the marginal band of the floodplain levels, likewise, has a fringe of late Pleistocene terraces—Riss and Würm glacial phases. These instances of late Pleistocene terraces, are common in the lower reaches of the Subarnarekha to the south of Ghatsila (23°34′N; 86°29′E; with 13 m depth of alluvium). A typological study with supporting evidence from the functional role of tectonics, both in its macro- and micro-dimensions also indicate to a polycyclic evolution of landscape of the parts of the Peninsula and Extra Peninsula Region in Eastern India. The superposition of the Subarnarekha drainage across the central ridges of Dalma lava area (22°56′N, 86°10′E, 300 m a.s.l.) has occurred on the adjacent quartzose-phyllite-schist level e.g. the Panch Pargana Plain.

The Tista basin with antecedent character is an uplifted and tilted terrain which has been divided into several large, elongated sub-parallel tributary valleys. Being an anticlinal valley the Tista to the south of the Takdah-Peshok-Ridge (27°10′N, 88°15′E, 3000 m a.s.l.), finds its way along the axis of a transverse anticline, and as a whole lineation in the topography has been largely controlled by structural and tectonic elements. The occurrences of transverse drainage throughout the hill or upland areas have provided channels of a wide range of drainage order within these basins.

Thus, the Tista and the Subarnarekha basins being sustained by the Sikkim-Darjeeling Himalayas and the Chotanagpur Plateau are examples of critical fields in structural developments and land evolution; these are classic areas in the study of neotectonics. The entire play of physical forces with the evolution of the polycyclic landscape-complex is revealed in all its facets in the Tista and Subarnarekha basin areas.

ACKNOWLEDGEMENT

The author is grateful to Prof. K. Bagchi, Head of the Department of Geography, Calcutta University, Prof. Dr. A.K. Banerjee, Head of the Department of Geology, Presidency College and Prof. Dr. A.K. Saha, Department of Geology, Presidency College, Calcutta for their valuable suggestions during the period of preparation of this paper.

TABLE
*Morphometric*

| Morphological Units | Absolute Relief (Hundred metres) | Relative Relief | Drainage Texture |
|---|---|---|---|
| 1 | 2 | 3 | 4 |
| I. Eastern Himalayan tributary valleys—(The Tista-Sikkim and Darjeeling Himalayas). | | | |
| A. The Great Rangit—Rangit Valley— | | | |
| (i) Great Rangit Valley (left bank tributary to the Tista) | 20-30 | High | Medium |
| (ii) Rangit Valley (right bank tributary to Rangit). | 21-32 | High | Medium |
| (iii) Rangbang Valley (right bank sub-tributary) | 25-42 | V. High | Fine |
| (iv) Rathakhola (right bank sub-tributary). | 29-45 | High | V. Fine |
| (v) Jhepijhora (right bank sub-tributary). | 28-45 | High | V. Fine |
| B. The Rangitkhola Valley : | | | |
| (vi) Rangitkhola Valley (left bank tributary to the Tista). | 25-40 | High | Fine |
| (vii) The Rangpo Chu Valley (left bank tributary to the Tista). | 26-45 | V. High | V. Fine |
| (viii) Rishi Chu-Rangbang Chu (left bank tributary to the Tista). | 27-48 | V. High | V. Fine |
| (ix) Gitkhola (left bank tributary to the Tista). | 27-50 | V. High | V. Fine |
| C. Loner short tributary Valleys :— | | | |
| (x) Rambi-Rayeng Valley (right bank tributary). | 6-10 | Mod. High | Medium |
| (xi) Kali Kohla—Sivok Valley (right bank tributary). | 4-8 | Moderate | Medium |
| (xii) Lish-Gish Valley (left bank tributary). | 4-10 | Low | Coarse |
| (xiii) Southern low rolling plains. | 2 and below | E. Low | V. Coarse |

5.1
*Attributes*

| Dissection Index | Average Slope | Roughness Index | Ruggedness No. | Morphogenetic outline of the terraces and flood plains |
|---|---|---|---|---|
| 5 | 6 | 7 | 8 | 9 |
| Moderate | Mod. Steep | High | High | 1. VM. Valley and meander terraces – a |
| Moderate | Steep | High | High | largely erosional flood plain which includes |
| High | Steep | High | High | the intramontane section of the Tista in |
| V. High | E. Steep | V. High | V. High | the Sikkim Darjeeling Himalayas (VM) VM |
| V. High | V. Steep | V. High | V. High | Characterizes the morphological units of IA, Bank major division of C from (X) |
| Moderate | Steep | High | High | to (XII) along with parts of II D to G. |
| High | Steep | High | V. High | 2. F A Flood Plain of accu- |
| V. High | V. Steep | V. High | V. High | mulation – including the alluvial fan of the |
| V. High | Steep | V. High | V. High | Tista in the piedmont segment and adjoining low rolling plains in Darjeeling and Jalpai- |
| Moderate | Mod. Steep | Moderate | Mod. | guri Divisions. FA is a typical feature of the sub-unit C (XIII). |
| Moderate | Mod. Steep | Moderate | Mod. | 3. REAF Rhythmical erosional- |
| Moderate Low | Moderate Gentle | Moderate Low | Mod. Low | accumulational flood plain associated |

*Table 15.2*

| 1 | 2 | 3 | 4 |
|---|---|---|---|
| II. Dividing Ranges of Eastern Himalaya— | | | |
| D. The Tista Tammur (Sapt Kosi) Divide— | | | |
| (xiv) Western Divide—Singalila Range | 30-60 | High | Fine |
| E. The Tista Mahananda Divide— | | | |
| (xv) Southern Divide-Darjeeling Range (Senchall hill) | 10-30 | High | Fine |
| F. The Tista-Rangit Divide— | | | |
| (xvi) Takdah-Peshok Ridge | 10-30 | V. High | Medium |
| G. The Tista Chumbi Divide— | | | |
| (xvii) Eastern Divide rugged-Dongkya Range | 30-50 | V. High | V. Fine |
| H. The Tista-Jaldhaka Divide— | | | |
| (xviii) South Eastern Divide-Paro-Darjeeling | 10-30 | High | V. Fine |

contd.

| 5 | 6 | 7 | 8 | 9 |
|---|---|---|---|---|
| | | | | mainly with the Tista section in part of the southern foothills and great North Bengal Plain, Darjeeling and Jalpaiguri Divisions. |
| Moderate | Mod. Steep | High | V. High | It is found largely in C (XII) and partly in I morphological unit. |
| Moderate | V. Steep | High | High | Here the alluvial fans and flood plains with meander terraces both of bedrock and |
| Moderate | V. Steep | V. High | V. High | alluvial types—are found as interwoven in spatial and temporal dimensions. |
| V. High | E. Steep | V. High | V. High | |
| High | V. Steep | High | High | |

## *APPENDIX I*

*Neotectonics* mean tectonic activities involved in moulding the earth's surface since Pleistocene time. Tectonic relates to the processes which tend to build up the various features of the earth's crust, and brings about complications to the 'base levels' specially in terms of interruptions' in the progress of cycles of erosion in the evolution of landscape.

*Ruggedness Number* is defined as the product of relative relief and drainage frequency for the basin concerned as a whole. This variable shows a combined expression of relief texture and slope. The number is directly proportional to the relative relief i.e. it increases with corresponding increase in relative relief.

*Drainage Frequency* is obtained as the total number of stream channels within a basin by the basin area.

*Dissection Index* is a ratio between a relative relief and absolute altitude. It ranges between 0.01 and 0.84 in the present study on geomorphological problems of two selected basins in eastern India.

*Slope-Average* (*Wentworth 1930*) C.K. Wentworth, (1930) has devised a method to obtain average slopes for a whole region with the following formula :

$$\frac{\text{Average No. of Contour Crossings per mile} \times \text{Contour interval}}{\text{3361 (a constant)}}$$

This method, one of rapid sampling, provides a considerable measure of accuracy, and it is expressed in degrees, minutes or seconds etc. unit. The average slope of the studied areas varies from 0° (zero degree) to more than 40° (forty degrees) as obtained by this method (Wentworth 1930).

*Roughness Index* has been adopted by J.C. Hoock (1955) in connection with 'Arid zone Research' UNESCO. It is defined as equal to $\frac{N \times M/4}{4} \times 10$ where N=total number of intersections of contour lines with two sets of perpendicular grids, M=distance in miles between grid lines 4 and 10 are constants. The index shows a variation from zero to about 170 specially in the rugged terrain of the upper stretches of the Tista basin while the maximum attainment of such index does not exceed 80 in the Subarnarekha basin. The studied area is covered by the topo-sheets 73E, F, I, J, for the Subarnarekha and 78 A, B etc. series for the Tista basin (Survey of India). Chayes and Chayes-Kruskal methods are applied for testing and interpreting the correlations between the modal variables in terrace-sequences. Ordinarily correlation coefficients of modal variables from different suites of terraces are calculated and tested for significance using these methods adopted by Chayes, F. (1960), Chayes F. and W. Kruskal, (1966). They give an equation as given below for which the null value ($\rho$) can be calculated approximately and observed correlation coefficient (r) between original proportions can be tested against the null value etc.

$$\rho_{12} \cong (p_1 - p_2\,\sigma_t^2 - p_i\,\sigma_2^2 - p_2\,\sigma_1^2)/\sqrt{[\{p_1^2\,\sigma_t^2 + (1-2p_1)\,\sigma_1^2\}\{p_2^2\,\sigma_t^2 + (1-2\,p_2)\,\sigma_2^2\}]} \quad \ldots(1);$$

and the open variance $\sigma_i^2$ for the $i^{th}$ variable can be calculated from the observed means and variances with the help of another equation No. 2,

$$\sigma_i^2 \cong [\tau^2/(1-2\,p_i)]\,\{s_i^2 - p_i^2\,(\Sigma[s_j^2/(1-2\,p_j)])/(1+\Sigma[p_j^2/(1-2\,p_j)])\} \quad \ldots(2)$$

Where $p_1$, $p_2$, $\sigma_1^2$, and $\sigma_2^2$ are respectively the mean proportions and the variances of the variables 1 and 2 in the corresponding uncorrelated open array X which on closure, yields a new array Y etc.

*Kolmogorov-Smirnov Statistic (Test)*: The Kolmogorov-Smirnov Statistic tests the hypothesis that two sample—cumulative frequency distribution were drawn from the population having the same distribution. For this purpose the maximum deviation $d_n$ is measured directly from the graph between two sample cumulative frequency distributions (Miller and Kahn, 1962, pp 464-470; Dixon and Massey, 1957, Table A-21, p 450). This test has the advantage of being non-parametric i.e. no assumptions need be made about the form of the population frequency distribution rather than only the mean values as in the case of 't' test (Fig. No. 15.4).

*'t' Tests* : A series of 't' tests amongst the sectors for each parameter (a, b, c, d, etc) can be found out :

$$t=\frac{\bar{X}_1-\bar{X}_2}{sp\sqrt{\frac{1}{N_1}+\frac{1}{N_2}}},$$

and $s_p{}^2$=pooled mean sq. estimate ($\sigma^2$) (which is expressed as

$$s_p{}^2=\frac{(N_1-1)s_1{}^2+(N_2-1)s_2{}^2}{N_1+N_2-2},$$

where $s_1$ and $s_2$ are the st. dev. of sector 1 and 2, $\bar{X}_1$ and $\bar{X}_2$ are mean for sector 1 and 2 (as the sectors concerned) respectively.

It should be noted that there is another 't' Test which refers the index and the degree of freedom as obtained from the 't' table or graph (Student's 't' Test). It tests the relationship between the difference of the mean and standard error of the difference relating to the variables e.g. the roughness index, drainage ruggedness number, stream lengths of various order drainage density, average slopes length and height etc. between two groups of the basins.

Percentage of total sum of squares accounted for by the computed curve ($Y_c$) is calculated by the formula :

$$\frac{\Sigma Y_c{}^2-\frac{(\Sigma Y_c)^2}{N}}{\Sigma Y^2-\frac{(\Sigma Y)^2}{N}}\times 100$$

The values of percentage of total sum of squares are obtained as 88.36%, 71.42%, 97.97% etc. for the areas drained by the Subarnarekha and its tributary rivers Kharkai, Sanjai, etc. respectively.

*The Exponential Curve* : It is also known as symptotic growth curve (modified). The exponential curve, $Y_c=ab^x$, involves a constant ratio of change and, therefore, a constant ratio of change in the amount of change. The straight line $Y_c=a+b^x$, on the other hand, describes a constant amount of increase or decrease which has also been used in the present paper. The equation of the modified exponential is $Y_c=ab^x+k$, where k is the asymptote as the curve also approaches an upper limit. Here, the *modified exponential curve* particularly of the 'form' when a is positive and b is greater than one (Croxton, Cowden and Klein 1969) has been applied fruitfully by the author (Fig. No. 15.2). The values of $Y_c$ are calculated as $\log Y_c = \log a + x \log b$, when a and b are constants and $Y_c$=computed y.

*Harmonic Analysis* : This method has been acclaimed as a powerful tool with whose aid 'noteworthy and unexpected features' of the temporal variation in a periodic phenomenon of the space-time characteristic etc. may be successfully worked out. It is capable of being used as an alternative to curve smoothening in a time series analysis. It also helps to solve the problem of exact and objective description of the natural behaviour both in temporal and spatial dimensions. It is based on Fourier series and consequently provides the advantage of being used in the exploration of both continuous and discontinuous variables like the annual march of rainfall over parts of Eastern India. (L.H. Horn, and R.A. Brysons 1960). This tool has helped much in the present attempt of manifestation of the peculiar features of the rainfall curves being marked by an annual trend.

**Morphometric Attributes**

(Scales of the measurements adopted in Table 15.1)

| *(a) Relative Relief (R)* | *Range Metres (Feet) Approx* |
|---|---|
| Extremel low (REL) | 0-30 (0-100) |
| Moderately low (RML) | 31-75 (101-250) |
| Low (RL) | 76-152 (251-500) |
| Moderate (RM) | 153-305 (501-1000) |
| Moderately high (RMH) | 306-458 (1001-1500) |
| High (RH) | 459-762 (1501-2500) |
| Very high (RVH) | above 762 (above 2500) |

| *(b) Drainage Texture [(Density (T)]* | *Km/Sq. Km Mile/Sq. miles* |
|---|---|
| Very Coarse (TVC) | 0-3.2 (0-2.0) |
| Coarse (TC) | 3.3-6.5 (2.1-4.0) |
| Medium (TM) | 6.6-13.0 (4.1-8.0) |
| Fine (TF) | 13.1-14.0 (8.1-9.0) |
| Very Fine (TVF) | above 14.0 (above 9.0) |

| (*c*) *Slope- average* (*S*) | *Degree-minutes* |
|---|---|
| Level (SL) | 0-2°0′ |
| Gentle (SG) | 2°30′-5°0′ |
| Moderate (SM) | 5°30′-10°0′ |
| Moderately Steep (SMS) | 10°30′-20°0′ |
| Steep (Ss) | 20°30′-30°0′ |
| Very Steep (SVS) | 30°30′-40°0′ |
| Escape Steep (SES) | above 40° |

| (*d*) *Dissection Index* (*D*) | *Range* |
|---|---|
| Low (DL) | 0-0.1 |
| Moderate (DM) | .11-.30 |
| Moderately High (DMH) | .31-.50 |
| High (DH) | .51-.70 |
| Very high (DVH) | above .70 |

| (*e*) *Roughness Index* (*RI*) | *Range* |
|---|---|
| Low (RIL) | 0-20 |
| Moderate (RIM) | 21-60 |
| Moderately High (RIMH) | 61-100 |
| High (RIH) | 101-140 |
| Very High (RIVH) | above 140 |

| *Ruggedness No.* (*RG*) | *Range* |
|---|---|
| Ext. low | 0-100 |
| Mod. low | 101-250 |
| Low | 251-500 |
| Mod | 501-1000 |
| Mod. high | 1001-2000 |
| High | 2001-3000 |
| Very high | above 3000 |

## REFERENCES

1. M. Pecsi, The Development of Foot Hill-slopes of the Hungarian Uplands, Studia Geomorphologica Carpatho Balcanica, Krakou, *Geog, Res. Inst. Hung. Acad. Sc.*, Budapest Vol. X, 1976, pp 15-24.
2. S.C. Mukhopadhyay, "Studies on the Terrace-Sequences and Neotectonics of the Tista-Subarnarekha Basin Areas," Eastern India. Univ. of Burdwan, Vol. I, 1978, pp 32-46.

S.C. Mukhopadhyay, "Geomorphology of the Subarnarekha Basin," Burdwan, Univ. Burdwan, 1978, pp 40-52, 56-65.

S.C. Mukhopadhyay, "Morphological Analysis of Land-forms of the Tista Basin Eastern India." *Geog. Rev. of India*, Calcutta (Press), 1978C.

S C. Mukhopadhyay, "Studies on Intra-Valley Variation in Terrace-succession, Dissection Roughness Index and Slope Angles within the Tista-Jaldhaka Basins, Eastern Himalayas." Occ. pap. Inst. Geog. Dept. of Geog., Burdwan Univ., Burdwan I, 1978C, pp 10.

3. C.K. Wentworth, A Simplified Method for Determining the Average Slope of a Land Surface. Amer. *Journ. Sci.*, 20, 1930, pp 184-94.
4. J.C. Hook, Roughness "Index of Terrain as Referred in Geomorphology, Chap. 5, "*Guide Book to Research Data for Arid Zone Development*, Arid Zone Research, UNESCO, 1955.

   M. Pecsi, The Dynamics of Quaternary Slope Evolution and its Geomcrphological Representation, *Symp. Int. de Geomorphologic* Liege-Louvain, Budapest, 8-10, 1967, pp 187-189.
5. R.L. Miller and J.S. Kahn, *Statistical Analysis in the Geological Sciences*, Wiley, New York, 1962, pp 464-470.
6. F. Chayes and W. Kruskal, An Approximate Statistical Test for Correlations Between Proportions : *Jour. Geology*, V 74. No. 5. pt. 2, 1966, p 693-700.
7. Budel, Julius, The System of Climatogenetic Geomorphology, 21st. Session, *Int. Geog. Cong. 1968, Proc. S-25, Erosion Surface*, pp 29-32.
8. S.C. Mukhopadhyay, "Erosion Surfaces in one Sanjan River Valley, (The Subarnarekha System), Bihar, 21st Session International Geog. Cong. Proc. 1970, pp 56-60.

   S.C. Mukhopadhyay, "River Terraces of the Subarnarekha Basin," Geog. Rev. of India, Calcutta, Vol. 35, No. 2, 1973, pp 153-169.

   S.C. Mukhopadhyay, "Geomorphology of a Part of Lower Kharkai Basin," *Geog. Rev. of India*, Calcutta, Vol. 39, No. 3, 1977, pp 249-57.
9. S.K. Acharyya, Geological Framework of the Himalayan Foot Hills of N. Bengal, West Bengal, Geog. Inst., Presidency College, 1970, pp 8-14.

   E. Ahmad, Bihar—A Physical, Economic and Regional Geography, Ranchi University, Ranchi, 1965, p 28-36, pp 310-312.

   K. Bagchi, Kalimpong—Its Land and People Cal. Geog. Rev., Calcutta, Vol 2, 1940, p10.

   K. Bagchi, 1944, The Ganges Delta, University of Calcutta, p 157.
10. G.U. Yule and M.G. Kendal, *An Introduction to Statistical Analysis* (London), 1980, p 100.
11. R.L. Miller and J.S. Kahn, op. cit.
12. F. Chayes, On Correlation ween Variables of Constant *Sum. Jour. Geophys, Res.* V. 65, No. 12, 1960, p 4185-4193.
13. R.L. Singh, "Morphometric Analysis of Terrain," Presidential Address, Geology and Geography Section, Ind. Sc. Cong. Asson. 54 Session, Hyderabad, 1967, pp 1-3, 6, 10, 14-19.

A.N. Strahler, "Statistical Analysis in Geomorphic Research," *Jonrn. Geol.* 62, 1954, pp 1-25.

14. M. Pecsi, Evolution of the Flood Plain Levels of the Danube and their Principal Bearings on the Geography of Agriculture. *Academic Scientiarum Hungarica-Institutum Geographicum*, Budapest, 1968, pp 215-219.

M. Pecsi, Factors Affecting Slope Evolution and Formation of Slope Sediments in Hungary, *Argumenat Geographica*, Band 12. Bonn, p 193-198.

# 16

# The Concept of the Continental Drift

—*Ram Bahadur Mandal*

THE CONTINENTS of today are not permanent. They have drifted about 200 million years ago, i.e., in the later part of the Carboniferous period from the original super continent of Pangaea surrounded by the ocean of Panthalasa which existed during the Archaean period. In Pangaea all continents have joined together of which the southern part was called Gondwana land comprising the continents of Australia, Peninsular India, Africa, Antarctica and northern part called Angara land comprising the continents of the Chinese massif, Russian massif and Canadian shield, and the Tethys sea in the middle (Fig. 16.1)

*The Continental Drift*

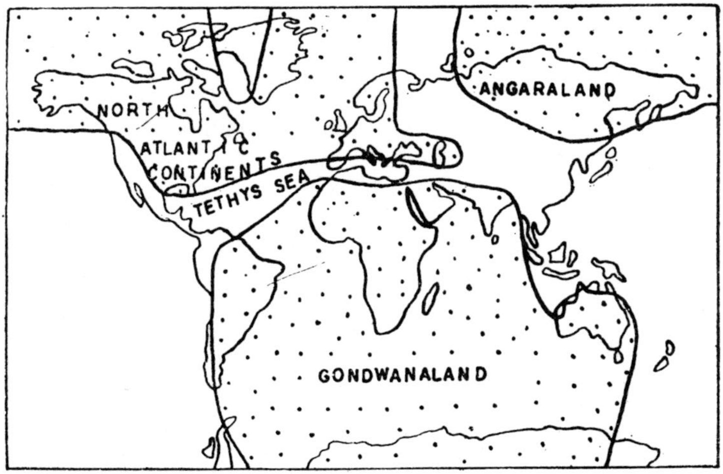


ANCIENT SUPER CONTINENTS CONNECTED BY LAND BRIDGES

FIG. 16.1

**Development of the Concept**

This dynamic idea which analyses the changing state of continents, ocean basins, climate, mountains, etc., is known as the theory of continental drift. It was devised by S.F. Bacon (1620) who said it was accidental. Further R.P.F. Placet (1658) suggested that the world became separated following the flood. T. Lilienthal (1756) gave the idea of continental drift and A.V. Humboldt (1800) considered that Atlantic ocean was originally a huge river basin which with the drift of continents became an ocean.

Further, with technological advancement, geological, geometrical and structural similarities of continents have been shown by A. Snider-Pelligrini (1858). In 1908 B. Taylor suggested the idea of continental drift in order to explain the origin of mountains while H.B. Baker (1928) explained the similarities of mountain systems on either side of the Atlantic Ocean.

The idea of continental drift was finally developed by A. Wegener (1912). He came to this idea by way of solving the ancient climatic conditions of the world. This is because, there are such places on the surface of the earth where the distribution of present climatic conditions did not actually exist in the past.

Alexander du Toit[1] supported the view of Wegener who discovered a large number of similarities between the west coast of Africa and the Brazilian coast along with the works of Henno Martin[2] who had advanced geological evidences in favour of the idea of continental drift. Further, the idea of radioactive dating of coastal rocks and the Palaeomagnetic investigations of Holmes (1927)[3] provided advance level support for the drift theory. The work of L.C. King[4] in South Africa, S.W. Carey in Australia and S.K. Runcorn[5] of Britain presented evidences of the continental drift from their respective areas of study. Recently, the work of the Plate Tectonic Theory as advanced by Ronald Schiller (1971)[6] favour that the continents are great rafts of rock supported from below by layers of denser plastic rock and the earth's crust is divided into a number of thin lithospheric plates. The internal energy of the earth moves the plate horizontally by thermal convection current of radioactive heating in the earth's crust.

**Direction of Drift**

The equatorward drift brought Africa and Eurasia closer and the formation of the Alpine folding took place from the material accumulated in the Tethys sea, while all other movements have relatively taken place from the African table land. This collision gave rise to the Pyrenees, Apennines and Alps in Europe. On the other side, the Americas, Africa and Australia moved towards each other across the Pacific where the piled up materials of geosyncline accumulated on the shore to form the Andes and Rockies of the Americas, the Island Arcs of the Aleutians, Japan and other Archipelagos of the westernside of the Pacific Ocean. Today the remaining part of the Tethys is present in the form of Mediterranean sea. Australia, Antarctica and the Indian Peninsula separated from Africa and the Indian ocean came into being. When India separated from Africa and Antarctica, it moved 8800 kilometres in 180 million years ago and with the dashing of the Lorentian shield of the north, the Himalayas were formed from the accumulated sediments of the Tethys sea. The north-westward drift separated North and South America from Europe and Africa respectively, and as a consequence of which Atlantic ocean formed.

**Evidences of Continental Drift**

The theory of continental drift of Wegener is not merely an assumption, but is based on logical evidences of physical, mineral, geological, biological and tectonic derived from the works of geographers, geologists, climatologists, geophysicists, plant ecologists and zoogeographers.

PHYSICAL EVIDENCES

**Structural Similarities**

There is too much similarity in the rocks of Brazil, whole of Africa, Peninsular India and western Plateau of Australia which are of the same nature and pattern. All these indicate that previously they were joined together. Similarly, in the northern hemisphere the rocks of the Canadian shield, the plateau of Scandinavia and the Silurian shield are of the same nature. Thus it is quite clear that the stable blocks of the world were joined together in the past.

### Jig-Saw-Fit

If we were to join the east and west coast of the Atlantic Ocean, they would fit like a jig-saw puzzle. The closer fitting is only observed along the Atlantic coast. Near Brazil and the Gulf of Guinea, the difference in fitting is about 15°. This difference may be due to the newly formed coastal plains. But similarity of the coastline of the Atlantic ocean is not a proof of the Pangaea. If Australia is kept in the Arabian Sea that may be the best fit but no one has claimed that Australia has come out from the Arabian Sea.

### Similarities in Rock Deposits

(1) Alexander du Toit has made a study of the South American coast and more especially of South African geology, and recognized a striking series of correspondence in rock deposit, earth movement and igneous intrusion of the two sides. Both of them have experienced the same geological history during the Palaeozoic and early Mesozoic periods.

(2) The Caledonian mountain folds of northern Britain and the Hercynian system of Europe continue into the North America. Other resemblances noticed are similar deposits found in Spitsbergen, North Greenland and the occurrence of pre-Cambrian intrusive rocks in Labrador.

### MINERAL EVIDENCES

(1) Backeth deposits are found in South Africa and similarly in Brazil diamond area. They revealed that once upon a time these two areas were single landmass and later on they have drifted.

(2) Marl studied the coral-reef of England and other parts of Europe and came to the conclusion that the North Pole had moved several times in the past.

(3) Dorothy Hill studied the huge potash deposits of Germany and came to the same conclusion that Germany must once have had a warm climate.

GEOLOGICAL EVIDENCES

(1) The presence of red sandstone on the upper part of the Vindhyan sandstone presents an idea that this region must facilitate warm climate in the past and because of this fact the continents had drifted.

(2) Vine and Methues of England studied the rock samples of Charlsberg Ridge on both sides in the Indian Ocean and found that rocks of similar distance from the mid-oceanic-ridge were similar in age, structure and this shows that the ocean is spreading.

(3) Gold veins are found in Brazil, and alluvium gold is found in Ghana. In case these two areas are brought together it will be obvious that the Brazilian gold veins had been eroded away and brought to Ghana. This proves that previously they must have been joined together.

(4) One hundred and twenty years ago W.T. Blanford, H.F. Blanford and W. Theobald claimed that the deposition of the boulder bed in Talcher series of Orissa (India) was about 200 million years old. Only glaciers could carry such huge boulders of erratics. A similar bed was found on the banks of river Penganga near Chanda in Madhya Pradesh and below this boulder bed "striated pavements" were found which cannot be produced by any other agency except glaciers.

Further, the till deposits of the upper Palaeozoic Glaciation are found in the Salt Range of Indiana (U.S.A.) Schuchert also found the occurrence of similar till deposits in the north-western part of Africa and at Boston in U.S.A. According to Wegener the South Pole at that time lay somewhere near the present position of Durban or the mid southern side of Pangaea.

(5) The distance between Greenland and North America is decreasing at a rate of 20 cm per year. This has been surveyed many times. This shows that the continental drift is on its way.

(6) At present North America and Europe have cool climates but they produce huge amount of coal. Coal is a mineral which forms in warm tropical climates only. But the present distribution of coal in North America and Europe contradicts this theory. This is possible when the warm

climate had prevailed over these continents. The flora and fauna in the coal area also prove that there was a warm climate in the past.

(7) Plant fossils have been found in the coal bearing Damuda series of Bihar and Bengal. Quite a large number of varieties have been reported, such as Glossopteris, Gangamopteris and Neoggerathiopsis. Similarly, the same plant fossils were reported from South America, under which the rock formations of Talcher beds of India are found. This supports the theory of the continental drift.

(8) According to Holmes the earth contained radioactive minerals inside which after disintegration produce heat. It was felt that conduction alone could not bring this heat to the surface. Besides this convection currents must exist inside the earth which could be moving the continents as raft on Sima.

(9) The minute observation of magnetised rocks shows that rock sediments contained minute quantities of iron which tended to be aligned parallel to the magnetic axis of the earth at the moment of coming to rest. The lavas of the Deccan Trap were an excellent example as they contained considerable amount of iron and at the time of solidification the ferruginous grains also came to be aligned parallel to the magnetic field of the earth. This favours the idea that the "rocks of a particular age in any continent were systematically aligned in a particular direction, but rocks of different ages were invariably differently aligned."[7] This idea was tested in the Pyrenees and the Rockies on the *Palaeomagnetism* principle and the same was found to be valid that the continental drift had occurred in the past.

## TECTONIC EVIDENCES

### Mid-oceanic Ridges

In support of the drift theory *mid-oceanic ridges* (Fig. 16.2) have been reported as a continuous chain of 65,000 km all along the middle of the ocean parallel to the opposite coasts in the 1950s. According to Bullard, convection currents were ascending in the area of mid-oceanic-ridges and the same were descending in the submarine trenches bordering the land notably in the Japan trench, Minadanao trench, South Australian trench, Philippine trench and Canadian trench in the west, etc. This also suggests that through the cracks of mid-oceanic ridges convection currents oozed molten lava from inside the earth to create these oceanic ridges and with this process the ocean began to expand and the continents became further and further apart.

*The Expansion of Oceans and the Continental Drift*

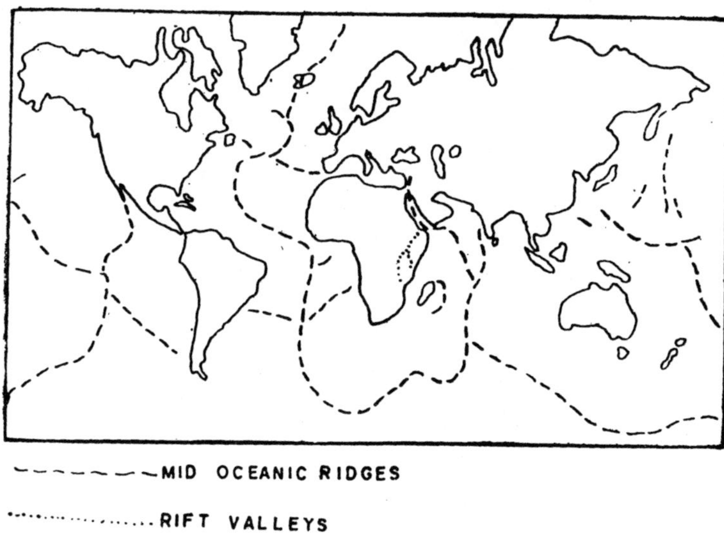


FIG. 16.2

**Global Plate Tectonics**

With the help of Glomars Challenger a team of ships, scientists are examining the sedimentary rocks of millions of years old from the bottom of the ocean through a nine kilometre long drill pipe. This favours the idea that continents are drifting on the "pliable mantle of inner earth". This is known as the concept of "*global plate tectonics*" and forwarded the view that 4600 million years ago oceans were open and closed like accordions and continents have been buffeted like hulks on the stormy sea. It answered all the questions regarding the formation of continents, oceans, mountains, islands, volcanoes, earthquakes and give reasons as to why fossils were found in the rocks of Himalayan mountains.

The following are the views advanced by this theory of continental drift :

(1) Instead of being solid, the earth's crust consists of ten separable major plates of varying size made of the rocks of 65 to 95 kilometres thick, which float on the hot viscous layers inside (Fig. 16.3).

*The Theory of Continental Drift*

(AFTER TARLING)
GLOBAL PLATE TECTONIC

FIG. 16.3

(2) The oceans and the continents of the earth rest on these plates, and some 200 million years ago they were welded together and later on the process of split began to form continents and islands like parcels on a moving conveyor belt.

(3) The edges of these plates are built with molten rock welling up from deep fissures at mid-oceanic ridges and slowly propelling the continents at various destinations and directions with a speed of 15 centimetres a year.

(4) Whenever plates jostle with one another for space phenomena like volcanic eruptions and earthquakes occur. In case the moving plate of land bearing granite meets an ocean bearing plate of dense and less-buoyant basalt, the former rides over the latter "like a titanic bulldozer, scraping up the sediment deposited on the

seafloor over millions of years, along with slices of crystal rock. The debris piled up along the edge of the land like a rumpled blanket to form mountain ranges. The ocean bearing plate, forced down at a steep angle into trenches under the land plate, melts from the heat of the friction forming deep underground pockets of white hot lava. The trapped lava is forced up through crevices, erupting on the surface as inland volcanoes. The collision, separation and shearing of the plates also create the seismic disturbances that we call earthquakes."[8]

## BIOLOGICAL EVIDENCES

(1) In the Scandinavian Islands a variety of birds travel in a westward direction every year, but ultimately they die after drowning in the Atlantic Ocean. This hereditary nature of birds proves that their forefathers used to go in that very direction when the land was united. But now the western portion (Canada) has separated away.

(2) The ideas of ice-age and land bridges suggest that there were ice ages during the Permo-Carboniferous and Pleistocene period (10,00,000 years ago) when the northern hemisphere was mostly covered with ice. During the Carboniferous period the whole of the northern hemisphere was warm except India. To explain this and the dissimilarities of plants in northern and southern hemisphere, the idea of a land bridge emerged as hypothetical super continent of "Gondwana land" which assumed this name after the ancient Gond Kingdom of Eastern India. Further *Glossopteris* was discovered in Antarctica from beneath the ice cape of the South Pole. This clearly shows that Antarctica was not covered with ice when coal beds were deposited in India, Australia, Africa and South America.

(3) The fossilised bones[9] of *Mesosaurus*, *Listosaurus* and *Denosaurus* had been found in South-West Africa, Argentina, Brazil and Antarctica. It was certain that they must have lived in marshes, feeding fish and it was also certain that they could not cross the sea. Thus the mystery is how the fossilized bones of those animals were found in different countries separated by the ocean. This proves that the continents have drifted.

## CONCLUSION

The idea of the continental drift is not new but only the evidence collected by various thinkers is truly new. Till now the credit has gone to Wegener who formulated this concept and showed the distribution of continents, oceans and mountains on the earth surface but the recent advancement of the idea of mid-oceanic ridges, palaeomagnetism, convectional current of radioactive minerals and the concept of Plate Tectonics are changing the idea fast.

## REFERENCES

1. A.L. du Toit, *Our Wandering Continents,* Oliver and Boyd ; 1937.
2. H. Martin, "The Hypothesis of continental drift in the light of recent advances of geological knowledge in Brazil and South-West Africa" *Transactions of the Geological Society of South Africa,* vol. 64, 1961, pp 1-47.
3. A. Holmes, *Principles of Physical Geology.* Nelson, 1965.
4. L.C. King, *The Morphology of the Earth*; Oliver and Boyd. 1967.
5. S.K. Runcorn (Ed.), Continental Drift, *International Geophysics Series,* Academic Press, Vol. 3, 1962.
6. R. Schiller, "Earth's Continents are Adrift." *Reader's Digest,* July, 1971.
7. F. Ahmad, "The Drifting Continents," *Science Today,* Vol. 1, No. II, 1967. pp 22-23.
8. R. Schiller, op. cit.
9. D.H. and M.P. Tarling, *Continental Drift, A Study of the Earth's Moving Surface* ; Penguin Books ; Suffolk, 1971.

# 17

# Palaeomagnetic Evidences and Continental Drift

*—Nageshwar Prasad, and Shukdeo Sahu*

WHATEVER may be the possible forces for continental drift (discussed either in the Theory of Thermal Cycles by Joly or Convection Current Theory of Holmes or Tectonic Plate Theory of Hess, Dietz, Vine, Mathews and many other recent views all leaning to the terrestrial forces, not celestial as invoked by Wegener), the displacement of the continents during geological time is widely accepted by geologists as a framework into which other geological findings are to be fitted. Among a number of supporting evidences like (i) palaeoclimate, (ii) mountain-building, (iii) structure of the ocean floors, (iv) volcanism, earthquakes and rift-valley formation, (v) faulting, (vi) juxtaposition of the continents, (vii) structural parallelism of the rocks of the continents, (viii) palaeomagnetism, etc. *Palaeomagnetism* is the evidence that has convinced most geologists that drift is the reality with which other apparent difficulties must be reconciled. This indicates that the continents have changed their positions relative to each other. Continental drift of some sort or another has taken place, resulting in a scattering of polar determinations for each geological period. Each continent indicates separate polar positions.

## PALAEOMAGNETISM

Palaeomagnetism is the determination of the magnetic poles and magnetic fields of the earth during geological past. This is regarded as the most scientific method by which the direction and amount of continental drift can be measured.

Every rock on the earth has a magnetism of its own. It was related to the magnetic field of periods when that rock was formed or solidified. It is stated that the main part of the earth's magnetic field is nearly equivalent to that of a powerful bar magnet (a dipole placed near the earth's centre and oriented along the axis joining the N and S magnetic poles). This geocentric axial dipole field accounts on an average for about 95% of the whole. Most of the balance is made up of feeble and irregular fields equivalent to a number of weak non-axial dipoles, not situated centrally. At present, the magnetic axis departs considerably from the axis of rotation. However, the departure is continuously changing and during the last hundred years the average divergence between the magnetic and geographical north poles has been very much smaller than the value at any given time. But to make it a reasonably sound assumption that the mean magnetic field is that of a geocentric dipole oriented along the axis of rotation, the period over which the average is taken should not be less than several thousand years (Holmes : 1966).

It is also stated that normally the coercive force and the intensity of magnetization decrease with temperature, the decrease being particularly rapid just below the curie point. Consequently, in the presence of the geomagnetic field, the lava becomes strongly magnetized as it cools below the curie point when its coercive force is low. On cooling to ordinary temperature the coercive force rises and subsequent changes in the direction of the geomagnetic field have no further influence on the magnetization (Runcorn : 1957).

It is notable that the direction of the permanent or stable magnetization of certain sedimentary and igneous rocks of many ages from various parts of the world, have now been determined. Most of the rocks studied have been well bedded red sandstones and basaltic lava (basaltic sills and dykes). These rocks often possess a high degree of magnetic stability, and have consistent directions of magnetization over considerable distances within one continent in any one geological period. Other rocks are known to be permanently magnetized, but have not yet been so extensively studied. Basaltic lavas are found to have a strong permanent magnetization, red sandstones a less strong magnetization. Except in Cainozoic times (the last 60 million years) these magnetizations are in different directions from the magnetization induced by the present geomagnetic field (Runcorn : 1957).

### BRIEF HISTORY

Apart from some earlier works in France, which were not systematically followed up probably because of the world Wars, the present repidly accelerating study of palaeomagnetism was begun about 1950, some 29 years ago, by P.M.S. Blackett, who soon attracted an enthusiastic team of co-workers; some of these, notably S.K. Runcorn have started teams of their own. Because of the obvious bearings of results on the geological problems of continental drift and the physical problems of the origin of the earth's magnetic field, the subject has rapidly spread throughout the scientific world, as one of the most fertile and actively developing branches of geophysics. The possibility that the magnetization of rocks could be used in the investigation of polar wandering and continental drift has long been recognized (Runcorn : 1957).

### INTERPRETATIONS

We may follow the above statement supposing that we find out the position of the pole during a particular geological era from the rocks in a particular continent. We take, for example, the Permian period. Now, the magnetic pole found for the Permian period, whether it is from the rocks of the continent of Europe or North America or some other land-masses should be the same pole. If on the other hand, for the same Permian time, we get different magnetic poles (as shown in Fig. 17.1), then it shows that since the Permian times the continents concerned have undergone relative movement by the amount suggested by the different poles of the same geological period.

In this way, the magnetic pole during the Mesozoic era was different from the present geographical pole. This shows that the magnetic pole in the past was far away from the present and this is what is known as the "Wandering of the Pole", *i.e.*, the pole in the Mesozoic was different from the present pole. Similarly the pole in the tertiary was different from the present pole. But there should be only one pole in a particular geological period, indicated by the ***direction of magnetization of the rock of the period***. Therefore, there should be

*Geomagnetic Poles*

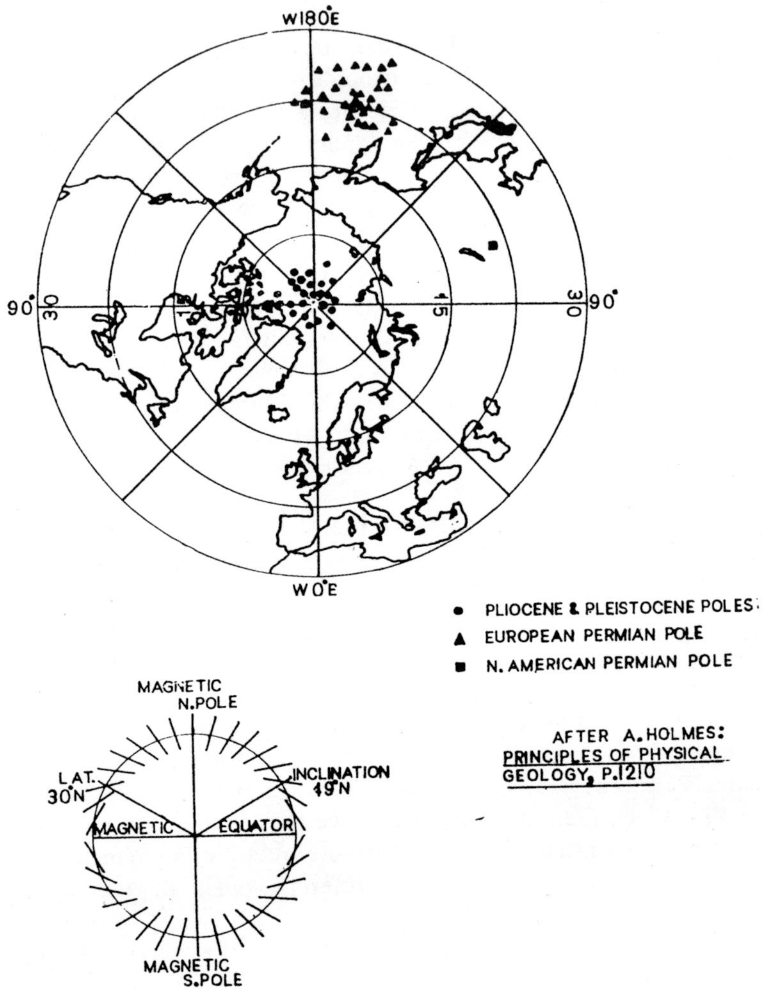


FIG. 17.1

only one magnetic pole related to the Mesozoic era, but from the diagram we see that the magnetic poles obtained from the palaeomagnetic data are numerous. This means that the rocks on which these magnetic poles are based have been moving or drifting along with the continents in which they occur since the Mesozoic time.

And then it becomes clear that if the magnetic poles obtained during a particular geological era, from the magnetic directions indicated by the rocks of that era in the different continents, do not coincide, but lie in different positions of the globe, it becomes an obvious conclusion that the rocks in question, *i.e.*, the continents themselves have moved away from the position which they occupied during that particular geological period.

### RESULTS OF PALAEOMAGNETIC SURVEYS

Palaeomagnetic observations have given the following results about the drifting of continents :

(1) The study of remnant magnetism in rock-masses shows not only that the continents have changed their positions relative to the geomagnetic poles progressively during geological time, but that they have also moved with respect to one another. As between Europe and North America, this amounts to a rotation of 25° or more corresponding to the opening of the North Atlantic Ocean. In other words, about 250 million years, the movement has taken place by a distance of 2000 km.

(2) For the South Atlantic, Creer (1958) postulates a clockwise rotation of South America relative to Africa of 22° (King : 1962). And results obtained in Australia, South America and South Africa lead one to suppose that a very considerable amount of continental drift occurred in the Southern Hemisphere in Mesozoic times.

(3) Australia has moved away from the South Pole since the Permian times by 60°, *i.e.*, 6000 km in 250 million years.

(4) Palaeomagnetic work on the Deccan plateau basalts shows that India drifted northward from 37°S to 13°S during the period of lava extrusion, *i.e.* from the late Cretaceous to the Lower Eocene. Since the end of the Cretaceous the northward drift has brought Bombay to its present latitude of 19° N, a displacement of 56°, *i.e.* about 5000 km in 70 million years. The movement involved not only a northward translation but also anticlockwise rotation of 25° or 30° which increased the angle of the triangular gap between Arabia and the West Coast of India (Holmes : 1966).

(5) Clegg concluded from studies of the British Trias that England must have rotated clockwise some 34° since the rocks were magnetized, which we assume provisionally was at their time of formation some 150 million years ago. Further, it is concluded that England was probably markedly nearer to the Equator than today. Creer suggests strongly that England must have been near the equator about 300 million years ago and possibly south of it still earlier (Blackett : 1961).

(6) From the palaeomagnetism of Cainozoic and Quarternary rocks from Europe, Hospers (1955) concluded that if Polar Wandering has taken place at all it has not exceeded 5°-10° since Eocene times.

(7) A palaeomagnetic survey of African Karoo sediments and lavas by Nairn (1960) also, after indicating a movement of the African continent, from polar to equatorial situation from late Carboniferous to late Triassic times, noted an anticlockwise rotation of nearly 90°.

Similarly various results have been put forward by the study of palaeomagnetism which renewed interest in the problem of Polar Wandering and continental drift.

In this way, from the preceding paragraphs, it becomes clear that the most important fact of these deductions is concerned with the use of rock magnetism to obtain experimental evidence for possible movements. During geological time the continents in relation to other geographical poles is highly controversial hypothesis of continental drift and Wandering of Poles.

## DISCUSSION

(1) Recently the effect of mechanical stress on the direction of remnant magnetization has been discussed. Graham has shown by laboratory experiments that the direction of magnetization of lavas and metamorphic rocks changes appreciably under uniaxial stresses of an order which might be produced by burial beneath some thousand feet of rock. Although he is not able to show that such effects have irreversible characteristics, it is probably true that, over long periods of time irreversible changes in the magnetization of rocks might occur in this way. In other words, the magnetized rocks by magnetic axis of one period might be affected by that of

another period. Similarly, if Polar Wandering has been taking place, it might be gradual. And then there might be gradual change in the magnetization of rocks during a long period.

(2) In case continents have drifted, it is clear that there might be no change in the axis of rotation or magnetic axis. On the other hand, if we agree to the hypothesis which is generally understood to mean that an outer shell of the earth, involving the crust and probably part of the mantle down to the low velocity zone has shifted as a whole relative to the axis of rotation, which remained almost fixed, relative to the stars, then there is, probably no explanation how the continents were departed (If they were not departed, it is difficult to explain how their rocks of magnetization vary in the same period).

(3) The only remaining explanation is continental drift, which is abundantly demonstrated by both geological and palaeomagnetic evidence. D. Van Hilten has put the present position very clearly. According to our present knowledge both processes may have played their part in the geological history of the earth. We are, however, completely at a loss about their rotations. Whether both processes occur contemporaneously or in some succession and is there any causal relation at all ? Palaeomagnetic evidence is not yet sufficiently accurate to produce the answers. In these circumstances, it seems wise to adopt Blackett's practical suggestion that the term continental drift should be provisionally allowed to include any Polar Wandering component that may eventually be recognised in such a way as to become separable from the total combined movements (Holmes : 1966).

(4) Accepting as England has rotated clockwise some 34° in the last 130 million years one is immediately faced with the question whether this rotation has been shared by all part of the continent of Europe. The discussion will have to await the results of systematic studies of continental rocks. Of special interest will be studies of European Trias, since these rocks seem generally to be rather specially suitable for magnetic measurements. Moreover, there have been suggestions by geologists based on purely geological evidence that relative rotations of part of Europe especially the Spanish peninsula and France have taken place. This hypothesis is now under

test by studies of the Spanish and French Trias.

(5) Atlantic margins of North and South America are probably characterized by a continuous filled trench. It might be argued that these trenches are rift on the back side of the continents, the space left when the continent moved away from the previously formed Atlantic ocean crust. However, these trenches are filled with enormous volumes of sediment, which could not be expected to have been deposited in any less than one or two hundred million years, so it seems unlikely that the filled trenches of the Atlantic can be used in such explanation. On the Pacific margin of South America and Central America, modern marginal trenches exist and if we were to explain marginal trenches as gaps left behind drifting continents, we would be forced to assume that Central and South Americas were drifting east rather than west (Runcorn : 1968).

(6) The second alternative that the extension has been distributed throughout the width of the ocean at any time, does not explain the apparent regional differences in age of the relief is the Atlantic. The youngest relief of the ocean appear to be along the axis of the oceans the relief becoming progressively older toward adjacent shares. This is proved by the recent information also. General Electric scientists recently dated four samples of rock from the ridge that has been dredged up by a Geological Survey of Canada Expedition. The rock from the centre of the valley separating the two ridges that form the mid-Atlantic Ridge is comparatively young, about 13 thousand years old. Seven kilometres away from the centre, the rock is 290 thousand years old. Rock samples from 16 and 60 kilometres away are 740 thousand and 8 million years old, respectively. The distribution of ages is consistent with the theory that the ocean floor is spreading out on both sides from the mid-Atlantic Ridge. The rock in the centre would be new magma forced up through the rift surface at the site of the ridge by convection currents deep within the earth. As the new rock appears, it thrusts the old rock farther and farther away. At the same time the spreading of floor slowly, drives the continents apart.

The spreading of the ocean is not a cause for immediate alarm. It is estimated that the outward movements is taking place at the rate of just 2.5 cm a year (Span : 1969).

The Indian Ocean floors are known to be moving outward as fast as two or three inches a year from the crack, constantly breaking and filling anew.

It seems most likely that the continents on either side of the Atlantic are moving apart by the successive addition of new crustal material along the axes of the ridge in the floor of the median rift valley, and that the continent and a large part of the crust of the Atlantic, Indian and South Pacific deep sea floors are moving away as units from the median line. Therefore, it is rather wrong to say that the continent of the Americas has moved westward from the continent of Africa and Eurasia as palaeomagnetic evidences have proved. It should also be proved that the continents of Eurasia and Africa have moved eastward.

## CONCLUSION

With these criticisms too it is widely agreed that palaeomagnetic evidences promise more precise and accurate date for the future. But this is no idle armchair evidence; it stands upon hard work in the field, upon literally thousands of facts garnered upon the rocks that can be seen, measured, sectioned, hammered and compared individually and in sequence, structurally and in age by any one qualified to do so (King : 1962).

## SUGGESTED READINGS

P.M.S. Blackett, Comparison of Ancient Climates with the Ancient Latitudes Deduced from Rock Magnetic Measurements, *Proc. of the Royal. Soc.* Vol. A 263, 1961, pp 1-30.

A. Holmes, *Principles of Physical Geology*, Thomas Nelson & Sons, London, 1966, pp 987-993, 1?04-1216.

L.C. King, *The Morphology of the Earth*, Oliver & Boyd, London, 1962, pp 26-34.

S.W. Mathews, "Science Explores the Monsoon Sea," *Span* Feruary Volume. 1969.

S.K. Runcorn, "Rock Magnetism" *Study of the Earth* (Ed. J.F. White) Prentice-Hall of India, New Delhi, 1968, pp 271-290.

C.R. Twidale, *Analysis of Landforms*, John Wiley & Sons, London 1976, pp 33-40.

# 18

# The Indian Ocean—Its Dynamic Evolution and Mineral Resources in Relation to Plate Tectonics

*—Dr. Anil Kumar*

ENOUGH RESEARCHES and studies have been conducted over Atlantic and Pacific oceans; but it is only recently that the Indian ocean has attracted the attention of marine geologists and geophysicists. Since the beginning of the Geophysical Year, information regarding the topography of the Indian ocean is being added every year to our knowledge. On the basis of these discoveries and even on the increasing amount of information regarding the complexity of its geology and topography, it is now possible to prepare a chronology of the evolution of the Indian Ocean under the conceptual framework of Global Plate Tectonics. The theory appears to explain many things and provides several practical dividends.

Several sets of ideas have been advanced in modern times, in the field of marine geology and geophysics such as continental drift, polar wandering, sea floor spreading and magnetic reversals, Benioff zones, transform faults, ocean topography and a better fit of continents. The concept which unified these separate ideas and concepts was inducted the subject as that of "Plate Tectonics," in 1968 by Sacks, Oliver and Sykes, "Tectonics" is a geological term pertaining to earth movements. The concept has created a revolution in earth science.

Plate Tectonic explains that the lithosphere is divided into seven major plates, each of which is bordered by oceanic ridges, subduction zones and/or transform faults. Plates are generated along ridge systems and destroyed in subduction zones. Continents, as they are lighter rest on denser oceanic crust and drift about in accordance with the movement and directions of the plates on an underlying plastic layer, the asthenosphere. The mantle convection produces drag effects along the base of the lithosphere and causes the plates to move.

**Magnetic Anomalies in Indian Ocean**

The rocks of the ocean floor carry a parmanent record of their own history. The magnetometer surveys conducted over the Atlantic, Pacific and Indian oceans, reveal that oceanic crusts are not uniformly magnetized. They consist of a series of parallel bands on the either side of the mid-oceanic ridges, which are strips representing additions to the present magnetic field (positive anomaly) alongside, which are subtracted from it (a negative anomaly). Vine and Mathews (1963) confirming the earlier views of Hess (1960), related these strip patterns to sea floor spreading where new materials

*India and the Indian Ocean : Evidence of Drift*

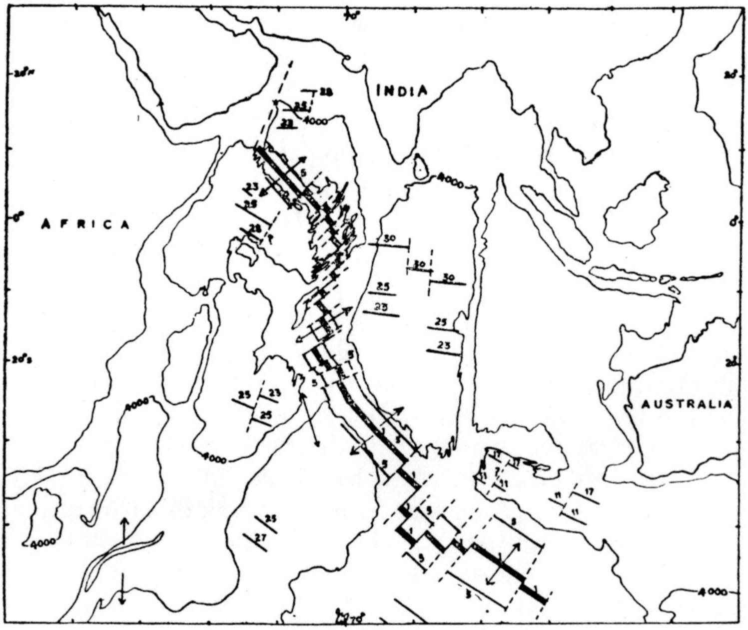


FIG. 18.1

are added to the crust along a ridge and new material will lock-in magnetic polarity at the time it crystallizes. There are about 33 magnetic strips floating on each side of the present ridges which have been distinguished. Reversals of the earth's magnetic field have been dated back about 80 million years.

By combining radioactive measurements of rocks with transform faults often offsets the pattern of magnetic anomalies. Much of the information regarding its geology comes from the works of Fisher Mckenzie and Sclater (1968). Their observations of magnetic anomalies of the Indian Ocean provide ample evidence of the drift of plate carrying India across the Indian Ocean. (Fig. 18.1)

The floor of the Indian Ocean reveals four large north-south ridges, two still actively spreading. The Ninety East Ridge and the Chagos-Laccadive Plateau are on the other hand, inactive ridges. The ocean floor also consists of numerous ancient continental fragments. Fisher's work shows that the plate boundary that runs between the Chagos-Laccadive Plateau and the Mascarene Plateau in the north-western part of the Indian Ocean is offset by numerous transform faults and none of them cross these two inactive ridges. According to Mckenzie and Sclater these faults are not the products of the original break. They tried to map the shape of the ridge axis that produced them using anomalies Nos. 23 and 29. The similarly numbered anomalies, south of Sri Lanka, are about 2500 km south of the east-west running anomaly in the Arabian Sea. At one time, these two distantly separated ridges had been joined by a huge transform fault, which divided the floor of the Indian Ocean into two large segments. Between 80 and 55 million years ago the segment carrying India travelled at the rate of about 16 cm per year. It is assumed that some 55 million years ago the relative motion south of Sri Lanka slowed from 16 cm per year to about 6 cm per year and probably stopped altogether in the Arabian Sea. The movement again resumed in the Arabian Sea, but its direction was no longer parallel to the ancient transform fault. The single north-south fault of the Chagos fracture zone evidently broke up into a number of ridge segments connected by short faults. A new sea floor was again generated and the boundary changed from a ridge oblique to the spreading direction to a series of ridge segments that are perpendicular to the direction of motion and are joined by transform faults. The direction of sea-floor spreading was rotated about 45 degrees from the original direction. (Fig. 18.2)

*The Indian Ocean Floor*

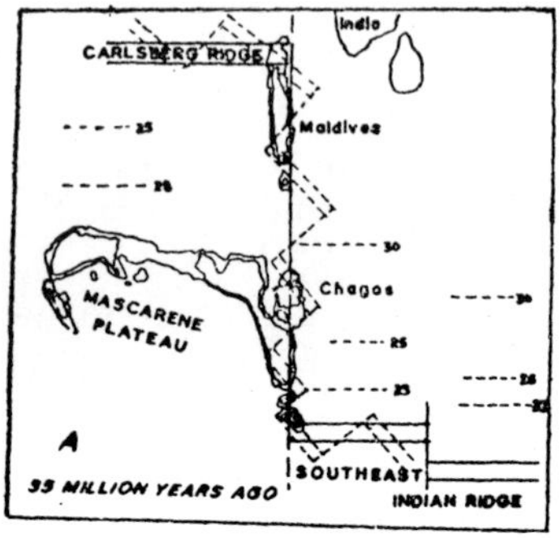


FIG. 18.2 (a)

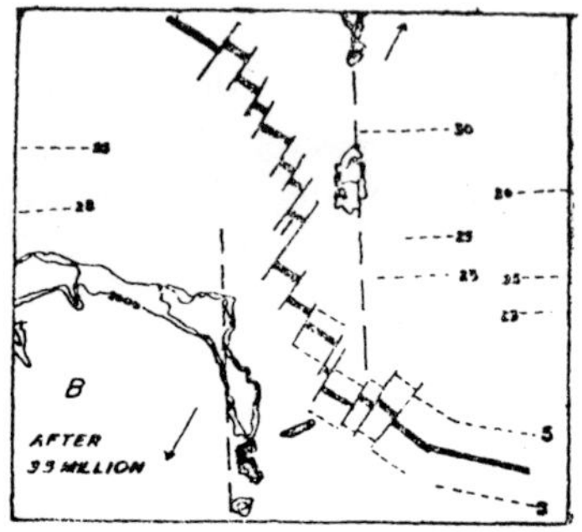


FIG. 18.2 (b)

Similarly anomaly nos. 21 to 32 formed between 75 and 55 million years ago, are found in the region south of Ceylon. Their matching anomalies on the other side of the ridge, particularly anomaly no. 32, are about 5,000 km apart near the Crozet island. It seems that after breaking away from Australia some 75 million years ago, India must have travelled this distance at the rate of 7.5 cm per year. In some cases the separation rate during this period was as high as 16 cm per year.

Magnetic lineations have been employed to reconstruct the relative position of the different plates underlying the Indian Ocean. Mckenzie and Sclater, employed palaeomagnetic observations from Australia to correct the position of the plates. As the motion of each plate can be described by a rotation, the closing of the Indian Ocean required the rotation of several plates that have moved at varying rates and in different directions.

**Evolution**

The palaeomagnetic evidence reveals that the continental mass of Gondwana land was at high southern latitudes during the late Palaeozoic (Greer, 1965) as indicated by the Permo-Carboniferous glaciation. The Gondwana land broke up in the early Mesozoic and drifted towards their present position. India drifted northward from middle latitudes crossing the equator. The study of the relative movement of Africa,

Australia, Antarctica and India from their ancient anchorage makes it possible to trace back to the Mesozoic the history of the opening up of the Indian Ocean.

The evolution of Indian Ocean can be reconstructed on the basis of various anomalies. The oldest anomaly is identified as anomaly no. 32, formed about 75 million years ago, when the plate carrying India started to move rapidly towards the north-east, after breaking away from the ancient primitive continent, Gondwana land. This was accompanied by the generation of new plate along the ridges. Long transform faults, east and west of these connected series of ridges provided unimpended movement to the Indian Plate. The ridge south of Ceylon which was spreading at that time, propelled India a little east of north. The northward slipage of the Indian plate was facilitated by an active transform fault now known as Ninety East Ridge. At that time the Asian plate was either stationary or was not moving as fast as India was approaching it. The Indian plate thus rammed and subsequently consumed in a trench spreading south of Asia.

*The Indian Plate*

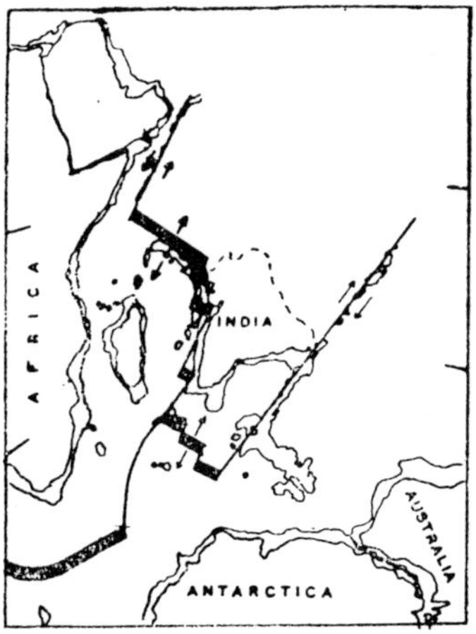


FIG. 18.3

The northward motion of Indian plate between two transform faults, namely Ninety East and the Owen fracture zone on the west, continued to the next 20 million years (Fig. 18.3). The relative movement between plates slowed down or stopped some 55 million years ago. As the new plates are now only generated along the ridge between India and Antarctica, the rate of separation between these two blocks thus slowed down from 16 cm to 6 cm per year. Australia which was still joined to Antarctica, become separated as a rapture developed between them, some 45 million years ago. At this period motion ceased on the Ninety East Plate boundary. The northern edge of the Indian Plate began pushing the layers of shallow water sediments that bordered the southern edge of Asia and the Himalayas were formed between these plates.

Indian Plate had travelled some 4,000 km northwards from its southern anchorage in 35 million years at rates that evidently varied from 16 cm to 7 cm or less per year. Soon afterwards all the plate boundaries in Indian Ocean became active again and direction of relative motion among several of the major plates changed quite drastically to produce the complicated sea floor geometry which we see today. Arabia has remained attached to Africa until now. Only 20 million years ago, Arabia split off from Africa and subsequently the Red Sea and the Gulf of Aden were formed (Fig. 18.4)

*The Formation of the Red Sea and the Gulf of Aden*

FIG. 18.4

The plotting of earthquakes recorded over recent years seems to cluster along the present ridge axis and on lines where the axis is offset by transform faults. The greatest number of earthquakes take place around the island arc between south-east Asia and Australia, where the Indian Ocean plate and the Pecific Ocean plate plunge into deep trenches and are consumed.

The plate tectonics based on a dynamic earth model, in which continents are constantly moving and ocean basins are opening and closing, have provided a useful conceptual framework towards a better understanding of global distribution of mineral deposits. It can be used to extend the limited knowledge of mineral resources of the ocean basin towards a realistic appraisal of their potential. Now it is possible to anticipate hidden mineral resources beforehand and this is surely saving a good deal of energy, time and cost of oceanic surveys.

**Plate Tectonics and Possibility of Mineral Resources**

The concept of plate tectonics provides clues in locating the economically important hydrothermal minerals along its three types of plate boundaries. Hydrothermal mineral deposits are formed by precipitated solutions along convergent and divergent plate boundaries. The most important types of hydrothermal deposits are sulphides in which various metals combined with sulphur precipitate from the hydrothermal solution. According to Frederick Sawkins (1972), sulphide deposits are located along present or former convergent plate boundaries where an oceanic lithospheric plate plunges under the margin of continents. Sampling from the floor of the Indian Ocean, particularly of the Red Sea and the Indian Ocean ridges, provides important information of the potential of metallic sulphide deposits at divergent plate boundaries. Taking two extreme examples from the Indian Ocean, the occurrence of minerals along divergent plate boundaries, can be examined.

The Red Sea provides an accessible natural laboratory for the study of mineral processes associated with the divergent plate boundaries. The richest submarine metallic sulphide deposits were found in three small basins along the centre of the Red Sea at a depth of about 2011.6 m below sea level. These minerals are disseminated in sediments that fill the basins to a thickness of 19.7 to 100.7 m. It is estimated that

10 m of sediment has a total dry weight of about 80 million tons, with average metal contents of 29% iron, 3.4% zinc, 1.3% copper, 0.1% lead, 0.005% silver and 0.00005% gold. The deposits are saturated with salty brines carrying the similar metals in solution. The salty brines are considered to be the hydrothermal solution from which the sulphide minerals are precipitated.

The divergent plate boundary of the Red Sea represents an early stage of the hydrothermal process while its advance stage is represented by the mid oceanic ridges. The sampling indicates that hydrothermal processes are actively concentrating metals from volcanic sources underlying mid-oceanic ridges. Sediments on active ridges are generally enriched in iron, manganese, copper, nickel, lead, chromium, cobalt, uranium and mercury with trace elements of vanadium, cadmium and bismuth. Though economic concentrations of minerals are universal, much higher concentrations exist locally. Under a continuous process of deposition, a mineral deposit, on a mid oceanic ridge would be expected to extend in a linear zone from the ridge across the ocean basin. The sea floor is supposed to have originated by spreading from the mid-oceanic ridge.

Samples taken from the Indian Ocean ridge are rich in metallic sulphides. The Deep Sea Drilling Project collecting samples at several sites, revealed small veins of pure copper in rocks. At the crest of the Ninety East Ridge, near the equator, veins of copper are found in volcanic rocks overlain by 438.9 m of sediment at a water depth of 2249.5 m.

Petroleum, which is a hydrocarbon derived from the remains of plants and animals, requires a natural reservoir to contain it and a trap to concentrate its fluid and gas constituents. Both convergent and divergent plate boundaries provide conditions favourable to the accumulation of petroleum. The convergent boundary provides a habitat favourable for accumulation of petroleum. The convergent boundary provides a habitat favourable for the accumulation of petroleum in small ocean basins and deep sea trenches marginal to continents. The divergent boundary on the other hand also creates conditions that favour the development and accumulation of oil extending from the continental shelf into deep ocean basin under the continental ridge. As the enclosed sea of the Indian Ocean widened and was created in response to the symmetric creation of new lithosphere by sea floor spreading from the mid-oceanic ridge, the organic matter and salt were buried under sediments. The Red Sea, the Gulf of Arabia, the Bay of Bengal provide examples of such basins. Petroleum found at Bombay High illustrate this point.

According to Rona (1973) petroleum may likewise be found in other ocean basins that have grown through the stage of a restricted sea by sea floor spreading. Such a possibility opens up immense opportunities for petroleum resources in the deep oceanic basins like the Bay of Bengal and the Arabian Sea.

The evolution of the Indian Ocean not only finds justification under the concept of plate tectonics but also explains various geophysical problems associated with the ocean floor and mountain buildings and earthquakes. The patterns of distribution in the Indian Ocean are emerging from the conceptual framework provided by the plate tectonics and this can be expected to accelerate the discovery of resources of the Indian ocean in near future.

## SUGGESTED READINGS

E., Bullard, "The Origin of the Ocean" *Scientific American*, Vol. 221, No, 3, 1969, pp 30-45.

K.M. Creer, "Palaeomagnetic Data from the Gondwanic Continents", in A Symposium on Continental Drift, *Phil Trans R. Socl*, 258A. 1965, pp 27-40.

E.T. Degens and D.A. Ress, : "The Red Sea Hot Brines"; *Scientific American*, 1970, Vol. 217, No. 1.

X. Le Pichon "Sea Floor Spreading and Continental Drift". *J. Geophys, Res.*, Vol. 73, 1968, pp 3661-3697.

D.P. Mckenzie and J.G. Sclater, "Evolution of the Indian Ocean", *Scientific American*, Vol. 228, No. 5, 1973, pp 62-74.

J.L. Miro, 'Minerals on the Ocean Floor, *Scientific American*. Vol. 203, No. 6, 1960, pp 64-72.

E. Orowan, 'The Origin of the Oceanic Ridges" *Scientific American* Vol. 221, No. 5, 1969, pp 102-119.

P.A., Rona, "Exploration Methods for the Continental Shelf: Geology, Geophysics, Geochemistry; *Nat. Oceanic and Atmospheric Adm. Tech. Report* ERL 238-AOML 8, U.S. Govt. Print Office., 1972.

——"Plate Tectonics and Mineral Resources" *Scientific American*, Vol. 229, 1973, pp 86-95.

F. Sawkins, "Sulfide Ore Deposits in Relation to Plate Tectonics" *Jou. of Geol.* Vol. 80, 1972, pp 377-397.

J.T. Wilson, "A Revolution in Earth Science" *Geotimes*, Vol. 13, No. 10, 1968, pp 10-16.

# 19

# Evolution of the Himalayan Drainage

—*Ram Bahadur Mandal*

MOST OF the rivers of northern India, originated from the Himalaya's as well as from those of the peninsula owing to the mountain building movement of the late Tertiary age. Thus the evolution of Himalayan drainage is a recent development and differs a lot in features and functions from that of peninsular India.

Most of the Himalayan rivers are older than the mountains they traverse. This is because due to the slow process of mountain formation by the folding-contortion, and upheaval of the rock leads the old rivers follow much to their respective channels, although working at an accelerated rate by receiving great stimulus from the uplift of the region near their source. This momentum has been acquired by the upheaval and this helped in eroding their channels on higher speed. Hence the elevation of mountains and the erosion of valleys proceeded *pari passu* and these two processes keeping pace with one another up to the end and ultimately a mountain chain emerged with completely developed valley-systems intersecting at deep transverse gorges. Such linear deep and precipitous gorges of the Himalayas cutting at right angles the line of the highest elevations, are the most characteristic features of the landscape, and are the results of the evolution of inconsequent drainage of this region. As such the Himalayan drainage is known as an antecedent drainage which means a system of drainage in which channels of the flow were in existence before the present features of the landscape were formed.

The whole area forming the tertiary mountain chain north of India including the Tibetan plateau and the Kunlun ranges in the North formed part of an ocean known as the Tethys sea (Fig. 19.1). This sea existed in the Cambrian period and received deposits of that time partly contributed by rivers of the Peninsular India which had a northerly direction.

*Past and Present Landscape of the Himalayan Area*

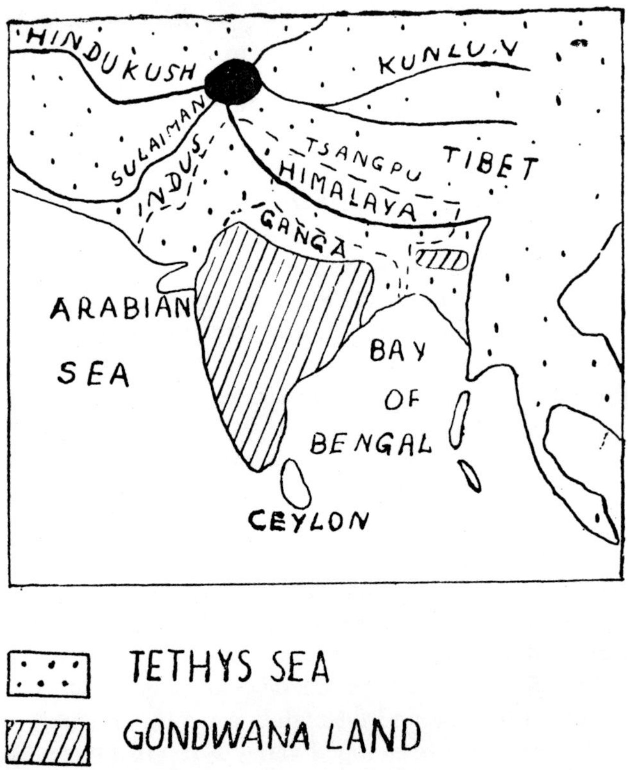


FIG. 19.1

The Tethys sea was extended up to the present Sinkiang basin and Thianshan, etc. in the north which formed a mountain chain during the Hercynian movement. The southern boundary is fixed by the peninsular land mass of the Gondwana land. The Tethys remained a sea and a basin of deposition of silt till the Upper Eocene period when the first Himalayan upheaval took place and part of the Tethyian geosyncline was uplifted. This is the sign of the beginning of

Himalayan drainage. The part of Tethys sea was raised into a land mass with a median mass of high Tibetan Plateau into the centres and two bordering ranges on either side i.e. the Kunlun and the great Himalayas, on the southern edge of the median mass or a "Geanticline" that the system of drainage started flowing south towards the fore-deep which had started forming since the first Himalayan upheaval for which the resistance was offered by the Peninsular land mass. The formation of east-west ranges created east-west valleys, the rivers flowed these valleys as indicated by the upper courses of the Indus, Sutlej and the Brahmaputra towards the southern flow. At this stage whole Tethys was not so enormously raised as to be a fully land surface. The patches of sea existed particularly on the marginal areas and the drainage lines were not properly developed.

The second Himalayan upheaval of the Mid-Miocene increased the altitude of the median mass and the bordering ranges. It also raised the remnant of the sea into land. A high relief and vast land surface meant greater and more invigorated drainage. But along with these general changes the bordering depression of the sea which had been fully sedimented was raised as the lesser Himalayan ranges. The major Himalayan streams e.g. the Indus, the Sutlej and the Brahmaputra which came into existence after the first Himalayan upheaval maintained themselves across on the rising lesser ranges and they are antecedent drainage.

The Sutlej, Indus, Kosi and other streams that have their courses in the interior valley of the Himalaya cut out deep gorges across the outer ranges also. As such all those rivers are cited as noteworthy example of antecedent drainage. As for example the Indus valley in Gilgit flows through a narrow defile, between enormous precipices, nearly 20,000′ in altitude while the dead of the valley is only 3000′ above mean sea level at Hyderabad near Indus delta. This elevation marks the successive levels of river bed. The example of similar gorges are numerous as for example those of the Sutlej, Gandak, Kosi Alknanda, etc., which have deep defiles ranging from 6000′ to 12000′ in depth having just 6 to 18 miles in width between the summits of the mountains.

The previous drainage lines of the older mountain belt

were entirely competent to cope with the new uplift. The interior valleys of the older mountains hold vast gravel accumulations, which are eroded to form terraces. This is the supporting evidence for the antecedent origin of rivers across the southern ranges. The lesser Himalayan ranges themselves gave rise to another set of consequent drainage lines which were emptying themselves into the fore-deep which formed along the whole length of the Himalayas on either side of the south during the Mid-Miocene upheaval. Now it is difficult to say whether this drainage reached up to the Arabian Sea through a single river (Indo-Braham or Siwalik) or through the inter-connected vast lakes and lagoons into both the Bay of Bengal and the Arabian Sea are the questions which are delineated here.

With the third upheaval in the Plistocene period the Siwalik fore-deep was folded and streams rising near the edge of the median mass across the rising Siwalik as antecedent streams. The last movement developed the final shape to the Indo-Gangetic trough or fore-deep which extended from the southern edge of the Siwalik to the northern edge of the Vindhyan Ranges and this has been jointly filled by the Himalayan streams from north and peninsular streams from south. The uplift of the Siwalik gave rise to the last sets of small consequent streams flowing from the courses of the Siwalik range and discharging their water into the larger and older streams.

It seems that the major portion of the Himalayan drainage is antecedent and only a part of it appears to be consequent formed at the time of that final Himalayan uplift.

It is not clear that the drainage from the extra-peninsular mountain wall collected into the Indo-Gengetic fore-deep. Although it is a problem how this drainage discharged water into the sea after the emergence of the Himalayas and the fore-deep as a consequence of the second Himalayan upheaval in Siwalik times. The older school consisting of Pascoe, Pilgrim, Oldnam and Evans held that the extra-peninsular mountain drainage was discharged by a single westward flowing large river into the Arabian Sea. Fox, Krishnan and Aiyangar, however, hold that during the Siwalik times the drainage followed both into the Arabian Sea and the Bay of Bengal.

*Evolution of the Himalayan Rivers*

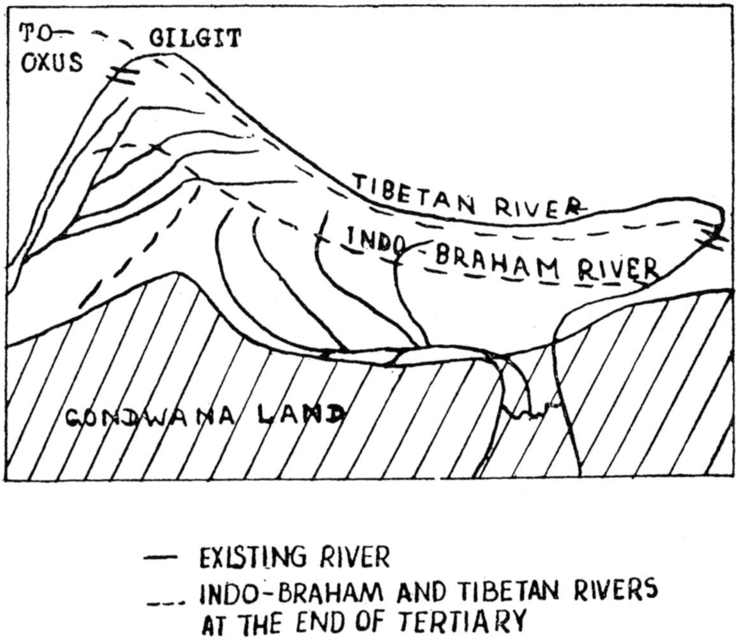


FIG. 19.2

A portion of sea extended into the Eocene from Sind towards Afghanistan and then eastwards up to about Nainital. It was gradually filled up and replaced by the great river originating from the north eastern corner of present Assam, which deposited a thick sediments of Murree and Siwalik rocks in between the early Miocene and the end of the Pliocene period. Post-Siwalik movements in the north-west Punjab brought about a dismemberment of the river system, which hitherto has flowed from the head-waters in Assam, throughout the whole breadth of India, to the Potwar and then to the succeeding head of the Sind gulf into these subsidiary streams: (1) The present Indus from North-West Hazara and (2) the five Punjab tributary rivers of the Indus which is the portions of the present Jhelum, Chenab, Ravi, Beas and Sutlej which have originated after the uplift of the top most stage of the Siwalik system and subsequent to the severance of the Indus from the Ganges.

Evans and Oldham in the discussion on Pascoes paper, pointed out that the alluvium in the Rajmahal gap, though relatively thin is of lesser effect without structural assistance. Further, they pointed out that earth movement was more likely an agent of such change in temporal dimensions. An important point is the occurrence of sharp V.S. in many of the rivers of the Northern flank of the Indo-Gangetic plain which points to the north-west direction of the Siwalik alluvium boundary. The Northern limbs of these V.S. are regarded by Pascoe as the remnant of the old right bank tributaries of the Indo-Braham. The middle Indo-Braham was being attacked from two directions i.e. from South-West the tributaries of the Indo-Braham itself from the south-east by the headstreams of the Ganges now diverted into the Bay of Bengal. The former streams may have been rejuvenated by the earth movement associated with the uplift of the Potwar plateau and similar events taking place in the sub-Himalayas itself. The limb of the upper Chenab, upper Jhelum was cut into sections and, hence, there is no doubt that the present small one is a misfit in its broad alluviated valley, while all this was happening on the Western front. The Rajmahal-Shillong watershed had been separated to form gap and the Ganges-Brahmaputra drainage diverted into the Bay of Bengal. The rejuvenated western headstreams of Jamuna thought to have flowed in the course now marked by the dry Ghaggar depression in Rajputana and in fact it was the river Saraswati of Hindu legend. The upper Sutlej have debouched into the Ghaggar bed by river *capture* which is activated by a tributary of the Sutlej river. The Indus confluence near Attock was a tributary of the Indo-Braham. This tributary captured a large river flowing into oxus in eastern Afghanistan, marking the east-west headstreams of the Indus.

This river flow into the Arabian Sea as indicated by the Siwalik deposits. Two separate rivers Proto-Ganga and Proto-Brahmaputra flowing into the Bay of Bengal cut across and captured the river. The eastern one captured the Assam portion to form the Brahmaputra which had captured the Tsangpo supposed to be then flowing westwards as suggested by its westward flowing tributary. The western one beheading the section of the Indo-Braham or Siwalik river between the present Jamuna and Assam.

**Single River Theory**

The point in favour of westward flowing indo-Braham by Pascoe's or Siwalik River by Pilgrim are as follows:—

(1) It has been suggested that Murree gulf from Kumaon to Mekran had been filled up by sediments containing coal, gypsum and petroleum and after the migration of sea from the land it is generally followed by a single river whose valley extends up to the Arabian Sea. This is true about the area of the Punjab where deposit of such minerals are found. Moreover, with the emergence of the Punjab beneath the Arabian Sea, the Indo-Braham river extended its course according to the initial slope of the land. For this fact a lower course of the Indus is developed as the consequent stream. As for example, Burma gulf by the Irrawadi and the Mesopotamian gulf by the Tigris and Euphrates.

(2) The thickness of the Siwalik deposit also prove that the materials has been deposited by single river. The Siwalik rocks have a thickness of 16,000′ mostly deposited by fluvial sedimentary rocks. Other rivers flowing from the north had the sedimentary embankments and not crystalline rocks. Somewhere the Tertiary and the Crystalline rocks would have alternated. Moreover, in case the Siwalik was deposited by an independent river the Siwalik regions would have numerous alluvial cones, terraces, fans and small valleys. But in this case the deposit would not have been so enormous. In the case of the single river it would have brought sediments from larger catchment areas. Hence the sinking of the basin to great gaps would have been possible. The vast thickness of the Siwaliks and the nature of their boundary can be more logically explained by a single river Indo-Braham.

(3) Most of the rivers of the Himalayan regions have their 'V' shaped upper courses. The upper limbs of 'V' are considered to be the remnants of the main courses of Indo-Braham River which open towards the North-

West as the number of river captures are still found, with connected and detached river into numerous pieces, which ultimately turned into big rivers of the North Indian plain. Hence the major rivers of the present are extended tributaries of the Indo-Braham river. The northern limb of the 'V' shaped valley gives a very important clue to solve the mystry of past Himalayan drainage. The meeting of tributary with the main river along the prevailing slope towards west are obviously the westward moving Indo-Braham River.

(4) The formation of the Rajmahal gap is a very important factor in the evolution of Himalayan drainage. Rajmahal is a plateau made of old crystalline rocks of mainly granite, Rajmahal quartzite and slates. Volcanic action has taken place in this area during the Jurassic period and hence the lava flow is found there.

According to Pascoe and Pilgrim the gap was formed only in the Pleistocene period. Before this period the Bay of Bengal was shut off by the extension of Peninsular table land of Assam which separate the channels of the Indo-Braham river. The only outlet left for the river was the Arabian Sea through which the Indo-Braham has to flow westward to cut a distance of about 3000 miles. So according to Rish there was no Rajmahal gap.

(5) The constituents of the Indo-Gangetic plain are similar to that of the rocks of the Siwaliks. The deposits of the Indo-Gangetic plain have been made by the similar river system thus why not the Siwaliks be possibly deposited by a single river.

(6) There is a similarity between the Dolphins and fishes of the Ganga and the Indus system of rivers which suggests that the two drainage lines were formerly inter-connected by the Indo-Braham river. These evidences have been presented here to prove the single river theory present throughout the length of the Himalayas.

**Multiple River Theory**

Krishnan and Aiyangar have discarded one river theory

and supported the view of Himalayan drainage in Siwalik times through inter-connected lagoons and lakes both into the Arabian Sea and the Bay of Bengal. The arguments forwarded are as follows:

(1) The Siwaliks are continuous beyond the Brahmaputra up to the Bay of Bengal. Hence if Siwaliks were formed of river deposits it will be inferred that the river flowed both into the Bay of Bengal and Arabian Sea.

(2) Estuarine conditions were found by Fox near the source of Indo-Braham. In case the source region was an elongated sea it would be obvious that the relief between the source region and the Arabian Sea was very little for rendering the flow of Indo-Braham from north-east Assam into the Arabian Sea.

(3) It has been enunciated that there was no continuous hill in the south of the Indo-Gangetic fore-deep during the Siwalik formation because Rajmahal gap has already been formed. In case this gap existed previously it is obvious that the Indo-Braham would have gone nearer the Bay of Bengal than the distant Arabian Sea. Now the arguments about the pre-Siwalik origin of the Rajmahal gap are as follows. The second violent Himalayan movement had taken place in the Mid-Miocene period when the Rajmahal gap was formed. The topography on either side of the 'gap' is relatively mature suggesting an older age than the post-Siwaliks.

(4) Keeping in view the westward pointing 'V' shaped valley in the course of the Himalayan drainage it has been argued that it is impossible to draw the courses of the Himalayan rivers during Siwalik times after viewing the present trends and the subsequent changes. This is because the westward 'V' shaped is not the rule in most of the rivers.

(5) The similarity of fauna of the Ganga and the Indus could be easily explained by assuming inter-connected lagoons and lakes.

(6) There is no evidence of recent uplift in Delhi-Ambala ridge. According to Pascoe and Pilgrim it is a dismembering fact for the Indo-Braham river. The Ghaggar which flowed even in historic times across

the divide have failed to do so as there had been uplift in the Siwalik times.

## CONCLUSION

In concluding the views presented by the few groups of geologists it can be said that the western geologists are in favour of single river theory, while the Indian geologists are of the opinion that the Siwalik rocks have been deposited by the joint efforts of small and big rivers originating from the Himalayas and from the extra Peninsular drainage lines.

## SUGGESTED READINGS

E. Ahmad, Evolution of Himalayan Drainage, *Geographical Outlook*, Vol. VI.

O.H.K. Spate, *India and Pakistan.*

W.D. Thornbury, *Principles of Geomorphology.*

Vonengles, *Geomorphology.*

D.N. Wadia, *Geology of India.*

# 20

# Weather Forecasting

*—Ram Bahadur Mandal*

THE COMMON man's concern is the hour to hour or day-to-day conditions of the weather. The weather is the condition of the atmosphere at any place at a specific time, with respect to the various climatic elements, i.e., temperature, sunshine, wind, cloud, fog precipitation, etc. Whether it will be rain or snow or hot summer or cold winter, only a meteorologist can give the correct answer of all these questions so as to know the conditions of probable weather. The purpose of this paper is to analyse the principles and methods of weather forecasting to the students of geography.

Man has always tried to probe the mysteries of nature and the elements. The moon, the sky, the wind and the clouds have been studied in the hope of being able to predict the weather. Empirical rules have been applied for its formulation but none of them is reliable.

Weather forecasting means predicting the probable weather conditions for short or long terms with the help of past and present weather maps and upper air observations. It has been a problem from centuries for aviators, sailors, businessmen and agriculturists who are concerned with accurate weather forecasting.

From the meteorological point of view the problems of weather forecasting are different for navigators, aviators and agriculturists respectively. For agriculturists the weather forecast indicates the force of wind, fog, temperature, rainfall and drought, etc. ; while navigators are concerned with the storms, gales and destructive cyclones so that safety measures can be adopted in advance.

In order to forecast the weather in the middle latitudes with some degree of accuracy, it is necessary to have available basic observations on temperature, atmospheric pressure, humidity and other weather elements for many places and over areas of considerable extent. Formerly weather forecasting was of importance primarily for shipping and agriculture. The development of a network of airlines has greatly increased the need for basic weather observations, made not only at the surface but also aloft, so as to foresee flying conditions and sea route planes whenever and wherever necessary. In turn, the aeroplane, the pilot balloon, and the radiosonde are furnishing increased quantities of observations at higher levels of the atmosphere, and these are indispensable for forecasting surface weather with higher precision.

## DEVELOPMENT OF WEATHER FORECASTING

This is one of the oldest occupations of man. Till the later half of the nineteenth century forecasting the weather was mere guess-work, without any scientific basis or explanation. The first scientific forecasting was made possible after 1844 as Samual Morse sent his first telegraphic message which makes easy the collection of data in the form of code from a larger area. The real progress in forecasting has been experienced during the last fifty years, specially after the first world war when the science of forecasting assumed a great significance. The methods of forecasting depend upon the techniques of weather analysis.

## PRINCIPLES OF FORECASTING

A weather analysis is a consideration of physical properties of the air in relation to space and time. In predicting the conditions of probable weather the principles involved are the following :

(a) Determination of weather elements;
(b) Determination of the movement of air mass and storm systems;
(c) Preparation of maps and synoptic charts;
(d) Forecast the weather.

## DETERMINATION OF WEATHER ELEMENTS

The elements of the weather which are studied separately and in combination for the purpose of weather forecasting are temperature, barometric pressure, direction and velocity of wind, humidity, cloudiness and sunshine, precipitation and others.

Various elements of the weather should be analysed properly to make a correct forecast of conditions of the atmosphere present three dimensional problem involving altitude, length and breadth of the climatic phenomena. These are hardly changeable and are, therefore, to be analysed properly. Most of the meteorological stations at which the data of the weather conditions are collected are stationed on the ground, obviously it is very difficult to obtain the date about the free air. They are obtained by means of balloons equipped with meteorological instruments which can go up to the required heights. They are known as aerological stations.

**Cloudiness** : The nature of the cloud is a good sign of a particular type of weather. The increase in cloud and low pressure is a good sign of frontal weather. Such clouds are good indicators of one's position in a cyclonic air-mass. The nature and amount of precipitation, the direction and speed of upper winds are indicators of particular weather condition.

**Wind** : Wind speed is related to the pressure gradient. The location, direction and the intensity of storms will depend on high and low pressure centres and fronts. A steep drop in barometric pressure means that a cyclone is approaching. A rise indicates frontal passage and the rise after cold front indicates fair weather. The sudden rise and fall of the barometric pressure are indicative of rapidly moving storms or high winds. "Bad Weather" often occurs with high pressure and not all lows have cyclonic activity.

**Temperature** : In a cyclone, the areas of high temperature may lead to thunderstorm with the contact of a cold air mass, which leads to the development of a front. The lowering of temperature after a cold front passage means cold air mass will persist for a longer period in terms of hours or days. This process of lowering temperature in the early hours of the evening will lead to the warning of dew, fog and frost in case the relative humidity of that area is high.

**Humidity** : Increasing relative humidity is an eminent sign of frontal weather. This is followed by fog, dew or frost formation under a clear sky. In case the high relative humidity comes in contact with warm surface temperature, thunderstorms may occur.

**Visibility**: Cold air mass has better visibility than warm and with the passage of fronts the visibility changes rapidly. Dust particles indicate dry air while the lower layers of dust or smoke give rise to stable air inversions.

**Sea : Swell** : At sea "the longer the swell, the stronger the wind" has been the reliable sign of navigators about the intensity of speed and the direction of wind.

## DETERMINATION OF THE MOVEMENT OF AIR-MASS AND STORM SYSTEMS

The direction and speed of movement and the characteristics of air-mass depend upon the characteristics of the source region. It is modified by the interaction of fronts and the surface over which the air travels. Some storms typify a particular region—with seasonal regularities. A knowledge of such climatological facts is necessary for forecasting. As for example, in some parts of the world the changing conditions of weather depend on the latitudinal shifting of pressure belts. The shifting of the pressure belts is meant for the north and south movement of wind systems specially the sub-tropical highs and polar lows. A movement of these belts occurs annually with the change of seasons, but if the amount of movement varies from year to year, obviously, there will be differences in the position and development of the trade winds, the prevailing westerlies, and in the paths of depressions and anti-cyclones. An abnormal position of these belts results in weather abnormalities of various kinds in many parts of the world. Where direct methods are not applicable, it is useful to calculate the chance of occurrence of certain types of weather on the basis of its past history. Specific determination about a particular storm system in future is based on developments in the past and the existing available conditions. Limited coverage and inaccurate data leads to an erroneous picture of weather forecast. The absence of good coverage over the oceans causes "blind spots" which hamper accurate weather analysis.

## PREPARATION OF MAPS AND SYNOPTIC CHARTS

The maximum difficulty in recording the data is felt in preparing weather maps or synoptic charts of an area. These maps are one of the essential tools for making a forecast. A correct weather forecast depends upon the entire analysis of the free air data and the synoptic chart.

Weather data is generally shown on a synoptic (a general survey) weather chart which represents the surface distribution of weather elements for a given time. For short range forecast sectional maps are generally drawn and the same may be used as the supplement of larger synoptic map. In this way larger maps include all major storm systems. Synoptic observations of weather elements are taken at a number of meteorological observatories throughout the world by an agreement with the World Meteorological Organisation (WMO). The observations are generally taken with the help of Greenwich Mean Time at 0000, 0600, 1200 and 1800 everyday. Besides these hourly forecasts are also made by the airways stations. This helps in short forecasts for flights and detailed analysis of the trends of the weather. The collection of weather data needs quick communications like telephone, telegraph, teletype, radio, radio-teletype, satellite, etc. For quick transmission of weather information, World Meteorological Organisation codes in numerical form are generally used. At the forecasting centres the coded messages are decoded and these figures are placed in the form of symbols on the weather map. The isobars on the map represent pressure distribution and fronts. The areas of rainfall are generally shaded, information about cloud, wind speed, isotherms, visibility and humidity are generally added. The upper air observations about air mass types, effect of jet streams are included in the map for which the data are generally obtained from distant stations.

## FORECASTING THE WEATHER

A forecast involves the following points :

(1) It requires careful enquiry about the surface and air charts;

(2) Calculations are generally done to determine the rate of movement and the future position of air mass, pressure systems and fronts;

(3) The subjective forecast with experience should not be used as it is not scientific;

(4) Recently high speed electronic computers are generally used to handle large quantities of data for an accurate forecast.

(5) Weather forecast maps can also be made with the help of past synoptic situations. Such maps are called analogous. In this situation the forecaster can easily select the historical weather sequence from analog map. This process is more scientific as not even a single map is like any other in all respects. Hence comparison of maps will be of greater use.

(6) Special forecasts are made for air flights, shipping, industry, port authorities, sailors and farmers regarding violent storms, hurricanes, tornadoes, typhoons, willey-willies, bursters, brick fielder s, etc. which involve great abnormal changes.

(7) With respect to the involvement of time, the forecast can be divided into :

  (i) Short-term forecasts for 24 to 72 hours ;

  (ii) Extended-period forecasts for several days or a month in advance ;

  (iii) Long-range forecasts covering several months or even a year.

  The long range forecast includes general information and that is susceptible to greater error, while short-range forecast includes a small area with great precision in assessing weather elements, and the results are more accurate.

(8) The forecast made by amateur forecasters for small areas and a short duration are valid for a few hours only without any sound judgment of future trends.

## STEPS FOR WEATHER FORECAST

As it has been noted, an effective weather forecast depends upon a careful analysis of weather conditions. It also requires certain examinations to be made, which are always in respect of weather conditions, which help in weather forecasting.

### Interpolation of Fronts and Air-Mass

The study of fronts and air-masses is quite necessary for making a correct forecast. It is on the fronts that the movement and the characteristic of the air-masses depend. The location and movement of air-mass influence the entire weather conditions. The fronts are actually drawn with the help of tendency of the fronts of a number of days in the past. Extrapolation needs the study of weather maps on synoptic charts of previous days, which must be kept in a logical or sequential order. This is known as the principle of historical sequence and gives a good idea of the location of fronts in future hours. The extrapolation of air-mass on the basis of fronts gives a full picture of the future weather conditions.

*Warm Front*

| *Ahead of the Front* | *At the Front* | *Behind the Front* |
|---|---|---|
| Cumulus or Ultra stratus cloud, Drizzling | Drizzle rain or shower, cloudy sky | Occasional rain but most often clear weather |

*Cold Front*

| *Ahead of the Front* | *At the Front* | *Behind the Front* |
|---|---|---|
| Fog, rain, snow or thunder storms with increasing intensity | Heavy rain, snow sometimes winter, blizzards | Occasional scattered rain but generally clear weather. |

### An Examination of the Differcnt Phenomena of Air-Mass

A detailed examination of different phenomena of air-mass is also needed for making a weather forecast. It is on the condition of the air-mass that the stage of frontozenesis develops upon which the entire weather conditions depend. Besides the life history of the air-mass, it should be studied on the basis of down or heaped conditions. For this, the source region and the inherent physical properties of the air-mass are also to be studied. A movement from warm to cold or from cold to warm area makes a lot of difference in the weather conditions associated with an air-mass. This study also includes whether the air is stable or unstable, because this tendency also influences much the overall weather conditions.

### Conditions along the Front

A detailed examination of conditions along the fronts is

also essential. This shows the extent of cloud, velocity of air, amount of changes in the physical properties of the air-mass, etc. All these depend upon the rate of movement of fronts.

## METHODS OF FORECAST

Various methods have been adopted to make a weather forecast but only some of them are important, i.e. :

### Path Method

This is the simple method which involves most careful study of the direction of the movement of air, cyclones, fronts, troughs, etc. In fact, the tendency of the movement of all these elements of weather from one place to another gives more or less a correct idea of the conditions happening in the future. This method, therefore, includes a historical study of the different types of weather conditions taking place in the past and makes a forecast about the actual location of cyclones, troughs, etc. (Fig. 20.1).

*Method of Forecasting*

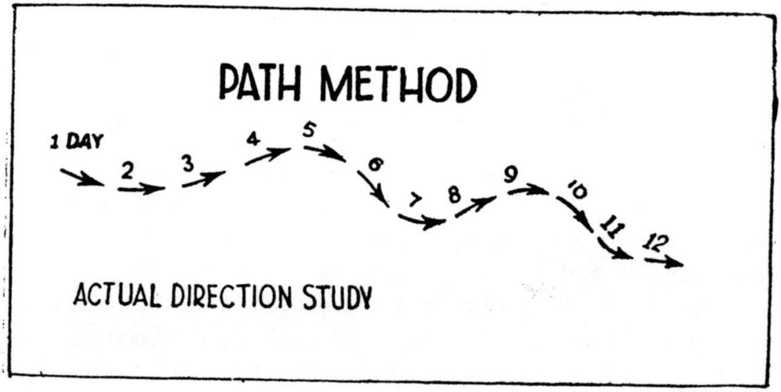


FIG. 20.1

### Tendency Method

There is a very small difference between the path and tendency methods. In the former, the actual location is given due to much accretion and in the latter there is a study of the tendency of displacement of the pressure system upon which other aspects of the weather are dependent. The tendency of displacement is computed from isobars. By this method of calculating the intensity of tendency it becomes easy to locate the pressure systems of the future hours. (Fig. 20.2)

*Method of Forecasting*

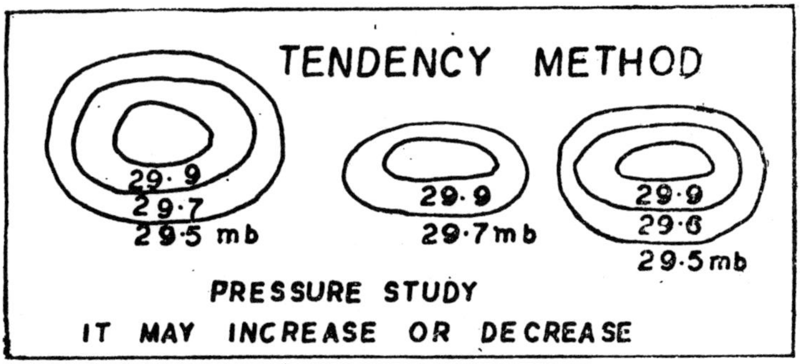


FIG. 20.2

**The Geostrophic Wind Method**

This method is helpful in determining the movement of cyclones. There is no difference between the speed of cyclones and fronts but fronts move in lesser speed as compared to geostrophic wind due to friction and vertical oscillations. The speed of cold front is 60% to 80% of the geostrophic wind, while warm front has 70% to 90%.

**Isollobaric Gradient Method**

In isollobaric gradient, the uniform pressure system denotes stationary wind. In a changing gradient generally troughs move from high to low pressure and wedges move in the opposite direction. These are proportional to the isollobaric gradient and inversely to the intensity of the system.

APPROACHES OF WEATHER FORECASTING IN INDIA

In India the Meteorological Department has a central forecasting office at Poona. This centre prepares weather maps and charts for India as a whole and its specific parts and issue forecasts twice daily. For the purpose of weather forecasting, India is divided into five zones. In each zone there is a principal forecast centre and one or more dependent centres. These forecast centres are located at Bombay,

Calcutta, Delhi, Madras and Nagpur. These centres prepare all India weather maps but forecasts are generally made for their respective regions only, besides providing meteorological informations for air flights. The forecast centres of Bombay, Calcutta and Madras are also providing informations to the shipping and port authorities. The secondary forecast centres are located at Trivandrum, Cochin, Jodhpur, Allahabad, Lucknow, Ahmedabad, Begumpet, Bangalore and Barrackpur. These are located under the command of principal forecast centres.

The Indian Meteorological Department makes two categories of forecast namely short range and long range forecast. Under short range forecast the weather conditions are made known for 24, 36 and 48 hours only. Short range forecasts are generally issued for general public, aviators and farmers. For air-flights, the forecasting centres make local forecasts, route forecasts, terminal forecasts and air field warnings of rainfall, thunder storm, fog, high winds, etc. For aviators the forecast made are generally related to the wind, cloud, visibility, air turbulence and ice accretion. For farmers the forecasts are generally related to the conditions of cloud, rainfall, high winds, conditions of temperature and hail storms. The forecasts of rainfall and temperature are made for the general public. The long-range forecasts are generally issued about the normal and seasonal departure of rainfall.

In India the first official forecast about the monsoon rainfall was made in 1886, when the forecasting about the world weather was scientifically started in Europe. It has been found that the monsoon rainfall in India is related to the local heating and cooling of the landmass and is further governed by the disturbances in Africa, Australia, Japan and the Himalayas, besides easterly and westerly strongly blowing jet streams over 20,000 feet above in the sky. A closer study of the data of a number of years has yielded results which can be applied to making forecasts of the season's rainfall in advance.

In order to forecast the winter rainfall during January to March, over North-Eastern India comprising Uttar Pradesh, the Punjab, Western Himalayas, Rajasthan, Gujarat, Saurashtra and Cutchh, the factors employed are :

(a) western rain during December;
(b) South American pressure during December;
(c) western upper winds at Agra from mid-September to mid-October; and
(d) rainfall at Port Blair during December.

EXPLANATION OF THE FACTORS OUTLINED ABOVE

(a) Elaborating the first winter rainfall in north-eastern, India is associated with the western disturbances which come from the west. The activity of the disturbances is noticeable at an earlier date in regions lying further west to India than India itself. Hence the rainfall received by these regions in December gives an idea of their likely effect on India in the subsequent months.

(b) Secondly, it has been found that the South American pressure gives us an idea of the magnitude of major oscillations between two hemispheres and this has an influence on the rainfall in north-western India.

(c) Thirdly, the western winds over North India during winter are a feature of the circulation of atmosphere and they appear at higher elevations at the end of September or the beginning of October. This appearance of air current measured at Agra during mid-September to mid-October, gives an indication of the likely intensity of winter rainfall.

(d) Fourthly, the retreat of the South-West Monsoon from the Indian waters influences the coming winter rainfall in north-western India, the late retreat affecting it adversely. The retreat of monsoon from Indian waters is indicated by the amount of December rainfall at Port Blair. Later if the monsoon withdraws higher than normal the December rainfall at Port Blair in North-Western India is likely to be poorer than usual.

Similarly for forecasting the monsoon rainfall over the peninsula comprising Gujarat, the Konkan, Bombay, Deccan, Madhya Pradesh, Hyderabad and North Tamil Nadu coast during June to September the variations of the following factors should be considered :

(i) rainfall from October to April in South Rhodesia;
(ii) pressure during April and May in different parts of South America;

(iii) rainfall in Java during October to February; and
(iv) the temperature during December to April at Dutch Harbour in Allution Islands, North Pacific.

In order to forecast the monsoon rainfall over North-West India comprising western Uttar Pradesh, the Punjab, Kashmir and Rajasthan during June to September, variations in the behaviour of the following factors should be considered :

(a) rainfall in South Rhodesia during October to April;
(b) pressure in different parts of South America during April and May;
(c) Accumulation of snow in western Himalayas in May;
(d) Equatorial pressure as measured at Seychelles during February to March, at Batavia during January to April and at Port Darwin during March to May; and
(e) temperature at Dutch Harbour during March and April.

For forecasting rainfall during August to September over the peninsula, the chief factors considered are :

(a) rainfall in Southern Rhodesia during October to April;
(b) distribution of pressure in India in June-July;
(c) pressure at Mauritius in June-July; and
(d) rainfall in Java from October to February.

For forecasting rainfall over North-West India during August-September, these factors should be taken into account :

(a) pressure in different parts of South America in June-July;
(b) rainfall in Southern Rhodesia during October to April;
(c) distribution of pressure in India in June-July;
(d) Indo-Gangetic river discharges during April-May; and
(e) departures of temperature range in the Punjab during April and may.

Forecasting for North-Western India comprising Uttar Pradesh, the Punjab, Western Himalayas, Rajasthan, Gujarat, Saurashtra and Cutchh, the extra Indian factors are :

*The Rainfall from January to March*

(1) western rain during December;
(2) south American pressure during December;

(3) western upper winds at Agra; and
(4) rainfall at Port Blair.

*The Rainfall from June to September*

(1) rainfall in South Rhodesia during October to April;
(2) pressure in different parts of South America during April and May;
(3) Accumulation of snow in Western Himalayas;
(4) Equatorial pressure; and
(5) Temperature at Dutch Harbour during March and April.

*The Rainfall from August to September*

(1) pressure in different parts of South America during June-July;
(2) rainfall in Southern Rhodesia during October to April;
(3) distribution of pressure in India June-July;
(4) Indo-Gangetic river discharge during April and May;
(5) departures of temperature range in the Punjab during April and May.

For forecasting the monsoon rainfall over the Peninsula comprising Gujarat, the Konkan, Bombay, Deccan, Madhya Pradesh, Hyderabad and North Tamil Nadu coast the following factors should be considered :

*Peninsular June to September Rainfall*

(1) rainfall from October to April in South Rhodesia;
(2) pressure in different parts of America during April and May;
(3) rainfall in Java during October to February; and
(4) temperature during December to April in Dutch Harbour in Allution Islands (Pacific ocean).

*Peninsular Rainfall during August to September*

(1) rainfall in Western Rhodesia during October;
(2) distribution of pressure in India during April, June, July;
(3) pressure at Mauritius in June-July; and
(4) rainfall in Java—October to February.

## RECENT ADVANCES IN WEATHER FORECASTING IN INDIA

In recent years it has been revealed that weather is an aggregated form of climatic elements which are distributed widely in time and space. For example, how much rain Japan gets in summer is intimately connected with the monsoon in India. Thus the global approach to forecast in any country is highly essential. The meteorological services of the world are trying to provide the First Global GARP (Global Atmospheric Research Programme) Experiment (FGGE)—which commences in 1978—of which Monex (Monsoon Experiment) forms a part.

India proposes at MONEX to augment considerably the weather observation systems over the Indian seas—the Arabian Sea, the Bay of Bengal and the Indian Ocean—over the Indian land mass and the adjoining countries affected by the summer and winter monsoons.

During the periods January to February and May to August of 1979, meteorological observations from land, sea and air will be increased manifold. These will be supplemented by observations from special floating balloons, a few polar orbiting satellites and a geostationary meteorological satellite (GMS) to be placed over the equator near longitude 70°E by the Soviet Union. In addition, a total of about ten ships will be employed in the Arabian Sea and the Bay of Bengal.

The objectives of Monsoon Experiment (MONEX) can be broadly divided into (a) large scale aspects of the Monsoon, and (b) regional aspects of the monsoon.

(a) It will cover the energetics and the gross characteristics of the seasonal atmospheric response to the presence of the vast-Europe-Africa "island" north of the equator. The data collectcd will be used to construct and verify ocean-atmosphere coupled with numerical models capable of reproducing seasonal and international variations of the monsoon. Such models are expected to be useful not only in the identification of the physical factors which control the monsoon, but also for making extended range forecasts.

(b) The regional aspects will deal with the summer monsoon, the winter monsoon and the effects of the mountain systems of south-east Asia on the monsoons. The studies on the summer monsoon will cover the air, sea interaction and other features over the Arabian Sea, the monsoon storms and depressions over the Bay of Bengal, the time of onset of the monsoon and variation in its activity, the interaction of the monsoon circulation with the wind systems elsewhere over the globe and the main energy sources which maintain the wind systems of the Asiatic monsoon.

India has had some experience in the collection of data similar to that of MONEX and its utilization. During 1963-65, India had participated in the International Indian Ocean Expedition (IIOE). Ships from India, Germany, U.K., U.S.A. and U.S.S.R. participated in the expedition. During the summer of 1974, the Indo-Soviet Monsoon Experiment (ISMEX) was conducted, in which additional data over the Indian seas were collected by four Soviet research vessels and two Indian ships. A similar experiment entitled Monson-77 was conducted during the summer of 1977.

Monsoon-77 was an international experiment in which many countries around the Indian Ocean area had collected extra meteorological data. The data were pertaining to winds, temperature, pressure, humidity at different levels of the atmosphere, cloud patterns, sea surface temperature incoming and outgoing radiations, rainfall and soil moisture. These provide valuable information for MONEX and for predicting the weather trends for a longer period in India.

## SUGGESTED READINGS

H.R. Byers, *General Meteorology*, McGraw Hill Book Company, New York, 1959.

J.J George, *Weather Forecasting for Aeronautics*, Academic Press, Inc. New York, 1960.

L.P. Smith, *Farming Weather*, Thomas Nelson & Sons, London, 1958.

R.D. Huschke, *Glossary of Meteorology*, American Meteorological Society, 1959.

# 21

# The Concept of Jet Streams

—*Ram Bahadur Mandal*

SINCE A.D. 1900, the successive use of kites, gas filled balloons, small unmanned balloons, balloon sonde, radio sonde, rocket sonde, meteors, rocket grenade methods and weather satellites have highly helped in the investigation of jet stream in the zone of upper troposphere.

The concept of jet stream is first of all introduced by Rossby[1] and his fellow workers[2] in the field of meteorological literature in 1947. According to the structural division of the sky in the upper troposphere, there blows a very high speed light wind over specific parts of the globe. Such wind in the latitude of westerlies are known as jet stream. The locations and intensities of jet stream are found to vary daily and seasonally both. According to the World Meteorological Organization jet stream is defined as a horizontal, flattened tabular current (Fig. 21.1) of wind blowing just below the tropopause. The axis of this wind always lies in the line of maximum wind speed followed by transverse wind shears. Thus any wind with a maxima of 60 knots and excess of wind speed embedded in weaker zonal currents is also known as jet stream. It is a guiding factor for streams in the middle latitudes as well as the means of transporting vast quantities of air across the continents. It is a wind of several thousand feet thick and 300 to 400 miles wide.

In this way jet stream is a more local high altitude air current of oval form concentrated in a narrow belt of above 30,000 to 60,000 feet above in the sky. During summer the average wind speed is about 60 knots, but in winter it may be stronger and the wind speed varies from 95 to 120 knots and

*The Jet Stream*

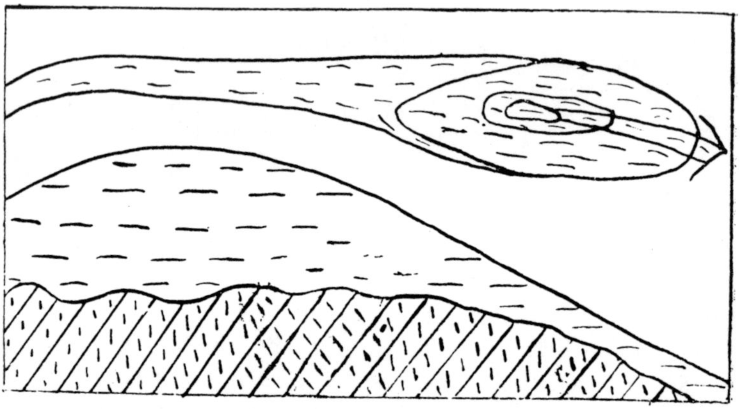

FIG. 21.1

on rare occasion winds of 200 to 300 knots have also been recorded.

The purpose of this study is to highlight the concept of jet stream,discovery, classification, distribution and effects on general climatic conditions of India.

### DEVELOPMENT OF THE CONCEPT OF JET STREAM

In the early twentieth century the observations of the drift of cirrus clouds indicated the presence of strong zonal currents of air in the upper troposphere. At the time of the eruption of Krakatoa volcano in 1883, the movement of noctilucent clouds revealed the existence of high speed winds at great height going towards west over the sub-tropical belt of middle latitude, and the same has been taken easterly direction over the equatorial region. Noctilucent clouds are those which remain sunlit at night due to their presence over higher altitude. The data collected by Bjerkness in 1933 also favoured the idea of the presence of 'strong zonal current' in the upper troposphere. In the same year of 1933 an experiment of a pilot balloon over India showed an evidence of strong wind concentration at an height of 35,000 feet on 25° N latitude. However, the concept of jet stream evolved by the meteorologist with the establishment of a large number of observatories. During Second World War of (1939-45) meteorologist recorded these strong wind current and they have used the same in high level jet aircraft operations. The meteorologists and aviators studied these air currents all over the world intensively after 1945 when American pilots operating over the Pacific Ocean near Japan came across exceedingly strong jet streams having maximum wind speed of 230 knots or 250 miles per hour (mph) in some of the exceptional cases. During the last three decades since 1945 the investigation and study of jet streams have increased several times.

## SOME CHARACTERISTICS OF JET STREAMS

In the way of jet stream the wind builds up strength slowly from the marginal areas up to the core of the jet stream. In the periphery the wind speed is just half in comparison with the core of the jet. A longitudinal cross-section of the jet stream is as follows in Fig. 21.2

At any time the jet streams have varying altitude which corresponds with the height variations of the tropopause level at the latitudinal belt and seasons concerned. On the basis of synoptic situation in a particular region there are minor variations in the altitude of the jet stream.

Jet streams are found in both the northern and the southern hemispheres. At one time the distribution of these winds was found over a specific latitudinal belt only. In accordance with the latitudinal shift of the sun in relation to earth. the positions of the jet streams undergo considerable shift towards north and south respectively. On global wind charts, the jet streams appear to be circumpolar showing considerable oscillations towards north and south simultaneously from the so called mean latitudinal position. They also suffer from discontinuities over longer distances. The observations of jet streams on the global chart, it looks like a meandering river of strong wind with some portions of dry segments (Fig. 21.3). In successive days jet core show a considerable variations in

*Cross-section of a Jet Stream*

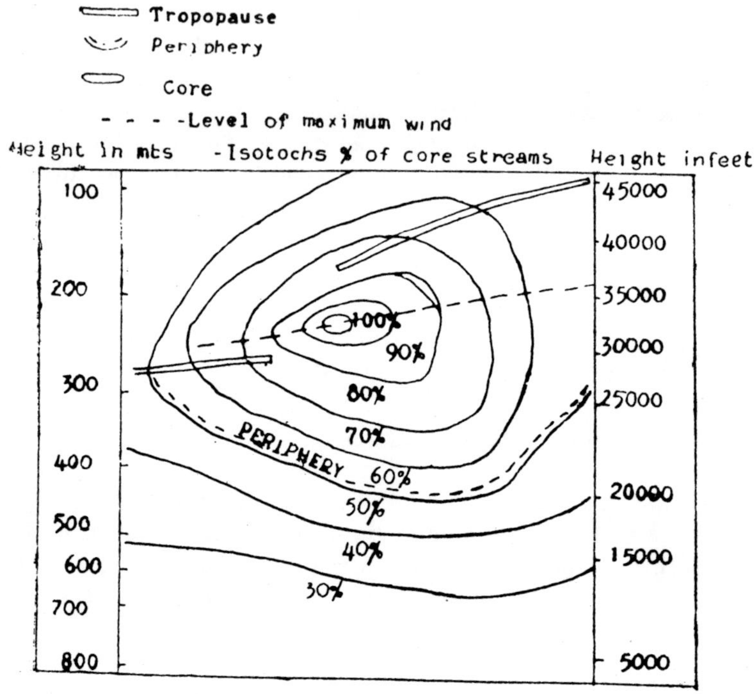


FIG. 21.2

intensity of wind speed, location, orientation along with height of such winds from the surface of the earth.

*Meandering Course of Jet Stream*

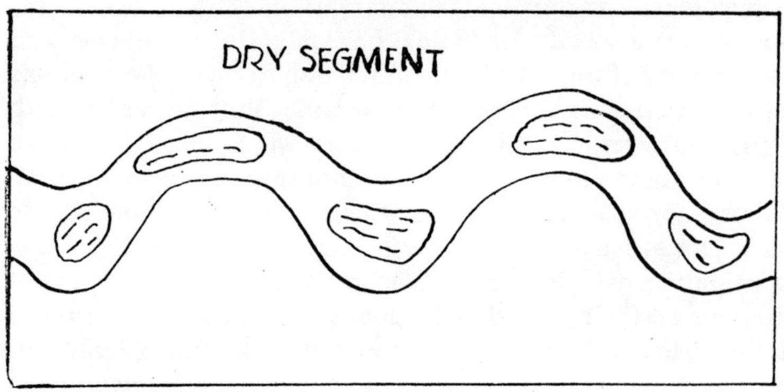


FIG. 21.3

**Classification of Jet Streams**

The jet streams which blow over different latitudes are classified as arctic, polar, sub-tropical and tropical. Among these the first three are westerly in which winds blow from west to east and the last is only easterly, in which winds blow from east to west.

(1) **Arctic Jet** : It extends from polar regions upto 60° latitude towards equator. The height of arctic jet is usually below 30,000 feet or 9 kilometres. The pressure of air is 300 mb. Arctic polar night jet stream recorded speed over 330 knots at a height of nearly 18 kilometres.

(2) **Polar Jet** : It extends from 70° latitude upto 30° latitude towards equator. The height is 30,000 feet to 40,000 feet or 9 to 12 kilometres. The air pressure varies from 200 mb to 300 mb. Polar night jet stream is a strong westerly current.

(3) **Sub-Tropical Jet** : It is also known as Ferrel westerlies which extends from 35° to 20° latitude in both north and south of the equator. The height of this jet varies from 33,000 to 46,000 feet or 10 to 14 kilometres respectively. Pressure level is just 150 to 250 mb.

(4) **Tropical or Easterly Jet** : It is also known as Krakatoa easterlies because they have carried the volcanic dust westward after the eruption of Krakatoa in 1883. It extends from 15° latitude upto equator in both hemispheres. The altitude of this jet is from 46,000 to 53,000 feet or 14 to 16 kilometres respectively. Pressure level is just 100 to 150 mb.

## DISTRIBUTION AND EFFECT OF JET STREAM OVER INDIA

Over the Indian sub-continent only sub-tropical westerly and tropical easterly jet streams have been observed by meteorologists. In Northern India the sub-tropical jet holds sway over a greater extent in comparison with the tropical jet.

During the period of south-west monsoon in India, the sub-tropical jet is not found. The intensification of sub-tropical jet is observed during winter (November to March) over the north. On Peninsular India the tropical jet can be observed from June to August. At this time sub-tropical westerly jet flows at the height of the Himalayan range. Sometimes, the splitting of sub-tropical jet stream is found due to the obstacle of the Himalayas. Here the sub-tropical jet is divided into two branches, one flow from the north and the other from the south. These two branches combine into one over China in the east. Generally, the core velocity of the combined jet stream is high in comparison with two divided branch of the north and south across the Himalayas (Fig. 21.4).

*Splitting of Jet Stream due to the Himalayas*

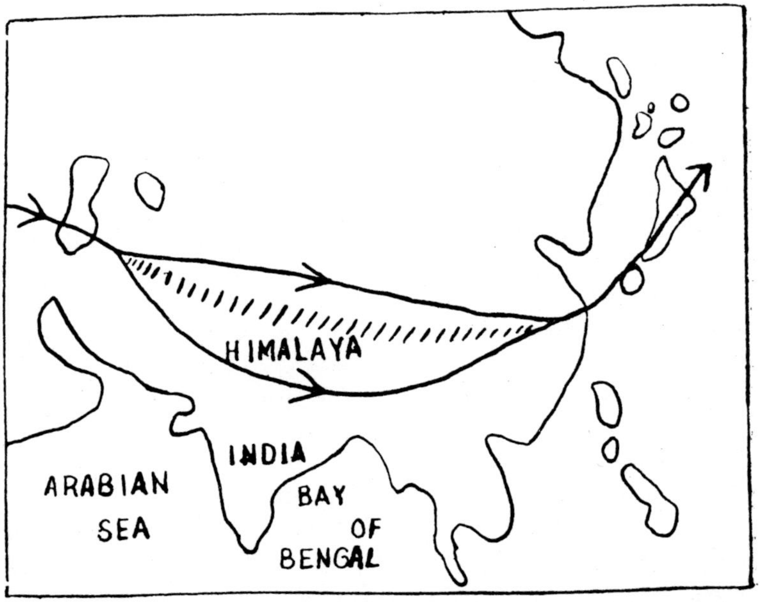


FIG 21.4

The easterly tropical jet is commonly found over the south-west monsoon and is located south of the sub-tropical ridge of high pressure Tropical easterly jet starts over the south coast of China (Fig. 21.5) and flows towards west roughly near 15° North latitude over Thailand, South India, South Arabia and finally loses existence over the world's greatest desert of Sahara. In the tropical jet stream the current accelerates from Malaysia to Peninsular India, where maximum core velocities of 100 to 120 knots maxima have been observed but this speed lessens as it proceeds further west over the Arabian Sea.

*Tropical Easterly Jet over South Coast of China*

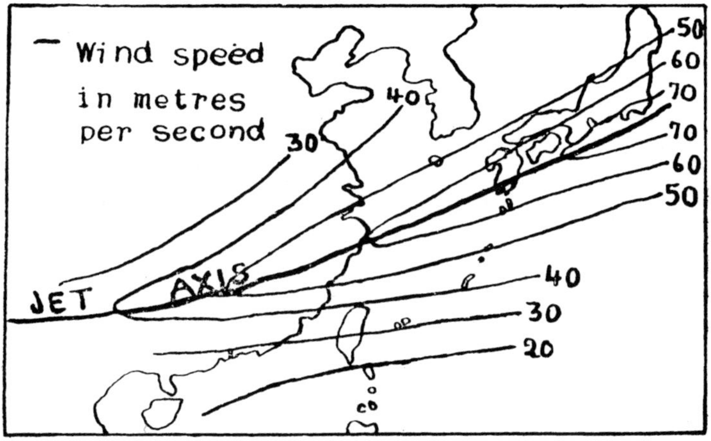


FIG. 21.5

SEASONAL VARIATION OF JET STREAM

Over the North Indian plains the sub-tropical jet appears in October at the time of the withdrawal of the south-west monsoon and changes its position southward with the increasing cold in winter. It reaches up to the Tropic of Cancer which is the transitional and southern-most position of the shift of sub-tropical jet. It reaches up to this position at the end of February and March.

After March the sub-tropical jet generally recedes from the north and it is non-existent over the North Indian plains at the end of May.

The tropical jet makes sway over Peninsular India beginning from June up to the end of August. In July and August the tropical jet stream is found to be located at the latitudinal extent of 12° to 15°. The observations show its non-existence from Peninsular India at the end of September. In sub-tropical westerly jet streams the lowest mean core velocity is 65 knots in October and May, and the highest velocity of 100 to 150 knots reached is in January.

DISTRIBUTION OF TEMPERATURE

Below sub-tropical jet the temperature increases from north to south below the core, while the revsree case has been observed above the core. The temperature gradient below the core is observed 1°C in between one degree of latitude, and the higher gradient is observed in well developed jet streams. The node of the sub-tropical jet stream is usually found above the tropopause. A normal temperature of 50°F is found there. The wind of maximum velocity is found at a height of 15,000 feet.

In sub-tropical jet streams, the horizontal shift is marked 30 knots more per degree of latitude in the north than to the south of the central core of jet stream. In the south the shear is just 10 knots per degree of latitude. The shear value of tropical jet is much lesser in comparison with the sub-tropical westerly jet.

In jet streams certain zones are liable to turbulent flow of air both horizontally and vertically with rapid fluctuations. This type of clear air turbulence (CAT) is generally found in upper troposphere and is not associated with the formations of cloud. It has no apparent signs in the upper sky as it is hazardous to the aircraft in clear air.

## EFFECT OF JET STREAMS ON WEATHER

The change of seasons indicate the structural change in the character of jet streams, as it is part and parcel of the general circulation of atmosphere. In India the appearance of tropical and disappearance of sub-tropical jet coincides with the onset of the south-west monsoon. The shifting of the sub-tropical jet in North India is associated with the withdrawal of the south-west monsoon. Some investigators tried to correlate the location and intensity of the westerly jet in India with the onset of the south-west monsoon, as the breaks in monsoon are closely associated with the northerly position of the tropical jet. Due to horizontal and vertical shifting, large velocity gradients of the jet stream usually give rise to convergence and subsidence of wind and they have some important bearing on the origin of dust storms, thunder storms besides intensification and dissipation of storms.

In order to maintain the planned flight of aircraft the analysis of turbulence and reconnaisance survey of the area is highly needed. The knowledge of jet stream is essential for the operation of jet aircraft, jet fighters and bombers in order to take advantage of the fast moving air current and to avoid alternative route and altitude of aircraft flight. It is also economical in the use of fuel, comfort to the crews, passengers and hence it is beneficial for commercial aviation.

## CONCLUSION

Jet stream is a strongly concentrated high speed air current blowing over certain parts of the globe in a meandering fashion over heights of 15,000 feet to 60,000 feet in the zone of upper troposphere. It affects the climate of the area over which blow and change their position, orientation due to oscillation in different seasons. It is helpful for jet aircraft and jet bomber flights.

## REFERENCES

1. G. G. Rossby, American Meteorological Society, Vol. 28, p 53.
2. S. D. Sarma, "The Jet Streams" *Science Reporter*, Vol. 6, No. 2, 1969.

# 22

# Paleo-climatic Development of Landscape in Hazaribagh Plateau

*—Anil Kumar and Ram Niwas Pandey*

THE PHYSICAL landscape is a cumulative function of three major casual factors, endogenic processes, exogenic processes and the local geology (Büdel, 1972)[1]. Apart from general geological and lithological features like structure and tectonic activities, the other important factor is climate. In small areas, various geological factors may appear dominant as climatically caused differences would not be easily detectable in the landscape. However, in large areas the geological factors are far outweighed in importance by those of climate particularly in relation to slope and nature of surface cover. The analysis of landscape necessitates the identification of relict or inherited landforms developed under some past climatic condition. The present day processes have either modified the inherited surface or have created new landforms in the region. The role of climate in moulding the physical landscape is very significant. It is always desirable to bring out the climatic characteristics of the area of study and correlate them with the evolution of landscape. Relief depends on climate almost as much as on the bed rock.

Various topographic forms, such as wide and almost plain undulating erosion surfaces at different altitudes, relict weathered profiles, residual hills, gullied surfaces, rock benches, terraces, flood plain deposits and deep soil covers confirm the changes of climatic condition which is believed to have played an important role in punctuating the processes engaged in the landscape evolution at different geological periods. There is a dearth of artifacts or fossils records of climatic changes in the Hazaribagh Plateau Region, but soil structure, depositional surface of broad valleys and the Gondwana basin in the south contain records of climatic changes

TABLE 22.1

Figures in brackets show the lapse of time from the beginning of the particular Era, period of systems. Figures without brackets show the total duration of the period or epoch)

| Geological Time Scale | | | Stratification of Rocks and Succession of Climate | | | |
|---|---|---|---|---|---|---|
| ERA | PERIOD | EPOCH | AGE (In Million of years) | CLIMATE | EARTH MOVEMENTS | Structural Growth of the Area |
| CENOZOIC 60 | Quarternary | Recent | | Warm | | Newer Alluvium |
| | | Pleistocene | 1 ( 1) | Cooling | | |
| | Tertiary | Pliocene | 11 (12) | Cool (Temperate) | | |
| | | Miocene | 13 (25) | Cold+Semi Arid | | |
| | | Oligocene | 15 (40) | Warm+Humid | Tertiary Earth Movement (Epeirogenetic) | Laterite lithomerge |
| | | Eocene | 20 (60) | Cold+Warm | | Deccan Trap Basaltic lava Dykes |
| | | Paleocene | 10 (70) | Humid+Semi Arid | | |
| | Cretaceous | UPPER | | Fluctuating | | Infra-trappean (Calcified-Silicified rocks) Grit |
| | | LOWER | 50 (110) | Very Warm | | |
| MESOZOIC 120 | Jurassic | UPPER | | | | |
| | | MIDDLE | | | | Rajmahal Volcananues Dolerites |
| | | LOWER | 40 (150) | Warm | | |
| | Triassic | UPPER | | Warm | | |
| | | MIDDLE | | Arid | Mahadeo | Dykes Mahadeo (Super-Panchet) Panchet Series Sandstone, Shale, |
| | | LOWER | 30 (180) | | | |

| | | | | | | |
|---|---|---|---|---|---|---|
| | Permian | UPPER | | | Damuda | Ranigani Series Sandstone, Shale. Middle Damuda, Grits, Sandstone Barakar Series (Conglomerate Sandstone, Grit, Shale) |
| | | LOWER | 35 (215) | Dry | | |
| | Carboniferous | UPPER | | Cold (Glaciation) | | Talchir beds (Boulder beds) |
| | | LOWER | 60 (275) | Hot + Humid | Earth Movement Eperogenetic | |
| PAIAEO-ZOIC 330 | Devonian | UPPER | | | | |
| | | MIDDLE | Arid + | | Damodar Trough | Unrepresented |
| | | LOWER | 50 (325) | Semi Arid | | |
| | Silurian | UPPER | | | | Chotanagpur Granite, Granite-Gneiss. |
| | | MIDDLE | | Arid + | | Dharwar, Phyllites, |
| | | LOWER | 35 (360) | Semi Arid | | Schists, Quartzitse Lime Silicate rocks and Basic rocks. |
| | Ordovician | UPPER | | | | |
| | | MIDDLE | | | | Unrepresented |
| | | LOWER | 90 (450) | Dry | | |
| | Cambrian | | 150 (600) | Warm Temperate | | Unrepresented |
| PROTE-ROZ OIC | Pre-Cambrian | | 1500 | Dry + Cold | Archaean Earth Movement | |
| | Archaean | | 4000 | | | |

Source : Krishnan, M.S. (1969) : *Geology of India and Burma*, p 92.

in its different layers or beds. The Gondwana stratigraphy shows that the period had experienced a number of cycles of sedimentation. Its organic fossils and sediments bear the evidence of the climatic condition in which they were deposited. The study of various stages of Gondwana formation reveals not only the sequential change in past climate from cold glacial climate to warmer humid to cold temperates and drier arid climate which ended with the return of moist condition of climate but also the source and nature of sedimentation and the history of the drainage system during the Gondwana era. (Krishnan, 1960).[2] The basin has lost its true character due to sub-aerial denudation in later geological periods. Elsewhere the presence of deep soil and laterite also bear the evidence of climatic change during some recent geological periods.

The fossils or dead erosion surfaces can be interpreted in terms of periodic climatic changes from arid to humid condition. There is compelling evidence in relict landforms and inactive deposits that past climatic conditions have differed from the present.

## PALEO-CLIMATIC CHARACTER

The Hazaribagh Plateau Region, as a part of the ancient Peninsular landmass has been continuously denuded since pre-Cambrian times. During the major part of the Palaeozoic era denudation of the plateau region continued in dry, arid and semi-arid climatic conditions with phases of sedimentation in primeval depressed zones leading to the formation of Dharwarian rocks. The records of the continuous denudation can be identified over the widespread erosion surfaces of Kodarma plateau, particularly in mica region. At present these areas are being dissected by contemporary processes.

The Gondwana trough in the Damodar valley records the sequence of climatic change from the Carboniferous period to the Jurassic period. About 275 million years ago, the Carboniferous period started with the formation of the Damodar trough separating the erosional surfaces of central Chotanagpur into north and south segments. The climatic condition during the upper Carboniferous was favourable and cold enough to support a continental glacial phase which seems to have been extended over the areas of Rajasthan, Madhya Pradesh as well as the Eastern Ghats. The Talcher boulder bed overlying the Cambrian succession contains pebbles and boulders derived at least in part from Rajasthan.[3] The Talcher boulder beds in the Damodar Valley contain large quantity of quartzites and gneissic rocks. The resemblance of these quartzites to the Vindhyan quartzites of the Sone valley, expresses the possibility that these materials may have been laid down by glaciers moving east or south-east from Vindhyan. At present the traces of Carboniferous glacier erosion are not available as they might have been erased by subsequent denudation in the areas.

The climate gradually warmed up during the close of the Carboniferous period as a vast amount of fluvio-glacial sediments and varved clays were laid down over Talcher boulder beds by melt water. "The prevailing greenish tints of the rocks indicate that the climate was still quite cold. Though there are plant impressions in the beds immediately overlying the boulder-beds, the flora was apparently still scanty during the Talcher period."[4]

215 million years ago during the Permian period, the temperature gradually turned warmer and more humid and favoured the growth of luxuriant vegetation. Streams carried down enormous amount of vegetation into the swamps and lake basins of the Gondwana trough. During this period the largest number of Genera and Species of the *glossopteris* flora developed and were deposited by streams to form the coal beds. The period of coal formation is known as Damuda period and the strata consists of sandstone containing kaolinised feldspars, shales and coal in succession. Each coal seam represents a sedimentary cycle. "There may have been as many as fifty to sixty cycles of sedimentation during the whole of Damuda period"[5], in a duration of only 35 million years. This again points out the possibility of continuous sinking of the basin and erosion by running water over the higher surrounding regions. The nature of flora and the presence of undecomposed or partly decomposed feldspar in the sediments suggests that the climate of this period was cold temperate.

The increasing number and thickness of strata from east to west and decreasing purity of coal seams towards the west, indicate that the source of sediments was somewhere in the west. In Damodar valley the drainage seems to have flowed from west to east. Its outlet was located somewhere north-east where there are several small coalfields in West Bengal and Santhal Parganas.

With the onset of the Triassic period, about 180 million years ago, climate became arid. This change in climate is indicated by a slight unconformity between Raniganj (Upper Permian) and Panchet (Lower Triassic) beds in Raniganj coal-fields. Absence of Carbonecious matter in Panchet beds indicates that sediments were laid down in flood plains and shallow lakes (Krishnan, 1969[6]. This was a period of gradual drying up of the sedimentary basins (as indicated by the presence of *Labyrinthodonts* fossils.)

From middle to upper Triassic age the climate of this region remained arid as the supra—Panchet and Panchmarhi beds of the Mahadeva series are definitely sediments of an arid climate. The *Glosopteris* flora practically died out during the dry period of Triassic.

The climate again changed from arid to more favourable and moist in the Jurassic age. A new flora, *Ptylophyllum* flora, gradually established itself (in Rajmahal) and continued well into the Cretaceous. The comparative humid condition seems to have continued with perhaps slight variations during the rest of the Gondwana era, spanning over about 50 million years.

The Tertiary period was marked by ever changing climatic conditions and earth movements leading to the successive uplift of the Chotanagpur plateau. The climate varied from cold to warm and humid from Eocene to Oligocene and again to cold in the Miocene period. No significant sedimentation took place in the area under study as the entire plateau experienced phases of uplift and consequently rejuvenation in erosional processes.

The Pliocene was humid enough to create conditions of sub-aerial alternation leading to the formation of laterite in Chotanagpur. It is now generally agreed that the conditions favouring the formation of laterites are a warm, humid climate with plentiful and well distributed rainfall and good drainage. With the establishment of monsoon climate in India the process of laterization began and is active even at the present day (Wadia, 1966)[7]. In Chotanagpur Highlands the continuous weathering of Basaltic lava, even Dharwar and Gneissic rocks have produced zones of clay minerals. The subsequent uplift of the plateau surface provided a good drainage condition for bringing about the formation of laterites and down migration of clay zones and formation of laterites of secondary origin. The deep weathering of Granite Gneissic surface into reddish brown soil is the result of the prevailing climatic condition of the Tertiary period.

The climatic fluctuation during Pleistocene was related to the glacial advances in the Himalayan region. Each glacial maxima corresponds with period of increase seasonal humidity (pluvials) in the Peninsular parts. Erosion and planation are related to these periods. The presence of some Himalayan plants particularly some species of rhododendron in the high hills of Peninsular India and in Parasnath hills, points out the possibility of Himalayan flora migration during glacial ages. It is certain that during the Ice Ages, the Peninsular parts experienced a mild temperate climate. Sedimentalogical and geomorphic evidence suggests that during pluvial condition heavy sedimentation was generally favoured in the lower basins and in the river valleys. During the last glacial period, siltation occurred in the middle courses of many tropical rivers i.e. the Ganges and Narbada, and the rivers appeared to have had insufficient discharge to move their load. Similarly chemical weathering processes were less intense in the late Pleistocene, possibly because of a reduction in rainfall.[8] Presence of 14 metres thick pebble and sand bed in the Belan river, west of Kaimoor is overlain by another 6.10 metres thick silt and clay beds. The human artifacts and animal fossils found in the lower bed establish its age of about one lakh years. This finding has been supported by similar archaeological evidences from Ajoy valley, Birbhum and some other parts of Bihar.[9]

The importance of these archaeological findings lies in the fact that vast amounts of sediment were deposited under conditions of heavy rain and during mild climatic conditions of post-glacial period, clay beds were deposited over pebbles and sand beds in river valleys. It also provided favourable conditions of deep chemical weathering and erosion of highlands and subsequent transportation and deposition of such materials in basins or river valleys.

Although climatic conditions have varied in terms of distribution, the basic climatic types have not altered drastically in the last few million years. Seasonal climate has always in a seasonal lessening of the ground cover and provided conditions for high erosion rates. In case of the tropical wet-dry regime, it seems that interference with plant cover may extend far backward in time. In general climatic shift, towards seasonal aridity probably results in accelerated upland erosion and stream aggradation. While shift to equable climate results in upland stability and stream incision. This has been supported by Langbein-Schumm concept (1958)[10] which considers vegetation-run off relationship as critical in controlling erosion rates.

It is not difficult to recognise 'fossils' form elements in the Hazaribagh Plateau Region, particularly, such 'Genetic' features of landscape which was greatly influenced by the climatic changes from the early Tertiary tropical phase to Pliocene temperate phase and to the severe fluctuations of the Quarternary with many repeated temperate (pluvial) periods. Some recent studies by Singh (1971)[11] have shown clearly that in India, the climate of post-glacial times has been both drier and wetter than that at the present. While analysing the landforms of such a complex region it should be borne in mind that giving much importance to these past phases of climatic changes as Büdel and others have tried, in terms of present conditions is simply exaggerating the ideas of climatic control on landforms. Recognising the true effectiveness of the present processes, Machatschek (1969)[12] expressed his views that it is illogical to measure the accumulated changes of the past with the alterations achieved in the brief period of post-glacial time. One must consider the existence of certain threshold values of many agencies in Post-Pleistocene denudation, especially, fluviatile ones, while making the analysis of landforms.

The landscape apparently reveals the control of climate through its general impression of the entire landscape and through the peculiar components of landforms. Landforms can be, thus, well judged "as the adjustment of forms to process in the denudational systems.[13]' Almost each component of landscape, such as vegetative cover, surface duricrust, drainage texture, pediments and tors reflect influence of climate. Weathering, both chemical and mechanical, and erosion, transportation of materials, run off, creep, wash, any or all of these processes active in the Hazaribagh Plateau Region, in reality, are closely related to climatic elements.

## DIFFERENTIAL WEATHERING AND LANDSCAPE

The many ways in which constituent minerals weather and the dominance of either of chemical and mechanical weathering cause resistant rocks to be etched into relief as a prominent landform. In the area under study the differential weathering of igneous and metamorphic rocks has produced various landforms such as residual hills or bornhardt, tors, domes, corestone, narrow ridges, etc., depending upon the variation of their composition and texture. In the Damodar trough resistant sandstone forms ridges while shales over low lying areas or intermediate low hills.

The most prominent and common surface accumulation of weathering products in the area are soils. Soils are both residual and transported. Most of the soils over upland are residuals directly derived from the weathering of the present rocks. The presence of decomposed granite or 'Gruss' provides visual impressions of the depth of such weathering through the tendency of rock breakdown which sometimes exceeds 15 metres in depth. In low lying areas and intermittent valleys soils of secondary origin are found. Their depth varies according to the gradient and distance from the uplands. The study of these soils proves to be of great help while reconstructing the physiographic history of the region.

In the Hazaribagh Plateau Region loose debris of rocks fragments in various stages of decay, known as regolith, overlie the bed rocks in the hilly areas. Most of the hillside slopes are covered with regolith which tend to migrate downhill under mass movement. The degree of decay of scree over scarps reveals that they are not older than Pleistocene. Lower foothills are the areas of coluvium derived from up slope scree or regolith covered areas and departed by sheetwash action of water. Undeveloped horizons of caluviums and their limited zonation around scarps further suggest that only after Pleistocene or particularly in latter parts of glacial phases conditions were favourable for their deposition and time was not enough for their decomposition into developed horizons.

The residual soils of the areas developed over granite, granite-gneiss and gneiss, mica schist are rich residual—brown to yellow in colour. The colour varies with the amount of the iron content in the soil. In the south-western parts of the area under study limited patches of secondary laterites can be observed as hard pan like features. The common occurrence of Murum bed of few centimetre thickness, in soil, in the western part of the region, is the result of the accumulation of iron oxide from upper laterite zones or iron rich soils, by the dominant action of downward leaching of rain waters. The Murum layer when exposed acts as capping on soil and crests of micro relief forms. The layer of Murum is absent towards the eastern half of the region.

## CLIMATE AND GEOMORPHIC PROCESS

The outstanding modern exponents like Büdel, Tricart and Coilleux, have argued for climatic geomorphological approach of landform study. Büdel (1973)[14] on the basis of distinct land assemblages tried to deduce the climate responsible for them. But his criterion of recognising 'landform style' of a region as the product of prevailing climate, as Tricart and Coilleux pointed out, was extremely general as the zonal criterion was not applied. Whereas Tricart and Coilleux (1973)[15] considering vegetation as an important geomorphic factor provided a regional division based on their variation in the process-complex consisting of physical and bio-chemical reaction. They put much emphasis on present day processes but also recognised the importance of paleo-geomorphic and paleo-climatic factors. Emphasis was on the interplay and denudational potential of linear erosion. Stressing the importance of climate De Martonne (1973)[16] has also singled out the forms of fluvial erosion as being particularly expressive of climatic variety.

His consideration of the effect of climatic change in the partial replacement of one landform assemblage by another to produce a polycyclic landscape, introduced the concept of climatogenetic geomorphology. The role of climate in dynamic equilibrium is further elaborated by Louis (1973).[17]

It is clear from the foregoing that both the erosional and depositional landscape of the Hazaribagh Plateau Region have resulted under multi bio-climatic regime. The rate of denudation has been influenced by vegetation cover. Under humid conditions of the geological past, it can be expected that vegetation would lead to the plentiful supply of carbon-dioxide and chelating agents to the soil. Under these situations deep penetration of rock weathering was favoured and many products of the weathering processes were recycled by the deep rooting trees, creating a condition of biostasy (Erhart, 1955)[18] over the Pre-Tertiary stable erosional landscape. According to Douglas (1969)[19] under such condition increased. "Complexity of the plant community is associated with increased soil development and a geomorphological trend towards a state of minimisation of work and even distribution of entropy throughout the drainage network. Under such conditions, most energy is used in the development of the plant community and not in erosion and removal of debris by streams." But positive movements of the surface during the Tertiary period had increased the basin relief and slope values of the Hazaribagh Plateau Region, disturbing the biotic equilibrium and accelerated the removal of weathered material. A condition of 'rhexistasy' or periods of natural disturbance prevailed. Again under Post-Tertiary changing climatic conditions, the condition of rhexistasy existed over the area under study.

It is also clear that the present day rates of denudation is also the outcome of the changes in vegetative cover brought about by human interference. The studies from other parts of the world has outlined that in Holocene, man himself was an increasingly potent agent of environmental change (Pennington, 1969)[20]. Early human occupation of the area under study is well authenticated and his effect may have been far reaching in terms of vegetation changes over a period of more than three to four thousand years, with the establishment of agricultural activities. Early human culture in the plateau region dates back to more than one lakh years (Sankalia and Nagar, 1972) but knowledge of their impact on the vegetation cover is not known to us. However, neolithic man did possess the axe, which could have been effective in forest clearance, but more importantly Mesolithic man may have utilized fire deliberately for driving game and clearing forest (Goudie, 1977). A number of Palaeolithic, Mesolithic and neolithic sites in and around the Hazaribagh Plateau Region, leaves little doubt that early man and his successors were capable agents of ecological change. We only know that gullying in most parts of the plateau can be safely related to the anthrogenic process introduced a few thousand years ago by the agrarian communities. However, if such substantial changes in environment have occurred only in the recent past then their influence over landscape in larger association must be tremendous.

In logical sequence a generalised account of the area can be thus: The area experienced humid conditions throughout the greater part of the Tertiary era. Erosion intensity initiated by the uplift was increased during Quarternary periods. The surfaces were exposed to intense denudation during climatic fluctuation and probably lost a high proportion of their former regolithic or soil cover. Some of the earlier surfaces may have remained in the interior forested zones. Widespread occurrence of tors, corestones, domes, gully erosion, etc. all record the changes in climatic and balance of geomorphic processes. The undulating plain and residual hills can be attributed to sub-tropical process of weathering and planation activities. It is not out of the way to accept the possibility of K. Cycle (Butler, 1959) in development of local surfaces as climatic variations during Post-Tertiary may have created landsurface instability (Thomas, 1974). The wide weathered mantled surfaces of the region under study may be dated to the Tertiary while stripping and dissection of planation surface may be associated with the Post-Tertiary climatic conditions and expeirogenetic movement experienced by the region. The recent denudation thus has greatly modified the Pre-Pleistocene landscape.

"The combination of palaeo-climatic landforms, current morphoclimatic processes and authropic effects goes a long way to explain the range of landforms." (Tricart and Coilleux).

## REFERENCES


1. J. Budel, The Climatic Geomorphic System, Climatic Geomorphology (Ed. by E. Derbyshire), Macmillan, London, 1972. p 104.
2. M.S. Krishnan, Geology of India and Burma, Higginbothames (Pvt.) Ltd., Madras, 1960, pp 324-343.
3. Ibid., p 325.
4. Ibid., p 325.
5. Ibid., p 325.
6. Ibid., p 325.
7. D.N. Wadia, Geology of India; English Language Bool Society and Macmillan, Madras, 1966, p 403.
8. A.S. Goudie, Environmental Change; Clarendon Press, Oxford, 1977, pp 83-84.
9. N.G. Sankalia and M. Nagar, Bihar in Prehistoric Periods, Dharmyug, May 7 and May 14, 1972, pp 24-26.
10. W.B. Langbein and S.A. Schumm, Yield of Sediment in Relation to Mean-Annual Precipitation ; Trans. Amer. Geophys. Un., Vol. 39, 1958, pp 1076-84.
11. G Singh, The Indus Valley Culture Seen in the Context of Post-Glacial Climate and Ecological Studies in North West India, Arch. and Authro in Oceania, 6, 1971, pp 177-89.
12. F. Machatschek, Geomorphology, Oliver and Boyd, Edinburgh, 1969, p 102.
13. D.R. Stoddort, Climatic Geomorphology in Water, (Ed. by R.J. Chorley); Earth and Man, Methuen, London, 1969, p 284.
14. J. Budel, The Climatic Geomorphic System in Climatic Geomorphology, Macmillan, 1973, pp 104-130.
15. J. Tricart and A. Coilleux, Morphogenic Systems and Morpho-Climatic Regions, in Climatic Geomorphology, Macmillan, 1973, pp 228-268.
16. E. de Mortonne, Climate : Factor of Relief, Climatic Geomorphology (Ed.), 1973, pp 61-75.
17. H. Louis, The Problem of Erosion Surfaces, Cycles of Erosion and Climatic Geomorphology, in Climatic Geomorpholoy, Macmillan, 1973, pp 153-170.
18. H. Erhart, Biostasie at Rhexistasic : Esguisse d'une Thiorie sur le role de la pe 'dogene' en tant que Phe 'nomina geologique, compt. Rend Acad. Sci. Francaise, Vol. 241, 1955, pp 1218-20.
19. I. Douglas, The Efficiency of Humid Tropical Denudation Systems, Trans. Geol. Soc., South Africa, Vol. 71, 1969, pp 131-134.
20. W. Pennington, The History of British Vegetation, English Univ. Press London, 1969.

23

# Contributions of L.C. King in Geomorphic Thought

*—G.N. Singh*

THE CONTRIBUTIONS of Lester C. King have heralded a new era in the history of geomorphic thought. Unlike other traditional geomorphologists, King does not belong either to Europe or to North America. His official position as Professor of Geology in the University of Natal, Durban, South Africa, has provided him an opportunity to become a sincere student and painter of arid and semi-arid landscape. If Davis was tempted to explain the bewitching scenery of the Appalachian, and Penck was taught lessons in physical geology by the young folded Alps, the age-long denuded plateau surface of South Africa provoked King to analyse its evolution. The humid temperate environment of North America made W.M. Davis to believe that the landforms fashioned under such conditions were "normal", and Penck's ideas of earth-movement dominated geomorphology were a byproduct of his keen observation of the rising Alps. In the same way the genesis of King's geomorphic thought can ably be conceived by the environmental knowledge of his area of study i.e. the semi-arid landscape of the South African Plateau.

Due to limited time and resources the present study is based on some of the important works of Lester C. King. King's ideas have appeared in geomorphic literature since 1930, but his major contributions appeared in the forties and fifties of this 20th century. His two text books, *The South African Scenery* (1942)[1] and *The Morphology of the Earth* (1962)[2] and a number of his learned research papers published in reputed international journals have enabled the assessment of some of his important geomorphic contributions. Various text books of geomorphology and research publications in geology and geography also refer to and comment upon King's signal contributions in this field. These studies have greatly helped in preparing a comprehensive note on the major contributions made by L.C. King in the advancement of geomorphic thought.

## THE PROBLEM OF SLOPES

The configuration of the earth is not flat. It is made of several slope segments. Therefore, the analysis of slope form, process, pattern, evolution and classification provide an important theme for current discussion in geomorphology. "The geomorphological attention has turned away from denudation chronology, quantitative measurement and analysis of slope have set new objectives."[3]

The problem of slopes has been studied by Gilbert,[4] Davis and others but their studies are mainly qualitative and based on the method of inference. So they talked about the "convexity of hill-tops" and flattening of the divides by down sculpturing. In the Davisian Geographical Cycle with the advancement of time "the valley sides are becoming more and more gentler than before. The slope is becoming fainter and fainter so that after some time the latest stage . . . the region is only rolling lowland."[5] Thus in the scheme for the normal cycle there is little room for straight erosional slopes, which are universally present in all parts of the world.

The most fundamental discussion of the present problem is that of W. Penck[6] (1924) who could recognise simple parallel retreat of the straight slope element for the first time. But it is essential to note that his slope elements were diastrophically or tectonically significant.

The starting point of modern hill slope analysis is the work of the British geomorphist Alan Wood (1942)[7], who distinguished four elements in a fully developed slope. King and Fair[8] (1944) while discussing slope development in Natal, South Africa, designated these four elements (Fig. 23.1) as : (a) Waxing slope (crest), (b) Free face (scarp), (c) Constant slope (debris slope), and (d) Waning slope (pediment). All these elements of slopes may not be present everywhere and locally one or more elements may fail to develop. 'This analysis applies to hill slopes all over the world in all stages of development in geological ages, as . . . . . hill slopes are the result of physical forces acting upon physical materials in ways that are now established as natural laws.'[9]

*Elements of a Slope (After L.C. King and T.J.D. Fair)*

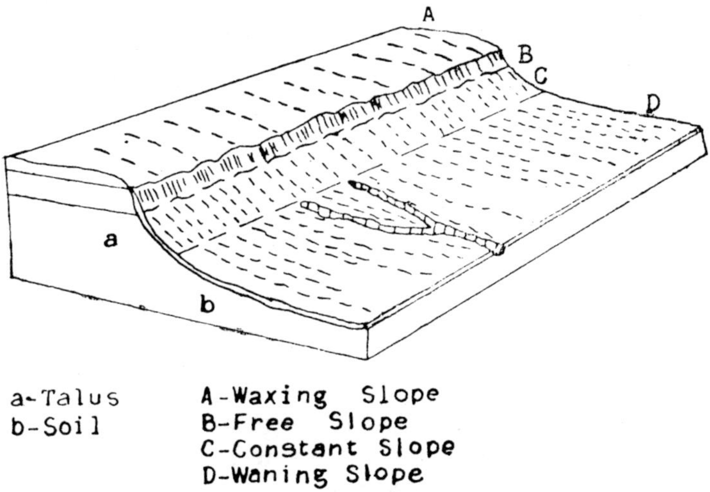


FIG. 23.1

ELEMENTS OF SLOPE

Before we analyse the "physical forces" or processes responsible for slope development, we will like to have an idea of elements of a slope, proposed by King. The waxing slope or the "Crest" is the summit area of a hill. It is limited in extent by the scarp and measures from a very small proportion to half of the total length of the hill-slope. This is the zone of weathering and soil creep. The profile is generally convex. It is longest in the areas of greatest surficial detritus.

The free face or "scarp" is the zone immediately below the waxing slope. Outcrops of bedrocks are usually exposed on the scarp. The steepness of the slope deter against any accumulation of sediments and soil formation. The disintegration of this scarp zone leads to tumbling down of fragments under the influence of gravity. This is the most active element of slope which retreats parallel. Thus back wearing, and not down wearing, of the slope as a whole is registered in this section. In areas of hard rocks such a free face is better developed. King has cited the example from Natal, South Africa, but such scarp faces have been reported by others in South East England, North America, India and other parts of the world.

The constant slope or "debris slope" originates below the scarp face. It consists of the debris slipped or fallen from the crest and scarp and accumulated at the free face. If the debris is weathered into fine particles it will be removed by erosion and consequently the debris slope extends backward with the retreat of scarp face above.

The waning slope or the "pediment" is a broad concave lower most slope extending from the base of debris slope to the bank of a stream or the limit of its alluvial plain. It is feshioned under the influence of running water. Usually a thin veneer of fine sediments covers this pediment slope. There is an abrupt change of slope between the hillside and the pediment. On the hill side the debris is coarse, while on the pediment fine debris mask its face. In areas of weak rocks, the break is smooth, but where the bed-rock is hard, the pediment and scarp face may be sharply angular (King, 1957).[10]

King has regarded the pediments as the most important of all landforms. They constitute over 60 per cent or more of the landscape in the eastern and southern districts of South Rhodesia. In King's view the pediment "is the fundamental form to which most, if not all, subaerial landscape tend to be reduced under erosion, the world over."

### Slope Processes

The landscape is moulded by the agencies of running water, wind, ice and mass movement operating on the earth's material. But except in glaciated regions where the land surface is covered with a sheet of ice, and in hot deserts where sand dunes are shifted by winds, all the world over, landforms are sculptured by the flow of water.

Different processes in different degrees are operative on all the four elements of a slope. The crest of a divide i.e. the area of waxing slope, is a zone of no-erosion. Flowing water cannot erode a slope until the erosive power exceeds the resistance of soil to erosion. Thus on all hilltops there appears a zone of no-erosion under water flow. The profile of this zone is convex. Weathering and soil creep are processes of land sculpture. Naturally in areas of greater infiltration of water into soil, this crest zone is wider. Thus on porous soil and bedrocks the convex hillcrest attains maximum width. For example the boulder clay country of midland Britain and the chalk country of Britain and France hillcrests exhibit abroad convexity. On the other hand, granite and quartzite terrain in Africa where the proportion of runoff is high, the minimal summit convexity is expressed everywhere.

In case the critical distance from water-parting is surpassed by erosion, the gullying rills steepen the profile. Horton[11] (1945) has shown how critical distance from divide may be calculated at which the hydraulic power of the flow is sufficient enough to overcome the strength of soil and vegetation cover and to erode a channel. "The critical distance from divide is controlled by surface flow intensity, the gradient, the hydrodynamic roughness of the surface and the strength of vegetation mat."[12]

The gullying of scarp face by rills deserves special mention. The erosive power of rills make the scarp face steeply concave. The closely spaced rills cut through superficial deposits to the bedrock which is often exposed as a scarp. Even where the bedrock is not exposed the soil cover is so thin and poor, that it is rapidly removed. Thus once the critical limit from water parting is surpassed, retreat of hill slopes and scarp becomes inevitable.

Below the scarps the removal of waste consumes the energy of rills and gullies. It is a fact that only the larger incision extends across the debris slope and pediment, carrying either seasonal or perennial stream; all the smaller gullies usually end at the foot of the steep slope. An essential feature of King's hypothesis is the contrasted model of water flow between the steep slope and the pediments. In view of Pugh (1966) "the change in flow appears to be the result, and not the cause, of the contrast in slope."[13]

On the flat zone of pediment generally the slope is less than 5°, though it may rise from 7°-13°, when the amount of water discharge is maximum. This discharge incorporates rainfall of the upper surface, minus infiltration and the precipitation upon the pediment itself. King has observed (1953)[14] that in such a case water will flow in sheets or as sheet flood. Turbulent sheet flood is perhaps the agent primarily responsible for the hydraulic profile developed by pediments. Further King has cited many instances where he has personally observed sheet floods, one to three inches thick and 30 feet apart. But where rainfall is insufficient, sheet water falling upon pediments may pass in rills only. "Pediments are best exemplified (King 1953) on hard rocks in semi-arid environment where the transport of land waste across the terrain is most efficiently conducted."

King's (1962) as the manner of water flow over land-surface is controlled by the physical laws. The four-elements or hillslope formations are important for landforms development all over the world in the regions of sufficient relief and under all climates where water flow is a prominent agent of denudation.

Table 23.1 shows the relation between slope elements and process observed by King.

TABLE 23.1

*Slope Elements and Their Relation to Slope Processes*

| Slope Element | Characteristics of slope | Slope Processes |
|---|---|---|
| Crest or waxing slope | Zone of no erosion, Convex profile | Physical and chemical weathering, Soil creep, rainwash. |
| Scarp or free face | Straight in profile with a slope of 30° or more fully developed in hard rocks | Mass movement in the form of slipping and sliding. Rill action, basal erosion. |
| Debris slope or constant slope. | Concave in profile, covered with talus and sand. | Transport by water, rainwash, mudflow. |
| Pediment or waning slope | Concave in profile, gentle sloping ground with a low angle (6°-7°), accumulation of slope waste. | Linear and sheet flow of water, minor creep and mass movement. |

Table 23.1 presents a summary of the slope elements with their characteristic forms and processes under the influence of

water flow and various forms of mass movements as gravity flow. King has stressed the role of water flow and mass movement but he has also remarked that in nature both the processes do not act independently : in fact, they act and abet each other.

Recent studies of slope owe much to the contributions made by L.C. King. Others who have contributed on slope analyses include Savigear (1960), Schumm (1966), Dalrymple, Blong and Conacher (1968), and Young (1963, 1972). A comprehensive summary of their works can be found in the work of C.A.M. King (1971)[15]. Their works are based on the findings of L.C. King but emphasis is placed on the measurement of slope profiles in the field leading to recognition of various slope elements and analyses of processes involved in their evolution.

## STANDARD CYCLE OF DENUDATION

L.C. King has expressed the view that all over the world denudation is run by flow of water, of course, with the exception of glaciated areas and the arid land covered with sand dunes. The normal cycle of erosion propounded by W.M. Davis is applicable only to the humid areas of the world. But King asserts that the term 'normal' should be deleted, and even if this term is applied, it should be applied for semi-arid regions. King's standard cycle of denudation is the cycle of pediplanation. In his earlier work, *South African Scenery*, King advocated only the cycle of pediplanation, but as this cycle operates all over the world in his later work ***The Morphology of the Earth***—he has preferred "the Standard Cycle of Denudation" and "the Standard Epigence Cycle of Erosion (pediplanation cycle).

Climate is a major factor in landsculpture. But it is not the fundamental controlling factor at all. King has expanded his vision to cover all the landforms of the world both past and the present under his scheme of pediplanation cycle. In his opinion the bornhardt or inselberg landscapes are not typical features of arid and semi-arid regions of East Africa or Southern Africa as Passarge regarded them. Such features have also been reported from maritime regions of Brazil, the sub-tropical coast of Natal, the humid regions of the Blue Ridge Mountains of Virginia, the cold temperate part of south-west New Zealand and Africa. King (1948)[16] has stressed that bornhardts are formed in areas of plutonic rocks, where plenty of relief is available. Similarly, pediments are not only confined to the arid and semi-arid regions, they have been reported even from humid Europe and North America (Dury op. cit. 1972). Even Davis (1930)[17] has mentioned concave rock floors occurring in abundance along the belts of the folded Appalachians. According to King these should be called pediments. Thus in broad view, the pediment should be taken as the fundamental landform to which landscape have tended to be reduced all over the globe. The difference in climatic elements such as temperature, rainfall, etc., are not fundamental; they differ in degree alone.

The influence of bed rocks upon the development and maintenance of scarp faces, an essential feature of King's pediplanation cycle have not been ruled out. As the hard massive rock types have their definite control upon the development of inselberg, the weak rocks also affect the landscape in a different way. In regions of hard rock as in Africa, scarp retreat and land surface with multi-concave is dominant, whereas weak rocks surface may show concavity and convexity as in parts of Northern Europe. In such regions convex upward slope, meet with the concave pediments smoothly. In such cases prolonged down wearing is more important, than backwearing recognised as dominant process in the areas of hard rocks.

**Pediplanation Cycle**

King accepts the general principles of cycle of erosion introduced by W.M. Davis, in which a series of denudational changes occur involving stream incision and increasing relief, interfluve reduction and diminishing relief, and finally planation. The chief defect in Davisian philosophy, in the eye of King lies in the fact that there is no provision for parallel retreat of hill slopes, which is perhaps the most important activity of the pediplanation cycle.

In the Davisian trio of landscape development structure, process and time, King regards the process most important.

"Landscape is dominantly a function of the processes operating to mould it." The processes responsible for cyclic development of landforms include (a) river incision, (b) scarp retreat and pedimentation and (c) soil creep in low relief. The dominance of these activities respectively and successively define the stages of youth, maturity and old age in the landscape. In the evolution of the continental landscape these stages are carried progressively inland. The progress of the cycle depends upon factors like the size and spacing of rivers, the nature of uplift, the rock type and the physical agencies involved.

**Youth** : The stage of youth is characterised by young streams cutting deep ravines and gorges. By rapid downcutting maximum relief is attained. Rainwash and weathering produce a stable angle of slope dependent upon the structure and processes acting upon it. This stage is of shorter duration in comparison with other stages of the cycle.

**Maturity** : The stage of maturity is dominated by activity on hillslopes. River incision practically ceases. The hill slopes and valley sides retreat at virtually constant gradient with a fine texture of dissection, interfluves are rapidly consumed. In case the relief is low the upper convex curve of the crest may meet the lower concave curve of the pediment. But in areas of high relief and hard rocks the reduction of interstream upland continues till the pediments extending back from two sides. Even in very advanced stage of the cycle the upper surface of the divide may remain without any alteration, thus preserving the remnants of ancient surfaces. King has identified on Shoshi hill in Shamva district of Southern Rhodesia an outlier of the main plateau.

This stage of maturity is characterised by a general angle of declivity upon a majority of hill side. Between steep retreating hillslopes and widening pediment a distinct break of slope is often apparent. In regions of hard bedrock break may even become sharper. The secrecy of mature landscape lies in the mode of development of hillslopes in relation to bedrocks. The hillslope may originate either as a valley side or as tectonic features due to faulting, warping and tilting. Once generated these scarps will go on retreating like a nickpoint of a river. The rate of retreat of the continental scarps has been calculated by King as about 1 foot in 100 years.

The scarps may not always remain clean and wall like vertical. Some scarps may degenerate into zones of dissected country, separating upper and lower planation surfaces.

As the scarps retreat the pediments widen. From either side of the stream twin pediments extend laterally. Coalescing pediments give birth to pediment plains with standing steep-sided residual hills. In southern Rhodesia, King suggests about 65 per cent of the land surface is constituted by pediments and the rest by residual hill masses.

The original declivity of about 10° or more is reduced to 1/2-1° only. Perhaps sheet flood and rill wash are responsible for this regrading. Even in the case of very flat pediments beneath a thin veneer of transported sediments, a smoothly cut unweathered rock surface occur.

**Old Age.** The senility of landscape has certain characteristic features. The hills are reduced to small rocky Koppies and the pediments stretch for miles, producing a multi-concave landscape. "The sum of all pediments together with residual nubbins (monadnocks) and any flood plain of major river constitute a pediplain". King has defined a pediplain as "a land surface of low relief, multi-concave upwards and resulting from prolonged erosion." At senility a continental erosion surface is a complex assemblage of closely related surfaces, chiefly pediments. Weathering and downwearing processes modify the pediments and introduce a broad convexity across the low interfluves.

## Importance of the Pediplanation Concept

The concept of pediplanation is the outcome of the works of W. Penck, Kirk Bryon and Jessen. L.C. King accepts the basic idea of cycle of erosional changes expounded by W.M. Davis, but he is very critical of his emphasis on universal down wearing which is a negative concept responsible for bringing sterility in geomorphic thought. Some of the special advantage of the pediplanation concept are as follows :

(i) Scarp production is recognised as an inherent property of landscape. Thus the abundance of scarps in continental landscape is explained.

(ii) As two land surfaces, one above and one below the intervening scarps are inherent and the polycyclic landscapes are explained.

(iii) The unaltered summit levels in landscapes of all continents receives explanation.

(iv) The frequent discordance of topographic and geologic boundaries are clearly explained. Probably, in all instances the cyclic scarps have transgressed from weak to resistant formation.

(v) The multi-concave forms of certain senile landscapes are explained.

(vi) A single cyclic regimen has been envisaged to fit all the non-glaciated landscape of the world.

(vii) Cyclic land surfaces independent of the bedrock and of the drainage pattern, is the natural result of pediplanation.

(viii) The features of multicyclic continental landscape are capable of measurement, comparison mapping and statistical analysis.

(ix) Inter-continental correlation of cyclic land surfaces becomes possible.

Thus to quote King (1962) "the consequences of pediplanation hypothesis and its application in the interpretation of global landscape form a major contribution to historical geology."

## A STUDY AND SYNTHESIS OF WORLD GEOMORPHOLOGY

L.C. King's greatest contribution lies in the synthesis of world landscape. For him geomorphology includes two functions[18] : (a) description of present day landscape, and (b) elucidation and explanation of their histories. He has himself accepted that the main theme of his classic work, the *Morphology of the Earth* (King 1962, p 163) is the synthesis of the continental landscape. He has faithfully painted the evolution of the landforms on the outer part of our planet principally from the Mesozoic era to the present day. He has travelled widely in different corners of the world for this study and has done intensive field work in the continent of

Africa. Thus one of the chief virtue of King's study is "its critical scrutiny of processes, gained by first hand field observation of slopes undergoing denudation."[19]

The glimpse of synthesis of world landscape appears in King's work as early as 1950 when he interpreted widely separated surfaces in Africa, Asia, North America, Europe, South America, Australia and India as ancient pediplains dating back to the Cretaceous period.[20] Referring to the evolution of the Indian landscape, he remarked that the Gondwana surface was heavily warped down in the Cretaceous time in Deccan, and "was covered by Deccan Trap, across which the Indian landscape was carved at a higher level during the early Tertiary."[21] In his next work *'Canons of Landscape Evolution* (King 1953, op. cit.) King has outlined the universality of his elements of hillslopes and the pediplanation cycle vigorously. In this way King has discarded the Normal cycle of erosion expounded by W.M. Davis.

In his monumental work *The Morphology of the Earth : A Study and Synthesis of World Scenery*, King has presented a full view of global landscape development. This book alone is sufficient to acquaint oneself with the major geomorphic ideas of L.C. King. A full review of the book is not possible in the short span of this paper, so only the salient points have been considered.

The seven hundred page book is divided into five parts, each one spanning over 100 to 125 pages, except part C which has about 170 pages and part F, the Review covering about 70 pages. Part A deals with the "Basis of World Scenery" including the structure of the earth's crest, the unity of Gondwana land, and the continental structure, dividing into two parts the shields and the "Mobile Belts" or the geosynclines and orogenic cycles. Part B is concerned with the "Development of Continental Scenery" both by degradation and aggradation. It is in this part that we conceive King's idea of standard hill slopes, standard cycle of denudation, landforms of aggradation, tectonics and landscape evolution, and the ages of land surfaces. Parts C, D and E depict the "Scenery of the Plainlands", "the Scenery of the Mountain Lands," and "the Geomorphology of the Ocean Basins", respectively. In the "Scenery of the

Plainlands" King first analyses the evolution of African scenery in various landscape cycles starting from carboniferous up to the present day, and then he examines the extension of "Cyclic land surfaces in the remainder of Gondwana land" "North American Plainlands" and "Cyclic Land Surfaces in Europe and Asia".

The scenery of the mountain lands have been discussed chronologically in three chapters covering "the Palaeozoic mountain chains", "Mountain Ranges of Mesozoic and Cainozoic Time" and the "Modern Mountain Ranges". The ocean basins, as they cover about 70 per cent area of the earth, deserve special attention in King's work, wherein the forms and features of the Atlantic, the Indian, and the Pacific Oceans have been dealt with in detail.

In part F of the book, a review of some recent concepts has been presented. Burning topics like Cymatogeny and Isostatic anomalies, time factor in landscape evolution, the sources of crustal activity, the nature of subcrust and the theory of continental drift have been examined.

Much has been said about the popularity of this book which is often quoted in geomorphic literature. The book does not treat geomorphology region by region in space; it deals with the geomorphic evolution historically and chapter by chapter in time. King (1962, op. cit., p 246) himself accepts that "the object of this book is not to produce a compendium upon African topography but a scientific geomorphology that may be compared with geomorphologies of other continents."

## CYCLIC LAND SURFACES OF THE WORLD

In the study of the world's scenery King has applied certain premises. Firstly, pediplanation is responsible for the development of continental landscape ; secondly, pediplanation acts cyclically in collusion with intermittent earthmovement; thirdly, the number of continental denudation cycles is small, and finally, a certain range of local variations in minor landforms is possible under the influence of local tectonics, heterogeneity of bedrock, and the changes of climates. Based on these premises the essential features of major denudational cycles can be recognised, and their imprints in different corners of the world correlated. King has initiated the study of cyclic land surfaces with Africa and his objective is not merely to afford a description and history of the African landscape but to deduce from this region a natural scheme on a global scale, with suitable modifications.

In Africa the major landscape cycles treated by King (1962, op. cit., p 246) in chronological order are as follows :

(a) The 'Pre-Karroo'' and "Intra-Karroo" landscape cycles (jointly known as Karroo Cycle after Karroo system) which extended from Carboniferous rolls to the Triassic period.

(b) The 'Gondwana' and 'post-Gondwana' landscape cycles (jointly known as Gondwana planation cycle) from Jurassic to Cretaceous. In the Cretaceous the fragmentation of Gondwana land occurred, which was followed by the post-Gondwana dissection.

(c) The early-Cainozoic "African" landscape cycle of extreme planation forming the most perfectly planed surfaces of Africa, much dissected now by later cycles.

(d) The Quarternary canyon cutting and scarp retreat following the plio-pleistocene cymotogeny.

(e) Coastal drowning has been recorded in recent times. Recent alluvium in inland valleys, and recent sand dunes in coastal areas characterise the present landscape cycle. The following Fig. 23.2 shows the relations of cyclic land surfaces in Africa.

*Relation between Cyclic Land Surfaces in Rhodesia*

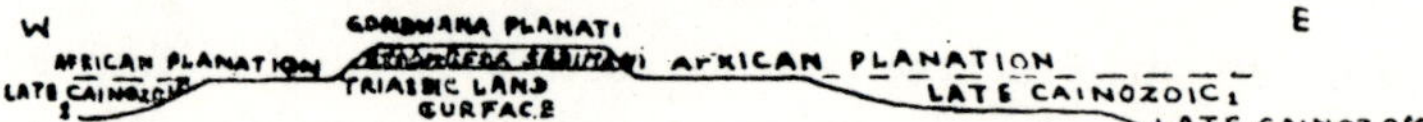


FIG 23.2

L.C. King feels the necessity of the late Mesozoic continental drift to explain and correlate the cyclic land surfaces of the Gondwana land. In his opinion right from late-Mesozoic up to the recent period of the earth's history the morphological development of all the fragments of Gondwana land have been identical. All the southern continents including India "have experienced similar type of tectonic displacement occurring at closely synchronous epochs." During the longer time intervals between tectonic episodes, cyclic landscapes on a continental scale were carved out by prolonged denudation.

Table 23.2 shows the identical character of cyclic land surfaces over different fragments of the Gondwana land.

The identification of widely separated cyclic land surfaces as a result of pediplanation is of course a sweeping generalisation. It has aroused wide criticism of King's interpretation. "Few geologists would draw such a sweeping conclusion either as to the origin or the age of the surfaces interpretated by King as pediplains.[22] But later workers from various parts of the world have supported King's contention. Singh (1969)[23] while working on the erosion surfaces of Chotanagpur Highlands, India, has supported the idea of King's planation surfaces. Similarly Bishop (1966)[24] agrees fully with Kings view of cyclic landscape development in eastern Africa. He states plainly "the continent-wide cyclic landscapes recognised by King under the headings Gondwana, African, or various Cainozoic stages, seem likely to prove the broadly correct relevance of events for many areas. The concepts involved to explain and date the cycles involved may be perfectly valid."

Recent works on continental drift (Tarling and Tarling 1972)[25] show that the disruption of Gondwana land was preceded by a violent volcanic lava flow 160 million years ago. The lava spread on various fragments of Gondwana land dating back to the Jurassic and are, therefore, contemporary. Examples of such lava are Tasmanian Dolerite (Australia), Rajmahal Traps [India, Yemen Traps (Middle East)], Ferrar Dolerite (Antarctica), and Karroo lava (South Africa). This simultaneous volcanic episode on the southern continents confirm King's view of cyclic development.

## OTHER CONTRIBUTIONS

Besides his important contribution in the study of world geomorphology, King has treated various topics of geomorphic interest. It has been stated that he is an advocate of

TABLE 23.2

*After L.C. King (1962 op. cit.)*

| Epoch | Brazil | Southern Africa | Peninsular India | South-eastern Australia |
|---|---|---|---|---|
| Recent | Raised beaches, drowning of lower courses of rivers. | Raised beaches and drownings of coastal lagoons. | Raised beaches and coastal lagoons. | Raised beaches, drowned river mouth, Murray Basin alluvia. |
| Quarternary | 'Paraguacu' cycle of canyon cutting in coastal hinterland. | 'Congo' cycle of deep gorge cutting in coastal hinterland. | Gorges and water falls of rejuvenated river systems. | 'Post-Kosciuko' canyon development of maritime slope. |
| Plio-Pleistocene | Epeirogenesis on earlier pattern of deformation. | Cymatogeny an old pattern. | Marked uplift and warping. | Kosciusko Cymatogeny |
| Pliocene | Second river-valley phase of late Cainozoic denudation coastal plain now 700-1000 feet at the inner edge. | Second phase of late-Cainozoic valley development. Zululand-Mozambique coastal plain-700 feet maximum. | Valley excavation | Valley excavation preceding "Newer" Basalls. Coastal plain rising to 700 feet in Gippsland. |
| Mio-Pliocene | Minor continental uplift. | Slight uplift. | Minor movement. | Minor uplift in south-eastern Australia. |
| Miocene | "Velhas" denudation cycle, reaching to the heads of many valley systems. | Broad valley-floor pediplains widespread into the early cainozoic planation. | Main dissection of early-cainozoic landscape. | "Monan Planation cycle. |

| | | | | |
|---|---|---|---|---|
| Mid-Cainozoic | Widespread, generally uniform uplift. | Widespread, regular uplift. | Widespread uniform uplift. | Widespread uplift with local depression (Murray Gulf). |
| Oligocene (with Eocene in Brazil and some places in Australia). | "Sul-Americana" landscape of extreme planation. | "African" cyclic landscape of extreme planation forming the most widely extended surface of Africa. | Extreme planation of plateau India and Ceylon, surviving only on smooth ridge crests. | "Great Australian pediplain" Fundamental surface of South-eastern and Western Australia. |
| Mid-Cretaceous | Mid-Cretaceous disturbances. | Mid-Cretaceous disturbances. | Cretaceous disturbances. | Late cretaceous disturbances. |
| Early to Mid-Cretaceous. | "Post-Gondwana" landscapes in high country along upwarpings. Planation not advanced, e.g., mountain-land of southern Minas-Gerais. | "Post-Gondwana" landscapes, usually in the vicinity of uplands incompletely planned, e.g., highlands of southern Rhodesia and Benguella. | "Post-Gondwana" highland landscapes of southern India and Ceylon. | "Post-Gondwana" landscapes not certainly identified unless making the high plateaus of the Australian Alps. |
| | Dismemberment of Gondwana land. | Dismemberment of Gondwana land. | Dismemberment of Gondwana land. | Dismemberment of Gondwana land. |
| Jurassic | "Gondwana" landscape of extreme planation and formerly wide distribution. | "Gondwana" landscape, attaining a stage of extreme planation over most of central and southern Africa. | "Gondwana" Summits of Nilgiri and Cardamom Hills. | "Gondwana" planation known certainly only as unconformity below continental Jurassic deposits |

continental drift, which arises from his necessity of the unity of Gondwana land. In his several papers he has examined various problems related with continental drift.[26] Recent studies of palaeomagnetism of rocks and plate tectonics have proved beyond doubt that the continents have drifted apart in geologic times.

King has also applied the principle of isostasy in explaining various earth movements. He is of the opinion that isostatic disturbances lead to the genesis of earth movements which restore isostatic balance.

King's treatment of "cymatogeny" deserves special mention. He has proposed this term (1961)[27] to denote "undulating agency." It is one of the important concepts in geomorphology. It involves only vertical displacement. The earth's surface is thrown into gigantic undulations, sometimes measuring hundreds of miles across, and with vertical displacement amounting to thousands or even more than ten thousand feet. King (1962, op. cit., p 638) has defined cymatogeny in the following manner : "cymatogeny is a mode of major vertical (radial) deformation of the earth's crest wherein the fundamental structure induced at intermediate depth is a steeply inclined tectonic gneiss, with or without minor igneous activity, which is usually basic only. At the earth's surface it is expressed by arching sometimes thousands of feet in height and hundreds of miles in width. Rift valleys due to tension, and wedge uplifts, due to unbalanced forces, are frequent minor attributes." In the evolution of the African landscape the role of cymatogeny as a characteristic tectonic activity of late Cainozoic and Quarternary times, has been highlighted by King. In his opinion perhaps the rift valley of Africa is the product of cymatogenic arching (Fig. 23.3).

*Opening of a Tensional Rift Valley along the Crest of a Cymatogenic Arch (After L.C. King)*

FIG. 23.3

## CONCLUSION

The contributions of L.C. King have evolved new thinking in modern geomorphology. His major contributions include the study of hillside slopes and processes, standard cycle of denudation (pediplanation), synthesis of continental landscape and cyclic land surfaces, the theory of continental drift, and cymatogenic arching of landscape. Some of his geomorphic ideas have stood the test of the time and they are now widely accepted.

## REFERENCES


1. L.C. King, *South African Scenery,* First Edition, and Oliver Boyd Ltd. London, 1942.
2. L.C. King, *The Morphology of the Earth,* First Edition, Oliver and Boyd, Edinburgh and London, 1962.
3. G.H. Dury, "A Partial Definition of the Term Pediment with Field Tests in Humid-climate Areas Southern England," *Institute of British Geographers,* Tran. No. 57, 1972, p 139.
4. G.K. Gilbert, "Convexity of Hill-tops ", *Journal of Geology* Vol. 17, 1909, pp 344-50.
5. W.M. Davis, The Geographical Cycle, in Geographical Essays, Ed. J.W, Johnson, Dover Publication, New York, 1954, (1899) p 255.
6. W. Penck, Die Morphologische Analysis, English translation—Morphological Analysis of Land Forms, (Macmillan, 1953), 1924.
7. A. Wood, "The Development of Hillside Slopes," *Proceedings of the Geologists' 'Association,* 53 128-40, 1942.
8. L.C. King and T.J.D. Fair "Hillslopes and Dongas" *Trans. Proc. Ged. S.S. Africa,* 1944, 47, pp 1-4.
9. L.C. King, *The Morphology of the Earth,* First Ed. Oliver & Boyd, Edinburgh and London, 1962, p 143.
10. L C. King "The Uniformitarian Nature of Hillslopes" *Trans. Geol. Soc.* Edinb., Vol. 17, 1957, pp 81-102.
11. R.E. Horton, "Erosional Development of Streams and their Drainage Basins : Hydrophysical Approach to Quantitative morphology; *Bull of the Geol. Soc. of America,* 56, 1945, pp 275-370.
12. M.J. Kirkby, "Erosion by Water on Hillslopes, in *Water, Earth and Man*—edited by R.J. Chorley, Methuen & Co. Ltd., London, 1969, p 231.
13. J.C. Pugh, The Landforms of Low Latitudes in *Essays in Geomorphology* Edited by G.H. Dury, Heinemann, London, 1966.
14. L.C. King, "Canons of Landscape Evolution, "*Bull. Geol. Soc. Am.* Vol. 64, 1953, pp 721-752.
15. J.C. Doornkamp and C.A.M. King, *Numerical Analysis in Geomorphology,* Edward Arnold, London, first Edition, 1971, pp 182-187.

16. L.C. King, "A Theory of Bornhardts", *Geog, Journ.* Vol. 112, 1948, pp 83-87.
17. W.M. Davis, "Rock Floors in Arid and in Humid Climates" *Journ. of Geol.* 38, 1930, pp 1-27.
18. L.C. King, *South African Scenery, A Textbook of Geomorphology*— 2nd Ed. Oliver and Boyd. Edinburg, London, 1951, p 1.
19. S.W. Wooldridge and R.S. Morgan, *An Outline Geomorphology : The Physical Basis of Geography*, 2nd Edition, 1959, p 244.
20. L.C. King, "The Study of World's Plainlands : A New Approach in Geomorphology," *Quart. Jr. Geol. Soc.* London, 106, 1950, pp 101-127.
21. Ibid. p 111.
22. W.D. Thornbury, *Principles of Geomorphology*, John Wiley and Sons, New York and London, First Edition, 1954, p 296.
23. R.P. Singh, "*Geomorphological Evolution of Chotanagpur Highlands, India*; *N.G.S.I.* publication Varanasi-5, 1969, pp 108-120.
24. W.W. Bishop, "Stratigraphical Geomorphology : A Review of East African Landforms, "in *Essays in Geomorphology*, Edited by G.H. Dury, Heinemann, London, 1966, p 140.
25. D.H. Tarling and M.P. Tarling Continental Drift, Penguin Books, 1972.
26. L.C. King, "Necessity for Continental Drift", *Bull. Am. Assn. Petrol. Geol*, Vol. 37, 1953, pp 2163-2177.
27. L.C. King, "Cymatogeny". *Tr. Geol. Soc. S. Africa*, Vol. 64, 1961.

# 24

# Climato-genetic Geomorphology : Review and Reassessment—A Case Study of the Barakar Basin

—*Nageshwar Prasad and Shukdeo Sahu*

A SIGNIFICANT trend of geomorphology in the past few decades has been the attempt to interpret an area under the changing climates during the geological past. The Davisian method was content with the beginning of a cycle—an initial surface and prediction of an end—a peneplain, "making full allowance for accidents, climatic or volcanic and interruptions, i.e., changes of base-level." It assumed "a period of standstill which permitted a cycle to nearly featureless condition." This idea fails to face the need of employing morphological evidence to elucidate earth-history. The weakness of Davisian geomorphology has, consequently, "prompted recent developments in three main directions, all overlapping to a certain extent. They may be designated as tectonic, climatic and quantitative geomorphology."[1-3]

Here our main aim is to discuss and examine the significance of climato-genetic geomorphology in the geomorphic interpretation of the Barakar Basin.

## CLIMATO-GENETIC GEOMORPHOLOGY

Climato-genetic geomorphology is one of the two distinct concepts of climatic geomorphology. It has been opined that within the present major climatic zones, there are some characteristic complexes or relief expressions, and the investigation of these features, influenced by the present climate is the business of '*climatic geomorphology*'. But the most important modern development, however is the "recognition that existing relief is not everywhere merely a synthesis of present structural features, but contains also older relief generations originating in earlier period of climatic and tectonic history.

The investigation of these varied generations in the evolution of the relief is the task of *climato-genetic geomorphology*."[4] Thus Budel[5] distinguishes the two distinct concepts of climatic geomorphology by the terms climatic geomorphology and climatogenetic geomorphology. As so defined, "climatic geomorphology *sens stricto* is largely a systematic study, and climato-genetic geomorphology becomes the historical analysis of particular areas."[6]

## THE CLAIMS OF CLIMATO-GENETIC GEOMORPHOLOGY

The emerging school of climatogenetic geomorphology has gained strongest support among continental writers (Bakker and Levelt, Budel, Carlston, Chorley, Common, Cotton, King, Louis, Martonne, Passarge, Pacsi, Tricart and Cailleux, Wilson and others)[7-32] on the following postulates :

(1) The claim has been made that climato-genetic geomorphology differs in aim, philosophy and method both from the traditional Davisian geomorphology and the other geomorphic philosophies.

(2) "Since climate has demonstrably changed during the Tertiary and Pleistocene, and continues to change, sets of climatically-controlled landform features are being continuously superimposed on each other, so that the landforms of any one area bear the imprint of climates no longer operative."[33]

(3) Thus "the present day landforms are for the most part not the result of present climates, but the product of Pleistocene processes. Consequently the present day active forces and their morphological expression depend on the climatic zones in which they occur."[34]

(4) The claim has been made by the climato-genetic geomorphologists that the significance of *petrovariance* has been over-estimated by the structural geomorphologists. They hold the opinion that "the ability of the rocks to withstand erosion does not depend on their physical but rather on their *morphological hardness*. This varies according to climate. Granite, for example, is one of the hardest rocks in the mid-latitudes whereas in the moist tropics, where deep

weathering prevails, it is soft and sand like."[35]

(5) Similarly *epeirovariance* is not, as a whole, a determining factor of the relief features. Epeirovariance governs the ultimate height and general steepness of slopes. Of course, "the form of an uplifted block or a step produced epeirogenetically appears in its pure form only in the relief of the ocean bed. On land, the effect of exogenic erosion processes begins at once, changing the 'raw block' into a greatly reduced, many-membered relief of flat surfaces, ridges, slopes and valleys which we see today."[36]

(6) "Often 95 per cent of the relief consists of remnants of older landform generations."[37]

(7) In King's opinion, pediplains are typical of semi-arid tropical zones, but may develop at a lower intensity also under moderately humid climatic conditions. He considers "the difference between the forms developed in arid, semi-arid and moderately humid climatic zones to consist merely in the stages of intensity of their development."[38]

(8) King[39] expressed the view that pediplanation is the general form of relief smoothing, substituting, as it were, the pediplanation concept instead of the Davisian peneplain concept.

(9) Since pedimentation is considered the general process of planation, the pediplains outside the semi-arid zones of today would all be fossil landscape forms.

(10) So, the climato-genetic geomorphologists hold the view that the zone of humid tropical climate of today must have had sub-humid, arid or semi-arid climates in the geological past.

(11) Since the exogenous relief shaping processes are related to only 5 per cent of the existing continental landforms, the emphasis has been given to the study of fossil erosion surfaces.

(12) Since climatic variance controls all dynamic processes, all over structural components, the climatic geomorphology will become the decisive part of geomorphology of the future.[40]

(13) Thus "the weaker the present-day processes are, and

the shorter the period of their operation in an area, the more evident are the older relief generations and greater is the part played by climato-genetic factors in their formation in comparison with that of tectono-genetic factors."[41]

Now in the light of these postulates, we have to establish whether the relief features of the Barakar Basin are climato-genetic relief generations or epeirogenetic relief generations, i.e., whether the 'fossilization' of the earlier elements has been caused by a different tectonic behaviour in former times or by a different climate.

## THE CLAIMS OF FOSSILIZATION IN THE BARAKAR BASIN

### The Barakar Basin

Lying between 23°41′30″ and 24°31′30″ N latitudes, and 85°10′00″ and 86°54′00″ E longitudes, the Baraker Basin is a purely geomorphic unit consists of the whole catchment area (7026.1 km$^2$) of the Barakar river in the State of Bihar (Fig. 24.1). The Basin is a part of the Chotanagpur Highlands which are the easternmost continuation of the Indian Peninsular upland. The Basin lies in the hilly country of Chotanagpur and occurs mainly in the 'Lower Chotanagpur Plateau (305 m Level) region.'

The Barakar Basin is a region of physical inequalities. It presents a rich panorama of topographical features, characterized by scarp-zones, erosional surfaces and residual hills, etc. The general slope of the region is in the south-east direction. Extensive areas lie between the 152 and 457 m contours, which present a variety of morphologic features. River valleys, most of them long, deep and terraced and a few of them with gentle slopes, alternate with low elongated ridges, sometimes hilly spurs, projecting from higher tableland or elongated into chains of hills and ridges.

The area under study has been a part of Archaean Basement Complex which has been subjected to many cycles of erosion. Therefore, this is a polycyclic region.

A large part of the region is covered with rich forest and scrub-jungle. These protect the area at places from severe soil-erosion. But great elevations and steep slopes facilitate quick run-off which renders the streams ephemeral and ponds empty scars in summer.

*Present Surface in Chotanagpur* (*Generalised*)

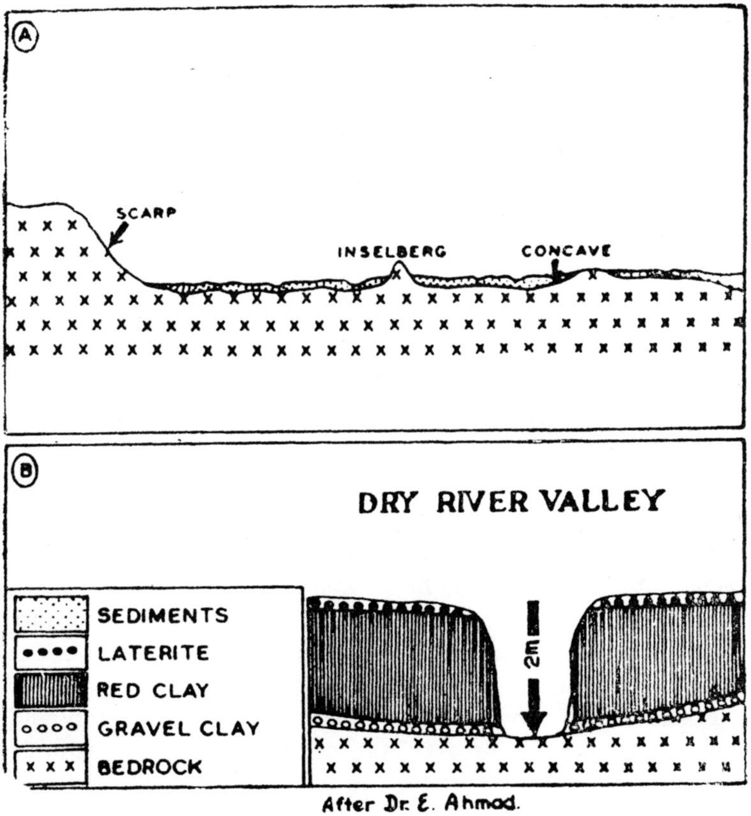


After Dr. E. Ahmad.

FIG. 24.1

Other features of morphological interest in the region are waterfalls and springs. Waterfalls and rapids have greatly influenced the topography of the area, causing the long profile to be one of abrupt break of slope.

Thus, in brief, it may be said that the area presents a variety of landscapes characterized by scarps, rugged topography long, deep and terraced river valleys; waterfalls and rapids residual hills (monadnocks, tors, etc.), meanders, gullies and other features of morphographic importance. The Basin has a tongue-like projection towards south-east and is roughly rectangular towards west.

The 305 m surface constitutes the core of Barakar Basin. On the south-eastern side, it gradually slopes down to about 100 metres at the confluence of the Barakar with the Damodar river. On the south-western corner, it ascends up 600 m which is a part of the Upper Hazaribagh plateau scarps. The transition from 305 m to 381 m surface is gradual on three sides of the plateau. Thus the general topography is level land except in the south and south-western parts which comprise of scarps and hills. But if we include the lofty Parasnath Hill, the relief accounts for about 1265 metres—the Parasnath being 1365 metres high and the lowest area near Chirkunda-Barakar being only 100 metres above the sea level.

The Barakar Basin has been characterized by multi-cycle of erosion. Observations have shown that an ideal cycle is rarely allowed to run its full course. The Barakar Basin has never been reduced to zero, i.e., up to sea level.

The Barakar Basin falls under the sub-tropical monsoon climate where precipitation comes in the form of rainfall with an average annual of 120 to 125 cm with alternate dry and wet periods of March to May; and June to September respectively. The mean maximum temperature is about 38°C to 45°C during May and the average minimum temperature during December-January ranges from 3.5°C to 7°C. The monsoon generally starts by the middle of June and lasts till the end of September, the wetter months being July and August, accounting for over 80% of the total annual rainfall. Winter rainfall in December to January is irregular and the amount is negligible. Much of the rain water is returned directly to the atmosphere by evaporation. Heavy storms average three or four per monsoon, mostly in July and August of about 3″ (76 mm) of rainfall in a day. These storms tend to centre over the northern or north-western sections of the Basin. Prolonged storms over 4 to 5 day periods occur every third or fourth year, the heaviest being 12″ (305 mm) in 5 days in 1913 and 10″ (254 mm) in 7 days in 1928 Intensity of rain per hour rarely exceeds 5″ (127 mm), but rains of 7.68″ (195 mm) per hour of 5 minutes' duration have been recorded. We know that unless the rain continues for an hour or so, no overland flow may be expected. Also rapid runoff is caused by uniform and heavy rainfall which the Basin lacks except for some weeks in the rainy season.

### FOSSILIZATION

Among the fossil erosion features is probably the bed-rock surface of the region which at present is out of touch with the subaerial processes, because between the bed-rock surface and the upper soil surface is a layer of sediment concealing the bed-rock. The bed-rock is designated as 'dead erosion surface' as it is not actively eroded by the subaerial agents. It is believed to represent an earlier drier climate. Subsequently the climate has become more humid so that the hot and humid climate of the region causes the weathering of so much sediment which cannot be transported under the physical conditions of present drainage and gradient, where the thick layer of sediments is concealing the bed-rock. This is the view of eminent geomorphologists like Marton Pecsi (Director and Academician of the Institute of Geography of the Hungarian Academy of Sciences), Paul Macar of the University of Liege (Belgium), C. Embleton of England and A.J. Wiggers of the Netherlands who attended the symposium on 'Erosion Surfaces' held at Ranchi University, Ranchi (Bihar) between 10-13 December, 1968.

They opined their views in the following observations :

(1) In the whole of Chotanagpur the bed-rock is practically absent. It is not more than one per cent of the surface or so. This is of course true outside the scarp-zone in relatively even surfaces like the Ranchi Plateau or Chaibasa Plain, etc. Elsewhere the solid rock of the substratum is covered with a layer of soil and sediments which may be as much as 5 to 7 metres deep in the Valleys. This layer of sediments which consists of hard rocks in Chotanagpur is one of the most important feature of its present geomorphology and has to be properly explained. It is obvious that the upper surface of bed-rock and the present ground surface are more or less parallel suggesting that the

present ground surface has been inherited from the bed-rock which was earlier at higher level than the present ground surface. Subsequent weathering and erosion of bed-rock have converted the upper layer into a mantle of waste and debris.

(2) They felt that during the last 10-12 million years, several changes in climate had taken place. The area was, in the geological past, under arid or semi-arid climate which shaped the region into the features of an arid region (Fig. 24.2). Present-day processes, conditioned by the climate of the post-glacial or Holocene epoch and covering no more than 10,000 years, are responsible for only 5% of the existing relief in the mid-latitudes, while 95% belongs to earlier generations. In Chotanagpur the arid and semi-arid climate of the past has changed into the more humid climate of the present times with the rainfall of 50″ to 60″ (1270 to 1525 mm) and a temperature of 65° to 90°F (18.5°C to 32°C). These climatic conditions cause an enormous production of waste and debris which cannot be all transported from the

*Location Map of North Chotanagpur Showing Basins*

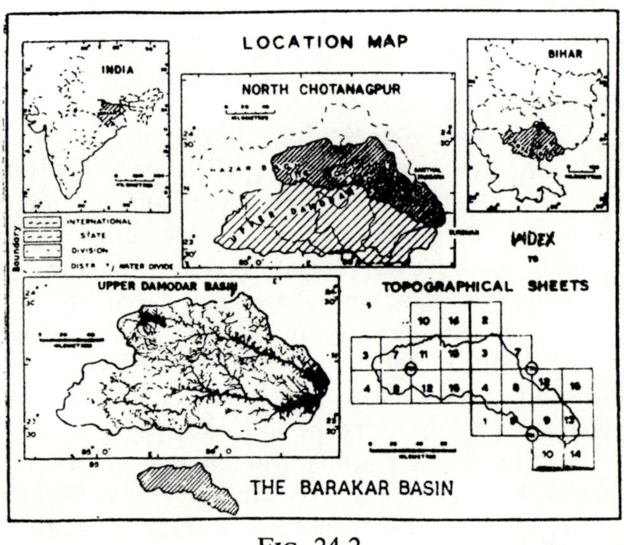


FIG. 24.2

area where these sediments are being produced from the parent rock.

(3) In Chotanagpur, there are two phenomena apparently related to two different types of climate. One of these is the surface formed by the bed-rock or hard rock which generally consists of gneisses and granites. This bed-rock surface, although for the greater part obscured from view, is of pediplain nature which consists of a series of pediments or concave slopes on one side ending in inselberg or bornhardt (i.e., a residual hill of unusual hard rocks) locally known as *tongaries*. These pediments end in some river-valley (Fig. 24.2).

(4) During the field trip to Netarhat, in connection with the December symposium, Pecsi drew attention to a feature near the road between Lohardaga and Ghaghara. Here a small dry river valley was cut into different layers of laterite, red clay and gravels, and then the bed-rock was observed at the bottom. Pecsi suggested that such a feature, i.e., a bed of gravels occurring directly over the bed-rock points to the existence of an earlier pediment surface, association with a more arid climate. Subsequently the climate changed to more humid conditions leading to the formation of the Red Clay and laterite.[42]

(5) Pecsi pointed out that "he had, a few days ago, seen an active planation surface in Tamil Nadu marked by pediments and extensive exposure of bed-rock, in marked contrast to what was visible in Chotanagpur."[43] He felt that on the Chotanagpur plateau now, erosion has ceased to operate on bed-rock which lies buried under a thick mantle of waste.

Thus the observations of the foreign delegates concluded that the surface under study was not an active erosion surface but a fossil and dead erosion level which now lay under a thick mantle of eroded debris. They also concluded that the dead erosion level was a pediment floor with the possibility of pedimentation in the region in the past.

## DISCUSSION ON AND VALIDITY OF FOSSILIZATION

The foreign delegates, Pecsi in particular, emphasized the role of the arid or semi-arid climate of the geological past on the fossil erosion surface. The prominent supporting factor, according to him, was the layer of gravels occurring directly over the bed-rock. This layer of gravels must be, at any case, allochthonous or drifted in nature. So we are faced with two sets of questions : (i) Is there any gravel-bed occurring directly over a bed-rock, and (ii) if the gravel-bed does exist, is it drifted or *in situ* in origin ?

(i) So far as the present landscape of the Barakar Basin is concerned, it is mostly formed of soil and sediments and there is no question of gravel-beds lying over the bed-rock. The report of the Damodar Valley Corporation[44], presented by its Soil Survey Section of the Soil Conservation Department reveals this fact. A number of soil-profiles have been drawn, which show that the region is characterized by red loam, representing an intermediate stage in the process of laterization. These have developed from acidic rocks like biotite-gneiss and granite-gneiss.

The red loam soils are highly leached, slightly acidic in reaction, usually deficient in organic matter, nitrogen and available phosphoric acid and have a low base exchange capacity, while the potash content is usually medium to high. The description of some of the soil profiles is given below :

### SOIL PROFILE—1

*Village: Maheshpur, Usri Catchment*

Slope: 1-3% Upland

| Horizon | Depth in cm | Description |
|---|---|---|
| $A_1$ | 0-17 | : Very pale brown, Loamy sand, single grained, loose dry, friable moist, very rapidly permeable, plenty of roots, clear boundary, pH 5.4. |
| $A_2$ | 17-30 | : Yellowish brown, sandy loam, granular to very weak sub-angular blocky, slightly hard, friable moist, rapidly permeable, plenty of roots, diffused boundary, pH 5.4. |
| $A_3$ | 30-47 | : Strong brown, Sandy clay loam, weak sub-angular blocky, slightly hard, sticky moist, moderately slow permeability, few roots, diffused boundary, few Fe concretions, pH 5.3. |

| Horizon | Depth in cm | Description |
|---|---|---|
| $B_1$ | 47-107.5 : | Yellowish red, clay loam, moderate sub-angular blocky hard, sticky moist, slowly permeable, few roots, few Fe concretions, pH 5.6. |
| C | 107.54+ : | Weathered granite-gneiss. |

*Range in Characteristics*

Texture varies from loamy sand to sandy loam in $A_1$, sandy loam to loam in $A_2$, loam to sandy clay loam in $A_3$, $B_1$ being clay loam.

| | |
|---|---|
| Relief : | : Undulating with gentle slope 1-3% |
| Drainage : | : Surface runoff rapid, internal drainage moderately slow. |
| Vegetation : | : Mohua and Grass. |
| Use : | : Occasionally cultivated for millet crops such as *Marua* and *Gondli*. Maize is also grown. |

## SOIL PROFILE—2

*Village* : *Koradih, Rajoya Catchment*

Slope: 0-*1*%/Upland

| Horizon | Depth in cm | Description |
|---|---|---|
| $A_1$ | 0-25 | : Yellowish red, loamy sand, coarse very weak granular, loose dry, loose moist, clear boundary, plenty of roots, pH 5.5. |
| $A_{11}$ | 25-65 | : Yellowish red, coarse sand, single grained, loose dry, loose moist, clear boundary, plenty of roots, pH 5.5. |
| $A_1$-$A_{12}$ | 65-75 | : Yellowish red, sandy loam, very weak sub-angular blocky, loose dry, loose moist, clear boundary, plenty of roots, pH 5.5. |
| $A_1$-$A_{11}$ | 76-84 | : Yellowish brown, fine sand, coarse single grained, loose dry, loose moist, clear boundary, plenty of roots, pH 5.5. |
| $A_1$-$A_{12}$ | 84-90 | . Strong brown, loamy sand, single grained, loose dry, friable moist, clear boundary, pH 5.5. |
| $A_1$-$A_{13}$ | 90-99 | : Yellowish red, fine sand, single grained; loose dry, friable moist, clear boundary; pH 5.5. |
| $A_2$ | 99-110 | : Strong brown, sandy loam, weak coarse sub-angular, blocky, loose dry, friable moist, few roots, clear boundary, pH 5.5. |
| $B_1$ | 110-152 | : Strong brown, clay loam, medium fine angular blocky, hard dry, firm moist, few roots, clear boundary pH 6.5. |
| C | 152+ | : Quartz gravel in clay matrix. |

These are developed on the weathered material of gneissic rocks. The colour of the soil is yellowish red on the surface but becomes darker with the depth. The soils are well drained and contain concretions of Fe.

Mango, *Banjan* and *neem* are the main vegetation, and the land is mostly used as grazing land.

The above-noted profiles have no gravel-beds occurring over the bed-rock as suggested by Pecsi and others. Almost all the bottom horizons consist of weathered or decomposed granite-gneiss, and the soils lying over them are residual or *in situ* in origin. It means that by erosion and weathering, the upper portion of bed-rock has been converted into sediments from hard rock, and the layers of sediments have come into existence above the bed-rock. These sediments have no alternate layers of fine or coarse sediments, because they have not been deposited by flowing water. Had these areas been characterized by alluvial deposits from some adjoining highland, they must have been characterized by a series of fine and coarse layers because of the alternate wet and dry seasons. But the findings show that this is not the case.

Thus the present dry soil horizons, observed in the profiles noted above, go against the opinion of Pecsi. The alternate wet and dry seasons (about 4 and 8 months respectively) along with the high temperature prevailing in the region bring rapid decomposition of organic matter, increased microbiological activity and weathering of parent material during the hot season while during the wet months intense leaching results in the removal of soluble constituents like calcium and helps in the development of soils rich in iron and alumina, The soils developed under this environment have the characteristic reddish tinge with a tendency to laterisation. On the other hand, the upper layers of the soils are subjected to erosive action by torrential rain. This is why the soil groups of this region have been classed as red loam, representing an intermediate stage in the process of laterisation. The other groups of soils, the calcareous soils, have developed from basic rocks like hornblende schists and these occur in patches scattered all over the area.

So the gravel beds do not exist over the bed-rock as claimed and explained by the foreign delegates to the field trip.

(ii) It is true that in many cases, the upper layers of soil profiles have been found to be truncated by erosion over which materials from the adjoining region have been deposited giving rise to different layers of soils without any genetic relation. Often the contrasting layers are intervened by 'Stone-lines'.[45] But these drifted materials neither occur immediately over the bed-rock nor are they the product of palaeoclimatic environment.

(iii) There may be confusion in the recognition of a gravel-bed. There does exist the horizon of sub-angular blocky materials in the Basin as noted in the soil-profiles. But these are not gravels. Had these gravel-like materials been deposited, transported from the adjoining highlands, they would be rounded, smooth or the like. Again they must occur over the bed-rock whereas the case is that the bed-rock is covered with weathered or decomposed granite-gneiss.

(iv) The deposit of drifted materials is the recent formation only, which form superficial deposits of the Barakar Basin. They are the alluvial sands of rivers and the residual soils derived from the weathering of bed-rock. Thick deposits of such recent alluvial sands occur adjacent to the banks of the Barakar rivers and the Khudia, Pusai and Usri streams.

Such deposits do not explain Pecsi's view that the bed-rock surface is a pediment which is usually topped by gravel layer in the areas of active pedimentation. In the absence of a gravel layer, the bed-rock surface is not a pediment or pediplain (of the past). They are the product of present-day environment.

The origin of the layer of clay and similar soils which represent the thickest part of the sedimentary cover, cannot be explained by the climato-genetic view.

Peltier's[46] work is to deal with gross morphometry on a world-wide basis. Peltier claimed that "with a constant mean slope greater than 5°, drainage density is related to climate in a complex manner. It is lowest in the desert (2 drainage ways per mile with mean slope 10°), rises to a peak of 4 drainage ways per mile in semi-desert areas, falls to 3 in temperate areas, and then rises to more than 5 in the humid tropics."[47]

Most of the area in the Barakar Basin is characterized by a fairly fine and a medium to fine texture ranging from 6 to

55 per sq. mile. It falls under sub-tropical monsoon-climate. Again, a number of features in the Barakar Basin have proved that the drainage is antecedent.

In case the claim of Peltier is correct, there arises no question of the existence of fossil erosion surface of arid or semi-arid climate in the geological past due to the occurrence of high stream frequency in the Basin during the past and the present.

The climatogenetic geomorphologists claim that in the Barakar Basin in particular and Chotanagpur in general, the upper surface of bed-rock and the present ground surface are more or less parallel. So, the slope-forms of the bed-rock should be almost similar to those of the ground surface.

The characteristics of slope forms in the Barakar Basin are not related to the pedimentation. Again the 'composite' nature of the slope cannot be explained by slope retreat which is an important phenomenon in the evolution of the arid landforms.

So, the view that the bed-rock represents an earlier geomorphic phase in the Barakar Basin characterized by more arid climate appears to have serious difficulties.

Many geomorphological features, e.g. incised meanders, drainage and a number of genetic features (waterfalls, springs, gullies, tors and stream capture, etc.) of the Barakar Basin are the result of epeirogenetic uplift. It is rather impossible to interpret these features by climatic changes).

Thus, the above discussion concludes that there seems to be no palaeoclimatic fossilization or dead erosion surface in the Barakar Basin as was claimed by the delegates during the symposium held at Ranchi in 1968.

## RE-ASSESSMENT OF TECTONOGENETIC CONCEPT

In contrast with the climato-genetic geomorphology, we go back to the tectonogenetic view for the interpretation of the geomorphic features of the Barakar Basin. This tectonogenetic concept has gained strongest support among regional workers (Dunn, Ahmad, Singh and others,

"According to accepted usage landforms that result from crustal movements are described as tectonic."[48] The characteristic landforms caused by crustal movements are folds, faults (fault-scarp or fault-line scarps), block mountains, the basin ranges, antecedent and anteconsequent gorges, tectonic valley forms, incised meanders and interlocking spurs and so on. Metamorphism of the basement rocks and non-uniformity of strata are other geological evidences of tectonic history. The investigation of these features in the evolution of the relief is the task of *tectonogenetic geomorphology*. In other words, it may be said that the role of geology, structure and tectonic movements in the evolution of landscape is greatly emphasized by tectonogenetic geomorphologists.

Now let us examine the significance of tectonic history in the geomorphic interpretation of the Barakar Basin.

**Tectonic History of the Barakar Basin**

Most geologists and geographers assume that the region has been subject to crustal movements, i.e., from the Cambrian to the Tertiary. Thus it has been shaped by the tectonic movements.

We know that the area consists of hard crystalline gneissic granites, and therefore, it has never been subject to crustal shortening as the Himalayas or the like. But the great earth-movements have, indeed, affected it epeirogenetically. It is supposed that Hercynian movement during Permo-Carboniferous period would have resulted into a down-faulted trough upon which the Gondwanas were laid. This is observed in the Giridih and Cis-Barakarian coalfields. During the Himalayan orogeny, it was uplifted epeirogenetically along with other parts of Chotanagpur.

In this long span of time, from the Carboniferous to the Tertiary it would have been sculptured by exogenic forces several times or probably several part-cycles would have characterized the area. This was perhaps due to tectonic disturbance.

**The Significance of Tectonogenetic Concept**

The tectonogenetic concept has the following aspects for its support :

(1) The steep scarps standing above planation level in the

Basin, their dissection by erosion into mountainous ranges of peaks and valleys, and the reduction of many parts of the Basin to undulating lowlands or plains prove the general uplift of the area, caused by tectonic movements.

(2) Metamorphism of the basement rocks and its alteration into banded gneiss may probably be due to Archaean tectonic movements.

(3) Thick accumulations of the Gondwana sediments in a downfaulted trough and formation of coal, and the erosion of the softer sediments recently can only be explained by tectonic disturbances during the Carboniferous period and the Tertiary period respectively.

(4) The northern edges of the scarp-zone above the general level are the older folded strata which have been epeirogenetically uplifted during the Himalayan orogeny.[49] These, probably, cannot be explained by interpretation other than tectonic.

(5) The epeirogenic uplift is "essential to the explanation of several features of rejuvenation"[50], e.g., incised meanders, sinuous courses of the streams, disregarding bands or joints of gneisses or other structural characteristics, waterfalls and rapids, discussed earlier.

(6) The gullies are the most striking feature of the Basin. Study and observation show that these may be interpreted geomorphologically only by uplift of the area.

(7) "River terraces are topographic surfaces which mark former valley floor level."[51] River terraces and nickpoints along the valley show evidence of rejuvenation. Similarly gorges on the plateau lying above 305 m can only be interpreted by regional uplift.

(8) We have discussed earlier that the Barakar Basin comprises antecedent drainage which has carved out the hard granitic surface without keeping pace with the structural lines. This eroding capacity can only be due to rejuvenation.

(9) In the Basin, a striking geomorphic feature is the summit plain. The summit plains and residual hills have been shaped by active erosion on gneissic and granitic areas with different geologic structures.

Because of the long duration of erosion, hard and comparatively soft structures are truncated alike. The erosive power and different geological formation may be interpreted by tectonic movement which has uplifted the area. There are, thus, many similar features in the Basin which show evidence of epeirogenic uplift.

(10) During the seminar on Himalayan Geology held in September. 13-17, 1976, many papers led to the line of thinking that "the Himalaya was built up out of the products of tectonisation of the frontal edges of the Indian Peninsula, sliced up as wedges, during the interaction of plates."[52] So, the Indian Peninsula is working as a plate which is moving northward. A recent exploration of the Indian Ocean has revealed that the crest of the oceanic ridges is moving outward due to tension at the rate of 5 to 8 cm per annum. These phenomena apparently show that the Peninsula is subject to tectonisation. This goes in favour of the tectono-genetic concept of the area.

## REVIEW OF THE VIEWS

Now keeping in view both the ideas—climatogenetic and tectonogenetic, we are led to the following conclusions :

(1) As regards the Barakar Basin, the significance of tectonic history in the geomorphic interpretation is more than that of climatic changes. Many morphological features may easily be intepreted by tectonic history.

(2) Delegates ignored all the evidences favourable to tectonogenetic view, and put only those points forward which supported their ideas. They probably had an observation of pediments and pediplanation from their work in arid parts of Central Africa.

(3) The arcuate layout of scarps may be formed by antecedent drainage over the former surface which is enforced to carve out after uplift.

(4) There may be some features which are now difficult

to be interpreted by tectonic history, but this does not mean that the area has not been tectonically affected. It requires intensive study so that the fact may come to light.

(5) The tectonogenetic geomorphology presents an explanation of many features of the area, while the climato-genetic geomorphology has limited scope. So, it will be better to work in detail on the former rather than the latter.

(6) The significance of *petrovariance* and *epeirovariance* has been underestimated by the climatogenetic geomorphology. In case of the Barakar Basin, their importance is greater than the *climatic variance*. This is clear from the discussion which concludes that "the degradational forces acting on the earth's surface are opposed by constructional forces.[53]

(7) Generally "the early German workers overstressed the erosive action of wind, and underestimated the action of running water."[54] It seems that "the rise of climatic geomorphology represented a revolt against a dominantly historical structurally-based geomorphology".[55] It is, no doubt, an unsound procedure.

(8) "Much of climatic geomorphology is (a) not new, being implicit in Davis's own work; (b) not well-established, especially in terms of the climatic control of process ; and (c) premature, in setting up generalised world-wide schemes, both of present climatic-morphologic regions, and of climatic-morphologic changes in the Pleistocene and earlier, before sound factual bases exist."[56]

(9) Lastly it had rightly been remarked by Ahmad that "climatogenetic geomorphology had only a minor and doubtful role to play, and real work still had to be done in the field of tectonogenetic geomorphology."[57]

## REFERENCES

1. S.W. Wooldridge, "The Progress of Geomorphology", *Geography in the 20th Century*, (Ed. by G. Taylor), Methuen, London, 1962, p 170.

2. W.D. Thornbury, *Principles of Geomorphology,* John Wiley & Sons, Inc., New York, 1954, p 11.
3. R.P. Beckinsale and R.J. Chorley, "History of Geomorphology", *The Encyclopaedia of Geomorphology,* (Ed. by R.W. Fairbridge), Reinhold Book Corporation, New York, 1968, p 413.
4. J. Budel, "Geomorphology—Principles'. *The Encyclopaedia of Geomorphology* (Ed. by R.W. Fairbridge), Reinhold Book Corporation, New York, 1968, p 417.
5. J. Budel, "Klima-genetische Geomorphologie" *Geographische Rundschau 15,* 1963, pp 269-285.
6. D.R. Stoddart, "Climatic Geomorphology : Review and Re-assessment" *Progress* in *Geography* Vol. 1, (Ed. by C. Board, R.J. Chorley, P. Hagget and D.R. Stoddart), Edward Arnold (Pub.) Ltd. London, 1969, pp 163.
7. J.P. Bakker, and T.W.M. Levelt, "An Inquiry into the Probability of a Polyclimatic Developments of Peneplains and Pediments (etch-plains) in Europe during the Senonian and Tertiary Period "*Publications dela Service Geologigue de Luxembourg,* 14 1964, pp 27-75.
8. K. Bryan and C.C. Albritton, "Soil Phenomena as Evidences of Climatic Changes "*American Journal of Science,* 241, 1943, pp 469 490.
9. J. Budel, "Climatic Zones of the Pleistocene" *International Geoolgy Review* 1 (9), 1959, pp 72-79.
10. J. Budel, *op. cit.*, 1963, pp 269-285.
11. J. Budel, "The system of Climatogenetic *Geomorphology*", *Proc. of Symposium* on Erosion Surfaces (Ed. by E. Ahmad and D.S. Gupta), National Committee for Geography, Calcutta, (1968a), pp 29-32.
12. J. Budel, "The Climatic Geomorphic System" *Climatic Geomorphology* (Ed. by E. Derbyshire), Macmillan, London, (1973a), pp 104-130.
13. J. Budel, "Climatogenetic Geomorphology, *Climatic Geomorphology* (Ed. by E. Derbyshire), Macmillam, London, (1973b) pp 202-227.
14. J. Budel, "Geomorphology—Principles *Encyclopaedia of Geom.* (Ed. by R.W. Fairbridge), Reinhold Book Corporation, New York, 1968, pp 416-421.
15. C.W. Carlston, "The Effect of Climate on Drainage Density and Streamflow, *Bull. of the International Assoc. of Scientific Hydrology,* 11, 1966, pp 62-69.
16. R.J. Chorley, "Climate and Morphometry", *Journal of Geology,* 65, 1957, pp 628-638.
17. R. Common, "Slope Failure and Morphogenetic Regions" *Essays in Geom.* (Ed. by G. H. Dury), Heinemann, London, 1966, pp. 53-81.
18. C.A. Cotton, *Climatic Accidents in Landscape Making*, Hafner Pub. Co., Inc., New York, 1969, pp 354
19. C.A. Cotton, "The Theory of Savanna Planation" *Climatic Geomorphology* (Ed. by E. Derbyshire), Macmillan, London, 1973, pp 171-185.

20. L. Holzner and G.D. Weaver, "Geographic Evaluation of Climatic and Climatogenetic Geomorphology", *Annals of the Assoc. of Am. Geographers*, 55, 1965, pp 592-602.
21. L.C. King, "The Uniformitarian Nature of Hill Slopes" *Irans. of the Edinburgh Geological Society*, 17, 1957, pp. 81-102.
22. L.C. King, *Morphology of the Earth*, Oliver and Boyd, London, 1962, pp 699
23. H. Louis, "The Problem of Erosion Surfaces, Cycles of Erosion and Climatic Geomorphology, *Climatic Geomorphology* (Ed. by E. Derbyshire), Macmillan, London, 1973, pp 153-170.
24. E. De Martonne "Climate : Factor of Relief", *Climatic Geomorphology, ibid.* 1973, pp 61-75
25. S. Passarge, "Morphology of Climatic Zones or Morphology of Landscape Belts *Climatic Geomorphology, ibid*, 1973, pp 91-95.
26. M. Pecsi and J. Szilard, "Principal Problems of Exploration and Nomenclature of Planated Surfaces, *Proc. of Symposium on Erosion Surfaces*, National Committee for Geog., Calcutta 1968, pp 17-28.
27. M. Pecsi, "Surfaces of Planation in the Hungarian Mountains and Their Bearing on Pedimentation", *Proc. of Symp. on Erosion Surfaces, ibid*, pp 33-39.
28. A. Penck, "Climatic Features in the Land Surface", *Am. Jour. of Science* Series, 4, 19, 1905, pp 165-174.
29. W.F. Tanner, "An Alternate Approach to Morphogenetic Climates" *Climatic Geomorphology*, (Ed. E. Derbyshire), Macmillan, London, 1973, pp 186-192.
30. J. Tricart and A. Cailleux, "Morphogenic Systems and Morphoclimatic Regions", *Climatic Geomorphology, ibid*, (1973), pp. 228-268.
31. J. Tricart and A. Cailleux, *Introduction to Climatic Geomorphology*, Longman, London, 1972, pp 295
32. L. Wilson, "Relationships Between Geomorphic Processes and Modern Climates as a Method in Palaeoclimatology", *Climatic Geomorphology, op. cit.*, 1973, pp 269-284.
33. D.R. Stodart, *op. cit.*, 1962, p 163.
34. S. Passarge, *op. cit.*, 1973, p 91.
35. J. Budel, *op. cit.*, (1973b), p 203.
36. J. Budel, *op. cit*, (1973b), p 204.
37. J. Budel, *op. cit.*, (1968a), p. 30.
38. M. Pecsi and J. Szilard, *op. cit.*, 1968 p 20.
39. L.C. King, *op. cit.*, 1962, pp 154-164.
40. J. Budel, *op. cit.*, (1968a) p 30.
41. J. Budel, *op. cit.*, (1968b), p 421.
42. E. Ahmad, "Report on Symposium on Erosion Surfaces", *Proc. of Symp. on Erosion Surfaces*, National Committee for Geography, Calcutta, 1968, p 8.
43. E. Ahmad, *ibid.*, 1968, p 13.
44. D. V. C., Hazaribagh, "Soil Conservation Working Plan, *Soil Conservation Department*, DVC Hazaribagh, 1972, pp 37-65.

45. D.V.C., Hazaribagh, *ibid.* 1972, p 36.
46. L.C. Peltier, "Area Sampling for Terrain Analysis", *Professional Geographer*, 14 (2), pp. 24-28 Reprinted in *Climatic Geomorphology*, 1973 (Ed. by E. Derbyshire), Macmillan, London, 1962, pp 193-201.
47. D.R. Stoddart, *op. cit.*. 1969, p. 194.
48. C.A. Cotton "Tectonic Landscapes", *Encyclopaedia of Geomorphology*, (Ed. by R. W. Fairbridge), Reinhold Book Corporation, New York, 1968, p 1109.
49. E. Ahmad, "Ranchi to Rajrappa—Some Geomorphic Observations", *Geographical Outlook*, Vol. VI, 1969, p vii.
50. E. Ahmad, *ibid.* 1969, p iv.
51. W.D., Thornbury, *Principles of Geomorphology* 1954, p 157.
52. G.S.I., "Himalayan Geology Seminar "Geology of the Himalaya—Appraisal of Status and Definition of Problem held in New Delhi, Sept. 13-17, *GSI* Calcutta, 1976, pp 32-33.
53. L.B. Leopold, M.G. Wolman, and J.P., Miller, *Fluvial Processes in Geomorphology*, Eurasia Publishing House (Pvt.) Ltd., New Delhi, 1969, p 3.
54. R.P. Beckinsale and R.J. Chorley, "The Growth of Climatic Geomorphology", *Encyclopaedia of Geomorphology, op. cit.*, p 414.
55. D.R. Stoddart, *op. cit.* 1969, p 212
56. D.R. Stoddart, *op. cit.*, 1969, p 212.
57. E. Ahmad, *op. cit.* 1968, p 3.

# Index

# RECENT TENDS AND CONCEPTS IN GEOGRAPHY

VOLUME TWO

## About the Editors

**Dr. R.B. Mandal** (*b.* 1948) teaches geography at R.D. and D.J. College Monghyr and at the Post-graduate Department of Geography Bhagalpur University. He had his education from graduation to post-doctoral work at the Patna University. He has also been associated with research projects sponsored by the University Grants Commission and the Rockefeller Foundation.

**V.N.P. Sinha** is currently on the faculty of the Department of Geography, University of Cape Coast, Ghana. He holds a Ph.D. degree from Patna University.

# RECENT TRENDS AND CONCEPTS IN GEOGRAPHY

## VOLUME TWO

*Edited by*
Ram Bahadur Mandal
Vishwa Nath Prasad Sinha

**CONCEPT PUBLISHING COMPANY PVT. LTD.**
**NEW DELHI-110059**

**ISBN-13: 978-81-7022-853-0 (Set)**
**ISBN-13: 978-81-7022-854-7 (Vol. I)**
**ISBN-13: 978-81-7022-855-4 (Vol. II)**
**ISBN-13: 978-81-7022-856-1 (Vol. III)**

First Published 2016

*Published and Printed by*

Concept
Since 1974

**Concept Publishing Company Pvt. Ltd.**
A/15-16, Commercial Block, Mohan Garden
New Delhi-110059 (India)
*T*: 25351460, 25351794, *F*: 091-11-25357109
E: publishing@conceptpub.com, W: www.conceptpub.com
**Editorial Office:** H-13, Bali Nagar, New Delhi-110 015, India.

**Cataloging in Publication Data**--Courtesy: D.K. Agencies (P) Ltd. <docinfo@dkagencies.com>

**Recent trends and concepts in geography** / edited by Ram Bahadur Mandal, Vishwa Nath Prasad Sinha.
3 volumes cm
Reprint. Originally published: 1980.
Includes bibliographical references and indexes.
ISBN 9788170228530 (set)
ISBN 9788170228547 (vol. I)
ISBN 9788170228554 (vol. II)
ISBN 9788170228561 (vol. III)

1. Geography. I. Mandal, R. B., 1948- editor. II. Sinha, V. N. P. (Vishwa Nath Prasad), editor.

G62.R43 2016 DDC 910 23

# Preface

GEOGRAPHY is a unique science in the sense that it integrates various disciplines, encompassing as it does the physical, economic, social, political, historical and medical sciences in relation to man, space and environment.

With the widening scope and increasing dimensions of geography, both in respect of its conceptual and systematic branches, today, there is available a huge amount of literature on this subject. Considerable changes have been made in the methods of study adopted by modern geographers who have spared no effort to make theirs as exact a science as possible.

In trying to make choices between old fashioned and modern geography, descriptive and quantitative geography, spatial analysis and interdisciplinary approaches and so on, geographers today reveal an eagerness to enlarge their horizons and to keep abreast with the latest developments in the subject.

There is also visible a good deal of enthusiasm for deeper research into topics already touched upon with the application of newer techniques to geographical findings such as statistical methods, model building, remote sensing, the use of data matrix, and a host of other sophisticated tools. All this adds up to what can be termed as a conceptual revolution in the field of geography.

What is important today is the dissemination of such knowledge so that both concepts and techniques can be studied and utilized by an ever widening group of geographers and not limited to a coterie of professionals and specialists. It is with this purpose in mind that the essays in these three volumes have been collected together and published. The readings suggested at the end of chapters as well as the references, provide a comprehensive bibliography of research materials.

The three volumes in this series cover the conceptual, physical, resources, agricultural landuse and human geography aspects in their broad outlines. Woven into the main fabric are the social, political, historical and allied problems which interact and provide complexity to the main themes.

All the three volumes contain brief introductions to each part which gives the reader glimpses into the topics dealt with in each of them.

It is hoped that this unique collection of essays will serve the purpose of enlightening the reader regarding the kind of research and the variety of topics which are being taken up by modern geographers today.

The extensive use of maps, figures and diagrams will help illustrate the problems that have exercised the minds of scholars and researchers in this wide-ranging subject.

These volumes have been made possible because of the unstinting cooperation of a host of people too numerous to acknowledge individually. Gratitude must be expressed, however, to Prof. P. Dayal, Head, Department of Geography and Dr L.N. Ram, from the same department at Patna University for their valuable suggestions and the latter for sponsoring this project on behalf of the Geographical Research Centre, Patna. The credit goes to the contributors who contributed scholarly papers on a variety of topics with great enthusiasm in spite of other urgent engagements. To friends and critics like M. Jha, G.N. Singh, P.N. Mandal and Suresh Kumar go special thanks for constructive suggestions and generous help at all times. Without the cooperation of all of them these volumes would not have seen the light of day.

MONGHYR
*18 May 1979*

RAM BAHADUR MANDAL
VISHWA NATH PRASAD SINHA

# Contents

## SECTION TWO

## AGRICULTURAL LAND UTILIZATION

# List of Contributors


1. *Barnwal, Shekhar Kumar,*
   Department of Geography, R.D. & D.J. College, Monghyr, India.
2. *Das, K.N.,*
   Reader in Geography, Bhagalpur University, Bhagalpur, India.
3. *Dey, B.,*
   Department of Geography, University of Sasketchwan, Canada.
4. *Jafri, S.S.A.,*
   Lecturer in Geography, North Eastern Hill University, Shillong, Meghalaya, India.
5. *Mandal, R.B.,*
   Lecturer in Geography, R.D. & D.J. College, Monghyr, India.
6. *Mishra, D.K.,*
   UGC Teacher Fellow, Centre for the Study of Regional Development, Jawaharlal Nehru University, New Delhi, India.
7. *Prasad, Manager,*
   Department of Geography, Magadh University, Bodh Gaya, Bihar.
8. *Sami, A.,*
   Department of Geography, Patna University, Patna, India.
9. *Satpathy, Debidutta P.P.,*
   Reader in Geography, Ranchi University, Ranchi, India.

10. *Saxena, H.M.*,

Department of Geography, Government College, Sri-ganganagar, Rajasthan, India.

11. *Shanmuganandan, S.*,

Lecturer in Geography, Kamraj, Madurai University, Madurai, India.

12. *Singh, B.B.*,

Department of Geography, M.M.H. College, Ghazia-bad, U.P., India.

13. *Singh, Tej Vir*,

Director, Institute of Himalayan Studies and Regional Development, Garhwal University, Srinagar, Garhwal, India.

14. *Sinha, Udai Prakash*,

Lecturer in Economics, C.B.R. College, Bariarpur, India.

15. *Suresh Kumar*,

Department of Commerce, R.D. & D.J. College, Monghyr, India.

16. *Uyanga, Joseph T.*

Department of Geography, University of Calabar, Nigeria.

# List of Figures

# SECTION ONE

## RESOURCE GEOGRAPHY

# Introduction

NATURE does not know what is a resource and what is not. Resources are the materials, broadly speaking, both natural and man-made which are utilized for the satisfaction of human wants. Human knowledge and ingenuity are essential for the tapping of these resources to their optimum extent. Thus man is the greatest resource of all. Increased human knowledge and technological development produce more wants thus leading both to a demand for more resources, and more importantly, for geographers, a change in the spatial dimensions of such resources. Recent trends in resource study are directed towards a search for new and alternative resources to replenish the older ones and also cater to the newer human needs that are increasing each day. Some of the studies directed towards these ends are, transportation, marketing, recreation, consumer behaviour pattern, the role of satellite imageries and remote sensing techniques.

Resources are the social, economic and political strength of a nation. The importance of a resource is mostly dependent on the capacity of man who commands, utilizes and conserves the natural as well as the man-made resources. With the advancement of time, the concept of resources is changing. In prehistoric civilization a permanent agriculture was felt unnecessary. At that time due to sparse population and free availability of land, people were nomadic. Nowadays, with the increase of population and decreased per capita income, people live in permanent village communities. Hence the concept of resources has a dynamic aspect in relation to space, time, region and the concerned inhabitants. Recently, it has been assessed and surveyed with the help of satellite imageries also.

The study on systems of resources explores the feed-back mechanism process of natural, biotic, human, economic, social and political resources which show how they exert their integrated influence on man—the greatest resource of the earth surface. All these interact with each other in between the welfare of man and the investigation of resources at the same time. The idea of the systems of resources revolve in a cyclic way in order to raise production for the functioning of the society. Here the cultivated land, agricultural labourer, production of foodgrains as supply of resource and human resources are analysed in a graphical manner.

Remote sensing techniques are applicable to land resource management along with land utilization survey. The Indian Space Research Organization (ISRO) and the National Remote Sensing Agency (NRSA) are engaged in conducting experiments on the application of airborne remote sensing techniques in land resource inventory. Based on the experience gained from those experiments in India, and also similar research and operational experience in Canada and U.S.A., it is suggested that remote sensing techniques could be used very economically in India to provide a land resource data inventory and also in updating landuse data which are essential for resource planning and development at regional and national levels.

In recent times, tourism land recreation have attracted the attention of modern geographers. Christaller, Clawson and Wolfe have made praiseworthy contributions. Indian geographers, however, have evinced only a peripheral interest though the tourist movement has appeared very boldly on the Indian scene. Geographers with their training in landscape perception, resource planningand development, environmental management and mapping techniques can greatly help decision-makers in sound tourist structures. The paper brings forward the status of the subject with possible areas of work to be undertaken within the purview of this newly emerging branch of geography.

Spatial organisation is a result of spatial structure and its interactions. It integrates the continuing flow of people, ideas, goods and services. It may be stated in this connection that the geographer's spatial expression of transportation geography is the study of transportation as an aspect of the spatial organisation of an area. This paper discusses the conceptual development of transportation geography by dividing it into different aspects. Firstly, it surveys the methodological developments in geography and then goes on to describe the progress of transportation geography, keeping in mind the concept of spatial organisation under various headings, viz., theory and essentials of transportation geography, spatial structure, spatial system or functional organisation, spatial dynamics or dynamics in the spatial system, spatial efficiency measures of space adjusting techniques, and finally, spatial organisation and spatial reorganisation or space fulfilling techniques. The paper lastly advances a method of consolidating the aspects of spatial organisation as discussed earlier in an effort to provide a conceptual frame-work.

The objective of the paper is to explain the developing concepts and trends in the field of marketing geography. Marketing geography deals with the geographical study of the trading activities and with the market and market place as a unit of space. The concepts discussed in the paper are those of spatial interaction, hierarchy, morphometric analysis, consumer behaviour, market area and periodicity.

While studying the geography of retailing, shopping activity as the main function in most of the urban centres of India has been discussed. This study aims to investigate the significance of retailing including factors of spatial development, identification and delimitation and hierarchy of functions etc. In most cases the approaches of study by different scholars have been delineated to analyse the problems of retail business research in social science discipline.

Retail business occupies a prominent position in the contemporary urban culture but the study of such a pursuit has been mostly neglected by geographers, economists and the business class of people. Hence this study has been placed here with a view to introduce the inventory aspects like meaning, objectives, scope, methods, approaches and principles of retail business structure, particularly with a view to aiding research students.

The paper entitled *The Concept of Consumer Behaviour*, contains discussions about the movement of consumers in and around the city of Monghyr for the purchaseof consumer goods. Consumer behaviour pattern is generally determined by the movement of consumers in the city for th epurchase of cloth, medicine, books and several other items, and the umland is generally determined by the incoming of dried cowdung for fuel, foodgrains, labour, essentials of living, vegetable and milk. Monghyr is a novel urban centre in respect of its location encircled from east, west and north by the river Ganges whereas the south is a low lying countryside with intervening hills, invariably liable to floods every year. Hence the consumer behaviour of the city of Monghyr is unique in nature in comparison with other urban areas of the State of Bihar.

The paper on the role of mineral resources in economic development deals with the use of minerals by man as the prime source of food, raw materials for industry, and power for machines. In view of the economic limitations and endowments of minerals, Drolet's diagram has been presented which gives a visual impression of the classification of mineral resources.

The paper entitled "Progress and Problems of the Coal Mining Industry in India" analyses the distribution of coal fields especially in the triangular area of the Damodar valley, the Godavari valley, the Chambal valley of Gondwana formation, while the coal of the tertiary sector is found in Kashmir, Assam, Meghalaya, Rajasthan and Tamilnadu, Further, the discussions about coal reserves, management, production during plan periods, impact of nationalization on coal production and distribution; problems of coal mines, e.g., accidents, flood havocs, fire hazards, the plight of the coal miners, lack of effective and speedy means to transport coal from pit heads, illegal coal mining, shortage of explosives, obsolete equipments, and out of date techniques of production, irregular coal supply to steel plants and power houses, inadequate port facilities, lack of finance, poor quality of coal, dearth of trained manpower, and inadequate power supply to the coal fields are emphasized.

Under varying geological conditions the availability and productivity of petroleum differ greatly. Because of this fact and due to the lack of technological know-how in the developing countries of the world, the oil crisis and the problems related to the extraction, refining, distribution and consumption of oil have increased enormously these days. Keeping this in mind, the occurrence of oil in different geological structures, increase in production, the distribution of oil fields and the development of oil refineries with special reference to the Indian sub-continent have been analysed here. The aim is to highlight the importance of oil for the welfare of Indian society.

# 1

# The Concept of Resources

*—K. N. Das*
*—Udai Prakash Sinha*

IN RESOURCE study the factors of natural environment control the number, energy and abilities of inhabitants. Functional activities are guided by the efficiency and mental abilities of neighbours. In this sense some areas offer little resources for human use while others do in abundance. As for example, the tribal people of the Nilgiris in South India still lead a primitive life because the topography is rugged, the climate unsuitable for raising crops and the forest environment creates favourable situation for hunting and forest gathering. In such an environment people spend much of their time and energy to satisfy a few primitive wants such as food, shelter, fuel implements and personal adornment. In such an extreme environment inhabitants have little choice in the selection of functional activities. On the other hand, the people of Madras city have multitudinous occupations, good accessibility of roads, railways, waterways and airways. The industrial development, location of educational and social institutions, administrative establishment, medical facilities, etc. have given rise to dense population per hectare of land in comparison with the Nilgiris.

The strength of any nation, whether it be social, economic or political, depends mostly on the resources available at command and their proper utilisation. The evolution of civilisation centres command their spectacular growth in the field of social, economic, cultural and political advancements. History bears testimony to the fact that ancient civilisations flourished because they utilised available water resources; and wherever the growth of civilisations has been slow it has been in areas with less potentialities using the resources and these areas come under the category of underdeveloped countries. It is clear that the capacity of tapping the resources must be developed for the advancement of a country.[1]

Today there is a race among the underdeveloped countries as to how to make the best use of resources at their command to keep the pace of growing necessities set by the growing population and unemployment. Unless these two major problems are solved properly, they will sap the vitality of the nation; and scientists and technologists the world over must find out ways and means to combat the crucial problems being faced by the underdeveloped countries and must try to make them at par with the developed countries.

It is clear that a heavy responsibility lies on the scientists and technologists to uplift the underdeveloped countries by teaching them how to exploit the natural resources and make the best utilisation for the development of the country. The scientists by virtue of their various discoveries must direct the world to foster brotherhood and live in harmony.

The point is that we have not attached importance to the geography of resources in the university curricula. We do not realize that the planning of natural resources cannot be determined without a systematic introduction to the subject. We must know about our natural resources and thereby develop the art of their proper utilisation. This is necessary as the economic development of various nations has reached saturation point and some positive work in this field is most necessary for the survival of the human race. The purpose of this paper is to shed light on some recent advances in the concept of resources.

## MEANING AND SCOPE

According to Jones and Darkenwald resource geography deals with "the productive occupations and attempts to explain why certain regions are outstanding in the production and exportation of various articles and why others are significant in the importation and utilisation of these things".

The demand for any item used by man is known as resources. In this sense anything in itself is not a resource but the need of such thing for a man makes them resources. In this way, an item is known as resource if in a certain time and space man has developed technological, cultural and physical abilities to use it. This shows the close relationship between the abilities of man and the resourcefulness of an item.

Generally hunting, fishing, forest gathering industry, trade, minerals, water, land and soils are known as resources. But all these are the items which supply food, clothes, shelter, fuel, tools, materials of industries and luxuries as the physical base of natural environment, hence all these are known as resources. Besides these, other items which have an indirect influence in obtaining the resources are health, wants, knowledge, social adjustment, political stability, economic prosperity, national organisation and international co-operation, etc. are also known as resources. Commonly coal, iron, gold and silver are known as resources but the same serve the purpose of a man with the advancement of technological knowledge only, without which the resourcefulness of gold or silver cannot be estimated. The value attached to them in terms of capital, marketing technological knowledge, sources of transport and favourable social, economic and political environment makes them resources.

To understand the resourcefulness of a thing, one has to understand the knowledge, habit and external and internal technological knowledge of persons including neutral stuff. There are two forces in physical environment: one is the natural element and the other the natural resistant force. The latter is generally harmful for the development of a nation. In this way, the progress of an economic landscape is determined by the above forces. Primitive people were surrounded by the neutral stuff of physical environment which has no direct influence on man. People in those days tried to fill their belly just like animals; and they had no courage to face the struggle actively. Such conditions are known as the passive adaptation of man which is just like the animal character of the present world.

Primarily the landscape was covered with forests but slowly with the occupation of land, exchange of little surplus with the neighbours and the growth of resources developed with the commercial activities of men. As for example, most of the early civilizations viz. Egyptian and Mesopotamian in the old world and the Aztec, Melan and Incan in the new world have developed with the growth of subsistence agriculture. In early civilizations the impetus for advancement came from European countries by occupying the professions of fishing, tending animals, tilling land, manufacturing, trading etc. With the decline of commercial powers of the Mediterranean, Phoenesian, Cretan, Gressian, Carthaginian and Roman (from A.D. 1096 to 1270) revived the resource development in most of the American, African and Asian countries of the world.

The resources are dynamic in nature, as for example, many years ago coal was regarded as a blackstone. But nowadays, with the technical advancement of a man, coal is regarded as an essential element of the present civilization. Thus man has developed his technical abilities with the adaptation of nature in comparison with animals and the same is known as the "active adaptation of nature" for the growth of nations. In a natural environment according to the activities of men, the forest clearings give rise to permanent agriculture and in place of wild animals, domestic livestock are reared slowly. Later on fields and settlements develop.

According to Zimmerman and Michell, the knowledge of man is the greatest resource because the elements of physical environment have developed different cultural characteristics such as multifarious knowledge, love, social urbanisation and political organisation etc. which help in determining an item as resource. The limitation and barrier of human abilities cannot generate energy in itself and the same cannot restore resources. This is because the technical and cultural abilities of man are dynamic which have changed the use of resources in different manner. According to Bowman, the geographical elements of environment are static but with the contact of man the characteristics of those elements have changed like cultural elements. In this way so far as the dynamic relationships of resource and cultural functions are concerned, there is no resource in itself but the same have been made by man as resource. Due to this fact the economic landscape of different nations are different in time and space which ultimately determines the potentialities of consumption and cultural advancement of the societies. Even in one country with the technical abilities of men and physical environment, the concept of resources is changing as for example the resources which are sufficient in peace time are not sufficient for a period of war.

The world is dynamic in its internal structure and so the age-old social, economic and political institution must change with the advancement of age and time. The factors which once guided the development and utilisation of resources now find themselves incapable of ushering in a golden era, and they are now considered to be outdated. For instance, the yield which is produced with old techniques of production to cater the needs of the people now cannot keep pace with the time as there has been a tremendous growth in population and therefore, yields must be trebled by using good manure and with the introduction of new crops with modern scientific methods, otherwise we will hardly be able to feed the hungry mouths and to cater the needs of the people. A change in the old order of society is a prerequisite for a more efficient and higher utilisation of resources at a place which will provide proper solution to the fast multiplying problems.

The underdeveloped countries produce a large quantity of the world's raw materials yet they control an insignificant fraction of international trade in finished goods because their resources are being exploited by the market mechanism of the advanced countries. The underdeveloped countries have to face many knotty problems such as illiteracy, limited capital, underdeveloped industry and a primitive agriculture system. Even with limited resources at our disposal there has been a tremendous development after independence in India which managed to steer the ship out of the rough water. India has managed to go ahead with the development strategies with limited resources at its disposal. Poverty stands in the way but it does not hamper progress.

It is most striking to note that the developed countries, overutilised the natural resources and this has had an adverse effect on the underdeveloped countries. On the other hand, the underutilisation of natural resources has kept the Indian economy in abject poverty. This unbalanced exploitation of natural resources may cause immense damage to the quality of human life and the life-supporting systems of ecology. In case the present trends of industrial and agricultural production, population growth and exploitation of non-renewable natural resources continue, this would cause immense damage to the earth's carrying capacity of human beings. The oil crisis also continues to be a sore problem in the economic affairs in most of the underdeveloped nations. The crisis of oil may diminish the rate of production of their natural resources. The value of oil production and prices has added new dimensions to resource development. The prices of raw materials in the international market have been artificially fixed at a lower level, whereas the prices of finished goods have been inflated to suit the convenience of the developed countries. The prices in most cases do not give the true economic value of the products. After all, self-reliance is the goal of the underdeveloping countries and a rational policy of international trade can help to remove the obstacle in the way of development.

The responsibility for jointly safeguarding the natural resources would rest with the developing countries. This is the only way to eradicate poverty and its consequences.

India represents the paradox of poverty in the land of plenty. The reason is not far to seek. We have not been able to utilise our resources in an orderly manner. There are so many complex problems facing the Indian sub-continent like housing, medical facilities and per capita consumption of food, clothing, energy supply, minerals and other natural resources. The per capita income is very low.

We have not yet gone through the process of raising production in fields and factories to banish poverty from the land of plenty. We have not utilised modern techniques to raise production. We are still using the age-old methods in spite of the fact that if we do not use the modern techniques we cannot feed everyone. The population problem is a baffling one. It has to be checked by using scientific methods. The illiterate masses must be taught that they have no moral right to increase the number of family members without finding resources for their proper maintenance. Illiteracy is the major cause which stands in the way of the progress of a country. It has to be fought at the war level and if we want to raise the status of the country, illiteracy has to be banished.

Lack of education, inadequate health care and the presence of irrational attitudes towards the toiling masses and traditional social institutions are detrimental to advancement. All these have threatened to thwart the forces of modernisation. Absence of knowledge about development, utilisation and conservation of our natural resources is the root cause of our unability to generate the powerful forces of modernisation. The central focus of this complex problem is that unless there is more for mass participations in the cause of advancement, there could be no prospect of rapid progress along a wide front.

## APPRAISAL OF RESOURCES

The concept of resources includes physical, biotic, human and agricultural resources. This also includes mineral and man-made resources of settlements, industries, transportation lines and the development of scientific pursuits by acquiring technical abilities by people. Among all the resources, the human resource has great value as it formulates resources according to its choice. In any region the vast reserves of coal and iron ore may have great economic value but the same is possible only with the help of technical abilities of man. Unless and until the inventions are made about any reserves by man, the same may not be regarded as resources because that is beyond the reach of man. A man as the creator and fashioner of resources occupies a dominant position in all economic and geographic studies. The relationship between man as the resource and environment can be divided into the following points :[2]

(a) Man utilising the resources of nature as the creator of natural environment;
(b) Man utilising the resources of nature by virtue of his working in the natural environment; and
(c) Man utilising the resources of nature because human resources are the product of the physical environment.

Man is the dominant factor of resource geography as he produces, manufactures, distributes, consumes and also destroys resources. From prehistoric forest economy up to the present agricultural-cum-industrial economies, man occupies most of the habitable parts of the earth and uses its resources to satisfy his needs.

Thus the utilisation of resources depends on the favourable conditions of natural environment where each and every variable competes with the other in shaping and determining the cultural advancement and economic prosperity of the region. As for example the production of wheat is not possible in highly wet areas. Similarly, the hot deserts are also not favourable for the cultivation of wheat in bulk. Only the reclamation of marshy land in wet areas and the provision of irrigation facilities in hot deserts can usher an era of favourable 'ecological balance' of environmental factors for the production of wheat in a particular region.

### RESOURCES AND CULTURE

The culture of a man is the product of his physical environment and mental abilities of man and the increase of these two have given rise to the cultural advancement of a landscape. According to Karl Sauer, man's deformation of the pristine is assumed as essential. In a real sense cultural landscape is the result of dynamic and deformation of cultural abilities. In this way culture in itself is the greatest influencing resource of a man. The table of other resources is also changing in time and space according to the cultural attainment which determines the intensity of the usefulness of an item as resource. In order to show the functional importance of this cultural hypothesis, Paul Seers has developed a formula of the physical resource of environment as land/population culture. According to this formula the base of resource is divided by population in a country which is the effect of a cultural landscape. As for example for primitive people coal, petroleum and other minerals are not regarded as resources but for a well-to-do person the same may be a resource.

### INDUSTRIAL RESOURCES

Physical environment is a great source of human living which becomes the base for cultural advancement and the growth of industrial civilizations. On this analysis three aspects have great influence : (1) technology, (2) social organisation, and (3) the knowledge of habitat and analysis. In any space and time, any level of industry fully depends on those natural resources which are utilised by the industry. Industries themselves are the product of the cultural landscape and the analysis of specific resources cannot be made keeping out of the cultural make-up of societies. The greater availability of resources, renewability and exploitability, etc. are the result of developed industrialisation. Underdeveloped, uncultured human society receives less resources from the physical environment; for example, petroleum was not known to primitive people as fuel oil but nowadays it is an essential commodity. In this way the societies which have advanced in industrialisation are a great source of the resource base. (Figure 1.1).

In any region the question of the availability of resources depends on the technological advancement and the amount of resources available in that region. As for example, an area having hundred square kilometres may be an area of under population if the fertility of soil and the production of crops are better in relation to the inhabiting population. Secondly, if the production of crops in that area is equal to the demand made by the people then it is said that the area is under optimum population. Thirdly, if an area has a higher number of people in relation to its productive capacity then it is said that the area is over populated. Fourthly, if the pace of population growth is rapidly increasing in comparison with the growth of production then it is said that the area is under a population explosion.

Malthus has studied the relationship between population and resources at the end of nineteenth century and he used the term "the means of subsistence" in lieu of resource. According to him, in any region population growth is always higher than resources. He suggested that the geometrical progression of population growth takes place in case "the means of subsistence" increase in arithmetical progression. In such a situation mal-nutrition will automatically bring famine conditions to region.

In order to show the relationship between population and resources, E.A. Ackerman has classified the whole world into five groups[3] :

(i) The United States Type; developed technology, a case of under population.

*Advanced Industrialisation and Resources*

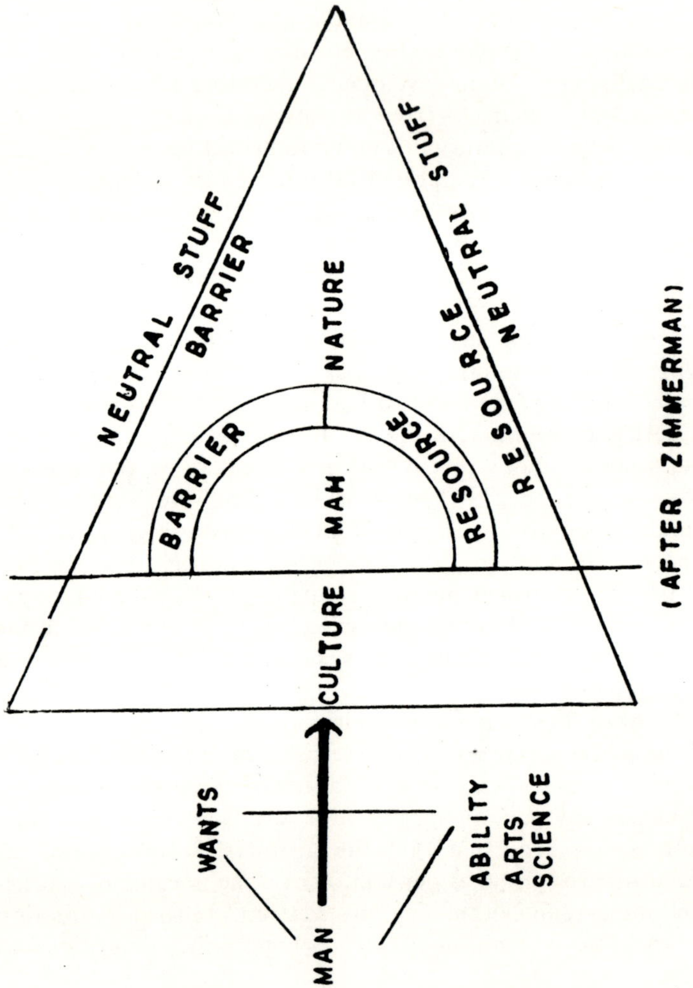


FIG. 1.1

(ii) The European Type; developed technology, a case of over population,

(iii) The China or Egypt Type; less developed technology, a case of over population,

(iv) The Brazil Type, less developed technology, a case of under population,

(v) The Arctic Desert Type, less developed technology, a case of intensive over population.

Resources are scarce in nature. Nowhere in this world is the desired level of resource development possible unless a contentment prevails among the people for the fulfilment of their desires. Wants are unlimited and it is practically impossible to fulfil the desire of all persons at the same time. In areas where the growth of industry, agriculture, urban expansion and accumulation of capital, mineral wealth takes place, it is said that the growth of civilization has reached its peak, e.g., the areas of the Eastern Sea Board of North America. On the other hand, the forested rugged terrain, unfavourable climatic conditions of heavy rains and hot weather in the equatorial belt of the world, bears some vulnerable diseases which make the environment uninhabitable for human beings and as such this belt is the area of under development from the point of view of resource development.

With the advancement of scientific knowledge, the dimensions of resources are increasing, as now-a-days, the search for tidal power, solar energy, thermal power, atomic energy are fast changing the scene of resource development. Now the training in agricultural practices is given on the television. The use of the tractor, combine harvester, thrasher, and the use of chemical fertilizers in the fields as well as the use of new dwarf varieties of seeds including the changed methods of cultivation have helped in the growth of agricultural resources beyond imagination. In some areas it is felt that the green revolution has taken place in the cultivation of wheat, rice and maize especially in northern India. Similarly, material culture has also changed because of the changes of food habits, clothing, advancement of technical abilities; and the old practices of cultivation through plough and oxen are now altered with the modern method of cultivation through scientific implements and the use of dwarf varieties of seeds, the use of synthetic fibres in place of traditional cotton cloth.

## CLASSIFICATION OF RESOURCES

The resources can be classified into two groups—natural and human resources.

### Natural Resources

This is the combination of element available in nature. Keeping this in mind any element, creature, oxygen, nitrogen, rocks, minerals, landscape, soil, water, vegetation, livestock etc. are regarded as resources due to their usefulness to man. The combination of a number of natural resources sometimes forms a composite resource as for example agricultural land, forest landscape, mining etc. The agricultural landscape is the integrated form of land, soil, temperature, rainfall etc. In this way, different elements and creatures of natural environment form resources both separately and in an integrated manner.

### Human Resources

Man as the creator, producer and consumer of natural resources, is a great resource in himself. Natural environment in itself is a dynamic element but without human use it has no importance. In this way the activeness of man is necessary in order to procure benefit from the unseen resources of nature. The activeness of man is dependent on three things :

(a) Biological hereditary which includes race, physiological behaviour, etc.

(b) Social, economic, cultural and political organisations which are acquired and adapted through organisational systems and traditional customs in different time and space as dynamic elements. In this sense living conditions, level of education, personal and family affairs, occupations and service position, social and economic status, citizenship and the consciousness of nationality, time, region etc. are dynamic elements which affect population migration; and density, numbers, and behaviour of international relations which in turn, are important factors which affect resources. The factors which affect the use of resources can be divided on the basis of several factors in which the following three are important :

(i) The influence of physical characteristics on natural resources,

(ii) The influence of cultural factors on resources, and

(iii) An integrated influence of natural and cultural elements.

The influence of physical characteristics on natural resources includes : (1) The total resource base and its influence; (2) The variability of the base of resources; (3) The characteristices and amount of each resource element; (4) The inter-relationship amongst different resource elements; and (5) The locational attributes of resources.

In any space and time the base of human culture is the product of natural resources. The regions which have the greatest endowment of natural resources are the areas of highest technological advancement. Similarly, the availability of resources in greater amount will help in economic development. Recently, in the trade and industrial economic landscape, due to heavy influx of people on food items, raw materials, agricultural land, forest, minerals, water, houses, sources of transport, sources of recreation are developing, keeping in view the demand. In countries like U.S.A., U.S.S.R., China, India and Brazil, the bases of natural resources are vast, and in these countries the economic attainments are also substantial. During the Second World War when international trade was almost stopped, the wheels of the economy of every nation mainly depended on the natural resources base. But in peace time for the countries which have greater potentialities of favourable climatic conditions, soil, vegetation, livestock, birds and mineral wealth, the chances of prosperity are enough. In areas of single type of climate, the types of vegetations and crops grown are limited in number while in areas of temperate sub-tropical and tropical climatic conditions, different types of crops and vegetables are available. Similarly, in a country where only coal is available, the chances of industrial development are less in comparison with the countries where different types of minerals, e.g., coal, petroleum, gas, iron and other minerals are available. With today's competitive economy, the demand for different resources is increasing according to the needs of the society (Fig. 1.2). Due to high technological

*Resources and Demand*

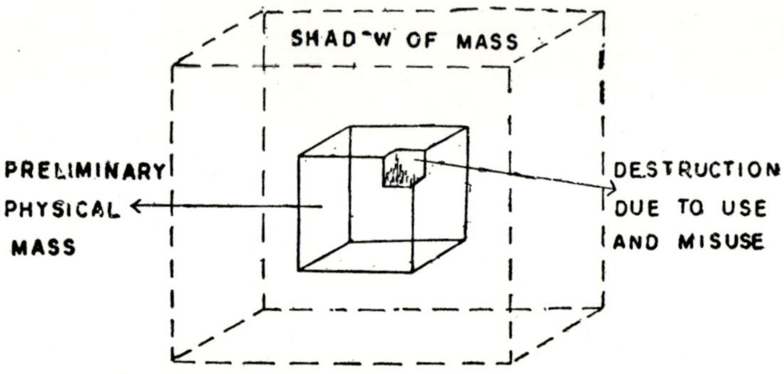


PHANTOM RESOURCE THEORY
(AFTER ZIMMERMAN)

FIG. 1.2

advancement, some items which are in demand these days, may not be so tomorrow or the demand may increase in place of the former.

### INFLUENCE OF CULTURAL RESOURCES

The priority of resources is generally fixed by man. In any country, the total population, standard of living, wants of society, fulfilment of demand of different resources, competitive demand, public opinion, political pressure, national and international political expediency, technological and scientific knowledge, availability of labour and capital, resource ownership pattern, projects and legislation of the Government are the factors which are not equally distributed on the earth surface. Their amount, variety, frequency of occurrence are distributed in a haphazard manner. Keeping this in mind, Zimmerman has classified natural resources into four types :

(a) Ubiquities available everywhere such as oxygen in the air,

(b) Commonalities which are generally available in most of the regions as for example soils suitable for agriculture, water etc.,
(c) Rarities : Those resources which are available rarely such as tin, gold, silver etc., and
(d) Ubiquities which are available in a few places very rarely such as the business of cryolite of Greenland.

Zimmerman has classified resources into four groups keeping other ideas in mind such as (i) unused, (ii) unusable, (iii) potential, and (iv) latent.

Those elements, materials, attributes, location and site which have as yet not utilised is known as unused resources. According to present knowledge and availability of other factors if any element or material is not utilised as resources then the same may be regarded as unusable resources. Those materials which have the chance to be utilised by man in the near future are known as potential resources. Those resources which have attributes of unusability but the same have not been utilised are known as latent resources, such as solar energy and the availability of carbon dioxide ($CO_2$).

## CLASSIFICATION OF NATURAL RESOURCES

Physical environment is an integrated and composite unit of different materials and biotic resources which have formed different niches in space and time. Thus resources are dynamic and their presence helps directly in the activities of natural environment as a necessary part and parcel of geographical locations, size, accessibility, climate, landscape features, soil, water, mineral resources, vegetation, livestock and birds including human resources.

The future demand of natural resources like flow and fund resources, water and mineral resources mostly depend upon the demand set by people on them for their judicious use. The flow and fund resources are always limited in quantity for a particular space and time. Thus the constant demand will pose a serious problem of population explosion on any kind of resources.

Water resource acts both as flow and fund resources. In case we allow the water to flow in channels, such water forms flow resources on which the navigation, rearing of fish and the sailing of boats for recreation purposes are possible up to the distant areas. The water resource is also utilised as fund resource if it rests in a reservoir. In such a situation the rearing of fish is possible but the generation of hydel power is possible only at the point of the waterfall. In any region the excess accumulation of water may create a problem of flood which might damage the resources in general. In case the required amount of water availability is just possible then it is said that the water has been used judiciously. In hot desert areas where the water is available in scanty quantities, it is regarded as the most precious resource amongst all the resources. In this way, according to the change of environmental conditions and the available amount of water, the concept of resources changes.

Among power resources atomic energy, hydel power, thermal power, tidal power, solar energy, petroleum, natural gas and gas from cowdung are important which vary from place to place and time to time for the economic development of a region.[4]

Mineral resources include a host of minerals, viz., iron ore, copper, zinc, sulphur, limestone, gypsum, mica etc., for the growth of material culture through the establishment of industries especially in areas where the power resources are available in bulk.

In this way it can be said that the concept of resources is changing from place to place, time to time, man to man according to the task set by nature and the demand put by the society on a particular kind of resource.

## PRINCIPLE OF RESOURCE ADEQUACY

In any country the efficiency of resources is measured by the number of population, the desired level of consumption and the total population etc. Population and their living standard determined the total demand of resources. Whether the demand on any level can be fulfilled or not can be determined by human knowledge, technology and the available physical resources. Man is not only the gatherer of resources; he also uses the scarce resources scientifically, increases their production and tries to lessen the pressure of human beings on a particular piece of land with a view to avoiding scarcity of physical resources. During the last two decades the growth of population and the economic development have taken place simultaneously but in order to raise the standard of people, family planning has been introduced in India. In this way, it can be said that the relationship between physical resources and population is a dynamic process.

The covariance analysis of these two variables has been presented by Ackerman in the form of an equation. Considering the demand and use of resources, the formula is as follows :

$$L = T + S + R \pm Sep - T_r.$$

where,

L = Living standard of population,

T = Total base of resource characteristics and attributes of natural and physical resources (physical technology factor, administrative technology factor, resource stability factor),

S = Scale of economies,

R = Resource added in trade,

Sep = Social, economic and political institutions and their characteristics, and

$T_r$ = Techniques of resource use.

## CONSERVATION OF NATURAL RESOURCES

The conservation of resources has attracted the mind of scholars mainly after the Second World War due to population explosion, technological-industrial revolution and in order to fulfil the increased wants for a good living standard. During A.D. 1600 only a few minerals had been discovered such as clay, whetstone, millstone, iron ore, glass, tin, mica, coal, salt, mercury, etc. But now-a-days more than a hundred minerals have been discovered including petroleum, potash, nickel, aluminium, chromium, etc. In 1750 the total population of the world was only 66 crores but in 1978 this rose to about 400 crores. Hence, the resource requirements of these teeming millions also increases with the same rate.

A man can live in a flexible manner in any environment which may be designated as an unspecialisation, but the evolutionary characteristics of human beings are of a different nature and the habitat and scope of his activities are also not restricted. Man as a rational being, utilises the natural environment of the active ecological system. In such a manner that should be the dynamic elements of organic and inorganic world which sometimes used as resources, while at the other create barrier. In order to extract natural resources which have some economic value, a man determines the systems of organisation and space actively. The theory of the conservation of resources considers the base of physical balance of population and living standard of people which not only develops the present social, economic and political activities of man but also controls the future use of resources by less destruction, systematic utilisation and unexpectable exploitation.

The conservation of resources does not mean the use of resources either frugally or preservation. Conservation means the purposeful and planned use of resources and the protection of soil, forest and minerals resources from erosion and restoration. In this way the conservation of resources always means the optimum use of resources which are intimately related with the population, living standard, technological advancement and cultural characteristics of the region which determines the productive capacity of resources and their utilisation. The conservation of resources also includes continuous, constant and sustained use of resources. However, the rational use of resources as when and how to use, allocation, depletion, wastage, state of optimum use, planned integrated production come under the conservation of resources.

For planning and conservation of resources the consideration of the following aspects are necessary :

(1) The consideration of a region as a unit is necessary so that the knowledge of total resource base and their types, amount, characteristics and attributes of resources can be fully estimated for proper utilization.

(2) In any space and time after considering the present resources need, the supply of the resources on lower prices should be immediately stopped in order to prevent destruction and misuse.

(3) For the prosperity and production of resources, all possible methods of conservation must be adopted.

(4) Those resources which are insufficient in quantity or exhaustible should be scientifically used.

(5) Such resources should be invented which are not exhaustible or take a long time such as solar energy in place of coal and petroleum.

(6) Resources complexes should be utilised in an integrated, balanced and in multiple ways.

(7) Keeping in view the amount and availability of resources the study of a social, economic and political relations are highly needed.

(8) By planning the national resources in a multitude of ways one should try to be self reliant personally, regionally and socially.

(9) In resources planning, full co-operation between the government, public and private institutions is necessary.

(10) The planning and conservation of resources are necessary in order to make the general people good citizens.

## CONCLUSION

Having gone through the important factors which affect the survey, conservation and utilisation of resources, it is suggested that the level of mass education and human health should be given due importance as without it no progress can be achieved. Proper attention to the development and right application of science and technology must be aimed to achieve this goal. State agencies, administrators and scientists should consider this problem of resources-conservation and planning on a strong footing. The natural resources should be tapped and modern technology should be mobilized to achieve success in harnessing resources for balanced regional development.

## REFERENCES

1. E.W., Zimmerman, *World Resources and Industries*, Harper and Row, New York, 1961, p 402.
2. R.B. Mandal, "Nature of Population Geography", *Dimensions in Geography* (ed. V.N.P. Sinha and R.B. Mandal), Associated Book Agency, Patna, 1978.
3. E.A. Ackerman, iu K.N. Singh and J. Singh, *Bhoogol Ke Mul Tatwa*, Tara Publication, Varanasi, 1973.
4. P. Dayal, "The Energy Resources of India," Geology and Geography Section, 64th session of *Indian Science Congress*, 1977.

# 2

# Resource System

—*Udai Prakash Sinha*

## INTRODUCTION

THE AVAILABLE endowment on the earth surface for the well-being of mankind is known as resource. It may be natural, biotic, human, economic, social, political and atmospheric and lithospheric in its dimensions. The system on the other hand, may be defined as "any set of objects and the linkages between them can be called a system". Men and their artifacts make up a man-made system. Natural phenomena and the processes which link them can be thought of as natural system.[1] Hence the system shows how resources exert their integrated influence on man—the supra resource of the earth surface—and how in configuration together they react on the earth by virtue of their functional structure in the environment. In any system, the process of feed-back mechanism is always at work, for example, if man needs a resource to be raised for his betterment and if it does, then it is man who has developed the resources for human consumption. Thus the systems of resources are always interacting with each other in between the welfare of man and the investigation, growth, production, planning and conservation of resources at the same time.

## OBJECTIVES OF STUDY

The purpose of the paper is to highlight some of the natural resources and their systems in relation to man who plays a vital role in shaping and utilising different kinds of resources as spatial linkages on economic, social and political grounds. The production and destruction of resources are the two extreme points, and in between these two, the same takes several forms in order to serve mankind within the particular climatic and topographic conditions of a region under study. This study has been designed to throw light on the forest and agricultural resources in the district of Monghyr.

## REVIEW OF THE LITERATURE

Resource utilization is the greatest function of the human society among which, according to Leonard Zobler, the resources can be grouped under three stages: firstly, when technological development is in the initial stage with poor technology and unskilled labour. At the second stage, the cultural advancement and technological change, abundant and proper use of resources are found. But at the third stage with the exhaustion of resources again scarcity prevailed for most of the resources of the world.[2] Water is an exception which is available ubiquitously. The complex system of acute scarcity may be expressed in Table 2.1.

Table 2.1 shows that the district of Monghyr has been faced with a vicious circle of poverty. This challenge has to be met with the economic development in the country. This major process requires mobilization of natural resources on a huge scale. All the frontiers of natural resources have been opened during the period of planned development in the district. Utmost precaution regarding resource-use policies should be taken. The resource-use policies should be so planned where no stone should be left unturned to utilize optionally the existing natural resources.[3]

Secondly, the study of Duncan shows that with the opening of vital resource fronts their best utilization is possible.[4] But only a few areas in the world have sufficient resources in a viable economy. This condition is available in the Mediterranean countries and China. According to Morgan and Munton, "Agricultural system cannot be meaningful unless it is associated with the production system along with the forest, soil, irrigation and systems of human resources".[5] Similar attempt has been made by Simmons who had successfully analysed the interdependence of resource in close connection with the ecological conditions of the region.

TABLE 2.1

| Limited production in terms of agriculture, industry and trade | Drain of resources : agricultural and industrial | Drain of capital through defective system of trade and agriculture | Ruin of industrial skill | Population explosion |
|---|---|---|---|---|
| 1 | 2 | 3 | 4 | 5 |
| | | STAGE OF SCARCITY | | |
| More land under industrial-cum-commercial crops. | Limited per capita agricultural production. | Negligence of scientific and technological training. | — | Misuse. |
| Export of food and industrial raw materials. | Lesser capital formation and investment in economic activities. | Limited Industrialisation. | — | Misuse. |
| Limited improvement in agricultural efficiency. | — | Limited trained hands | — | Wastage. |
| More mouths to feed. | Lesser capital for mobilization of resources. | Scarcity of skilled labour. | — | Agricultural scarcity of agricultural industrial raw materials. |
| Per capita scarcity of all resources. | Scarcity of capital. | — | — | — |

*Source* : P. Verma : *History of Resource Use*, p 135.

## IMPORTANCE OF RESOURCE SYSTEM

It has been observed that resources act as factors of production when these are used in the economic process. Entrepreneurial ability, labour, capital and natural resources counted as factors of production are known as human resources and the others are called property resources. Technology teaches us to make effective use of scarce resources in different ways. The other capital resources—machinery, plants and the like—have to be adopted and adjusted to other uses. Sometimes due to technological advancement, it has been observed that some resources are still unused. The four factors of production, e.g., land, labour, capital and entrepreneurship become scarce because they have alternative uses. If natural resource like land is used for a new golf course, the society provides space for an industrial park or a residential development. Similarly, a capital resource and material resource building can be used to produce something new in resources. Thus the most scarce factors of production rotate behind choice and pricing.[6]

Population as resource is the great creator, fashioner and destroyer of resources. In this way, systems of resources always revolve round a mechanism of integrated push-up in raising the production for the functioning of society. A man needs nearly twenty-two hundred calories of food for sustenance in a healthy manner in the Indian environment. The gain of these calories may be possible with the eating of prepared food whose preparation requires coal, oil, edible items, utensils, home, defence, amenable social stratification, amicable environment, energy, water and several other products of nature. It proves that only those items come under resource system which are useful for man.

The function of a system may be exemplified with the help of the production of food item. A man produces foodgrains to eat. The foodgrains give energy in the form of proteins and calories to mankind, and on the other hand, the same is used as seed to produce foodgrains. But here the system is not a meaningful one, because once a particular foodgrain is consumed, the same cannot be replenished, time and again whether one uses water in a steam engine or in a hydraulic turbine or the same is operated in the air, but ultimately after doing its job the water again comes into the form of water on the earth surface. Thus water balance in a man, in a field, in a machine and environment is a complete system of its regulation, and cannot be the food item. The production of food items in this sense gives an incomplete system wherever the replenishable capacity of water gives a full fledged and complete system.

### NATURE

The materials used by man come from the physical, biological, social and economic components of the earth. They interact and the same is known as eco-sphere in which a complex collectivity, alternation of quantity and the behaviour of elements and environmental components are followed by the adjustment of one with the other. The satellite imageries taken from the space craft tell us that the earth has sub-planet and limited dimensions. The rate of the flow of solar energy on the earth surface is at the rate of 3400 K cal/$m^2$ day and the same becomes the basis of light and energy of the biotic resources on the ground. The importance of solar energy lies in the proportion which is observed by the plants and its conversion into an organic matter to make it available to the other organism in the form of food through insolation. This process is known as photosynthesis and the world over yearly 2% of the solar energy is converted into the organic matter in the form of food. Thus unless and until this proportion is received by technological means, it may not form a limit in raising the resource level of an organism in this world.

There is a series of stages by which the energy flows through a system of human and biological organism (Fig. 2.1).[7]

In the above system at each transfer of energy, there is a loss of heat back to the atmosphere. Such systems are also in need of the factors of inorganic elements which are available to the plant from the soil. Animals derive them from plants in case the same is herbivora and by animal prey if it is carnivora. In nature these elements are recycled within the biological system which is susceptible to loss as far as it is possible. Such systems of living resources and their inanimate habits which affect man's general characteristics is known as eco-system. Here we are interested in discussing only one or two

*Resource Systems*

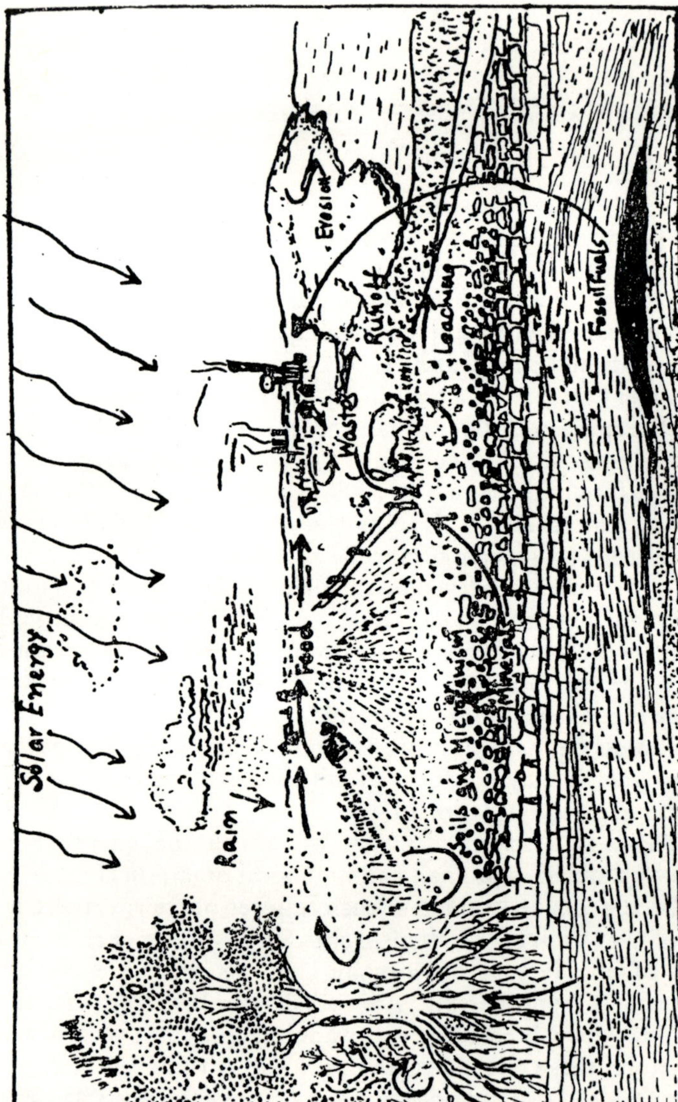


FIG. 2.1

components of eco-systems. In the domestication of wild animals the number of energy path can be reduced for the preparation of the structure of a system. Hence, per unit area of crop may be greater under cultivation than under natural conditions, but the general level of organic matter in the form of crop and the rate of energy flow for the greater production of resources is lower. At the time of harvest of all the crops, the essential elements of the system can be removed and the same can be reinstated artificially if the cropping is to be maintained. This is one of the ecological reasons for the use of fertiliser in the farm. Ultimately, man's modification of eco-system removes their natural barriers and balances, and makes them prone to instability as in the case of the soil erosion which suddenly changed the patterns of cropping.

## CULTURE

Among all the resources man's idea is the greatest storehouse of resource because unless a man perceives a thing as a resource potential, the same can never be one. In the western world the cultural attitudes are dominating over nature due to their intellectual superiority. In those places a man does not care about nature and for this fact the need to conquer nature is not a necessity. Somewhere, the limitations set by nature may change the behavioral patterns of people as for example Malthus sensed a widespread trouble created by nature in maintaining the number of persons of an area. In Asia, man and nature were philosophically viewed as one because both of them do not set out to destroy the ecological systems set by nature for the rapid growth of population. Recently man viewed himself as a part of the total community of the planet. The people of western world said that in the composition of world community, a series of commodities emerge as balancing forces. Recently cultural attitudes are dominated by the economic principle of demand and supply and the system of price mechanism. In nature, the demand is set by consumers for environmental resources, spatial arrangement of wilderness and pure air for breathing, but the price of industry quite often restricts, the supply of natural resources in bulk to the inhabitants.[8] In this sense, it is desirable to develop a reformed economics in which all external diseconomics are fully taken into account. Again, economics presupposes a rational man, but studies of environmental perception show the objective knowledge given by science through imagination, experience and prejudice, all of which play a vital role in resource use.[9]

Resource use can also be measured on a scale of rational to individual and immediate or unthinking selection of an action. In most cultures, only correct modes of behaviour persist which include the modification of eco-sphere in garnering resources. The idea of duty to posterity is strong in the western world especially in the present use of the planetary systems which may not be set to endanger future generations and resources as well. In religious writings, the concept of stewardship of man as the temporary guardians is also set in motion to make this planet as a useful storehouse of resources for the moral and perceptual well-being of mankind.

## SYSTEM ANALYSIS

In a resource system man considers natural system because it is also the resource producing system. Every system has to receive energy for movement from its natural state according to the use set by man because man is dominant in the system.[10] The system is also the response of economic and behavioral aspects of man, cultural values but an ecological aspect of the system; man and society have no power to limit nature.

## SYSTEMS OF FOREST RESOURCES

The district of Monghyr is rich in forest resources where forest lands are found in Sono, Chakai, Jhajha, Kharagpur, Sheikhpura, Jamalpur, Lakshmipur and other anchals of the district. In system analysis forest resources come under the close system with special reference to the tepographic, climatic, hydraulic and general environmental conditions. In forest environment rugged topography, higher amount of rainfall, lesser density of population, unproductive ground, the menace of wild animals, etc. are the rule. On the other hand, from forests as the resource, mankind derives wood by forest cutting, wild life with the prey of animals for meat, tending of trees for glue and several other forest products as raw materialsfor paper industry, bidi industry and so on. Thus forest as the marginal ground of human resource provides a firm base to the economy of the region, directly with the forest products and indirectly with the industrial production giving employment to the people. Cultural advancement to the society and the modification of environmental conditions it brings rain, protect soil from leaching, increasing the fertility of land, excretion of leaves as nitrogenous element and humous content to the soil including the construction of houses, highways, bridges, industries, railway wagons, etc. On forest resource, the primary function of human society lies with the forest cutting, secondary function is the manufacturing of goods from wood, tertiary function with the sale of manufactured goods by the shopkeepers, and building in urban areas. Quarternary function is provided by the forest rangers, etc., who teache the society the ways of preservation of forest resource and the virtues of resources for shaping mankind, whereas the quinary function is provided by the chairman, and ministers of the Forest Department who direct other persons on telephone for the protection, conservation, planning and extensions of forest resources. Its regeneration is not possible because once the wood drawn from the forest, the same never be replenished. Thus the cycle of forest resource in itself cannot be a fullfledged system but the incentive of man to grow and protect forest provides incentive for raising the forest resource including their growth and wood is an example of fund resource. Wood itself is the form of manufactured resource material. The presence of tree in nature is a natural resource. The utilisation of wood in the construction of house acts as a basic resource. Thus the wood of forest resource and the desire of man fulfils and in turn acts as a catalyst for the development and conservation of forest resources on the earth surface.

In resource systems with the utilization of solar energy and moisture by rain, the growth of plants on the ground takes place where micro organisms, minerals, leaching of soil, erosion by water run of, and industrial waste are the ecological cycle which runs from one resource to the other. The growth of plant, production of food and agricultural development system in between the food production and raw materials along with fossil fuels, of soil fertility by leaching, the growth of grass diet for animals, deposition of alluvial soil by meandering courses of the river Ganga and again the evolution of the initial cycle of landscape on newer land are the various cycle of natural and biotic resources which form the base for economic development, cultural advancement, and strategic position over different parts of the district of Monghyr.

In the district of Monghyr, the distribution of forest resource, net sown area and production of foodgrains are given in Table 2.2.

TABLE 2.2

*Distribution of Forest, Net Sown Area and Production of Foodgrains (1974-76)*

| Name of blocks | Percentage of forest | No. of registered craftsman industries | No. of workers engaged in craftsman industries in '000 | Net sown area in '000 acre | Total production in '000 M.T. |
|---|---|---|---|---|---|
| Sadar | Nil | 156 | 31 | 27 | 123 |
| Jamalpur | Nil | 98 | 22 | 24 | 93 |
| Dharhara | 42 | 25 | 14 | 19 | 85 |
| Kharagpur | 33 | 102 | 22 | 43 | 214 |
| Tarapur | Nil | 43 | 23 | 27 | 108 |
| Sangrampur | 7 | 30 | 11 | 21 | 89 |
| Lakhisarai | 10 | 97 | 24 | 55 | 200 |
| Surajgarha | 5 | 78 | 22 | 50 | 198 |
| Barahiya | Nil | 37 | 19 | 47 | 164 |
| Sheikhpura | Nil | 22 | 26 | 51 | 191 |
| Ariari | Nil | 16 | 13 | 23 | 102 |
| Barbigha | Nil | 10 | 15 | 33 | 181 |
| Jamui | 0.4 | 88 | 12 | 27 | 61 |
| Lakshmipur | 51 | 34 | 24 | 42 | 183 |
| Sono | 32 | 41 | 10 | 13 | 89 |
| Jhajha | 48 | 38 | 12 | 14 | 65 |
| Chakai | 33 | 16 | 13 | 25 | 83 |
| Khaira | 24 | 19 | 16 | 23 | 84 |
| Sikandra | 7 | 12 | 31 | 46 | 172 |
| Halsi | Nil | 13 | 19 | 38 | 181 |
| Khagaria | 0.3 | 68 | 24 | 39 | 121 |
| Alauli | Nil | 24 | 20 | 31 | 100 |
| Chautham | Nil | 32 | 13 | 33 | 100 |
| Beldaur | Nil | 22 | 12 | 32 | 101 |
| Gogri | Nil | 14 | 10 | 29 | 92 |
| Parbatta | Nil | 16 | 11 | 37 | 101 |

## AGRICULTURAL SYSTEM

Agricultural system provides the sustenance almost 80% people in the district of Monghyr. It is a traditional society of agricultural production where intensive subsistence farming is the rule due to larger number of people engaged in agricultural production and tiny plots of land owned by the farmers under permanent settlement of land tenurial system. For agricultural system the concept runs through land, labour, capital, suitable environmental condition and the incentive of local people for the success of such a system. In the district of Monghyr only 50% land is available for cultivation where the density of population is about 400 persons per square kilometre and the per capita income is only Rs. 200 in the area. The level of resource development has been achieved by the production of food crops in four different seasons of Aghani, Garma, Bhadai and Rabi. Most of the years the failure of natural rain, flood havoc, the spell of heat, cold waves, dissipation of crops by frost in the winter and the presence of innumerable natural hazards are responsible for present agricultural system of the district of Monghyr. Agriculture is a vital resource from which local people derive food, capital, materials of personal amusement, industrial raw materials, and in turn agriculture provides employment, energy and shelter for living to man. In any region the mechanisation of agriculture, use of higher number of agricultural labourers are examples of employment avenues. In the district of Monghyr the relationship between net sown area and agricultural labourer is very nice. In terms of the availability of agricultural labourers for almost all net sown area is relatively more than 800 per anchal.

Similarly, the correlation between total population and production of food is also high but it is curvi-linear in the graph (Figure 2.2). This shows that some areas are over populated with lesser production of food, while some areas are highly productive with lesser number of population in group III. Some areas have both lesser population and lesser food production in group I. In the district of Monghyr, the total net sown area and production of foodgrains is given in Table 2.2.

*Total Population and Food Production*

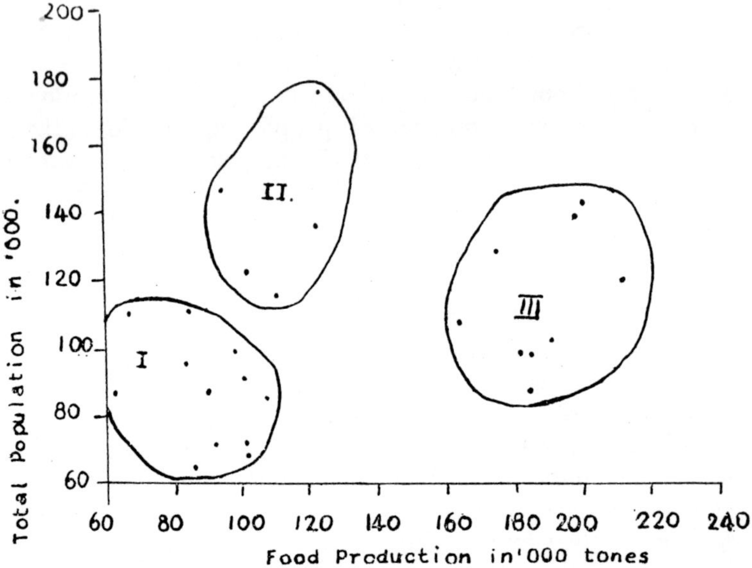


FIG. 2.2

## CONCLUSION

With the study of systems of resources, it has been observed that various resources available on the earth surface to their usability for man, it has been found that the resources e.g. forest, water, mineral, agriculture, population, livestock etc. are interlinked for the well-being of man and on the other hand, modifies the usability of a particular resource in different places according to their need and hence act as catalyst for resource development in a systematic manner. This is determined by the man himself. In this study with the justification of close relationship between forest resource and cultural and economic advancement of people shows the existence of a variable resource system. Similarly, the close relationship between total population of anchals and food production with net sown area and agricultural labourer justified the wholeness of the system.

## REFERENCES

1. F. John Kolars & D. John Nystuen, *Ways to Weigh the World: A Preface to Man Environment Systems Geography The study of Location, Culture and Environment*, 1964, p 247.
2. L. Zobler, "An Economic Historical view on Natural Resource Use and Conservation," *Economic Geography*, vol. XXXVIII, 1962, p 191.
3. P. Verma, "History of Resource Use." *Applied Geography*, 1968, pp 135-36.
4. C. Duncan, "Resource Utilization and Conservation Concept", *Economic Geography*, Vol. XXXVIII, 1962, p 119.
5. R.J.C. Morgan & Munton, "Agricultural Activity and the Physical Environment", *Agricultural Geography*, 1974, p 39.
6. A. Ronald, Wykstra, *Introductory Economics*, Harper & Row Publishers, New York, Evanston, San Francisco, London, 1971, p 15.
7. C. Simons, *Resource System : Aspect of Geography*, 1974, p 7.
8. Ibid., pp 8-9.
9. Ibid., p 10.
10. Ibid., p 13.

# 3

# The Application of Remote Sensing in Land Resource Management with Reference to India

*—B. Dey*

## INTRODUCTION

THE DEMANDS of a rapidly increasing world population coupled with the interaction of economic growth and resource depletion are creating a growing awareness of the need for sound land resource management. The management of land resources usually involves a three-step process : inventory, analysis and planning. At present, there is no systematic inventory of existing landuse and changes in land for agriculture, forests, water resources and urban areas at national or international levels. The major problems with the existing data sources are inconsistency and the age of the data. Moreover, many developing countries who urgently need a resource data base for national planning cannot afford such an inventory due to high cost and lack of trained personnel. On the other hand, the preparation of landuse maps in the developed countries take so long a time that the maps are out of date when they are published. Satellite remote sensing which offers a synoptic view and repetitive coverage of the earth's surface at regular intervals provides the means for data inventory, updating landuse data and removes the inconsistency of different data sources.

The use of satellite imagery in the inventory of resource data has several advantages. Firstly, satellite data is less expensive and faster than conventional ground observations. Secondly, satellite imagery provides a daily view of our resources over large areas. Such coverage is difficult and expensive to obtain by conventional aerial photography. Thirdly, satellite imagery itself provides a data base on which data of different land resources are stored. The operation of remotely sensed data for solving problems of land resource management is truly multi-disciplinary and involves geography, hydrology, forestry, agriculture, landuse, cartography and regional planning.

## REMOTE SENSING AND LAND RESOURCE DATA

The term "remote sensing" relates to acquiring information about features on or near the surface of the earth through the use of aerial cameras or other sensing devices. These are operated from an aircraft, helicopter or space craft that is at a distance from the area being sensed.

Until a decade and a half ago, aerial photographs were the most familiar form of remote sensor data employed in most studies of resource management. However, the world has received remotely sensed meteorological data with the launch of the first weather satellite Tiros-1 on 1 April 1960. Unfortunately, in the early stages, environmental data extraction from the imagery was very poor due to limitations of sensors and poor ground resolution. Today, a wide variety of environmental satellites equipped with various sensors are transmitting data on land resources to the earth at unprecedented rates. The Earth Resource Technology Satellite (ERTS—now known as LANDSAT) provides very high quality resource data due to the high ground resolution.

LANDSAT-1 and LANDSAT-2 were launched in July, 1972 and January, 1977, respectively. The satellites are sun-synchronous with near polar orbit at an altitude of 915 kilometres above the earth. The playloads of both satellites are three Return Beam Vidicon (RBV) cameras consisting of three bands : Band 1 (0.45—0.57 $\mu$m), Band 2 (0.58—0.68 $\mu$m), and Band 3 (0.69—0.83 $\mu$m); and a Multispectral Scanner (MSS) with four spectral bands: Band 4 (0.5—0.6 $\mu$m), Band 5 (0.6—0.7 $\mu$m), Band 6 (0.7—0.8 $\mu$m) and Band 7 (0.8—1.1 $\mu$m). LANDSAT has the advantage of offering small-scale photography with an area of 34,225 sq. kilometres (185 km $\times$ 185 km) while maintaining relatively high ground resolution (80 metres).

LANDSAT data are available either in computer compatible tapes, glossy prints or transparencies. Black-and-white images for each band (Bands 4, 5, 6 and 7) and two types of colour composites ($C_1$ and $C_2$ in Canada; and $C_1$ in the United States) are also available. $C_1$, a colour composite which simulates false colour aerial infra-red photography, is the combination of two visible bands (MSS Bands 4 and 5) and one infra-red band (normally MSS Band 6). $C_2$, is a combination of the visible red (MSS Band 5) and two infra-red bands (MSS Bands 6 and 7). Each type of image emphasizes different aspects of the scene and hence is more useful for some observations and for measurements than for others. A generalized listing of uses of each waveband is presented in Table 1 (Gregory and Moore, 1975). Bands 5, 6, 7 and $C_1$ and $C_2$ give better views for land resource information.

LANDSAT (1 and 2) resource satellites operate continuously, and with two satellites currently in orbit, scans are repeated every 9 days. The uniform scale and synoptic view of each MSS image has provided an important view of the world. The image is not interrupted by tonal contrast present in a conventional photographic mosaic.

## MULTIPLE DATA AND LAND RESOURCE MANAGEMENT TECHNIQUES

The field of remote sensing has been characterized by development of multiple data acquisition capabilities and multiple data analyses techniques. The concept is significant in relation to the usefulness of LANDSAT imagery augmented as necessary by aerial photography and ground observations. Supporting this approach in the field of data extraction and interpretation, it is found that more information is usually obtained through multi-enhancement of photography and also multi-disciplinary analysis.

Two methods, digital and analog, can be applied to land resource inventories. With the digital method, the remotely sensed data are processed for geometric and radiometric corrections, calibration and noise filtering. These are done to improve the aesthetic quality of the image for photo interpretation and for digital feature selection and classification. The advantages of digital processing are accuracy and rapid processing of

TABLE 3.1

*Generalized List of Uses for each ERTS Waveband (for well-illuminated, snow-free scenes)*

| Feature | Band 4 | Band 5 | Band 6 | Band 7 | $C_1$ | $C_2$ |
|---|---|---|---|---|---|---|
| 1. Well-exposed soil and bedrock, beaches, salt flat, desert drainage | F* | E | F | P | G | F |
| 2. Geological structure, surficial textures, lineaments | F | G | E | G | G | G |
| 3. Landforms | F | E | G | F | E | G |
| 4. Soil moisture | P | F | G | G | F | G |
| 5. Rivers and streams | P | F | G | E | G | G |
| 6. Boundaries of water bodies and wetlands, flooding | P | F | G | E | G | E |
| 7. Water turbidity, submarine topography | G | E | F | P | G | F |
| 8. Water penetration | E | F | P | P | F | P |
| 9. Vegetation coverage and stress (including logging, forest fire scars, etc.) | F | G | G | G | E | G |
| 10. Anthropogenic elements (urban development, transportation infra-structure) | G | E | P | P | E | G |
| 11. Rural landuse, agriculture | G | E | E | E | G | G |
| 12. Atmospheric transmission (through smoke, haze and dust) | P | F | G | E | F | G |

*Arbitrary ratings of Excellent, Good, Fair, Poor from experience by Gregory Geoscience Limited.

Object resolution averages about 200-300 feet, with higher resolution for objects of high contrast.

(After : Gregory and Moore, 1975)

large amounts of data. However, the bulk processing of LANDSAT data is not economical at this stage of computer hardware development.

The analog method is based on simple photo interpretation. Tonal differences in the image between water, forests and agricultural areas can be identified by visual interpretation with the help of magnifying lens, stereoscope, or by applying density slicing. The density slicing is done by an optical enhancement device which attempts to separate different features on the basis of arbitrarily selected reflectance levels.

The normal application of LANDSAT data in land resource management surveys is the preparation of a resource data base which could be a series of thematic maps showing the location and distribution of water, forests and agriculture. Other applications related to the development of a resource data base are landuse inventories, and land resource evaluation (land capability)

Landuse maps are very useful for land resource management planning. As we know, land resource potential (land capability) affects basic decisions concerning the reservation of land for its best use. A particular parcel of land could be well suited to recreation or agriculture based on its resource potential. A landuse classification system that provides a standard approach for use with remote sensing data has been developed by the U.S Geological Survey (Anderson, 1973).

Thematic landuse and land resource potential information can be obtained by visual photo interpretation of black-and-white images and colour composites, by density slicing or by digital computer processing.

Remote sensing scientists have made great progress in land resource inventories in regional and national levels. LANDSAT imagery has been used to update landuse and resource inventories in several areas of the United States, Canada and other parts of the world (ASP, 1975; Colwell, 1973; NASA, 1973, 1975; University of Michigan, 1975, 1977). The success of LACIE (Large Area Crop Inventory Experiment) in the United States and GAPE (Global Agricultural Productivity Experiment) in Canada indicates the feasibility of making uniform inventories of land resources (MacDonald and Hall, 1977; Mack *et al*, 1977). Indeed, many remote sensing scientists are considering the development of globally uniform inventories of renewable resources.

TABLE 3.2

*Remote Sensing Aerial Survey Projects*

| No. | Year | Test Site | Area Surveyed (sq. km.) | Equipment Used | Film, Filter Spectral Band (μm) | Format Scale | Platform | Purpose |
|---|---|---|---|---|---|---|---|---|
| 1 | 2 | 3 | 4 | 5 | 6 | 7 | 8 | 9 |
| 1. | 1970 | Kerala | 10 | Hasselblad camera | CIR, W12<br>.5—.9 | 70 mm<br>1:5000 | Helicopter | To detect coconut wilt disease. |
| 2. | 1972 | Kadgaon (Poona) | 80 | Bank of 3 Hasselblad cameras | CIR, W12<br>.5—.9<br>IR B/W, 89B<br>.7—.9<br>Pan B/W, W12<br>.5—.7 | 70 mm<br>1:2500<br>1:6500 | Helicopter | Identification of agricultural crops with CIR Imagery |
| 3. | 1973 | Karjat (Bombay) | 100 | Cluster of 4 Synchronized Hasselblad cameras | CIR, W12<br>.5—.9<br>IR B/W, 89B<br>.7—.9<br>Pan B/W, Red<br>.6—.7<br>Pan B/W, Green<br>.5—.6 | 70 mm<br>1:10000<br>1:50000 | Aircraft DC-3 | Identification of agricultural landuse with CIR and Multiband Imagery |

| 1 | 2 | 3 | 4 | 5 | 6 | 7 | 8 | 9 |
|---|---|---|---|---|---|---|---|---|
| 4. | 1974 | Anantapur District | 4000 | Cluster of 4 Synchronized Hasselblad cameras | CIR, W12 .5—.9<br>IR B/W, 89B .7—.9<br>Pan B/W, Red .6—.7<br>Pan B/W, Green .5—.6 | 70 mm 1:50000 | as above | Crop inventory and landuse classification. |
| 5. | 1975 | Patiala District | 4500 | as above | as above | 70 mm 1:30000 1:75000 | as above | as above |
| 6. | 1976 | Panchmahals District | 9000 | as above | as above | 70 mm 1:30000 | as above | as above |

After : Dhanju, 1977.

## REMOTE SENSING ACTIVITIES IN INDIA*

In India, various agencies are engaged in the application of remote sensing techniques for gathering data in the areas of agriculture, forestry, hydrology, meteorology, geology and oceanography. The activities connected with the research and development aspects of the applications of remote sensing are now coordinated by the Indian Space Research Organization (ISRO) through its divisions of Remote Sensing and Meteorology Applications (RSMA) of the Space Application Centre (SAC), Ahmedabad. The Survey of India and the Geological Survey of India have been using black-and-white aerial photography for mapping purposes. The Indian Photo Interpretation Institute extends the use of black-and-white imagery for forest inventory (Tiwari, 1974). The remote sensing activities in India expanded with the establishment of National Remote Sensing Agency (NRSA) for operational remote sensing surveys (Dhanju, 1977). The detailed activities of ISRO can be described in terms of data acquisition, data processing, data interpretation, and data utilization and resource management.

The airborne remote sensing with multiband and infra-red false colour photography have been used from various platforms like helicopters, aircraft and balloons for crop identification, landuse classification and inventory of resources (Table 3.2). The present airborne data acquisition system consists of a few Hasselblad cameras with different film filter combinations and an IR scanner. The IR line scanner is a single channel instrument operating in the 10-12 $\mu$m region. The infra-red false colour photography was used for its better identification of different crops and landuse patterns.

Commercial organizations and Government Research Laboratories have been processing photographic films including false colour infra-red films and also providing prints, slides and enlargements. The RSMA is setting up a fairly well equipped photographic facility primarily for precision processing and high quality reproduction. Elaborate support facilities for automated data processing have been developed by using an IBM

* Part of the information in this section is extracted from Dhanju, 1977 and Hariharan, 1974.

360/44 computers. Computers facilities are also planned for a computerized image processing facility which will handle digitation, image rectification, preprocessing, classification and pattern recognition for aerial and space acquired tapes and imagery (Hariharan, 1974).

Simple photo interpretative techniques are used extensively for data interpretation because they do not require any sophisticated equipment. This also gives the developing countries the opportunity of employing their own nationals and building up their own know-how.

The airborne surveys for resource studies in India are still in the research and experimental stages. It would be very costly for a large country like India to make land resource inventories with airborne remote sensing. For that purpose, the use of LANDSAT imagery is essential. LANDSAT imagery taken over India is available at the SAC of ISO, Ahmedabad; and EROS Data Centre, U.S. Geological Survey, Sioux Falls, South Dakota, U.S.A. Several studies in the areas of hydrology, forestry, agriculture and geology are in progress using LANDSAT imagery. A few scientist in India and abroad have used LANDSAT imagery for Indian resource and landuse studies (Es. 1974, 1976; Sharma *et al*, 1977). A study is underway by the author to examine the possibility of using LANDSAT imagery for agricultural resource inventories in India.

## FUTURE POTENTIALS AND LIMITATIONS

The feasibility of making national inventories of land resources in India is high due to a good remote sensing institutional set-up, research and experimental experience, and the ensuing launch in early 1978 of the second Indian satellite called Satellite for Earth Observation (SEO). Broad landuse classifications have been mapped and seasonal changes may be monitored by using LANDSAT imagery (Sharma *et al*, 1977).

Indian interpreters have developed a wide range of skills for using LANDSAT, airborne imagery and air photos. It is important that users take areas that are well known with well documented physical environmental data to establish signature limitations of LANDSAT imagery. Much experimental duplications can be avoided by utilizing overseas techniques. There is extensive current remote sensing literature which records these techniques (ASP, 1975; NASA, 1973, 1975; University of Michigan, 1975, 1977).

The limitations of satellite remote sensing for resource inventories in India are : (i) satellite imagery has low resolution as compared with aerial photographs. The individual species of mixed crops and small fields (common in rice producing areas) cannot be separately identified on LANDSAT imagery; (ii) the cloud cover has precluded the procurement of high quality LANDSAT imagery of large areas (i.e. imagery with less than 10% cloud cover); and (iii) well trained personnel. Cooperation should occur between Governments and overseas institutions for remote sensing training.

## CONCLUSIONS

Remote sensing is a very useful data gathering technique applicable for many resource inventory problems. LANDSAT data, a major input into any natural resource information system, also provide the perspective for viewing regional problems and the repetitive coverage for forecasting seasonal changes. Both are necessary when making decisions about the use of renewable natural resources from the data inventory. However, this inventory is to be supported by airborne surveys and by field investigations. The amount and reliability of data obtained depend on the competence of the user and his knowledge of the potentials and limitations of various techniques. Comparisons of LANDSAT colour composites between two seasons provide much additional useful information for data inventory.

Studies in the United States, Canada and a few other countries have demonstrated the important role LANDSAT imagery provides in land resource inventories and management. LANDSAT imagery would be useful to developing countries for providing a wide range of thematic mapping at a scale of 1 : 250,000 or smaller.

Remote sensing is also multi-disciplinary. Integration and sharing of research and analytical resources is vital in India. Many scientists from the disciplines of physics, geology, forestry and hydrology are engaged in remote sensing research in India.

Unfortunately, the geographers in India have not yet come forward to accept the challenge of using satellite remote sensing techniques in their research activities.

## SUGGESTED READINGS


American Society of Photogrammetry (ASP), *Manual of Remote Sensing*, 2 vols., Falls Church, Virginia, 1975, p 2144.

J.R. Anderson, *A Landuse Classification System for Use with Remote Sensing Data*, U.S. Geological Survey Circular 671, Washington, D.C. 1973.

R.N. Colwell, "Remote Sensing as an Aid to the Management of Earth Resources", *American Scientist*, vol. 61(2), 1973, pp 175-183.

M.S. Dhanju., "Remote Sensing Activities in India", Paper presented at the U.N. Regional Training Seminar on Remote Sensing Applications at Karachi, Pakistan, January 17-22, 1977.

E. Van Es, "Interpretation of Forest Vegetation of the Doon Valleys on ERTS-1 Imagery", *ITC Journal*, vol. 5, 1974.

E. Van, Es, "The Dry Deciduous Forests of Bastar, Central India on LANDSAT-1", Symposium Paper for ISP Commission VII, Helsinki, 1976.

A.F. Gregory and H.D. Moore., "The Role of Remote Sensing in Mineral Exploration with Special Reference to ERTS-1," *Canadian Mining and Metallurgical Bulletin*, vol. 68, 1975, pp 1-5.

T.A. Hariharan, "Techniques and Applications of Remote Sensing in India", Paper presented at the Institute de Pesquisas Espaciasis, S. Paulo, Brazil, June 1974.

A.R. Mack, J. Schubert, C. Goodfellow, P. Chagarlamudi and H. Moore : "Global Agricultural Productivity Estimation from LANDSAT Data," *Fourth Canadian Symposium on Remote Sensing*, Quebec City, May 16-18, 1977, pp 8-18.

R.B. MacDonald and F.G. Hall, "The Large Area Crop Inventory Experiment—A Major Demonstration of Space Remote Sensing." Invited paper presented at COSPAR/IAMP, Tel Aviv, Israel, June 8-10, 1977.

National Aeronautics and Space Administration (NASA), *Symposium on Significant Results Obtained from the ERTS-1*, GSFC, NASA SP-327, NASA-356, NASA-357, X-650-73-127, 1973.

———*Proceedings of NASA Earth Resources Survey Experiment*, 4 vols, NASA TM X-58768, June 1975.

R.D. Sharma, B N. Raina and M.S. Dhanju, "Studies Utilizing Orbital Imagery of India for Geology and Landuse", in *Remote Sensing Applications for Mineral Exploration*, (edited by William L. Smith), Dowden, Hutchinson & Ross Inc., Stroudsburg, Pennsylvania, 1977, pp 334-362.

K.P. Tiwari, "Forestry, Aerial Photo Interpretation and Resource Inventories in India, *Proceedings of the Symposium on Remote Sensing and Photo Interpretation*, International Society of Photogrammetry, Commission VII; Banff, Alberta, October 1974, pp 177-189.

University of Michigan, *Proceedings of Tenth and Eleventh International Symposium on Remote Sensing of Environment*, Environmental Research Institute of Michigan, Ann Arbor, U.S A. 1975, 1977.

# 4

# Tourism and Recreation—The New Frontier of Geography*

*—Tej Vir Singh*

> Not labour but leisure will be the great problem in the decades ahead. That prospect should be accepted as a God given opportunity to add dimensions of enjoyment and grace to life.
>
> *—David Sernoff*

THE SPIRIT of wanderlust and nomadism seems to have possessed modern society so much that more and more people are on the go in search of "new pastures" and "new lands", more often than not, by way of escape from their own environment. Some trot the globe by crossing international boundaries, others rest content with their own resorts, sea-sides and national parks while many more can do no better than invading their country-sides.[1] City parks and open spaces can be seen full like beehives on evening sessions, work-breaks or weekends. This is more true of the affluent societies, often afflicted with urban excesses and artificialities of life, subscribing to Wordsworth's poetic murmurings : "What man has made of man".

Since all these short-term migrants are characterised by bohemian moods they demand sophisticated recreational infra-structure, better recreational transport and a home away from home with perfect hospitality. All these services have to be procured by the host countries because tourists pay handsomely, often bringing jam to their breads. Nor can 'he' be ignored by the developer or the planner and least by an economist and a geographer, the last being committed to the concept of balanced growth and space organization. A geographer's. involvement in the recreational behaviour of the people should

* First published in *Tourism Recreation Research, June, 1978* and adapted here as desired by the author.

be all too obvious particularly in the context of recreational landuse in regional or urban planning where the problem has become almost threatening. Clawson comments[2] :

> The growth of outdoor recreation and the attendant squeeze on recreation facilities is just about the most open secret of our time. What is happening is clear enough. The question is : What does it mean ? Are we suffering a natural healthy spell of growing pains or are we working up to a crisis in outdoor recreation. To me the evidence suggests that we are approaching a grade—a crisis, unprecedented both in size and character.

The picture of Indian landscape is even worse where the recreational zones, in most cases, lie in distress : riversides wantonly polluted, greens fast disappearing and the city-lungs infiltrated irreparably.

## TOURISM GEOGRAPHY

Until recently geographers have had surprisingly little to say on this aspect except a peripheral discussion in their urban investigations. While tourist geography* is well established as a branch of broader field of recreation geography nonetheless it did not go further than studying resort settlements or at best qualitative assessment of scenic resources of countries like Switzerland or Norway. In fact, it did not make much headway till the last decade or so when some of the serious minded American and European geographers gave their best thoughts to it. Christaller, Clawson and Wolfe are some of the pioneering names in this field. The establishment of a non-profit research organisation : The Resource for the Future, Inc., in U.S.A. (1952), was a turning point in the history of recreation studies. The publication of Outdoor Recreation Resources Review Commission (ORRRC) study reports** (Washington, 1962) was

* Tourist geography focuses attention mainly on vacation industry at the international and national scales while outdoor recreation studies are largely home-based.

** From the beginning, one of the strongest currents in American thought has been the idea that the Outdoor Recreation is a right of Americans. And President Kennedy's contribution in procuring these facilities can hardly be underestimated.

*The Key Elements in Tourism and Outdoor Recreation*

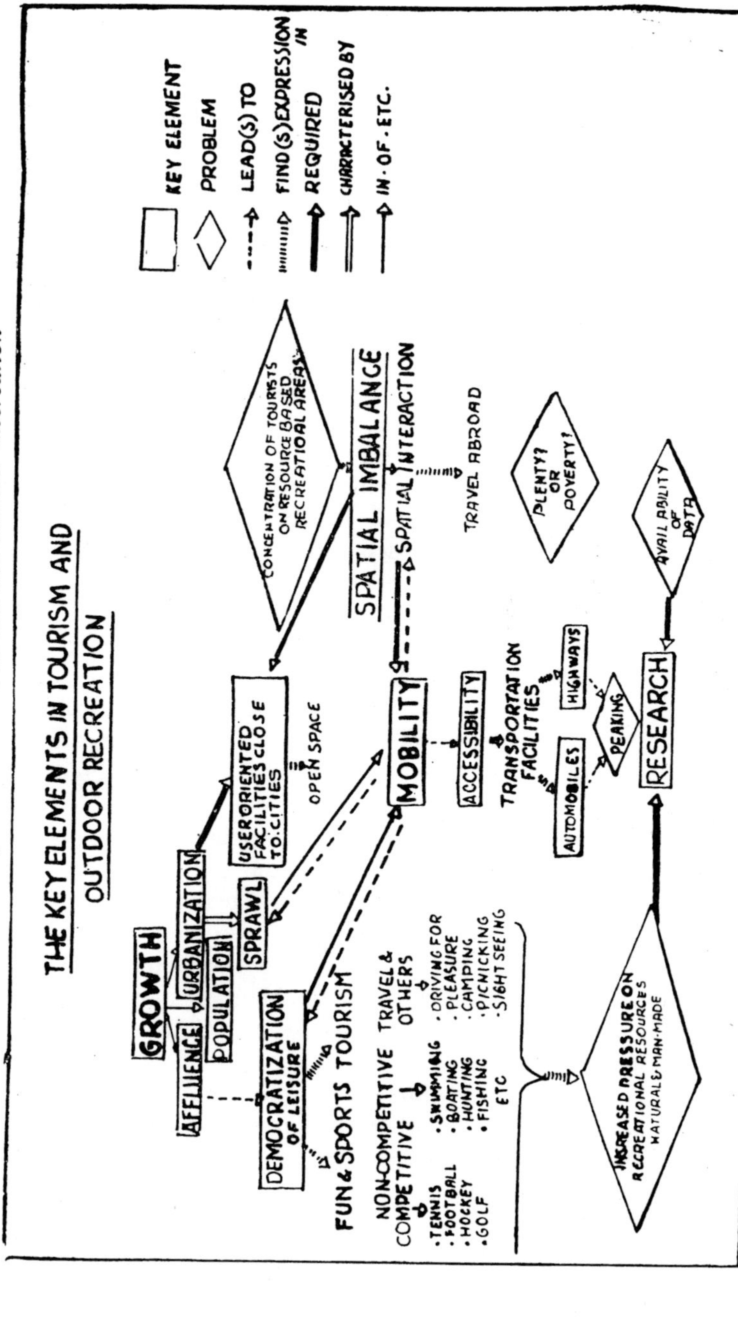


FIG. 4.1

an American magnumopus. Marian Clawson enviably emerged as a sensitive recreation-geographer. In Europe, French and Swiss geographers made handsome contributions to the tourism industry while German tended to look at the recreation pattern in its entirety. Christaller wrote voluminously after micro-geographical examination of his own region. He not only produced primary data on tourist traffic but applied his Central Place Theory in reverse, i.e., tourists gravitate to peripheral regions from the Central places. "Christaller brought recreation into the mainstream of modern geography."[3] Unfortunately, recreation geography did not thread its way in India. Except a few non-descript papers and all too few research-monographs, that leave everything to be desired, not much has been attempted in this country; much less has been published. It is indeed a sad commentary that this fascinating theme has failed to attract matured geographers of India[4], who perhaps, don't wish to turn away from their classical pursuits or as Wolfe puts it, "their discipline offers them such a multitude of specialities to choose that recreation to them is only a little more than a study of fun". The other reason to my mind is that our geographers don't find it quite worthwhile to study what people do in their work-free leisure time—and how many have leisure, indeed ! Some of them that do have, don't know what to do with it. May be the world of "economic man" is all too public and amenable to study whereas the world of "leisured man" is traditionally difficult of access. Hence data collection on factory production is easy to come-by but statistics on discretionary income and expenditure on park visitations or recreational travel is a challenging task. Customary geographical interrogatives may provide the key to the problem, viz., *why* do people seek outdoor recreation in such large number ? *What* kind of natural resources are most in demand and *how* are these resources used for this purpose? Answers to these pertinent queries would not only unfold the nature of the subject but would also emphasise how deep a geographer's involvement can be. To quote Murphy "Geographical research in outdoor recreation includes aspects of nearly every major division of systematic geography".[5]

## LEISURE CIVILIZATION

An explanation to the 'why' query is *prima facie* very simple that people are outdoors because they have sufficient leisure, wherewithal and mobility. This need further elaboration. While this is true that recreation and leisure are intertwined, the phenomenon of "mass leisure" is the direct outcome of the industrial revolution. We are now out of demand of Weberian "iron cage" and can reap the "ethic of salvation" by work, and enjoy the fruit of our hard labour.[6] Thus recreation is no longer the prerogative of a few "leisured elite"* or the parasites of society. Veblen's "industrial elite" can now also display worth, wealth and status in society and can indulge in "conspicuous consumption" and "conspicuous leisure" to a certain degree. Thanks to this Brave New World that has brought about a vast range of technical, economic and social changes, some of which are certainly dramatic. As the automation of production processes would lead to higher output and wages, leisure would increase only to go hand in hand with recreation and tourism. Since this curve is likely to go upwards we shall have problems of "leisure not labour". To quote Robert Cam[7] :

> In relation to the present, society may then become comparatively inactive: only a part of the population will work, tourist operators in particular, and the rest will have to be paid, in fact, for doing nothing.

It is worthwhile recalling in this connection some remarks of Jean Fourastie[8] :

> It is generally expected at present that in fairly near future the average citizen of economically developed country will be able to meet his needs by working 30 hours a week. Thus, the time that an individual will devote to productive activities in future, will take up about 6 per cent of his existence.

* The ancient Greeks granted leisure to their elite. For the slaves, work was an end in itself.

Fourastie's forecasts do not differ from projections made by H. Kahn of the Hudson Institute in which he observed that post-industrial society will be largely pleasure-oriented, working only 1100 hours a year, that is 147 working days and 218 days of free time. This, indeed, is a big challenge of leisure, of course with awfully bright prospects, to all those engaged in the leisure industry. Trends have already been set in; working hours have been shortened, weekdays reduced and weekends lengthened. Hodidays are now recognised as social needs, providing relaxation to compensate the strain of living in an industrial society that saps physical and mental energy produced by work. Legislation for paid-holidays to workers exists in all countries of the world and the number of paid vacations are constantly increasing.[9] With this transport facilities are vastly improved and the rural population is shrinking nearly everywhere in the economically developed countries, resulting in the change of occupational structure of society which has better social mobility and of course more genuine leisure.

Now, one can ask what relevance this immense leisure and vast discretionary income do have with Geography or Geographers. To answer the question, let us proceed to take up other two unexplained interrogative *what* and *how* these tourist/recreational products are made and used. This is just a question of "demand" and "supply" in tourism resources—virtually a geographer's realm, for in them emphasis would always be on those space-consuming activities that take place out of doors.

## RESOURCE SUPPLY

Man of today afflicted with urban cruelties, deafening noises and maddening crowds, seek in nature refuge and solace.[10] It is for this benign (rather divine) influence that Hindus placed their "Dhamas" (sanctums) in the heart of the wonderful Himalayas or along the shores or junctions of the rivers, lakes, sea-shores, spas and hill resorts go full with recreationists during tourist cycles and rhythms. Evidently, the physical characteristics of these natural elements of the landscape affect their use for outdoor recreation but as Zimmermann sums up : "Resources are not, they become."[11] In fact, it is the combination of natural qualities and man's desire to use it which gives it an attribute of resource out of what otherwise may be more or less "a meaningless combination of rocks, soils or trees". To quote Clawson "in this respect outdoor recreation resource is no different from farming, forestry, grazing, mining or any other natural resource.'"[12] Some of them can be created artificially, viz., reservoir, area dams, mounds and artificial lakes or canals, Based on management problems, the ORRRC (USA) made the following six major classification :[13]

(i) High density recreation areas
(ii) General outdoor recreation areas
(iii) Natural environment areas
(iv) Unique Natural areas
(v) Primitive areas
(vi) Historic and cultural sites

This list makes it sufficiently clear that outdoor recreation is a user of natural resource in a big way. Of them land forms, water bodies, climate, forest and wild life are the foremost. Man affords each of these a value in accordance with his outlook. Even a primitive wilderness area is a cultural feature because it is legally defined and an artificial boundary drawn around it (Taylor, G. D., 1961).[14] Thus, the resources range, as Mercer says,[15] along a continuum from underdeveloped wilderness areas to the highly artificial environment of urbanised resorts or landscape parks (Mercer, 1970). Clawson's classification of resources—user-based, intermediate and resource-based takes into account physical recreation resources and spatial imbalance; the last one (resource based) is demanded by tourists and the first-two are home-based resources for the consumption of outdoor recreationists.

Of all the factors, climatology of recreation deserves attention. Ullman[16] was among the firsts to recognise the importance of climate amenity (Ullman, E. L., 1954). Certainly climate is a vital recreation resource, motivating or dissuading tourist traffic. The concept of a destination area is, indeed, inalienably linked up with climatic comfort regimes; an obvious fact that recreationists and tourists seek besides other things, a sparkling bracing quality of weather, makes it a precious tourism merchandise. It is perhaps for this reason that Mediterranean lands in winters turn into vacation lands and playgrounds of Europe. People like to relax, revitalize and repair after prolonged exposure to urban life, fatigue and continued sedentary pre-occupations at centres which have invigorating climates with strong uplifting effect.[17] John Keats in his inimitable style observed : "Nothing cheers up my spirit than a fine day". To dwellers of micro-climates, sunshine is a tourist commodity whereas the residents of meso- or mega-thermal zones seek hill resorts for a spell of cool weather. "The burgeoning of ski-resorts of Chilean Andes, of the Australian Alps and of Scotland are a tangible expression of the changed appraisal of this climatic resource".[18] All these areas of outdoor recreation resource supply are areas of genuine geographic search and research. Efforts should be made to relate resort-growth to physiological climatic factors, comfort and discomfort regimes. Terjung's[19] Physiological Climate of Coterminous U. S. (Annals, Vol. 56, Part 1) and Singh's "Nainital : Centre Resource Based : A Study in Climatology of Recreation"[20] and "Climatology of Recreation : A resource-input for Indian Tourism" (TOURISM RESEARCH, Vol. 2, 1977) are thought-provoking papers worth going through. Flora and fauna recreation resources must also be a geographer's indepth study. Of late, forest recreation has been surging forward rapidly as a major use of forest areas in countries abroad: ranging from losing of oneself in the trackless wilderness to merely viewing the forest backdrop or enjoying the 'non-contact feel' of the forest-waters or bird-watching anything can be sought. In any case, physical geographers have rich pastures to graze. Some of them may be mentioned here for guideline: "Land assessment and land scale value for space preference", "Carrying capacity of various recreation areas", "Water resource of a region for recreation", "Aesthetics of physical landscape for resort growth", "Recreation resource inventory of region", "Climatology of recreation", Anthropo and Ethnic studies : "People resource, Primitive folk culture, Recreational desire line of an area", "Potential recreation areas', "Problem of Geomorphological mapping", "Impact of leisure on land and man",'Non-economic influence of recreational activities and many other problems pertaining to conservation and ecological balance, viz., "how can we reconcile Access with Beauty" or how do we balance developmentand conservation; "Is recreation in conflict with quality of environment", "quantitative measurement of capacity or area-attraction"—its beaches, mountains, fishing waters for human use without or least damaging the resources are some of the problems abegging solution. American geographers have assuredly showed the way and their work can serve as model. Dower[21] cites the case of Huron Clinton Authority at Detroit where they know the capacity of each of its regional parks, element by element (beaches, for example, can take 25 people per acre, golf courses 10, nature study area 3). "The car parks are fitted to this capacity with overspill spaces onto grass areas for peak periods, and the roads are fitted to peak traffic and the flow of cars to each park is checked" (Dower, 1970). Cannot we take such fertile though intriguing research projects. For that matter, I would like to cite the case of Kumaun Recreation belt in U. P. Himalaya where Nainital explodes with vacationists at the expense of other micro-tourism regions, posing problem of peaking in seasonal tourist rhythms. Or Badrinath bound pilgrim go stranded amass and the problem of environmental pollution infests the divine destination every pilgrim season; the Ghats of Varanasi also invite attention. This seems so ridiculous that our planners have initiated space-action in the sensitive pilgrim zone of Himadri, (most of it remained a protected area till June 1974) without adequately assessing its recreation-resource supply,[22] fixing typology and quality of resource for distribution. Even the famous Bhyundar Valley, popularly known as the Valley of Flowers[23] has been exposed to human-erosion with little planning and no preparedness for resource protection. Cosgrove quotes Wibberley's[24] investigation in England and Wales where it was made possible to assess acres of recreational land per 1000 population by region (Wibberley, 1968). Similarly Perloff's study on Urban Growth and the Planning of Outdoor Recreation provides the keystone for whole structure erected by the ORRRC, U.S.A. With the help of a devoted band of geographers and social scientists the Commission could adequately evaluate U.S. optimal outdoor recreation supply up to the year A.D. 2000. I can hardly resist the temptation of mentioning some of the observations made in 1970 London Conference on "Countryside" on recreation as "the top defiler of the environment by way of trampling of vegetation, pollution of rivers and lakes, destruction of wild life and spoilation of scenic area by unsightly development".[25]

This reminds me of the prophet Jeremiah :
But when ye entered ye defiled my land and made mine heritage an abominable place.

I do agree that impact-assessment studies are tough and challenging and often insidious in character for the reason that environmental deterioration goes imperceptibly, and more often than not, goes hand in hand with the take off phenomenon. But this area of research can hardly by ignored by tourism geographers. Wager,[26] however, was able to demonstrate the possibility of evaluating direct recreational effect of man on a wilderness area. Singh* presents an interesting study; "Impact of Tourism in the Himalaya: A Partial Analysis of Uttarakhand Himalaya".[27]

India is immensely rich in historico-cultural resources. The spirit of religion pervades all through her physical and cultural landscapes; there flows Gita in her rivers and there is, indeed, a sermon on the mount Himalaya; monuments and monsolea, caves and temples throw soft invitations to lovers of art and architecture. In fact, they consitute India's paramount tourism resources. Geographers with their Argus eye and sharp tools of analysis can ably interpret this queer intermix of culture and geography, so subtly interwoven in the fabric of her landscape aesthetics.

## DEMAND ASPECTS

If tourism and recreation offer so much to physical geographers along the supply side, there is a lot worthwhile to do by human geographers on the demand side. In any area of tourism, mobility and spatial imbalance would constitute core-studies, for the people must move to the place where the

* The paper was discussed at the National Seminar on Resource Development and Environment, organised by the National Committee of Environment, New Delhi, 1978.

resource exists. Thus transport holds the key to tourism. The extraordinary growth of tourism in modern society would not be conceivable without the remarkable development of means of transport. To a certain extent transport and tourism are complementary. Considering tourism manifests itself at the outset through a demand for transport services, it is understandable that the size of tourist demand has expanded greatly as a result of the increasing need felt by man to travel."[27] Wolfe in his study "The Geography of outdoor recreation" emphasised this aspect and added fresh dimensions to the subject in his thought provoking report "Travel : the New Migration".[28] Cambell (1966) pleaded for more organised studies around highway oriented pattern of movement. Tourist traffic flows were constructed in Europe by Defert and Todt. Some of the geographers did good cartographic exercises in traffic patterns along recreational and non-recreational highways. Clawson's designation of three-tier-facilities, user-oriented, intermediate and resource-based take into account time-element and spatial imbalance. Time comes to us in chunks of varied size. The pattern of these chunks of free time is constantly changing. Time rhythms of leisure as utilised by people according to space have to be investigated for better results. More worthwhile work can be done on vacation and weekend extra-urban nodes than on predominantly intra-urban travel flow lines. Traffic-volume-patterns are important to discover the recreation hinterland and for augmenting facilities. Mercer's Melbourne recreation study revealed that day trips in metropolitan areas are barrier to movement and they were close to nearest artery. Others concluded that the car is used for flexibility in timing and route and also for optimal use of the tourist belt. This area of work needs better attention from committed geographers.

Accommodation and tourist dwellings are more important areas of tourism research. A primary study of Human Geography is dwelling and hence human geographers can considerably contribute to decision structures—a neglected field so far. Vacation homes, hotels and tourist residences are spatial expressions of a region and they should reflect the cultural mosaic in respect of architectural blend, house-keeping, cuisine and indigenous entertainment. Hotels, particularly star complexes, occupy an important place in the morphological structure of a resort town. Case studies can be piloted to assess the impact of such hotels on the economic/social growth of the region. Except some travel bodies and non-profit tourism-agencies, viz, IUOTO, PATA and tourism corporations who have conducted in-depth researches in the economics of tourism, rarely did a geographer think it his domain to take up a project of this nature.

In fine, demand studies in recreation ask for meticulous data-collection, processing and measurement for mapping and decision-making. Properly processed data-mass can provide a meaningful recreational, pattern of planning. Some other components can be demographic studies of size, sex and age-group of the population, the amount of leisure and discretionary income available with the people. High standard motivational and attitudinal studies would be needed for building a tourist profile of the area. There is assuredly an opportunity and a challenge to geographers if they can accept it.

## REFERENCES

1. T.V. Singh, "Planning Indian Tourism Promotion" : Paper Presented at the First Asian Conference on Regional Planning and National Development, Mysore University, (Mysore 1974). Published in *Tourism Research*, Vol. 1, No. 1, Lucknow, 1976.
2. M. Clawson, Study Reports ORRRC SR-57, Washington.
3. R. I. Wolfe, "Perspective on Outdoor Recreation", *Geographical Review*, A. G. S., Vol. LIV, No. 2, New York, 1964.
4. T. V. Singh, "Tourism and Tourist Industry, U.P." : *New Heights*, Delhi, 1975.
5. R. E. Murphy, "Geography and Outdoor Recreation : An Opportunity and an Obligation", *The Professional Geographer*, Vol. XV Part 5, 1963.
6. I. Cosgrove and R. Jackson, "Geography of Recreation and Leisure", *Hutchinson University Library*, London, 1972.
7. Robert Came, "The Future of Tourism : Long-Term forecast and objectives", Travel Research Journal, IUOTO, Geneva, 1969.
8. Fourastie, Cited in Robert Came, "Future of Tourism" IUOTO, *Travel Research Journal*, Geneva, 1969, p. 23.
9. J. J. Taylor, "Social Consequences : Past, Present and Future" In Sales, H. P. (Ed.) *Encyclopaedia*, Travel and Tourism World, London, 1959.
10. T. V. Singh, op cit.
11. E. W. Zimmermann, "Introduction to World Resources", in Hunker, H. L. (Ed.), *Harper ond Row*, New York, 1964.

12. M. Clawson, "The Crisis in Outdoor Recreation", American Forest, Vol. 65, Part 3, 1959.
13. M. Clawson, "Recreation Resources". In Smith, G. H. (Ed.), Conservation of Natural Resources, John Wiley and Sons, Inc. New York, 1971.
14. G. D. Taylor, "An Approach to the Inventory of Recreation" *Canadian Geographer*, Vol. 9, Part 2, 1965.
15. D.C. Mercer, "The Geography of Leisure—A Contemporary Growth-point", *Geography*, journal of the Geography Association, No. 248, Vol. 55, Part 3, Sheffield 1970.
16. E. L. Ullman, "Amenity as a Factor in Regional Growth", *Geographical Review*, Vol. 44, 1954.
17. T. V. Singh, "Exploring Climate Recreation Resource for the Tourism Industry of India With Accent on Uttar Pradesh", paper read at the first Indian Geographical Congress, Published in *Tourism Research*, Vol. 2, No. 1, Lucknow, 1977.
18. D. C. Mercer, "The Geography of Leisure" op cit.
19. W. H. Terjung, "Physiologic Climates of the Coterminous, United States". *Annals*, Vol. 56, Part 1, 1966.
20. T. V Singh, "Nainital : Centre Resource-based : A Study in Climatology of Recreation", Nainital, Souvenir Volume Committee (Autumn Festival) and U.P. Directorate of Tourism, Nainital, 1974.
21. M. Dower, "Leisure—Its Impact on Man and Land", *Geography* (Journal of the Geog. Association) Vol. 55, No. 248, Part 3, Sheffield, 1970.
22. T. V. Singh, "Opening Garhwal for Tourism : Towards Research-based Planned Development", *The Himalayas* (Vol. No. 1), Institute of Himalayan Studies, Garhwal University, Srinagar-Garhwal, 1977.
23. Jagdish Kaur, "The Valley of Flowers ; An Exploratory Survey of High Altitude Flora", *Tourism Recreation Research*, Vol. 2, No. 2, Lucknow, 1977.
24. G.B. Wibberley, "Pressure on Britain's Rural Land", (proceedings of the 22nd Oxford Farming Conference), cited in Cosgrove. Op cit.
25. Report. The Countryside in 1970, H.M.S.O., London, 1964.
26. J. A. Wager, "The Carrying Capacity of Wildlands for Recreation", Society of American Foresters, Forest Science, Monograph 7, Washington, 1964.
27. T. V. Singh, "Impact of Tourism in the Himalayan Region, Proceedings of the *National Committee for Environment Protection* (NCEP), New Delhi, 1978.
28. R. I. Wolfe, "Recreation Travel : The New Migration", *Canadian Geographer*, Vol. 10, Part 1, 1966.

5

# Conceptual Development in Transportation Geography

—*S. Shanmuganandan*

## GEOGRAPHY AND ITS METHODOLOGICAL IMPLICATIONS

GEOGRAPHY has acquired a professional identity and states of its own during the course of its development in order to build up a platform for sound methodology. The development of this professional identity in geography has two aspects—the future development of the theoretical study of spatial distributions ; and a reappraisal of the overriding problem recognised by the discipline (Ackermann, 1953).[1] In this manner geography seeks to organise its field of scholarly procedures of investigation and presentation (Davies, 1972).[2]

The mosaic of human communities and their inter-relations in multifarious activities is in a continuous process of change and this has resulted mainly because of the forces which shape or re-shape the sense of problem, perspectives and concepts of the discipline.

### Nature of Perspectives in Geography

The essence of the dominant perspective of geography are intersect of place, space and time. The clusters of these approaches or themes exhibit a good deal of shifting emphasis and the change is apparent. The different approaches include (i)Theoretical approach, (ii) Behavioral approach, (iii) Histo-genetic perspective, (iv) Ecological approach, and (v) Regional perspective.

### Shaping the Conceptual Frame

The conceptual elements are location, spatial distribution, spatial association, spatial interaction, and spatial systems.

Terms denoting allied concepts and characteristics with location-site, situation, place and locality ; with spatial distribution continuity, discreteness, contingency, pattern and density; with spatial association—integration, interdependence, coherence, ensemble, coincidence and complementarity, with spatial intersection—accessibility, linkage, and flows; and with spatial systems—connectivity, order and hierarchy are associated with the above and form an integral part.

The theoretical deductive reasoning is quickening as recognition of the applicability of systems concept for all of which geography is mainly concerned with the spatial attributes of systems. This has paved the way toward a flexibility of research thus allowing geography an almost endless variety of subsystems significant to the over-riding problem. Thus the systems approach in geography characterises the discipline in a unique bridge position between the natural and the social sciences. The interest is in the identification of hierarchy of problems with the over-riding problem of understanding the man-natural environment system. Geography has thus acquired for itself a significant place in satisfying man's scientific curiosity. Geographic studies will be irreplaceable components of the scientific support for an efficient space management.

Geography has, in the past, witnessed different levels of generalisation which further foster geography in its dynamic way of expounding theories through a conceptual revolution. Three major recent concepts considering the studies of spatial structure, organisation and processes have placed geography in three levels of theoretical inquiry. Such a conceptual development of science ejected out in geography in very recent times developed the science of geography in every aspect and placed geography as a modern science. Geography now stands on the verge of developing man-machine systems through which it distinguishes itself as a distinct discipline as far as its process of learning is concerned.

At present the methodology of geography is largely prequantum. A discipline like geography will always be needed for the formulation and testing of significant hypotheses at all levels in order to make the conceptual revolution gain momentum.

## QUANTITATIVE REVOLUTION AND THEORETICAL GEOGRAPHY

The movement which led to the revolution in geography was begun by non-geographers like physicists and mathematicians. The quantitative revolution was inspired by a genuine need to make geography more scientific and by a concern to develop a body of theory. The quantification in geography made a new era which has as its main concern more rigorous hypotheses testing, interest in statistical regularities which ultimately result in the formation of a highly sounded theory and also in furthering the methodological revolution in geography. The development of theoretical, as well as model-building in geography is likely to be the major consequence of the quantitative revolution. The building and testing of theories and models require precision. Such precision could be only achieved through the language of mathematics and it is because of this reason that geography, like other social sciences has become involved with "quantification" the use of mathematical and statistical techniques. All these are basically important to provide a better understanding of spatial patterns and spatial processes. Thus the focusing of the importance of quantification adds to the need and offers a technique whereby the theory may be developed and improved. Geographers have felt the imperative need of quantification which is inextricably intertwined with theory and the importance of theory is thus felt because it provides a measure for the recognition of exceptional and unusual events.

## TRANSPORTATION GEOGRAPHY

### Emerging Discipline

The three great parameters for any scientific problem although in varying dimensions and attributes are space, time and composition of matter. It is also emphasized that the study of spatial distribution and spatial inter-relationships and the spatial dynamism contributes a better treatment to any great problem of scholarship. In this concern it may be pointed out that the man-environment system in geography is primarily concerned with space and time and in this context it is essential to state the Geographer's organising concept explaining the framework of study of spatial distributions and space relations. Location theory studies, a new research cluster has faced a tremendous development in its diversity of aspects, starting from a traditional subject at its root to the involvement of extensive application of mathematical methods to facilitate refinement of theory.

In elucidating the essential aspects of location studies, it is important to analyse the following aspects :

1. The structure of spatial distribution and its variation over different scales bring it to the position of a regional-topical continuum. Emphasis has been also laid on the static aspects of pattern as location, spread, density and geometry.
2. Analysis of actual linkages and flows between places should be essentially considered in terms of its various phenomena-study of spatial systems or Functional Organisation in terms of connectivity, accessibility, dominance and hierarchy.
3. Identification of the temporal dynamics of spatial structure and spatial systems.
4. Application of normative models to derive efficiency solutions. The potential geographical contributions to the understanding of the social problems in transportation geographic research have been geared to the improvement in transportation techniques and penetration particularly has increased the interdependence and interaction, the major distinguishing features of all systems. It may be said in this connection that transportation requires spatial relations analysis.

**A Spatial Perspective**

Transportation geography in a spatial context is stressed by so many geographers (Taaffe, 1973;[3] Alan Hay, 1973)[4]. It deals with the measuring and mapping of the traffic flow of goods and passengers volume, speed, origin and destination and all these, with various environmental conditions affecting movement. According to Ullman (1954)[5] "transportation is closely related to the concepts of geography as a whole, and such studies contribute to the formulation and evolution of geographic theory.[5]

Transportation is a measure of the relations between areas, and is therefore, an essential aspect of geography (Ullman, 1954). From this we can understand the basic geographical concepts of spatial and areal association.

As a matter of fact, to a greater extent, the study of transportation is the understanding of the spatial organisations. The goal of the spatial organisation is to determine the optimum structure of movement. Spatial differentiation is a direct result of movement and the overcoming of distance i.e. distance is basic to geography. The spatial differentiation is explained by the French geographers in a single word called "Circulation". Further geographers would also share a decision in determining the location of an industry, which ultimately leads to a definition of clear spatial expression. This spatial expression my be represented in the form of spatial structure which is directly resulted from the cost of distance (economic or social) between people and activities. Though the distance has lost its significance (due to the advancement of technology) the indirect role played by distance for the progress would have never been ignored.

The recent decades have stimulated the quantitative stream of geography including the introduction of location theory, spatial analysis, regional science and varied techniques such as linear programming, dynamic programming and advanced mathematical and statistical applications. All these have fostered the progress of transportation geography and placed it as a distinguishing discipline among the other branches of geography.

## INTERDISCIPLINARY ATTITUDES

An explicit spatial expression in the case of transportation shows that the geographer studies transportation as an aspect of the organisation of an area, whereas the economists may study the cost characteristics of different modes of transportation and so on. Hence the study of transport is not a unique prerogative of any one discipline but is shared by many fields of inquiry each with its own viewpoint; above all the geographer focuses his attention on the spatial structures formed by these modes and attempts to understand the processes that have created them. In addition to this the nature of transportation planning education has broadened the disciplines such as engineering, operations research, economics, sociology, geography and political science. The nature of any transport problem necessitates the essential of interdisciplinary research which reflects the real world environment for a transportation planner.

## METHODOLOGICAL AND PHILOSOPHICAL FOUNDATIONS

The present study has cited major review works from a considerable amount of literature in order to bring out a clear exposition of the progress of transportation geography. The capsule of research presented here is to shed light towards an understanding of the conceptual revolution in transportation geography. The emphasis is given here on a verbal analysis of the ways in which transportation geography has emerged. This will enable us to identify areas where present knowledge is fragmentary and where future research may be concentrated.

The methodological and philosophical foundations of modern transportation geography have been established by a growing body of substantive literature and all these have appeared only in the last twenty years.

Reviews of literature in transportation geography have been done by so many persons (Ullman[6] Ullman, 1956)[7]. Most of these statements in transportation geography have basically attempted to synthesize the development in the field rather than call for new directions.

## PHASES OF DEVELOPMENT

The development of transportation geography is a recent phenomenon. The theoretical-quantitative revolution has resulted in the consolidation of the field of transportation geography. Notably, it is significant that the subject was sparked off by the pioneering introduction of research identity by Edward Ullman on spatial interaction models, and the spatial movement theory supported by the employment of rigorous statistical analysis by William Garrison which put a new dimension in the spatial analysis of transportation geography with special reference to transportation networks. A remarkable number of substantial works and increasingly sophisticated analytical literature form the supporting evidence to build up a conceptual framework for the spatial movement theory and a refinement to the growing influence of movement in which the analysis of transport networks and flows present a major task to distil the whole range of research activity.

Later phases of development in transportation geography have concentrated on the socio-economic reality which leads to finding out a hope for a broader socio-economic horizon followed by mechanistic analyses of transportation. These studies seek to analyse and to develop a set of principles of network interdependence with the economic causes of the changes in the network components.

Muller (1974)[8] in his review stresses the lack of an organising framework in transportation geography. He has also emphasized the following : (1) transportation development in terms of spatial interaction of Ullman triad; (2) types and causes of movement as embodied in migration, trade and diffusion theory; (3) areal specialisation in relation to input-output analysis; (4) explanation of interaction theory through gravity and potential models; and finally, (5) analysis of transportation systems using the techniques of movement geometry, network analysis and flow analysis.

The goal of transportation geography is the elevation of the subdiscipline to the level of an objective science. The phases of development of transportation geography marked the adoption of scientism on which the geography has become concerned with major processes of conceptual revolution i.e. law seeking, model building and articulation of theory. Scientism has made the progress of transportation geography emerge as a comparatively easier one to describe the complex patterns of network in mathematical term. Further it has also geared up transportation geographers in pattern identification which is usually beyond the competence of ideographic or nomothetic approaches. In this connection the transportation geographers may be said to be technique-oriented.

Geographers' interest in understanding transportation geography rests on the scientific empiricism in which the geographer considers geography as a frontier of analysis rather than as a field. This indicates the spatial organisation of an area. The relation and connections between areas are frequently reflected in the character of transportation facilities and in the flow of traffic. All these are explained by three concepts, viz., spatial interaction, areal association and spatial differentiation. At all levels of research, the interest of geographers lies in identifying the intensity and degree of spatial interaction, which by its nature considers the organisation of space, thus establishing the hierarchies of interaction. These hierarchies of interaction are the basis of spatial organisation whereby it enunciates the immense prospect to the future researchers to explore further to understand the dynamism in spatial systems. At this stage it would be highly relevant to mention the importance of transportation in evolving the dynamic spatial systems.

> . . .Transportation and relative location have dominated our thinking and explanations up to now because they were indeed some of the important elements of social, economic and political life. An economy and society which gains its livelihood by producing goods and services must possess integrated transportation systems. Transportation is literally the lifeblood of the society which engaged people in secondary and tertiary activities . . . . (Abler, *et al*, 1971).[9]

## METHODOLOGICAL FOUNDATION AND CONCEPTUAL INTEGRATION

The entire concerns of future transportation geography will create the spatial organisation and spatial interaction of the future and the geographers will continue to be vitally concerned with the spatial organisation of structures and processes. The present study focuses the methodological and conceptual integration of transportation geography under various broad headings namely:

(1) Theory and Essentials of Transportation Geography,
(2) Spatial Structure—A Prelude,
(3) Spatial Systems or Functional Organisation,
(4) Spatial Dynamics or Dynamics in Spatial Systems,

(5) Spatial Efficiency Measures or Space Adjusting Techniques, and
(6) Spatial Organisation and Spatial Reorganisation or Space Fulfilling Techniques.

It may be, however, important to consider the methodological issues that have been developed in transportation geography for the conceptual integration of knowledge in the discipline. Most of the works that have been done in this particular fields are invariably to ascertain the importance in the following fields of research :

(1) Analysis of network structure and its characteristics,
(2) Analysis of accessibility measures to observe the spatial integration and spatial efficiency,
(3) Finding out an optimal structure or least cost routes both from a locational and functional point of view,
(4) Works have also been done though not widely in analysing the relationship between the landuse transport system,
(5) Analysis of flows of goods, person and information to study about the functional organisation or to delimit the zones of influence in a spatial context, and
(6) Introduction of models such as Gravity model (pertains to the study of movement and provides a conceptual base for spatial interaction theory etc.) and some simple linear programming to dynamic programming approaches to solve the various problems associated with the process-pattern analysis.

The significance of transportation geography calls attention to the following :

(i) Attempts have been made by persons (Vasilevsky, 1963;[10] Wheeler, 1971[11]) to classify the branches of transportation geography in view of the relationships between the transport sector and economy; the study of transport networks and its impact on the socio-economic system.

(ii) In recent years the Interdisciplinary locational analysis has gained momentum (Garrison, 1960).[12]

(iii) Followed by Christaller's theory the inter-city and intra-city linkages brought new approaches in developing a conceptual frame for gravity model (Warntz, 1960).[13] The interdependence of transport with economy has been the main focus for such approaches (Berry, 1959).[14]

(iv) The modern era has not only witnessed the data explosion in the field of transportation geography but has also geared the quantitative revolution in terms of hypotheses testing and model building (Haggett, 1965)[15] Network analysis utilising graph theory (Kansky, 1963).[16] systems analysis, behavioral analysis and the application of probability concepts have further enlightened the research in transportation geography by behavioral assumptions and decision-making.

**Theory and Essentials**

The earlier study of the theory of transportation has stressed the character of transportation as an organised activity which is determined by its inter-relationships with physical and social forces and their conditions. In addition to this he has also focused the importance of transportation for social development which in turn forms a tool for socio-economic-political development. The study also elaborated the certainty of transportation in affecting every part of society as far as its interdependence, specialisation and organisation is concerned. However, the development of transportation geography as a distinct discipline has emerged when geographers have started to analyse the transportation network as a system dealing with the theory of spatial structure and pattern with a specific note in relation to organisation of an area (Taaffe and Gauthier, 1973).[17] The development of transportation may be approached by utilising the theory of spatial interaction of Ullman which stresses the dynamic spatial spectrum of transportation geography in a broader perspective dealing with different components of the transportation system. The earlier approaches i.e. before 1970 were concerned with the transportation geography as a spatial analysis based on its derivatives on a theoretical-quantitative one whereas after the 1970s, due to the growing changes and wider research interests in the social dimensions of humanism, a humanistic transportation geography began to emerge. These developments are the harbingers of the metamorphosis of the quantification-theory paradigm of the 1960s into a new paradigm of the 1970s. In the early 1970s, in accordance with this, a spatial transportation analysis has emerged pertaining to the social and related policy development applications (Eliot Hurst, 1973).[18]

Most of the latest works in the field of transportation geography have proposed a number of research frontiers and agendas in relevance to the societal framework actually where it has set the stage for the opening up of new avenues of research in transport geography. It must be important to mention that certain geographers (Appleyard, 1971;[19] Muller, 1974[20]) have considered the social goals of transportation and its impact on the environment. This has further enlightened and widened the horizons of transportation research especially in the case of metropolitan area studies.

In brief, the recent progress of transportation geography has witnessed the emergence of two major research clusters of which one is the production of a humanistic transportation geography and the other is the exploration of the new type of applied and policy-oriented research reported in 1973 by a special issue of *Economic Geography* ("Transportation Geography : Societal and Policy Perspectives." Wheeler, 1973)[21]. Noteworthy in the progress of transportation geography is the definition of a more refined theory of interaction, the specification of inter-relationships among quantitative models (such as the gravity model and linear programming) and model building of transportation within the regional and urban systems (Wilson, 1974).[22]

## Spatial Structure—A Prelude

Geographers are very much concerned about the internal organization of a distribution, the location of the elements of the distribution with respect to each other. The spatial structure refers to the internal locational organisation of a distribution in space and these are produced by the mechanisms of spatial processes. Hence in this context it may be pointed out that both the spatial structure and spatial process are circularly causal. The concept of spatial structure can better be explained by means of understanding the processes of both static distributions and dynamic distributions. It may be even understood that the spatial structure has the explanatory power for understanding the underlying dimensions of spatial processes. Broadly speaking, the spatial structure compresses the following : (i) Patterns of spatial distribution, (ii) Production of economic spatial structures by means of understanding the varying dimensions of socio-cultural-economic significance of distance, (iii) Distance decay relationships, (iv) Layout of Transportation networks, (v) Distance and market relationships, and (vi) Locational aspects.

The progress of transportation geography has started its development from the descriptive study of transportation and communication with special reference to a particular region. These studies have invariably described the general transport pattern formed by different modes of transportation (Singh, 1966).[23] Thus resulting in adequate spatial structures in some places and some inadequate spatial structures in some other places because those might have been overburdened with transportation problems (Kayastha, 1960).[24] Some of the studies have attempted to explain the transportation system of both the urban (metropolitan) and rural areas in connection with the transportation problems (Mukerji, M. 1969).[25] Of these studies certain studies have based their importance in studying the transportation systems by way of analysing their characteristics and patterns of network, the incidence and structure of traffic flow (Kayastha and Singh, 1972).[26] Emphasis has also been laid to study the quality of the surfaced roads (Adrashannawar, 1968) and also the investigation of the physio-economic factors which retarded the development of roads (Prasad, 1968).[27]

Later studies have explained the spatial structure of transportation networks and its potential use by way of analysing the commuter's travel (Aangeenbrug, 1968;[28] Wheeler, 1970[29]) with specific emphasis on the metropolitan areas (Wood, 1976).[30] Studies have also appeared to deal with the impact of technological development for the improvement of transportation (Hilling, 1969)[31] and analysing the built-up transportation network structure in relation to locating the multiple facilities (Robertson, 1974).[32]

Attempts have been also made to analyse the expansion of transportation network in keeping pace with the economic development (Grimshaw, 1976).[33] Of which certain studies have thrown light in giving much more consideration to the locational factors such as the location of market centres and their influence in evolving the spatial integration of efficient transportation routes (Saxena, 1975)[34] and some of the studies have been carried out to explain the transportation factors affecting the location decisions and economics of location (Barr, 1970).[35]

Models have also been utilised to evaluate the optimum location of public facilities such as hospitals in rural areas (Gould and Leinbach, 1966)[36] and development of an heuristic model to assign or allocate a set of hierarchicalfacilities to nodes on the network (Scott, 1971).[37] Some of the studies by their emphasis explain the growing economic development along with the growth of suburbanisation which makes the areas face the improved mobility with an exacerbration of the transport crisis, because of suburban employment (Breckenfeld, 1972)[38] and also some studies were concerned in explaining the spread of the road network by way of examining the effect of spatial structure of economic activity and the physical environment (Hubhard, 1974).[39] Most of the works of recent origin have employed widely the application of graph theory in analysing the spatial structure of the network (Ramaswami, 1977)[40], and also the analysis of the above at different district levels was also carried out to understand the differing spatial structures (Bradnock, 1976).[41]

Attempts have been made in analysing the role of distance and its importance in determining the characteristics of different spatial structures in relation to polynomic population (Matela, *et al* 1976)[42] and also the performance of four different distance functions such as polytropic, exponential, square root exponential and Log-normal (Black *et al* 1972).[43]

**Spatial System or Functional Organisation**

Geographers have based their emphasis in treating the world organisation in theories of landuse, location, transportation and trade. Hierarchies from the basis of most spatial organisation. In this context, the role of geographers is mainly concerned with the spatial interaction and with hierarchies of interaction.

Peter Haggett (1965)[44] has remarked that the built-up systems involve,

> Movements, that (produce) the channels along with the movements occur, the network structured around nodes, organised as a hierarchy with the interstitial zones, viewed as surfaces.

From the above, it is explicit that hierarchy is a fundamental concept organisation and interaction. It may be important to stress the principles of the fundamental system dealing with the organization, interaction, hierarchy and growth.

In analysing the application of the systems approach in transportation geography. It is worthy to mention the following remarks which were made by the U.S. National Academy of Sciences and National Research Council in 1965.

> .... The development, testing and refinement of location theory, related to the studies of geographic organisation of economic life or urban and transportation systems, have been fundamental .... The applicable body of theory includes .... abstract concepts concerning spatial distributions and space relations .... Very recently a new synthesis has begun to emerge :
> (1) The identity of spatial concepts and principles developed in (several) subfields of geography; and (2) emphasis upon the interaction of economic, urban and transportation phenomena in inter-dependent regional systems that are the material consequences of man's resource-converting and space-adjusting techniques. This emerging synthesis results from a concerted application of systems theory within geography.

Focusing the important goals of geography, it should be clear and necessary to understand the functioning of the spatial system, where the geographer is concerned with the use of resources, the determination of locations, patterning of space and the growth of particular regions. In this way the systems approach has enhanced its importance by way of providing a conceptual framework within which it analyses the whole range of different answers in relation to a wide range of questions. This ultimately provides a complete and concrete understanding of the structure and operation of the other systems. Hence in this direction, the systems analysis has provided new insights and many dimensions of the problem(s) in locational studies with particular reference to transportation geography.

The present paper is an attempt to analyse the handful of work in a systems approach manner. In this connection transportation geography is concerned with (1) the particular linkages and flows that comprise a transportation network, (2) the centres, or nodes connected by these linkages, and (3) the entire system of hinterlands and hierarchical relationships associated with the network (Taaffe and Gauthier, 1973)[45] :

> . . . . increasingly, too, the planning of large urban agglomerations—in particular transportation planning is being formulated in systems terms. Lloyd and Dicken, 1972, pp 3-5).[46]

The analysis of spatial system or functional organisation is carried out in different divisions which are as follows :

(i) Transportation networks and its structure into the analysis of spatial distributions and space relations,
(ii) Interaction and interdependent spatial systems in terms of different structural phenomena,
(iii) Spatial behaviour and travel preferences,
(iv) Accessibility measures including the degree of its connectivity and regional specialization,
(v) Movement studies (flows of goods, people, and information to study the functional organisation),
(vi) Organisation of spatial systems.

**Spatial Structure of Transportation Network and its Attempt to Describe the Spatial Distributions and Spatial Relationship**

It may be important to state that geographers are interested in accounting spatial facts and the inter-relationships by means of spatial association and spatial inter-action. Thus the long term goal of the discipline can be achieved by having a careful and gainful analysis of the comprehension of the spatial domain of problems.

Analysis of the spatial structure of the network has been activised by the introduction of graph theory. The basic aim of the technique is to provide measures of the structural properties of real world transportation networks by abstracting the system as a set of points (nodes or vertices) and routes (line segments or edges). Progress in the application of graph theory has been instrumentalised after 1960. In addition to this the computation of aggregate measures of graph theory (dealing with connectivity and structure) provides a consistent basis for the analysis of spatio-temporal variation of the network structure and their comparison at various levels hold a strong emphasis in explaining the spatial aggregation and spatial association of network structure with measures of economic and regional development.

A pioneer work was carried out by Garrison (1960)[47] in applying the graph theory to identify the spatial structure of transportation network. Actually this has introduced a substantial body of research using the mathematical tool as an analytical technique. His earlier work examined an aspect of network geometry in analysing particular places on a route system. He has used a more refined measure of accessibility (the Shimbel-Katz index). By way of explaining the relative accessibility of 45 places in the south-eastern part of the United States, he was able to predict the unexpected pattern of high accessible areas by visual examination. Followed by this, numerous attempts have been made to describe the spatial structure of transportation network by utilising graph theory. (Shanmuganandan and Ilengovan, 1978).[48] Attempts have been also made to compare the networks structure and its growth at various levels for example, with the measures of urbanisation (Mukerjee, 1974),[49] agricultural landuse (Shanmuganandan, 1977)[50] and economic and regional development (Kansky, 1963).[51]

Kansky (1963) has attempted to describe the spatial structure of transportation network by deriving the basic graph theoretic measures and used the same in constructing simulation models of network generation in analysing economic development. The increasing trend of analysing the spatial structure of transportation network has grown to the level of introducing matrix algebra. It may be used to manipulate from a simple incidence matrix (showing the accessibility measures of individual nodes in the network) and the same may be replaced by a recent one i.e. topological distance matrix which gives the measure of the distance between nodes in actual mileage or travel time so that more refined measures of nodal accessibility result. Such procedures have been widely dealt in network analysis.

James et al, (1970)[52] explained the commonly used measures of graph structure which are often poor indicators and also they have proposed a new measure which appears to discriminate the currently used methods of analysis.

The spatial organisation as reflected by the structure of the transportation was dealt with by Garrison and Marble (1964) and Gould (1967). In this study the authors have analysed the spatial dimension of the Venezuelan and the Ugandan network. The analysis was based by means of a reduction of a binary connection matrix to a characteristic equation and illustration of the vector which contrasts the "inter-connectedness with peripheralness".

The characteristics of the network were analysed to show the significant aspects of network theory and the various techniques which are potentially useful in overcoming the geographical problems. Works have also been carried out to deal with the general definitions and descriptive measures, geometric structure and Steiner problems network flow problems, etc. related to the network (Dalton, 1973).[53] Solutions have been arrived at using the minimal tree concept in network analysis which has offered a theorem and proofs relevant to the geometry of minimal trees in the Euclidean plane (Black, 1970).[54] It has emphasized that a network must have at least one circuit with an odd number of linkages (edges) to have a solution matrix (Werner, et al, 1968).[55]

Investigation of real world networks and their degree of close association tend through time towards the ideal linkages of centres proposed by Christaller and Lösch, accomplished through a measure of entropy level. A statistical index of redundancy is obtained in comparison with the real world network and an ideal network. This results in the fact that the networks associated with service centres in areas surveyed according to a grid system tend towards the complexity of the square linkage pattern and much less complex than the more efficient linkage pattern associated with the hexagonal network structure (Semple and Wang, 1974).[56]

The introduction of stochastic techniques have further sharpened the discipline. An analysis was carried out on the patterns of two graph theoretic representation of empirical evidence in which an attempt was made to examine the several situations which arise in link-node problems. Probability theory and the properties of certain statistical distributions are then used to develop solutions for a non-directed/directed edges in graph theoretic analysis (Cummings et al, 1973).[57] Tinkler (1973)[58] has made an attempt to use the connectivity matrix as a transition probability matrix. This stochastic matrix may be used to describe the probability of movements over the graph and finally using a binary association coefficient, a comparison of connectivity matrix is made to utilise direct and principal component analysis. The essence of the research says that a particular network implies a set of equilibrium processes, by way of concluding that the connectivity matrix may be viewed as a model of these processes.

**Interaction and Interdependence of Spatial Systems in Terms of Different Structural Phenomena**

The development of model building is a significant progress in the field of transportation geography. An active influence on the spatial form of the economic system is the cost of overcoming the friction of distance. In this direction the gravity model brings forth the ideas of the specific geographical complementarities and overcoming of the friction of space. Thus the gravity model has captured a great deal of attention because of its pervasive simplicity (Abler et al, 1971).[59]

*Development of Gravity Model* : An analogy was recognised in the mid-nineteenth century by Coney (1858). Later both Ravenstein (1855)[60] and Yound (1924) used the same ideas in their migration studies, the former in the 1880s and the latter in the 1920s. Shortly after the study by Yound had appeared, Reilly (1931)[61] further applied concept to the delimitation of retail trade areas, while Bossard (1932)[62] discussed it in a sociological study dealing with the choice of marriage. The new formulations made by the astronomer, Stewart (1958)[63] have developed as a result of similarities, and ideas inherent in the Newtonian Law of Gravitation (Olsson, 1965).[64]

In all these, it can be pointed out how Nystuen and Morrill (1959) indicated the lack of a theoretical basis. The increasing number of studies using the gravity model demonstrates the necessity for such a device. Since the gravity model can be made simple and easy to work with, there may be sufficient reason to continue its use. The gravity concept explains that the amount of movement or the amount of interaction between any two places will be directly proportional to the products of their populations and inversely proportional to some of the power of the distance separating them.

Alcaly's (1967)[65] paper on aggregation and gravity models analyses the effect of aggregation over models for travel. It may be pointed out that the crude gravity model as formulated in the equation explains travel by all nodes better than it explains travel by an individual mode with the possible exception of automobile travel. The results, however, offer support for the underlying presumption to minimise the objectives when the model is applied to aggregate data.

The utility of the gravity model was evaluated (Black, 1971)[66] to add a new shape to the model. Black (1971) has evaluated the utility of the model in examining 24 sets of regional commodity flows for the U.S. in 1967. The study summarised the variations in the definition of inter-regional distance which has a significant effect on the absolute value of the exponents and also confirmed the importance of the following first two hypotheses through regression analysis and the third one as insignificant one using the stepwise model. The hypotheses were as follows :

(1) The lower the exponent of distance the greater the regional specialisation for a particular commodity.

(2) The greater the inter-regional consumption the greater the exponent of distance.

(3) The greater the per unit value of a commodity, the lower the exponent of distance.

Niedercorn and Bechdolt (1969)[67] has examined the gravity model's qualitative or theoretical substance. An inquiry made in this direction shed light on the construct and the derivation of the gravity model within the framework of the utility theory. Exciting works have been also carried out by Wilson.[68] In this connection it may be important to focus the pedagogic review of the Wilson monograph (Gould, 1972) an indispensable one to understand the analytical stimulus of the work.

Black (1973)[69] has extended his study of gravity model distance exponents to indicate that the exponent will be low if either production or consumption of a commodity is geographically concentrated. To facilitate the study, a multiple regression model was used to estimate the exponents which leads to a re-estimation of the flows. The difficulties in estimating a gravity type function in a spatial context was attempted by Curry (1972).[70] He emphasized the need for parameters and for the introduction of the calibration method and estimation of flows without considering weighting of areas, while for those places he has used a spatial regression operator which serves to adjust the pure gravity allocation procedure by providing a trip distribution problem. The operator can be viewed in a certain sense as describing the average minimum movement to a point such as the place of work.

In elucidating the importance of both linear programming model in comparison with the gravity model, Mrea (1971)[71] has explained that for homogeneous commodities (wheat, rice and cement) linear programming gives a superior prediction while for diversified commodities the gravity model performs better. Later approaches in the gravity model had lead from the parameter estimation for the constrained gravity model (Stretzer, 1976)[72] to a family of spatial interaction models that correspond to a wide variety of transport flow situations.

The application of the gravity model is widely felt in the analysis of traffic studies. Since landuse and transport combine to form the traffic, the study brings various facts to understand the entire functional organisation of the region. The functional aspect of a region is a product of spatial interaction of areal economy and lines of circulation. In this context, the gravity model contribution is immense. In the present paper the traffic studies carried out the following aspects.

There are a wide range of models in explaining the spatial interaction through the trip distribution theory, model and route split, inter-regional commodity flows and locational problems in which they involved computational complexities. These are good in describing the real world situation. Attempts were widely made to describe the structure of traffic of a region in relation to landuse, such as in metropolitan area and the impact of traffic on agricultural landuse. Wheeler (1970)[73] in his studies on the structure of metropolitan work-trips, described the findings based on aggregate work-trip data to an improved understanding of elements of the journey to work as a backdrop for more detailed planning analysis. The approach used factor analysis in order to bring out a good understanding of metropolitan linkages. An attempt has also been made by Wheeler and Stutz (1971) in analysing the spatial dimensions of urban social travel. Two hypotheses were derived of which one deals with the frequency of social trips and distance and the other one was on the socio-economic status of an individual and the social travel. Factor analysis was employed to analyse the effects of distance bias and residential status on the social interaction in the city through a social ties matrix.

Studies on trip purposes and urban activity linkages made by Wheeler (1972)[74] explain the presence of a high degree of similarity in the pattern of flow for various purposes because of a common dependence on the spatial arrangement of a single trip, in transport inputs which are also related to the areal pattern of activities and most frequently involve clustered business and commercial landuses. Whereas Leinbach (1973)[75] in his studies derived modernisation scores for two hundred and sixty five individual hexagonal cells each measuring 210 square miles. The scores were derived from a principal component analysis of current data on commerce, education, health facilities, communications and transportation. He has also analysed in his separate studies regarding automobile ownership, its utilisation and the social interaction.

Studies have also been carried out in analysing the travel time in competition (Chapman, 1968)[76] and also the classification of service centres based on available public road transportation services (Carrothers, 1957).[77] Studies carried out in describing the structure of different modes of transportation networks indicate not only the traffic densities and patterns but also explain the locational dimensions of the region in relation to transportation costs (Barr, 1970)[78] with particular reference to industries (Singh, 1968).[79] There was also an attempt to describe the structure of journeys and the nature of travel patterns (Hanshore, 1976).[80] An empirical framework in this connection was drawn out to explain the theory and measurement for the demand for travel (Quandt, 1970).[81]

The structure of passenger transport and its importance has been dealt by various persons. Williams and Zelinsky (1970) focused on international tourist flows. The analysis suggests that tourist flows do not satisfy a pattern that does exist. Cost, distance inertia and the presence or absence of other types of national connectivity are among the explanations offered for the patterning. Two areas of investigation are also noted by the authors which can broaden our understanding of tourist flows, tourist perception and the evaluation of various destinations and feedback effects of tourism upon national image, migration flows and the diffusion of cultural traits.

Hay (1969)[82] presents an inquiry to describe the major patterns observed in Nigeria followed by the general explanation in terms of economy and society. He has also delineated that passenger transport in a tourist industry which felt the imperative need to distinguish between the reason for the trip research and the occasion of the trip research on seasonal fluctuations. Emphasis has also been laid on passenger transport because it allows a free mixing of capital between different sectors apart from recognising its role in social and commercial life. The role of passenger linkages in the commercial core and the metropolitan nodality was also carried out (Snyder, 1962).[83]

The techniques of flow analysis on a single data set was evaluated by Britton (1971).[84] He has analysed the origin and destination of road transport and freight flows at the city and regional level on England and Wales. Original data matrix as well as matrices which are standardised by transaction flow analysis and the gravity model (relative regression residuals) are also analysed to identify the largest flows and hierarchical ordering with the application of R-mode (Column vector correlations-scores on origin regions are evaluated) and Q-mode (row vector correlations-scores on destination regions are evaluated) 'Factor analysis', 80% of the variance of the flows were collapsed into 14 factors. The author has arrived at his conclusion that the "largest flow analysis" and hierarchical ordering are one of the dubious values as a standardised technique and also viewed that "factor analysis" as a successful technique for determining similarities in internal relationships.

The concepts and methods in commodity flow analysis were reviewed by R.H.T. Smith (1970)[85] and suggested three central concerns in commodity flow analysis :

(1) The diagnosis as 'more than' or 'less than' expected for volume flows over individual routes.
(2) the efficiency of a given flow pattern; and
(3) the structure of commodity flow patterns.

Researchers in these areas have availed themselves a wide range of descriptive indices and statistics, simple and multivariate inferential statistics, and activity analysis techniques. The major diagnostic tool in the study of freight movement is afforded by the application of regression formulation of the gravity and potential models. The spatial efficiency of the flow patterns is determined by the degree of the node in relation to its operations cost i.e. least transport cost or minimum distance pattern of the distribution of network. Structure in a flow pattern can be easily absorbed and recognised by R-mode and Q-mode factor analysis.

Various studies have been carried out in the flow analysis of freight movement (Chisholm and Sullivan, 1973).[86] Its pattern (Black, 1973)[87] and descriptive studies giving a particular emphasis to the regional analysis of flow structure (Perle, 1964).[88]

Chisholm and Sullivan (1973) in their studies describe the inter-regional freight flows in Britain. The authors have emphasised the two most suitable frameworks for modelling the freight flows, which are the gravity model and linear programming approaches. It is also identified that the latter hold good for predictive purposes with a probabilistic element. The study has also analysed the spatial variation in gravity model distance exponents. Golledge (1963)[89] has made an attempt to study the geographical pattern of rail-freight traffic in Newcastle.

In describing the pattern of freight movement Bhaskara (1964)[90] in his thesis, analysed the present pattern of freight movement through Haldia Port and the possible co-ordination between road and rail transportation in its hinterland. Whereas the same study was carried out by Lakshmi (1967)[91] in identifying the patterns of railway traffic flow in Madhya Pradesh. Her study gains importance by way of comparing the state's railway system with its economy. The study of Black's (1971)[92] on the commodity flow patterns in U.S. is a distinct approach to the above. He has experimented with inter-regional commodity flow patterns in the U.S in order to explore the utility of dyadic factor analysis. This study has discussed the industrial or agricultural specialisation in a regional economy and it may be viewed as a directed spanning tree, i.e., a single production system forms the source for a set of K-1 directed flows to sinks, which are consuming regions. The dyadic factor analysis has proven to bring out the results that if a region is less developed in fact, it does not dominate in the shipment of any commodity to other regions. Five factors, were drawn from the above analysis and it revealed the close approximation of those significant factors to confirm the directed spanning tree model. Further it has also indicated that the technique of factor analysis has been applied over a wide range of flow situations and proven to be useful probing and synthesizing technique. O'Sullivan (1971)[93] has applied Wilson's entropy maximising model in his studies (specifically the production and attraction constrained gravity model) to the forecasting of road freight flows among seventy-eight zones in Great Britain. A friction of distance coefficient was obtained for each origin, and the mapping revealed high values in rural peripheral areas and low values for urban-central areas. A final exposition out of the correlations however between actual and expected were poor suggesting that even at a crude level of aggregation the gravity model lacks accuracy. Studies dealing with the freight and transportation system described the general traffic pattern of movement.

*Spatial Behaviour and Travel Preferences* : Behavioral geography is of recent origin dealing with issues of growing concern for the nature of decision-making. This is mainly because of the widening role of government in the evolution of spatial systems, increasing concern for overcoming the spatial and social inequalities and maintenance of the environmental standards to the satisfactory level and other such public issues. It must be important to point out that the spatial system is born out of individual decisions and behaviour. In this aspect Eliot Hurst (1973) has categorically stated a new research cluster of spatial transportation analysis as a major thrust into the social and policy related development applications. This may be strengthened by the introduction of a special issue of Economic Geography (Wheeler, 1973)[94] entitled "Transportation Geography : Societal and Policy Perspectives". Followed by the Wheeler's introductory overview stresses the setting up of the stage for new exploration into social transport geography and various aspects in connection with the operation of social system in relation to space economy.

Berry (1968)[95] has outlined a general field theory of spatial structure and spatial behaviour. In his studies a comparison between formal and functional regionalisation through time has been made in order to understand the relationships between spatial structure and spatial behaviour. Through the construction of the attribute and interaction matrices, he has assessed the importance of the theory to integrate attributes of places as well as interactions among places through time. Finally with the help of the canonical correlation analysis, he has drawn out the meaningful dimensions in the behavioral and structural matrices. There are four kinds of environments in the behavioral classification viz., geographical, operational, perceptual and behavioral. The objective dimensions of the behavioral environment determines it as a complex subset of broader geographical environment when it is analysed in its subjective perceived dimensions.

The behavioral environment can be defined as that part of the environment of which the individual is aware and also elicit a behavioral response or toward which a behaviour is directed, such as results in a conscious utilization or transformation of environment. The behavioral environment need not be an alternation of an environment but involve a change in the individual's relationship with or exposure to that environment. An attempt was made to describe the spatial behaviour of hospital patients (Earickson, 1970).[96] This study endeavoured to analyse the individual's (patient) choice of hospital and the difference in their choice behaviour varies over space.

Later research sprang up in dealing with the interaction patterns of the blacks by way of explaining the spatial behaviour of the group. In this direction Bernstein's (1973)[97] study has shed additional light on the problem of the Central Brooklyn Ghetto in New York. He has analysed social travel with the gap in the ghetto perception between transport need and existing service. Additional information collected from this survey exhibited the same level of dissatisfaction with transport for shopping and medical services. These findings are of considerable interest by way of illustrating the growing tendency to allow for the input of local communities in urban planning decisions which affect their decisions.

Travel behaviour studies have also been brought into the limelight by way of understanding the mobility problems of urban social groups. Herr and Fleisher (1968) have attempted to study the deficient in mobility amongst the poor. It was also pointed out that though deficient in mobility, the travel behaviour of the poor is surprisingly like that of the total population. An empirical analysis for the Boston the metropolitan area dealing with the trip frequency and movement decisions explored that the interaction deficiency resulting in strengthening the claims i.e. poverty is employment-related in that spatial separation and transport shortcomings between home and potential work places form a barrier, which greatly limits the range of job opportunities for the working poor.

Golant (1972)[98] has presented his study to describe the spatial behavioral of the urban elderly. Analysis has shown that the mobility problems of the aged are to maintain access to necessary facilities in a metropolitan environment undergoing steady decentralization an auto-oriented activity locational shifts. It was also found out that the mobility is acute in the case of vital medical facilities which continue to realign themselves in outlying specialised function districts. In all, he has talked about the increasing accessibility problems and the impact of rapidly changing social organisation resulting in the environmental stresses.

In order to reduce the problems of urban, metropolitan planners keep suggesting new programmes. In this connection an attempt was made in introducing the demand-responsive (Myers, 1968)[99] flexible route concepts. This included a recently endorsed DOT which may include the dial-a-ride minibuses, taxi-jitney and personal rapid transit (PRT) cab systems. Towards a further understanding on the personal demand-responsive transportation experiments, a study was made by Morlok (1971).[100] He has introduced as fare reduction programme, mainly meant for the urban poor to show considerable promise for alleviating certain mobility problems. The study has also covered detailed information on hourly transport demand patterns to be processed and translated in the form of more effective routing and scheduling.

An active influence on the spatial form of the economic system is the cost of overcoming friction of distance. The role of accessibility is felt pertinent in giving new spatial dimension to the landscape by way of achieving its maximum utility. In most of the studies on accessibility, the connectivity matrix was formed as the base. Nystuen and Dacey (1961)[101] have made an attempt to study the nodal accessibility through connectivity matrices. The analysis of nodal accessibility through an analysis of telephone message flows in indirected and directed graphs provided a better understanding. Gauthier (1968)[102] in his study used a connectivity matrix powering design for analysing a portion of the Brazilian highway network. Cost values were used to provide after powering and summing up the accessibility values for individual nodes.

An a tempt was made by Reed (1970)[103] to show the powering technique and its application to a connectivity matrix (weighted or unweighted) which may be modified to identify the relative accessibility provided by urban activities linked in flow patterns. Reed has also identified a hierarchy of four levels of connectivity—national, regional, subregional, and route endings associated with cities in the Indian Airlines system. The utility of this method is the attainment of generality and the measure of average graph distance may be used to any flow system, taking into account both direct and indirect flows over the shortest path from an origin to destination.

The broader, perspective of Reed's paper is that the measure may provide information for examining planning problems related to both connectivity and the specific locational roles of activities which require different degrees of accessibility to full or regional components of a system.

There are certain studies dealing with nodal accessibility. The studies carried out merely describe the powering of about the nodal accessibility measures derived from the powering of connectivity matrix to the diameter level. Whereas in the case of Stutz (1973),[104] it differs from the above. He has experimented his study with a multiple linkage measure of nodal accessibility using a scalar sensitivity analysis. He has explained the resulting nodal accessibilities when scalar weights of powered connection matrices are allowed to vary over a given interval. It has also been found that the increasing scalar values indicate nodes with inaccessible location possesses a higher absolute accesssibility. Studies have been also carried out in explaining the changing patterns of nodal accessibility through the expansion of domestic air transport. (Hubert and Murphy, 1971).[105] The authors have based their analysis through explaining a powering of a connectivity matrix. They have concluded that their results warrant an acceptance of the original hypothesis, namely that as the domestic U.S air transport network evolves through time, levels of accessibility of individual cities on the network will increase, but in a disproportionate manner. Certain cities emerge in the growth process as more dominant than others and the main regional flow of the nation clearly pulled and warped the accessibility surface over time. Investigations were also carried out in connection with the historical analysis of accessibility.

Lal's (1968)[106] study merely described about the influence of transport and accessibility in the lower Ghaghara-Gandak doab whereas the study of Pitts (1965)[107] explained the tool used in historical analysis to compute the relative accessibility of Moscow during the twelfth and thirteenth centuries in an attempt to shed light on the city's growth and centrality.

Time-distance studies appeared as a part of the later methodological pursuit in computation of accessibility. Alao (1970)[108] in his study pointed out the additional conditions which guarantee the existence of the solution matrix. He has also noted that an adjacency matrix possesses a solution matrix which is an element of the set of eigen values of the matrix, although for large networks irreducibility is the necessary and sufficient criterion for the existence of the solution matrix. Marchand (1973)[109] has followed a similar theme in examining the time-distance matrix among cities in Venezuela. The author calculated and mapped a measure of the "rentability" of road investment over time based on the mean speed between cities and the population of those cities. Using the multi-dimensional scaling technique, the dimensionality of the network has been analysed. Of these, seven dimensions were to indicate the homogenization of network conditions in 1950. The usefulness of this study rests in mapping of relationships among nodes on a network in a manner and the adoption of MDS techniques helps to shed additional light on the nature of changing spatial organisation and relationships.

Considerable attention is paid in understanding the crisis in accessibility and several worthwhile proposals have been derived to solve the problems that face the situation of the centres lying outside of the activity spaces of the city. Leinbach (1974)[110] has undertaken a study to describe the social influences of accessibility at a broader scale are of critical importance in the diffusion of modernisation across the rural areas of Third World nations and have an enormous impact on migration flows, the planning of new growth centres, and spread of such innovations on agricultural technology, birth control information, health care delivery systems and communication facilities.

Wilbank's (1972)[111] study has emphasized that the technological change in agriculture can be expected to be influenced by the geometry of networks of personal communication. Data from a sample of villages in India indicate a consistent association between accessibility to transportation facilities and the rate of technological change. Personal communication networks tend to correspond with transportation networks but correlates of road transportation accessibility suggest that innovation is more closely related to information-flow variables than strictly economic factors.

## Movement and its Problems

Spatial structures are born out of movements and if once the structure is established, the increasing spatial arrangements will influence subsequent movement. Here it may be said that the spatial arrangements are reciprocally tied to movement processes. It is also found that the increasing movement structure gives new dimension to the landscape thus facilitating it into a complete spatial system.

The movement problems are very acute in the growing metropolitan areas. The most immediate and far-reaching impact of the restructuring of the metropolis has been upon the movement behaviour of its population. The existing metropolitan transport network is ill-equipped to serve the changing needs of its residents in the mid 1970s.

Mapping of the movements can be helpful in accurately delimiting the spatial extent of the ghetto. The study by Wheeler has accounted for the spatial extent of the movement of black neighbourhoods. Activity and action spaces of this social group tend to be highly localised. Further he has brought out that the blacks are highly subjected to the movement biases resulting from poor transport facilities and services of their residential neighbourhoods. Their decision in the selection of purchasing sites are only along the peripheral main streets rather than those in the heart of the city. It has been also stated that the rise in the mobility problems of blacks had reached a vantage point where there is a stress on the transportation planning which may be able to direct its attention to the different problems.

The movement problem results out of social segregation. This perpetuated the increasing mechanisms of dual housing markets. The clustering process of this results in cleavages amongst different social groups at all economic levels since it constitutes the characteristics of the contemporary metropolis. In this connection, the problems of the ghettos have been separately analysed to have a fuller understanding of their problems. The study mainly emphasized the interaction patterns of the blacks through the analysis of their spatial behaviour.

Kraft and Moencich (1962) presented their arguments about the minimal value in relieving the mobility difficulties of those lying in poverty. This is mainly because for the inner city poor, cost is far less of a problem than is the unavailability of adequate forms and services of public transit.

Studies have also been carried out in order to investigate the distributional characteristics of the elderly blacks in Philadelphia and their social problems (Cybriwsky and Meyer, 1973),[112] In this connection it focused its importance to understand the location of facilities vis-a-vis the distribution of senior citizens then alone can it be useful for planning better transportation facilities. Whereas Perle (1965)[113] has made an analysis in identifying the urban mobility needs of the handicapped. In addition to this, little attention has been paid to study the movement plight of this group in the design of vertical and horizontally sprawling metropolitan environments.

Attempts have been also made to study the geographical dimension of the currently intensified mobility problems of the rural regions (McKelvey and Bueker, 1974). This study deals with the similarity of the central movement crisis in both the city and country, the autoless segment of society (especially the aged, the poor and disabled) and the most transportationally deprived. It is highlighted to relieve the mobility problems of the rural elderly by means of telephone generated-fixed-flexible minibus system focused on new activity centres, in order to meet the travel demands of the aged in successful central city experiments. In this context, a new comprehensive regional transport development programme has been suggested to improve the accessibility of the region's far-flung population and the new road network was designed to link many of the isolated valleys with the few major urban nodes and centres of economic activity in the region. It is hoped that in this manner a more equitable distribution of income and social services could be achieved.

Gauthier (1973) has attempted to show in his review that remote rural areas depending upon links to cities for improving their economic lot were passed over as those urban centres pulled in and used up most of the available highway funds for constructing bypasses and making other road investments in their immediate vicinities. Whereas Leinbach (1974)[114] in his studies explained that the social influences of accessibility are of critical importance in the diffusion of modernisation across the rural areas of Third World nations and have an enormous impact on migration flows, the planning of new growth centres and the spread of such innovations as agricultural technology, birth control information, health care delivery systems and communication facilities.

## Systems Growth and Spatial Dynamics

The importance of the spatial system and its dynamism is determined by its growth and development. In distinguishing growth and development, growth refers to the expansion of the system as a whole, through either addition of new parts or the increasing size of existing parts; whereas development indicates a particular kind of change in which there is substantial restructuring or reorganisation of the parts. For example, this may be referred to the increasing functional specialisation or the growth of new types and levels of interaction and organisation (Berry, 1976). According to the geographer's interest there are four types of systems, viz., closed no growth system (changes taking place are deterioration and decay); self-maintaining systems (in which neither growth nor change is taking place, but in which a sufficiently high level of interaction and interchange with their environments is maintained to ensure that the existing size, composition and spatial organisation of the system is preserved); growing open systems (the form may change and in this part must grow at different rates to ensure that the enlarging system remains functional as a whole) and open developing system (growth is accompanied by substantial restructuring of both internal organization and the nature of parts that are so analysed).

It is quite explicit that the spatial organisation is a result of spatial structure and its interactions i.e. that spatial organisation would integrate the continuing flow of people, ideas, goods and services. Invention, innovation and diffusion are the three interrelated processes which link the growth of spatial systems and the development of spatial organisation. Invention is central to the internally generated process of structural transformation. Innovation and diffusion decide the inducement of growth, change and development by external means through the transmission process of inventions to other places. Indeed such are the various means by which a great number of developments in spatial organisation have occurred at a variety of scales resulting in changes in the spatial system.

In a spatial system, it is really a subject of interest to analyse the relationship between transport and development. Geographers in this aspect are mainly concerned with the changing spatial structure of transport networks in relation to other aspects of the landscape, and with the factors affecting changing patterns. Transport is explicitly forms a factor of fundamental importance in all economic activity, and the development is a matter of vital concern because it is an imperative need which is felt to maintain an awareness of the multi-dimensional nature of the problem—the economic, social, political and spatial dimensions of transport. It is also widely felt from the literature of transport economics and of the transport geography of less developed countries in understanding the importance of transport and in the efficiency of transport improvements in accelerating the development process. The transport/development relationship is essentially a two-way interaction process, and the results of the interaction depend upon the economy involved and upon the level of development at which transport improvements are effected (Hoyle, 1973, p. 11).[115]

Numerous researches have been carried out to analyse the transport/development relationship in a spatial economy (Hay, 1973; Garrison, 1960). It is stressed that transport improves in any economy as a concomitant of a economic development, not a precondition for it (Hunter, 1965).[116] In this attempt, Hunter stressed that the functioning of an economy requires the use of transport and it is also suggested to analyse the economies with varying structures and levels of development, and finally to identify the variations in its use in relation to other determinants of spatial variation. An attempt to relate to the causal possibilities of transport development was made by Wilson (1966).[117] The study indicated the three causal possibilities using Storey's categories such as a positive effect (wherein new, directly productive activities are the direct result of providing facilities); a neutral effect (0) (wherein transportation facilities do not themselves call forth directly productive activities and subsequent increase in the level of economic growth); and a negative effect (—) (wherein the presence of transportation facilities eliminates directly productive activity and effectively reduces the level of economic growth). Wilson categorises these as a positive stimulus to further development. He has concluded that transport is no more an initiator of growth than any other form of investment or deliberate policy.

Wheeler in his attempt (1971)[118] described the significance and impact of networks and flows on the space economy. Attention was paid to study the economies of transportation as an independent variable to determine the city size and the hierarchy of facilities (Bharadwaj, 1962).[119] Considerable attention was paid towards the study of the general problem of transportation development and economic change, where it examined the highway transportation and urban growth in the region of Sao Paulo, Brazil and area that is undergoing a rapid economic transformation. The objective of this analysis is to investigate for the period 1940 to 1960, the inter-relationships between changes in accessibility to the study of highway network and the growth of urban centres. He has found out that the change in network structure may have an effect on economic development. The significant aspect of this study is the application of canonical correlation analysis to consider whether the relationships between nodal accessibility and urban growth have been balanced or unbalanced. It is also understood that the increase in the network accessibility have benefited by the increase in their rates of economic development. More recently there has been an increasing recognition by the geographers to study the value of economic models and generalisations. Berry (1959) has attempted in his study to describe the role of transportation in the spatial economy and the approaches such as dealing about the spatial structure of the economy arising out of the impact of transportation and the provision of a systematic framework for the descriptive analysis and explanation of the spatial patterns in transport phenomena are noteworthy. Hay (1973) in his study briefly discussed the non-spatial aspects which directly provide an understanding of the overall spatial patterns and finally outlined the problems of planning for transport and identified the immediate need of some analytical tools in their solution.

Emphasis on the permissive element considering the locational aspects in relation to spatial economy has led to the formation of a framework of locational dimensions of spatial economy. It is also important to mention the empirical study of transport network and economy in North Bihar (Singh, 1968).[120] Hoyle's (1970) paper on transportation and economic growth in developing countries was based on the simulation model of network development. This paper attempted to set railway and port systems serving East Africa within the broader context of ideas relating to transport developments and economic growth. He has concluded in his paper that all transport media are vital factors in the economic growth of less developed countries. In this aspect, it is to say that transport media are permissive rather than stimulative in themselves. An attempt was made by Hance (1967) to study the great power of improved transport to quicken the economic pulse of a region and also in this direction he made it clear that all transport media are vital factors in the economic growth of the less developed countries.

A wide range of material was provided by Brookings Institution (Washington D.C.) through its "Transport Research Program" which during the 1960s yielded a number of important studies of the role of transport in development: two of these are the Wilfred Owen studies in the importance of mobility in developing countries and also a study of transport problems of India (Owen, 1968)[121] in a multifarious manner. The author of these studies sought to explore and to illustrate economic aspects of the transport/development relationship. Hoyle (1973)[122] introduced a variety of a body of extensive literature available on the role of transport in the development of less-developed countries. Emphasis has been laid to cover the various aspects of the problems and areas; material based on economic theory and on spatial analyses has been included; discussions are also made in analysing the role of specific transport modes, especially roads, railways and seaports. Problems such as intermodal integration, technological change and passenger transport are also presented. Issues related to agriculture, industry and urban growth are also discussed.

Geographers have long recognised that economic development is closely related to the growth of efficient transport systems and how the variation of results in studies of transport and development may be explained by the differences in the creation of economic opportunity and the response to economic opportunity. The first depends upon the quality and quantity of resources in the regions served, whereas the latter depends upon an awareness of opportunity and what may be broadly defined as attitude towards economic change. The extent of awareness and their potential non-economic nature is a direct function of the degree of accessibility brought out by the implications of transportation with respect to the economic analysis of development.

A distinguished work in transportation geography is the one that studies its relevance in the spatial economy (Gauthier, 1970). Gauthier (1970) in his study has given importance in reviewing some traditional ways of viewing transportation in the development progress. He has further argued the necessity of considering transportation as a spatial system in regional development and also suggested some problems that arise and in the same way the same be related to the spatial impact of transportation to goals of regional integration. In his concluding remarks he has necessitated the role of transportation in the development process as a precondition or prerequisite for economic growth. Airways have often assumed a particular predominance in the transport systems because in many cases they have acted as a major initiator of development and modernisation and have tended to dominate subsequent patterns of growth in the transport sector. The paper discussed the relationship between the installation of air transport services and economic growth in the underdeveloped north-west of Australia. Railways have also gained an importance in the development process of the region both as a initiator of development and modernisation. Long (1965)[123] has made his attempt to deal with a considerable expansion programme affecting railways, lake services and ports. Whereas Black (1971) has devised network replication models utilising historical data on the development of marine railway network. A discriminant analysis estimation is also devised which allows distance to a node, potential traffic and linkage angle measure to predict the edge construction. Black's model deals exclusively with network growth in which the idea of the network growth is a deterministic diffusion process. His study provided a framework to take proper decisions by the decision-makers to allocate their funds accordingly. Leinbach (1974) has presented his study to explain the transport development in Malaysia through a regression approach. The road network growth is modelled as a contagious diffusion. The results indicated the evidence which pointed out the importance of the diffusion process in transport forecasting. The study of structural changes of economic regions was made based on the railroad commodity flow pattern with the help of factor analysis. Werner's flow approach has been used to find out the probable flows over links together with the location of sources and sinks to determine the flows converging on a primitive network and in order to fit the geometrical scheme to a group of points, assignment of flows are made to network links. Hence, in this manner, assignment also serves as an aid in constructing the network. The usefulness of this approach is strengthened by means of its admission of bivariate probabilities when it has elastic flows to associate with the network nodes. Attempts have been also made to analyse the role of transportation in the urbanisation process, based on the dominant streams of work trips, the interaction patterns were determined as against physical distance among residential work place locations. The techniques employed to accomplish these objectives are transaction flow analysis, factor analysis and simple Markov chain analysis.

Considerable attention was paid to study the interactions between transportation system and urban development (Reichman, 1970).[124] Three aspects deserve particular attention viz., (1) the relationship between the localization of transportation facilities and the spatial elements of the urbanisation process, (2) the role of specific transportation factors in the evolution of urban form and activities, and (3) the inter-dependence between transportation networks and flows and the development of urban systems. In his concluding remarks he stated that the urban systems were an inseparable part of the functional organisation or space and stressed the need to investigate the three levels such as (1) urban field which includes urban-rural interactions; (2) a system of central places within a given area generally comprised of the whole national territory; and (3) the supra-national level where metropolitan areas acquire specific functional dominance over urban systems. It is mentioned that each of these levels, particularly elements of the transportation system have contributed to the organisation of space. In describing the levels of importance, the lower level is meant that the main contribution of transportation is that of the cost reduction at the same time allowing for a more efficient utilisation and mobilisation of resources; and at the higher level, it assumes greater importance namely spatial differentiation due to network location.

Studies were also made to analyse the impact of technological change in the transformation or plantation and transportation systems (Auty, 1975)[125], the changing patterns of oil transportation and refining in West Germany (Waller, *et al* 1977).[126] Another well-known attempt to analyse the relationships between the highway development and the process of geographic change is studied by Garrison *et al*, (1959).[127]

Taaffe, *et al* (1963) have attempted to study the ideal sequence of transport network development in a less developed country and he has provided a basis to identify factors helping to explain spatial variations in transport network densities which is an important pointer towards a model stimulating the economic growth process and enables the achieving of the effective evaluation of alternative transport development strategies. The foremost priorities of transport policy of the many developing countries are the improvement of rural road systems involving the application of development capital at the grass roots level of the transportation which in turn will make an attempt to introduce at least one form of modern transport within reach of a high proportion of the population. Transport development at this level would certainly assist the agricultural activity and also promote the rural industries at an intermediate technological level. Further, it may be also pointed out that the improved transport links between rural areas and urban industrial growth centres tend to increase the polarisation and reduce the problems of the towns.

Stanley's (1970) other study attempted to interpret the changes in maritime and land transport systems and to note events peculiar to Liberia by examining the broad stages of transport expansion. In his concluding section, he has analysed the critical role of ports in the development of inland transport. Lloyd *et al* (1972) have presented their approach starting with an introduction to economic geography and the essentials of model building and systems theory. He has organised this in three parts: (1) Building and testing of a simplified model of economic landscape in which all but one variable (distance) among the complex variables operating in the real world are constrained, (2) the impact of spatial variations on the quality and accessibility of resources and market outlets on the tendency for production units to agglomerate and perhaps the most significant of all, the nature of the location decision process itself are all introduced. In the third part, time dimension is added to examine the underlying causes of economic development in space and how variations in economic growth from place to place create stress situations. Other common studies include the growth of transportation and communication and the influence of railways in changing the characteristics of towns. Miller has undertaken an investigation to estimate the increase in agricultural production which could be gained from highway improvements. There are a considerable number of studies which account for the influence of transport change and its impact on the agricultural specialisation (Leaman and Conkling, 1975) as well as the highway improvements on urban land (Garrison, *et al*, 1958).[128]

Leaman and Conkling ware concerned to study the Brinkmann thesis to test that local comparative advantage assumes increasing locational importance as transport costs decline under non-homogeneous conditions. The results confirm the hypothesis that transport costs are inversely associated with the degree of specialisation and trade.

Studies were also carried out to analyse the competition in terms rail-road competition (Koilsen, 1968) and the competitive significance of different ports in the interplay of trade (Kenyon, 1970).[129] An attempt was made to analyse the shifting magnitudes and composition of general cargo freight among selected ports of the United States. It has also been pointed out in the study that great inertia (or perhaps momentum) characterises the ports of U.S. and the success is partly determined by their regional location, the size of their local and nearby economy and their proximity to overseas trading areas. In addition to this, it is also important to mention the predictive nature of the extent of their commercial financial and organisational structure. Koilsen (1968) has presented his study in dealing with the economies and control of rail-road competition.

In explaining the role of transportation network it is important to recognize the following: maximise freeways, automobile and profit yielding freight links; minimize public transit, rail links to isolated rural communities and social and community welfare.

In an operating environment there are many stages in understanding the broader set of utilities influenced by individual and group aspirations, life-styles, political motivations and socio-economic relationships. The essential of conceptualisation of dimensions of space is felt an imperative need to give shape to the terms movement space. The other functional aspects of different stages include, (1) to channel the movement where the modes and stocks interact over space and time, competing or complementing, but following through the channels in a variety of interacting paths. The degree of movement depends on the degree of concentration and specialisation in the economy, (2) decision-making in a variety of movement options, (3) determination of the 'level of service' which includes the establishment of the routes and sehedules, pricing, and reliability and other factors, (4) demand estimation, (5) the influence of direct and indirect factors over movement decisions. Considering all the above, it can easily be recognised about the changing concepts of transportation geography and newly emerging research avenues in societal and policy perspectives in an environment/transport system. In this, the conceptualisation of space has diverted the attention of the geographers and then the general area of perception which can never be ignored. Three major approaches reviewed by Roger Downs in his bibliographic survey are, *Structural*: concerned with the identity and structure of space perception. It also consists of storage of information. Nature and structure of its storage and the relationship between the image and the real world. *Evaluative*: People's consideration of ideas in their perceived worlds and how these factors affect decision-making. *References*: Consideration of a set of spatially differentiated objects and the assessment of their choice between them. These three avenues fall within the frameworks of the behaviorist to measure rigorously and evaluate these perceptual dimensions.

Since movement space is defined as perceived felt and conceived part of the build environment, the need arises to examine such a space conception which could possibly lead to a better understanding of travel patterns. The literature and concepts of the above mentioned suggest that both a philosophical and a theoretical foundation for a re-orientation of movement studies away from a description of habit patterns to an analysis of the causes behind movement behaviour.

Attempts were also made to develop a teaching model which discussed about the inquiry-discovery learning of processes involved in determining transportation patterns (Wheeler and Pannell, 1973). The purpose of this paper was to develop the exercise of a generated model of route network growth, followed by explanatory models of causation and theory formulation. These suggested hypotheses were then tested against a real-world example, the evolution of road network in Taiwan. The model promotes an understanding of human behaviour and geographic change through systematic scientific inquiry and cross-national application, and its emphasis is sufficient for use at several teaching levels. There is another well-known attempt to analyse the relationships between transport provision and economic development and the growth of modern transportation facilities in several underdeveloped countries through the processes of spatial diffusion. The analysis focused the following inferences :

> As population increases in an area, the demand for transportation is intensified; as new transport lines are built into the area, a greater population increase is encouraged, which in turn calls for more transportation. The models used here hold two separate effects; the ideal typical sequence considered transportation expansion as though it were independent of population distribution; the regressions treated transportation as though it were caused by population. By and large, the residual maps provided intuitive evidence of the lag-and-lead nature of transport development. Apart from the above, studies were also made in analysing the behavioral model selection (Craig, 1973)[130] and the freight rates and regionalism in a behavioral context.

The Landuse-Transportation relationship bring forth an important concept known as 'Static' and 'Dynamic'. Once the region becomes dynamic, it is characterised by the functional aspects. The functional aspect of a region is generally a product of spatial interaction of areal economy and lines of circulation. The landuse-transport forms a closed loop and leads to the study of the effect of traffic and its imposing changes, one on the other. If both the landuse and transport do exist the traffic that comes into occurrence is determined by the landuse potential and transport capability (Blunden, 1971).[131] One of the important objectives of planning a landuse-transport system is to ensure that there is an efficient balance between landuse and transport capability. A unique property of landuse is its ability to generate traffic. The study of traffic in a wide sense that it is a measure of socio-economic activity definitely saying, that the traffic that arises determines the scale of the landuse and transport activity. It is also quite explicit that the analysis of the study of landuse/transport system in a socio-economic environment would focus that if there is any balance or imbalance between the landuse potential and transport capability. Studies of landuse/transport system were of recent origin that have created a new link between the environment and transport in a wide spectrum of problem-oriented researches.

An attempt was made to establish a conceptual framework for the relationship between landuse and travel movement by utilising traffic movements (Eliot Hurst, 1969a)[132]; and it allowed both landuse and traffic as two facets of the same general problem. Discussions were also held thus by explaining that movement becomes a function of the nature of demands and supplies (desires and fulfilments) and of the character of the available transport systems. The amount of flow (the quantity of traffic) is a function of the character of the available transport systems and their environment. The relationship between the aforesaid has been strengthened by way of analysing the positive potentials (landuse and population distribution in a particular functional entity have particular travel needs) and negative potentials (the spatial imbalance of the desire and the fulfilment identified factors such as (1) household size, (2) income or socio-economic status, (3) location and distance (with the corollary of density), and (4) the potential surface. The techniques used to identify the aggregate travel behaviour and travel linkages in a spatial context are to account for the frictional effect of distances between origin and destination (methods such as factor analyses, gravity method and probability method). The study has set out an Osotsky's Modesto model to calculate the theoretical vehicle movements from the regression analysis thus by way of giving assurance that the study of landuse and traffic can no longer be thought of as separate, but related fields of study. Suffice it to say in this particular study that land can only be effectively utilised only when it is accessible. An attempt relate to the travel decisions at the more personal level without the constraints was made to distinguish the growing area of transportation geography to spill over into the more general area of behavioral geography. In his attempt, he gave an emphasis in dealing with the individual's perception and conception of transport alternatives.

The landuse-traffic generation relationships of non-residential landuses were analysed to account for the functional relationships between the landuse and the volume of urban travel. The study analysed the location and intensity use of the land with important variables in establishing such a relationship. It is found that the economic growth depends upon a soundly developed transport network. The study focused its importance in describing the changing landuse and transportation pattern in Sierra Leone. The following factors were outlined to discuss the role of transportation such as (1) running down of the national railway, and (2) expansion and modification of the road network and the increase in motor vehicles and finally the improvements of the port. Blunden (1971) in his study wrote about the relationships between the landuse transport system through traffic. This forms a very good account in investigating the degree of closeness between the two.

The spatial interaction which forms the basis of interaction is defined by the Ullman's triad (Ullman, 1954)[133] viz., transferability, complementarity and intervening opportunity. This is a situational concept and is mainly intended as a more positive and dynamic concept both in the situation relations and even in circulation relations. Focusing research on interaction may well provide an infrastructure for the fuller understanding of the spatial structure and the dynamism in spatial system. The approach is of a geometrical one to deal interaction in which all quantities are treated as continuous distributions over space rather than concentrations, at nodes in a network. The model produced an allocation of trips from any origin to destination area based upon a derived trip density function. The outputs from the model are spatial distribution of the accessibility of locations of jobs and residences as well as traffic flow. Olsson's concept assumes that the whole of man's experience can be evaluated on a linear scale of finite and readily perceptible units. The reference made here suggests that traffic along a given route is a normative function of the physical distance. Other studies of note are the estimates of the parameter of spatial interaction and the study of freight rates in selected aspects of uniform and nodal regions.

The study of the dynamics of spatial systems have poured much of its potential in raising the status of transportation geography. Wheeler (1974)[134] has analysed the changing spatial properties of urban communities as defined by local social networks. The city neighbourhoods extent was accurately delineated by the application of the growing tendency of automobile decentralisation of social contacts and personal activity spaces. The study of traffic characteristics in comparative analysis of ports and transport freight complex structure in relation to the metropolis and the analysis of inter-relationships of the changing structure of transportations in relation to making changes in industrial locations are of considerable importance in explaining the inertia of the spatial organisation and the dynamic role of spatial systems. Attempts were also being made to assess the energy uses and air pollution in transportation systems through the application of weighted graph models (Roberts, 1975).[135]

**Spatial Efficiency through Space Fulfilling and Space-Adjusting Techniques**

Movements of phenomena through space result in the establishment of movement systems. These in turn facilitate further movements through space which is termed as space fulfilling. In describing the importance of the expanse of the others, the relationship between movements (process) and transportation systems (structure) is not one-way but is reflexive and circularly causal. The movement systems or space-adjusting techniques are very powerful determinants of the location of human activities. The influence of the operating of the powerful system invents spatial organisation in which the interaction between process and structure is a fundamental aspect of any locational problem. In order to establish an efficient functional organisation and effective spatial organisation, it is important to determine the best possible or optimal, transportation pattern from every centre to every other centre. Such a profoundly established mode which would allocate flow of goods between producing and consuming areas in a way so as to minimize transport costs, is for builders and one for users. A model which determines such an optimal pattern is known as a "normative model".

To strengthen the spatial efficiency of transportation it has to consider two important aspects viz., primal and dual of the transportation problem. The objective of the primal is to be minimized and dual is to be maximized (the economic return). Attempts were made to develop the various efficient algorithms or set of rules for solving the general transportation problem.

Several algorithms have been developed to find out the minimum length circuit. The basic aims of these algorithmic approaches are to reduce the long and expensive computations needed. A continuous research in this field of inquiry produced more efficient algorithms (e.g. the "Cascade" algorithm of Farbey, Land and Murchland and Shen Lin algorithm). Studies were also made to design and optimal route through 'n' cities so as to start at city 1 to visit each of other n-1 cities only once, and then return to city 1 by applying travelling salesman problem approach. In describing this, a network connecting set of points can assume the form of a tree-like structure of connected lines without any complete loops or it can be a single continuous path connecting many places. The path can also be closed in a circuit or loop. The former two are typical of collection and delivery systems; the last is called the travelling salesman problem because a salesman's itinerary takes him away from home to make his calls but brings him home at the end of his trip. It might also seem that the travelling salesman problem could be solved by measuring all the possible paths and then picking the shortest one, to make one approximation of the shortest travelling salesman's random. The solution to the travelling problem depends on the kind of surface over which movement between centres must take place. Lin (1965)[136] presented his approach to the approximation solution method for the travelling salesman problem. The basic idea behind the problem is to establish a set of arcs linking 'n' vertices so that the sum of the shortest distances between each pair of vertices is a minimum.

The first algorithm starts by establishing the minimal spanning tree (a tree whose arcs are of minimum aggregate length) and gradually arcs to the tree. The solution process approaches the objective function value from above. It is understood that there is an indirect relationship between arc addition and the value of the objective function; whereas the second algorithm begins with a completely interconnected set of vertices, and there works towards the optimal solution by deleting arcs. Although the forward and backward heuristics involve lengthy computations, it can treat the optimal network problems of moderate size rapidly and while the uses loses rigour and exactness. The utility of such programming is the provision of flexibility and practicability. An attempt was also made to arrive a solution for the two centres location-allocation problem by using an efficient algorithm. Two algorithms were developed to solve certain important transportation problems which have gained importance in recent research potential. The first one is a branch and bound tree searching programme and the other one is an heuristic algorithm which allows the locations and capacities of the facilities to vary while a search is made over the population surface for the locations which minimize cost. A full explanation of input information along with practical examples, as well as data and sensitivity experiences are provided by a monography (Tornqvist, 1971).[137]

The heuristic algorithm was attempted to minimize the distances in a spatial context. Mackinnon and Hodgson (1970)[138] adopted an heuristic algorithm from Scott (1969) in determining to minimize inter-nodal short path distances over a portion of the highway network for Ontario and Quebec subject to a budget constraint. The creation of the minimal spanning tree answers the solution for an optimum network. In addition to the above, he has also presented the gravity model based on the objective function. The authors have also noted where comparing the above two, the Shimbel model (minimizing inter-nodal distances) provides a good replication of the network as a gravity model. It also implied that the highway structure is determined by the relative location of the cities rather than by relative size. The authors have also emphasized the ability to create new roads in network generation and to develop solutions that are less than optimal. Mackinnon and Barber have also developed an heuristic algorithm and the method developed is quite analogous to a regression problem where a line is fitted to a set of points so that the sum of the squared vertical deviations is minimized. The problem is very similar to the location allocation problem and in addition, it may be important to mention the partitioning algorithm as an initial stage of optimal network generation. The preliminary stage of development, emphasises the research which has offered an imaginative approach to study the problems in association with the evaluation of different spatial patterns.

An attempt was made to construct a mathematical model (Werner, 1969)[139] to provide the geometrical construction of a minimum length network if the graph theoretic structure is given. It is also dealt the problem of selecting a network structure which leads to a network with the absolute minimum length. An estimation technique is provided for the MLN (minimum length network) and emphasis has been laid in the construction of a general topological function which would facilitate the development of an optimizing procedure to derive the topology associated with the MLN. Werner has also extended the application of the law of refraction to the identification of a transport route that minimizes total cost when a region is partitioned into a number of total cost regions. The success in better utilising the transport problem in linear programming as a predictive model was attempted by O'Sullivan (1972).[140] It is found that the correlations between the minimum cost and actual solutions were much higher than those for the gravity prediction and actual values. A study was also made (Lösch, 1944)[141] to identify the optimal route between two sub-regions by a straight line boundary.

Kissling (1969)[142] in his study has thrown light upon the 1966 highway network in the province of Nova Scotia. He has calculated expected average costs for operating a 30,000 pound loaded vehicle over each link from an analysis of the physical properties of the highways, annual average daily traffic statistics, and the extent of legally controlled zones. He then modified the Shimbel method of assessing accessibility to provide an alternative method which requires that minimum value routeways between all pairs of network nodes may be calculated only once. Once when a detailed inventory of highway linkage characteristics is at hand, an algorithm of such type has considerable utility for planning and development purposes.

A study was made to establish (Sen, 1971)[143] an optimal network configuration that connects a node with a portion of its hinterland by the addition of secondary routes to the trunk route while minimizing the total movement cost from all points in the hinterland to the node. The study identified the significance of individual parameters through sensitivity analysis such as the width and breadth of the rectangle defining the area under study, the optimum angle at which traffic flow off the network enters the network, the spacing interval between parallel secondary routes and the main line. Vector solutions were provided to the problem of optimum location of interior nodes which in turn will determine the minimum length network. Network analysis using the minimal tree concept has offered a theorem and proofs relevant to the geometry of minimal trees in the Euclidean plane.

Flow assignment problem through location-allocation was made to discuss the given 'n' points and a set of 'm' centroids without predetermined locations, find locations for the 'M' centroids which allocate each point to some centroid in an optimal fashion. The pedogogic review of Wilson monograph provides an indispensable one to grasp the analytical keys by way of establishing the transportation models. The utility of Wilson's work in introducing transportation models are of immense value and it should not be ignored. Based on Wilson's probabilistic frame work, an attempt was made to apply the Morkov Chain model. This deals with the generation of flow conditions within a fixed node-link network with inputs defined and under varying degrees of user uncertainty about the system. In signifying the importance of model development in transportation geography, it is important to mention Kolar's and Malin's (1970)[144] development of a post-dictive simulation model of the Turkish railway system based upon population and topographic structures. In addition to the identification of optimal routes, the network model is generated using a gravity model to compute potential interaction and incorporating a parsimonious route length notion while allowing for physical features. Summing up, the model clarifies the additional support for the current theory of network growth and thus lets one have a new insight in feasibility studies of link construction.

The growth of urbanisation continuously puts for enormous problems especially in the transportation sector. Hence the need for transportation planning is felt urgently to alleviate the increasing problems. Emphasis has been laid on a more integrated and comprehensive approach to transportation planning and analysis. Principal deficiencies were identified in this study and suitable recommendations were drawn to carry out planning priorities. In addition to the above, a model for the evaluation of transport alternatives have been introduced and the purpose of the model rests in evaluating for particular combinations of economic and geographic conditions, (i.e. a country) alternative transportation plans and the associated long-term investment requirements. Along with these, the time period concept has been introduced into the model. An attempt was made to apply the linear programming technique to facilitate the flow assignment problems and the same was applied in the spatial planning of school districts.

A considerable number of works have been done to cover the salient features of transportation planning and perspectives on regional transportation planning. Two research avenues in transportation planning have emerged to discuss the planning and development to roads for rural areas. The attempt covered the various aspects such as opening up rural areas the need for amendments, new rural roads traffic on rural roads, bases for the design and the need for rural roads. Traffic forecasting is the principal axis in transportation planning. A study was devoted to deal with the traffic problems prevailing in the Netherlands. The study was initiated by the Ministry of Transport and Waterstaat and it had to bundle the ideas to an integrated view on the problems of Ministry of Transport and Waterstaat for the future. An attempt was made to describe traffic restraint as part of transport planning. Studies were also made to account for the impact of environment on transportation policy assessment and the role of pedestrial planning in the perception of the urban environment (Marihand, 1975).

Since the geographer's primary concern is with location decisions, the whole set of decisions that a firm must make, must adapt to a constantly changing environment. In order to avoid the spatial repercussions arising out of the relationship between the firm and environment, decision-making is essential because this will allow us to examine certain basic elements underlying decisions as a whole before looking specifically at location decisions. The actual decision-making is a complex one, but for easy understanding it can be divided into three stages: (1) the perceived stimulus to make a decision and definition of the nature of the problem; (2) search for a satisfactory solution; and (3) evaluation of the consequences of alternative courses of action and choice of solution. In this context, it may be important to point out that one of the many challenges facing the geographer today, therefore, is to build a healthy body of location theory based on sound behavioral principles because in the final analysis, the economic landscape is the cumulative expression of the decisions made by individuals at different points in time and under a variety of conditions.

In the above, an attempt was made to assess the role of decision-making within the operational milieu of the society in Thailand. The author has concluded that the decision processes are an amalgam of *ad hoc* approaches, many *laissez faire* as far as the government is concerned, but still far from optimal, even allowing that it is a nation dominated by the capitalist ethos.

## SPATIAL PERSPECTIVE IN GEOGRAPHIC RESEARCH

In transportation, the structural elements in spatial organization are linkages and nodes. The linkages and nodes are organised into hinterlands, systems of hinterlands and hierarchies. The mosaics characterising the uniform region and the cobwebs of linkages characterising the functional region can both be considered aspects of a broader system of spatial organisation. The changes in the functional organisation ultimately bring out change in the landuse mosaic representing its spatial uniformity. This indication of close relationship between spatial uniformity and functional linkages form parts of the region's over-all spatial organisation. In this context, the spatial pattern based on minimum-distance paths shows a stronger spatial organisation. The spatial pattern of organisation is determined by the spatial dominance of flows between centres. Highway interchanges have formed new focal points in changing the patterns of spatial organisation. Studies have been made in a elaborate manner to discuss the various aspects starting from order, science and geography, to measurement, relationship and classification, location and spatial interaction, spatial diffusion, meshing space and time and finally spatial organisation and the decision process. These studies have invariably considered the inter-relationships of both the uniform and functional aspects of spatial structure, spatial system and spatial interaction all of which are contained in one term called "spatial organisation".

An attempt was made to study the relationship between spatial organisation and the temporal evolution of the transportation system through a holistic approach (Janelle, 1969).[145] (Fig. 5.1). The study mainly considered examining the role of

*Spatial Reorganisation—Model-1*

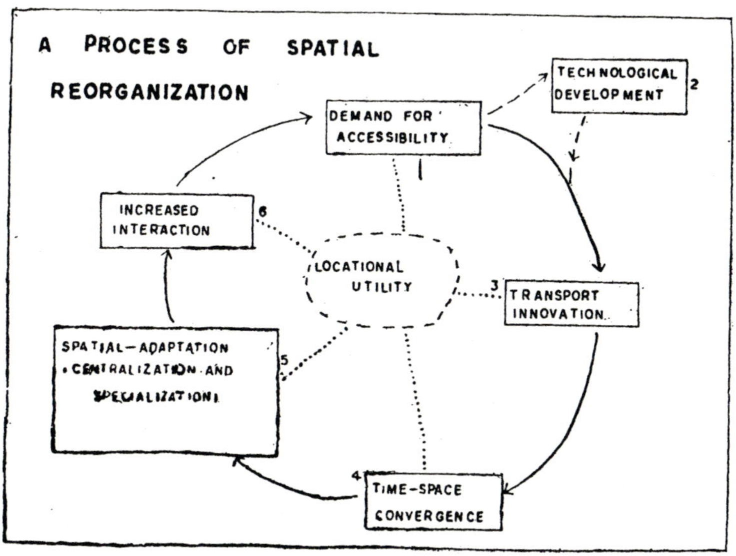


FIG. 5.1

transportation technologies in the spatial organisation of human activities, the changing spatial pattern over time and the spatial variation occurring between different areas on the earth's surface. The author has used the travel time matrix as the key factor in defining a process of the spatial reorganisation of man's functional establishments. The concept of spatial reorganisation includes the following aspects such as demand for accessibility, transport innovations, time-space convergence, spatial adaptations—centralization and specialization and spatial interaction. The application of the concept of relative advantages holds good in the process of spatial reorganisation in the form of centralisation and specialisation. Apart from the above, the time-space convergence is a significant factor leading towards spatial adaptation and that spatial reorganisation will accelerate most rapidly at those places which stand to benefit most from increasing accessibility. In short, the understanding of the process of spatial reorganisation may rest upon one's knowledge of the decision-making processes. Dynamic programming techniques may hold some promise for the solution of these problems (Fig. 5.2).

*Spatial Reorganization—Model-2*

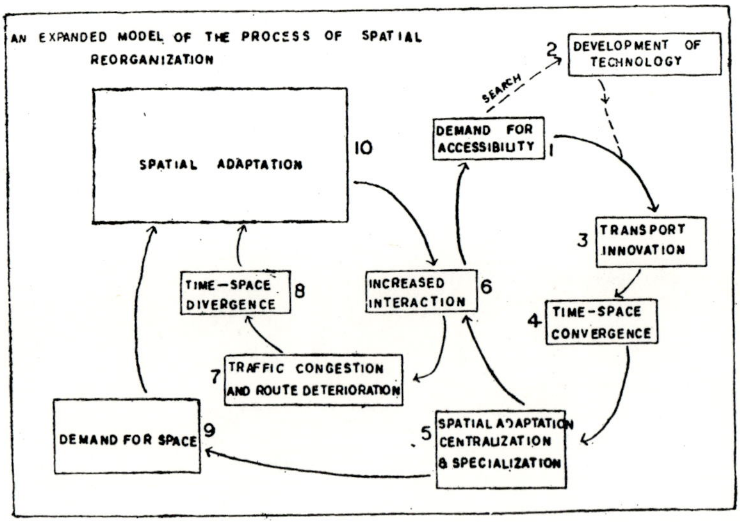


FIG. 5.2

## CONCLUSION

The progress of the conceptual revolution in transportation geography has thus gained a professional identity of its own during the course of its development whereby it has built up the platform for a sound methodology for future researchers. The geographer's contribution in considering transportation as a prime variable in the organisation of area should not be ignored. Contributions by geographers in the field of transportation geography both particular and general reveal the existence of spatial inter-dependencies which is the foundation stone for future research. The basis of their contribution is a better understanding of the geography of transportation. The imposing challenges have to be met only by way of taking present spatial organisation and the geographical pattern of the world's system.

## REFERENCES


1 E.A. Ackerman, "Regional Research—Emerging concepts and Techniques in the Field of Geography", *Economic Geography*, 29, 1953, pp 189-197.

2. G.S. Davies, "The Reverse Commuter Transit Problem in Indianapolis", in H M. Rose, Editor, *Geography of the Ghetto : Perceptions Problems, and Alterations*, De Kalb, III, Northern Illinois University Press, 1972, pp 169-96.

3. J. Taaffe, et al; *Modern Transportation Facilities in Underdeveloped Countries in Transport and Development*, Edited by B.S. Hoyle, Macmillan, London, 1973.

4. A. Hay, *Transport for the Space Economy*, Macmillan, 1973.

5. E.L. Ullman; "Transportation Geography", *in American Geography, Inventory and Prospect*, Ed. PE. James and C.F. Jones, Syracuse, 1954, pp 311.

6. E.L. Ullman, "Geography as Spatial Interaction", *Studies in Regional Development : Cities and Transportation*, University of Washington. Press, 1973.

7. E.L. Ullman, "The Role of Transportation and the Basis for Interaction", *Man's Role in Changing the Face of the Earth*, Edited by O.L. Thoman, Chicago University Press, Chicago, 1956.

8. P.O. Muller, "Toward a Geography of the Suburbs", Proceedings of the *Association of American Geographers*, 6, 1974, pp 36-40.

9. R. Abler, John S. Admas and Peter Gould, "*Spatial Organisation : The Geographer's View of the World*", Prentice Hall Inc., New Jersey, Chap. 8, pp 236-239, 1971.

10. L.J. Vasilevsky, "Basic Research Problem in the Geography of Transportation of Capitalist and Underdeveloped Countries", *Soviet Geography : Review and Transportation*, 4, 1963, pp 36-58.
11. J.O. Wheeler, and F.P. Stutz, "Spatial Dimensions of Urban Social Travel", *Annals of the Association of American Geographers*, 61, 1971, pp 371-86.
12. W.L. Garrison, "Spatial Structure of the Economy", *A.A.A.G.*, 49, June, 1959, pp 232-39, December, 1959, pp 471-82, Sept. 1960, pp 357-73.
13. W. Warntz, and D. Neft, "Contributions to a Statistical Methodology for Areal Distributions", *Journal of Regional Science,* Vol. 2, 1960.
14. B.J.L. Berry, "Recent Studies concerning the Role of Transportation in the Space Economy", *A.A.A.G.* 9, 1959, pp 328-42.
15. P. Haggett, *Locational Analysis in Human Geography*, Edward Arnold, London, 1965.
16. K.J. Kansky, "Structure of Transportation Networks" Research Paper No. 84, Department of Geography, The University of Chicago, Chicago. 1963.
17. J. Taaffe, and H.L. Gauthier, *Geography of Transportation*, Prentice Hall, New Jersey, 1973.
18. M.E. Eliot, *Transportation Geography : Comments and Readings*, McGraw Hill Book Co., New York, 1973.
19. D. Appleyard, "Social and Environmental Policies for Transportation in the 1970s", University of California, Berkley, Institute for Urban and Regional Development, Working Paper 161, 1971.
20. Muller, 1974, op. cit.
21. J.O. Wheeler, and W.P. Colffton "A Teaching Model of Network Diffusion : The Taiwan Example", *Journal of Geography*, National Council for Geographic Education, Florida., Vol. 72, May, 1973, No. 5, pp 21-31.
22. A.G. Wilson, *Urban and Regional Models in Geography and Planning*, London, John Wiley, 1974.
23. R.B. Singh, *Transport Geography of Uttar Pradesh*, Varanasi, 1966.
24. S.L. Kayastha, "Transport and Communication in the Himalayan Beas Basin", *National Geographical Journal of India*, 1960, 6 (2).
25. M. Mukherjee, "Transport Town of Bihar", *The Indian Geographical Journal*, Vol. XLIV, July-Sept. and October, 1969, No. 524, pp 42-51.
26. S.L. Kayastha, and D.N. Singh, "Some Aspects of Transportation in Dhanbad", *The National Geography Journal of India*, Vol. XVIII, Part 2, June 1972, pp 64-79.
27. A. Prasad, "Roads in Chota Nagpur", Abstracts of Papers of the 21st International Geographical Union Congress, New Delhi, 1968.
28. R.T. Aangeenburg, "Automobile Commuting in Large Suburbs : A Comparative Analysis of Private Car Use in the Daily Journey to Work", in the North Western University, *Studies in Geography*, No. 16, 1968, pp 1-19.

29. J.O. Wheeler, "The Structure of Metropolitan Work Trips", *The Professional Geographer*, Vol. XXII, May 1970, pp 152-158.
30. C. Wood. "Railways and Resources in British Columbia", *Geography*, Journal of Geographical Association, Sheffield., No. 270, Vol. 61, Part I, January, pp 31-33.
31. D. Hilling, "Container Potential of West African Ports" in *Transport and Development*, Edited by B,S. Hoyle, Macmillan, London, 1973.
32. I.M.L. Robertson, "Road Networks and the Location of Facilities", *Environment and Planning* Vol. 6, No. 2, March-April, 1974, Pion Ltd., London, pp 199-206.
33. P.N. Grimshaw, "Stream Railways : Growth Points for Leisure and Recreation, *Geography*, Jl. of the Geographical Association No. 271 Vol. 61, Part 2, April 1976, pp 83-88.
34. H.M. Saxena, "Geography of Transport and Market Centres" S. Chand & Co. (Pvt.) Ltd., New Delhi, 1975.
35. B.M. Barr, *The Soviet Wood Processing Industry* : A Linearprogramming Analysis of their Role of Transport Cost in Location and Flow Patterns, in University of Toronto Press, 1970.
36. P. Gould, and T.R. Leinbach, "An Approach to the Geographical Assignment of Hospital Services", *TESG* 57, 1966, pp 203-206.
37. A.J. Scott, "A Model of Nodal Entropy in a Transportation Network with Congestion Costs", Transportation Science 5, 1971, pp 204-II.
38. G. Breckenfeld, "Downtown had Fled to the Suburbs", in *Fortune* 86, 80-87, 156, 158, 162, October, 1972.
39. R. Hubbard, "The Development of the Jamaican Road Network By 1846, *The Journal of Tropical Geography* Vol. 38, June 1974, pp 31-36.
40. G. Ramaswami, "Road Network in Tiruchi-Thanjavur Planning Region : An Analysis", *Indian Geographical Society*, Golden Jubilee Volume, pp 150-160, 1976.
41. R.W. Bradnock, "Measurements of Accessibility and their Relationship to Development in Tamilnadu" in *Indian Geographical Society*, Golden Jubilee Vol. 1976, pp 129-149.
42. R. Matela, and Ellin O'Hare, Graph-Theretic aspects of Polyominoes and related Spatial Structures, *Environment and Planning* Vol. 3 No. 1, June (1976), pp. 79-110.
43. R. William, Black, and L. Robert, "A Comparative Evaluation of Alternate Friction Factors in the Gravity Model", in *The Professional Geographer*, Vol. XXIV, Nov. 1972, No. 4.
44. P. Haggett, Op cit, 1965.
45. Taaffe and Gauthier, Op cit, 1973.
46. P.E. Lloyd, and P. Dicken, *Location in Space : A Theoretical Approach to Economic Geography*, Harpe International Edition, London, 1972.
47. W.L. Garrison, Op cit., 1960.
48. S. Shanmuganandan, and P. Ilengovan, "Transport Network of a Drought Prone Area : A Case Study of Dharmapuri District in

Tamilnadu", Symposium on all-India Symposium on Drought Prone Areas, Department of Geography, Venkateswara University, Tirupati, 1977.

49. A.B. Mukherji, "Road Transportation Network Structure and Levels of Urbanization in Rajasthan", *The National Geography Journal of India* Vol. XX, Part I, March 1974.
50. S. Shanmuganandan, "Developmental Aspects of the Transportation Network", Paper submitted for publication, Deccan Geographer, 1978.
51. Kansky, op cit; 1963.
52. G.A. James, et al.; "Some Discrete Distributions for Graphs with Applications to Regional Transport Network", *Geografiska Annaler* 52B, 1970, pp 14-21.
53. R. Daltan, et al., *Networks in Geography*, George Philip & Sons Ltd., London, 1973.
54. W. Black, "A Note on the Geometry of Minimal Trees", in *Geographical Analysis* 2, 1970, pp 410-14.
55. C. Werner, et al., "Research Seminar in Theoretical Transportation Geography", Northwestern University, *Studies in Geography* 16, 1968, pp 128-70.
56. R.K. Semple and L.H. Wang, "A Geographical Analysis of Changing Redundancy in Inter-Urban Transportation Links", *Geografiska Annaler*, 55B, 1971, pp 1-7.
57. L.P. Cummings, et al; "Measuring Association in Link-Node Problems", *Geoforum* 4, 1973, pp 43-51.
58. K.J. Tinkler, "The Physical Interpretation of Eigen Functions of Dichotomous Matrices", *Transactions of the Insitute of British Geographers* 55, 1972, pp 17-46.
59. Abler, et al; 1971, op. cit.
60. Ernest George, Ravenstein, "The Laws of Migration", *Journal of the Royal Statistical Society*, Vol. 48, 1855, pp 167-233.
61. W.J. Reilly, *The Law of Retail Gravitation*, New York : The Knickerbocker Press, 1931.
62. James H.S. Bossard, "Residential propinquity as a Factor in Marriage Selection", *American Journal of Sociology*, Vol. 38, 1932, pp 219-244.
63. J.D. Stewert, Discussion : Population Projection by Means of Income Potential Models", Papers and Proceedings of the Regional Science Association, Vol. 4, 1958, pp 153-154.
64. G. Olsson, "Distance and Human Interaction", A Review and Bibliography, Regional Science Research Institute, 1965.
65. R. Alcaly, "Aggregation and Gravity Models : Some Empirical Evidence"; *Journal of Regional Science*, 7, 1967, pp 61-73.
66. W. Black, "The Utility of the Gravity Model and Estimates of its parameters in Commodity Flow Studies", in proceedings of the Association of American Geographers, 3, 1971, pp 28-31.
67. J.H. Niodercorn, and B.V. Bechdelt, "An Economic Derivation of

the Gravity Law of Spatial Interaction", *Journal of Regional Science*, 9, 1969, pp 273-82.

68. A.G. Wilson, "A Family of Spatial Interaction Models and Associated Developments.", Environment and Planning, 3, 1971, pp 1-2.
69. W. Black, "An Analysis of Gravity Model Distance Exponents", Transportation 2, 1973, pp 299-312.
70. L. Curry, "A Bivariate Spatial Regression Operator", *The Canadian Geographer* 16, 1972, pp 1-14.
71. K. Mrea, "An Evaluation of Gravity and Linear Programming Transportation Models for Predicting Inter-Regional Commodity Flows", In J.R. Meyer, and M. Strazheim, Editors, Technigres of Transport Planning, Washington, D.C., 1971, pp 297-308.
72. F. Stetzer, "Parameter Estimation for the Constrained Gravity Model : A Comparison of Six Methods", *Environment and Planning*, Vol. 8, No. 6, Sept. 1976, pp 673-·84.
73. Wheeler, op cit, 1970.
74. J.O. Wheeler, "Trip Purposes and Urban Activity Linkages", *A.A.A.G.*, Vol. 62, December, 1972, No. 4, pp 641-654.
75. Thomas, R. Leinbach, "Distance, Information Flows and Modernisation : Some Observations from West Malaysia", *Professional Geographer*, Vol. XXV, Feb. 1973, No. 1, pp 7-11.
76. A.S. Chapman, "Trans Europe Express : Overall Travel Time in Competition", *Economic Geography*, Vol. 44, Oct. 1968, pp 283-295.
77. W.I. Carruthers, "The Classification of Service Centres in England and Wales—An Analysis of Public Road Transport Services", *Geographical Journal*, 122, 1957, pp 371-385.
78. Barr, op cit, 1970.
79. D.N. Singh, "Transport Network and Economy in North Bihar", Abstracts of Papers of the 1st International Geographical Union Congress, New Delhi, 1968.
80. D.A. Hensher, "The Structure of Journeys and Nature of Travel Patterns 2, *Environment and Planning*, A, Vol. 8, No. 6, Sept., pp 655-672, 1976.
81. R.E. Quandt, "The Demand For Travel : Theory and Measurement", Lexington, Mass, D.c. Health and Ci., 1970.
82. A. Hay, "The Importance of Passenger Transport in Nigeria", in *Transport Development*, Edited By B.S. Hoyle, Macmillan, London, 1973.
83. D.L. Snyder, "Commercial Passenger Linkages and the Metropolitan Nodality of Montevides", *Economic Geography*, Vol. 38, No. 2, 1962, pp 95-112.
84. J.N.H. Britton, "Methodology in Flow Analysis", *East Lakes Geographer* 7, 1971, pp 393-403.
85. R.H.T. Smith, "Concepts and Methods in Commodity Flow Analysis", *Economic Geography* 46, 1970, pp 404-16.

86. M. Chisholm and P. Sullivan, "*Freight Flows and Spatial Aspects of the British Economy*", Cambridge, University Press, 1973.
87. Black, op cit, 1973.
88. E.D Perle, "The Demand for Transportation : Regional and Commodity Studies in the United States", Research Paper No. 95, Dept. of Geography, University of Chicago, 1964.
89. R.G. Golledge, "A Geographical Analysis of New Castles Rail Freight Traffic 2, *Economic Geography* Vol. 39, 1963, pp 60-73.
90. B. Bhaskara Rao, "Prospects of Freight Movements Through Italdia Port and its Hinterland 2, For the M. Tech. Degree Thesis, I.T.I., Kharagpur, 1964.
91. B. Lakshmi, "Pattern of Rail Traffic Flow in Madhya Pradesh", *National Geographical Journal of India*, 1967, 13(2).
92. Black; op cit; 1971
93. O.P. Sullivan, "Forecasting Interregional Freight Flows in Great Britain". M. Inchisholm, et al, Editors, *Regional Forecasting*, London, Butterworth, 1971. pp 443-50.
94. J.O. Wheeler "Transportation Geography : Societal and Policy Perspection", *Economic Geography*, 49, 1973, pp 95-184.
95. B J.L. Berry, "Interdependency of Spatial Structure and Spatial Behaviour . A General Field Theory Formulation", Papers and Proceeding of The Regional Science Association, 21, 1968, pp 205-27.
96. R. Earickson "*The Spatial Behaviour of Hospital Patients : A Behavioral Approach to Spatial Interaction in Metropolitan Chicago*", University of Chicago, Department of Geography, Research Paper, 124, 1970.
97. S.J. Bernstein, "Mass Transit and the Urban Ghetto", in *Traffic Quarterly*, 27. 1973, pp 431-49.
98. S.M. Golant, "The Residential Location and Spatial Behaviour of the Elderly : A Canadian Example", University of Chicago, Department of Geography, Research Papers 143, pp 1-109.
99. S. Myers, "Personal Transportation : Conference on Poverty and Tansportation American Academy of Arts and Sciences," INK-in, J.F: et al., National Technical Information Service, 1968, p 25.
100. E.K. Morlok, "The Effects of Reduced Fares for the Elderly on Transit System Routes", The Transportation Centre, Northwestern University, 1971.
101. J.D. Nysteun, and M.S. Dacey, "A Graph Theory Interpretation of Nodal Regions", Papers and Proceedings of the Regional Science Association 2, 1961, pp 29-42.
102. H.L. Gauthier, "Least Cost Flows in a Capacited Network : A Brazilian, Example", In Horton, F. editor, Geographic Studies of Urban Transportation and Network Analysis, Department of Geography, Northwestern University, 1968, pp 102-27.
103. W.E. Reed, "Indirect Connectively and Hierarchies of Urban Dominance", *A.A A.G.*, 1970, pp 770-85.
104. F.P. Stutz, "Distance and Network Effects on Urban Social Travel

Fields", *Economic Geography* 49, 1973, pp 134-44.

105. B. Herbert, and E. Murphy, "Evolution ot an Accessibility Surface : the Case of the United States Air Network", proceedings of the Association of American Geographers 3, 1971, pp 75-79.
106. R.S. Lal, "Transport and Accessibility in Lower Ghaghara-Gandak Doab", Abstracts of Papers of the 21st International Geographical Union Congress, New Delhi, 1968.
107. F.R. Pitts, "A Graph Theoretic Approach to Historical Geography", *Professional Geographer*, 1965, pp 15-20.
108. N. Alao, "A Note on the Solution Matrix of a Network", in *Geographical Analysis* 2, 1970, pp 83-88.
109. B. Marchand, "Deformation of a Transportation Surface", *A.A.A.G.*, Vol. 63, 1973, No. 4, pp 507-521.
110. Thomas R. Lainbach, "The Spread of Transportation and its Impact upon the Modernisation of Malaya, 1887-1911", *The Journal of Tropical Geography*, Vol. 38, December, 1974, pp 54-62.
111. T.J. Wilbands, "Accessibility and Technological Change in Northern India", *A.A.A.G.*, Vol. 62, Sept. 1972, No. 3, pp 427-436.
112. R.A. Cybriwsky and K.C. Meyer, "Distribution of Elderly Blacks in Philadelphia : Some Research Findings", *The Pennsylvania Geographer* 11(4), 1973, pp 26-32.
113. M.D. Perle, "Urban Mobility Needs of the Handicapped : An Exploration", Geographic Studies of Urban Transportation and Network Analysis, Northwestern University, Dept. of Geography, *Studies in Geography*, 16, 1968, pp 20-41.
114. Leinbacks, op. cit, 1974.
115. B.S. Hoyle, "Transport and Economic Growth in Developing Countries", in *Transport and Development* Edited by B.S. Hoyle, Macmillan, London, 1973.
116. H. Hunter, "Transport in Soviet and Chinese Development", *Economic Development and Cultural Change*, Vol. 14, 1965, pp 71-84.
117. G.W. Wilson, "The Impact of Highway Investment on Development", Washington, D.C., Brookings Institution, 1966.
118. J.O. Wheeler, "An Overview of Research in Transportation Geography", East Lakes Geographer, Vol. 7, 1971, pp 3-12.
119. U.S. Bharadwaj, "The Poone City and its Economics of Transportation", For the M. Tech. Degree, Abstracts of the Theses, I.I.T., Kharagpur, Vol. 1, 1955-66.
120. J. Singh, "Transport Network Traffic Patterns and Location of Industries in the Iron and Steel Belt of India", Abstracts of Papers on the 21st International Geographical Union Congress, New Delhi, 1968.
121. W. Owen, "Distance and Development Transport and Communication in India", The Brookings Institution, Washington D.C., 1968.
122. B.S. Hoyle, *Transport and Development*, Macmillan, London, 1973.
123. A. Long, "Internal Transport Developments in East Africa", Geo-

graphy, *Journal of the Geographical Association* No. 226, Vol. L, Part I, Jan. 1965, pp 78-81.

124. S. Raichman, *Transportation and Urban Development in West Africa : A Rcview*, 1970.
125. M. Richard, "Technological Change in Plantation Transport Systems : A Caribbean Example 1930-70., in *The JI of Tropical Geography*, Vol. 41, December 1975, pp 1-8.
126. P. Peter Waller and S. Harvy Swain, "Changing Pattern of Oil Transportation and Refining in West Germany", *Economic Geography*, Vol. 43, 1967, pp 143-156.
127. W.L. Garrison, et al, *Studies of Highway Development and Geographic Change*, University of Washington Press, Seattle, 1959.
128. W.L. Garrison and M. Marts, "Influence of Highway Improvements on Urban Land : A Brief Summary", University of Washington, Seattle, 1958.
129. J.B. Kenyon, "Elements in Inter-Port Competition in the United States", *Economic Geography* Vol. 46, No. I, 1970, pp 1-24.
130. Thomas Graig, "A Behavioral Model of Model Selection", *Transport Journal* 12, 1973, pp 24-28.
131. W.R. Blunden, *The Landuse Transport System*, in Pergamon Press, Oxford, 1971.
132. M.E. Eliot Hurst, "Landuse Travel Movement Relationships", *Traffic Quarterly* Vol. 23, 1969, pp 263-274.
133. Ullman, op cit., 1959.
134. J.O. Wheeler, *The Urban Circulation Noase*, North Scituate, Mass; Duxbury., 1974, pp 137.
135. F.S. Roberts, "Weighted Digraph Models for the Assessment of Energy Use and Air Pollution in Transportation Systems", *Environment and Planning* Vol. 7, No. 6, Sept. 1975, pp 703-24.
136. S. Lin, "Computer Solution of the Travelling Salesman Problem", *Bell System Technical Journal*, 44, 1965, pp 2245-69.
137. G. Tornqvist et al; "Multiple Location Analysis", in Geography (Land Studies), Series C. *General and Mathematical Geography*, 12, Lund; Gleerup, 1971.
138. R.D. Mackinnon and M.J. Hodgson, "Optimal Transportation Networks : A Case Study of Highway System", *Environment and Planning*, 1970, pp 267-84.
139. C. Werner, "Networks of Minimum Length", *Canadian Geographer* 13, 1969, pp 549-67.
140. O.P. Sullivan, "Linear Programming as a Forecasting Device for Inter-Regional and Urban Economics, 1972, pp 383-96.
141. A. Lösch, *Die Raumliche Ordnung Der Witschaft*, Jena, 1944.
142. C. Kissling. "Linkage Importance in a Regional Highway Network", *Canadian Geographer*, 13, 1969, pp 113-27.
143. H. Sen, "The Geometric Structure of an Optimal Transport Network in a Limited City Hinterland Case", *Geographical Analysis* 3, 1971, pp 1-14.

144. J. Kolars and H. Malin, "Population and Accessibility: An Analysis of Turkish Railroads", *Geographical Review* 60, 1970, pp 229-46.
145. D.G. Janelle, "Spatial Reorganisation : A Model and Concept", *Annals of the Association of American Geographers*, Vol. 59, No. 2, 1969, pp 348-364.

6

# Marketing Geography—Concepts and Trends

*—H. M. Saxena*

THE GEOGRAPHICAL study of markets, market places and marketing is a result of increasing specialization in the various branches of geography. The need for marketing geography like agriculture, industrial and transport geography, etc. was first pointed out by William Applebaum. He argued, "the study of marketing had been neglected by geographers up to that time, in spite of a large section of working population engaged in marketing functions, the large part of urban landscape devoted to structures of wholesale and retail trade, and the complex channels of distribution leading from producing to consuming areas".[1] With the growing interest among geographers it was claimed during the sixty's of the present century that marketing geography had "come of age". Its separate status was recognised by IGU only in 1972 at the time of 22nd IGU Congress at Montreal, when a working-group on "Market Distribution Systems" (now Market Place Exchange Systems) was constituted and its first symposium was held in Moscow at the time of the 23rd IGU Congress in July, 1976.

The objective of the present paper is to explain the developing concepts and trends in the field of marketing geography.

## SCOPE OF MARKETING GEOGRAPHY

Marketing geography deals with markets and market places as units of space and with the process of marketing. A market is an authorised public gathering of buyers and sellers of commodities meeting at an appointed place either daily or at regular intervals. While market places are sites with social, economic, cultural and other referents where there are a number of buyers and sellers, and where prices offered and paid by each are affected by the decisions of the others.[2] On the other hand, marketing is a process or a function of a market centre which involves three sets of inter-related activities of production, consumption and exchange.

It is rightly observed by McCarty and Lindberg: "Geographical studies dealing with trading activities usually appear under the rubric of marketing geography, whose concern has been almost exclusively with the location of retail and wholesale establishments, their sales, facilities, market territories and similar facets of the trading institutions".[3] In brief, "marketing geography is concerned with the delimitation and measurement of markets and with the channels of distribution through which goods move from producer to consumer".[4]

## CONCEPTUAL FRAMEWORK AND TRENDS

Geographical phenomena are described through the varied nature of postulates and concepts which ultimately act as the basic postulates for theory building. In spite of rapid development in geographical literature, geographers have not developed a definite set of concepts for explanation. Therefore, there is an urgent need to develop a conceptual framework for scientific geographical study.

As pointed out earlier, marketing geography is a developing branch of economic geography and requires a framework of concepts not only to explain its nature but also to provide a solid base for further research or in other words for the development of hypothesis and theories. Some developing concepts in marketing geography can be listed as follows:

### Spatial Interaction

The study of markets and marketing required both spatial as well as temporal analysis and their relationships. Markets, which are also centres of population and economic activity are elements of landscape thus form a spatial organization and their spatial behaviour requires explanation in marketing geography because markets have (a) a regularity in size and spacing, (b) a tendency for concentration of activities to take advantage of location, and (c) a degree of connectivity which forms a base for spatial interaction.

The spatial arrangement of markets requires analysis of the distributional pattern in their regional setting. Such a distribution pattern of markets can be described by either nearest neighbour analysis or by Chi-square method. By adopting these methods one can explain the distributional pattern both quantitatively as well as cartographically.

Since marketing activity is dominantly performed by urban centres, the models used by urban geographers have also been adopted by marketing geographers. These models are used to explain the linear, cluster and a uniform pattern of spatial distribution. In this connection models developed by Christaller[5], Lösch[6], Isard[7], and Bunges[8] are notable. Prof. Berry feels that the geography of retail and service business displays regularities over space and through time, that the central place theory constitutes a deductive base from which to understand these regularities, and that the convergence of theoretical postulates and empirical regularities provides substance to marketing geography.[9]

**Hierarchy**

The concept of hierarchy is very common in geographical study and widely used for determining the categories. In marketing geography it simply means ranking of market centres into successive groups or orders on the basis of a single or a set of variables. It is a fact that in every region there are fewer larger markets than medium and smaller ones and their characteristics can be explained properly when hierarchical orders are determined. Such hierarchical order is fundamental to central place models of Christaller and Lösch. Similarly Berry and Garrison[10] demonstrated how hierarchies arise on the basis of "threshold".

Market hierarchies can be determined on the basis of threshold size, number of retail establishments, size of the market's sphere of influence, percentage of population engaged in trade and commerce activity, etc. The pattern of market hierarchy that emerged out of a detailed analysis will provide a base for planning the market centres by avoiding duplication and wastage in inputs. In brief, the spatial form of the marketing process consists of a hierarchy of market centres and their trade areas. Each level of hierarchy contains a set of uniform size of market centres, and are spatially arranged on the plain either in a triangular or in a hexagonal pattern.

The concept of "rank-size rule" developed by G.K. Zipf[11] for explaining the rank and size of the urban centres is also applicable in market study for determining hierarch on the basis of rank-size.

**Morphometric Analysis**

Morphometric analysis provides a framework in which shapes and forms in space are examined by geographers. The morphometric analysis in marketing geography is, although a part of urban morphology considerably different from it in the sense that urban morphology deals with all aspects of morphological landuse, while marketing geography deals only with the commercial landuse, i.e., land occupied by retail and wholesale traders, financial establishments, insurance, goods storage etc., thus providing an intensive base for analysis and planning.

The morphological pattern of a trade centre is a complex areal phenomena which consists of the features of market places. In a classic topology Proudfoot[12] describes the five components of commercial structure of the principal cities of the U.S.A. Brown and Davidson[13] also offer a similar fivefold type of market morphology. Other contributors in this field are Hoyt[14], Kelley[15], Berry[16], Alam[17], Gopal Krishan and Agarwal[18] and Saxena[19].

A classification of the components of market morphology is suggested below, for a detailed analysis specially for the markets in developing countries.

*Typology of the Components of Market Morphology*

| Primary Components | Secondary Components |
|---|---|
| I. Central Business Area. | (a) Mixed Retail Zone.<br>(b) Wholesale Zone.<br>(c) Regional Centres. |
| II. Secondary Business Area. | (a) Retail shops.<br>(b) Wholesale shops. |

| | |
|---|---|
| III. Principal Business Arteries. | (a) Retail Strings. |
| | (b) Service-cum-saleshops. |
| | (c) Service shops. |
| IV. Connecting Business Streets. | (a) Grocers' Row. |
| | (b) Furniture Row. |
| | (c) Vegetables & Fruit Row. |
| | (d) Automobile Repair Row. |
| | (e) Other specialised Rows. |
| V. Isolated Clusters. | (a) Commodity Stores. |
| | (b) Service Stores. |
| | (c) Mixed Stores. |

**Consumer Behaviour**

In marketing geography consumer behaviour is studied as a pattern that links household units to retail structure. It also requires a detailed study of "least-effort shopping trip behaviour" i.e. behaviour in which consumers patronize the *nearest* place offering the goods or services. The above hypothesis is not a universal one, since shopping trip behaviour is not always based on least effort.

Geographers have developed several models in their study of consumer travel behaviour. The models developed by Lowry[20], Harris[21] and Harvey[22] are notable. Similarly Reilly's law of retail gravitation also explains the "breaking point", "approximately in direct proportion to the population of two cities and in inverse proportion to square of the distances from the two cities to intermediate town".[23] Some of the other models are developed by Baumol and Ide[24], Nystuen[25], Marble[26] and Huff[27] : There is a need to test these models in various types of economy of scales.

**The Location of Retailing**

The concept of retail location is a prime one when we study the marketing pattern of urban centres. Retail locations in towns are the product of (i) the density of demand for a particular product or service, and (ii) the market demand of a particular product or service. In brief, retail location is directly related with (i) size of the urban place, (ii) magnitude of its trade area, (iii) population of sales area, and (iv) distance to a place offering a similar number of functions.

**Market Area**

A market centre cannot live in isolation. The origin of markets lies in the surplus of the countryside which allowed special activities other than growing food. Market area or sphere of influence is that territory from which it draws its customers or in which it sells its products. One of the significant aspects of marketing geography is to trace the movement of goods and people between markets and their countryside. The measurement of market area "is the basis of presenting marketing data, evaluating store or shopping centre performance, studying changes due to new competition, or estimating the locational advantages and potential earning of new sites for stores or centres."[28]

Several methods have been adopted by geographers to delimit the market area, some of them are based on traffic flow, while others are on population and consumer behaviour. Empirical methods based on detailed field work are still important in which spotting of the consumer origin place is required for mapping and analysis. In fact, market areas are "nodal regions" in which the surroundings are tied to the focus by lines of transportation.

**Periodicity**

"Periodicity is an essential element of local indigenous market structure of most under-developed countries as it was of medieval Europe.[29] Periodic markets provide a mechanism for satisfying the limited demands for goods and services of the mass of rural consumers with self-dependent economy." The market functions are performed by mobile agents moving from place to place. Such spatial behaviour of individual or aggregate suppliers gives rise to the weekly market or the fair.[30] These "periodic markets play a key role in the internal trading process, mediating exchanges between farmers, craft manufacturers, forestallers, and itinerant traders on the one hand, and urban wholesalers on the other hand."[31]

The literature dealing with the various aspects of periodic markets (weekly market and fair) is a voluminous one. For a detailed bibliography one may consult the bibliography prepared by R.J. Bromley.[32] Some of the concepts developed in the geographical study of the periodic markets are (i) the concept of temporal periodicity, (ii) the concept of locational spacing, and (iii) the concept of market cycles and marketing rings. All the above concept of the periodic marketing require detailed analysis specially in developing countries.

## CONCLUSIONS

The analysis of the important concepts indicate that there has been a remarkable growth of interest among geographers to study periodic and daily markets, resulting in a separate field of study known as "marketing geography". An inter-disciplinary approach is necessary to handle the problems of research in marketing because the subject has received considerable attention from economists, sociologists, anthropologists and, of course, geographers.

The conceptual framework as discussed in the paper is still a suggestive pattern because its theoretical base is also in a developing stage. The various concepts described in the paper i.e. concept of spatial interaction, hierarchy, morphometric analysis, consumer behaviour, market area and periodicity will definitely provide a base for research in the field of marketing geography.

## REFERENCES

1. W. Applebaum, "Marketing Geography" *American Geography: Inventory and Prospects*, Syracuse Univ., 1954, pp 245-51.
2. C.S. Belshaw, *Traditional Exchange and Modern Markets*, Prentice Hall, 1965, p 8.
3. H.H. McCarty and J.B. Lindberg, *A Preface to Economic Geography*, Prentice Hall, 1966, pp 133-34.
4. W. Applebaum, "Marketing Geography" *American Geography: Inventory and Prospects*, Syracuse Univ., 1954, p 246.
5. W. Christaller, *Central Places in Southern Germany*, Englewood Cliffs, New Jersey, (Translated by C.W. Baskin), 1966.
6. A. Lösch, *The Economics of Location*, New Haven (Trans. by W.H. Wogeom and W.F. Stoper) 1964.
7. W. Isard, *Location and Space-economy*, New York, 1956.
8. W. Bunge, "Theoretical Geography"; Lund Studies in Geography, Series C, *General and Mathematical Geography, I* 1962.
9. B.J.L. Berry, *Geography of Market Centres and Retail Distribution*, Prentice Hall, 1967, p vii.

10. B.J.L. Berry and W. Garrison, Recent developments of Central Place Theory, Regional Science Association, Papers and Proceedings, 4, pp 107-120.
11. G.K. Zipf, *Human Behaviour and Principle of Least Effort*, New York, Addison, Wesley Press, 1949.
12. M.J. Proudfoot, "City Retail Structure", *Eco. Geog.*, 1937, 13 (4).
13. P.L. Brown and W.P. Davidson, *Retailing Principles and Practices*, Ronald Press, New York, 1953, p 75.
14. H. Hoyt, Classification and Significant Characteristics of Shopping Centres, *Appraisal Journal*, April, 1958.
15. E.J. Kelley, Retail Structure of Urban Economy, *Traffic Quarterly*, Vol. 9, 1955, pp 411-30.
16. B.J.L. Berry, "Commercial Structure and Commercial Blight", Deptt. of Geo. Research Pap. No. 85, University of Chicago, 1963.
17. S.M. Alam, "Commercial Structure of the Cities of Hyderabad and Secunderabad", *Journal of Osmania University*—Arts, Commerce, Law and Educations, 1963.
18. Gopal Krishan and S.K. Aggarwal, "Commercial Pattern of a Planned City—Chandigarh", *NGJI*, Vol. XIX, 1973.
19. H.M. Saxena, "Kota.—A Study in Market Morphology", *Geog. Review of India*, Vol. 36, No. 1, 1974.
20. I.S. Lowry, "A Short Course in Model Design", *Journal of the American Institute of Planners*, 1965, 31, pp 158-166.
21. B. Harris, "Quantitative Models in Urban Development : Their role in Metropolitan Policy Making" in H.S. Perloff and L. Wingo, Jr. (Eds.) *Issues in Urban Economics*, Baltimore, 1968, pp 363-412.
22. D. Harrey, *Explanation in Geography*, 1969, London : Edward Arnold.
23. W.J. Railly, *The Law of Retail Gravitation*, New York, 1931.
24. W.J. Baumol, and E.A. Ide, *Variety in Retailing, Management Science*, 1957, 3, pp 93-101.
25. J.D. Nystuen, "A Theory and Simulation of Intra-Urban Travel," in W.L. Garrison and D.F. Marbal (Eds.), *Quantitative Univ. Studies in Geog.* Evanston, No. 13, 1967, pp 54-83.
26. D.F. Marble, "A Theoretical Explanation of Individual Travel Behaviour," in W.L. Garrison and D.F.Marble (Eds.) *Quantitative Geography*, Part I, 1967, pp 33-53.
27. D.L. Huff, A Topographical Model of Consumer Space Preference, Papers and Proceedings of Regional Science Association, 1960, pp 154-173.
28. B.J.L. Berry, *Geography of Market Centres and Retail Distribution* Prentice Hall, 1967, p. 127.
29. B.W. Hodder, "Distribution of Markets in Yorubaland", *Scottish Geographical Magazine*, Vol. 81, No. 1, April, 1965, p 48.
30. B.I. Garner, "Models of Urban Geography and Settlement Location," in Chorley and Haggett (eds.), *Models in Geography*, Methuen, 1967, p 320.

31. R.H.T. Smith, "West African Market Places : Temporal Periodicity and Locational Spacing," in Cloude Meillassoux (ed.), *The Development of Indigenous Trade Markets in West Africa*, p 319.
32. R.J. Bromley, *Periodic Markets, Daily Markets and Fairs : A Bibliography*, Monash Publications in Geog. No. 10, 1974.

# 7

# The Geography of Retailing—A Conceptual Review

—*A. Sami*

## INTRODUCTION

THE STUDY of intra-city market centres is a comparatively recent outcome of Urban Geography, which has not attracted much attention of social scientists in India so far. An urban centre is generally studied from all aspects of urban landuse. But now the retail nucleations in towns, cities or metropolitan centres have proved to be of greater importance in all parts of the earth in urban life. Shopping activity is the nerve centre of most of the cities of the world, and hence the study of this particular aspect deserves greater attention and emphasis by urban geographers, especially in India, where urbanization is developing in its full swing. In each census year, new towns and cities are added to the old list with greater potentiality, of commercial activity in them. At the same time the existing towns witness an unprecedented expansion of their commercial activities leading to a diversification of their functional character.

The purpose of this paper is to illustrate the urban centres with special reference to the significance of retailing including factors of spatial development, hierarchy of functions, hierarchy of market centres, classification of functions, land utilization and rent values.

## SIGNIFICANCE OF MARKET CENTRES IN URBAN AREAS

Every urban centre, large or small, is in some degree the headquarters of traders and commercial institutions, as there are provisions for goods and services for tributary areas. A city is characterised by the greater variety and quality of its commercial services and act as a centre for collecting and marketing of the products of the surrounding areas. There are towns which are established only for commercial purposes, e.g., the market where the country folk come to exchange their produce and the port where merchandise from abroad is collected. But even towns in which commerce is not the main activity, such as industrial or administrative centres, become commercial in due course. This is because market develop there for the convenience and well-being of the local inhabitants. Grocers, drapers and general merchandise vendors set themselves up for the convenience of these inhabitants, who could not do without them.

The geography of the retail markets is concerned with the market as a fixed point on the earth surface. The marketing function takes place at specific centres distributed in an area with facilities for collecting and distribution of goods as well as availability of services. Geographically, it is a well defined area in human habitations varying from a petty village to a metropolitan centre. In the present context market or commercial centre is a place for the sale and purchase of different types of commodities generally brought from outside the town and sometimes produced locally. There are world markets for cotton, rice, wheat, jute, tea and coffee. Most types of exchange, however, involve a specific market place. The Chicago wheat pit and London metal exchange are dramatic examples of market places of international importance, yet equally important are the thousands of small market centres that are spread all over the world. Thus market centres are of different orders, grades or ranks which vary from the rural market of a hamlet to the urban market of a metropolitan centre.

The village market must have been the first form of market. The village economy of the Middle Ages could not be entirely closed, certain levies had to be paid in kind, and so certain products were commercialised. And in this way market towns were born. These often come to life only once a month or perhaps once a week. Markets were developed in contact with different regions which traded their goods, e.g., towns of the voyages were built at the junction of sandstone and limestone regions. Gradually market places developed at trading posts, at different sea, water, road and railway traffic spots and sometimes as a commercial centre in the form of fairs. In the olden days, large scale trade was principally carried on at fairs. The fairs temporarily transformed small local markets into important trading centres. Smallest market centres or major metropolitan centres not only store specialities but also wholesale, the products they have assembled to relatively smaller surrounding urban places, which ship them to even smaller places or the market centres where consumers visit to purchase the goods and services they need from retail business.

The essence of the geography of commercial centres is the clustering of establishments with a wide range of units. The relative importance of town and city is to be judged on the basis of types and quality of goods available there. It is obvious that the larger centre will have more establishments and business types, offer more goods and services and do greater volume of business than small centres. Small centres generally provide only goods such as foodstuffs which are purchased frequently with minimum consumer travel.

Retail shops are the last link in the chain of activities by which goods are transferred to the consumers. Shops to be successful, require good contact with the public and are highly sensitive in location. The importance of shops as distributive organs can be measured by the rent they manage to pay. Similarly, since shops prefer level streets of moderate width, the chief shopping street coincides with chief pedestrian thoroughfares in a town or city.

## SPATIAL DEVELOPMENT OF THE SCIENCE OF RETAIL CENTRES

The geography of commercial centres carries the retail interest of the geographer into practice within metropolitan areas. The national economy is essentially related to the process of marketing and market places where this phenomenon actually takes place. In Britain, for example where retail sales account for more than one-half of total consumer expenditure, employment in the retail trade is the largest component in many British towns.[1]

The importance and need of geography of retailing was initially spelled out by its outstanding practitioner, William Applebaum[2] who pointed out that the well developed topical field of economic geography was at that time, with the exception of transportation geography concerned with the production of material goods. But goods must not only be transported from production to consumption areas, they must also be transferred from the hands of producers, by collection and subsequent distribution, into the hands of consumers. This is the function of marketing. He argued that the study of marketing had been neglected by geographers up to that time, in spite of the large section of working population engaged in marketing functions, the large part of urban landscape devoted to structure of wholesale and retail trade and the complex channels of distribution leading from producing to consuming areas. The objective of this field should be the delimitation and measurement of markets and study of channels of distribution through which goods move from producers to consumers.

About ten years later, it was claimed in the introduction to a collection of papers that marketing geography had "come of age". The papers had several characteristics in common, and were concerned with actual business operations of firms. This dealt with the problems of selecting new store location based on measurement of market areas, present and future, and sale potentials. Much of the emphasis was on how to conduct such studies, leading to better store location science. Store-site evaluation and store location research, however, were recognized as being more of an art than a science.

## SCOPE OF MARKET GEOGRAPHY

Geographers have paid little attention to this subject and most of the early literatures was published by economists and people in allied disciplines. Schwartz, pointed out that the study of marketing geography had taken place approximately in A.D. 1900 when this study produced a "Plethora of facts that seem to add up to very little."[3] In the aftermath, a body of marketing theory emerged so that one could in fact begin to talk of a marketing science. Marketing geographers realises the drawbacks in their work, after Applebaum's 1954 plea, within the framework of marketing science. It is surprising that the geographers did not pay adequate attention to the pioneering work done by Christaller and extension by Lösch who presented theoretical models for delineating trade areas, selecting retail location and evaluating markets. This work had been there within geography for at least twenty years before Applebaum prepared his paper on marketing geography.[4]

## OVERVIEW OF LITERATURE

In spite of the importance of retail business in geographical studies, it has remained until recently a comparatively neglected field of research. Commenting on the economics of retailing, Henry Smith said that the economists seemed to think that retail trade was unworthy of their attention in comparison with other scholarly subjects. This view seems to have been shared by geographers and sociologists as well.

During the twentyfive years, the contributions in this field came mostly from town planners, and geographers continued to neglect various aspects of spatial analysis of retailing. More recently (1933) Christaller's theory of central places was a significant contribution[5] which was brought to the attention of the English-speaking world by Ullman.[6] This was followed by the praiseworthy attempt by Berry and Pred[7], to catalogue and analyse the literature available on the subject which testified to the interest aroused in the subject.

The theory advanced by Lösch (1954) is at variance with Christaller.[8] According to Lösch, the centres of similar size do not necessarily perform the same function, and the size of a centre does not depend upon the size of its market area. He suggests that centres ranked by size may sometimes appear to be a continuum and may not necessarily form a hexagonal distribution with uniform size of market areas. Berry and Garrison[9] have presented a modification of the central place theory of Christaller in which they have developed a hierarchical structure of markets without assuming uniform distribution of purchasing power so necessary for the creation of the hexagonal patterns.

The balance of opinion among scholars seemed to support the concept of hierarchy in central places. But there can be areas where the stepped hierarchical spatial order may not be distinguished, and the pattern may conform, at least partly, to the conclusion of Lösch. For example, Schell[10] has found that higher order centres are not necessarily more complex in functions than lower order centres, and that market areas need not increase with increase in functional complexity or form hexagonal patterns. Johnston[11] has also found that there are marked areal variations in the market size of each class of centres. Kenyon[12] asserts that in his study he has found scant evidence of descrete size of groupings. All these seem to lean towards a modification of the Löschian system.

The central place theory also fails to take into account the importance of the rent theory which determines the internal site and location of a particular function within a given market.

Different aspects of geography of retailing have been studied by different scholars in their respective study areas, while reviewing the literature, however, only a few aspect of the study of different scholars have been mentioned here.

## IDENTIFICATION AND DELIMITATION OF RETAIL CENTRES

"Market Geography carries the retail interest of the geographer into practice within metropolitan areas, in the service of private business enterprises."[13] The very preliminary task before the market geographer is to demarcate and delimit the different shopping centres within the metropolitan area of city. Researchers have relied upon self perception and extensive field work. However, delimitation of shopping centres should be done on a uniform basis. The early studies of central places and shopping centre systems seem to have avoided this problem. The 1961 Census of Britain defined the limit of a central shopping area as the point "where the ratio of shops to all properties fell below one in three" (Leeming, 1959).[14] Lomas describes this as "a fairly liberal definition"[15] (Lomas, 1964), but Davies used "a slightly broader definition", a shop being included in a centre "if it lay within a distance equivalent to the frontage of four terraced houses"[16] (Davies, 1967). A notable and precise definition is a "contiguous group of retail stores, usually more than four or five in number, having at least two kinds of retail types."[17]

All the above definitions have their limitations, and not even a single of them is applicable to commercial ribbons.

Such development needs to be systematically surveyed, and none of the definitions above provide a means. Similarly, major shopping streets are often separated from each other by non-retail landuse and no solution has been offered to this problem either.

A number of other experts interested in urban geography have contributed towards the delimitation of commercial activity within a city. Notable among them are Lomas (1964), Murphy and Vance,[18] Mattingly[19] and Bohnert and Mattingly.[20] But their studies deal with the delimitation of Central Business District (CBD) only.

Though the Murphy and Vance technique has a fairly general acceptance when applied to medium-sized grid iron cities in newer western countries, it cannot be used without arbitrary modification for cross-cultural comparison, historical analysis and for smaller, particularly one street centres.

Any technique for delimiting a commercial centre or its core will depend on the circumstances of its use; and without deep testing in relation to the specific condition, it is not possible to say which method is most appropriate.

## HIERARCHY OF COMMERCIAL CENTRES

Many geographers have established the hierarchy of market centres or business centres in their areas of study. However, the majority of the work has been done in the field of regional geography. The study of retail centres within a city is a relatively newer branch of the subject and has a recent history. Among the early descriptions of the business structure of a city, the work of Proudfoot is supposed to be the pioneer literature. He discovered on the basis of detailed study of the cities of Chicago, Philadelphia, Cleveland, Atlanta and Des Moines, that the majority of the cities of the U.S.A. possess five types of retail structure : (a) the central business district; (b) the outlying business centres; (c) the principal business thoroughfare; (d) the neighbourhood business street; and (e) the isolated store cluster.[21] Although this classification has a sound base, Proudfoot has nevertheless ignored the importance of the number of shops at different shopping centres, as well as the range of functions available.

Carol distinguished three major levels of market centres (Central Places) : (a) Neighbourhood business district, (b) regional business district, and (c) the Central business district in the city of Zurich. He classified the market centres on the basis of two factors : (a) inventory of the amount and variety of goods and services, and (b) extent of the service areas. He found that only the common or convenience goods and services were available at neighbourhood centres, a slightly wider variety being found in regional level centres and exclusive choice of goods being offered only in the central business district located in the heart of the city.[22]

It is logical to assume that the number of functional units would also increase with a rise in the rank of a centre, but this fact seems to have been totally ignored by Carol in his study which takes into account functional types only.

Johnston used type of establishment and number of units as the chief criteria while classifying shopping centres in the Melbourne Metropolitan area. According to Johnston : "Grouping of centres into hierarchy when two variables are known is commonly achieved by graphical analysis. However, when the 718 centres in this study are plotted on a graph with number of functions and number of establishments on the axes they form an almost unbroken continuum of centres by size" (Johnston, 1966). The hierarchical groups identified by him, however, are purely subjective as admitted by Johnston himself.

A hierarchy of retail centres was devised by Simmons in the study of growth and blight of Toronto. He used the term in progressive order as neighbourhood, community, shopping goods, regional and super regional (C.B.D). The order suggested by him is also mainly based on functional types rather than number of stores.[23]

Berry has applied factor analysis[24] of ten variables to obtain a hierarchical pattern in the city of Chicago. On this basis four classes of retail centres, major regions, other shopping goods, community and neighbourhood centres have emerged.

Garner has identified a hierarchy of business centres in the city of Chicago.[25] He utilised mainly two variables : number of shops and type of shops. A long linear relationship between the number of establishments and the number of business types distinguished three centres designated as regional, community and neighbourhood.

Calcutta, one of the important cities of India and a trade centre, has been studied by Dutta from a commercial point of view. He adopted a methodology for identifying different levels of hierarchy which is based on two factors: number of customers served by the central place (market centre) and their catchment area. By adopting a formula, he finally finds out the composite hierarchy level (CHL) for each market centre except New Market which he maintains is actually CBD of Calcutta and has almost one-half of the city as its tributary area. He plotted the CHL values of all market horizontally without any vertical scale, and distinguished three levels of hierarchy on the basis of clustering of the CHL values designated II, III and IV respectively (I position was given to New Market).[26]

Commercial activities of Hyderabad and Secunderabad have been ranked into four hierarchical levels by Alam.[27] He called them central business district, local, neighbourhood and sectional markets. The classification achieved by Alam is mainly based on types of functions and quality of goods available at different levels of centres. He ignored the number of functional units which is bound to be in large numbers in the higher level commercial centres.

## COMMERCIAL FUNCTIONS AND THEIR CLASSIFICATION

A large number of geographers have studied the commercial structure and functional characteristics of a city. But so far as classification of functions is concerned the number of functional types recognised by each of them are totally different. This is because of the fact that commercial functions generally differ from place to place; while large city can have a variety of functions, a small town has relatively less functions. It is also noticeable that the functional list presented by different scholars follow mainly their purpose of study. During the study of the structure of retail trade of a Swedish town, Bengtsson selected only eleven types of business.[28] But Vance while comparing the commercial structure of an American city in 1929 and 1954, listed 55 types of establishments.[29] Similarly, Carol selected only 17 central functions to investigate shopping habits of people and how consumers make use of the central function of various levels of the business district. Ratcliff, in the course of the study of internal arrangement of landuse of Philadelphia, classified the commercial functions into ten generalised groups.[30] On the other hand, Berry and Garrison have selected a more detailed list, i.e., 52 types of business.[31] Arther Getis and Judith Getis presented 19 groups of store types based on three digits of standard industrial classifications to find out the spatial associations of retail stores within business centres of a few cities.[32]

This difference in the list of business types in different studies is due either to the purpose of study in which some geographers believe in generalisation and have taken broad and major functional groups or in some cases due to the availability of some and non-availability of certain other functions in different cities. Johnston while determining the intra-city central place hierarchy in the city of Melbourne (Australia), identified 91 types of business with the help of Sand and McDougal's Directory of Victoria, 1962. Simmons listed 135 types of retail units on the basis of Standard Industrial Classification which have been grouped into 11 broad categories during the study of Blight of Toronto. Similarly, Garner recorded 64 types of functions in the study of internal structure of Chicago, based on standard industrial classification.

The time period is also one of the important factor responsible for the difference of business types in different cities. Berry is also of the same opinion when he says there is difference in the time period between the two studies (between Spokane and Cedar Rapids). In the eight years between the studies, variety store, meat markets, fruit and vegetable stores have been diminishing in significance and the super-market has spread rapidly.[33]

So far as the classification of business types adopted by Indian Geographers is concerned, they have used a broad classification and have categorised the functions according to availability of business in their own study areas or according to the purpose of study. Very little work has so far been done on the commercial structure of a city in India.

Venugopal identified only 6 major categories of shops during the study of commercial structure of Warangal City.[34]

This is a very generalised grouping. Thakur has selected 21 types of business in the city of Patna in the course of delimiting the CBD of Patna.[35] His classification is also generalised which may be useful for the delimitation of CBD only but is unsatisfactory for a detailed study of functional structure of the smallar market centres of the city. Singh has identified about 40 types of functions in Kanpur city.[36]

Alam and Khan have made the most valuable contribution in the field of commercial structure of a metropolitan centre. They have listed fifteen functional types for Hyderabad-Secunderabad (twin cities) on the basis of detailed field survey.[37]

A very small list of functions results in broad groupings which fail to bring out the variations between small retail centres. Whenever a generalised group of functions is taken into account, the speciality or individuality or peculiarity of a single market would no longer exist. The typical function of the centre will merge into the broader group of functions, which obviously has no importance for that particular shopping centre. In the present study a detailed grouping of functional types has been preferred as a basis for the analysis of functional character of commercial centres at different levels of hierarchy.

## FUNCTIONAL HIERARCHY AND MARKET TYPES

On an intra-city level, Berry and Garrison and Vance have established a ranking of business types in their respective study areas. Vance while comparing the commercial structure of the American city in 1929 and 1954, ranked the retail trade on the basis of total number of each type of establishment. But his ranking of functions is unsatisfactory because, according to his ranking list, food, drugs and hardware stores appear in the higher order group, and office-store, machine equipment, department store, camera and radio and T.V. store are listed in the lower order. Similarly fruit and vegetables, second-hand goods, men's and boy's clothes and jewellery are placed in the middle order. It is obvious that food stores cannot be of higher order than jewellery; on the other hand, office equipment cannot be lower in rank than hardware.

Berry and Garrison have also established a ranking of functions on the basis of population and establishment ratio,[38] which is not applicable in an intra-city study.

The method of Garner for ranking of functions for the city of Chicago seems to be convincing because, "Since the number of establishments of each business type decrease as threshold size increases, the lowest threshold type will at the same time be the most ubiquitous in the urban area." Thus, following Garner, the ranking of functions in the present study has been determined on the frequency of its occurrence at different retail centres.

Because of the difference in classification used in this study, it is not possible to compare the present rank order of functions to that recognised by Garner. However, it is broadly found in both studies that the convenience goods (mostly food articles) are placed in the lower order group; and the shopping goods are listed in the list of higher order functions. But a marked difference is observed in a few functions. Thus in Garner's study, financial services like insurance and bank are placed in middle and lower order groups respectively, but in the present study they have a high rank. Similarly, in this study bakery has been identified as a higher order function but in Garner's work, it is listed in the lower order function.

Alam and Khan have also ranked the goods in the city of Hyderabad and Secunderabad into three orders, higher, middle and lower on the basis of frequency by which goods of various order are sought by the consumers.

Alam's method seems to be logical and convincing and his schemes is not much different from the present study except that in Patna the jewellery shop is classified as middle order function, while in Hyderabad-Secunderabad it is placed in the higher order group. Such small changes are bound to occur in all studies.

The functional characteristics of different levels of markets have been widely analysed by a number of urban geographers in their respective study area.

According to Rolph who has identified four levels of retail centres in the city of Baltimore : (1) the Central Business District is characterised by general merchandise, apparel, jewellery, furniture and household facilities; (2) 'retail subcentres' are signified by food and apparel group; (3) 'string streets' are observed to be highly specialised in having traffic arteries, i.e., automobile, besides furniture and household, general merchandise, and apparel groups; and (4) 'Neighbourhood' centres are facilitated by mainly food, automobile, lumber and building materials.[39] This scheme has been criticised by Berry and Garrison on the ground that "the precise functional character of centres was marked by the higher aggregative groups of business types utilized".[40]

Proudfoot on the basis of study of five cities, i.e., Chicago, Philadelphia, Cleveland, Atlanta and Des Moines as well as a reconnaissance study of Washington, New York, Baltimore, and Knoxville, has identified five levels of commercial centres within a city and has described the functional structure of each. To him central business district is the retail heart of the city with complete range of shops including the most specialised urban functions, such as cigar, department store, drug, jewellery, men's clothing, variety and women's clothing shops. The next outlying business centres are considered as smaller form of the central business district and are generally located at the main point of the intra-city transportation system. They draw customers from a wide area and have business establishments of mainly apparel accessories, cigar, department store, drug, fancy goods, grocery, general merchandise, household appliance, jewellery, men's clothing, millinery, shoe, variety and women's clothing shops. The third type of centre identified by Proudfoot was principal business thoroughfare possessed of widely spaced convenience and shopping goods, such as auto accessories, tyre and battery, farmer's produce, filling station, milk depot, new cars, used cars and grocery. The fourth centre of retail structure was the neighbourhood business street where convenience goods dominate, e.g., bakery and catering, combination grocery and meat, drug, fancy, grocery, meatmaker and variety shops. The isolated store clusters are the final and individually the least significant type of retail centres. These centres usually comprise two or more complementary rather than competitive convenience goods such as drug store, grocery store, a meat market, or a fruit and vegetable stores.

Canoyers is of the opinion that at the central business district all types of goods are sold while in other lower order centres the business is generally restricted to convenience and service goods, mostly dependent upon the immediate neighbourhood residents for support.

Garrison has studied a portion of the metropolitan area and has classified the business centres into different levels. The community business centres of Garrison have a concentration of business of food, cleaning and laundry, eating, medical, apparel, fuel and automobile facilities. The major neighbourhood centres have food, fuel, cleaning and laundry, furniture, household, radio sales, automobile service, beauty and barber shops. The minor neighbourbood business centres possess mainly food, eating, fuel, beauty, barber, cleaning and auto service establishments. The principal business thoroughfares enjoy the facilities of fuel, automobile service, eating, building materials, recreational services, auto dealers and beverage sales. The isolated business establishments generally have food stores which are usually located with respect to local access streets. Moreover building material, moving and storage facilities are also associated with isolated business establishments.

Berry and Garrison while analysing the business types of Phoenix city describe the functional characteristics of different levels of market. According to them Phoenix centres fall into two levels. Two lower level includes certain lower business types such as grocery, barber, cleaner and laundry, drug, beauty, clothing, variety and bakery. Jewellery, department store, bank, restaurant, shoe repair, furniture, appliance, gift and novelty are associated with the higher level planned nucleations.

Carol has distinguished three major levels of markets. They are central business district, regional business district and neighbourhood business district. According to him each higher order of market was found to have all functions represented in the lower as well as special functions of its own.

Simmons identified five levels of markets in Toronto on the basis of functional characteristics at each level. Markets of "super-regional level" possess two or more department stores; those of "regional level" have one department store and shopping goods; "shopping goods" level is recognised by having one or two chain variety and shopping goods store; "community level" is characterised by chain variety plus some shopping goods store; and the "neighbourhood" market is identified by the presence of super-market.

Garner is of the opinion that lower order retail centres offer only low level goods (convenience goods) while higher order retail centres not only have higher order goods (shopping goods) but also have all order goods.

Alam has identified four types of business centres in the city of Hyderabad and Secunderabad. Functional characteristics of each business centre clearly reveal that CBD includes all important functions, e.g., banking, administration as well as wholesale while the local business centres dealt only in the convenience goods. Neighbourhood centres consist of sweetmeat, general merchandise, ready-made dress, bakery coupled with a restaurant, and a maternity and general clinic. Besides these functions, Alam includes all the municipal markets (twenty-one, mostly of vegetables) in this group. Sectional business centres, according to him "performed on a minor scale, the same services for their market which the central business district does for the whole city and the surrounding region".

## COMMERCIAL LAND VALUES AND RENT

It has been found by several research workers that the Central Business Districts of almost all cities in the world are characterised by the highest land values in the whole city. Most of the urban geographers have arrived at a common understanding that the CBD always enjoy the highest land values. The basic cause of this peaking tendency of land values in the CBD is the keen competition among the retailers for purchasing space for the location of their outlet, which has naturally resulted in the shortage of land, ultimately raising the price of land as well as the rent and premium values of shops. But as we go away from the core of the city in any direction, the land value gradually decreases. In Berry's words, each CBD is "characterised by a cone of land value which is highest at the main intersection and declines away from it".

Land value (commercial), rent and premium reach a peak level in the central area of the city rising towards the central locality of the retail nucleation. The decline takes place with increasing distance from that point. The decline in values, however, is not characterised by a smooth curve. Rather it comprises an irregular surface which declines in general with distance from the peak area but is broken by minor elevations or projections. Harris and Ullman have suggested a hetero-nucleated urban structure[41] and Mayer[42] has provided evidence of the peaking of land values at major traffic intersections, differential peakings according to relative importance of the intersections, and the association of the peaking land values with business use of the land around intersections.

Garner had access to detailed official data of each retail nucleation by type of functions which is not available for any Indian city. This is why Alam and Khan while analysing the commercial street of Hyderabad and Secunderabad have not gone into this aspect. Alam, however, in his earlier work has given a generalised picture of land values in the city of Hyderabad and Secunderabad indicating that highest land value is associated with commercial areas and the shopping centres of the city.

One difficulty in determining land and rent values in Indian cities is the use of unaccounted money transaction. The sale and purchase of land is seldom registered at the real monetary consideration which changes hands—the actual figure is usually two to three times the officially disclosed price. Similarly in the case of rents, the amount disclosed in the rent receipt is generally much less than the actual rent paid. The cash premium paid to the landlord at the time a shop is taken on rent runs from Rs. 1,000 to Rs. 1 lakh, but is never mentioned in the account books. However, where the premium paid is high, the rent paid is generally low, as the latter has been partly capitalised in the form or premium paid. Thus we cannot rely on the official records for this purpose, and recourse had to be taken to filling up a large number of questionnaires in the course of field work for gathering primary data for the purpose. It is believed that the oral statements by a large number of people about these facts prevailing in their locality can be averaged to give the true picture, even if they would be reluctant to admit the same in their own cases. It may be pointed out that this fact is amply proved by the very similar replies furnished by most of the people for particular localities even when they lived at far off places and were engaged in different activities. The range to be averaged was, therefore, found to be small, giving a reasonably accurate mean value for land value, premium or rent as the case may be.

## CONCLUSION

There is a dearth of work by Indian geographers in the field of market geography. This is mainly due to the non-availability of relevant data. Most of the studies of market geography in India have been attempted on a regional basis.

The study of intra-city retail structure of an Indian city reveals that there is a distinct pattern in the spatial distribution of both markets and their functions. Higher order of retail centres are generally located in the centre of the city, surrounded by smaller centres of decreasing hierarchy towards the rural-urban fringe.

The functional structure at different levels of hierarchy of business centres reveals that an increasing functional complexity is associated with the higher level market. On the other hand, a decreasing trend in the number and type of functions is noticeable in the lower level centres. Moreover, lower level centres are characterised by convenience goods while in higher level centres, both convenience and shopping goods are sold.

The study of land values and rent of commercial premises of a city also tend to support this conclusion. Land value, rent and premium increase with the hierarchical level of markets, decreasing outwards from the city centre (CBI). There is an inverse relationship between rental, premium and as the former tend to rise, while the latter declines and *vice versa*.

## REFERENCES

1. P. Scott, *Geography of Retailing*, London, Hutchinson University Library, 1971, p 12.
2. W. Applebaum, *Marketing Geography in American Geography—Inventory and Prospect,* Edited by James, P.E. & C.F. Jones, Syracuse, 1954, p 245.
3. G. Schwatrly, *Development of Marketing Theory.*
4. H. Smith, Retail Distribution : A Critical Analysis, London, 1937.
5. W. Christallor, *Central Places in Southern Germany*, Englewood Cliff, 1966, New Jersey (Translated by C.W. Baskin).
6. E. Ullman, "A Theory of Location for Cities, *American Journal of*

*Sociology*", 1941, pp 835-64.

7. B J L. Berry, and A. Pred, 1961, Central Place Studies : A Bibliography of Theory and Application, Regional Science Research Institute Bibliography Series No. 1, 1961; Supplement through 1964, by H.G. Barnum, R. Kasperson and S. Kiuchi, 1965.
8. A. Losch, *The Economics of Location*, New Haven (Translated by W.H. Woglom, and W.F. Stopter), 1954.
9. B.J.L. Berry, and W.L. Garrison, "A Note on Central Place Theory and the Range of Goods" *Economic Geography*, 34, 1958, pp 304-11.
10. E. Schell, "Change in Bostons Retail Land Scope : A Geographical Study of Fourteen Year's Change in Retail Distribution Patterns", New York, 1964, p 14.
11. R.J. Johnston, "The Distribution of an Intra-Metropolitan Central Place Hierarchy" Australian Geographical Studies, 1966, 4, pp 19-33.
12. J.B. Kenyon, "Annals of the Association of American Geographers, 1967, 57, pp 736-50.
13. B.J.L. Berry, "Geography of Market Centres and Retail Distribution" Prentice Hall INC, New Jersey, 1969.
14. F.A. Leeming, An Experimental Survey of Retail Shopping and Service Facilities in part of North Leeds" Transaction and papers of the Institute of British Geographers, 26, 1959, p 140.
15. G.M. Lomas, "Retail Trading Centre in Midlands. *Journal of the Town Planning Institute*" 50, 1964, pp 104-9.
16. W.K.D. Davies, "Centrality and Central Place Hierarchy" Urban Studies, 1967, pp 61-79,
17. E. Schell, "Aspects of Central Place Theory Unsubstantiated by Store Facilities in Boston's Retail Landscape" Abstract of paper, 20th International Geographical Congress, London, 1964, p 202, (Edited by F.E. Ian Hamilton).
18. R.E. Marphy and J.E. Jr. Vance, "Delimiting the CBD" *Economic Geography* 30, 1954, pp 189-222.
19. P.F. Matting, "Delimitation and Movement of CBD Boundaries through Time : The Harrisburg example" *Professional Geography*, 16; 1964, pp 9-13.
20. J.E. Bohnert and P.F. Mattingly, "Delimitation of CBD through time, *Economic Geography*, 40, 1964, pp 337-47.
21. M.J. Proudfoot, "City Retail Structure, *Economic Geography*, XIII, 1937, pp 425-28.
22. H. Carol, "The Hierarchy of Central Functions within the City" *Annals of the Association of American Geographer*, 1960, Vol. 50.
23. J.W. Simmons, Toronto's Changing Retail Complex, No. 104, University of Chicago, 1966.
24. B.J.L. Berry, "Commercial Structure and Commercial Blight : Retail Pattern and Process in the City of Chicago", University of Chicago, Department of Geography Research paper, 1963.
25. B.J. Garner, *The Internal Structure of Retail Nucleations*, Department of Geography, Northwestern Univ., Evanston, Illinois, 1966, p 29.

26. A.K. Dutta, *Intra-City Hierarchy of Central Place*: Calcutta as a Case Study, *The Professional Geographer*, Vol. XXI, No. 1, 1969, pp 18-22

27. S.M. Alam, *Hyderabad and Secunderabad : A Study in Urban Geography*, Allied Publishers Private Ltd., Osmania University Press, Hyderabad, 1965, pp 101-107.

28. R. Bengtsson, "The Structure of Retail Trade in a Small Swedish Town" Lund Studies in Geography, Lund, 1960, p 300.

29. J.E. Jr. Vance, "Emerging Pattern of Commercial Structure in American Cities, Lund studies in Geography, No. 24, The Royal University of Lund Sweden, Deptt. of Geography, Lund, 1960, p 495.

30. R.U. Ratcliff, Internal structure of Landuse, Urban Land Economics, McGraw Hill Book Co., New York, 1949, pp 386-97.

31. B.J.L. Berry and W.L. Garrisson "The Functional Basis of the Central Place Hierarchy" *Economic Geography*, 1958, pp 145-54.

32. A. Getis & J.M. Getis, "Retail Store Spatial Affinities," Urban Studies, Vol. 5, No. 3, 1968, pp 317-332.

33. B.J.L. Berry and W.L. Garrison, Studies of High way Development and Geographic Change, University of Washington Press, Washington, 1959, p 86.

34. V. Venugopal, "Commercial Structure of Warangal City in Andhra Pradesh, India, 21st, International Geographical Congress, Proceedings of Symposium on Urban Landuse; Hyderabad, National Committee for Geographers, Calcutta, 1972, p 13.

35. B. Thakur, "Central Business District of Pata" 21st International Geographical Congress, Proceeding of Symposium on Urban Landuse, Hyderabad, National Committee for Geographer, Calcutta, 1972, p 88.

36. H.H. Singh, Kanpur, A study in Urban Geography, Banaras Hindu University, 1972, pp 55-68.

37. S.M. Alam and W. Khan, Metropolitan Hyderabad and its Region : A Strategy for Development, Asia Publishing House, New York, 1972, p 170.

38. J.W. Simmons, "Toronto's Changing Retail Complex" Development of Geography, Research paper No. 104, University of Chicago, Illinois, U.S.A., 1966.

39. I.K. Rolph, The Location Structure of Retail Trade, Vol. 80, Domestic Commerce Series, U.S., Bureau of Foreign and Domestic Commerce, Government Printing Office, Washington, 1929. Vol. 80.

40. H.G. Canoyer, Selecting a Store Location (U.S. Department of Commerce, Economic Services, No. 56) Government Printing Office, Washington, 1946.

41. C.D. Harris & E.L. Ullman, The Nature of Cities, A.A. Academy of Pol. and Soc. Science, CCXLII, 1945, pp 7-17.

42. H.M. Mayer, Pattern and Recent Trends of Chicago's Outlying Business Centres, *Jour. of Land and Public Utility, Econ.* 1942, pp 4-10.

# 8

# Spatial Dimensions in Retailing

—*Suresh Kumar*

GEOGRAPHY is a science of discipline in distance.[1] Keeping this view in mind, spatial relationship of the retail business structure in the market, its spatial relationship with the consumer on the one hand, and the wholesaler on the other, creates a wider dimension for its locational decision.[2] Nowadays it plays a dominant role in our daily life. We all have to use either a big shop or a small one as the source of the last link in the chain of supply. It has become the life blood for the present world civilization as a primary link in the society. The purpose of this study is to throw light on some of the recent trends of retail business study with special reference to the city of Monghyr.

Retail business is the product of civilised life. During the prehistoric period when man was leading a wandering life, retail business was unnecessary but now with the cultural advancement, increasing population pressure on land and agricultural development lead a man to occupy the parcels of land in the nearby areas for business pursuits. The need was also felt for the product of one region to be exchanged with another where the same was not produced. Slowly the barter system, the exchange of 'surplus goods' in lieu of grains and food items, was enforced in society. With the invention of metals the exchange of goods was done with the help of metal money. Here were the beginnings of the retail business. This gave an impetus to locate varieties of shops at convenient points from the point of view of the consumers and transportation linkages.

## MEANING AND SCOPE OF THE STUDY

The term 'retailing' is derived from two French words, prefix *re* and verb *taillor*, is a science of purchasing and selling of goods at a particular place and time to cater to the needs of the consumer's demand and seller's profit. A retailer generally purchases goods from a wholesaler either from the nearest point (own city) or from higher order centres. Normally such centres depend upon the surrounding areas of produce which may be a centre of agricultural goods or industrial items. On the other hand, retailers sell the goods profitably at such market centres where the consumer or the surrounding areas seeks them. The market centre may be a rural service centre or an urban centre of the city rank. The place and time are relative terms for retailing because they vary from place to place, in temporal dimensions from city to city, at the junction area of rail and road, and at even intra-city location of business establishments. The word location is a very pertinent aspect of retail business because the selection of the site, approachability from the hinterland, flow of goods, capital, availability of labour, and human knowledge take place with the help of transportation and communication lines. All these make an environment of "ecological balance" under the control of a number of variables as the push and pull mechanism of the location of business establishments. A successful retailer offers the customer a geographic convenience. A household customer do not want to travel more in order to find produce and services for their use. Most of the characterised stores are within the reach but buyers and sellers must seek each other and this searching may make available the retail shop at a house length or thousands kilometres apart, with the help of minimum travel cost involved in communication.

The scope of the study of retailing includes both basic and non-basic functions and income for the town. Retailing is intimately related with the wholesaler, wholesale market, producer, consumer and their availability, approachability and mobility, pricing, demand cone in relation to price feasibility of the sellers to sell the goods at a particular price, consumer behaviour patterns, means and modes of advertisement, influence of site and situation, effect of climatic conditions and "ecological balance" within the parameters available in the environment. These are some of the basic factors which are studied in the retail business structure of a city. On the other hand, the accumulation of capital, banking facilities, productivity in the hinterland, means of communication and transport and the economic status of the consumers are potential factors to a retail business of an area, wherever they may be situated.

## SPATIAL SYSTEM OF RETAILING

Generally retailing is established where any degree of demand is placed upon it from the surrounding areas. For this fact retailing is the ultimate result of the spatial arrangement of shops in relation to consumers' demand and favourable location. The spatial system of retailing has been a problem since its origin. Location and site are not, however, always responsible for success and failure of any retail structure. Retail business may be located in a village, town, city, downtown of a metropolis or industrial centres but there must be a perfect relationship with the consumers with greater purchasing capacity. Thus retail business is usually confined under physical, economic, social and political systems in a mixed fashion. The fiscal policy and good economic condition of a retail may be harmful or helpful under the said manners. Hence the system may be closed or open but there must be the process of "feed-back mechanism" between the producer, manufacturer, buyer, seller and ultimately the consumer. Under this process of "feed-back mechanism" retail business is performed due to the support of producers of the surrounding areas and the far away higher order central places. On the other hand, sustenance is also provided by the surrounding hinterland as ultimately the consumer is meant for the retailing in a city and *vice versa* where he may be located in the surrounding areas or the city itself. But the hierarchy of such centres must exist from the point of view of hexagonal arrangement of market centres both inside and outside the city.

The system of Monghyr retail business structure is believed to form an open system restricted by the second order central places located around it. The retail shop of marginal and poor people are situated on the outskirts of the city and such shops are beneficial as the neighbourhood store for the poors, whose income, cost of travel and demand of particular goods on lower price generally finds a place in marginal shops only. On the other hand, the shops of high standard having fresh and fashionable goods with high capital investment are available only in the central business district like Chowk Bazar of Monghyr. Only well-off people with good income dare to visit such shops for the availability of choicest varieties of costlier goods on a particular location.

In such a situation of retail location and consumer behaviour the central business area generally derives cheap labour from the fringescape of the city, whereas the finance for the outskirt retail business comes from the shopkeepers located in the Central Business District in most of the cases.

## CONCEPT

The concept of retailing can be stated on three points. Firstly, spatial relationship deals with the location of retail shops on space and their variations from place to place due to the facilities available there. The shop itself is an item of commerce and economics but its spatial variation gives it a place in geography. The present study aims to investigate the geographical, commercial as well as the economic aspects of retailing wherever the same is located.

Secondly, the concept of environmental relationship determines why and how of a particular place the retail business has flourished because only those places are fit for retailing, where, at least the optimum threshold of consumers must be received by a centre. In the case of under threshold of consumers, a shop may fail or begin to die slowly. In the case of maximum turnover of consumers at a particular shop for a particular kind of goods, it determines that the shop is in the over threshold of consumers. Threshold means the required number of consumers for a shop on a particular day or a month or a year with the help of which a business concern may run profitably. Besides these items like capital, incentive, transport, etc. generally attract consumers.

A shop is on the whole a product of the cultural advancement of people in the process of economic development of the society in order to maintain a happy and prosperous life by purchasing and selling of goods. In this way the shop is an emblem of retailing. It is an example of the getting together geographical, commercial as well as economic aspects of the society at the same time even if the retailing is a non-productive function.

The hierarchy of market centres is generally determined by the massing of shops on a particular location and the workers engaged in functions like trade and commerce. Thus the hierarchy of retailing may be used for differential locations of the retail business of the different areas of the city. In such a situation, the intra-city central place hierarchy will certainly emerge as a basic tool to study this aspect of the city.

Thirdly, the concept of retailing includes the size of an urban place and functional magnitude of the centre including population served and area served. In such a situation the size may be determined by units of shops, number of population, areal expansion of a city and magnitude of trade area determined with the help of the demarcation of the umland of the city with the help of certain goods which generally come from the surrounding areas of the city. The population served and area served can also be determined with the help of the functional magnitude of the city and demarcation of the hinterland. Under the concept of periodicity, the retail business generally runs under a shade or inside house in front of a road for 24 hours only if the site is basically defensible from the onslaught of thieves, anti-social elements, hazards of climate and the availability of workers and consumers. The concept of retail trade zones includes wholesale shops, retail shops, regional centres mixed with retail shops, furniture shops, vegetable, automobile and other specialised shops.

Under the concept of consumer behaviour the trip behaviour of consumers available from the urban and the rural areas for the purchase of consumer goods is considered along with the distance travelled area served, productivity of the region and the demand of particular goods, including accessibility. Generally with greater accessibility the consumers trip behaviour is found to be maximum whereas in an area of lesser accessibility the consumers try to purchase a number of items from the market in a single trip. In the retail business generally the floor space occupied by a shop and location of the same in areas of particular land value determine the rent of a shop. All over the world the "Central Business District" is the area of higher sale where the shops are generally let to the shopkeeper by the houseowner according to the land values and the construction cost of building and maintenance costs. Besides this, whenever the previous business man tries to vacate, the prospective comers will have to pay an exhorbitant amount in the name of *pagri* or gratuity to the previous shop owner.

## APPROACHES OF STUDY

Five approaches have been considered to study the retail business structure. These approaches are the product of time, space and principles of the function concerned.[3]

(1) **Sample Approach** : This approach may be called commodity approach or case study approach of a retail business establishment and its neighbourhood in all details including floor space, general turn out, input, height of the building, workers engaged in service facilities, and daily sale of goods, etc. It may also be regarded as a case study approach in which a representative item is generally chosen for detail analysis through the stochastic process. Here the study of house plans, front space and function of the shop along with other details are considered.

(2) **Regional Approach** : In the regional approach even one market is divided into a number of zones and the integrated study regarding its entirety is known as the regional approach of a retail business as the case may be, e.g., in one area of the city there may be a location of medicine shop, grocery shop, cloth shop, betel shop, tea stall, hotel, embroidery shop, readymade shop, grain shop, book-stall and so many others. In this way, the integrated study of all these items in one part of the city or a region is known as regional approach of the study.

(3) **Principle Approach** : Under this approach the principles of location of a shop are generally determined by the number of variables available in the surrounding environment and in this way the various kinds of functions are generally located on a particular place in the city. This happens as the consumers know that fresh goods at competitive prices are only available in the core area of the city, e.g., all vegetable shops are found to be located in one corner of the city, while all medicine shops are located near hospitals, all book-stalls are found to be located in the area of the university or college or school location. On the other hand, grocery shops are generally

located near residential localities in order to serve all old errands and the cloth shops, *purtune* shops and some others are found to be located in the central market areas of the city. All these indicate that certain principles are always there behind the location of the business structure of the city and such a study is known as the principle approach in order to find out the general regularity of such locations.

(4) **Genetic Approach** : Under this approach an area of retail business may have a particular type of size and shape in order to accommodate its establishment considering the river side location, location around a tank or a temple, mosque, sea side or lake side. The business structure may take a linear, circular, semi-circular, rectangular and grid iron patterns, radial patterns, herringbone patterns, checkerboard patterns and so on.

(5) **Remedial Approach** : Any scientific study is generally directed towards the solution of a problem by the way of advice, planning, projecting the future goals and highlighting of the problem of the market for improvement of the lot of the toiling masses of the traders and businessmen.

## PRINCIPLES OF STUDY

Principles involve the location, siting, types, patterns and maximization of profit and employment of a market place in the retail business system with special reference to the city level centres. The following principles are adopted here for the study of retail business structure at a particular location of a market :

(1) maximising profit,
(2) satisfying the customer,
(3) maximisation of the opportunity of employment and food,
(4) cultural advancement of people, and
(5) planned development of society.

Profit is the ultimate goal of a business, which may satisfy a man through the location of a shop, industry, etc. In retailing a shop is generally located at a convenient point where the access to sellers and buyers finds a suitable place for the maximisation of profit.

As a function, retail business gives opportunity to satisfy a man with an employment in a shop or in an industry. Normally, retailing is treated as a non-productive function carried on by lazy persons, but sometimes it seems to be essential for the maximisation of employment for the unemployed people of society in order to have a happy life with sufficient means of food and lodging. Actually the business class people are rarely affected by floods or any natural hazards like in the farming business. Hence the sufficiency of food available to the business class people seems to be an assured gift of the nature.

Retailing thrives in an urban culture where the goods are manufactured in industrial establishments and in such a situation more people can be accommodated on a small piece of land due to their technological superiority, accumulation of capital, and labour force and all these lead to the cultural advancement of a society and a stable social order.

Retailing gives an opportunity for the location of industry in the nearby areas which generally attract administrative functions, transportational services, establishment of the offices of a number of companies and many other items for economic progress and the planned development of a society.

## REFERENCES

1. J. W. Watson, "Geography : A Discipline in Distance", *Socttish Geographical Magazine*, Vol. 71, 1955, pp 1-13.
2. R. B. Mandal, "Role of Transportation Linkages in Rural Community Development", *Geographical Outlook*, Vol. XIII, 1977-78, p 69.
3. R. B. Mandal, "Nature of Population Geography", *Dimensions in Geography*, Associated Book Agency, Patna, 1978.

# 9

# The Concept of Consumer Behaviour

—*Suresh Kumar*

CITIES do not grow by themselves. The countryside sets the tasks that must be performed in the city. This concept is first of all enumerated by Mark Jefferson (1931).[1] Christaller, while studying the hierarchy of central places, found that the smaller towns have lesser number of consumers and the consumer behaviour is simpler marked with single trip, lesser purchasing capacity and short distance travel. The range of goods as suggested by Christaller is also much smaller in such markets. In comparison with this, in larger centres, the massive turnover of consumers with multipurpose trips, long distance travel and bulk purchase are the point of difference which is also helped by the range of goods at the same time. It has been suggested that all goods of lower order centres are found in the larger central places and hence the complexity of functions, morphological structure and wider zone of consumers' travel are the rule of Christaller's hierarchy of central places both for intra-city central places and the inter-city central places.[2]

Lösch on the other hand, favoured the idea of Christaller but he suggested that all the goods of lower order centre should not necessarily be found in higher order centres. On the basis of the consumers' movement and the delimitation of the umland of a city the hierarchy of central places can easily be achieved.[3]

Recently the ideas of Christaller and Lösch have been extended by many investigators after giving graphical and mathematical solutions to the problem. Among these the works of Berry and Garrison[4], Beavon and Mabin[5], Clark[6], Baumol and Ide[7], Jhonston[8], and Tarant[9], etc. are noteworthy. In India, this study has been developed by N.R.Kar[10], in some areas of West Bengal, A.K. Dutt[11], in West Bengal, A. Sami[12] for the intra-city consumer behaviour pattern of Patna, B. Thakur[13] for Ontario province of Canada and R.B. Mandal[14] in North Bihar Plain. All these works are empirical in nature and their problems are also narrower in the sense that they have not developed the concept of consumer behaviour as the basic issue of investigation.

In the increasing complexity of consumer behaviour, technologically and psychologically, a person does not necessarily make all his purchases within area of residence. On the other hand, the size of retail market area depends on feeding centres, the number and types of highways serving the umland, the mobility of people, the wealth and standard of living and proximity of rival market centres. Recent techniques in retailing is also directed towards the capture of consumer through newspaper, radio, magazine and billboard. Thus, the purpose of this study is to throw light on the range of goods and the zone of central business which fluctuate by the following parameters :[15]

(1) transportation has facilitated the buyers to travel five times more than on foot;
(2) the improved types of communications and highways have reduced the irksomeness of long distance travel;
(3) increased motion picture attendance, the increased number of radio receiving sets and wide reading of newspapers all of which has helped in increasing knowledge and stimulating interest in new merchandise goods;
(4) the expansion of rural free delivery routes, now make it unnecessary for the rural dweller to go to the town for his mail; and
(5) the tendency of wide areas to become still wider, due to the concentrated trading in larger urban centres. The scattered location is economically practicable to establish chain stores and mail order to branch retail concerns, which in turn draw more people from the feeder nearby towns, to the larger more distant towns.

## OBJECTIVES OF SIUDY

The movement of customers in and around in an urban centre is guided by the purchasing capacity of the consumers, availability of goods in the market, the price of goods sale and the distance which a consumer covers in making his trip to a trade centre. Such an area from where the consumers came to the market is known as hinterland, zone of influence, umland, tributary area and the zone of urban influence, all having similar meaning but with differential terminology. These ideas have developed a meaningful concept which is known as Consumer Behaviour. Under this idea the trip behaviour of a consumer is generally studied, keeping in view, the distance covered and time taken and the purchase made. Generally a consumer of just nearby areas makes a multipurpose trip to a market and the item of purchase is very limited and their quality is also very simple or of general use. But the consumers of distant areas have very few selected trips to the larger urban centres but they purchase bulk goods and the trip behaviour is also selective to a particular apartment which is customary in nature. Hence the trade areas of the principal urban centres are becoming complex integrated systems, each with a major diversified centre and numerous smaller more specialised and slower growings nodes at the outskirts. As the rural people over a wide area travel greater distances to reach the larger food drug, clothing or specialised stores of the complete shopping centre, they have the choice of travelling greater distances in other directions, to smaller or less diversified centres, as clients of a particular seed dealer, appliance service, etc.[16]

Recently a large number of consumers in Indian cities have migrated into relatively affluent and secure positions as their purchasing power has increased, the opportunity to embroider upon basic needs with a sense of individual taste and creativity, due to their keen interest in a new style of life rather than for security of trades. This idea is supported by the fact that the growing mass of suburban trend, high speed electrified trains, fast moving single and double storey transport vehicles along with a host of the sources of transport and communication have endangered the environmental conditions as the same has been polluted extensively in recent years of transport development,

This is again nullified by the humdrum reality of the urban area and the heat generated by the transportation and communication system and massing of people. In urban areas the style of life refers to the consumers who cerefully develop sets and patterns of personal consumer behaviour. In some areas income, leisure and time saving devices have permitted types of personal explorations that have raised a large number of industries devoted to agricultural industrial and sport, products etc. On the other hand, the political ideology based on conflicts between economically well-off people appeared in minority groups which looked upon with much interest at the innovations and consumption along with the political ideas. The potential variations in consumer behaviour are the outcome of economic differences and because of this fact the consumers for a personalised style of the life have been necessitated.

## RESEARCH APPROACHES IN THE STUDY OF CONSUMER BEHAVIOUR

In an idea of perception and behaviour analysis, the consideration of environmental factors attracts the mind of social scientists because one may expect much employment of experimental analysis to identify regularities in behaviour within the area. However, there is still a lack of general theory about the consumer behaviour. Katona[17] has developed a theory of the role of consumer expectations or confidence upon the levels of spending and saving the economy.

Nowadays the consumers are seen more powerful than expected and that has prompted the study to identify the trip behaviour of consumers living in the tributary area of an urban centre. In the consumer behaviour study six different approaches are as follows :

### General Approach of the Consumer Behaviour

The role of consumers and consumption level as investments and consumption in the order of economic development of region as a whole comes under this approach of study. Here the analysis of different groups of people in a city and their differential character of trip behaviour and responses to the call of a local shop is determined in the region or city as a whole.

**Income Approach** : Changes in income distribution, credit and expenditure by various classes of people are studied in this approach. This category accounts for a large extent of past studies in temporal dimensions and the income distribution of different groups of people in urban areas in accordance with their social status and working force.

**Population Study Approach** : Population trends and shifts in life cycle influences urban demographic classification which is included in this study.

**Life Style Factors** : This factor influences the style of living such as purchase, status symbols, social class, images, needs, leisure, convenience, etc. which are included in this approach. Nowadays in the study of consumer behaviour this approach is gaining ground in most areas of the world.

**Consumers as Decision Makers** : This studies who makes what decision, what decisions are planned, who is influenced by this decision, who is under the influence of this decision, who is influenced in various situations keeping in view the household as a decision unit and the behaviour of family members.

**Specific Purchase and Consumption Approach** : Studies of purchases and determinants of purchases for specific products such as coffee, dry cleaner, gasoline, refrigerator, ladies wear, soaps, toys, etc. At present the study of consumer behaviour is associated with the main body of the behavioral sciences.

**Influence of Changing Consumer Wants** : With the changing social condition in space and time, most of the consumers of the city of Monghyr like change. They are irresistibly attracted by whatever seems new and different. This trait has made shifts in the fashion of great importance in almost every line of merchandise—not only in the area of women's clothing but in the sports fields also. By studying consumer likes and dislikes, the shopkeeper may be able to keep his product abreast of changing market differences, on the one hand, by changing his product, while on the other, they may actually be able to modify and alter consumer preferences to their own advantage. With the growth of self service, retail store packaging has become of critical importance and continuing surveys are carried on to reveal consumer changing preferences in the packaging of such goods as soaps, breakfast, food items and napkins. The nature of the store in which the particular commodity is successfully marketed is also subject to change. The supermarket operator, e.g., insists on customers buying products that were once in the cherished line of druggist, the hardware merchants, and the other types of retailer. As an illustration, a manufacturer of camera flash guns may not only sell it to photographers but also to research scholars, students, housewives and people in general too. As a result of this discovery the manufacturer has been able to incorporate these retail outlets into his marketing programme. Rich dividends accrue to those who are first to discover the changing habits of consumers.[18]

## METHODS OF STUDY

The methods of study will be variegated in nature considering the principles of investigation in social science studies. It may be a descriptive analysis, quantitative measurement by drawing inferences, graphical study or mathematical computations and system analysis, etc. In view of this, first of all the notion of consumer behaviour as given by Baumol and Ide is presented in order to explain the basic method of consumer behaviour. After this the measurement of the zone of influence of a town as the delimitation of the area of consumers' attraction of a city, the law of retail gravitation given by Reilly are also analysed here subsequently. At the end, the process of investigation in this study will follow with special reference to the city of Monghyr and its environs.

## NOTIONS OF CONSUMER BEHAVIOUR

It is interesting to discuss the theory of tertiary activity and shopping behaviour of consumers. Baumol and Ide have developed a simple model displaying the choice variables in the selection of an urban centre for retail shopping trip by a consumer. A particular consumer always tries to visit a centre when his demand function is such that

$$f(N.D) = wP(M) - v\left(vdD + cn\sqrt{N + ci}\right)$$

is positive. f (N.D) is a measure of the expected net benefit of the consumer from entering into a store. It varies with D, his distance from the store, and N, the number of items offered for sale at the store. Assumed costs are cd, a cost of transport assured proportional to distance, Cn $\sqrt{N}$, the assumed costs of actual shopping, and ci, the opportunities foregone. P(N) is the probable satisfaction functions, and W and V are the subjective weights assigned by the consumer when he evaluates the size of each element in the equation.

The economic implications from this statement are many. For example, the minimum number of items necessary to induce a customer to shop at a given store will increase with D. Maximum shopping distance is given by the equation of indifference curve which offer the consumer zero net benefit from shopping at a store, and is obtained by setting (N.D)=0 and solving for D to yield.

$$Dn = \frac{W}{VCD} P(N) - \frac{1}{CD} (Cn/N + Ci)$$

Given the hierarchical spatial system of the theory of tertiary activity.

P(N) = will be a step function related to levels of goods available at each rank of the hierarchy.

Cn $\sqrt{N}$ is a like step function.

Ci dependent upon the spatial characteristics of the hierarchical system.

Hence, only solution Dn will be a step function related to levels of the hierarchy. The Baumol and Ide system is entirely compatible with the theory of tertiary activity. The statement is also true for the development of a total retail sales model since such model is based upon the previous one of maximum profits (which are of course based upon N). Not only do consumers discriminate among centres hierarchically and spatially, but varieties and sales levels and the extent of potential profits are likewise determined.

### THE LAW OF RETAIL GRAVITATION

The regional influence of functional types of a market are also measured around an urban centre with reference to the other urban centres in the neighbourhood. This can be estimated by applying the law of retail gravitation developed by Reilly. Ultimately one may arrive at the same conclusion after considering a number of shops at a centre because there is a close connection between the number of business types and the number of shops at a nodal point inside the city. In a real sense, the use of a number of shops gives better results whereas the quality of shops may distort the result in comparison with the number of shops which give a better evaluation of the market centres in terms of the consumer's influence, quality of shops and the number of shops at a centre. Reilly had used population as an index of size, while Berry preferred business type as the fundamental measure of the attractiveness of a market centre inside a shopping locality. In this study it has been found appropriate to consider a number of persons engaged in business activities as the basic measure of the importance of retail trade at a given places and consumers, and population served in the area as the regional influence on a particular centre.

According to the break point equation of retail gravitation, the trade area boundary between two centres A and B is, in kilometres from B, equal to

$$\frac{\text{Kilometres between A and B}}{1+\sqrt{\dfrac{\text{size of persons engaged in business activities at the centre A}}{\text{size of persons engaged in business activities at the centre B}}}}$$

where, A and B are two trade centres and their relative size is determined by the number of shops, i.e., retail outlets that they possess. Berry has pointed out, however, that this equation cannot be applied indiscriminately.[19]

Berry suggested that it must be applied to cities and larger regional centres to determine market areas for goods of the city. Further, it may then be added to derive market areas for lower order of goods inside the market and the goods of city order. In this study, in order to distinguish the market areas of Monghyr, the intra-city central places have been included to determine the market areas for goods of all varieties available in the city. In order ro determine the zone of influence of consumers in between Monghyr and Sultanganj in which the latter is a town, selling goods of a higher order, the break point is determined by the following equation :

Distance from A to B=24 km

Size of A=4,442 persons engaged in trade and commerce

Size of B=491 persons engaged in trade and commerce

The distance from Sultanganj (B) to the break point of market area is

$$\frac{24}{1+\sqrt{4442}}=\frac{24}{1+3.0083}$$

$$=\frac{24}{4.0083}=6 \text{ km from B Sultanganj.}$$

Similarly for goods of lower order the break point between Purabsarai and Chowk Bazar is found as follows :

Distance between A and B=1.5 km

Size of A=718 persons engaged

Size of B=433 persons engaged.

The distance from Purabsarai (B) to the break point of market area is

$$\frac{1.5}{1+\frac{718}{433}}=\frac{1.5}{1+1.66}=\frac{1.5}{1+1.29}$$

$$=\frac{1.5}{2.29}=.65 \text{ km from Purabsarai.}$$

It has been found that the Central Business area of Monghyr forms a large retail market zone which attest the regional importance of that centre.

### AGE GROUPS AND CONSUMER BEHAVIOUR

Much has been done on the age groups for consumer behaviour. Davis[20], Lasson, Weigand and Wright[21] have drawn attention regarding the spatial pattern of consumer behaviour. In categorising the types of shops, the author has categorised three main types which are as follows :

(1) Hotel, restaurant and entertainments,

(2) Food, clothing and groceries, and
(3) Medicine and cultural purpose.

The consumer behaviour, as observed in the city of Monghyr in January, 1979 produces the following percentage of market trips (Table 9.1).

TABLE 9.1

| Serial Number | Age groups | Categories of Shopping visited in % | | | |
|---|---|---|---|---|---|
| | | Type 1 | Type 2 | Type 3 | Total |
| 1. | 14 to 23 years | 50 | 20 | 20 | 90 |
| 2. | 24 to 50 years | 30 | 50 | 30 | 110 |
| 3. | 50 to onward | 20 | 30 | 50 | 100 |
| | Total | 100 | 100 | 100 | |

The above Table 9.1 shows that group 1 although it covers a large percentage ultimately the total percentage figure is lesser, whereas the third group is able to maintain its status. The second group which is high in shopping behaviour proves that the earning members who are generally of this age group dominate in the family and hence they also dominate in the system of consumer behaviour.

## ANALYSIS OF CONSUMER BEHAVIOUR IN THE UMLAND OF MONGHYR

The study of consumer behaviour pattern has been contemplated through field work in the umland of Monghyr in December 1978. In this investigation it has been found that the supply of vegetable, milk, dried cowdung and other essential items produced in the vicinity of urban areas are generally meant for sale for the city residents.

The zone of supply of cloth, medicine and other items of higher order goods by the city of Monghyr is larger in comparison with the above. This is due to the attractive market, cheap goods, chances of the availability of higher order goods in abundance and purchase of a large number of consumer goods in one trip make the zone of influence larger in comparison with the above (Fig. 9.1). This map shows that north of the Ganga villages of Mirailipur, Samastipur, Sanha, Chauki, Raghunathpur, Kailashpur, Taufir, Tikarampur, etc., are the pet villages from where the consumers generally come to the city of Monghyr for the purchase and sale of their daily requirements and produce of the land such as vegetables, grain and milk from domestic livestock, etc. Similarly, south of the Ganga where the primate city of Monghyr lies the service of villages received are those of Sitakund, Nauagarhi, Binda Diara, Bariarpur, Jamalpur, Surajgarha, Kiranpur, Khawatajpur, Sunderpur, Singhiya, Parham, Farda, Indrukh and Herudiara, etc.

*City of Monghyr, Milk, Vegetable Supply, Cloth and Medicine Consumers Zone (1979)*

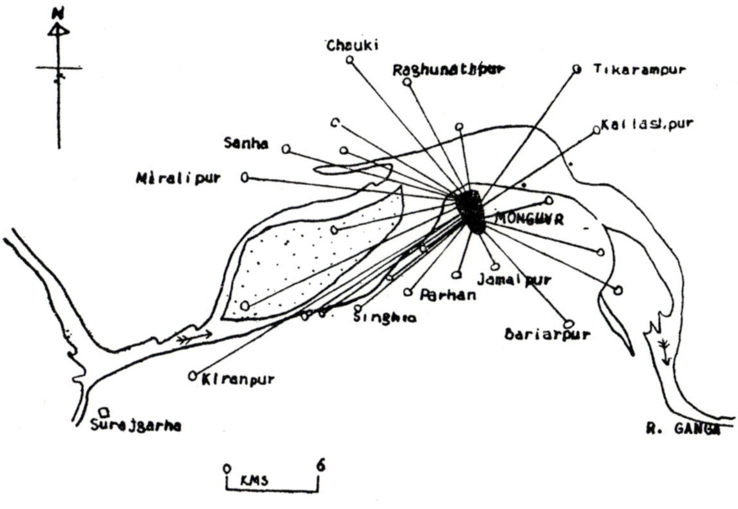


FIG. 9.1

With the help of the above rural settlements and the number of shops located in the city of Monghyr, the population threshold has been calculated. Population threshold has been defined as the minimum number of people required to support a functional unit.[22] Considering the number of shops in the city of Monghyr and the consumers of the zone of influence of

Monghyr the population threshold has been calculated and the same is given in Table 9.2, Fig. 9.2.

*Relationship between Number of Shops and Population Threshold (1979)*

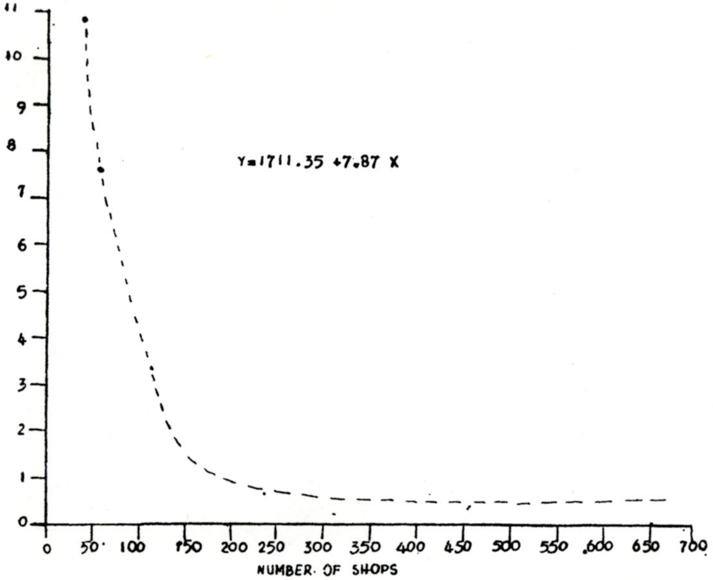


FIG. 9.2

TABLE 9.2

| Quality of shops | Number of shops in the city of Monghyr | Population Threshold |
|---|---|---|
| | X | Y |
| Medicine | 52 | 7,601 |
| Hotel | 654 | 604 |
| Cloth | 116 | 3,407 |
| Tailor | 169 | 2,339 |
| Vegetable | 148 | 692 |
| Manihari | 185 | 2,136 |
| Portune | 306 | 335 |
| Fruit | 36 | 10,979 |
| Meat | 92 | 1,114 |
| Total | 1.758 | 29,215 |

After considering the number of shops and the population threshold a graph has been constructed in which X-axis represents the number of shops while Y-axis is meant for population threshold for medicine, hotel, cloth, tailoring, vegetable, *manihari, portune,* fruit and meat. For these functions the correlation value is very insignificant which comes to $Y=1711.35+7.87\,X \pm$ sey. It has been observed that the regression is not perfect because the point in the graph represents a curvilinear situation (Fig. 9.2).

## CONCLUSION

The notion of consumer behaviour is not a new concept in the field of social sciences but it is the outcome of olden days since the beginning of urban culture. Recently due to the overwhelming growth of urbanization and rise of population in the hinterland of urban centres the pattern of consumer behaviour is a matter of great concern for the sellers, buyers and the scientists too. The concept of the range of goods, purchasing capacity of the consumer, productivity of hinterland, the amount and the quality of market are the subject matter for the investigators of consumer behaviour study. In this context the work done by Clark, Baumol and Ide, Reilley and several others are noteworthy. So far as the spatial ordering of the concept of consumer behaviour is concerned it has yet not attracted the scientist to investigate the behaviour of the consumer in purchasing the goods from urban areas much. In this direction this study has been done with special reference to the city of Monghyr and the close connection with the people residing in its hinterland and for this the population threshold has been calculated, trip behaviour has been analysed and the conclusions have been drawn on personal experience.

## REFERENCES

1. M. Jefferson, "Distribution of the Worlds City Folds", *Geographical Review*, Vol. 21, 1931, p 453.
2. W. Christaller, "Central Places in Southern Germany", translated by C.W. Baskin, Englewood Cliffs, N.J. Prentice Hall, 1966.
3. A. Lösch, "The Economics of Location", translated by W.H. Woglon and W.F. Stolper, Yale University Press, New Haven, 1954, pp 109-134.

4. B.J.L. Berry and W.L. Garrison, "The Functional Bases of the Central Place Hierarchy", *Economic Geography*, Vol. 34, 1958, pp 145-54.
5. K.S.O. Beavon and A.S. Mabin, "The Lösch System of Market Areas Derivation and Extension", *Geographical Analysis*, 1975.
6. W.A.V. Clark, "Consumer Travel Patterns and the Concept of Range", *Annals of the Association of American Geographers*, Vol. 58, No. 2, 1968, pp 386-396.
7. W.J. Baumol and E.A. Ide, "Variety in Retailing", Management Science, 1957, Vol. 3, pp 93-101.
8. R.J. Jonston and P.J. Rimmer, "A Note on Consumer Behaviour in an Urban Hierarchy", *Journal of Regional Science*, 7, 1967, pp 161-166.
9. B.J.L., Berry, H.G. Barnum and R.J. Tennant, "Retail Location and Consumer Behaviour, Papers of the Regional Science Association, 9, 1962, pp 65-106.
10. N.R. Kar, "Urban Hierarchy and Central Functions Around Calcutta in Lower West Bengal", Lund Studies in Geography, No. 24, 1962, p 250.
11. A.K. Dutt and S. Banerjee, "Transportation Index in West Bengal : A Means to determine Central Place Hierarchy", *The National Geographical Journal of India*, Vol. XVI, Part 3-4, 1970, pp 199-207.
12. A. Sami, "Retail Structure of Patna", Indian Geographical Studies, No. 3, 1974, pp 69-78.
13. B. Thakur, "The Changing Functional Make-up of Central Place Hierarchies in South-Western Ontario between 1871 and 1971, M.A. Thesis, Dept. of Geography, University of Waterloo, 1972.
14. R.B. Mandal, "Central Place Hierarchy in Bihar Plain", *The National Geographical Journal of India*, Vol. XXI, Part 2, 1975, pp 120-126.
15. W.C. Waite and R. Sassady, "The Consumer and the Economic Order", McGraw Hill Book Co. Inc. 1949, pp 252-256.
16. John R. Borchert, *Upper Midwest Urban Change in the 1960's* Minneapolis, June, 1968, p 16.
17. G. Katona, *Psychological Analysis of Economic Behaviour*, McGraw Hill, New York, 1951.
18. V.A. Musselman and E.H. Hughes, *Introduction to Modern Business*, Anglewood Cliffs New Jarsey, Prentice Hall, 1959, p, 386.
19. W.J. Reilley, *The Law of Retail Gravitation*, New York, 1931, Quoted by B.J.L. Berry, The Geographical of Market Centres and Retail Distribution, Prentice Hall Inc. N.J., 1969, p 40.
20. R.L. Davis, *Marketing Geography*, Methuen & Co. Ltd., London, 1976, p 218.
21. C.M. Lasson, R.E. Weigand and J.S. Wright, *Basic Retailing*, Prentice Hall, 1976, p 84.
22. M.H. Yeats, *An Introduction to Quantitative Analysis in Economic Geography*, McGraw Hill Book Co., New York, 1968, pp 104-107.

10

# Mineral Resources and their Role in Economic Development

*—D. K. Mishra*

## CONCEPT OF MINERAL RESOURCES

A NATURAL resource is anything discovered by man in his natural environment that he may in some way or other utilise for his own benefits. What one finds on the earth's surface is made up of either plant life or minerals. But ultimately all of it is derived from the minerals of the earth. Thus, the mineral resource is defined as a concentration of naturally occurring solid, liquid or gaseous materials in or on the earth's crust in such forms that economic extraction of a commodity is currently or potentially feasible.[1]

A higher level of industrial utilisation of mineral resources is an indicator of an affluent society. It also marks the stages of human civilization from prehistoric times to the present day. Different eras of civilization are named after the names of minerals and their products such as the 'stone age', the 'bronze age', the 'iron age', the 'nuclear age', etc. The Palaeolithic people were one step behind the Neolithic people in the non-technological era, while the bronze age marks the beginning of technology to utilise one of the known resources to produce metallic implements. The present sophisticated age is the result of enhanced use of minerals, whether it be as fertiliser for food, coal, petroleum, natural gas and atomic energy as source of power, or countless other necessities of life are derived from the use of minerals. So mineral resources are given by nature, but they cannot be known as resource unless they are used or put into useful forms by the knowledge of technology.[2]

Search of minerals in the old time was more a matter of intuition and it was chiefly owing to prolonged association of man with a particular mineral belt, and attempts were limited to surface of a few mineral deposits of known use. As industry and trade were restricted and future needs were unknown, the situation of demand did not arise. The rapid increase in the rate of consumption of mineral resources started during the industrial revolution; and also the emerged developing nations have created a tense situation of increase in demand for the mineral resources.

### THE ECONOMIC LIMITATIONS OF MINERAL RESOURCES

The mineral resources, ever since their economic potential was known, displayed some distinct characteristics which distinguish them from other natural resources. Apart from many geological and technical peculiarities, they inherited some economical aspects of their distinct physical characteristics which are worth-mentioning :

(1) Mineral wealth is exhaustible. Once it is mined and consumed, it is lost for ever. Only the thrown out materials out of these can be consumed afterwards, by improved technology. All mining areas have, thus, distinct life cycle; youth-development stage, maturity-established stage, old age-declining stage of ore reserves, and finally they lose their importance due to depletion of their reserve.

(2) "Mineral resources are there, where they are."[3] The occurrence of minerals is localised and limited and the total area occupied by them is far less than that by populated settled areas or agricultural lands. The distribution of the majority of minerals is sporadic. Some of the mining areas, lie in a place which is highly inaccessible. So the mines closer to settled areas are worked earlier.

The depth of occurrence of minerals is another specific characteristic of mining. The cost of exploitation is directly involved with the depth of occurrence of mineral beds.

(3) Mineral occurrences know no political boundary. There is no nation or State which has with its own territory all mineral resources needed for industrialisation. The unequal distribution also makes trade of mineral inevitable and also makes some countries acquire monopolistic rights in respect of certain minerals; and some of them make a common cause to exert

pressure on the international trade and marketing of the minerals.

(4) The quality of minerals differs from mine to mine and also within the same mine itself. And this creates difficulties in pricing of any particular mineral. The pricing of a mineral is fixed considering only the commercial grade for specific industries. So the mined minerals if is not of the specifications of the order of the industry or industries, is thrown and dumped useless. This action of wastage tells upon the final pricing of the minerals.

(5) Mineral resources have to be discovered and even after discovery "the actual content of resource is difficult to be assessed until the deposit is depleted".[4]

(6) "The exploration of mineral resources constitutes the highrisk element of any project, and in most cases expenditure is made without proving up a viable deposit".[5] Much exploration has to be done and many trial pits have to be developed before deposits of economic value are found. "A systematic geological work needs to be followed to find a deposit and even then, there may be certain unforseen geological factors like faulting, pinching out of the ore bodies, quality deteriorating at depths, etc. which may ultimately effect the life of a mineral deposit".[6] So the risk in investment in mineral development is much higher than other manufacturing industries because of the uncertainty of the result of the exploration investment.

All above mentioned special characteristics of mineral resources along with high investment with higher degree of risk, though with high return of profit also, have made mineral resources sector different from other natural resource development projects and industries.

## MINERAL RESOURCES AND RESERVES

The increased rate of consumption and further potential demand of mineral resources by the developing and developed economies have necessitated a rational approach to the definition of mineral resources and reserves. With the standardised definition and classification of mineral resources the national and international mineral policies for development can be well understood. A resource classification has been evolved jointly by the United States Bureau of Mines and the United States Geological Survey. This classificationcategorically differentiates different types of deposits for the consideration of study and working. Karunakaran[7] has drawn a simpler form of mineral resource classification out of the United States Bureau of Mines and United States Geological Survey classification which is as follows :

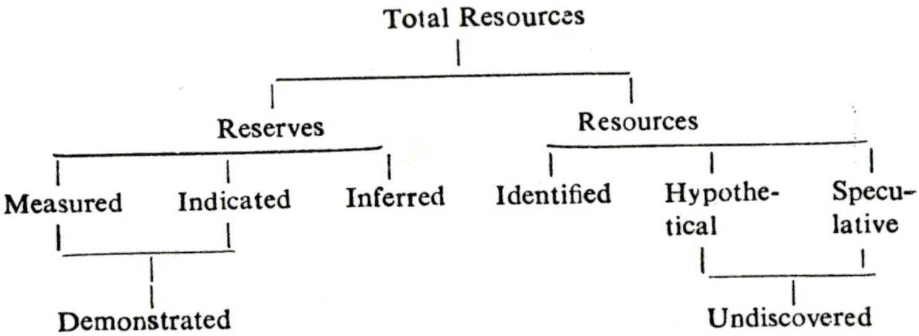


Resource is already defined as the total of materials available in or on the surface of the earth which is used or can be brought to usable form currently or in future is feasible. The total resource under the above mentioned scheme of classification includes all minerals already identified together with those presumed to exist from the geological evidences. "The proportion of an identifiable resource which can be produced at a profit at the time of classification, constitutes reserves of the mineral".[8] "A mineral resource is not classified as a reserve until there is sufficient indication that its size and grade are sufficient for it to be economically exploitable".[9]

The interest for investment on any mining project lies with the amount of measured mineral in a mineral reserve. These reserves are that which are delineated with estimation of size and grade of the deposit with reasonable degree of certainty by sufficient sampling and other measuring techniques. But in practice, there will generally remain an element of uncertainty, because the estimate of quality and quantity with a margin of error of 20 per cent requires a large initial investment for exploration and measurement. Therefore, mostly the decision on the working of any mine is taken on the basis of overall information available at the time of decision making.

Indicated reserves are also demonstrated type of deposit and they are partially delineated, though unlike measured one, quantity and quality of ore is drawn out of sampling results. These indicated reserves can be brought to the measured category only by increase in the bulk of sampling and enhanced measurement methodology. With limited information from the sampling a deposit may be categorised as inferred reserve. A known reserve, which is established by ordinary sampling activities, can be inferred of its approximate quantity and quality by using previous knowledge of extension of continuity of rock structures from the measured or indicated reserves. But here the degree of certainty about the grade and quantity is very low.

Again mineral resources are sub-grouped into three types : identified sub-economic, hypothetical and speculative. Hypothetical and speculative resources are undiscovered and are not explored on the site of the deposit. The identified resources which are sub-economic are so because they are not exploitable under the present knowledge of technology. By research and development of technology further these resources may be brought to the reserve category. The hypothetical resources are nothing of expectation of a reserve of known material in the same geological setting as they occur in other places. They are mostly expected in the established mining districts whose geological settings are previously well studied. By subsequent confirmation of the ore body they may be further classified either as identified sub-economic resource or a reserve. Speculative resources are those containing materials which are yet to be recognised as economic and expected to occur in the known types of deposits in a favourable geological setting. So in a broad sense this can be told that speculative resources are additional resources, undiscovered and not yet economic.

The following diagram (Fig. 10.1) prepared by Jean-Paul Drolet[10] gives a visual impression of the classification of mineral resources.

Jean-Paul Drolet has drawn the diagram taking a generalised form of classification from the original classification of resources prepared jointly by United States Geological Survey and United States Bureau of Mines. According to Drolet, from the unknown reserve of total mineral reserve only a part of it is

*Drolet's Diagram showing Classification of Mineral Resources*

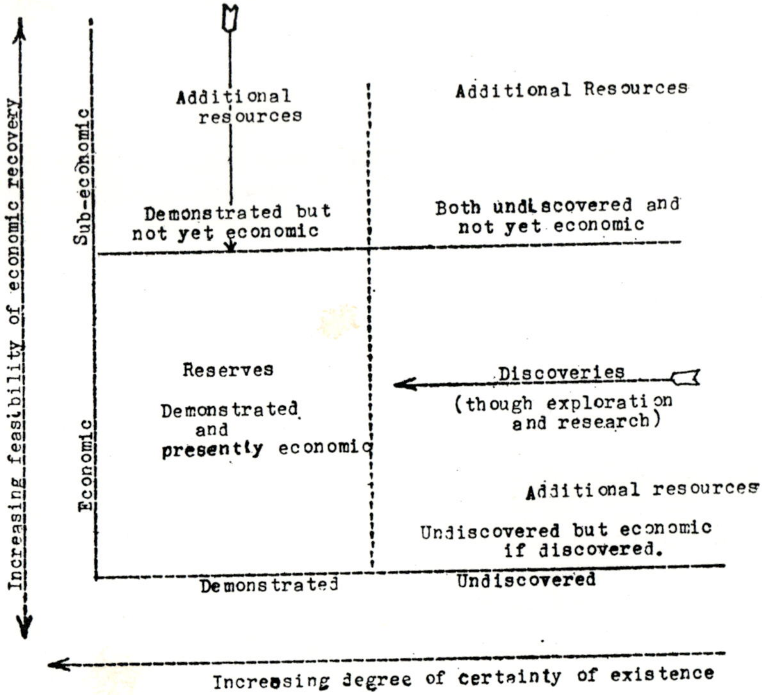


FIG. 10.1

discovered and demonstrated and considerd to be economical, whereas there are economic deposits which are yet to be demonstrated. With the present knowledge of science and technology a major part of deposits in the earth are considered to be uneconomical, out of which a portion is demonstrated and another is not.

## THE CONCEPT OF ECONOMIC DEVELOPMENT

The meaning of economic development has become increasingly clear in the modern world. Since 1945, economic development has become a major social and political issue in the world affairs. This is not due to sudden discovery of the extent of poverty in the world, but rather to a change in social attitudes towards its existence. The fact of poverty and its distribution have always been available. A century and half ago, the economically advanced countries of Europe and North America produced as much as two to three times more per capita output than other less favoured regions of the world. But now the gaps have increased from ten to twenty times than ever before in the case of some countries. This led to the development an area of vastly productive urban, industrial and technologically oriented society.

This diverse pattern of development was of course the product of a number of historical, cultural and circumstances. But the consciousness to alleviate poverty from these regions were not accepted until after the Second World War, when most of the colonial territories were freed and achieved independence. After becoming free of external control, these emerging nations immediately set out after higher standard of living; and most of these countries designed their political and economical set-up for the economic development. Also at this point of time economists started devoting their attention towards analysing the problems of developing countries. The emerging leaders of these countries desired to promote rapid economic development and developed nations realised that "poverty anywhere is a threat to prosperity everywhere".[11] As Meier and Baldwin have remarked, "A study of the poverty of nations has even more urgency than a study of the Wealth of Nations".[12]

However, the study of poverty of nations and methods of removing them cannot be based on the experience of the developed countries, because for the developed, rich nations it has been a tendency to consider economic development for granted—"as something that takes care of itself and to concentrate on short term oscillations of the economy".[13] Therefore, Myrdal[14] advised that, "the under-developed countries should not accept our inherited economic theory uncritically, but remould it to fit their own problems and interests so far economic theory has not concerned itself with the problems of under-developed countries, nevertheless if it is uncritically applied to their problem, it becomes wrong".

## THE VICIOUS CIRCLE OF POVERTY

In discussions of the problem of economic development, a phrase crops up frequently is 'the vicious circle of poverty'.

It is generally debated as "something too obvious and implies a circular constellation of forces tending to act and react upon one another in such a way as to keep a poor country in a state of poverty".[15] It is very easy to imagine such a situation in a poverty-stricken economy. For example, a poor man has not enough to eat; being under-fed he is weak; being physically weak he must not have a good working capacity; being his working ability is low his ability of production must be low; so he is a poor man; and being poor against he is weak, working capability is low and so on. A situation of this kind relating to a country on the whole may be summed up thus : a country is poor because it is poor.

Perhaps the most important circular relationship of this kind are those that effect the capital accumulation in economically backward regions. The supply of capital is controlled by ability and willingness to save. But there is a small capacity to save because of the low level of real income in these regions. The real income is low because of low productivity which is certainly due to lack of capital and the lack of capital is due to lack of saving. So this side of the capital formation also falls into the same type of circular causation.

On the other hand, the inducement to invest in production cycle may be low because of the small buying power of the people of the region, which is also due to their scanty real income. The low real income is due to low level of production which is again due to lack of investment, so a circular relationship exists on the both sides of the problem of capital formation in poverty-ridden areas.

So the supply and demand of capital formation is the important aspect of economic development and so the concept of 'big push' comes into the picture. Usually the trouble in the supply side receives all emphasis. "But the possible block on the demand side once one becomes aware of it, is also fairly obvious, though it may not be so serious, or so difficult to remove, as the supply deficiency".[16]

Though a major emphasis is given on the capital, but it should be remembered that capital is not everything. In addition to the capital problem there is another problem which can keep a country poor, for instance, lack of mineral resources, water resources, land resources and other bunch of natural resources. Some of the poorer countries of the world today are so due to lack of natural resources. So on the whole it can be said that the conscious undeveloped regions can come out of the vicious circle both by over-coming the problem of capital formation and identification and efficient utilisation of natural resources. All that one can say generally that, no country with any extent of poverty can come out of the vicious circle starting from the bare minimum of utilising primarily their natural resources.

## ROLE OF MINERAL RESOURCES IN ECONOMIC DEVELOPMENT

The importance of any natural resource in relation to the economic development of the region is well recognised. The natural resources together with the human resources constitute the principal economic asset to any economy. It might be true that, in the case of already developed countries, where they have already a good amount of accumulated capital and a large store of technology, the presence of natural resource under their territory is not necessary. But in the case of developing countries like India, the economic development revolves around largely the exploitation and the rational use of their natural resources.

Looking back into history one can conclude all the prosperous countries of the present world are either highly endowed with natural resources or they have taken advantage of historical events to exploit other regions which were rich in natural resources. Still one cannot forget their successful combination of labour, skill and knowledge of technology to utilise these resources they have drawn so rationally to bring them to effluence. There are also examples of countries with rich endowment of natural resources but still lying low in the process of economic development. The obvious lack of entrepreneurial ability to harness these resources in the same manner as few countries could do while they were leaping forward in the economic standard.

If we take the conditions of the developing countries, one can foresee the type and immediate limit of economic development considering its resources.[17] So again now we can say that the developing countries which are on the stride of development should start with the locally available resources and so as they can increase their standard of living and purchasing power of more sophisticated technology to make the pace of development fast.[18]

So looking into the different facts regarding natural resources, one can conclude that, for the economic development of any region, the natural resource it possess will play a determining role. In the debate of agriculture and industry, one finds many interesting points both in favour and against of one or the other. But keeping in mind the Indian condition one can say a balanced growth strategy would have a more welcoming effect on the economic field. But again looking at the per capita output from agriculture, one has to recommend cutting the size of labour force from that sector and putting them in industries. It can be readily appreciated that while the development of agricultural resources is essentially a matter of steady growth over a longer period, the exploitation of mineral resources for the industries can be a relatively short-term process. There are instances where the exploitation of mineral resources and industrialisation has given a dramatic effect on the economic development of the region.

## CONCLUSION

All this suggests that mining has been the cutting edge for economic development, the touchstone for advancement of backward regions. The economic development of a country depends to a great extent on the availability of mineral resources and their utilisation for the national purpose. "Industrialisation is not an end in itself; the development and utilisation of all available natural resources, as part of the total economic development process must increasingly improve the socio-economic conditions of the people of the region".[19] Industries based on mineral resources are always the key growth point in the economics of the developing regions, and these regions obviously become the growth centre to give impetus to grow other industries surrounding it. So mining and mineral based industries are considered to be one of the prime activities in any economy which is taking a rapid upward move in the economic development.

## REFERENCES

1. A.A. Archer, "Progress and Prospects of Marine Mining", *Mining Magazine*, March, 1974, p 35.
2. United Nations, *Natural Resources of Developing Countries*, New York, 1970, p 7.
3. R.K. Sinha and N.L. Sharma, *Mineral Economics*, New Delhi, Oxford and IBH, 1976, p 8.
4. It should be remembered that the amount of reserve is in all cases no more than simple estimates, even when the characteristics of a deposit have been examined as completely as possible, and with all desired competence, subsequent production does not always confirm them exactly and it may rightly be said that no mineral deposit is perfectly known before it has been completely worked out. United Nations, *Mineral Resource Development of Developing Countries*, p 6.
5. R. Bosson and B. Varon, *Mining Industry and the Developing Countries*, New York, Oxford University Press, 1977, p 8.
6. Op. cit., p 12.
7. C. Karunakaran, "Some Problems Related to Development of Mineral Resources in India", *Proceedings of Seminar on Science and Development Cammittee on Science and Technology in Developing Countries*, Bangalore, 1975, pp 71-72.
8. Op. cit., p 36.
9. Op. cit., p 215.
10. Jean-Paul Drolet presented a paper at the International Symposium on "Canada's non-renewable resources", Toronto, 1974 on "The Demand for Canada's Mineral Resources", The Introductory part of the paper contained this diagram to show the classification. (Memeographed).
11. M.L. Jhingan, *The Economics of Development and Planning*, Delhi, Vikas Publishing House, 1975, p 3.
12. G.M. Meier and R.E. Balowin, *Economic Development Theory, History and Policy*, Bombay, Asia Publishing House, 1976, p 12.
13. R. Nurkse, *Problems of Capital Formation in under-developed Countries*, New York, Oxford University Press, 1953, p 12.
14. G. Myrdal, *Economic Theory and under-developed Regions*, London, Methuen, 1969, p 99.
15. R. Nurkse, "The Theory of Development and the Idea of Balanced Growth", *Developing the under-developed Countries*, (ed.) Mountjoy, A.B., London, Mac Millan, 1975, p 115.
16. Op. cit., p 116.
17. W.A. Lewis, *The Theory of Economic Growth*, London, Allen and Unwin, 1955, p 52.
18. J.I. Fisher, *The Rule of Natural Resources in Economic Development, Principles and Patterns*, in H.F. William and J.A. Buttrick, (eds), *Economic Development*, New York, Allen and Unwin, 1954, p 32.
19. Kenneth Warren, "*Mineral Resources*", Harmondsworth 15, Pelican Books, 1973, p 3.

# 11

# Progress and Problems of Coal Mining Industry in India

—*Debidutta P. P. Satpathi*

## INTRODUCTION

THE COAL mining industry is one of the oldest industries of India, initiated at Raniganj, West Bengal in 1774. "By 1860, nearly 50 collieries were producing about 282 thousand tonnes of coal per annum in Raniganj area."[1] Production started in the coalfields of Madhya Pradesh, Andhra Pradesh, Upper Assam, etc. during the last decade of the nineteenth century. The Jharia coalfield went into production early in the twentieth century. At the beginning of the century India was producing 6 million tonnes of coal of which 5 million tonnes came from the coalfields of Raniganj, Jharia and Giridih. The First World War gave impetus to the coal mining industry and during that period new coalfields were developed at Bokaro, Panch Valley and Chanda Valley. At the beginning of the First World War, India was producing 16 million tonnes of coal of which 15 million tonnes were contributed by Jharia and Raniganj coalfields. In between the I and II World Wars the coal mining industry passed through periods of depression, slump, and uncertainties. Selective minings were adopted to maintain production. Some uneconomic coal mines were abandoned. Since the Second World War, coal has been in ever-increasing demand, and has had no occasion to look back.[2] During the Post-Independence period the production increased from 3.2 crore tonnes in 1950 to well over 10 crore tonnes at present.

## DISTRIBUTION OF COALFIELDS

The god of Geology was not a democrat. He favoured certain areas of this country with abundant coal and deprived others who are without that contemptible fuel. Almost all the important coalfields of India are located in a triangular area between the Damodar Valley in the East, the Godavari Valley in the South, and the Chambal Valley in the West covering almost entire M.P., Chotanagpur Plateau of Bihar, Western Orissa, South-western West Bengal and North-eastern Maharashtra (Fig. 11.1). Within this vast geographical area lie about 80 million tonnes of coal in the Gondwana beds. Tertiary coal is obtained in Jammu and Kashmir and Assam and lignite is found in Tamil Nadu. The Gondwana coal is good while Tertiary coal is inferior both in quality and quantity.

*India : Showing Coalfields*

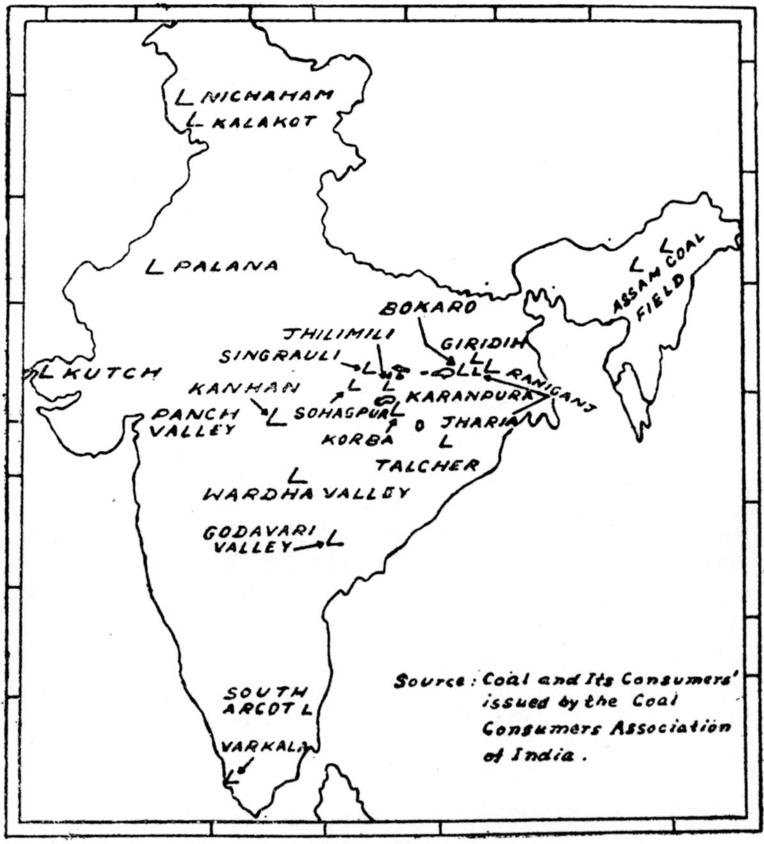


FIG. 11.1

## COAL RESERVES

The gross coal reserves of India are estimated at over 82,000 million tonnes of which nearly 2,000 million tonnes are of coking variety. Besides, there are also reserves of lignite estimated at about 2,000 million tonnes. "In other words, this amounts to nearly 2,000 tonnes of the nation's oil reserve in terms of heat content."[3] The reserves of both coking and non-coking coals might increase due to the discovery of a few new coalfields. The Geological Survey of India has discovered an additional reserve of 35 million tonnes of medium coking coal, spread over 5 square kilometres in the Parez block of West Bokaro, in Hazaribagh district. The area was previously not known to be coal-bearing. This discovery is quite important considering the country's dwindling coking coal reserves. As estimated 100 million tonnes superior quality non-coking coal has been found in the Moira-Khandra area of the Raniganj coalfields. The G.S.I. has established a proven reserve of 450 million tonnes at Ramagundam in Andhra Pradesh. In Sohagpur, M.P., 16.8 million tonnes of "Selected Grade" non-coking coal has been found. The area was previously not known to be coal-bearing. Besides, Bankura district of West Bengal is reported to have a great deposit of coal of inferior quality. The position in regard to non-coking coal is not unsatisfactory. But "the coking coal reserves are a matter of concern for the future."[4]

## COAL MANAGEMENT

All the coking coal mines were nationalized in 1971. Two years later all the non-coking coal mines were also nationalized. Coal, therefore, today is entirely in the public sector. The only important mines which have not been nationalized are the captive coking coal mines of the two private sector steel companies."[5] The production of coal in public sector is organised mainly by Coal India Limited through its subsidiary companies. There are Eastern Coal Fields Ltd., Bharat Coking Coal Ltd., Central Coal Fields Ltd., Western Coal Fields Ltd., and Coal Mines Planning and Design Institute Ltd. The National Coal Development Corporation (NCDC) formed in 1956 manages a large number of coking and non-coking coal mines.

## COAL PRODUCTION

We have the fourth largest deposits of coal in the world after U.S.A., U.S.S.R. and China, but we are seventh in production. "A small country like Poland produces 161 million tonnes, a year with reserves of 13 billion tonnes only.[6] Between 1868 and 1978 the production of coal has been increased by 2,000 tonnes which is illustrated by the following figures :

TABLE 11.1

*Production of Coal between 1868-1978*

(in million tonnes)

| Year | Production | |
|---|---|---|
| 1868 | 0.05 | |
| 1900 | 0.61 | |
| 1930 | 23.00 | |
| 1946 | 26.00 | |
| 1950 | 32.00 | |
| 1960 | 52.00 | |
| 1971 | 75.00 | |
| 1975 | 95.89 | |
| 1976 | 100.87 | |
| 1977 | 75.55 | (up to Sept. 1977) |

From Table 11.1, it is evident that coal production up to the end of the nineteenth century was quite nominal. The First World War gave much impetus to the industry. As a result, the industry registered a phenomenal growth during that period. During the period of planned development in the post-Independence era the production has not been phenomenal although we need more production for our economic development.

### Coal Production during the Plans

For an overall economic development, India started five year plans after Independence. During the First Five Year Plan (1951-1956) the target of coal production was 39 million tonnes but the actual production at the end of the Plan period was 38.23 million tonnes (Private Sector : 33.67 million tonnes,

Public Sector : 4.56 million tonnes) resulting in a shortfall of 0.69 million tonnes (Fig. 11.2). At the end of the Second Five Year Plan the total coal output was 55.72 million tonnes (Private Sector : 45.05 million tonnes, Public Sector : 10.67 million tonnes) against the target of 60.00 million tonnes resulting in a deficit of 4.28 million tonnes.[7] During the Third Plan the target was 75.93 million tonnes (Private Sector : 54.09 million tonnes and Public Sector : 13.65 million tonnes) and at the end of the Fourth Plan coal production amounted to 77.87 million tonnes against a target of 106 million tonnes. The Fifth Plan target was 135 million tonnes of which the share of the non-coking coal would be 103 million tonnes. The estimated requirements of coal during the Sixth Plan has been envisaged by the Fuel Policy Committee at 225 million tonnes (Fig. 11.2).

*Coal under Planning* (*in million tonnes*)

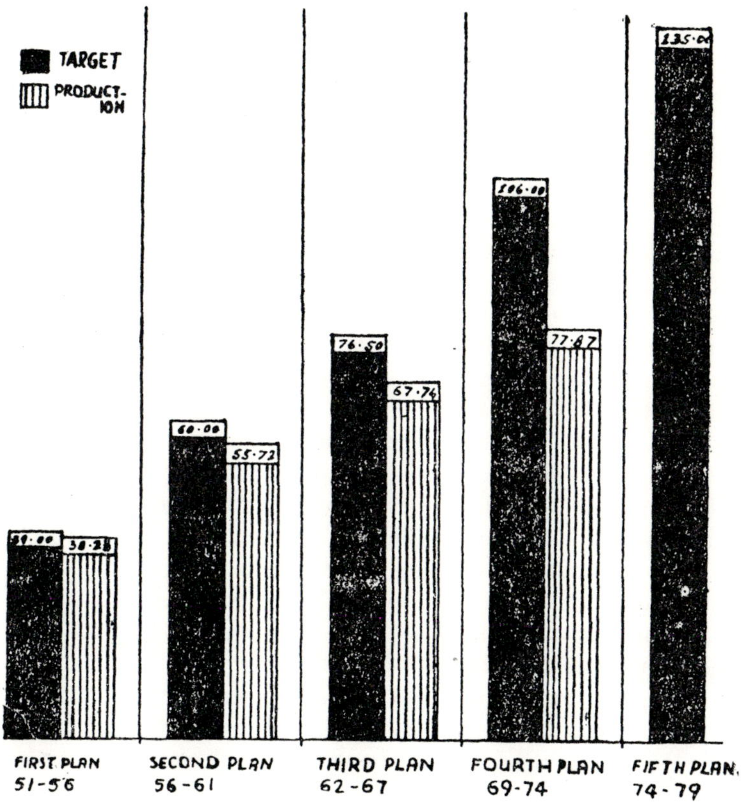


FIG. 11.2

From the above facts it is clear that during the earlier phases of planned development of coal mining industry, India reached very near to its production target. But in the later phases especially the targets for the Fifth and Sixth Plans seem improbable under the present conditions. The estimated coal production during 1973-74 is about 80 million tonnes. The increase in production, envisaged during the Fifth Plan was 45 million tonnes which meant an annual growth rate of 11.03 per cent. Between 1951 and 1972 the coal production in India registered a growth rate of 2.7 per cent and the best was 6.9% during the Second Plan period. The growth rate envisaged for the Fifth Plan, therefore, is almost double the best growth rate achieved so far. "Even in the advanced coal mining countries a growth rate of 11 per cent has not been achieved except in the case of China during the period 1950-60."[8] The European countries where the growth rate has been higher than that of India, have the advantage of a sound industrial base, efficient power supply, speedy movement of coal due to efficient facilities of transport, less absenteeism and strike of the labour force, etc. But the Indian coal mining is functioning under adverse conditions. It is constantly plagued by power shortage, lack of transport facilities, dearth of trained manpower, lack of efficient equipments, backwardness in mining technology, absenteeism apathetic attitude of the labourers towards work, frequent strikes, etc. Under these conditions a growth rate of 11 per cent is too ambitious by any standard.

According to a report, the average daily production of coal in the Coal India mines from 1 April to 17 June 1978 was 263,000 tonnes against the target of 303,000 tonnes.[9] The average raising in the corresponding period in 1977 was 289,000 tonnes and in 1976—"the best year for the CIL"—292,000 tonnes. When the demand for coal has been mounting, a daily shortfall of 40,000 tonnes in production but several factors are responsible for the low production. As against the actual raising of 88.96 million tonnes in 1977-78, the CIL target for the current year (1978-79) has been set at 100.5 million tonnes. But it will be difficult to reach the target unless the constraints on production and despatch of coal are overcome. The factors contributing to the shortfall in production are unsteady and erratic supply of power, high absenteeism in the mines, especially in the eastern coalfield area, a deplorable land and order situation, bad industrial relations and paucity of railway wagons leading to despatch bottlenecks.

The Energy Ministry of India has decided to launch a 25-crore crash programme to increase coal production in the western region to feed the power stations. Under the scheme, four opencast mines will be developed to yield a monthly production of 100,000 tonnes. Three of these will be at Shahdol in Madhya Pradesh and the other in Chandrapur district in Maharashtra. The Ministry has decided that the pit-head power stations of this region such as Koradi/Satpura and several others should get coal from the pit-head mines so that the pressure on wagons is somewhat eased. This will be possible with the opening of the opencast mines under the crash programme.

Against a daily allotment of 11,000 wagons, the Railways have been providing on an average only 9,200 wagons while the actual need is for 11,000 wagons. The Railways are unable to provide more wagons because imported fertilizer is being received at Tuticorin due to congestion at Bombay port, and this had to be carried by rail to other regions. The movement of foodgrains is also adding to the pressure on wagons.

Notwithstanding the problem of distribution, the Ministry is confident that the production target of 113 million tonnes of coal for the current year will be reached. The Western Coalfields Limited is all set to increase its production from 21 million tonnes to 25.5 million tonnes. The Central Coalfields and the Eastern Coalfields which are lagging behind are also expected to pick up.

Faced with the gigantic task of augmenting coal production from 100 million tonnes in 1977 to 180 million tonnes in 1978 the Government is also contemplating to switch over to new techniques in coal mining. The Central Mining Research Station (CMRS) at Dhanbad has developed expertise on strata control in long wall mining through intensive underground observations backed by laboratory investigations. The ever increasing target of coal production can be achieved at least partially by the method of mining. The coal industry could adopt the new system of mining known as long wall caving to achieve this end. It was safe and at the same time gave high productivity and large production from a single mine. Most of the major mining countries were either following this method or were trying to switch over to it. Adoption of this system in the East Katras, Ningarh and Moonidih mines has resulted in three tonnes face output per panshift in place of less than two tonnes obtained by using the board and pillar method which was the only method so far in vogue in underground coal mines.

### IMPACT OF NATIONALISATION ON THE COAL PRODUCTION AND DISTRIBUTION

The Government of India nationalised the coking coal mines in 1971 and the non-coking coal mines in 1973. Prior to nationalisation there were 884 coal mines in India—670 non-coking coal mines situated in West Bengal, Bihar, M.P., Gujarat, Orissa, Assam, Andhra Pradesh and Tamil Nadu and 214 coking coal mines concentrated mainly in Jharia, Barakar and West Bokaro in Bihar and Sohagpur in Madhya Pradesh. According to another report the Union Government nationalised 711 non-coking coal mines in 1973 of which 184 were vested in the Bharat Coking Coal Limited (BCCL) on the basis of their continuity and convenience of operations. Among the remaining mines 71 were small, isolated and seasonal of which only 30 were coking mines. CMA has, therefore, 297 mines. Taking into account 44 mines of National Coal Development Corporation (NCDC) already in the public sector, the total number of working mines with CMA at present is 341. An official review of the performance of these mines for 1973-74 shows that the production increased from 46.27 million tonnes in 1972 to 48.41 million tonnes in 1973. During 1973-74 as a whole, the total production achieved until the end of Feb. 1974 was 48.41 million tonnes. The original target of coal production for 1974-75 was 90 million tonnes for the country as a whole.[10] This consisted of 62 million tonnes for CMA, 10 million tonnes for BCCL, 6 million tonnes for Singareni and 3 million tonnes for TISCO and IISCO. For that period, therefore, the following division-wise targets were fixed :

| | (in million tonnes) |
|---|---|
| Eastern Division | 24.00 |
| Central Division | 18.99 |
| Western Division | 19.05 |
| Assam Division | 0.64 |

The production increase envisaged for 1974-75 is as high as 9 million tonnes corresponding to 90 million tonnes.

Needs of conservation and scientific development of coal resources have led to the nationalisation of coal mines. Prior to nationalisation bulk of production came from the private sector and the public sector which played only a secondary role in coal production. "When the well-intentioned and sisterly warnings by the Government of inadequate wages, inadequate capitalisation for development and the unscrupulous exploitation of mining resources were ignored by the private sector, it was decreed that coal be nationalised."[11] It was arrived at ensuring rational and coordinated development of coal production and for promoting optimum utilisation of coal resources consistent with the growing requirement with the economy. With the total nationalisation of coal the situation has radically altered. The leaseholds have lost the meaning. The size of the collieries can be substantial. With the large reserves available for each block, heavy investment and greater concentration of production are possible. For the first time, planning is no longer confined to the small leasehold areas held by different lessees but could be viewed at coalfield levels. "The shift is from micro to macro, insofar as coal mining is concerned."[12]

The Government claims that within a year of nationalisation coal production in India increased by 10 million tonnes which is no mean achievement. The Government has also made some qualitative changes in the approach to the coal mining industry. Coal has already been put into the core sector. The Government decided to lay all emphasis in the development of the coal mining industry during the first year of the Fifth Plan. There has been increase from the initial proposed outlay of Rs. 737 crores on coal development to Rs. 1200 crores.

In spite of all these, the critics feel that nationalisation has done harm to the industry instead of improving its condition. They say that subsequent to the nationalisation there was a "Coal famine" in which both the industries and domestic consumers suffered equally. Supply position deteriorated sharply and the coal marketed was of inferior quality. One of the former presidents of the Indian Mining Association remarked that "Nationalisation of Colliery is essentially a political move and has no economic relevance". According to Das[13] India produced 80 million tonnes of coal in 1969-70 but it declined to 75.25 million tonnes in 1971-72 and in West Bengal alone production declined by about 4 million tonnes during the corresponding period. The real position regarding coal production, stock and railway transport is illustrated in Table 11.2.

TABLE 11.2

| Year | Production | Pit-head stocks | Daily average loading |
|---|---|---|---|
| | (Figures in million tonnes) | | (No. of wagons per day) |
| | *All India* | | |
| 1969-70 | 80.00 | 7.08 | 8137 |
| 1970-71 | 76.34 | 9.58 | 7461 |
| 1971-72 | 75.28 | 7.83 | 7750 |
| | *Bengal-Bihar fields* | | |
| 1969-70 | 55.67 | 5.22 | 6378 |
| 1970-71 | 52.43 | 7.97 | 5774 |
| 1971-72 | 49.17 | 6.47 | 5606 |
| | *West Bengal* | | |
| 1969-70 | 20.29 | 1.35 | 2178 |
| 1970-71 | 18.61 | 2.20 | 1935 |
| 1971-72 | 17.01 | 2.06 | 1810 |

It is evident that coal production declined during the post-nationalisation period. But paucity of railway wagons to carry coal from pit-head sites and shortage of inputs like power contributed much to the slump in production. The unprecedented power shortage of 1972 seriously hampered coal production in West Bengal. But during that period coal production in other states considerably increased as a result of incentives given by the State Government like Andhra Pradesh, Madhya Pradesh, Maharashtra, etc.

**Problems of the Coal Mining Industry**

There is many a disquieting feature in the coal mining industry of India. The coal mining industry is bedevilled by a large number of problems like accidents, hazards of fire, flood havocs, absenteeisms, shortage of inputs like power, explosives, skilled manpower, finance, sophisticated mining equipment, paucity of railway wagons to carry coal from pit-head, lack of port facilities for export of coal to foreign countries, etc.

**Accidents in Coal Mines** : Although working conditions inside our coal mines are far away from being called "ideal" mining conditions in our coal mines are relatively easy. But the worker is often confronted with several mining hazards like occurrence of explosive gases, caveing-in of the roofs, wall collapse, flooding, haulage breakdowns, ventilation system, faults causing suffocation and death, etc. The recent and catastrophic accident occurred in the Chasnala Colliery in December 1975, flooding of the mine by the subsidence of the surface and release of enormous water stored in previous mining. That accident took a toll of 375 miners inside the mine. The previous superficial mining created a big underground cavern which was converted into a reservoir of water (Fig. 11.3). No subsequent mine owner tried to pump out water from that reservoir. Instead two shafts were sunk to reach lower levels of the seam which has a dip varying from 30° to 70°. "A width of 80 feet of the seam was left as a barrier between the old and the new mines—between the water-filled hollow and the new excavations".[14] This barrier was perhaps not wide enough to contain the weight of an unestimated amount of water. The overburden on the coal seam consisting of shale, clay and sandstone subsided and the water gushed into the galleries underground like a tidal wave from the top. It is not certain whether the subsidence was due to the vacuum created by the abandoned mine or the strata in the seam slid along a natural line of cleavage. It seems to be a man-made accident. Because, prior to this incident an expert advised the then owner of that coal mine not to work there without pumping out the water accumulated in the abandoned mine. But nobody did pay any heed and as a consequence so many people lost their lives.

*Position of the Coal Seams and Water Reservoir during Accident*

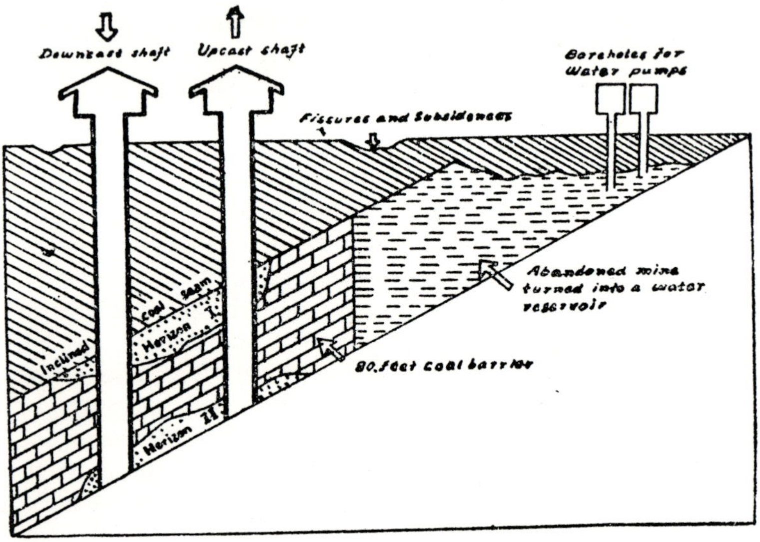


FIG. 11.3

Natural accidents occurred in Dhori Colliery in 1962 where explosive gases appeared all of a sudden killing hundreds of working labourers. Ten persons lost their lives in Saunda Colliery. Prior to these, accidents occurred in Ramgarh Colliery of IISCO, Barra Dhenod Colliery, Chinna Kusi Colliery, etc. but the loss of human lives was not great as in Chasnala or Dhori. In spite of rationalisation and use of modern techniques, we are yet to make our mines accident-proof.

**Flood Havocs** : During the monsoon months a major part of Eastern India especially lower West Bengal and Bihar regions which contribute about 60-70% of the total coal production of the country remains flooded. Due to unprecedented flood this year coal raising in the mines of West Bengal almost ceased for a month or so. In 1978, 45 out of a total of 116 mines in the

Eastern Coalfields Ltd. (ECL) had been flooded. A similar number of mines belonging to the Bharat Coking Coal Ltd. had been affected. Altogether, 26,500 workers of the two companies had been affected due to rains and floods. The production in the ECL might have come down by 4-5 million tonnes from the targeted 26 million tonnes for 1978-79. The BCCL production might have suffered to the extent of 3 million tonnes. These two companies have incurred an approximate loss of rupees 20 crores due to flood.

The flood caused a serious accident and serious damage to coal mining in the monsoon of 1976 in the Saunda colliery. Before the accident in that colliery, the Damodar and its tributaries were already in high spate and about 145 million gallons of water rushed inside the mine from a nearby tributary of the Damodar which was in high spate. Flood waters rose high and covered the surface area having cracks caused by earlier extraction of pillars by caving methods, entered the mine and trapped ten miners who lost their lives. We are yet to make any fool-proof scheme to save the mines from being inundated during the monsoon. As a former Irrigation Minister of India remarked "we have to learn to live with the floods", we have to be prepared for such eventualities.

**Fire Hazards** : The half-century old underground fire blazing in the Jharia coalfields is estimated to have already destroyed over 35 million tonnes of valuable coking coal. In other words, as many as one hundred-odd known coal seam fires, both surface and underground, covering an area of about 500 sq km of the Jharia Coalfield, the store house of the country's coking coal have been consumed in terms of value coal worth about Rs. 300 crores. Jharia, a thickly populated town in the heart of the coal-belt of Bihar can be said to be situated on a cauldron. The underground fire which has come out to the surface in as many as 23 places is threatening the very existence of this mining town. Flames often leap up through crevices on the ground near Jogra, Lodhna, and Fulitar. The extensive pot-holes observed on the roads which indicate that the ground may suddenly sink. Near the Junama Colliery 47 workers' houses and a two-storey building collapsed when 51,000 squre kilometre surface area subsided in March, 1975 killing two persons and injuring many. The fires which exist within a radius of 20 km from the town have affected the rail tracks and the roadways, especially the Jharia-Sindri Road, the plying of buses on which is fraught with great danger. It is believed that some illicit liquor makers initiated the fire which set the seams ablaze in the Jharia mines.

The BCCL which took over 214 coking coal mines as a result of nationalisation, has studied in depth the background aetiology of the Jharia fires and launched preventive measures for conservation of the limited resources of this valuable fuel essential for steel plants. It has divided the Jharia Coalfield into three major fire areas—Jogta, Lodha and Rajapur and launched a series of quenching-cum-salvaging measures in accordance with the nature of the fire. The mine-point method envisaged among others direct excavation after quenching massive sand stowing of the affected areas. Experts have suggested "blanketing and water pooling" on the surface but it is also found necessary to partly or fully flood the underground working to quarry out coal from the blazing seams. But nothing seems to have been done so far to save the valuable fuel.

In the coming years the requirement of stowing material would increase fast. "Sand availability is extremely short in the Jharia Coalfield. Alternative to sand like crushed material and blast furnace slag are being experimented with. At the same time investigations about the possibility of transporting sand from Damodar and Sone rivers are also in hand."[15]

**The Plight of the Coal Miners :** In 1970 our coal industry employed over 3,85,000 workers—the second largest labour intensive industry after the railways with an output per man shift of 0.67 tonne whereas the corresponding figures for Canada, Australia and the U.S.A. were 9.74, 9.91 and 17.81 tonnes per manshift respectively. At present about 6,00,000 coal miners are working in various coalfields of the country. The colossal human force is engaged in "energy producing industry second in importance perhaps only to the defence of our borders."[16]

Although according to the Government version after nationalisation the "days of ruthless exploitation" of the miners under the unscrupulous mine-owners are over and they are said to have ushered in an era of affluence, comfort and security. Yet the coal miner's lot does not seem to be any better at the present moment. Besides being subjected to the hazards of frequent accidents inside the mines he is perhaps the most exploited labour in India—in the clutches of both the recruiting agent and the money-lender. According to a report on money-lending problems in Dhanbad Coalfield "the sudden increase in the wages of coal miners after nationalisation has helped the growing of money-lending. . . .The money-lenders have great criminal power at their command. They operate as mafia gangsters. . . .they have virtually created a mini empire of their own in the colliery areas". Although the coal miners' wages have increased and they have got more leisure yet they are helpless before the human sharks—the trade union bosses and their musclemen who take away a major part of their wage and their condition is as miserable as before nationalisation. Besides, due to inhalation of poisonous gases and coal dust inside the mines many of the miners become victim to respiratory diseases. "Of the 3,91,536 coal miners engaged in 781 coal mines scattered in mine coal producing states in the country, 21,000 suffer from tuberculosis, 40,000 from pneumoconiosis (a disease caused by constant inhalation of coal dust and which has the same fatality rate as that of T.B.), 1,60,000 from anaemia, 1,55,500 from filarial infection and 9,000 from chronic bronchitis."[17] From the above figures it is clear that only a few fortunate persons in the coal mines are free from disease. Due to lack of proper medical facilities most of the miners die prematurely.

Very few coal miners have homes. Some of the collieries have "dhawras" (where miners live) which are without ventilation. Thus, providing every miner with well-ventilated quarters and amelioration of the mining ecology are an immediate necessity.

In spite of nationalisation 'bonded slavery' in coal miners is still prevalent in interior coal mines of Bihar. The inhuman practice should be abolished forthwith.

"The added security and higher wages for labour in the public sector adds another dimension of the labour-intensive industry which employs 3.85 lakh workers and expects to double this force by 1980. The basic thing is to provide an effective leadership and an awareness of the responsibility involved and not allow it to become a happy breeding ground for immature and harmful political thinking and thereby affect the production adversely."[18]

**Lack of Effective and Speedy Means to Transport Coal from Pit-heads** : "A rise in production also calls for a corresponding increase in transport capacity and an effective distribution system to ensure optimum utilisation of resources."[19] Due to shortage of wagons raised coal accumulate at pit-heads causing space problem, hampering production and causing inconvenience to consumer industries. According to a report of October 1978, the following 8 collieries were overburdened with extra-ordinary stock of raised coal :

TABLE 11.3

| Name of the colliery | Pit-head stock (in lakh tonne) |
|---|---|
| Kathara | 2.87 |
| Giddi 'A' | 2.57 |
| Sirka | 1.49 |
| Bhurrkunda | 1.46 |
| Jagarnath | 1.42 |
| N.S. Dhori | 1.41 |
| Dhakra 'B' | 1.00 |
| Dawarkand | 1.00 |

The simple reason accounting for the huge stocks of coal at the pit-heads is that the overall production has gone up and the railway could not lift it as envisaged. Some amount of coal is also sent by road. But still there was a demand of 58,000 wagons by the Central Coalfields Ltd. (CCL) per month after despatching 4-4.5 million tonnes of coal by road. Against this requirement CCL receives 30,000—35,000 wagons per month.

The inability of the railways to load the coal at the desired level is due to circumstances beyond its control. The blame must surely rest on Coal India Ltd., the collieries, the steel plants and some thermal power stations which receive coal on priority but hold up the wagons. The steel plants have been holding about 2,300 box wagons per day, thus depriving other consumers of wagons. Some of the thermal stations are to be blamed, as about a total of 300 wagons are being held by them. According to the railway authorities at present they are loading

10,000 wagons per day "which is a very high rate of loading" and which if sustained throughout the year can enable them to lift 87 million tonnes of coal in 1978-79 against estimated 84 million tonnes during the previous financial year.

Besides detention the railways also have to carry food grains, fertilizers, etc. to the deficit areas on the priority basis. Thus the railways cannot lift as much coal as is desired because they have to lift other goods as well. "But no plan of coal production will succeed unless adequate capacity is available to transport it to required consumptions".[20] Thus the Ministry of Railways should make sufficient arrangement of wagons to clear the coal stocks from the pit-heads because accumulation of coal besides causing immense loss to the thermal power plants, iron and steel industry, coal washeries, etc. also encourages theft and pilferage at the site.

**Illegal Coal Mining** : A number of persons have been allegedly mining coal illegally in contravention of the provisions of law, specially in the districts of Hazaribagh and Giridih, causing a huge loss of revenue to the Union and State Governments. Such unauthorised mining, "smuggling and illegal trade" of coal should be stopped forthwith.

**Shortage of Explosives** : Shortage of explosives has been seriously disrupting coal mining operations in 1978-79. The production from underground mines has already come down and the prospects of easing the shortage are none too bright because of sight in importing explosives. Against its daily requirements of 67 tonnes of explosives, CIL has been getting about 55 tonnes from two indigenous sources—the Gomia Explosives in Bihar and the Rourkela Unit of Detonators Ltd., Orissa. The former has an annual capacity of 10,000 tonnes whereas the latter can produce 8,000 tonnes per annum. But in both the units production has suffered in 1978-79 due to prolonged labour trouble.

Apart from the existing two units producing explosives, there is also a third unit with 5,000 tonnes capacity which has gone on steam—the Korba unit of Indo-Burma Petroleum. But its production has yet to be stabilised. The Bhatinda unit being jointly promoted by CIL and Director General of Ordnance Factories is expected to start production sometime next year. The feasibility report for production of 5,000 tonnes of nitroglycerine based on permitted explosives by utilising the available surplus capacity of the Ordnance Factory in Maharashtra is under preparation on the basis of the studies made by the erstwhile CMAL and the Director General of Ordnance factories. Another project based on ammonium nitrate is to be set up by Fertilizer Corporation of India (FCI) at Sindri to produce slurry type explosives. It will have an annual capacity of 15,000 tonnes.

The demand for explosives by the end Fifth Plan is expected to be about 41,000 tonnes per annum. With the existing indigenous production, it will be difficult to meet the requirements of coal mines. Explosives manufacturers, however, maintain that the shortage is not so much due to lower availability of explosives as to an increase in requirements of CIL which has stepped up mining operations. Whatever might be the cause of shortage, the position at present is that the requirements of the CIL cannot be met without import. It is reported that CIL has floated global tenders for importing about 5,000 tonnes of permitted explosives to replenish its dwindling stocks. But there will be clearance difficulties as no country is willing to allow the planes carrying the explosives to pass through its territory. The explosives may also be imported by sea, but it is time consuming. In the event of sea import, it may not be possible to maintain normal productions from underground mines as the explosive stocks have come down to a precariously low level.

**Obsolete Equipments and out-of-date Techniques of Production** : In order to maintain a fast tempo of progress we have to change the out-dated equipments and replace them by modern ones. "Employment of heavy machinery and latest mining techniques is the only way to maximise production and achieve self-sufficiency in coal."[21] We have the technical know-how to produce most of the mining equipments indigenously.

"The Government expects to utilise the idle plant capacity of MAMC, Durgapur, and HEC, Ranchi, to turn out the necessary mining equipment."[22] But it may be necessary to import a substantial part of the equipments from the foreign countries. "With a certain amount of balancing equipment for immediate increase in production, concerted efforts must be made to convert the pick coal into blasted coal and the matching transport capacity must be created."[23] The strategy of modernising our coal mining methods will be the following :

(1) progressive replacement of pick mining by partial mechanical coal mining by blasting or by solid blasting;
(2) providing balancing equipment to improve the utilisation of the underground transport and winding capacities;
(3) modernising pit and incline tops and bottom arrangements;
(4) rationalising underground workings to concentrate and hence improve production from selective productive districts; and
(5) provide better environment and working facilities to workers and staff.

Employment of heavy machinery and later mining techniques is the way to maximise production and achieve self-sufficiency in coal. The coal industry could adopt the new system of mining known as "longwall caving" to achieve higher production. It is safe and at the same time give high productivity.

**Irregular Coal Supply to Steel Plants and Power Houses** : The iron and steel factories need coking coal. The daily coal requirement by our steel plants is about 41,000 tonnes. But CIL has failed to supply required coal to our steel plants due to the shortage of wagons. But other factors also partially contribute to this short supply. The coal stocks in steel plants have been falling down since April 1977.

A declining trend in coal supplies and low stocks of coking coal at the steel plants has led the steel industry to ration coal consumption commensurate with coal receipts. According to the Steel Authority of India Ltd. (SAIL) sources coal stocks with steel plants on 1 September, 1978 was 277,000 tonnes against 456,000 tonnes on 1 September, 1977 and 714,000 tonnes on 1 September, 1976.

In order to achieve the pig iron and steel target during the Plan periods we have to keep the ovens constantly burning. Therefore, every effort should be made to supply coking coal—the prime fuel, to the steel plants. Due to their expansion Bhilai and Bokaro Steel Plants will have greater demand of coal. The Coal Department has worked out plans for regular supply of over 47,000 tonnes of coal every day to steel plants.

We have established several thermal and super thermal power stations which are solely dependent on coal. Without a regular coal supply production of electricity is hampered. In the summer of 1978 the Corba thermal power station in Mirzapur district of U.P. (the biggest thermal station in India) which feeds almost one-third of Uttar Pradesh faced the worst crisis of coal. Some units of this power station had to be temporarily closed due to coal shortage.

**Inadequate Port Facilities :** We have a great prospect of marketing our non-coking coal abroad and thereby earn some foreign exchange. Japan, Hongkong, Nigeria, Germany and our coal-hungry neighbouring countries with whom we have good diplomatic relations, are the traditional buyers of our coal. Table 11.4 shows the quantum of exports during the past years :

TABLE 11.4

*Coal Export from India*

| Year | Quantity exported (in million tonnes) |
|---|---|
| 1900 | 0.6 |
| 1910 | 0.9 |
| 1921 | 0.27 |
| 1931 | 0.44 |
| 1937 | 1.33 |
| 1945 | 0.46 |
| 1952 | 3.35 |
| 1959 | 1.43 |
| 1964 | 1.29 |
| 1970 | 0.47 |
| 1975-76 | 0.50 |
| 1976-77 | 0.60 |

Although our export reached 3.35 million tonnes in 1952 but in subsequent years it dropped abruptly due to the following causes : (1) the inability to supply coal on a sustained basis, (2) the foreign buyer's choice in favour of oil, and (3) development of power from other forms of fuel.

Japan is an important buyer of Indian coal. The requirements of non-coking coal by Japan were earlier indicated to be as follows :

TABLE 11.5
*Export of Indian Coal to Japan*

| Year | | Million Tonnes |
|---|---|---|
| 1975-76 | | 3.0 |
| 1980-81 | (Projected) | 4.7 |
| 1985-86 | (Projected) | 10.2 |
| 1990-91 | (Projected) | 15.0 |

(Import quota given by MITI, Japan)

"In the long-term plan for exports, Coal India is banking on the export possibilities mainly to EEC and Far East countries, particularly Japan, besides our neighbouring countries. Bangladesh, Burma, Bhutan and Nepal are the traditional importers of Indian coal. Exports to neighbouring countries are being increased to over five lakh tonnes per annum, 15,500 tonnes has been exported to Pakistan for the first time in recent years."[24] In June 1976, about 16,000 tonnes of coal was shifted to Belgium *via* Paradeep port. This was followed by shipments to France, Taiwan, Denmark, Japan, Holland, etc. Although Japan imports non-coking coal from Australia, USSR, China, etc., our coal competes favourably with those coals. As a result Japan is much more interested in importing Indian coal. "Being oil-hungry and conscious of pollution hazards, both Europe and Japan seem to be willing to accept the low grade Indian coal because it is sulphur-free."[25]

But lack of adequate port facilities affect our export of coal adversely. Because of shallow sea coast large vessels cannot reach the ports. Except Calcutta which handles 1.5 million tonnes of coal per year there is a general lack of mechanical loading facilities in most of our ports. Among the new ports Haldia in West Bengal and Paradeep in Orissa are the most important regarding coal export. "Haldia port is being developed as one of the most modern ports with fully mechanised arrangements for loading with an available minimum draft of 35 ft. now rising up to 40 ft. by 1980."[26] Considering the total port facilities already available and the planned facilities available in the near future, coal shipment, port-wise will be as follows :

Calcutta and Haldia—5.0 million tonnes per year. Paradeep—0.5 million tonnes per year. Besides, the inner harbour of Vishakhapatnam port may be utilised for coal shipment if there is further order of coal.

**Lack of Finance** : Coal mining is a capital-intensive industry. A substantial investment is necessary in the coal industry to maintain the tempo of production and to develop the new mines. The coal mining industry is known for its long gestation periods. It usually takes 5 to 10 years to develop a new mine and put it on stream. It is, therefore, essential that development programmes for the production of both coking and non-coking varieties of coal are formulated well in advance, so that the actual production and supply of coal from the mines can synchronise with the work schedules of industries consuming coal."[27] Thus, the coal plan is not of five years' duration—it is to be essentially a 15-year plan. It involves a huge outlay of resources. According to a rough estimate nearly 1,200 crores were needed to achieve the Fifth Plan target of coal production. Government after nationalising the coal mines is confronted with a Herculean task involving a huge capital outlay to achieve production of a gigantic magnitude.

**Poor Quality of Coal** : More than 50 per cent of our coking coal contains high ash and needs beneficiation before use for manufacture of metallurgical coke, but in washing about 45 per cent is left out as middlings and rejects."[28] Thus we have to develop some techniques to use our low-grade coal in steel making and various other purposes. In order to lower the ash contents in metallurgical coals and to blend the weak and semi-coking coals with the coking coals for use in the steel factories, we have established washeries at Dugda, Bhojudih, Patherdih, Durgapur, West Bokaro, Jamadoba, Lodna, Kargali, Nowrozabad, etc. We need more such washeries to supply coal to the iron and steel industries. The steel industry can use beneficiated non-coking coal known as Formed Coke. At a cost of 3.5 crores of Rupees a Formed Coke plant is being set up at Talcher in Orissa to produce a type of fuel which according to R.G. Mahendru would "profitably be used in blast furnaces, replacing thereby metallurgical grade hard coke to the extent possible."

**Dearth of Trained Man power** : Although there is an abundance of unskilled labourers in coal industry, the number of skilled personnel is very limited. The BCCL has an acute shortage of trained managerial, supervisory and skilled operators. Their number is to increase for an efficient handling the coal mining activity. Besides, training schemes for the supervisory and skilled workmen should be arranged. 5 training schools located at Talcher, Kargali, Kurasia, Giridih and Bhurkunda Collieries are at present functioning under NCDC.

**Inadequate Power Supply to the Coalfields** : Power constitutes one of the most important inputs of the coal mining industry. "The demand of power by the end of the Fifth Plan would be 250 MVA against current demand of 125 MVA. Continued and assured power supply is very essential for the safety of the mines. To ensure continuous power supply, actions are on hand to instal standby diesel and steam generating sets with an aggregate capacity of 45 MVA."[29]

Besides the above factors there is an erratic demand for coal from its major consumers—iron and steel factories, railways and thermal power stations. "There has been no scientific use of coal. We have been burning a most valuable mineral in boilers, for steam raising and for thermal power. We allowed western oil interests to swamp the country with cheap imported crude and petroleum products, rather than ourselves evolve a national plan to make the most profitable use of an indigenous product."[30]

In spite of the poor quality and uneven geographical distribution, coal is the most reliable source of energy for India. During the 60s it was swamped by oil and the latter as a source of energy replaced coal in various fields of economic production. But the age of oil—the Black Gold—seems to be drawing closer, while that of coal—the Black Diamond—appears to be making a debut for the second time."[31] In spite of great allurements of the non-traditional sources of energy like the solar energy, wind power, geothermal power, tide energy, atomic energy, fusion power, etc. We, perhaps, cannot afford to forget our old and reliable friend—"Mr. Coal". Perhaps during the next two decades or more, coal will be the dominant source of energy in India, possibly yielding place to nuclear energy. It is estimated that our non-coking coal reserves will last for another 150 years.

The problem lies with our coking coal reserves. If a growth rate of 6 per cent per annum for the iron and steel industry is envisaged, the present known reserves of coking coal should sustain us up to A.D. 2027 assuming that coke requirement of foundry industry remains constant."[32] Very recently the Steel Ministry has taken decision to import 1 million tonne of coking coal from Australia keeping in view the precarious condition of our coking coal reserves. It is likely that a long term order for the next five years will also be placed to take advantage of the price offer. The decision to import coal has been taken against stiff opposition by the Energy Ministry which repeatedly assured timely supply of coking coal to the various steel plants. But the Steel Ministry did not find these assurances convincing. The Steel Ministry does not believe that indigenous availability would be able to meet the increased requirements of 16.8 million tonnes in 1978-79.

According to its own calculations, there would be a shortfall of about 1.5 million tonnes.

Secondly, even if the supplies from indigenous sources are just enough, any disruption would affect the steel plants "unacceptably". Import would also give some relief to the washeries working at present under tremendous pressure which is affecting the quality of coal.

Thirdly, world coal prices are lower, because of the recession in the steel industry. Therefore, it is argued, this is the best time to take advantage of competitive prices rather than make distress purchases in a crisis situation later.

Also, import, even if only for a short period, will give the country necessary experience in dealing with the logistics of imported coal movement, handling and evaluate the actual plant trial and the economic and operating parameters of blending. The imported coal is proposed to be moved in separate rake leads from ports of Vizag, Haldia and Paradeep to the steel plants Rourkela and Bhilai.

A major problem that coal mines face is their own inability to utilise their machinery to full capacity. It will take some time to correct the handicap, a major one in an industry which deploys much of the investment on equipment.

The Government has rightly rejected the idea of using coal after its conversion into synthetic oil. This was not because the potential for producing synthetic crude from coal was not known, but because the likely costs were too high; there were also technological uncertainties and environmental problems which have yet to be fully resolved. Yet the idea continued to be canvassed, presumably because of the greater ease and efficiency with which oil can be used as fuel. Coal is more messy, but must remain India's principal conventional source of commercial energy. Conversion of coal into gas and other smokeless domestic fuels is, however, a wholly sensible technological pursuit, which needs to be substantially extended. The oil import bill had risen from Rs. 60 crores in 1950 to Rs. 1200 crores in 1974. Consequent to the recent price hike in oil the bill will be further inflated. In order to save our precious foreign exchange we have to use more coal and less oil. There are many areas in which oil-based systems could be replaced by coal-based ones or adapted to coal, *e.g.*, preference of river transport to oil-consuming road transport where such facilities exist. We have to augment our coal production by any means. It would be unfortunate if new oil discoveries generated false optimism.

"Wherever development of coal industries has been neglected in the past, reassessment of coal's role is now taking place and some of the countries which were, previously using oil have now started looking back upon coal in preference to oil."[33] India too has not lagged behind in attempts to ensure the development of the coal industry. The production and consumption pattern of coal within the country between 1972 and 1976 are given below :

TABLE 11.6

*Production and Consumption Pattern of Coal in India (1972-76)*

| Year | Production (in million tonnes) | Increase over previous year | Consumption | per cent increase |
|---|---|---|---|---|
| 1972-73 | 77.2 | — | 70.0 | — |
| 1973-74 | 78.1 | 1.2% | 77.0 | 10.0% |
| 1974-75 | 88.4 | 13.1% | 86.9 | 12.8% |
| 1975-76 | 99.3 | 12.3% | 91.1 | 4.8% |

"Unfortunately, while demand for electricity is increasing at 20 per cent annually and petroleum products at 13 per cent coal is increasing only at 4 per cent per annum. Steps should be taken to reverse this system."[34]

Recently our Coal Secretary has stressed the paramountcy of coal in a context where oil is gradually pricing itself out of reach. He has told the industry to increase its output without worrying excessively about the finance.

The first-two months of 1979 have already seen the record production of 13 million tonnes, which represents a healthy spurt. The target of 112 million tonnes for the year is not difficult to achieve if the pace is maintained. The goal for next year, however, has been set variously, at 118 million tonnes by the Energy Minister and more ambitiously at 152 million tonnes by the Coal Secretary. There is little doubt that the Energy Minister's target will not only be met but exceeded. The surplus, however, will not be large enough to meet the Secretary's expectation.

But the coal mining industry of India has no reason to look back. Only six years ago, Mohan Kumara Mangalam described coal as "the heart of India". But beset with numerous problems it is a 'sick heart' now. It is expected to regain vitality in the near future with timely and adequate supply of inputs like power, explosives, sand, plant and equipment, adequate serving by the railways and other organisations, and considerable improvement in the coal miner's lot.

## REFERENCES


1. R.N. Sharma, "*Coal : No Reason to Look Back*"; *Business Standard* Year-End survey, 1974, p 74.
2. Ibid.
3. "Coal India," *Business Standard* Year-End Survey, 1975, p 107.
4. "Coking Coal Mines Taken Over," *Business Standard* Year-End Survey, 1971, p 5.
5. India 1977-78, p 276.
6. B. Krishna, *Coal* : "*Our Major Source of Power*," *Illustrated Weekly of India*, Vol. 99, p 1 [illegible].
7. Surender, Mohan, "Coal After Nationalisation," *Illustrated Weekly of India,* Vol. 94, p 26.
8. R.N. Sharma, Coal : "No Reason to Look Back," *Business Standard* Year-End Survey, 1974, p 75.
9. *Hindustan Times*, June, 22, 1978.
10. *Hindustan Standard*, April 24, 1976.

11. Surender Mohan, "Coal After Nationalisation", *Illustrated Weekly of India*, Vol. 94, p 28.
12. R.N. Sharma, "Coal: No Reason to Look Back", *Business Standard, Year-End* Survey, 1975, p 75.
13. M. Das, "Climate of Confidence Absent in Coal Industry", *Business Standard Year-End* Survey, 1972, p 14.
14. *Hindustan Standard*, Jan. 4, 1976.
15. R.N. Sharma, "Coal : No Reason to Look Back", *Business Standard Year-End* Survey, 1974, p 76.
16. B. Krishna, "Coal : Our Major Source of Power", *Illustrated Weekly of India*, Vol. 99, p 15.
17. D.K. Ganguly, "The Agony of a Miner's Life." *Illustrated Weekly of India*, Vol. 94, p 29.
18. Surinder Mohan, "Coal Nationalisation and After," *Illustrated Weekly of India*, Vol. 94, p. 31.
19. Ibid.
20. Ibid.
21. Ibid.
22. Ibid.
23. R.N. Sharma, "Coal : No Reason to Look Back," *Business Standard Year-End* Survey, 1974, p 75.
24. M.P. Narayanan, "Coal Exports Look Up, *Business Standard Year-End Survey*, 1976, p 135.
25. B. Krishna, "Coal : Our Major Source of Power" *Illustrated Weekly of India*, Vol. 99, p 13.
26. M.P. Narayanan, "Coal Exports Look Up", *Business Standard Year-End Survey*, 1976, p 136.
27. "On Coal India", *Business Standard Year-End Survey*, 1975, p. 107.
28. H. Krishna, "Coal : Our Major Source of Power", *Illustrated Weekly of India*, Vol. 99, p 11.
29. R.N. Sharma, "Coal : No Reason to Look Back," *Business Standard Year-End Survey*, 1974, p 76.
30. Ibid.
31. B. Krishna, "Coal : Our Major Source of Power", *Illustrated Weekly of India*, Vol. 97, p 15.
32. R.N. Sharma, "Coal : No Reason to Look Back," *Business Standard Year-End Survey*, 1974, p 75.
33. M.P. Narayanan, "Coal Exports Look Up," *Business Standard Year-End Survey*, 1976, p 135.
34. D.P. Satpathi, *New Sources of Energy : Potentiality of their Development in India. Environment, Man and Society.*" Indian Institute of Asian Studies 1979 (Unpublished).

# 12

# Recent Trends and Prospects of Oil in India

*—Shekhar Kumar Barnwal*

PETROLEUM, as the base of modern scientific revolution especially in the field of transport, industry and agriculture, is a chemical substance composed of carbon, hydrogen, oxygen and nitrogen. It is known as hydro-carbcn and is generally expressed by the formula :

$$C_n + H_n + 2$$

where the value of n determines the form and quality of petroleum. It is available in solid, liquid and gaseous forms. The solid form is known as bitumen for paraffin wax and the gaseous form is known as methane. Petroleum has optical properties which are generated by chemical reactions. There also exists porphyrin which has originated from the green colouring matter of plant known as chlorophyl.

Petroleum is known as "liquid gold" which provides life to the industry and transport, and acts as a catalyst to the process of economic development. It is vital for defence, for lighting homes, for repairs of boats, furniture, for treatment of leather and even for enamelling. Thus oil has a greater role in warfare. Marcopolo has indicated the use of oil as medicine. Even as late as the nineteenth century, "a small trade was established in USA for the utilisation of oil to alleviate and cure all ills of the human body."[1]

There is an exhaustive list of oil-based chemicals, such as phenol, DDT, detergents, glue, moulding powder, linoleum, P.V.C. leather cloth, polythene, dry batteries, etc. Petrochemicals have a tremendous impact on life today. Through petro-chemicals, we get plastics, synthetic rubber, fibre, detergents, cosmetics, etc. Plastics alone have thousands of uses from home furnishings to satellites. Synthetic detergent has become an effective substitute for soap. The synthetic fibre has changed the sartorial habit of man.

Fertiliser produced from petroleum has brought about a revolution in agricultural production. Scientists have even extracted protein from oil which is a source of animal nourishment.

There are thousands of products which do not resemble oil, yet are made of petroleum, like ink, explosives and shoe polish. The wax extracted from petroleum serves in many ways. Petroleum gas has taken the housewives drudgery out of the kitchen. For want of lubricants, the wheels of man's progress would have taken thousand years to cover the distance which it has done in the last one hundred years.

Petroleum has served as a medicine an illuminant, fuel oil and raw material for the manufacture of many products. Today although electricity lights millions of homes, people on every continent still use kerosene extensively as the fuel for family lamps. A tremendous quantity of fuel oil runs steamships and locomotives, furnishes power in factory, and heats buildings.[2]

## OCCURRENCE OF PETROLEUM

Petroleum occurs in areas of shallow margin of epicontinental sea that existed during Tertiary, Mesozoic and Palaeozoic times. The other limiting factors of the occurrence of petroleum are as follows :

(1) Petroleum occurs in sedimentary rocks of marine and fluviatile conditions lying with unconformity in relation to the impervious bed-rock and cap-rock. In this sedimentary rock petroleum accumulates due to the chemical transformation of plants and other biological elements.

(2) Petroleum forms at comparatively low temperatures. The reservoir temperature is rarely 212° F but in some cases it may go up to 300° F also.

(3) The occurrence of petroleum ranges from Cambrian to Pleistocene sedimentary rock but the Cambrian petroleum is non-existent. This indicates that petroleum cannot resist destruction due to heat either by change in condition and long age.

(4) In high temperature with the presence of oxygen generally the porphyrin available in petroleum cannot exist in maximum temperature.

(5) Petroleum occurs only for a short time as only the Pleistocene rocks bear petroleum in abundance.

(6) Petroleum occurs in two forms: one as surface deposit and the other as sub-surface deposit. Surface deposit can be divided into two, i.e., seepage and mud-flows. From commercial point of view surface deposit is unimportant but generally it gives a clue about the larger sub-surface reservoir and the nature of the source rock. Sub-surface deposits are available in the form of petroleum dykes, plunge pools and oil series.

From the point of view of accumulation of oil the rock structure can be divided into several forms which are as follows:

1. **Domal or Anti-clinal Structure** : The petroleum along with gas in the upper part and water in the lower part sealed by impervious rocks as bed-rock and cap-rock, petroleum accumulates in domal structure (Fig. 12.1). Such situations are available in the oil fields of U.S.A., Middle East, Burmah and Assam oil fields of India.

2. **Faulted Structure** : Oil never accumulates in the horizontally laid sedimentary rocks. In case of compression of rocks from two sides the faulted structure takes place in the landscape. In this structure the same series of rock breaks into two pieces and moves up and down relatively which may be a normal fault or a reverse fault. In such a situation in the middle portion of petroliferous bed oil accumulates in an anti-clinal structure near the fault plane (Fig. 12.2).

*Domal Structure*

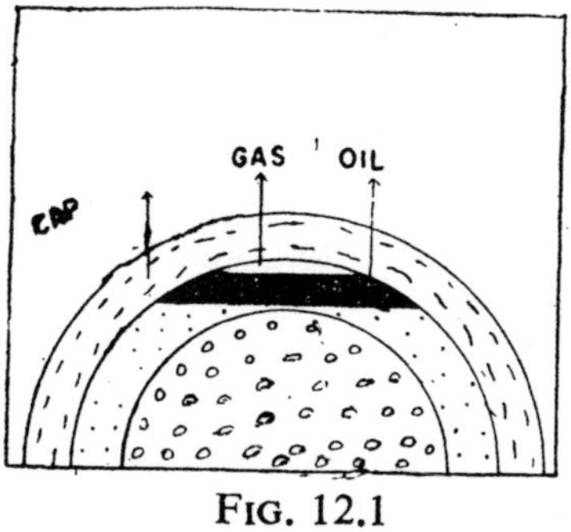


FIG. 12.1

*Faulted Structure*

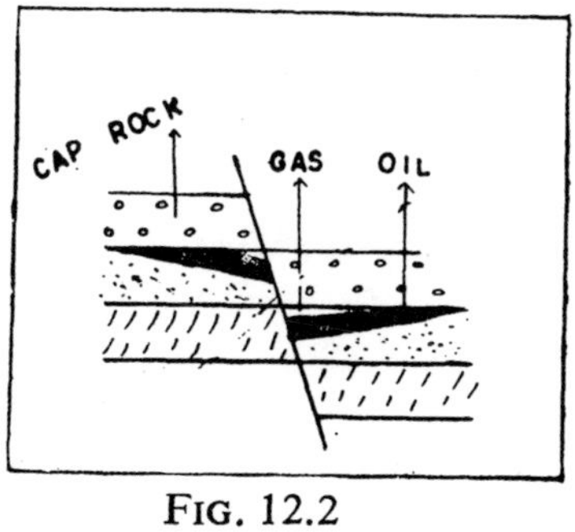


FIG. 12.2

3. **Lenticular Structure** : Sometimes the lenticular structure characterised by pervious and petroliferous in the form of lenses surrounded on all sides by impervious rocks gives some accumulation of petroleum. Such structures yielding petroleum are, of course, very few in number (Fig. 12.3).

4. **Igneous Intrusion** : The igneous intrusion, sometimes provides a condition for the accumulation of petroleum. Due to warping, tilting in surface structure the petroleum accumulates in anti-clinal part of the sedimentary rock surrounded by impervious sandstone (Fig. 12.4).

*Lenticular Structure*

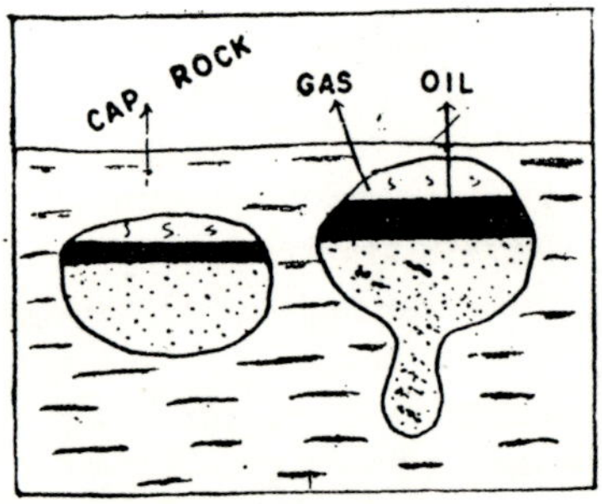


FIG. 12.3

*Igneous Intrusion*

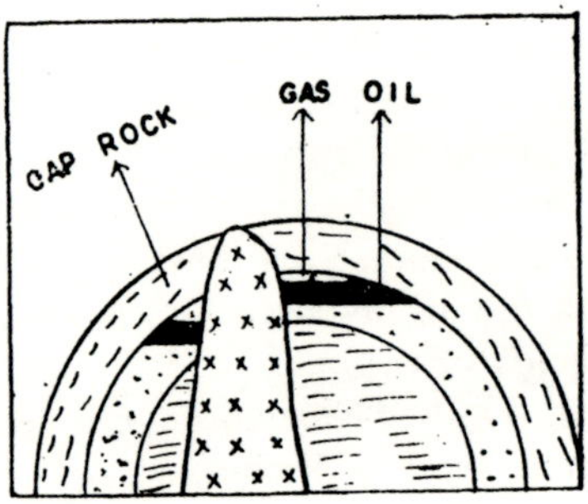


FIG. 12.4

5. **Salt Dome** : Salt dome structure is just like beehives and is subjected to heavy pressure in shallow gulf. During the geological recency the salt beds are formed, where oil accumulates, i.e., in the Mississippi river basin of U.S.A. Due to the pressure of over-lying sediments the compressed salt pushes up through the weaker part of the sediment in the form of salt dome. This salt dome leads to the formation of petroleum reservoir if one of the beds of the overlying sediment is petroliferous (Fig. 12.5).

6. **Unconformity** : There are various forms among which the unconformable bed favours the accumulation of oil due to impervious capping of rock which obstructs the seepage of petroleum from the reservoir (Fig. 12.6).

7. **Reef Structure** : Coral reefs are generally composed of calcium carbonate which contain petroleum if they are surroun-

*Salt Dome Structure*

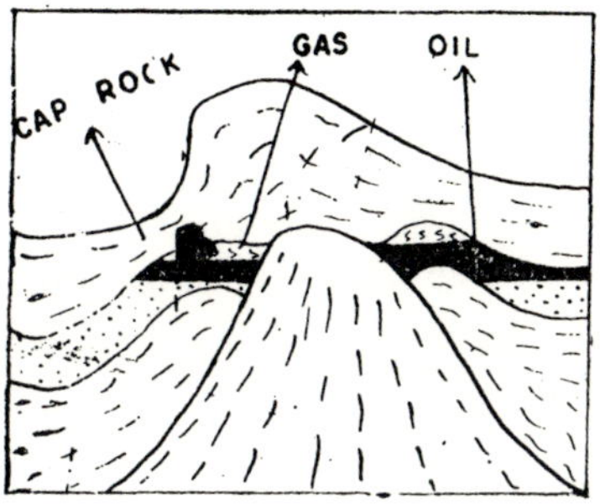


FIG. 12.5

*Pitching Structure*

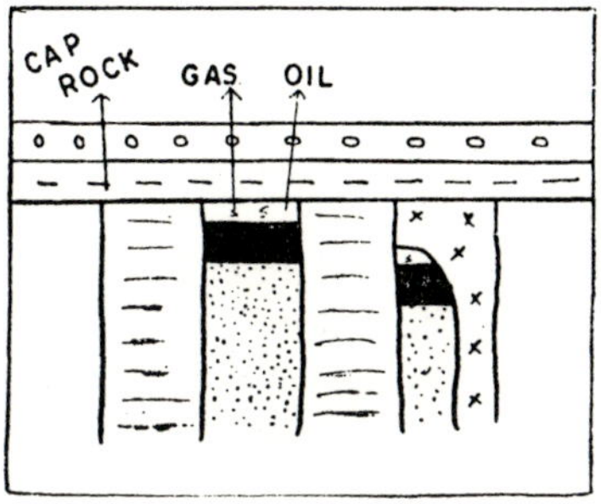


FIG. 12.6

ded by impervious strata, such structures are found in U.S.A. (Fig. 12.7), and the eastern sea coast of Australia.

*Coral Reef Structure*

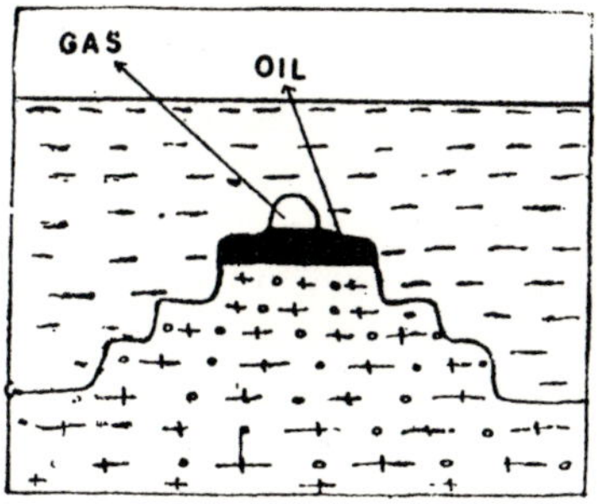


FIG. 12.7

8. **Types of Oil** : Keeping in view the percentage of moisture, silt content, presence of sulphur and other impuritiescrude petroleum can be divided into three types :

(1) Paraffin Base Oil.
(2) Naphthelic Base Oil.
(3) Asphaltic Basc Oil.

Generally paraffin base has been considered as a good quality of oil due to lesser excretion of impurities at the time of refining. Light oils which are used in aeroplanes and war tanks are prepared from paraffin base.

Naphthelic base provides a medium quality of oil which is generally used in transport vehicles in the generation of electricity. The presence of higher quantity of nitrogenous content in this oil is known as Naphthelic Base.

In the Asphaltic base oil impurities of sulphur, silt content, higher percentage of moisture are available. This oil is considered as heavy oil from which diesel oil, coal tar, kerosene oil, etc. are generally prepared.

## THE PROBABLE OIL BELT OF INDIA

The sedimentary structure of rocks which are found in extra-Peninsula areas of India bear rich petroliferous bed. In India one lakh sq. km. of area is seen to be petroliferous. Such areas are distributed as :

1. **Mountainous Region** : In this region eastern Himalayas, western Himalayas, eastern sub-Himalayas, western sub-Himalayas and north-eastern hills are included.

2. **River Basins** : Among river basins the Brahamaputra Valley, entire Ganga Valley, the Sutlej Valley, and the delta areas of the river Ganga, the Mahanadi, the Krishna, the Cauveri, etc. are included.

3. **Coastal Areas** : Among the sea-coast the Runn of Kutch, Malabar coast, Konkan coast, delta area of Indus, Coromandel coast, Uttri Sarkar coast, Minicoy, Amindive, Laksha Divi and Andaman and Nicobar islands including the Bay of Bengal and the continental shelf areas of the Indian sub-continent are included.

## DISTRIBUTION AND PRODUCTION OF OIL IN INDIA

In India the production of oil is limited in only two areas: Assam Oil fields and the Oil fields of Gujarat (Fig. 12.8).

The modern history of oil in India began in Assam. The successful drilling of oil wells, in fact the first in Asia, drilled by steam-powered machinery was completed at Makum in upper Assam on 26 March 1867. It produced just eight barrels of crude oil per day. However, this proved a mere flash in the pan and the reserves exhausted soon. Twenty-two years later, while the Assam Railway and Trading Company was building the first railway line from Dibrugarh to Margherita (where incidentally the first refinery was set up to refine the Makum crude, and the same closed with the closing of the oil well), its engineers came through oil seepages. The AR & T company decided to drill for oil and thirteen months later in 1890 oil was struck at a depth of 662 feet in the vicinity of Digboi (Digboi, dig).

*Digboi Oil Field*

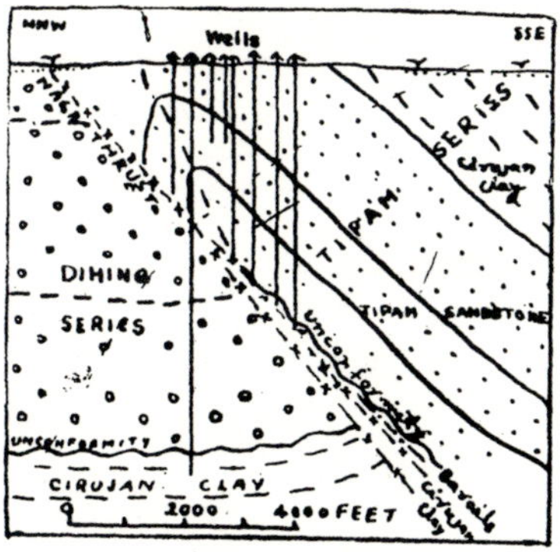


FIG. 12.8

There are two kinds of oil fields available in India. First old oil field and the second new oil fields. Among old oil fields comes the oil fields of Assam, i.e., Digboi, Nahorkatia, Moran, Hugrijan, Dumduma, Bardarpur, Patharia, etc.

**Old Oil Fields**

**The Digboi Field** : In Lakhimpur district of the Surma Valley of Assam with the help of Burmah Oil Co., oil wells have been discovered along the crest of folded anti-clinal part of the Naga thrust. In Digboi oil field oil occurs in the Tipam sandstone stage where twenty separate horizons have proved to be oil bearing over 3600 feet thick marine sandstone. It is also believed that the oil has migrated to this belt from Barail series by the way of the Naga thrust (Fig. 12.9).

**The Nahorkatia Field** : Some twenty miles away from Digboi within the seismic survey of Brahmaputra alluvium the promising oil field of Nahorkatia came to the light of geologists where in the Tertiary strata oil occurs in faults. Here the Tipam sandstone lies about 9000 feet deep beyond which oil occurs from the Barail series of over 9,500 feet depth where the depth of the Barail series is 3,000 feet in the Eocene Rocks of North Kachar Hills (Fig. 12.10). The annual production is about three lakh tons of oil.

*Nahorkatiya Oil Field*

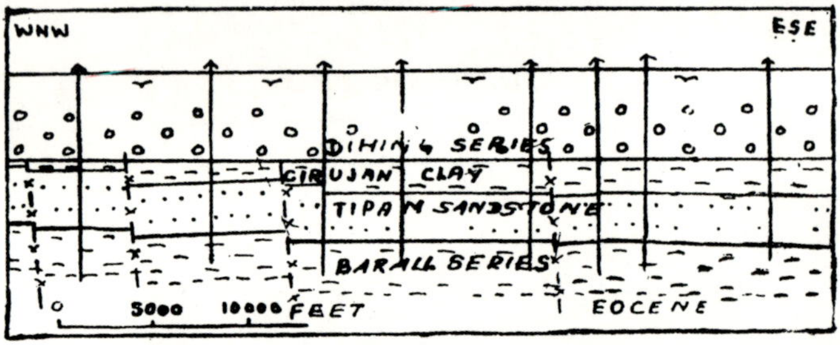


FIG. 12.9

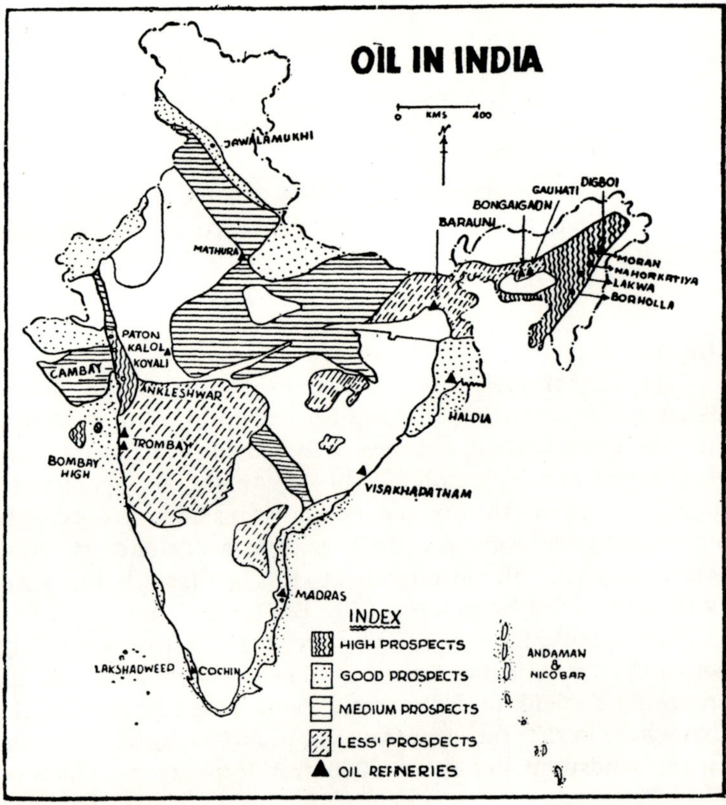


FIG. 12.10

**The Modern Oil Field** : At a depth of 11,000 feet on a faulted domal structure oil bearing strata were found from Barail Series which lies 25 miles south-west of Nahorkatia.

**Badarpur Oil Field** : In Badarpur Oil Field oil occurs from the Barail Series of rocks above the anti-clinal belt of Badarpur. Here Masimpur is a promising area but now-a-days the production of oil has gone to only 20 000 metric tonnes in 1977.

Among other oil fields of Eastern Region Makum, Shivsagar, Lakwa, Gelleky are very important.

## New Oil Fields

Among new oil fields the post-Nummulitic sediments of the Runn of Kutch, petroliferous rock occurs from faulted surface trap structure. This basin extends from Baroda district to the Gulf of Cambay and Bhawnagar of Saurashtra. The probable reserves of petroleum are estimated to be huge from here at least two to four million tonnes could be produced annually. At present the production is about five lakh tonnes annually.

**Ankleshwar Oil Fields** : The first well drilled on the crest of the Ankleshwar anticline in 1960. Oil bearing sand in Ankleshwar is found at a depth of 950 to 1250 metres. It encompasses an area of 30 sq. km. and lies south of Baroda and Broach. Out of 200 wells drilled in the area, 170 are oil producing and 13 are gas producers. The production is 83 tonnes a day.

**Kalol and Sanand Oil Fields** : With the Geophysical Survey of the 1960s in the northern part of Cambay Basin large prospective structures of alluvium were detected in Kalol and Sanand areas, which are highly petroliferous. In Kalol structure the oil occurs at a depth of 1400 metres. Oil and gas were also found there. These oil fields also include Bakrol, Mehsana, Kadim, Detroj, Becharaji and Tharad 240 km north of Ahmedabad. So far 170 wells have been drilled in this area out of which 48 wells are producing gas and oil, and the remaining are dry. At present the production is 1200 tonnes per day.

**Dholka Oil Field** : The Dholka oil structure is located at 8 km from Dholka town. Here the drilling was completed at the end of 1970. 81 wells were drilled here out of which 51 wells are oil producing and 15 are dry. At present per day 500 tonnes of oil is produced in this area. The oil is brought to Gujarat Refinery through the tankers during the initial period. But now the Kalol, Navgaon, Quali Trunk Pipe Lines has started operating in this area. Per day 6,000 cubic metres are produced from this field.

**Mehsana Oil Field** : In Mehsana Oil Field of north Gujarat, the production of oil comes into effect from 1969 where the oil fields are located at Detroj, North Kazi, Asjol, Becharaji, Mehsana, North Kalol and Tharad. At present, annually seven million tonnes of oil is produced from this field at a depth of 2200 metres in Mehsana structure. 26 wells have been drilled in this area, out of which 6 are oil producing and 10 are gas producers.

**Bombay Off Shore Oil Field** : The first off shore oil operation started in March 1970 in the alluvial structure of Cambay. At present the oil is producing from six oil wells where 50,000 tonnes are produced per day. The oil is brought to the Gujarat Refinery with the help of Jawahar Tanker. Bombay High seems to be a promising oil bearing structure in India.

Among other important oil fields of the western region are Badsher, Lunej, Shobhasan, etc.

**The Punjab Oil Field** : In the Punjab the Jwalamukhi Oil Field was discovered by Rumanian experts in 1957. Gas has been available at a depth of 800 metres but now oil is also being produced in the same locality.

**Oil Fields of West Bengal** : In 1960, three oil wells were completed at Jalangi, Burdwan and Galsi in West Bengal but none of them have a petroliferous bed from the commercial point of view. Very recently in the coastal area of the Bay of Bengal near Orissa oil wells are being investigated by American Natomas Company at a depth of 6,500 feet which lies 30 miles east to Bhuvaneshwar.

**Oil Fields of Arunachal** : In Arunachal with the help of Burmah Oil Company, Oil India has discovered a promising oil well at a depth of 2,850 metres in Kharaun. Here 21 metres petroliferous bed is available. The daily production available could be about 1500 jars. This will add 50 thousand tonnes of oil per year to the total Indian production. In Nagaland vast reserves of oil have been discovered near Damulgaon.

In Jammu and Kashmir vast reserves of oil have been struck at Surinsar. In Bihar near Subha Tola of Bhagalpur District vast reserves of gas have been detected in which 88.8% methane, 4.3% nitrogen and 4.5% carbon dioxide and carbon monoxide are found.

The progress of oil production in India has been stepped up specially after independence (1947). In 1869 only five thousand gallons of oil have been produced from Digboi area of Assam especially near Margherita. In 1951, the production was 2.64 lakh metric tonnes of oil, the same has gone up to 78.00 lakh metric tonnes in 1977.

TABLE 12.1

*Progress of Oil Production in India*

| Year | Production in lakh metric tons |
|---|---|
| 1951 | 2.64 |
| 1956 | 3.84 |
| 1961 | 5.16 |
| 1964 | 22.01 |
| 1965 | 30.24 |
| 1969 | 67.02 |
| 1970 | 68.01 |
| 1975 | 74.00 |
| 1977 | 78.00 |

Source : Statesman Year Book, 1978.

### OIL AND NATURAL GAS COMMISSION

The Oil and Natural Gas Commission started operation in 1956 and it has now entered its 22nd year. The difference between a private oil company and the Oil and Natural Gas Commission is essentially this that a private company restricts its operation to certain individual areas where they consider good prospects of profit, whereas the Oil and Natural Gas Commission has been making efforts to find oil wherever there is a theoretical possibility of locating oil, with this objective they have conducted geological surveys, gravity surveys, magnetic surveys and seismic surveys. These are all preliminary operations. These have been carried out wherever there appeared the slightest possibility of finding oil. For instance, in the entire foothill region of the Himalayas, extending from Jammu and Kashmir in the north-west up to eastern Uttar Pradesh and in the entire adjoining plains of Jammu and Kashmir. The Punjab, Uttar Pradesh and Bihar, the western part of Rajasthan in the so called Jaisalmer area (Desert of Rajasthan), the West Bengal area and coming to the north-east, in the Brahumaputra Valley and the Surma Valley of Kachhar district extensive survey operations are going on. O.N.G.C. has also done surveys in the Tripura area and in the Garo Hills south of Shillong. Most important of all O.N.G.C. explorations are conducted in the extensive area of Gujarat from Surat in the south to Mehsana in the Runn of Kutch. In the Cauveri Basin in the south and the Godavari and Krishna deltas in Andhra Pradesh, the Mahanadi delta and in the coastal area of Kerala also O.N.G.C. has done surveys. These are called sedimentary areas where theoretically possibilities of oil exist.[3]

The proved reserves of oil in India are at present exceedingly small. A great deal of survey and exploration work, however, remains to be done in the areas in which deposits of oil or natural gas might be found. The total sedimentary area of India onshore and offshore comprises 27 basins. These cover an area of 1.41 million sq km on land further 0.26 million square km lie within the 100 metres depth of the continental shelf zone in the Indian offshore. The basins are at present grouped by the Oil and Natural Gas Commission in their prospective value. The basins with high prospects comprise Cambay and a large part of Assam—Arakan, covering an area of 3,00,000 sq km or 18 per cent of the total sedimentary area including offshore up to 100 metres isobath line. These are areas with thick sections of marine sediments of relatively younger age, and contain all the existing productive wells. The second category with medium prospects comprises a further 15 per cent of the total sedimentary area, and includes small parts of Assam, Arakan, Cauvery, West Bengal, Jaisalmer, Tripura, Cachar and Andaman Nicobar. Here the Basic geological conditions are similar to those in Cambay and Assam, but so far no commercial oil or gas fields have been found. The remaining two-thirds of the sedimentary area consist of basins where the prospects are assessed to be either fair or low. There the sedimentaries are older and either wholly or mostly non-marine.[4]

As a result of exploration and drilling activities of the past 15 years an estimated 185 million tonnes of proved and indicated crude petroleum reserves have been discovered, all in the two areas of Gujarat and Assam. In addition, oil has been discovered in the offshore basin of Bombay High. An encouraging feature of the Bombay High structure is that oil has been found in limestone formations which are good producers in the Middle East. Recently seismic surveys and drilling have suggested commercial oil and gas deposits in limestone about 70 km west of Bassein.[5] The potential for new discoveries through High is uncertain but sometimes appears to be quite hopeful.

The current production (1975-76) of crude is about 8.28 million tonnes per annum of which about 5.2 million tonnes are from the Gujarat and Assam oil fields of O.N.G.C. and about 3.08 million tonnes from Assam oil fields of Oil India Limited. As against this the imports of crude in 1975 amounted to 13.67 million tonnes. The total indigenous production from O.N.G.C.'s oil fields both onshore and offshore is expected to be around 12 million tonnes per annum by 1978-79, while Oil India Limited is expected to maintain its production of the current level throughout the Fifth Plan period.[6] Total production of crude is thus expected to be of the order of 15 million tonnes by 1978-79.

The most promising way of reducing dependence on petroleum products, which is now being actively pursued, is the replacement of fuel oil by coal for steam raising in industry and those thermal power stations which use oil as primary fuel. Consumption of fuel oil in thermal stations in 1972-73 was about 1.4 million kilo-litres. Most of this was used as primary fuel in a relatively small number of power stations, and conversion of these to coal firing would, if feasible, probably save about one million tonnes of fuel oil per year. Conversion to coal is also technically possible in a number of major industries such as iron and steel, cotton textile and cement. It seems likely that out of a total consumption of fuel oils of about 5.5 million tonnes a year, about one-half could in time be replaced by coal without undue technical difficulty.[7]

During the Fifth Plan period oil exploration is being stepped up to intensify activities in the most promising oil bearing areas where the pay-back is likely to be quicker. Resources are, therefore, mainly being channelised to the development and production of oil from offshore and selected onshore areas, with a time bound programme to develop production from Bombay High to a rated capacity of 10 million tonnes per annum by 1980-81.

The draft plan envisaged a set-up of crude oil production from 7.2 million tonnes in 1978-79. The production target for crude oil is now placed at 14.18 million tonnes. Taking into account the enlarged programme, the outlays for O.N.G.C., during the Fifth Plan, have now been revised to Rs. 1,056 crores as compared to Rs. 420 crores in the Draft Fifth Plan.

### CRACKER BASED ON BOMBAY HIGH-BAY GAS

The Centre has agreed to entrust to the State the construction of a Cracker based on Bombay High Gas. Pipe line for supplying Bombay High Gas to a shore point in the State is, however, behind schedule and may not be completed as originally scheduled. The O.N.G.C. has yet to select a site for a terminal station which can also link gas supply from the Tapti structures. O.N.G.C. is also considering a proposal for linking associated gas of Vasai and the Tapti gas. A power station and a fertilizer unit are planned based on the Tapti gas in Saurashtra.[8]

After the above study on oil and natural gas in the light of the petroleum industry we proceed further towards the small-scale industries. The refining industry has helped a number of small-scale engineering industries to come up in Bombay and other Centres as ancillaries to the refineries and petroleum based industries. They produce small fittings, heat exchangers process equipment, instruments, pumps, compressors, pipes, tubes and various other components.

The calculated coke at Gauhati is being used for electronic industrial furnaces and is promoting the growth of the aluminium industry, arc lamps, battery cells and smelting of various metals. Barauni fertilizer is supplying fuel for heating fuel oil, for heating in steel plants and for the generation of power. Naphtha from refineries provides the feed-stock for the fertilizer plants at Trombay, Ennore and Baroda. The Fertilizer Plants under construction of Madras, Kanpur, Gorakhpur, Vishakhapatnam, Kota, Goa, Cochin and Barauni will also require Naphtha as their base stock.

## THE PROGRESS OF OIL REFINERIES

The Indian Refineries together are processing nearly 28 million tonnes of oil at present, of which the Indian Oil Corporation share comes to 16 million tonnes. The location of refineries are selected generally on the basis of crude oil available in the vicinity, marketing facilities, easy transportation for crude and petroleum products, sound structural base, security and availability of specialists and technical personnel. The refineries at Trombay, Vishakhapatnam, Haldia, Madras, and Cochin were located because of good harbours and well developed business centres. Refineries at Koyali (Baroda), Digboi, Noonmati and Bongaigaon were selected because of crude availabilities.

TABLE 12.2

*Refinery Capacity and Crude Sources*

| | Refinery | Capacity (Million tonnes) | Crude sources |
|---|---|---|---|
| 1. | A.O.C. Refinery at Digboi | 0.52 | Assam Valley. |
| 2. | ESSO, Trombay | 2.55 | Gujarat and other. |
| 3. | Burmah Shell, Trombay | 3.75 | Gujarat and Aghajari. |
| 4. | Caltex, Vishakhapatnam | 1.50 | Mines (Sumatra) and Aghajari. |
| 5. | Gauhati Refinery | 0.80 | Assam. |
| 6. | Barauni Refinery | 3.00 | Assam, Burmah and Sumatra. |
| 7. | Gujarat Refinery | 3.00 | Gujarat. |
| 8. | Cochin Refinery | 2.50 | Agha-Jari. |
| 9. | Madras Refinery | 2.50 | Darius (Iran) |
| 10. | Haldia Refinery | 2.50 | South East Asia. |
| 11. | Bangaigaon Refinery | 1.00 | Assam Valley. |
| 12. | Mathura Refinery | 6.00 | Gulf Kutch. |
| | Total capacity in 1978 | 29.62 | |

There are twelve oil refineries in India at present out of which Bongaigaon and Mathura are under construction. These refineries have the capacity to refine 30 million tonnes a year. Four refineries namely the Digboi, ESSO, Burmah Shell and Caltex are in the private sector.

**Digboi Oil Refinery** : Digboi is the first oil refinery of India constructed in 1923 after the Burmah Oil Company took over the management from Assam Oil Corporation in 1921. Generally it fulfils the demand of the people of Assam, Arunachal Pradesh, Tripura, Manipur, Meghalaya, Sikkim and Bhutan.

**Gauhati Refinery** : Gauhati Oil refinery was developed by the Government of India with the help of Rumanian Government technical assistance. This utilises the oil of Moran and Nahorkatia Oil fields and the main processing units are crude refinery unit, kerosene treating unit, coking unit and captive power plant. It was established in 1959 and the refining capacity is about 2.50 million tonnes at present.

**Barauni Refinery** : After Gauhati this is the second refinery developed in the Public Sector and it was commissioned in 1964 with the technical assistance of U.S.S.R. It has the capacity to refine 3 million tonnes of oil per annum but at present 3 million tonnes are refined for the market. Crude oil is available from Moran and Nahorkatia oil fields through pipe line. Barauni has so many units under its command such as atmospheric vacuum distillation unit, kerosene treating unit, coking unit, phenol extraction unit, dewaxing unit, contact filteration unit, lubricant oil complex including bitumen plant.

**Burmah Shell Rfinery of Trombay** : It was constructed in 1953 and it came into operation in 1955. The present refining capacity is 3.7 million tonnes per annum. The crude oil is available from Persian Gulf and Ankleshwar Oil fields of Gujarat.

**ESSO Refinery of Trombay** : This refinery came into operation in 1954 and its producing capacity is 3.66 million tonnes at present. The crude oil came from Persian Gulf and Lunes oil fields of Combay.

**Caltex Refinery of Visakhapatnam** : It has the capacity to refine 0.5 million tonnes oil per year. It has six different units e.g. kerosene unit, crude distillation unit, dewaxing unit and contact filteration unit etc.

**Gujarat Refinery** : This is also known as Koyali Oil Refinery. It depends upon the crude petroleum of Kalol, Ankleshwar, Mehsana and Dholka Oil fields. It was commissioned in 1965 and the assistance was provided by the U.S.S.R. Government. The refinery is processing 4 million tonnes of oil in a year.

**Cochin Refinery** : Cochin Refinery is constructed with the help of Government of India and Barless Ville Co. of U.S.A. and it went on stream in 1966. Annually it refines 2.5 million tonnes of crude oil and its capacity is being raised to 4 million tonnes per annum. In 1976, the refinery processed nearly 2.5 million tonnes of oil.

**Madras Refinery** : With the help of the Iranian Oil Company the Government of India entered into an agreement in 1965 to set up an oil refinery near Madras. For this refinery, oil is available from the fields of Iran and Java and Sumatra.

In India this is the first industry which is producing high quality lubricating oil and elemental sulphur. It has the capacity to refine 2.5 million tonnes of crude petroleum.

**Haldia Refinery** : In the public sector Haldia is the sixth oil industry under India Oil Corporation established south of Calcutta near Haldia Port. It has the capacity to refine 2.5 million tonnes of oil annually and the oil comes from Java and Sumatra oil fields, Burmah oil fields with the help of tankers. Now with the construction of a pipe line from Calcutta to Haldia this refinery has the facilities to send refined oil to Calcutta directly. In the setting of this industry financial and technical assistance were available from French and Rumanian Companies.

**Bongaigaon Refinery** : This refinery has started production in 1977. The refining capacity is one million tonnes at present but it is hoped that the same will be increased up to 2.5 million tonnes up to 1979. It is situated near Bongaigaon Railway Station in the Goalpara district of Assam and this would be one of the biggest petro-chemical complexes in India in the near future.

**Mathura Refinery** : This Refinery has the capacity to refine 6 million tonnes of crude oil annually. It would receive crude oil through a pipe line from the Cambay Gulf, Bombay offshore region and Jawalamukhi fields of the Punjab. It is one of the biggest oil refineries in India. It will help in shaping the economic growth and the requirements of oil in the north-western regions especially Delhi, Chandigarh, Agra, Mathura, Jaipur, Udaipur, etc.

## PETRO-CHEMICAL COMPLEX

The Government has accepted a policy of setting up petro-chemical complexes around the refineries in India. The end products of these petro-chemical industries will be synthetic fibres, plastics, synthetic rubber, detergents and the like. A beginning has already been made with the National Organic Chemical Industry Ltd. (N.O.C.I.L.) at Bombay in the private sector and a naphtha cracker plant will soon be constructed to adjoin the Gujarat Refinery. As the problems of technical and financial and for these capital intensive projects are solved, a number of petro-chemical complexes can be expected.

The first major petro-chemical complex at Baroda will be completed during the fifth plan period. The aromatic project of the complex has already been commissioned. The down stream units have been commissioned in April 1978. The Polyester filament yarn project of Petrofils Co-operative Ltd. is also expected to be commissioned in phases. A provision of Rs. 349 crores has been made for the above mentioned scheme.

A refinery-cum-petro-chemical unit at Bongaigaon has also been included in the Plan. In fine not only will every refinery in the country process petroleum to satisfy our growing energy needs but it will set off a chain reaction in the country's industrial progress.

## CONCLUSION

Apparently the recent trends and prospects of oil in India have passed through intensive investigation. In spite of long and arduous efforts, the present oil crisis has cast a gloom over the world. India having meagre crude resources is facing a threat of economic recession. The price of crude oil has shot up to the astronomical figure of nearly Rs. 700 per tonne (at the time of writing) as against Rs. 67 per tonne in 1971 and Rs. 300 per tonne in October 1973. In 1973 our foreign exchange expenditure on import of crude oil and petroleum products was to the tune of only Rs. 250 crores. This figure, according to present indications will now rise phenomenally to anywhere between Rs. 1200 to Rs. 1500 crores. As against this our foreign exchange earning from all our exports put together will be not more than Rs. 2000 crores. No wonder the chairman C.R. Das Gupta, in his open letter to the I.O.C. refineries in the year 1973 said, "We are facing today the severest test on our national resolve since the attainment of independence".

With the setting up of new refineries and the expansion of the existing ones, the demand for crude oil is likely to rise to 45 million tonnes by 1980, and by that time the production from country's proved reserves is likely to dwindle to a mere four million tonnes. This would create a tremendous gap between demand and supply in terms of foreign exchange. The gap is likely to be filled by the intensive inshore and offshore drilling, better utilisation of share of crude oil exploited in the concession areas abroad, and by imports. In the words of the Government, "Highest importance is being attached to the intensification of exploration and other activities with a view to locating new sources of crude oil and optimise the indigenous production from the oil field already discovered. The exploration activity is also being extended towards more promising, even though difficult structures such as Bombay High and Tripura." Results of test drilling of the offshore side in the Bombay High have raised hopes high.

In the field of petroleum products too, by 1980 there will appear a gap of nearly 10 million tonnes between the supply and demand with additional refineries and the augmentation of the existing capacity, the total production will be around 34.1 million tonnes as against a requirement of 40 million tonnes. Thus in the years to come Indian Oil will have to play a more vital role and shoulder a greater responsibility both in the production and distribution field. The maturity and skill with which this very young concern has hitherto been handling the task assigned to it have given rise to the hope that it will not fail the people where their energy requirement is concerned. Yet India has a long way to go to acquire self-sufficiency in oil production both crude as well as finished products and Indian Oil is destined to play its role in the country's efforts.

India has made considerable progress in fabricating equipment for refineries and is in a position to obtain indigenously a large proportion of equipment materials needed for the construction of oil refineries. A major public sector units established for this purpose are Bharat Heavy Plates and Vessels, Bharat Heavy Electricals India Ltd., Triveni Structurals, etc. Even though practically all types of refinery equipment can be fabricated indigenously, it may not be possible to fabricate pieces of larger size machinery than has been provided for in the indigenous fabricating plants. Such equipment would continue to be imported but would form lower percentages of the total cost of the refineries in future, as the indigenous fabricators extend their range of manufacture. Further development of indigenous capacity to achieve maximum self-sufficiency in this field is being pursued continuously.

## REFERENCES

1. K.K. Srivastava, "Story of Oil"—*Bulletin*—Public Relations Department, Indian Oil Corporation Ltd., Barauni Refinery—1974, p 3.
2. Plough M. Emerson, "Salient Characteristics of Petroleum Pipe Line Transportation"—*Land Economics*, XXVI(1950) pp 27-39.
3. "The Big Gamble", The Text of an interview Chief Editor, YOJANA, with Chairman, ONGC, YOJANA, Vol-XVIII, No. 1, 26 January, 1974, pp 41-42.
4. "Oil and Natural Gas"—Section of Geology and Geography, 64th Session, Indian Science Association, Calcutta, pp 9-10.
5. Report 1975-76, Ministry of Petroleum, Government of India, 1976, p 7.
6. Ibid., p 7.
7. P.D. Henderson,—*India, The Energy Sector*, Delhi, Oxford University Press, 1975, p 116.
8. "Energy Cracker Based on Bombay High Gas"—*Data India*-34. Aug 21-27. 1970, New Delhi, p 542.

# SECTION TWO

## AGRICULTURAL LAND UTILIZATION

# Introduction

IN THE Third World affairs, the field of agricultural geography is changing fast with the advancement of technical knowledge and the use of sophisticated techniques like computer programming, model building and hypothesis testing through quantitative methods. All these have led to the change of concepts and developed a more systematic knowledge of geography. Hence the meaning, scope, methods, approaches and principles of study are changing with the increasing demand for a particular use of agricultural land and the growth of population. In the field of geography a dynamic view has been developing fast led by the conceptual revolution especially in the fields of agriculture, resources appraisal, aerial photo-interpretation, remote sensing, land utilization and integrated regional development.

In most of the above-mentioned study samples, regional principle, genetic and remedial approaches have been employed including the complicated problems of physical and cultural landscape in relation to man. In almost all studies the descriptive-cum-analytical view has been adopted for the spatial differentiation of agricultural innovation, diffusion, typology and pattern analysis. Recently perception and behavioral approaches have almost revolutionized the scientific thinking of the rational of agricultural development and injudicious use of the land.

Agricultural geography is a systematic science which deals with the production, location and patterns of crop distribution in terms of cause-effect and functional-areal relationship at any level ultimately aiming at preparing a plan fot an over-all agro-economic development of the region. In brief, the main characteristics of agricultural production such as (i) many producing units, (ii) multiplicity of products, (iii) biological nature of production, (iv) location decision in agriculture, and (v) self-sufficiency in production have been critically analysed here. With the increasing complexities and problems of agricultural growth, the importance of aerial photography has been felt by many scholars in order to analyse the problems of agriculture of a particular region through the process of air-photo interpretation.

In a paper on some aspects of land utilization, the author has discussed the significance, scope and the review of literature on world landuse, British landuse and American landuse. Further the approaches of the works of Indian geographers have also been analysed.

Primarily, it was felt that model building was a recreation for the child. But recently it has encouraged scientific thinking in almost all branches of science in a diagrammatic, mathematical manner and by representation of real world situations. The scientific expression of real world situations and the generalization of land utilization in rural areas have been systematically analysed here in the models of landuse. A review has also been made of the models advanced by Von Thunen, McCarty Lindberg and Hiroshi Ishida. In landuse studies a new model of polar and anti-polar concepts of land utilization has been forwarded.

Aerial photo-interpretation is not a new cartographic technique for map analysis but in a developing country like India, the introduction of this technique in geographical studies and resources appraisal has only been recently made. In comparison with human labour, this technique has the ability to handle a large number of maps in a shorter period. This paper has been analysing the types of aerial photographs, their scale, principles of interpretation, quantitative analysis, pattern, shape, colour, texture and shadow analysis including image displacement method and stereoscopic vision analysis, etc.

The stereoscopic vision of aerial photographs has added a new dimension to the study of land utilisation and the features of landscape which are utilised by man in various forms according to his choice and need. The air-photo interpretation technique has the advantage of utilising large scale maps with greater accuracy in a shorter time. In this study the methods and techniques of landuse interpretation with the help of aerial photographs have been presented which would help the student of geography, geology, land economics, regional planners and in resources appraisal of a region.

In the spatial diffusion of agricultural innovation, the dimensions of the spread of any item may be responded from a common centre or from varying locations. Hence, the diffusion of an information is dependent on the force generated by the place of diffusion, while the distance covered in space is the outcome of the mean information field of an adopter in the form of distance-decay function.

Kostrowicki has developed a new concept of "agricultural typology" of land utilization. He aims at providing world typology of agricultural landuse following the type, patterns and colour schemes of 'world landuse survey'. With the help of 22 variables, 33 model types have been developed which are grouped into a number of main types. Kostrowicki has also given an indication of the regional occurrences of various types. The characterisation is also made by the construction of "typograms" for each land utilization type just like the typograms of settlements.

"A Rural Decision-Making Model and Its Research Implications in Agricultural Geography" analyses the fundamental concepts of decision-making and feed-back mechanism that a given representative farmer exhibits within this system. The farmer acts on decisions, generally in a disequilibrium situation based on limited information and on a severely circumscribed understanding of the total environment. The environment, consisting of all other farms, markets, political, cultural, and institutional organizations, and nature is affected to some degree by the decisions.

It can safely be said that in agricultural land utilization, study methods and concepts have changed substantially in recent decades with the advancement of technical knowledge and changing social needs.

# 13

# Nature of Agricultural Geography

*—Braj Bhushan Singh*

IT IS essential to study some of the fundamental themes of agricultural geography primarily in the interest of research students who often feel the paucity of such material; and secondly, to understand the growth of the discipline in terms of definition, subject matter, approach and methods.

The word 'agriculture' has been derived from the Latin words 'Ager' or land and 'cultura' or cultivation of land or raising of crops and the consequent culture. Agriculture also refers to soil husbandry and domestication of animals. Agriculture is defined as the science and practice of farming by Watson.[1] F.A.O. report[2] elaborates agriculture as a raising of crops including plant and grass lands (natural and other pastures). Fowler and Fowler[3] interpret it as a science of soil husbandry and concerned activities as gathering, animal rearing and ploughing. Zimmerman[4] used the term 'agriculture' in the widest sense and defined it as all humanised landscape related to stock raising, crop raising, planting, irrigation, fur farming and fish farming. The definition could not be accepted because of its extensive nature. Buchanan[5] used the word 'Portmanteau' for agriculture and treated it as a way of living in the first phase, followed by the way of getting a living and ultimately an enterprise in which all the activities are purposeful and deliberate. Therefore, McCarty[6] defined 'agriculture' as the purposeful tending of crops and animals. Leslie[7] too preferred to include only purposeful activities to transform the natural environment. Benneh[8] made it clear by defining it as a work, the purpose of which is to harness the force of nature to meet the human needs. Agricultural geography is the geography of cultivation for Brookfield[9], while for Reeds[10] it is to describe and explain areal differentiation. This idea of Reeds was quite similar to Bernhard.[11] According to him agricultural geography strives to bring light to the spatial variation in agriculture and reasons for them. Thus, agricultural geography seeks to interpret and understand the functional relationship of spatial patterns of crops. Leslie[12] treated agricultural geography as a branch of economic geography and defined it as the geography of man's husbandry of the land. It was similar to the definition of economic geography as the geography of man's work, put forth by Buchanan.[13] Benneh[14] felt that spatial analysis of the cropping systems is the main contribution of geographers. He qualified former definitions by mixing the role of multi-factors of the environment, both physical and cultural. Agricultural geography in his words, may be defined simply as that branch of geography that deals with the spatial distribution of agricultural phenomena, such as systems of farming and crops and also seeks to interpret such distribution in terms of the multi-factors of the environment. Agricultural geography is a systematic science which deals with the production-location and patterns of crop distribution (spatial and temporal) in terms of cause-effect and functional-areal relationship at any study level ultimately aiming at preparing a plan for an over-all agro-economic development of the region.[15]

## SUBJECT MATTER

In discussing the nature of agricultural geography the view of Hettner (1905), Bernhard (1915), Waibel (1933), Cholley (1946), Faucher (1949), George (1956), Otremba (1960), Haggett (1965), and Harvey (1968) have been examined. According to Hettner,[16] the nature of agricultural geography is based on three aspects : (a) systematic science, (b) historical science, and (c) geographical science. Bernhard's[17] agricultural geography is both agricultural science and geography; from the former it takes its object, from the latter its approach. He suggested three basic aspects : (i) systematic view of agricultural technology (a) plant production field, (b) animal production field, (c) allied agricultural technological field and agricultural economics; (ii) historical view (agricultural history) and geographical view (agricultural) geography. Bernhard realised that agricultural geography provides the agricultural science the missing link for the creation of a methodological whole. Waibel[18] stressed on : (i) statistical agricultural geography (ii) ecological or functional agricultural geography, and (iii) physiognomic agricultural geography. Cholley[19] pleaded the significance of "agrarian structure" in the study of agricultural geography while Faucher[20] differentiated agricultural science from agricultural geography on the basis of regional nature. He further classified statistical and qualitative agricultural geography or agrarian geography. George[21] emphasised human-cultural aspects in agricultural geography and suggested three basic relationships : (i) between physical environment and agricultural activities, (ii) between population density and agricultural land, and (iii) between historical processes and agricultural stages. Otremba[22] stressed the validity of (i) historical, (ii) economic, and (iii) social aspects in the study of agricultural Geography. Buchanan[23] expressed concern about the problem of scale in agricultural regionalisation. Haggett[24] and Harvey[25] pointed out the basic problem of scale at different levels. At present, the following aspects are treated to be the main themes of agricultural geography which synthesize the subject matter already viewed by different geographers : (i) Production-location and decision-making, (ii) Agricultural landscape and physico-cultural and technological environment, (iii) Dynamic aspects of agricultural landscape and the study of process, stage and structure, (iv) Spatial-functional inter-relationships, (v) Areal differentiation and organisation in distribution pattern and cause-effect analysis, (vi) Typology, regional agricultural development and planning.

Singh[26] has attempted a critical examination of the view perspectives in agricultural geography which, of late, has virtually emerged as an independent discipline with much closer links with agricultural science, economics and sociology.

## APPROACHES

Three basic approaches adopted in agricultural geography are similar to economic geography. These may be mentioned as : (a) systematic, (b) regional, and (c) theoretical. System approach deals with the general characteristics of the agricultural phenomenon considering the area as a single unit. Sometimes, sub-regions are studied under this approach under sub-heads. A regional approach has been quite common. The area is divided into regions and sub-regions in first step followed by a systematic analysis. Theoretical approach is concerned with the formulation of theories which is tested and experimented in the areal perspective. These approaches are, however, undertaken together in an integrative form in dealing with the agricultural aspects.

Generally speaking, the study of agricultural geography is conducted in three stages : (a) production-location and distribution pattern, (b) study of characteristics, and (c) study of relationship between physico-cultural elements and production systems. The first question concerning any aspect of agricultural geography is attached to the distribution of crops and their spatial location. It is essential to know the areal variations of different crops for further analysis. Attention is focused on the farming system, intensity and efficiency in the second stage. The study of interaction between physico-cultural elements and agricultural operations is conducted in the third stage which is usually analysed in three forms : (a) cause-effect relationship, (b) functional relationship, and (c) areal differentiation. Cause-effect relationship is studied in terms of location and distribution of crops. Production of rubber in Malaysia is the result of physical (temperature and rainfall) environment. Similarly, there is a functional relationship between corn farming and hogging in U.S.A. A slight change in one element results in relative change in another. It is realised that economic activity never develop in isolation. It is invariably inter-related to other elements belonging to internal as well as external surroundings. Thus, the study of regional inter-relationship becomes necessary.

## METHODS ADOPTED

The methods of agricultural geography has been mainly related to description, classification and analysis of the phenomenon which, in other words, may be termed as "Inductive method". Observations and experience form the base of description. In brief, features are to be explained on the basis of empirical enquiry. Another method is known as "Deductive method" in which hypothetical models are developed in the first instance. Later on, these are tested in the light of actualities. It is of immense value in forecasting phenomena and adopted recently in agricultural geography. Deductive models are usually developed in optimal allocations. Pred[27], Chorley-Haggett[28] presented optimizer models. Harvey[29] preferred a satisfier model in its place. Many geographers suggested the sub-optimal model based on simulation technique. Donaldson and Webster[30] are first to develop this type of model. Heady Egbert[31], Henderson[32] and Shaw[33] introduced linear programming model in forecasting optimal allocation and production. Such models are more valuable from the academic point of view as compared to its experimental utility. Kendall[34] and Henshall[35] used factors analysis models in place of satisfier model in order to enhance its applicability. Thus, inductive method is studied to understand and analyse the existing characteristics whereas deductive models are used to forecast the optimal conditions for the future.

## MODELS USED

All the models used so far can be divided into two broad classes : (a) normative model, and (b) descriptive model. The aim of the former model is to ascertain the future possibilities based on existing and hypothetical situations. Descriptive models are used to explain and analyse the existing characteristics of agriculture. Models, based on systems, are divided into three classes : (a) single system model, (b) total system model, and (c) sub-system model. Experimental models fall in the first category while landuse and farming system models belong to the second category. Subsistence farming system is an example of the sub-system model. All the models of agricultural geography studied so far are grouped[36] as such : (i) experimental model, (ii) conceptual model, (iii) environmental model, (iv) historical model, (v) taxonomic model, (vi) production-location model, (vii) decision-making model, and (viii) land potential model.

All the *experimental models* may be sub-classified into two : (a) model farm, (b) theoretical model farm. Blaut[37], Heller[38], Henshall[39] and many others adopted the model farm approach whereas Carter[40], Day[41] and others used the theoretical model farm approach. *Conceptual models* deal with fundamental themes of various aspects of agriculture. Such models were developed to analyse landuse, efficiency, agricultural productivity cropping pattern, crop concentration and diversification and crop location etc. *Environmental models* were studied mostly in the initial stage of the discipline. The model centres round the concept that agricultural characteristics of any region are the function of environmental condition. Mainly two approaches were adopted in such models. The first may be termed as the evolutionary model. It is concerned with a historical sequence and cultural dispersion in the evolution cycle of the plant, animal and other species. Saur[42] initiated this model. In the second model, an emphasis is laid on the changing relationship between population density and agriculture in the historical evolution of the system. Boserup[43] championed it. *Taxonomic models* explain and analyse classification systems. Recently the study of agricultural typology[44] has been of utmost significance. Studied conducted of *Production location models* are based on three aspects : (a) theoretical analysis of the location of agricultural activities, (b) rural and urban settlements and landuse zoning, and (c) transport and landuse zoning. Such models are grouped into two categories : (i) single unit spatial model, and (ii) inter-regional equilibrium model. Landuse locations are determined keeping in view the centrality of a single market in first model. Von Thunen[45] was the first to initiate this model. It was improved and advanced by Hoover[46], Lösch[47], Dunn[48], Garrison[49], Grotewold[50], Horvath[51], Kellerman[52], Harvey[53] hypothesized that area of various agricultural activities would be like points whereas producer, production factors and consumers would be located in discrete form amidst which transport input would be zero. Location of activities at different points in the region would be the result of comparative advantages. Such models are further sub-divided into two : (a) input-output models—agricultural locations are determined on the basis of net returns. Such models are the gift of agricultural economists. Loentiff[54] is said to be the father of such models. Carter and Heady[55] and Heady and Egbert[56] strengthened it; (b) spatial equilibrium model—optimal location of crop and area is determined on the basis of spatial equilibrium. Fox and Tanber[57], Judge and Wallace[58], Egbert and Heady[59] introduced this model. Optimal conditions were determined with the help of linear programming models. Many worked on *decision-making concept* while determining the role of socio psychological factors on landuse. It is further classified into two : (a) diffusion model and (b) game theory model. The contributions of Rogers[60], Hagerstrand[61], Wolpert[62] and Found[63] are significant. *Landuse potential models* have been presented by Stamp[64], Hockensmith[65], Stallings[66], and Beguin[67]. Beguin presented separate models of land potentials, population density potential and regional diversities. These models are statistical.

There appears to be an obvious shift from topical to regional and also from regional to typological studies. Typological studies form a base of spatial organisation in terms of a set or association of internal as well as external characteristics. Now, transformation of agriculture is not considered as an isolated phenomenon. Thus, the discipline requires an integrative approach in studying varied sets of assemblages in terms of internal processes (social-cultural, organisational and technical) as well as external processes[68] (locational-transportational and market conditions). The integration of agriculture, industries and services is of utmost significance in ensuring rural prosperity. It is, therefore. desirable that the scope of agricultural geography be widened in establishing diagnostic linkages with all allied aspects in regional economic development. This model has been currently[69] initiated under the title of 'Spatial Organisation'. The need is also felt to remodel[70] the traditional concept of general landuse models.

## REFERENCES

1. S.J. Watson, *Chamber's Encyclopedia*, 1950.
2. F.A.O. Report 1962, Agriculture in World Economy, Rome, p 3.
3. H.W. Fowler and F.G. Fowler, *Oxford English Dictionary*, Ed. 1964, p 26.
4. E.W. Zimmerman, *World Resources and Industries*, 1951, pp 47-48.
5. R.O. Buchanan, "Some Reflections on Agricultural Geography, *Geography*, 1959 XLIV, pp 1-8.
6. H.H. McCarty, and H B. Lindberg, *Preface to Economic Geography*, Prentice Hall, 1966 INC, pp 204.
7. L. Symons, *Agricultural Geography*, G. Bell & Sons Ltd., London, 1970, pp 1-4.
8. G. Benneh, "The Role of the Geographer in the Agricultural

Development of the Tropics with Special Reference to Ghana" in *Proceedings of Symposium on Landuse in Developing* Countries, edited by Prof. M. Shafi, 1972.

9. H.C. Brookfield, Questions on Human Frontiers of Geography, *Economic Geography*, 1964. 40, (4), p 259.
10. L.G. Reeds, *Agricultural Geography : Progress and Prospects Canadian Geographer* 1964, 8, p 51.
11. H. Bernhard, *Dier Agrargeographicals wissen Schaftliche Disciplin*, Petermanns Mitteilungen, 1915, 61, p 12.
12. L. Symons, op cit, 1970. pp 1-4.
13. R.O. Buchanan "Approaches to Economic Geography," *Indian Geographical Journal*, Silver Jubilee Souvenir 1951, pp 1-8.
14. G. Benneh, *op cit*, 1972, pp 180-191.
15. B.B. Singh, *Agricultural Geography : Definition. Approach and Methods*, Geographical Observer, 1978, 14, pp 16-19.
16. A. Hettner, *Die Geographie ihre Geschichte, ihr Wesen und ihre Methoden* (Breslau : Hirt) 1905, pp 114-17.
17. H. Bernhard, *op cit*, 1915, 12.
18. Waibel, *Probleme der Land Wirfschafts Geographie (Brcslau: Hirt)*. 1933.
19. A. Cholley, Problem de Structure Agraire et Deconomie Rural, Annals de Geographie, 1946, 60, pp 81-101.
20. D. Faucher, *Geographic Agrarie* (Paris) 1949, p ii.
21. P. George, *Lan Compagne* (Paris : Press Uni. de France), 1956, pp 7-9.
22. E. Otremba, *Allgemeine Agar—Und Industrie, Geographie*, 2nd ed. rev. (Stuffgart : Franckh), 1960, p 25.
23. R.O. Buchanan, 1958, op cit, p 5.
24. P. Haggett, *Locational Analysis in Human Geography*, London, 1965.
25. D.W. Harvey, "Pattern Process and Scale Problem in Geographic Research." *Trans Inst. of Brit. Geogrs.*, 1968, 45, pp 71-78.
26. L.R. Singh, "New Perspectives in Agricultural Geography". *National Geographer*, 1976, 11, (2), pp 113-122.
27. A. Pred, *Behaviour and Location : Foundations for a Geographic and Dynamic Theory*, Pt. I, Lund. studies, Geog. series, B 27, 1967.
28. R.J. Chorley and P. Haggett, *Models in Geography*, London, 1967.
29. D.W. Harvey, *Theoretical Concepts and the Analysis of Landuse Patterns*, Trans. Inst. Brit. Geogrs., 1966, 33, pp 123-44.
30. G.F. Donaldson and J.P.G. Webster, *An Operating Procedure for Simulation Farm Economy—Monte Carto Method*, Wye College, (1968).
31. E.O. Heady and A.C. Egbert, "Regional Programming of Efficient Agricultural Production Patterns" *Econometrics* 1964, 32, pp 374-86.
32. J.M. Henderson, "The Utilization of Agricultural Land : A Regional Approach", *Pap. Proc. Reg. Science Ass.*, 1957, 3, pp 109-117.
33. J.R. Shaw, "The Location of Main Crop Potato Production in Bri-

tain—Application of Linear Programming", *J. of Agr. Eco.*, 1970, 21, pp 267-82.

34. M.G. Kendall, "Geographical Distribution of Crop Productivity in England", *J. Roy. Stat. Soc.*, 1939, 102, pp 21-62.
35. J.D. Henshall, "Models in Agricultural Activity", Chapter XI, 425-458 in the book, *Socio-economic models in Geography*, 1967, Edited by R.J. Chorley and P. Haggett.
36. J.D. Henshall, Ibid, 1967.
37. J.M. Blaut, "Micro-Geography Sampling : A Quantitative Approach to Regional Agricultural Geography", *Economic Geography*", 1959, 33, pp. 79-88.
38. C.F. Hetter, "The Use of Model Farms in Agricultural Geography", *Professional Geographer*, 1964, 16 (4), pp 20-23.
39. J.D. Henshall, "The Spatial Structure of Barbados Peasant Agriculture", *Unpub. M.Sc. Thesis*, McGill Uni., Montreal, 1964.
40. H.C. Carter, "Representative Farms—Guides for Decision-making", *Journal of Farm Economics*, 1963, 45 (5), pp 1449-55.
41. L.M. Day, "Use of Representative Farms in Studies of Inter-regional Competitions and Production Respouse," *Journal of Farm Economics*, 1963, 45 (5), pp 1438-44.
42. C.D. Saur, "Agricultural Origins and Dispersals, American Geographical Society", *Bouman Memorial Lectures*, 1952, p 2.
43. E. Boserup, The Conditions of Agricultural Growth, London, 1965, p 124.
44. J.T. Coppock, "Crop, Livestock Enterprise Combinations in England and Wales", *Economic Geography* 1964, 40, pp 54-81; J. Kostrowicki, 1972, "A Preliminary Attempt at a Typology of World Agriculture", *International Geography*, 22nd Intern. Geog. Congress. Montreal, Uni. of Toronto Press, 1097-1100, *The Typology of World Agriculture —Principles, Methods and Model Types* (Warszawa, 1974), p 4; V.R. Singh, 1977, "*Taxonomy of Indian Agricultural Typology*", *Uttar Bharat Bhoogol Patrika*, 13, 182, pp 17-25.
45. J.H. Von Thunen, *Der Isolierte Statt in Bexiehung auf Landwirtschaft Und National Akonomie*, Hamburg, 1957.
46. E.M. Hoover, *The Location of Economic Activity*, New York, McGraw Hill, 1948.
47. A. Lösch, *The Economics of Agricultural Locations*, New Haven, Conn. Yale, Uni. Press, 1954.
48. E.S. Dunn, *The Location of Agricultural Production*, Gainsville, Uni. of Florida Press, 1954.
49. W.L. Garrison and D.F. Marble, "The Spatial Structure of Agricultural Activities", *Annals af the Ass. of Amer Geogrs.*, 1957, 47, pp 137-44.
50. A. Grotewold, "Vanthunen in Retrospect", *Economic Geography*, 1959, 35 (4), pp 346-355,
51. R.J. Horvath, "Von Thunen Isolated and State and Around Addis

Ababa, Ethopia", *Annals of the Assn. of Amer. Geogrs.*, 1969, 59, pp 308-323.

52. A. Kellerman, "Determinants of Rent From Agricultural Land Around Metropolitan Areas", *Geog. Analysis* (1978), 10 (1), pp 1-12.
53. D.W. Harvey, *op cit* 1966, pp 365-67.
54. W.W. Loentiff, *Studies in the Structure of American Economy*, New, York, 1953.
55. H.A. Carter and E.O. Heady, "An Input-Output analysis Emphasizing Regional and Commodity Sectors of Agriculture, Iowa," *Agri. and Eco. Exp. Stu. Res. Bull*, 1959, p 469.
56. E.O. Heady and A.C. Egbert, "Regional Programming of Efficient Agricultural Production Patterns", *Econometrica*, 1964, 32 (3), pp 374-386.
57. K. Fox and R. Tanber, "Spatial Equilibrium Models of the Livestock Feed Economy", *American Eco. Review*, 1955, 45, p 584-608.
58. G.G. Judge and T.D. Wallace, "Estimation of Spatial Price Equilibrium Models", *Journals of Farm Economics*, 1958, 40, 801-02.
59. A.C. Egbert, E.O. Heady and R.F. Prrokken, Regional Changes in Grain Production; *Iowa Agriculture exp. Stn. Res. Bull.* No. 521, Iowa, 1964.
60. E.M. Rogers, *Diffusion Innovations*, New York, 1962.
61. T. Hagerstrand, *Innovations for Loppert Ur Korologisk Synpunkt* (Lund), 1953.
62. J. Wolpert, "The Decision Process in Spatial Context", *Annals of the Assn. Amer. Geogrs.*, 1964, 54 (4), p 537-58.
63. W.C. Found, *A Theoretical Approach to Rural Landuse Patterns*, Edward Arnold, London, 1971, p 165.
64. L.D. Stamp, "Fertility, Productivity and Classification of Land in Great Britain", *Geographical Journal*, 1940, CXVI, (6), pp 397-401.
65. R.D. Hockensmith, "Classification of Land According to its Capability as a Basis for a Small Conservation Programme", Proct. Inter. Amer. Conf. on Conservation of Natural Resources, Colorado, 1949.
66. J.H. Stallings, *Soil Conservation*, Prentice Hall, INC. Englewood Cliff, N.J. 1957, p 424.
67. H. Beguin, *Models Geographiques pour Loespace Rural African*, Brussels, 1964.
68. M.T. Katzman, "The Von Thunen Paradigm, the Industrial-Urban Hypothesis and the Spatial Structure of Agriculture", *American Journal of Agriculture Economics*, 1974, 56, pp 683-96; C.R. Bryant, *The Anticipation of Urban Expansion Geographica Polonica*, 1973; 28, pp 93-115; G.J. Lewis and D.J. Maund, "The Urbanization of the Countryside : A Framework for Analysis", *Geografiska Annaler*, 1976; 58, 1, pp 17-27; Thomas Lunden, "Landuse Decisions in a Time-space Framework—Some Stockholm Examples", *Geografiska Annaler*, 1977, 59, 1, pp 1-13.
69. D. Jenelle, "Spatial Organisation : A Model and Concept", *Annel.*

*Assn. of Amer. Geogrs.*, 1969, 59, pp 348-64; J. Singh, Spatial Organisation : "A Geographic Concept", *Uttar Bharat Boogol Patrika*, 1977; 13, 182, pp 1-9.

70. A.M. Coleman, "A Geographical Model for Landuse Analysis", *Geography*, 1969, 54, pp 43-55; "The Quest for a General Landuse Model", *Essays in Applied Geography*, S.M. Ali Memorial Volume, Sagar, 1976; S.K. Srivastava, "General Landuse Model : Method and Experiment," *Uttar Bharat Bhoogol Patrika*, 1977; 13, 1 and 2, pp 26-34.

# 14

# Some Aspects of Land Utilization

*—Manager Prasad*

THE SUCCESS of national planning has to be measured by certain objective tests such as scrutinising whether the varied needs of the people are being satisfied. The needs have been classified by Stamp into six basic categories : the need to work, home, food, transportation, communication, defence and recreational.[1]

To provide all these needs the planner has to look towards land. The need of food means conservation of good agricultural land for production and at the same time the development of poorer land for production of food and other raw materials such as sugarcane and cotton, etc. for the industry. The same land has to satisfy the needs of transport and communication for the railways, highways and airports. Increasing acreage of land should also be provided for recreational purposes such as parks, playgrounds, gardens and clubs, etc. which is not a luxury but a a prime necessity for health, pleasure and mental balance of mankind. Apart from these basic claims a hundred different and apparently irreconcilable claims could be listed. The factories are located without any regard to the agricultural value of land or the growth of cities takes place, in most cases, on rich agricultural land. This tendency leads to grave dangers. Land that has once been withdrawn from agricultural use seldom returns to the farmer, and if it does, it will only be after a long lapse of time.[2] In the words of Osborn, the industrial complex of the country will gradually weaken unless we not only maintain but substantially improve our agricultural processes.[3]

Then again, there are conflicts within agricultural itself and between the use of land for cash crops and food crops. The increasing number of human heads makes the conflicts of land-use more serious. The rising population should be accompanied with a rise in food production, but land is limited and, therefore, has to be used judicially.

If the food growing fields are intensively cultivated without proper manuring, the food obtained from such fields would be deficient in mineral matters which come from the soil. We need minerals such as calcium, phosphorus, iron, copper, iodine, fluorine, and a whole array of 'trace elements' and vitamins with the carbohydrates, fats and proteins for our physical and mental well-being.[4] Even leprosy is connected with the use of rice grown in old and wasted paddy fields.[5]

Some day in our country a planned programme will determine the pattern of landuse and then not only crops and tame animals, but indirectly the occurrence of all living things will be determined by man's conscious planning and use of the land.[6]

## THE WORLD LANDUSE SURVEYS

With the increasing population and price of the land it is our duty to study the every use, misuse, overuse and underuse of land within because it is helpful in goading the planning of resources in the region. To feed the world is the problem for which a world landuse inventory is necessary. It was an awareness of this fact that S. Van Valkenburg of Clark University, Massachusetts brought forward a project at the International Geographical Congress held in Lisbon in April, 1949 to form a commission to formulate the programme of World Landuse Survey. The UNESCO supported the move and a commission of four[7] met at the Clark University in December, 1949. It drew up a scheme on the objectives and procedure of the World Landuse Survey. It was proposed that landuse publications should be on a uniform scale of 1 : 1 million and each map should be accompanied by an explanatory text. Later on in 1951 an organisation called World Landuse Survey came into being with the establishment of the Old World Division with headquarters in London. When the commission met at the International Geographical Congress in 1953 under the chairmanship of Van Valkenburg, it was proposed to carry out pilot surveys in as many parts of the world as possible. Van Valkenburg was to be in charge of the Americas and Stamp of the countries of the Eastern Hemisphere. An impressive record of the pilot surveys on different scales of various parts of the world were presented.

### THE BRITISH LANDUSE SURVEYS

The pioneer in respect of landuse survey has been Great Britain. Britain is the only country in the world which completed a national landuse survey over the whole country. This was done under the able guidance of L.D. Stamp.[8]

The credit for benchmark landuse survey goes to Patrick Geddes who evolved the concept of regional survey with study and mapping of land utilization as an essential basis.

The real work of landuse survey in Great Britain was organised by L.D. Stamp in October, 1930. It was conducted with the idea of finding out exactly what use was then made of every acre of the land of Great Britain. This pioneer work, to a large extent, was crowned with success in 1930s on account of voluntary exercise from the students of schools and colleges of Britain. The work was organised on a nationwide level and no less than a quarter of a million school children actively participated in it. Though many of the teachers organised it most skilfully, there were, of course, parts of the country where survey by the school students were not possible, and parties of university students and graduates were encouraged to cover these. "To cut short the long story, the bulk of the country was actually surveyed, the landuse recorded in the years 1931, 1932 and 1933. There remained small parts which had to be completed later but the whole was virtually finished before the outbreak of war in 1939, and before the beginning of the plough-up campaign which made such an enormous difference to land-use in the country.[9]

### AMERICAN LANDUSE SURVEY

The beginning of land utilization survey in the U.S.A. can be traced from O.E. Baker's published article entitled "Land Utilization in the United States: Geographical Aspect of the problem". He depicted the trends in land utilization and emphasised the need of land classification and surveys. The country was vast with many virgin lands to be opened. The

U.S.A. had set another example of landuse survey mainly for planning on a comprehensive scale. The idea of country landuse planning[10] was fostered by the U.S. Deptt. of Agriculture out of the need of coordinating the depression in agricultural aid programme, and the interest in securing local participation in planning. Although the programme of landuse survey was launched in the summer of 1935, it was properly executed only after 1938. Some organisations for planning had started functioning in 1891. It is estimated that over 1,000 communities were organised and 1,40,000 individuals, mostly laymen, took part in the work.[11] Studies of landuse including map making were undertaken and their analyses led to recommendations as to changes in "the kind of farming and sizes of farms in each landuse area, what land should be recovered for forestry, for wild life or for recreation and in what area settlement should be encouraged or discouraged".[12] The problems thus studied suggested plans in every branch, viz., farming and farm sizes, land recovery and settlement, etc. The development plans were later on worked out through proper authorities.

The concept of American landuse planning can be summarised in one sentence : "What is proposed in landuse planning is as broad as country life itself and organisation should, therefore, presumably begin with the most general functional organisation that can be discovered in an area."[13]

Thus in the U.S.A., landuse survey on countrywide scale was taken up mainly for the purpose of landuse planning which embraced the planning of both the physical and human side, i.e., all round development of the community.

### LANDUSE SURVEYS IN OTHER COUNTRIES

Probably no country has taken up the work more enthusiastically than Japan. The crucial problem there is how to support a huge population of some 90 million on the small area of cultivated lands, even when intensively used. It is not surprising that landuse survey has been given the highest priority.[14]

China with the largest population, badly needed a land utilization survey to improve its economy. Sample landuse studies suggested some landuse problems. *Land Utilization of China* by L. John Buck published in the year 1938, amplifies China's land utilization problems.

P.P. Karan[15] has recently made a study of Nepal and has prepared a land utilization map using air photographs.

Landuse surveys are in progress in various countries of the world, particularly in highly crowded countries, where land-use planning on scientific lines is vital, e.g., Pakistan, Canada U.S.S.R., Cyprus, etc.

## LANDUSE SURVEYS IN INDIA

An achievement in landuse study in India is probably due to the pioneer work done by the late L.D. Stamp in Great Britain. It was at the 1940 session of the Indian Science Congress held at Madras that S.P. Chatterjee pointed out the necessity of undertaking landuse survey in India on the lines of the landuse survey of Great Britain. Considering the vast area of India and the huge population who utilize land in various ways, there is no wonder that the problem of landuse would baffle easy and speedy solution. Land is our most valuable asset.

P. Dayal[16] prepared a thesis on agricultural grography of Bihar. Many geographers and economists started writing some accounts of landuse of a village or small area. Different types of landuse makes an intricate pattern, which needs geographical investigation in the context of their physical environment. This point has been emphasised by S.P. Chatterjee. He suggested undertaking a landuse survey by geographers on an all-India basis. His work in the field appeared as a published document under the caption *Land Utilization in the District of 24 Parganas Bengal.*[17] This work, profusely illustrated with diagrams and photographs, revealed that the district was poorly drained though nature endowed it with big rivers capable of draining it most efficiently. He further discussed the influence of soils and climatic elements on land utilization, the pressure of population on land and the nature of land utilization. A more detailed land utilization survey was conducted in Howrah District and 1200 landuse maps at the scale of 1 : 3960 covering 813 villages were prepared.[18]

V.L.S.P. Rao has emphasised the techniques of soil survey that he adopted for his analysis of landuse in the Godavari region, referred to the aims, objectives and classification of categories of landuse of the world Landuse Commission set up by the International Geographical Union.[19]

The sampling technique was suggested by M. Shafi[20] and he preferred purposive sampling to the other four types—random, stratified, cluster and systematic. A.K. Sen carried out landuse survey of part of the central Mayurakshi basin of the Birbhum District on the basis of existing landuse pattern. U. Sen and M. Guha (1960) made the detailed landuse study of a typical village in the Bankura District. R. Lahiri (1950) studied the major landuse types and agricultural problems for four typical villages, Kanjisar, Dhaulini Koradih and Koiridih near Jasidih, Deoghar.[21] S.B. Goswami (1960) selected one village in Ranchi district, Misirgonda and used it as a model for the study of the general pattern of the agricultural landuse in central Ranchi plateau. A. Sharan (1962) made a study of landuse changes in Salgawana village in the upper Damodar basin. S.M. Karimi (1950) made a study of landuse in Dinapur Ghusahra village near Patna. R.N.P. Sinha received his Ph.D. degree on the study of landuse in the canal irrigated area of Patna district. A commendable work on landuse was done by K.N. Das while preparing a thesis for the doctoral degree of Bhagalpur University.

The landuse of Delhi State was studied by M. Ameen (1956) and R. Ellefsen (1960). O.P. Bhardwaj dealt with landuse in the low lands of the Beas river in the Bist-Jullundur Doab with particular reference to two villages Mohalla and Akitunda.[22] A.N. Raina made an attempt to present basic factors affecting type and success of landuse in the floating garden region of the Kashmir valley.

Landuse study in much more extensive areas has been attempted in Uttar Pradesh. B.N. Ganguli (1955) has conducted a sample survey in a village Burki near Gyanpur. U. Singh (1955) surveyed five villages near Sarnath and found considerable changes in the crop pattern during the past 70 years. V.R. Singh (1950) studied land utilization in the Sonpar region. He also traced changing landuse pattern around Mirzapur town since the beginning of the present century. B.K. Roy

(1961) made a sample study in land utilization of five villages of Ballia District. He also contributed a paper on measurement of rural landuse in Azamgarh, Middle Ganga Valley.[23]

The pioneer work on landuse was done in South India by V.L.S.P. Rao. M.S. Honro (1953) studied the landuse in a typical village Halge in the coastal area of the North Kanara District of Karnatak. He subsequently (1962) brought out another paper on landuse patterns in the lower Kali basin in the same district. N. Anantapadmanabhan (1956) made an attempt to show the relationship between landuse and rural population density in 7 representative taluks of Tamilnadu. C.B. Joshi (1961) has made an intensive study of the landuse and settlements in the Ratangiri taluka on the western coast of Maharashtra. A.T.A Learmonth (1962) made an intensive field study of 60 sample villages in Karnataka State.[24]

B. Jha (1965) studied land capability classification of the Kosi region. K.Z. Amani (1968) brought out a paper on land utilization in village Golagarhi.[25] B.N. Sinha (1968) made an attempt to study agricultural efficiency in India.[26] Jasbir Singh (1971) dealt with the optimum carrying capacity of the land in the Punjab.[27] V.M. Deshmukh (1975) studied rural landuse of Lonkhede.[28] Manager Prasad studied the changing pattern of landuse in Bodh Gaya Block (Gaya)[29] between 1914-15 and 1974-75. The author has selected seven villages of the block on purposive sampling method and studied their landuse in detail. The work is profusely illustrated with maps and diagrams. He found considerable changes in the crop pattern during last 60 years.

## REFERENCES

1. L D. Stamp, *The Land of Britain and How it is Used*, Longmans, Green Co. Ltd., London, 1948, pp 74-77.
2. R.G. Stapledon, *The Land : Now and Tomorrow*, Faber and Faber, London, 1944, p 32.
3. Fairfield Osborn, *The Limits of the Earth*, Leonard Hill, London, 1953, pp 66-67.
4. E.H. Graham, *Natural Principles of Landuse*, London, 1944, p. 222.
5. R.K. Mukherjee, *The Rural Economy of India*, Longmans, Green and Co. Ltd., London, 1926, p 196.
6. K.H.W. Klages, *Ecological Crop Geography*, New York, 1947, p 10.
7. S. Van Valkenburg of Clark Uni. (Chairman), Hans Boesch of

Sorich, Leo Waihel of Brazil and L.D. Stamp of Great Britain.

8. L.D. Stamp, *The Land of Britain : Its Use and Misuse*, Third Ed. Longmans, Green & Co. Ltd., London 1962, pp 22-41.
9. L.D. Stamp, *Applied Geography, A Penguin Book*, Richard Clay & Co. Ltd., Bungay, U.K., 1964, p 39.
10. Arthur Hillman, *Community Organization and Planning*, Macmillan, New York, 1957, pp 44-48.
11. Ibid.
12. Country Landuse Planning Pamphlet, Country Planning Series No. 1, March, 1940, p 9, Quoted by Hillman in *Community Organization and Planning* op cit, p 46.
13. Quoted by Arthur Hillman in his book Community Organisation and Planning, op cit, p 48
14. L.D. Stamp, Op. cit., pp 48-50.
15. P.P. Karan, Landuse Reconnaissance in Nepal by Aero-Field Techniques and Photography, *Proceedings of the American Philosophical Society*. Vol. 104, No. 2, April, 1960.
16. P. Dayal, *The Agricultural Geography of Bihar*, London, Ph.D. Thesis (unpublished), 1947.
17. S.P. Chatterjee, "Land Utilization in the District of 24 Parganas," Bengal, 1945, B.C. Law Vol., Part 2, Calcutta.
18. S.P. Chaterjee, "Land Utilization Survey of Howrah District, *Geographical Review of India*," Vol XIV, No. 3. 1952.
19. V.L.S.P. Rao, "Soil Survey and Landuse Analysis, *Indian Geographical Journal*," Vol. XXII, No. 3., 1947.
20. M. Shafi, "Land Utilization in Eastern Uttar Pradesh," Deptt. of Geography, Muslim University, Aligarh, 1961.
21. R. Lahiri, "Land Utilization in Some Villages near Jasidih, *Calcutta Geographical Review*" Vol. XII, No. 1, 1950.
22. O.P. Bhardwaj "Landuse in Bist-Jullundur Doab, *National Geographical Journal of India*," Vol. VII, Part 4, 1961.
23. B.K. Roy, "Measurement of Rural Landuse in Azamgarh, Middle Ganga Valley, *The Geographer*, Special Number, Landuse, Vol. XV. Nov. 1968.
24. S.P. Chatterjee, *Fifty Years of Science in India*, Indian Sc. Congress Association, Calcutta, 1964, pp 145-51.
25. K.Z. Amani, "Land Utilization in Village Golagarhi, *Geographer*, Special Numbers, Landuse, Vol XV, Nov. 1968.
26. B.B. Sinha, *Agricultural Efficiency in India*, Ibid.
27. Jasbir Singh, "Optimum Carrying Capacity of Land in Punjab," *N.G.J.I.*, Vol XVII' Part I, March, 1971.
28. V.M. Deshmuk, "Lonkhede : A Study in Rural Landuse, the Deccan *Geographer*" Vol. XIII. Nos. 1 and 2, Jan-Dec., 1975.
29. Manager Prasad, *The Changing Pattern of Landuse in Bodh Gaya Block (Gaya)*, 1978.

# 15

# Models in Land Utilization

—*Ram Bahadur Mandal*

THE SIMPLIFIED representation of the real world situation is known as a model.[1] It is also called an analogy, an ideal, a description, an abstraction, a generalization, a form of reference, a programme of research, an explanation of how a system works, a theory, and a demonstration in the process of simplification of complex situations in a diagrammatic or mathematical form. Model building is seen to be a descriptive-cum-nomothetic approach in geography, which explains the complex perception of any spatial distribution in a logical and simple manner. The purpose is to analyse the models used in landuse study and their formulation.

Landuse varies according to the needs of the society in changing space and time. The generalization of the real world situation of land utilization either from category, natural regions of socio-economic models is generally helpful in formulating the concept of landuse. Earlier, Von Thunen had made generalizations about the landuse intensity in rural areas. Later on general landuse models, diagrammatic models, models of optima and limits, the polar and antipolar model of landuse and gravity models are considered here for analysis.

**The Category Model** : The determination of types of landuse in a region is known as category model. For example, five main categories of landuse are found in India, i.e., net sown area, fallow, pasture, forest and waste. Similarly, the first landuse survey of Britain conducted by L.D. Stamp records seven categories of landuse while the second landuse survey conducted by Alice Coleman recorded seventy categories, which have little use in map analysis. This shows that fewer categories are helpful for drawing of maps in landuse studies whereas larger number of categories are helpful in data analysis.

**The Natural Regions Model** : This model is helpful in order to highlight the regional patterns of land utilization which are the product of relief, climate and inherent structure of rocks. On the eastern coastal plains of India, for example, the cultivation of rice and coconut are instances of prominent land utilization of the area; whereas on the western seaboard of India, the northern side especially is covered with dense forest as the western ghats extend up to the sea coast leaving only a small part as sea beach. Such a manner of land utilization should be expected to prevail all over the world in characteristic regions but due to variations of climate and relief features of landscape, it does not always do so. This needs some kind of man-made modification of the observed discrepancies, whatever they may be, for the best use of land.

**Socio-Economic Model** : This model includes the influences of caste, income of the society, technical abilities, transport connections, settlement types, power stations, etc. on the land-use of an area. But without considering the structure, relief and climatic conditions of an area the cultural features play a little role in the shaping of land utilization types. In this sense the socio-economic model is regarded as a partial model, rather than a full fledged one.

## VON THUNEN MODEL OF LANDUSE

In determining agricultural landuse, the environmental control of agricultural land in relation to transportation, location of cities and consuming rural settlements in the surrounding areas play an important role in ascertaining the farm production to be raised in different areas. Von Thunen, a German economist, analysed the spatial distribution of farming systems, the kinds of produce to be raised at a given place and the influence of transportation lines in the localization of agricultural production. Von Thunen then considered a city in the centre of an isotropic plain.[2] In this way, he proposed his concept over a region which had no trade connections with the outside world and was surrounded by uncultivated wilderness. The soil was of uniform fertility, where the city was located centrally surrounded by rural settlements which had no improved means of transport. On the basis of production of goods, distance travel to the market for particular price, and calculated economic rent on each type of landuse from the central city to various distances, he obtained an ideal distribution of production zones as a series of concentric circles arranged from the city (Fig. 15.1).

*Von Thunen's Original Model of Landuse Types*

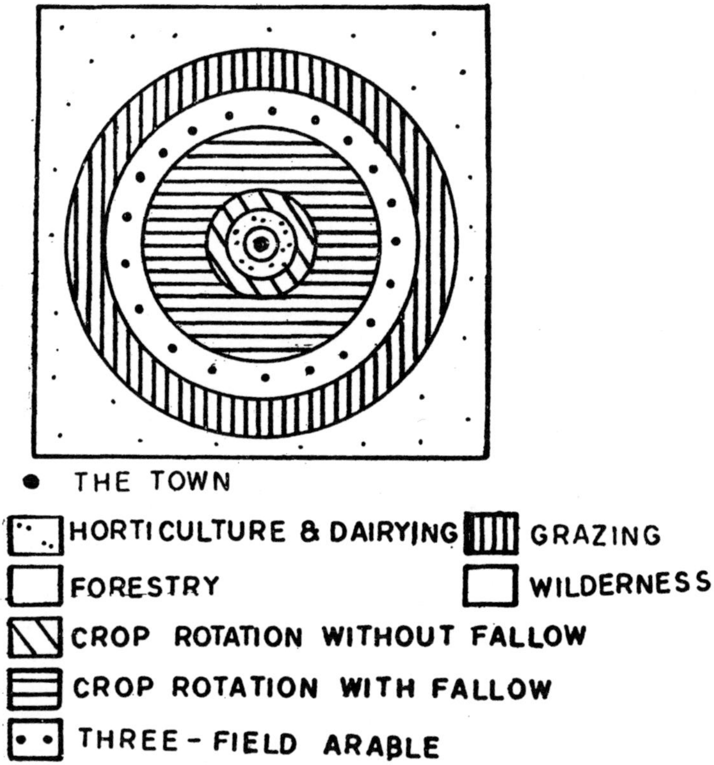


FIG. 15.1

He had also conceived a system that the land near the market shall have large inputs of manure, labour, etc., with intensive methods of cultivation, while it will be extensive when away from the market. There are two exceptions to this rule, which are as follows :

(1) When per hectare production is large despite small inputs as in the case of forestry;

(2) When a large quantity of inputs yield a small amount of products as with butter at a considerable distance from the market.

Von Thunen expressed his view that in zones close to the market on any settlement, vegetables, berries and milk may be raised, while in outer zones butter, root crops and grass for the livestock may thrive well. Thus Von Thunen placed forests close to the city. This is because forests supplied fuel and building materials which are bulky. At that time only wagons were available for transportation and it was because of this fact that forests had been placed near the market. In reality the natural forests have no such importance to be placed in the second zone except that forests were man-made and deliberately placed in the second zone.

**Landuse Competition**

When land is vacant then there is no competition in any use of the land. Landuse competition arises with the superiority of demand for a particular use on a particular piece of land. This is generally judged by (a) economic rent of the land, (b) superiority of demand for a particular use of the land, and (c) the character of land for a particular use.

Transport costs exerted an influence upon the spatial pattern of landuse with the help of its rent mechanism. Rent represents the primary production of the land. Hence the spatial pattern of agricultural production is the process of competitive bidding among landuse occupancy of a given place and time. That should yield the highest net return per unit of land and that will make the highest bid for the use of a particular plot of land. This kind of rent to which Von Thunen referred was more specific—"economic rent, which is the surplus of income accruing to a unit of land above the minimum income required to bring a unit of new land into use at the margin of production".[3] Of course, this difference is due to spatial variation of two places in terms of fertility and distance from the market in Von Thunen model.

Economic rent can be estimated with the help of the following formula :[4]

$$R = E(P - a) - Efk$$

where, R = rent per unit of land,

k = distance to market,

E = output per unit of land,

P = market price per unit of commodity,

a = production cost per unit of commodity (including the farmers' payments to himself for his own labour),

f = transport rate per unit of distance for unit of commodity.[5]

With the help of the original assumptions of Von Thunen, Figure 1, illustrates the ring formation by means of a hypothetical case in which five landuses compete for space around a settlement.

On the basis of techniques used in Figure 15.1 to the landuse type commonly found in the north Germany of his days Von

*Von Thunen's System of Landuse with Navigable Waterway and Satellite Town*

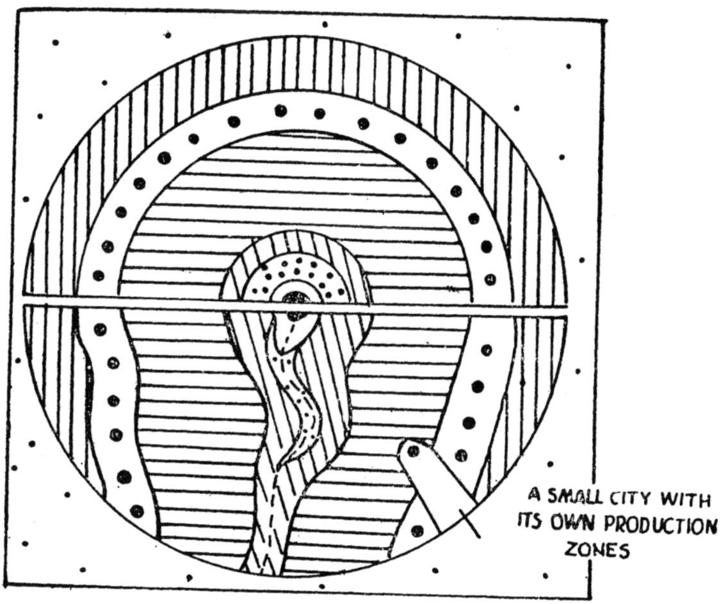


FIG. 15.2

Thunen found that the above theoretical pattern of the figure would result if wholly isotropic plains are assumed.

Subsequently, assumptions have been made about the change in the original concentric pattern if the initial assumptions of a uniform transport system and a single urban centre is relaxed (Fig. 15.2). The addition of a transport route would be beneficial for farmers closest to that route only. Hence the initial landuse pattern would change parallel to the transport routes. The location of a competing market would result in the formation of landuse zones around that centre. In the case of a river-side location of an urban centre, the effect of a navigable waterway can be easily visible (Fig. 15.3).

*Von Thunen's Model with Improved Transport Routes and Several Competing Centres*

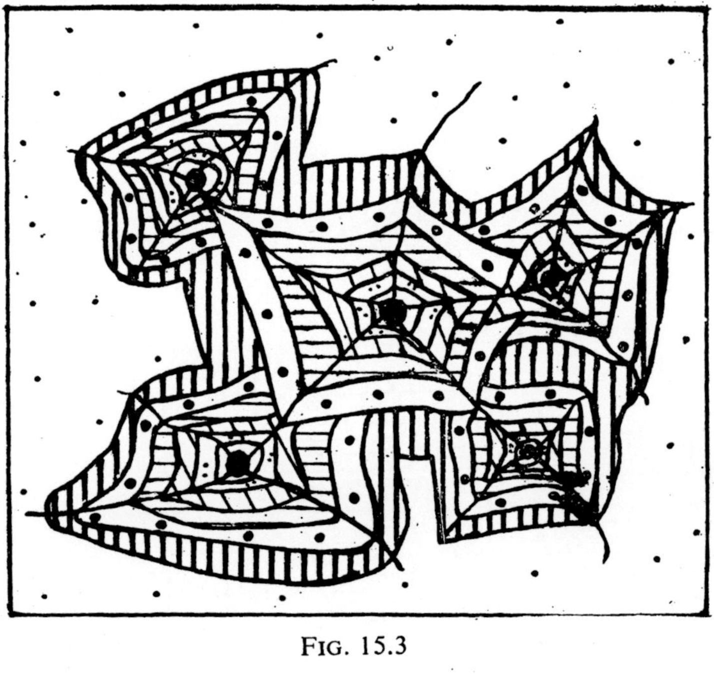

FIG. 15.3

Von Thunen also assumed a river flowing through the city across the plain. In effect, these new transportation lines extend the concentric zones in both upstream and downstream, but in upstream it will be more marked because downstream transportation would be easier and faster (Fig. 15.3). The extension of this type of landuse is more clearly marked along the southern bank of the Ganga in Patna, Buxar, Varanasi and Allahabad in Northern India. In modern times, however, due to steam engines, upstream river transportation is nearly as fast and efficient as downstream transportation. Still further in modern economics, streamlined railroad passenger service, e.g. paved highways, motor truck lines and airways extend the zones of the market along with these transportation networks. In this way "modern transportation and refrigeration techniques have overcome perishability and bulk limitations"[6] and hence, Von Thunen's concentric ring of circles no longer restricts the market areas which provide goods and services to the modern market centres.

Similarly, if several larger satellites had appeared on the plain the resulting pattern of agricultural landuse would probably have looked more like that of (Fig. 15.4). In this example, the satellites seem to vary in population size, reflecting considerable difference in the arrangement of landuse directly followed by transportation lines. Further, with the relaxation of Von Thunen's initial assumptions of uniform production costs, the area at the lower left (Fig. 15.4) has acquired some production cost advantage over the area at the upper right. This makes it possible for the farmers in the more favourable area to absorb a greater amount of transport cost in comparison with the producers of a poorer area. In view of this, agriculturally better landuse area expands in size at the expense of the other.

## GENERAL LANDUSE MODEL

This model has been suggested by Alice Coleman[7] in 1969 in order to explain the 'Second Landuse Survey' of Great Britain. Through this model Coleman attempted to find out five categories of landuse, i.e., farm scape, urban scape, wild scape, urban fringe and marginal fringe[8] (Fig. 15.5). In this way it is felt that conventional landuse categories must have been of a lower hierarchical order as the fundamental territories of landuse are generally found continuously up to

*Agricultural Landuse Model*

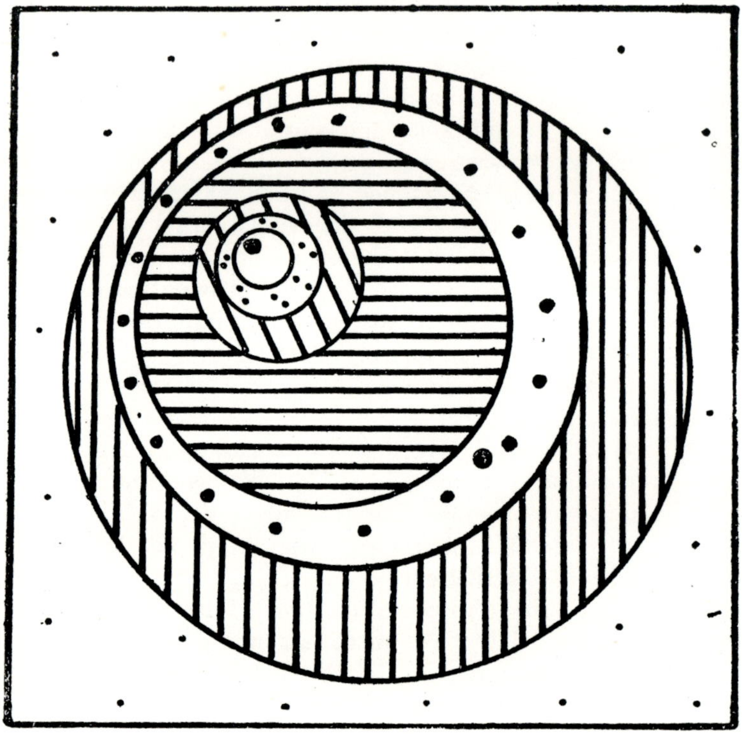

FIG. 15.4

*General Landuse Model (After Rawson and Sealy)*

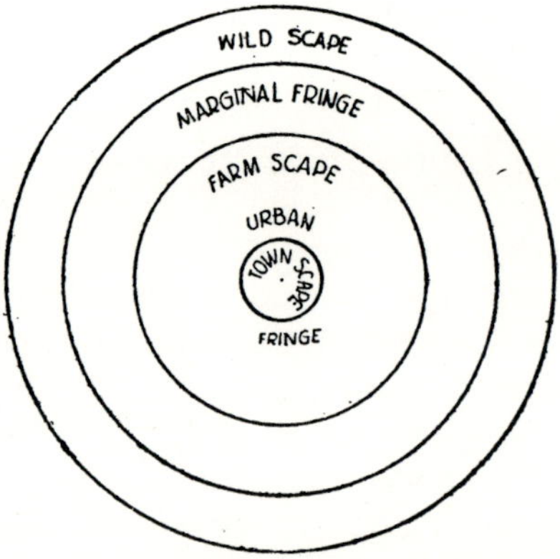


FIG. 15.5

the marginal fringe. The variations are due to the physical and socio-economic influence. The interacting zones are found in advancing, stable and retreating stages and it is hoped that these stages and the reasons for their formation should really be capable of wider application all over the world.

For a well structured general model, according to Coleman, the presence of the following characteristics should be a desiretum i.e.,

(1) a landuse model must be comprehensive,
(2) the model must be integrated,
(3) the model must be simple in order to give added insight,
(4) the model shall be hierarchical,
(5) the model should be practical,
(6) the model should be of a visual-spatial quality,
(7) the model should be flexible, and
(8) the model should be genuinely general in its application.

The characteristics of comprehensiveness seeks that all categories of land must be considered in the model or the relegated item should be there in some other form of integrated framework. The integrative character does not require the formation of categories in isolation but it should include the association of variables dynamically in 'landuse territory'. The attribute of added insight reflects that the model should be comprehensive as well as simple. The word hierarchical denotes that the model should be prepared keeping in view the varying objectives, i.e., territorial, partial, sample, regional, principle, etc. Later on, it is expected that the model should be problem-oriented such as planning the area, classification of an area, and must have a high level of academic value. The idea of visual-spatial quality requires that the model should be really geographical and one which should study the landscape territories and not just mathematical complexities. The model should be flexible enough so that the frequent change of landuse in the present situation could easily be included in the model and finally it should be wide in its application to any part of the world.

DIAGRAMMATIC MODELS

Hiroshi Ishida while delineating overseas comparisons of the models of land utilization in relation to field pattern in the Middle Ganga valley presented several diagrammatic models.[9] Under this model of landuse analysis, the hypothesis is that a ring-shaped territorial structure of the rural settlements was noted with inner, intermediate and outlying lands. Variations in fertility, however, from region to region are found due to varying conditions of land improvement. These have been the key criteria of land classification for the collection of revenue for a long time. It is also felt that the evolution of rural settlements and the growth of the agricultural system in India are the products of these ring-shaped arrangements of the customs of landuse and field pattern. Some cases of the variations in the zonal arrangement of land utilization are as follows : (Fig. 15.6).

(a) **Models of landuse in Eastern Uttar Pradesh**

*Bara* : Means encircled land : outer fringe of the settlement, rich soil, vegetables are cultivated.

*Goind* : Intermediate land around village, *i.e.*, land between *Palo* and *Bara*.

*Palo* : Outlying unmanured land around villages beyond the inner and intermediate zones, and is the least productive land.

*Gauchar* : Village common pasture.

*Dhankhar* : Paddy field.

(b) **Model of Landuse around Rural Settlements of Patna District, Bihar.**

*Dih* : High land as village site used for vegetable crops.

*Dhankhar* : Lowland, used both for *kharif* and *rabi* crops.

*Tal* : Outlying land, only *rabi* crops are cultivated due to flooding in summer season, *kharif* crops may not be cultivated.

*Models of Landuse in Villages of Eastern Uttar Pradesh*

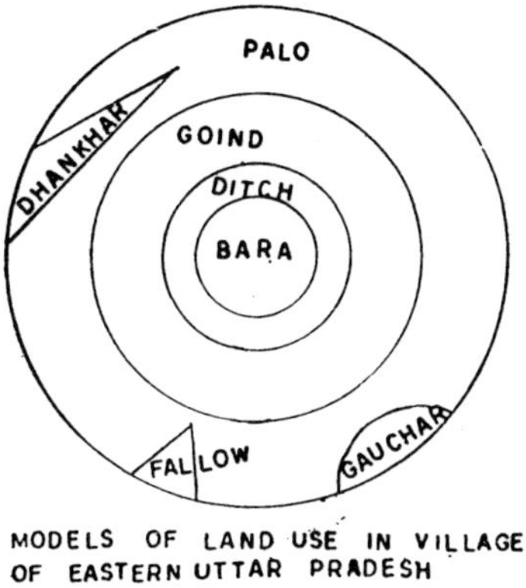


FIG. 15.6 (a)

*Diagrammatic Model of Landuse in a Village of Patna District, Bihar*

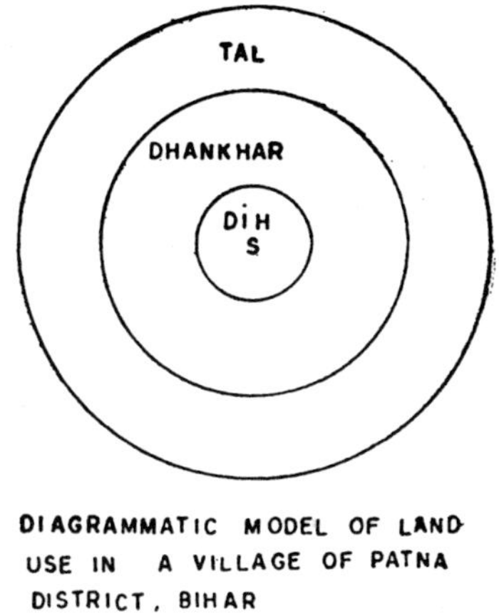


FIG. 15.6 (b)

*Landuse Model for the Village of Punjab*

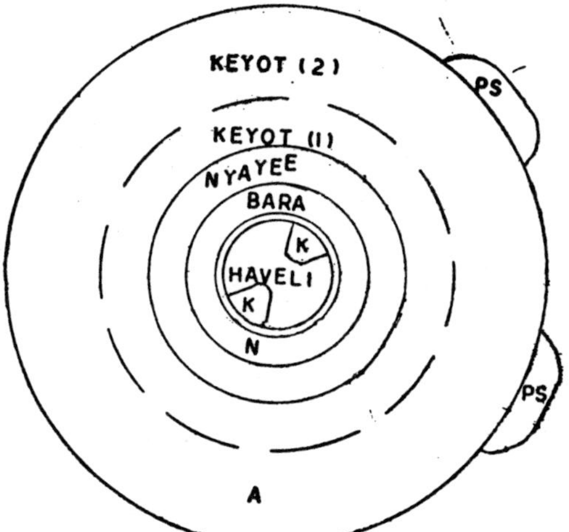


FIG. 15.6 (c)

**(c) Models of Landuse of the Punjab Village after Consolidation**

| | |
|---|---|
| *Haveli* : | Settlement site. |
| *Awal* : | First rate land. |
| *Bara* : | (Richest soil), vegetables are cultivated. |
| *Nyayee* : | (Very rich soil), vegetables are also grown, newly brought under the plough. |
| *Kayot* : | (Also *Kheyot*) Periphery. |
| *Keyot 1* : | (Rich soil), sugarcane, cotton, wheat, barley are cultivated. |
| *Keyot 2* : | Gram, bajra and jowar are cultivated and the area of fallow is large. |
| *PS* : | Pasture. |
| *Phirni* : | Circular road. |

MODELS OF OPTIMA AND LIMITS OF MCCARTY AND LINDBERG

In the formulation of models of agricultural localization they have considered physical and economic constraints and their limitation on the particular territorial space of agricultural land.[10]

Under the study of localization of agricultural crops one has to choose the best suited crop of a locality among other expected crops. For the localization of crops Von Thunen had adopted the aggregate analytical method for explaining the environmental factors of land utilization. Both in the industrial location of Weber and the agricultural location of Von Thunen, the theory of comparative advantage is involved. In case two different crops are grown in two different fields where only one crop could be profitably grown in each field, it is said that a particular field has got specialization in producing a particular crop. Let us suppose A and B are two plots which produce wheat and maize respectively as given below :

| | A | B |
|---|---|---|
| Wheat | 20 tons | 15 tons |
| Maize | 40 tons | 20 tons |

The production of maize in A and the production of wheat in B is profitable. In this way the specialization of crops gives the idea of localization of crops.

**Physical Optima and Limits**

Botanists and agronomists have determined the minimum requirements of temperature, humidity, nutrients of soil and other essential ingredients, through the model for the production of different crops. For the production of a particular crop, it is difficult to fulfil all these minimum conditions at all the places. In this way, with the availability of these minimum requirements, it is possible to demarcate such an area. In areas of the availability of these minimum requirements, many such smaller areas are available having an admixture of minimum requirements which are conducive for the production of so many crops. Those areas are known as limits of a physically optimum area for the localization and production of any crop. The physical limits of the production of crops and areas of favourable conditions are not always static; but due to technological advancement, the land resources, productivity of the soil and changing inputs, the optimum physical limits for the productivity of crops are changing. For example, with today's technological advancement, the dwarf variety of wheat can be grown in colder regions as well.

In the localization of agriculture in any region, the required amount of temperature and humidity create a favourable area for the production of crops, because these favour less input and high output per acre of land. From such centres outwards with the decreasing relationship of these two variables, the production of crops also goes on decreasing and, finally, there comes a point beyond which production is not possible. (After McCarty and Lindberg).

Figure 15.7 shows the physical optima and limits of cotton production in the cotton belt of U.S.A. This physical limit is determined by the amount of rainfall and time taken in the ripening of the crops. But in this vast country of the cotton belt, favourable conditions are found only in smaller areas where the productive soil is available. In this way the localization of agriculture is found only in such physical optima and limits. In the same way such optimal physical limits can be determined for any crop.

*Physical Optima and Limits*

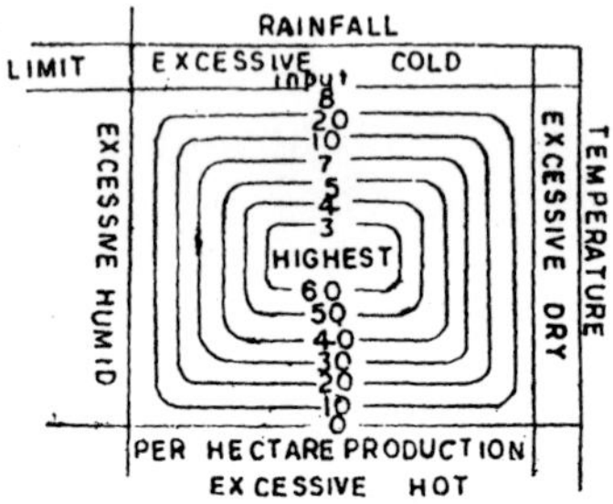


FIG. 15.7

**Economic Optima and Limits**

The production of crops is not only dependent upon the natural elements and stages of technological advancement, it is also determined by the percentage of price in the market and the cost involved in production. This limit is determined by a line which connects all those points where the total cost of production per unit is equal to the per unit production cost of the market. Under total cost the cost involved for land, labour, transportation and irrigation are also included. The determination of such types of limits is very difficult because it involves the determination of local variations of cost involved in each element, estimate of production, concession of the substitute of different essential factors and the future rate of sale in the market. But this optima and limits show such places where economic rent is zero and beyond this belt one begins to feel the disadvantages in production. Within this economic limit in any direction, per acre production costs will be lesser and such areas may be available where in the market, the per unit price and per unit total production costs will differ greatly. Such an area is called an area of economically profitable condition (Fig. 15.8).

**Competition in between Different Crops**

For any one crop, economic limits and determination of the area of favourable conditions, the probability of the production of other crops and the analysis of their effect could be made after the economic analysis of different crops. In between two or more crops the cost-benefit analysis and its comparison with the economic rent could be easily made. In areas of favourable conditions only such crops are grown which give more production per unit of area.

*Production in Relation to Optima and Limits Model of McCarty and Lindberg*

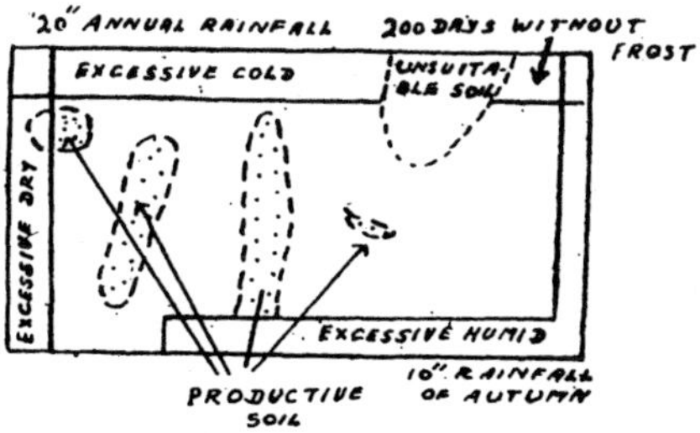


FIG. 15. 8

Generally speaking for a number of goods there is only one market and the necessary cost of investment is also the same. Usually market forms the areas of dense population, and because of this fact, for several crops, the areas of favourable economic conditions is also only one. There emerges thus a competition for the localization of agriculture. In such a situation those crops which help in gaining more economic rent would be produced in areas of favourable economic conditions. Those crops which are produced in smaller areas due to the restriction posed by transport, physical and economic limits, could be grown only in areas of favourable conditions because they provide higher economic rent.

## POLAR AND ANTI-POLAR CONCEPTS OF RURAL LANDUSE

The polar concept of landuse is based on the assumption of idealised polar conditions of the earth and this concept is mainly used here for the understanding of world rural landuse in the context of a village. This climate is the sole criterion which is mainly responsible for controlling the world's landuse (Fig. 15.9). Other factors are soil condition, topography of the land and slope, and the third is the density of population in a certain temperature which controls the landuse of a certain area.

*Polar Concept of World General Landuse*

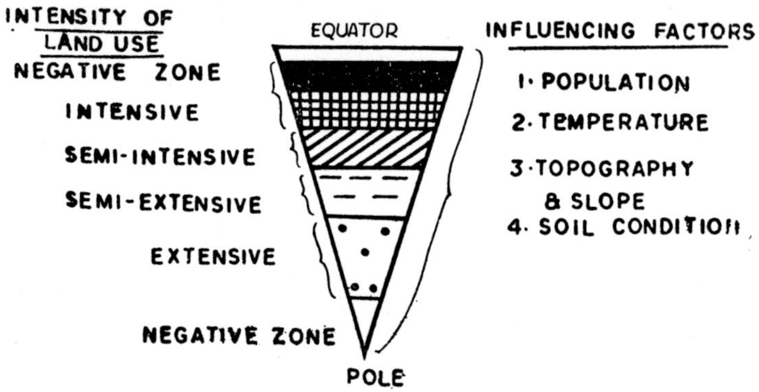


FIG. 15.9

It follows that in the polar region due to intense cold, the soil becomes frozen with moisture and that barren land is of negative use for human existence because no production makes the place uninhabitable. The second zone is of lower landuse because here the condition is better than in polar areas and a few people live in this region leading a wandering life based on the prey of animals and fish in lower land. The third zone of medium landuse is characterised by lesser cold than at the Pole with improving soil condition for agriculture and habitation of the people; thus the people of this region are laborious and sturdy because they are always struggling in anti-environmental conditions for their existence. The people of this region raise a single crop. The next zone is of high landuse. Here, because of timely rainfall and better soil conditions due to humus from the leaves of the forests, amicable temperature and double crop pattern, the inhabitants of these temperate lands have achieved a great miracle for the development of human civilization. The fifth zone is of higher landuse with a variety of crops, good rain and fertile soil which are favourable for the sustenance of the life of a greater number of people with 70° to 100° F temperature. This zone can be compared with that of triple cropped landuse zone of the former model (Fig. 15.10). The next zone is occupied by settlements which control the landuse from the negative equatorial area to the distant polar regions according to their mode of accessibility and lesser security of crops and human life. Thus this concept is known as the polar concept of landuse as propounded by the author. The only limiting factor for this concept is the presence of physical barriers because such topographic features may deviate the landuse pattern.

## ANTI-POLAR CONCEPT OF LANDUSE AND RURAL SETTLEMENT LOCATION

At the two Poles of the earth, the lines of the meridians converge like the spokes of a wheel; similarly do transportation lines which converge at the centre of the village. The consideration of the centre of the village as a pole is (Fig. 15.10) not

*Anti-Polar Concept of Landuse*

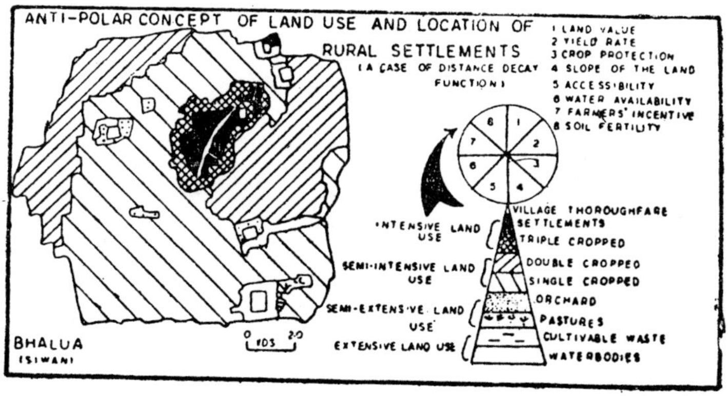


FIG. 15.10

appropriate because the conditions in and around the village are not polar but anti-polar in climatic condition.

These polar and anti-polar concepts are controlled by eight variables, whose intensity over landscape is higher near the settlement to high, medium, low and negative on the outskirt of the village, like that of the equator to the polar regions of the world. The variables are the price of the land, the yield per acre, security of crops, the topography and slope of the land, the fertility of the soil, accessibility, availability of water and the incentives of the peasants. This concept of landuse should be termed as the "anti-polar" concept of landuse.

This new concept dictates that in general the intensity of landuse is decreasing in the form of concentric rings from the centre of the village to the outskirts. This is due to the higher input, better irrigation and triple crops of village *bariland*, while on the other hand, the features on the outskirts of the village are single crop and less productive ground. This concept has a lesser degree of similarity with Von Thunen's concept in varying assumptions. The extensive farming, increased area of cultivable land on the outer margin is against the smaller areal extension of the pole of the earth and this theory is rightly named as the anti-polar concept of landuse.

For the validity of this concept in Bihar, a village Bhalua of Saran district has been considered here. In this village, settlement area and triple cropped land have been considered as high landuse, double cropped and single cropped land as medium landuse, orchard and water bodies as low landuse and others in the category of very low landuse, beginning from the centre of the village to the outer margins.

In modern times, the scientific way of cultivation on the outer margins of the village suggests that now this concept is losing its ground. But this proves the validity of this concept that the previous condition of landuse around the village was of an anti-polar type from the centre of the village to the outward; and the same is still continuing for most villages in India. But nowadays in some localities, the facilities of irrigation, intensive cultivation, deeper harrowing of fields by tractor, use of chemical fertilizer and compost coupled with green manure even on marginal fields make possible multiple crops and naturally the security of crops is necessary. This gives birth to smaller hutments and pumping set stations even outside the village, and this trend is universally discernible today. In this way, the dispersal of settlements is going on with the construction of flood production embankment and hence, planned development of rural settlements and landuse should be introduced by the government at this opportune moment of the changing economy and the present state of living.

MATHEMATICAL EXPRESSION OF THE INTENSITY OF LANDUSE

The above models can also be expressed in terms of the gravity model which is a measure of the interaction of forces and has been taken from Newton's law of gravitation. In the present context the difference in the intensity of landuse from the centre of the village to the outskirts is available only due to the variation in distance. Thus the intensity of landuse between the two places (i and j) is the product of the mass (p) of input of the two places divided by the distance (d) between them. This can be expressed as :

$$I_{ij} = \frac{P_i p_j}{d_{ij}}$$

In course of time it can be hypothesized that the intensity of landuse will increase in view of the increasing population and decreasing distance from the centre of the village to the outer margins. Thus in the second stage the situation for the two places will be as follows :

$$I_{ij} = \frac{p_i p_j}{d_{ij} b}$$

Thus the determination of landuse from the centre of the village towards outer zone of $n^{th}$ places will be equal to :

$$\frac{p_j\, p_i}{d_{i1} b} + \frac{p_i\, p_2}{d_{i2} b} + \frac{p_i\, p_3}{d_{i3} b} + \ldots + \frac{p_i\, p_n}{d_{in} b}$$

$$= \sum_{i=1}^{n} \frac{p_i\, p_j}{d_{ij} b}$$

As $p_i$ is common throughout the whole summation, it can be expressed as :

$$j = \sum_{j=1}^{n} \frac{p_i\, p_j}{d_{ij} b} = V_i \text{ (Potential)}$$

This division of landuse intensity from the centre of the village to the extensive farming area has been designated as potential (V) for different zones or places. This may be calculated and the quality can be expressed as : $V_i = j = 1 \frac{p_j}{p_{ij}\ b}$ for the number of zones of different distance and places in question.

## CONCLUSION

The representation of a real world situation in a diagrammatic or mathematical or structural form is known as a model. In landuse analyses in each and every part of the world, the location of settlements in the central part of the village surrounded by the vegetable gardens, agricultural fields and pastures is common with varying exceptions, apparently related with the local physical, climatic, social, political and economic conditions etc. This proves the validity of the modification of Von Thunen's model several times with some increasing scientific precision and regional ecological conditions. It is expected that the new model of "Polar and Anti-Polar" concept of landuse and location of rural settlements will help in understanding the problems of land utilization especially in rural areas. The better arrangement of the sectors of land in different parts of the village could be planned better with the help of a model developed in this study in order to have a good harvest from farm, farmers and the limited means of cultivation available, especially in Indian villages.

## REFERENCES


1. R.J. Chorley and P. Haggett, *Physical and Information Models in Geography*, Methuen, London, 1970, p 22.
2. M. Chisholm, *Rural Settlement and Landuse*, Hutchinson University Library, London. 1962, pp 21-35.
3. R.S. Thoman, E.C. Conkling and M.H. Yeats, *The Geography of Economic Activity*, McGraw Hill Book Company, New York, 1968, p 153.
4. Ibid., p 154.
5. Edgar Sand Dun Jr., *The Location of Agricultural Production*, University of Florida Press, Gainesville.
6. R.R. Renne, *Land Economics*, Harper & Brothers, New York, 1947, p 198.

7. A. Coleman, "A Geographical Model for Landuse Analysis", *Geography*, 1969; No. 242, Vol. 54, Part-1, pp 43-55.
8. V.K. Shrivastava, "General Landuse Models : Method and Experiment" *Uttar Bharat Bhoogol Patrika*, 1977; vol. x iii, Nos. 1-2, p 33.
9. H. Ishida, "Village and Field Pattern in the Middle Ganga Plain : A Diagrammatic Model and some overseas comparisons", (Edt. R.L. Singh) *Rural Settlement in Monsoon Asia*, National Geographical Society of India, Varanasi, 1972; pp 173-174.
10. H.H. McCarty and L.B. Lindberg, *An Introduction to Economic Geography*, Prentice Hall, N.J. 1966.

# 16

# Aerial Photo Interpretation Technique in Agricultural Geography

—*S.S.A. Jafri**

AERIAL photography and its interpretation is not new, but in developing countries like India, it has been a recent media of information. Though initially it is a costly affair, later on, its fruits are immeasurable. In fact, photographs are the substitute of manual surveys of the ground, which give the whole ground a sizeable shape in hand at once. Manual surveys are limited up to the requirements only and also without accuracy while photographs not only depict the required features, but also cover the entire ground in detail and with great accuracy.

In India it has been observed that manual surveys were conducted for the Master Plans of the cities; and these took such a long time that the acquired information later on became almost obsolete. Aerial-photography have been used for city planning, dam construction, bridge construction and flood control because manual surveys have proved outdated and lengthy in process.

At present the Survey of India is rendering a great service to the nation in trying to produce the topographic sheets, based on aerial photographs of 1 : 50,000 scale. In the same way, if agricultural researchers acquire the same aerial photographs of their study areas, they can carry out highly successful agrarian plans, on which the entire nation's fate is based. Here a brief aerial photo- interpretation technique is attempted to be explained which may be helpful to solve the problems, especially related to agricultural geograpy.

*Paper submitted for discussion in the "Aerial Photography Appreciation Seminar, October 1978", organized by Survey of India, New Delhi.

We have many agricultural problems which may be tackled effectively only by aerial photographic interpretation. For example, agricultural land consolidation schemes may be effectively made successful after interpreting the soil types and drainage pattern with the help of aerial photographs. Through aerial photographs, it becomes quite easy to mark out the land of different qualities and to cover the areas under different irrigational schemes. This process ultimately makes the agricultural development programmes economical and that too within a short time. These are the fundamental requirements of Indian agriculture. Otherwise if economy permits, we may use aerial photographs to forecast the agricultural production, to identify the agricultural diseases etc. For agricultural landuse planning, aerial photographs are yet to be utilized with greater emphasis.

## TYPES OF AERIAL PHOTOGRAPHS

There are three types of photographs i.e. (a) aerial Photographs; (b) Terrestrial photographs and (c) Extra-terrestrial photographs. Aerial photographs may generally be used for urban and agricultural landuse planning etc. with accurate scale. Aerial photographs are generally of the size of $9'' \times 9''$. Whereas terrestrial photographs (photos taken from two points on the ground) are used to plan a small patch of city for building purposes etc. Terrestrial photographs have a less accurate scale, because they are oblique and the scale is distorted after some distance. Extra terrestrial photos may be used for planning of national highways, railways and to prepare even small scale maps etc.

## SCALE OF AERIAL PHOTOGRAPHS

Before aircraft takes photographs of an area, it is demarcated on the topographic sheet to make sure that no area would be left unsurveyed. Also the mean height of the given area is noted because the height of the aircraft above the ground and the focal length of the camera fitted in the plane would, if combined, determine the scale of the photographs. The scale of the aerial photograph is determined by the formula as given on next page.

$$\text{R.F.} = \frac{\text{Focal length of camera fitted in the aircraft}}{\text{Height of the camera or aircraft above the ground}}$$

For aerial surveys the time between 11.00 A.M. and 2.30 P.M. is best suited, as long shadows do not appear and the objects are not hidden. When the pilot becomes sure that he has to make sorties at a certain height, he flies from west to east and then east to west and thus gradually covers the area from north to south. As these aerial photographs are to be utilized for three-dimensional purposes, the pilot takes pictures in such a way 'that each photo of the ground should have 60 per cent forward overlap. In the same way, when he makes the first trip from west to east and then east to west and so on to cover the entire area he has also to make a lateral overlap of 30 per cent. This is to make sure that each trip is connected with the previous trip and no piece of land is left unsurveyed. Thus the entire area is covered by the aerial photographs and in sequence, pair-wise photographs may be interpreted with the help of a mirror stereoscope.

## PRINCIPLES OF AERIAL PHOTO INTERPRETATION

Aerial photo-interpretation technique is based upon two principles : (a) Qualitative Analysis of Photographs, and (b) Quantitative Analysis of Photographs. In the qualitative analysis we screen and guess through the size, pattern, shape, colour and texture and shadow on the photographs. In the quantitative analysis, we measure the height, length and breadth of the photographs. Before scanning the photograph pairs to get the desired information, one should delineate the required areas on the photographs of a pair.

Photographic interpretation can be done with the help of few instruments, stereoscope, mirror stereoscope or stereoscope fitted with binoculars to see all the three dimensions i.e. length, breadth and height. The parallax-bar is the instrument, which is used to quantify the heights (see Fig. 16.1).

## QUALITATIVE ANALYSIS OF AERIAL PHOTOGRAPHS

In aerial photographs, while doing the qualitative analysis for identification, one has to make one's own assessment with

# MIRROR STEREOSCOPE

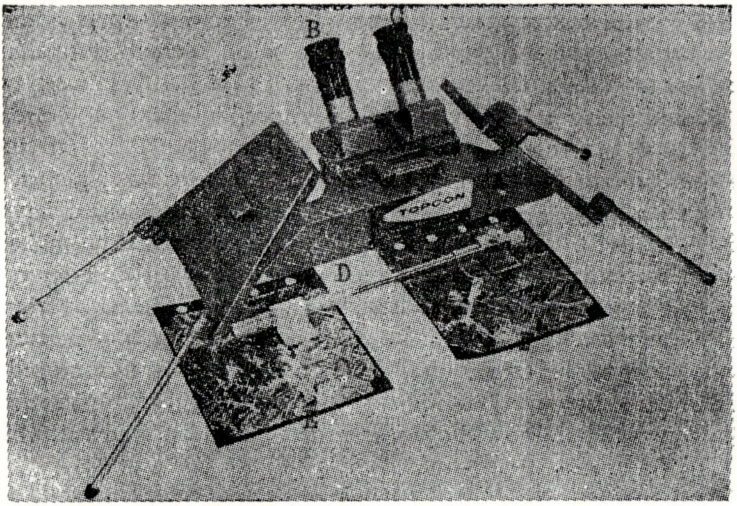


A Mirror Stereoscope
B, C Binocular
D Parallax Bar
E, F Aerial-Photo Pair

FIG. 16.1

the help of various available images. The following images which may be useful for agricultural geography, have been explained:

**Size**

This is an important image to differentiate between the hutment or cattle shed. In villages, cattle sheds are constructed in a typical way, which differs from hutments and houses. For the proper identification of cattle sheds the scholar sometimes has to assess the concentration of cattle. This would be an easy method to pin point all the cattle-sheds distributed in the given region. Sometimes in the grape producing areas, it may become difficult to identify a vineyard, because it is rectangular and of the same height. For such difficult objects one has to have ground verification. The Size of standing crops like bajra, jowar and maize is clearly distinguishable and thus may be marked on the photographs to analyse the agricultural combinations. The types of forests and other plantations can also be identified with the help of their sizes.

**Pattern**

Pattern is one of the important clues to identify various aspects of agriculture. Irrigated fields may be identified with the help of several parallel man-made gullies, while fields depending on rains do not have such gullies. Grazing land and agricultural lands may be distinguished with the help of boundaries, around the agricultural lands. Certain types of crops like potato or sweet-potato cultivation needs a certain type of field formation which can easily be identified by screening the pattern. Mechanized farming can be recognized by looking at some formation of crops on the fields. The pattern of trees gives an idea of afforestation and this may be distinct from the natural forests. Thus pattern is an important method to interpret various agricultural information available on the aerial photographs.

**Shape**

Identification, with the help of shape is again an important criterion to get information related to agriculture already available on the photographs. Rivers and road sometimes look similar, but their shape gives clues to distinguish them. A river does not look the same when the road is parallel and uniform and sometimes there are trees along its sides. Shape is also important to identify man-made water bodies like water reservoirs, tanks, natural water bodies like lakes, ponds or flood areas, etc. It also gives an idea of rural settlements and urban settlements by examining the shapes of house roofs and the shape of the settlement agglomeration. Agricultural fields of different shapes give a different picture, whether large land holdings or small ones. Also, one can get an idea whether the land has been consolidated or not. If there are low lying fields that may be important for rice cultivation while fields in upland areas may be important for wheat, millet etc.

**Colour and Texture**

Colour and texture variations on aerial photographs distinguish various agricultural features. Sometimes such variations are not easy to recognize, but if once they are verified from the ground as a key sample, one can go on to distinguish that particular feature from the entire region. For example, a big water tank looks similar like other agricultural fields. If it is minutely screened, one can observe that this rare water body has a different colour-texture than the surrounding areas. Sometimes roads and canals are difficult to recognise, but road and water surfaces radiate differently, so the colour on photographs is different. For soil classification, colour-texture is an important method to recognize, while standing crops demarcate areas under different crops, because, wheat, sugarcane, mustard, gram plants etc. give different shades on photographs. If coloured films are used in the camera for coloured photographs, one can easily distinguish all such important features such as diseases damaging these crops. It can easily be demarcated and remedial steps taken within a short time. Variations in colour-texture on photographs are quite meaningful for the aerial photo interpreter.

**Shadow**

Short distance shadow is an important phenomenon in mid-day light and it helps the interpreter to identify certain objects. Sometimes farmers plant fruit trees or trees for fuel in scattered farms which appear confusing on the photos. They look like black spots, but if properly viewed by mirror stereoscope, one can identify them by imagining their lengths with the help of their shadows. Most of the wells on agricultural fields may be identified, if they have their mud walls or some fencing. These are some examples to find out agricultural information with the help of shadows. Shadows are also helpful in measuring heights of several objects.

## QUANTITATIVE ANALYSIS OF AERIAL PHOTOGRAPHS

In aerial photo interpretation, the quantification of heights of different objects is the most important technique and needs very careful handling. For several pieces of information, related to agriculture, height measurement becomes unavoidable. Height of crops may be useful to understand for irrigation, weeding or if any deficit in fertility etc. To estimate the age of forests, one can easily calculate the height. One can also estimate the relevant heights of different areas to suggest the irrigational facilities or tarrace cultivation etc. Heights of different objects may be measured by the three following methods :

(1) Image displacement method
(2) Stereoscopic parallax displacement method
(3) Shadow method.

## IMAGE DISPLACEMENT METHOD

$$\text{Formula to measure the height} = \frac{\left(\text{Height of the flight from the ground}\right) \times \left(\text{Length of top from PP}\right)\left(\text{Length of bottom from PP}\right)}{\text{Length of top from PP}}$$

The height of the aircraft is generally given on every aerial photograph. If at times the height of the aircraft is not given, then it may be calculated with the help of the focal length (means the length of the camera base) and the scale of the photograph. If the focal length is 6 inches and the scale of the photograph is 1:13500, then the height of the aircraft would be:

$$\frac{6 \times 13500}{12 \text{ (inches)}} = 6750 \text{ feet}$$

Length of a tree top and bottom from PP (Principal Point) may separately be measured in centimetres. Principal point is exactly on the centre of the aerial photograph, which is generally given or demarcated. On the principal point the scale of the photograph is accurate; on this the camera of the aircraft is exactly vertical.

For example :

Length of a tree top from PP on photograph =8.60 cm
Length of a tree bottom from PP on photograph=8.35 cm
Image displacement =0.25 cm
Height of the aircraft from ground =2000 feet

$$\text{Height of the tree} = \frac{2000 \times 0.25}{8.60} = 58.1 \text{ feet}$$

STEREOSCOPIC PARALLAX-DISPLACEMENT METHOD

$$\text{Formula to measure the height} = \frac{\begin{pmatrix}\text{Difference of parallax}\\ \text{bar reading from datum}\\ \text{mean observation}\end{pmatrix} \times (\text{Height of flight})}{\begin{pmatrix}\text{Mean distance of photo}\\ \text{base}\end{pmatrix} + \begin{pmatrix}\text{Difference of parallax}\\ \text{bar reading from datum}\\ \text{mean observation}\end{pmatrix}}$$

Mean distance of photo-base would be measured from the principal points of photo pair. First the photo pair is fixed after obtaining the stereoscopic view. Then we put the left hand dot of the parallax bar on PP and then move the right hand dot till it floats on the left hand, enabling us to read on its scale.

**Difference of Parallax Bar Reading from Datum Mean Observation**

In case we calculate the height of the tree, the parallax bar measurements of the base and top of the tree would be done separately on photograph pair. Thus the difference in parallax bar readings of tree base and top would be calculated.

For example:

| | |
|---|---|
| Parallax bar difference of top and bottom of the tree | =1.32 mm |
| Flight height | =3000 feet |
| Mean distance of photo base | =88 mm |

$$\text{Height of the tree} = \frac{1.32 \times 3000}{88 + 1.32} = 44.3 \text{ feet}$$

SHADOW METHOD

$$\text{Formula to measure the height} = \begin{pmatrix}\text{Photo shadow}\\ \text{of an object}\end{pmatrix} \times \begin{pmatrix}\text{Sun angle in terms}\\ \text{of tangent}\end{pmatrix}$$

This method is quite easy if once the sun angle in terms of tangent is calculated. Then the height may be calculated by multiplying the photo shadow length of any object. This method does not have any complication and only with the help of scale, so many valuable pieces of information may be achieved.

$$\text{To calculate the sun angle in terms of tangent on a particular photograph} = \frac{\text{Actual height of any object on the ground}}{\text{Shadow length on the photograph}}$$

Once the angle of the sun in terms of the above object tangent is calculated for a particular photograph, then this value should be considered as a constant to calculate the heights of several objects.

## SUGGESTED READINGS

American Society of Photography; *Manual of photographic Interpretation*, Washington, 1960.

G.C. Brock, *The Physical Aspect of Aerial Photography*, New York, 1967.

G.C. Dickinson, *Maps and Air Photographs*, University of Leeds, 1970.

R. Leeder Donald, *Aerial Photographic Interpretation, Principles and Applications*, McGraw Hill Co., INC 1959.

S. Way Dougles, *Terrain Analysis, Community Development Series, A Guide to Site Selection Using Aerial Photographic Interpretation*, 1973.

Eugene Avery T, *Interpretation of Aerial Photographs*, second edition, Burgess Pub. Co., 1962.

Paul R. Wolf, *Elements of Photogrammetry* (with Air Photo Interpretation and Remote Sensing), McGraw Hill, 1974.

P.B. Shahni, *Advanced Surveying*, 2nd Ed.. Oxford & IBH Pub. Co., 1976.

Survey of India, *Photogrammetric Survey*, 3rd Ed., 1972.

# 17

# Air-Photographs in Landuse Analysis

*—Ram Bahadur Mandal*

DEPARTMENTS of Geography all over the world have begun to take an active interest in the interpretation of aerial photographs in geographical and economic studies. This is due to rapid progress in spectro-photogrammetry. In regional planning resource appraisal, landuse mapping, geology, geomorphology, forestry, pedology and agronomy, the students and research scholars, have found it helpful in their research. In developing countries the non-availability of air-photographs makes it difficult to undertake such studies. In India, the Air Photo Interpretation Centre, Dehradun, 'Air Force Agency, Calcutta' and several other agencies providing air-photographs and training for survey and interpretation, to the various institutions and scholars.

Air-Photo interpretation has the advantage of being able to handle large scale maps with greater accuracy in depicting the various types of landuse of wide spaces in less time. It has the advantage of depicting field boundaries; and thus making it convenient for use in assessing land tenure and providing a base for preparing regional plans and landuse analysis. Hence air-photographs can be used both as map and data. Even a cursory view of such photographs by Stereoscope provides a lot of information regarding the landscape. In a stereoscopic view the familiar scenes of the ground are transformed into lines and tones of varied faces where the true picture of a landscape could be viewed. The technical advancement of panchromatic emulsions, use of polarised filters, infra-red and colour photography have made the techniques of photo-interpretation easier and provide a scientific base for landuse analysis. Such accuracy is not humanly possible with the same team of people for many months of costly field work. In view of the need for land utilization maps through air-photographs, it is felt that methods, which have been applied for Cyprus[1] in the London School of Economics would yield useful results in a shorter period. This study aims to describe the materials required for survey, the stages by which the landuse maps are prepared and interpretations made.

## METHODOLOGY

Air-photographs are generally examined with the help of a stereoscope. A lens stereoscope with a good magnification (X-4) is particularly helpful in the identification of vegetation types, settlement types etc. Transitional areas between different types of landuse are drawn with the help of air-photographs and are checked at doubtful and crucial point. A preliminary landuse map is prepared in this manner. At this stage a field check and ground studies are helpful when a few sample checks are made for each of the different types of landuse. The purpose is to correlate photographic characters of each type of landuse with ground observations, and to check complex areas of land utilization. When mapping has been completed, all these boundaries are transferred with the help of a draughtsman to a 1 : 63,360 base. A detailed study of particular landuse can also be possible with the help of air photographs, which includes settlement types (rural or urban), type of houses, roads, lanes, footpaths, etc. For water-bodies which are being used as fisheries depth of water (approximate) and character (perennial or seasonal) are generally determined. Only doubtful issues are checked in the field. In this way, nearly 80 per cent of the information is collected from air-photographs, and the field study is reduced to a minimum. It also helps in covering a wider space within a short time.

## MATERIALS REQUIRED

### Air-Photographs

Photographs are generally taken in several trips in which turning, twisting, tilting and warping of flights quite often distort the scale of the photographs. Such photographs should be vertical with at least 60 per cent fore and aft overlap providing stereoscopic cover of the whole area. The large scale photographs are better in quality and highly accurate in providing a huge amount of landuse data. The air-photographs of Cyprus available in the London School of Economics ranges from R.F. 1 : 10,000 to 1:13,000 and as such, for any area prints of a smaller scale would be of little use. In areas of simpler landuse pattern such maps are useful, e.g., in the eastern coast of India where the cultivated land is mostly paddy and orchards of mango and coconut, etc. On the other hand, in Meghalaya where shifting cultivation is practised, interpretations have to be made minutely for patchy cultivation in bushes, and in such a situation photographs of larger scale i.e. 1 : 10,000 are necessary. Such photographs should be clear without shades of cloud and oblique cases.

**Topographic Maps**

The landuse data from air-photographs can be plotted on the topographic maps such as Cadastral maps or Village sheets on which field boundaries are clearly marked. In India the best map is 1 : 7920 Anchal level which record settlements, roads, tracks and water bodies clearly, but no field pattern. In Meghalaya the problem is intensified as no larger than a 1/63360 map is available due to the rugged terrain.

**Stereoscope**

This is of many types among which the pocket stereoscope is quite handy for air-photo interpretation in comparison with larger stereoscopes with binocular attachment.

## STAGES IN THE PREPARATION OF LAND UTILIZATION MAP

**Plotting of Photographs**

The air-photo should be placed on a table in runs of the aeroplane covering one side of the area under study to the other. This plotting should be made on the topographic or Anchal level maps. In case the number of first and the last prints are arranged in each run, the subsequent insertion makes the process easy. In case the runs are more than one, the only photographs of clear images should be selected for interpretation.

**Interpretation of Photographs**

This requires a good grounding in air-photo interpretation technique and the type of landscape available. The principal task of the interpreter is to establish a relationship between actual landscape features with that of the photographs and the interpretation of photographs with that of the topo sheets, in order to find out the error due to lack of acquaintance with the ground features.[2] In 1951, G.M. Howe published an account of his work on photographs of north-west Cardiganshire.[3] Due to easy access, he personally checked the critical areas of interpretation by the ground. Field work is not essential in case the interpreter is well acquainted with the ground. Otherwise, such studies must be supported by some other source of first-hand information.

With the help of densitometers, now it is possible to have a quantitative evaluation of air-photographs. The tonal variations as measured by this instrument have overwhelmingly reduced the chance of subjective lapses to a greater degree. To convert the land imagery into digital units generally a perception type of automatic spot scanner is used. The stimulation of this data gives correlation, regression for more than one variable dispersion and congregation of crops, etc.

The crops are interpreted with the help of textural contrast, tonal variations and differences in crop density through standard air-photo inventory procedure. In the interpretation of air-photographs for preparing landuse maps the following sixteen standard codes are generally in use (Table 17.1).

TABLE 17.1

*Standard Code for Landuse Analysis through Air-photographs*

| Code | Colour | Particulars of Landuse |
|---|---|---|
| 0.1 | Yellow | Cereals |
| 0.2 | Pink Saffron | Weeds (other than fodder) |
| 0.3 | Medium Saffron | Fibres and oil seeds |
| 0.4 | Saffron | Legumes (beans) |
| 0.5 | Light green | Weeds used for fodder |
| 0.5 | Medium green | Pastures (temporary) |
| 0.6 | Light grey | Fallows |
| 0.7 | Green | Pastures (Permanent), barn |
| 0.8 | Dark green | Moors, routes (Productive) |

| | | |
|---|---|---|
| 0.9 | Violet | Vineyards |
| 1.0 | Red | Orchard |
| 1.1 | Orange | Market gardening |
| 1.2 | Dark red | Gardens |
| 1.3 | Pink red | Other crops (tobacco) |
| 1.3 | Dark green | Moor and fallows (unproductive) |
| 1.4 | White | Buildings and courtyards |
| 1.5 | Green | Forest |
| 1.6 | Black | Uncultivable, rocks and waste lands |
| 1.6 | Blue | Water bodies. |

Source : After A.K. Tewari, *Geographical Review of India*, 1977.

In short, the whole process of aerial photo-interpretation for landuse analysis may be divided into the following sections:

**Recording the Landuse Data on Topographic Maps**

Data for individual fields cannot be collected from air photography, because :

(1) the field boundaries do not appear on the 1/50,000, scale photographs; and
(2) the fields are so small that on reduction to 1/250,000 two adjacent fields of different landuse could not be shown.

**Preparation of Land Utilization Map**

In the London School of Economics, while preparing the land utilisation map from air-photographs, it was thought that the final map prepared on a scale of 1/250,000 would be most convenient in handling. The traces reduced photographically, to the above scale, especially the air-photographs of Cyprus. For shading purposes eleven landuse categories have been evolved and combined with each other to record mixed categories. These maps are prepared in black and white in order to minimise the colour printing.

**Landuse Pattern Analysis**

The broad categories of landuse are generally determined on medium scale maps (1 : 25,000) followed by adequate ground checking. The areas of net sown, pasture, fallow, garden, road railway, graveyard, forest, water bodies, settlement site, harvesting and threshing ground, etc. are marked on the base-map with the help of the cartographer. On the air-photo, the cropped areas are marked by dark patches while light grey is for current fallows. This saves the time and energy of the investigator. The landuse areas engaged in different categories are generally determined by the Planimeter.

**Identification of Crops in Different Seasons**

Although the identification of seasonal variations in crops is difficult, the help of 1 : 10,000 scale photographs and its superimposition on cadastral maps is extremely useful for identification of crops. The density of crops and tonal variations, which are primarily determined by densitometers are useful aspects of photo interpretation. The National Institute of Geography, Paris uses the scale of density of crops as follows :[5] 0 : white, 1-2 : Grey—very clear, 3-4: Grey—clear, 5-6 : Medium Grey, 7-8 : Dark Grey, 9-10 : Very dark grey, 11-12 : Black.

The interpretation of farm crops, crop rotation, land classification and determination of land capabilities are a recently developed practice of photo-interpretation technique mainly developed in Zaire, Uganda, New Guinea, France, U.S.A. and U.S.S.R. With the help of this technique, France has recently started a census of agriculture and published a complete set of coloured landuse maps in alternate years.

**Analysis of Crop Productivity**

Measurement of agricultural efficiency and per acre yield are normally not possible but experts of agronomy can easily gauge the health, growth rate, yield and other items of crop productivity with the help of infra-red false colour imagery. This also facilitates the assessment of soil nutrients and moisture, duration of sunshine on particular land, plant mortality, immunity to pest and disease, quantity of inputs, type of seeds used and yield rate, etc. Infra-red radiation and heat pattern techniques are widely used in the interpretation of forest resources.

**Analysis of Landuse Change**

In U.S.A., Switzerland and Canada, air-photographs are extensively used to measure the landuse change. The photographs of different time periods are valuable assets in order to measure the advance and retreat of cultivated land, loss or forward marching of desert areas and sand dunes, etc.

**Interpretation of Input Parameter**

In the field of agricultural science and land utilization, the assessment of damages caused by insects and pests, soil erosion, degeneration of soil nutrients, use of a particular type of fertilizer, chlorophyll content of a plant, extent of mixed farming, intensity of irrigation, etc. are easily measured with the help of infra-red air photographs. The measures of agricultural innovation and operation research in the field of agriculture are easily possible with the help of air photographs.

**Landuse Planning**

Air-Photographs present the total view of the environment towards the planner in the briefest possible manner under a stereoscopic vision. The sharp perception of the regional planner clearly visualizes the ensemble of the whole in the miniature likeness of the landscape.[6] In India village level planning is generally tried but due to lack of the minutest detail of landuse data, meaningful planning is hampered to a great extent. Now it is hoped that the interpretation of satellite imagery and air-photographs will usher in an era of purposeful planning for land resource inventory, judicious use of land and integrated rural development through better land management.

REFERENCES

1. R.R. Rawson and K.R. Sealy, "Air Photographs and Land Utilization Maps" *Geography*, Vol. XLII, Part I, No. 195, 1957.
2. R.R. Rawson and S.H. Beaver, "Aerial Photography and Geographical Studies." *Geography*, Vol. XXXII, 1947, p 131.
3. G.M. Howe, "A Note on the Application of Air-Photography to the Agricultural Geography of North-West Cardiganshire" *Geography*, Vol. XXXVI, 1951, pp 15-20.
4. R.R. Rawson and K.R. Sealy, op cit.
5. A.K. Tewari, Air-Photo and Rural Landuse Analysis", *Geographical Review of India*, Vol. 39, No. 2, 1977, pp 158-164.
6. Ibid.

# 18

# Spatial Diffusion of Agricultural Innovations—An Empirical Analysis

*—Ram Bahadur Mandal*

DIFFUSION is a process of spread of any item in space. Agriculture relates to the production of crops from land with the help of human labour, mechanical implements, fertilizers, irrigation and pesticides, whereas innovation is a change in the established methods of agricultural practices, culture, ideas, etc.

In spatial diffusion, the spread may be responded to from a common centre or from varying locations. In this sense, the diffusion of information is dependent on the force generated by the place of diffusion, while the length of spread in space is the outcome of the mean information field of an adopter in the form of distance-decay function. Diffusion of any item is also the outcome of usability, popularity, ease of handling, source and force of information from where the same is received, level of income and the cultural traits of the adopter who inhabits the area. Hence the diffusion and innovations of the ideas for the adoption of newer agricultural practices by the agricultural population has long been of concern to people involved in transportation, marketing and business, communications, advertisement, agricultural extension programmes, rural sociology and several other fields of research. In each and every field, scientists approached the problem in terms of the individual, social, economic and behavioral characteristics of people in adopting any new agricultural practices. The geographers interest in diffusion lies in the inherent spatial implications and pattern analysis.[1]

The significance of diffusion and innovation studies in agricultural economy needs no emphasis. Because of the present vicious circle of population growth without access to latest techniques of agricultural production, it is impossible to solve the food problem of our teeming millions. Keeping this view in mind, it is hoped that the present study will answer the following questions.

(1) What are the processes through which innovations in agricultural implements spread from one individual to another at varying locations?
(2) What are the inherent characteristics of innovators and adopters?
(3) Why are some innovations readily accepted by the farmers while certain others are rejected?
(4) What are the roles of agents and opinion leaders in the diffusion of innovations?
(5) What are the spatial patterns of diffusion of innovations?
(6) Which method should be adopted to analyse diffusion of innovations?
(7) Is it possible to predict the spatial spread of innovation?

The study of the diffusion of agricultural innovation has great importance in the Indian agricultural sector where the traditional society is in the process of transformation as the change in agricultural production has led to economic growth and industrial progress. Any change from the traditional agricultural practices to scientific ones involves two processes: (a) agricultural research and innovation which may remove the barriers of higher production and (b) the level of the adoption of agricultural innovation by the farmers according to the technological know-how and socio-economic conditions. In the field of agricultural research there are scientists who are capable of developing innovations but on the other hand, it is not certain that our farmers will readily accept it. In order to make available the result of scientific research and innovations to our ill-fed, ill-clad and ill-sheltered farmers, it is our duty to make them ready to accept new ideas and innovations.

## REVIEW OF LITERATURE

The diffusion study in geography dates back to the early twentieth century. Among the pioneers, the work of Cook[2] presents the study of the diffusion of a number of geographic phenomena, i.e., plants, animals, birds and others. The original work of Hagerstrand on Monte Carlo Simulation Models of spatial diffusion of innovations acted as catalyst for a number of studies dealing with diffusion problems. To make the idea clear, Hagerstrand pointed out that "a start is made by a rather concentrated cluster of carriers. This cluster expands step by step in such a way that the probability of a conversion always seems to be higher among those who live near the carriers than among those who live farther away."[3]

Hagerstrand (1952)[4] presented a four-stage model for the passage of innovation waves for dispersal of settlements. In the first stage, the origin and evolution of initial agglomeration has been considered. In the second stage, diffusion is marked by centrifugal effect and creation of new centres in distant areas. In the third stage of condensing process, actual relative increase in the size of all locations is contemplated while in the saturation stage a symptotic increase towards maximum dispersal is analysed.

Later on, Nystuen (1963)[5] studies the simulation of travel pattern in urban areas. The effect of permeable and semi-permeable barriers on spatial patterns arising from diffusion process are studied by Yuill (1964)[6] on urban growth and development the works of Morrill (1963)[7] and further the simulation analysis of urban fringe and the growth of minority group ghettos in the city have been analysed. In model after nations of landuse at highway interchanges (Eighmy, 1964) and along major arterial routes leading out of a small centre (Colenutt, 1966) simulation methods have been employed. The review of Monte Carlo Methods and geographical research in spatial diffusion have appeared after the 1960s.

A Swedish geographer Bylund[8] (1960) has presented a four-stage diffusion model of settlements in Inner North Sweden. This is an important contribution to the dispersal process. The only difference in his four-stage model is in the number and size of settlements increase and the location of parent settlements. His A and B models assume spread from a coastal location while C and D go from an inland location.

The simulative structural model for diffusion considered by K.N. Singh[9] (1968), first of all presented all occupance of the territory through immigrational accretion, then sub-division of clan groups into 'tappas' and 'parganas' boundary and finally with the clan migration and competition even distant areas have been filled up by the growth of different-sized settlements.

Similarly, Hudson (1969)[10] has investigated a new area of dispersal process in U.S.A. while he was studying the settling process of Iowa. He has considered three processes : colonization, spread and competition from the growth and dispersal of settlements.

## PROCESS OF INNOVATION DIFFUSION

There are two processes of innovation diffusion i.e., relocation type and expansion type [Fig. 18.1 (a) and (b)]. In reloca-

*Relocation Type Diffusion*

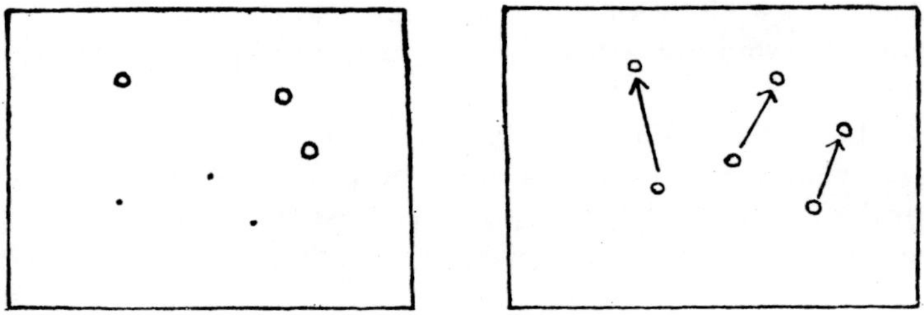

FIG. 18.1 (a)

*Expansion Type Diffusion*

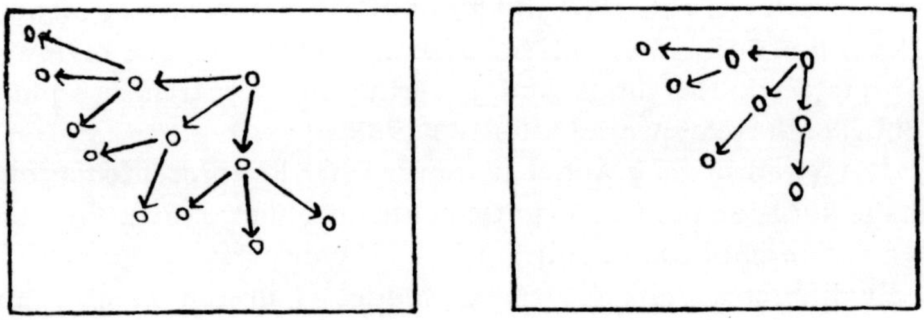

FIG. 18.1 (b)

tion type diffusion some members of the population change their location from time to time t+1 in space, while under expansion type diffusion new members are added to the population between time t to t+1. Changes in the basic spatial characteristics of diffusion depend upon six basic elements i.e., an environment, a temporal dimension, an item being diffused, nodes of origin of an item in space, nodes of destination of an item at the end of time t and the path of movement in between origin and destination points.

**Factors and Forces of Diffusion**

Diffusion of agricultural crops focuses attention on the clustering process of any item through innovation. This depends on the efficiency of the communication system, technological and economic advancement of the society and the involvement of time in the propagation of innovation waves for diffusion.

In this way, the factors of diffusion are innovation of the idea of a particular crop or implement, channel of communication, social system, spatial process and the component of time.

**Innovation**

Among a member of innovations introduced in an area only a few get diffused. The causes of this variation are several i.e. economic, cultural and perception of the innovation by the adopters. The agricultural innovation can be adopted by the farmers of an area under the following condition.[11]

(1) Geographic feasibility;
(2) Cultural compatibility;
(3) Technological simplicity;
(4) Experimentability;
(5) Communicability;
(6) Economic feasibility;
(7) Reliability; and
(8) Socio-technical compatibility.

**Geographic Feasibility**

In any physical environment and given state of technology a geographer can advise which particular crop can easily be raised. With a few exceptions of rugged land relief features, forests and cold climate of polar regions of the world, location, landforms, soils, climate, flora and fauna put some barriers on the type of crops and animals that can be raised in a given locality. Hence the same agricultural innovation is not equally suitable in every part of the world and to the same extent. In some areas the suitability will be greater in comparison with the physical and cultural elements of the landscape while in others the suitability will be lesser.

**Cultural Compatibility**

In judging the amount of agricultural innovation perceived besides physical elements, cultural features also play a significant role. As for example, in India the physical environment is suitable for the growth of beef animals i.e. cow like U.S.A. and Canada but the culture of India cannot permit the Hindus to eat beef. Hence it is not possible to include an innovation in an agricultural system if the same lies out of the boundaries of physical and cultural perimeters.

**Technological Simplicity**

This includes mechanical as well as mental advancement of the adopter to perceive an innovation. A simple technology used in the plough can be easily acceptable to an Indian farmer rather than a complex tractor. The proper utilization of natural resources also depends upon the technical abilities of people because the agricultural implements which have a complex device cannot be introduced easily to an uneducated and poor farmer in comparison with the educated and the rich.

**Experimentability**

For the proper acceptance of an innovation, experimentability and divisibility are essential e.g. in a small plot of land the experiment about the productivity of a crop could be easily ascertained after the division of a large plot into smaller one while it is difficult to plough a smaller plot of land with the help of a tractor as it is not divisible.

**Communicability**

Communicable innovations are easily get diffused at wider locations than the stable and newly introduced items.

**Economic Feasibility**

Unless an innovation is economically gainful, there is less of a chance to get it diffused in space. As for example irrigation is more economical for the Indian farmers in comparison with chemical fertilizers. This leads to the diffusion of irrigation quickly than fertilizers.

**Reliability**

Reliability of the diffusion of an idea is more helpful in the growth and development of agriculture. As for example in North Bihar BAC-32 a red and hard variety of sugarcane will thrive well even in areas of lesser rain, flood conditions and against the menace of flood coupled with higher production of cane for crushing. In such cases of the adoption of innovation, risk is always involved there as the investment may go waste in the purchase of seed, fertilizer and human labour if the physical environment is not suitable. Because of this fact many farmers in India are not desirous of adopting new variety of seeds as they are not in a position to take the risk which is involved in experiment.

**Socio-Technical Compatibility**

In general, the peasants do not adopt all innovations which are economically gainful, but at the same time that should be desirable. Quite a large number of innovations do not spread in space as they are out of the socio-technical and political make-up of the society. Big agricultural implements are not readily adopted by the peasants of India. The causes are not only ignorance of the machine but poverty and lack of repairing stations as such machines are almost absent from the traditional methods of farming in India. In this way there are two powerful forces e.g. economic advantage and the past experience of traditional society which reject the diffusion of agricultural innovation.

## COMMUNICATION

In the diffusion of agricultural innovation, the mean information field of a person or item is highly helpful. According to Shanon and Weaver there are five basic elements in the communication process :[12]

(a) Source of information;
(b) Transmitter;
(c) Channel;
(d) Receiver; and
(e) Destination.

The source of information may be a person or an organization. In case the source is a reliable regarding the innovation of a crop or implement, the same may get diffused in a lesser time with greater certainty, while in case of suspicion about the authenticity of piece of news the transmitter as well as the receiver will watch the situation before accepting it. Normally information passes through the transmitter, channel, receiver and ultimately arrives at a final destination of a place or an individual farmer.

**Social System**

The transmitter and receiver all help in the diffusion of an innovation under different social structures and economic status of a society. This is guided by the social norms, values, birth of a person and achievements in various fields of agricultural operations. In an educated and rich society an innovation will have to cross a few barriers while underdeveloped areas marked by illiterate persons put hindrances in the diffusion of innovations.

**Spatial System**

Adorned with physical and human resources, the different parts of the earth's surface have differential rates of attraction of agricultural crops or implements innovation. As for example, on the sea coast fish oriented occupations will thrive easily while an unbridged embankment may hamper the process of diffusion equally because the physical conditions on two sides are not equal.

Two types of effects of a barrier in transmitting the message have been realised in space : (a) due to socio-spatial distance some messsage may increase or some may decrease, and (b) the time involved in transmitting the message. These are the causes of maldistribution of the diffusion of an information, with increasing distance and time in between the source of origin and the destination.

## RELATIONSHIP OF ADOPTION WITH DIFFUSION

"If an innovation fails to get adopted then neither adoption nor diffusion has occurred." In this way it can be said that diffusion is not possible without adoption. According to Wilkening, "adoption is a process, composed of learning, deciding, and acting over a period of time. The adoption of a specific practice is not the result of a single decision to act but of a series of actions and through decisions".[13] Beal and Copp investigated five stages in the adoption process :

(a) Awareness, (b) Attention, (c) Evaluation, (d) Trial, and (e) Adoption at the final stage.

Adopters are of several types e.g. innovators as the pioneers of an idea, early adopters as local opinion leaders, inquisitive followers who wait for the early adopters to demonstrate the usefulness of innovations before they adopt; sceptical followers, or those who adopt an innovation under certain circumstances, and laggards are those farmers who are best concerned about the happenings which are going on in their surroundings.

The spatial aspects of adoption include five stages such as induction or beginning of an idea, clustering or nucleation, expansion, condensation and the last saturation.

## MEAN INFORMATION FIELD

The mean information field denotes the knowledge about an information of a person on particular space and time. This is generally determined by the accessibility of a person, transport linkages, efficiency of communication lines, technological know-how of adopters of a news, rate of diffusion of an innovation, etc.

The data for the mean information field were collected from fifteen farmers of Rosera anchal of Samastipur district by random selection. The farmers were asked to give a list of persons with whom the contacts were generally made along with their frequency. In this way, it has been observed that most farmers know each other quite well within a radius of 5.5 kilometres of their own agricultural fields, but beyond that distance the pattern of contact varies from one individual to another. Usually these contacts decline from the centre of the farm to all directions uniformly while in some places this rule may not follow. The reasons for such a decline may be due to the residence of the former on the outskirts of the village, illiteracy, poverty, caste barrier, religious superstition, etc., which would focus the interest of the farmers in a particular direction. The data collected from interviews show a decline of social contact with distance from the centre of the farm which is summarized by the equation:

$$P(D) = aD - b$$

where,

P stands for number of contacts,

D for distance from farmers home, and

a and b are constants.

In any social system the behaviour of an individual is always probabilistic rather than deterministic. Similarly, in order to know the diffusion of agricultural innovations in space as a random process we need a model to represent normally observed tendencies through an equation as written above. In diffusion analysis as the distance between two cells increases the number of people who are known to a person decreases and this decrease is much faster than increase in a linear fashion. In this study the probabilities assigned to the different cells in probability field are derived from the Pareto gravity potential function.

The main principle underlying this model is that human being have a close contact only in the nearby areas of a social system. In this sense we know that their relations in the form of close contacts are limited near home and that can be easily analysed in a probabilistic model. Keeping the above views in mind, let us assume that in an area of 25 square kilometres there emerges a transmitter of a given agricultural innovation. In this process of face to face transmission of information in the adjoining cells the number of people diffusing the information progressively increases because each additional adopter turns into a new transmitter also. The conditions under which the diffusion takes place are as follows :

(1) the area studied has an evenly distributed population;

(2) the transmitters do not go beyond five kilometres for pairwise meetings and tellings;
(3) the potential adopters adopt the innovation as soon as they know of it;
(4) all those who adopt the innovation become transmitters;
(5) within a given generation, all persons adopt the innovation at the same time;
(6) the transmitters move to any direction they wish for pairwise talks; and
(7) there are no physical or cultural barriers against the movement.

Under these conditions the simulation of the movement of people for pairwise talks is possible, in order to develop sets of artificial data. This may be compared with the data observed from a real world situation. This will indicate the success of the model and parameters used.

The above could be written as an equation :

$$\log P(D) = \log a - b \log D$$

Further, this relationship gets transformed into the following linear form after taking the logarithm of number of contacts and distance on both sides :

$$Y = a - bX$$

where,
$$Y = \log P\ D$$
$$X = \log D$$
$$a = \log a$$

TABLE 18.1
*Mean Information Field Data*

| Distance from farmers Home A in Kilometres | Mean Number of contacts per square kilometre for sampled farmers | | | | | |
|---|---|---|---|---|---|---|
| X | Y | log X | log Y | log X.log Y | log $X^2$ | log $Y^2$ |
| 0.80 | 5.97 | 0 | 0.778 | 0 | 0 | 0.605284 |
| 1.07 | 4.38 | 0 | 0.602 | 0 | 0 | 0.605284 |
| 2.30 | 3.25 | 0.301 | 0.477 | 0.143577 | 0.090601 | 0.362404 |
| 3.52 | 2.20 | 0.602 | 0.301 | 0.181202 | 0.362404 | 0.227529 |
| 4.97 | 1.92 | 0.700 | 0.301 | 0.210700 | 0.490000 | 0.090601 |
| 5.20 | 1.33 | 0.700 | 0 | 0 | 0.490000 | 0.090601 |
| 6.33 | 0.87 | 0.778 | 0 | 0 | 0.605284 | 0 |
| 7.21 | 0.63 | 0.778 | 0 | 0 | 0.605284 | 0 |
| 8.00 | 0.20 | 0.903 | 0 | 0 | 0.815409 | 0 |
| | | 4.762 | 2.459 | 0.535479 | 3.458982 | 1.376419 |

$\bar{x}=0.529, \qquad \bar{y}=0.273$

Thus in solving the regression equation y=a+bx the following process are involved

$$b=\frac{N\Sigma XY-\Sigma X.\Sigma Y}{N\Sigma X^2-(\Sigma X)^2}$$

$$=\frac{9\times .535479-4.762\times 2.459}{9\times 3.458982-22.676644}$$

$$=\frac{4.819311-11.709758}{31.130838-22.676644}$$

$$=\frac{-6.890447}{8.454194}=-0.8150329$$

$$a=\bar{x}+b\,\bar{y}$$

$$=0.529+.8150329\times 0.273$$

$$=0.529+0.2225039$$

$$=0.7515039$$

Therefore, log P(D)=0.7515039−.8150329 log D,

where log D=0.7515039 and b=0.8150329

P(D)=.56234 D—.8150329

with the help of P(D) value, the number of contacts for each cell of a five by five square grid of one square kilometre each was estimated according to its distance from the centre of the all. For the central cell, actual number of contact (3) was used (Fig. 18.2).

Probabilities were obtained by summary the estimated contact values of each cell and dividing each by the total (11.66). Thus,

$$P_i=\frac{Y_i}{\sum_{i=1}^{25} Y_i}$$

where, $P_i$ is the probability of contact between cell i and the central cell (A), and $y_i$ is the estimated number of contacts for cell i (Fig. 18.3).

It is assumed that in all the above cells the number of potential adopters are the same. After considering Hagerstrand model, it could be said that the probabilities of transmitters

*Estimated Number of Contacts*

| | | | | |
|---|---|---|---|---|
| ·192 | ·275 | ·317 | ·275 | ·192 |
| ·275 | ·544 | ·562 | ·544 | ·275 |
| ·317 | ·562 | A | ·562 | ·317 |
| ·275 | ·544 | ·562 | ·544 | ·275 |
| ·192 | ·275 | ·317 | ·275 | ·192 |

FIG. 18.2

*An Estimate of Simulation*

| | | | | |
|---|---|---|---|---|
| ·016 | ·023 | ·027 | ·023 | ·016 |
| ·023 | ·046 | ·048 | ·046 | ·023 |
| ·027 | ·048 | A | ·048 | ·027 |
| ·023 | ·046 | ·048 | ·046 | ·023 |
| ·016 | ·023 | ·027 | ·023 | ·016 |

a

FIG. 18.3 (a)

An Estimate of Simulation

| | | | | |
|---|---|---|---|---|
| 16 | 17–39 | 40–66 | 67–89 | 105 |
| 108-128 | 129-174 | 175-222 | 223-268 | 269-291 |
| 292-318 | 319-366 | A | 367-414 | 415-441 |
| 442-464 | 465-510 | 511-558 | 559-604 | 605-627 |
| 643 | 644-666 | 667-693 | 694-716 | 732 |

FIG. 18.3(b)

| | | | | |
|---|---|---|---|---|
| 1 | | | | |
| 2 | 4 | 7 | | |
| 3 | 5 | 9<br>A | 6<br>C | 4<br>F |
| | | 5<br>B | 4<br>D | 3<br>G |
| | | 3 | 2 | 1 |

C

FIG. 18.3(c)

getting the contact with receivers located in different cells will decline on an average at a rate which can be measured as reciprocal of the square of the distance. The probability field given for this purpose is given in Fig. 18.3. Due to probability distribution, it is necessary to add up the assigned values of different cells.

Later on in order to have a random selection of numbers, a range of discrete numbers is put in place of these probabilities [Fig. 18.3(b)]. Now it is assumed that transmitter A is located in one of the cells of the probability field. In $g_o$ (generation zero). A makes a move for pairwise meeting in all directions around him. Now in the selection of random number of three digits from left to right move, suppose we get 417 and the same is out of our cells, so it should be ignored. Suppose our next number is 558 which is within the field and is located one cell below the A's cell. Now one more person knowing the message may be called B; and here is end of generation one ($g_1$). In this generation both A and B are transmitters, and now both of them convert one more person as transmitter, so up to $g_2$ there will be four transmitters i.e. A, B, C and D. This process may continue in order to see the further situations. In this analysis the transmitters were permitted to move any direction within the field.

## TRENDS IN DIFFUSION STUDIES

Innovation and diffusion processes can be analysed with inter-disciplinary approach of study. The sociologists analyse the social structure, group behaviour and farm technology with particular reference to the impacts of community norms, social customs and cultural values, etc. The psychologists focus the problem of human perception, regional scientists study about the spread of disease and the zone of influence of hospitals, while the economists analysed the spread of innovation from one profitable location of a firm to another. So far as geographers are concerned, they have adopted spatial dimensions of technological innovations, impact of new transportation facilities near settlements, limits of population migration, seeds, fertilizers, agricultural implements, cultural values, spread of disease and epidemic, limits of marriage connections and human perception are important for diffusion of innovation studies. The geographers are interested in the differential processes of the spread of different items over space and in finding out the nature of barriers which obstruct the communication of ideas or innovations in order to predict the rate, so that the other innovations or ideas may spread over space under similar conditions. In this way geographers have a distinctive approach to study diffusion problems in comparison with sister disciplines. The geographers' contribution have mostly been directed towards inter-disciplinary approaches in solving the diffusion analysis.

The geographers approach to diffusion may be summarized as :

(a) the diffusion of any item tends to wane with the increasing distance as it travels from one place to the other and one individual to another;
(b) spatial diffusion is determined by the force generated by a message after propagation from its origin;
(c) distribution of any item or its location is partly determined by a chance factor;
(d) some areas lag behind others in the adoption of innovations; and
(e) the barriers in the movement of various elements are functionally variable according to the nature of the terrain, climatic conditions and the social make-up.

## REFERENCES


1. E.M. Rogers, *Diffusion of Innovations*, Free Press, New York, 1962.
2. M.T. Cooke, *The Spread of the European Sterling in North America*, U.S. Department of Agriculture, 1928.
3. T. Hagerstrand "On Monte Carlo Simulation of Diffusion" (Ed. W. Garrison) *Quantitative Geography*, Department of Geography, Northwestern University, Evanston, 1963, pp 1-32.
4. T. Hagerstrand, *The Propagation of Innovation Waves, Land Studies in Geography*, Lund (Sweden), 1952.
5. J.D. Nystuen, "An Approach to the Direct Measurement of Community Mean Information Fields" *Papers and Proceedings of Regional Science Association*, 1963, Vol. II, p 99.
6. R.S. Yuill, "A Simulation Study of Barrier Effects in Spatial Diffusion Problems", *Spatial Diffusion Study, Technical Report, I*, 1969.

7. R.L.Morrill, "The Distribution of Migration Distances", *Papers af the Regional Science Association, II*, pp 75-88.
8. E. Bylund, "Theoretical Considerations Regarding the Distribution of Settlement in Inner North Sweden", *Geografisca Annalar*, Vol. 42, pp 225-231.
9. K.N. Singh, "The Territorial Basis of Medieval Town and Village Settlement in Eastern Uttar Pradesh, India", *Annals of the Association of American Geographers*, 1968, Vol. 58, No. 2, pp 203-229.
10. O.C. Hudson, "Diffusion in a Central Place System", *Geographical Analysis*, 1969, Vol. 1, p 46.
11. R.P. Misra, *Diffusion of Agricultural Innovations*, University of Mysore, 1968.
12. C.E. Shanon and W. Weaver, "*The Mathematical Theory of Communication*, Urbana, University of Illinois Press, 1949.
13. E.A. Wilkening, "Social Isolation and Response of Farmers to Agricultural Programme", *American-Sociological Review*, 1951, 16, pp 836-37.

# 19

# The Concept of Agricultural Typology

—*Ram Bahadur Mandal*

In order to classify agricultural land in relation to social and ownership characteristics, organizational and technical characteristics, production and structural characteristics Kostrowicki (1972)[1] and Beek and Bennema (1972)[2] have developed agricultural typology of land utilization. Kostrowicki aims at providing "World Typology of Agricultural Landuse" following the pattern, process and colour schemes (Fig. 19.1) of 'World Landuse Survey'. The classification of Beek and Bennema is suitable for the study of land evaluation.

The variation in the attributes of agriculture in different regions is known as agricultural typology. The typology should not be confused with regionalisation. Both types and regions are systematic concepts. Thus typology is defined as similarities between units in terms of 'agricultural characteristics'. Such sets and associations of similar characteristics may occur repeatedly in time and space and the same types can be found in various periods or areas[3]. These characteristics are quite often distributed in space in a mosaic-like pattern, hence the distribution of resulting types does not necessarily form a contiguous area, but units of a certain type are usually dispersed in nature and intermingled with some other types. On the other hand, a region is a spatial concept which has some territorial limits and differences between places estabilized on a particular part of the earth's surface characterized by a peculiar association of environmental factors. Typology is a hierarchical concept in which a lower order may be grouped into a higher order; besides, lower orders often form territorial parts of higher order regions. Thus hierarchy is a relative term ranging from small to bigger vertically, while horizontally the same may be smaller areal extension to a larger in size. In an agricultural typology,

## *The Legend Proposed for mapping World Types of Agriculture*

Comment: colours - principal four types of world agriculture; horizontal lines - crop emphasis; vertical lines -, livestock emphasis; colour intensity - intensity of agriculture; asterisks - irrigation.

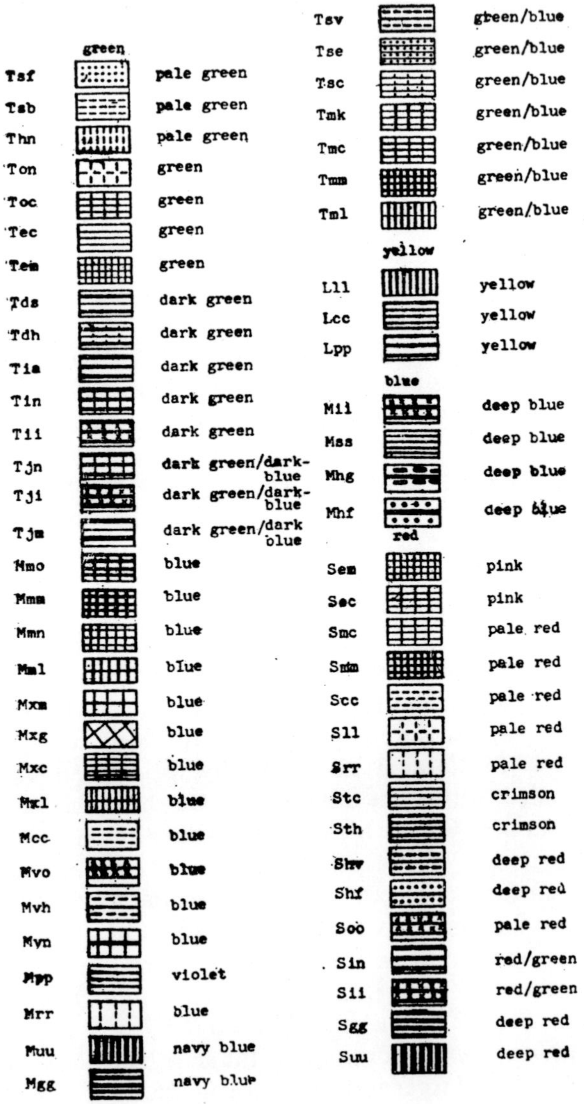


FIG. 19.1

the determination of agricultural regions could be made on the basis of the dominance and co-existence of individual characteristics of agriculture.

An individual plot of land is said to be the best operational holding for the determination of agricultural typology. However, on a macro-level the village, administrative units of anchal panchayat, subdivision, district and State level data may also be used to determine the agricultural typology. But the non-availability of data on each level is the greatest problem in a study of all levels. A sample study is very helpful in obtaining data on all levels, especially nested sampling procedure. Whatever be the territorial space, the uniformity of variables considered reflects uniformity of types and situations in an areal unit, even if the intensity of characteristic features are either non-existent or available scarcely or abundantly.

In the case of a micro-level study, a detailed differentiation about the smaller area is required only in such a situation; in the case of the identification of agricultural types, additional variables of local importance can be introduced. In any case, the variables irrelevant for one area may be significant for others. Because when comparability of the result is important, the common criterion should not be omitted without sufficient ground.

The classification of agricultural typology should be made with the help of internal and inherent characteristics of agriculture. The external characteristics should be used as why, when, how, where and with what process an individual type of agriculture has developed, but this should not be used as the basis for determining agricultural typology. The simultaneous use of internal and external variables is not permissible as such associations can be accurately shown with the help of correlation analysis and analysing the conditions of their development. The external variables such as location, transportation, market conditions, prices, supply and demand of agricultural products, play a dominant role in the formation of agricultural typology but their individual conditions, change, with the change of environmental conditions, time and space as well. A study about these facts will help in understanding the development and distribution of agriculture types only. A change in the agriculture type is essentially followed by a change of the external parameters necessary for agricultural development and in such a situation this provides a basis for planning the spatial organization and rational types of agriculture. It is also affected by the factors such as input and output, organizational and technical characteristics and the use of fertilizer and irrigation facilities as all these determine the production of crops. The efficiency of agriculture is determined by the fertility of the land, the amount of labour involved, the implements and seeds used and the system of land tenure which may foster or hamper the agricultural production keeping in view their applicability and intensity. Thus their identification is essential in determining the agricultural typology.

Agricultural characteristics can be grouped into the following:

(a) Social and ownership characteristics,
(b) Organizational and technical characteristics,
(c) Production characteristics, and
(d) Structural characteristics.

The first point answers the query as to who is the land owner and whether the scale of operation is small or large. The second point gives the characteristics of labour and material inputs along with the operation of land holdings. The third point gives the amount of production and the process of consumption. Finally, the fourth point answers the points regarding purpose proportion, combination of enterprise, livestock breeding, and gross agricultural production in terms of landuse. The parameters representing the characteristics of social, land ownership, organization, structure and input-output help in the identification of agricultural typology. In reality agricultural statistics are never correct and hence they are insufficient to provide the material for agricultural typology. Hence the scarcity of data, their unreliability and the estimation of agricultural statistics are responsible, for the inability produce some good results. The variables selected for agricultural typology should be expressed in quantitative terms. Although a scholar with a better knowledge about the area may produce a good result of agricultural typology and the same would be more reliable. In most cases, however, the same is not possible without proper agricultural statistics and methodology used so that the comparisons can be made and the line of thinking need not be repeated for all the places and all the times. Quantitative methods have the advantage of yielding quick results instead of labour-absorbing, traditional and lethargy techniques.

## METHODOLOGY

The method derived from the fifth meeting of the IGU Commission on Agricultural Typology held in 1972 in Hamiltan, Ontario, Canada recommended one specified method for all studies in agricultural typology irrespective of their scale, region and time. From a number of studies, it has been clarified that (a) the study conducted by the variables as indicated by the commission does not affect the results much in any part of the world and (b) as the comparability of results in time and space is necessary for agricultural typology. All methods which do not guarantee such a comparability cannot be recommended whether they are traditional, quantitative, highly precise and statistical.

For these factors that could not be expressed quantitatively, symbols for individual categories have been selected. It has been decided thus as those categories do not differ statistically and hence on this basis, it is not possible to include them in the same category. Such an area is denoted by a capital letter O meaning absence of any typology. Thus agricultural typology has been represented after considering the following code :

$$T = S \frac{O}{P} C$$

where, T is type of agriculture,

S is the social and ownership characteristics and size of holdings,

O is organizational and technical characteristics,

P is production characteristics, and

C is structural characteristics.

## REPRESENTATION OF DIAGNOSTIC FEATURES AND VARIABLES

The IGU Commission on Agricultural Typology has selected 22 variables with the verification of a number of individual cases. The first-two non-measurable variables are marked by symbols, while the other variables are shown by indices and their position within the prescribed characteristics has been aranged. For the typology of world agriculture, five classes were considered in a uniform way as (1) very low, (2) low, (3) medium (4) high, and (5) very high.

The following are twenty-two variables which represent the most significant agricultural characteristics applied :

**Social and Ownership Characteristics**

(1) Land ownership,
(2) Land operation,
(3) Size of holdings in terms of the number of actively employed per holding,
(4) Size of holdings in terms of the number of hectares of agricultural land per holding,
(5) Size of holdings in terms of the number of livestock measured in the number of conventional (large) animal units per holding,
(6) Size of holdings in terms of gross agricultural production per holding,
(7) Inputs of labour measured by the number of actively employed in agriculture per 100 hectares of agricultural land.

**Organizational and Technical Characteristics**

(8) Inputs of animal power measured by the number of draught animals in conventional (horse) units per 100 hectares of cultivated land,
(9) Inputs of mechanical power measured by the number of tractors and self propelling machinery in the number of HP per 100 hectares of cultivated land,
(10) Chemical fertilizing measured as the amount of chemical fertilizers used per hectare of cultivated land,
(11) Irrigation-measured as the percentage ratio of irrigated to total cultivated land,
(12) Intensity of crop landuse expressed as the ratio of harvested to cultivable land (including fallow),
(13) Importance of the system of perennial cropping measured by the percentage ratio of perennial crop

land to total agricultural land,

(14) Importance of the perennial grass land system measured by the percentage share of permanent grass land to total agricultural land,

(15) Intensity of livestock breeding measured as the number of farm animals in conventional (large) animal units per 100 hectares of agricultural land.

**Production Characteristics**

(16) Land productivity measured in gross agricultural (crop and animal) output in conventional (grain) units per hectare of agricultural land,

(17) Labour productivity measured as above per actively employed in agriculture,

(18) Degree of commercialization expressed as the percentage ratio of commercial to gross agricultural production,

(19) Level of commercial production in commercial agricultural production in the same units as above per hectare of agricultural land,

(20) The degree of specialization measured by the method described by Kostrowicki.[4]

**Structural Characteristics**

(21) Orientation in gross production measured by the percentage share of livestock products in gross agricultural production.

(22) Orientation in commercial production measured in the percentage rate of livestock products in commercial production.

TABLE 19.1

*Land Tenure Classification as Measurable and Non-measurable Variables*

**1. Land Ownership (Variable No. 1) : Symbols of Land owned :**

| | |
|---|---|
| As non-measurable variable in common (by a tribe, class, village community, etc.) under customary rights | ...C |
| by a chief, feudal landlord and allied under customary rights | ...F |
| by a religious or other non-profit organization | ...R |
| by a private owner | ...P |
| by a corporate (business) organization | ...U |

| | |
|---|---|
| by a cooperative organization | ...V |
| by a collective organization | ...K |
| by a State (government, local authorities etc.) organization | ...G |

*As measurable variable*

| *Class* | *Percentage rate* | |
|---|---|---|
| 1 | below 20 | Under traditional customary tenure, share tenancy, private or joint property. |
| 2 | 21—40 | |
| 3 | 41—60 | |
| 4 | 61—80 | |
| 5 | over 80 | |

**2. Land Operation (Variable No. 2) As non-measurable Variable :**

*Land is operated* :

| | Symbols |
|---|---|
| in common (by a tribe, class, village community members) | ...C |
| by an extended family | ...B |
| by slaves | ...S |
| by serfs under any form of landed bondage involving compulsory labour | ...F |
| by servile tenants (in exchange for services) | ...E |
| by sharecroppers | ...M |
| by tenants under fixed rent | ...T |
| by squatters | .. Q |
| by rent from tenants | ...N |
| by landowners | ...P |
| by part-time farmers | ...W |
| by hired workers | ...H |
| jointly by cooperative members | ...V |
| by members of a collective farm | ...K |

*As measurable variable*

| *Class* | *Percentage rate* | |
|---|---|---|
| 1 | below 20 | Agricultural land operated by the consciously planned collective or State enterprises. |
| 2 | 21—40 | |
| 3 | 41—60 | |
| 4 | 61—80 | |
| 5 | over 80 | |

In case of two or more farmers of equal importance, it should be treated as mixed farming marked by two or more symbols, e.g. $\left(\frac{\text{P}}{\text{H}} \text{ or } \frac{\text{T}}{\text{H}}\right)$.

*Classes of Measurable Variables based on their World Ranges*

(Number of variables are the same as listed above)

| Class No. of variables | 1 | 2 | 3 | 4 | 5 |
|---|---|---|---|---|---|
| 1 | 0-20 | 21-40 | 41-60 | 61-80 | 80 |
| 2 | 0-20 | 21-40 | 41-60 | 61-80 | 80 |
| 3 | 0- 2 | 3- 8 | 9-50 | 51-200 | 200 |
| 4 | 0- 5 | 6-20 | 21-50 | 51-200 | 200 |
| 5 | 0- 2 | 3-10 | 11-50 | 51-200 | 200 |
| 6 | 0-100 | 101-900 | 901-10,000 | 10001-80000 | 80000 |
| 7 | 0- 3 | 4-15 | 16-40 | 41-150 | 150 |
| 8 | 0- 2 | 3- 8 | 9-16 | 17-30 | 30 |
| 9 | 0- 6 | 7-15 | 16-36 | 37-90 | 90 |
| 10 | 0-10 | 11-30 | 31-80 | 81-200 | 200 |
| 11 | 0-10 | 11-25 | 26-58 | 51-80 | 80 |
| 12 | 0-0,1 | 0,2-0,3 | 0,4-0,7 | 0,8-1,3 | 71,3 |
| 13 | 0-10 | 11-20 | 21-40 | 41-60 | 760 |
| 14 | 0-20 | 21-40 | 41-60 | 61-80 | 80 |
| 15 | 0-10 | 11-30 | 31-80 | 81-180 | 180 |
| 16 | 0- 5 | 6-20 | 21-45 | 46-100 | 100 |
| 17 | 0-40 | 41-100 | 101-250 | 251-800 | 800 |
| 18 | 0-20 | 21-40 | 41-60 | 61-80 | 80 |
| 19 | 0- 3 | 4-12 | 13-30 | 31-80 | 80 |
| 20 | 0-0,1 | 0,1-0,2 | 0,3-0,4 | 0,5-0,8 | 0,8 |
| 21 | 0-20 | 21-40 | 41-60 | 61-80 | 80 |
| 22 | 0-20 | 21-40 | 41-60 | 61-80 | 80 |

In areas of great variations from the proposed twenty-two variables for the determination of agricultural typology some additional variables of local character may be added as for example[5] :

(1) Percentage of hired workers to the total number of employed in agriculture

(2) Percentage share of selected size groups of holdings in the total number of agricultural holding or in the total agricultural acreage

(3) Degree of farm fragmentation

(4) Total inputs of draught power in conventional units for unit area and the percentage of mechanical power in the total draught power

(5) Organic manuring (expressed as the number of farm animals multiplied by conventional manual unit and by the percentage coefficient of manure used for agricultural purposes for 100 hectares of agricultural land
(6) Total fertilization (chemical, organic, etc.) in conventional units per unit area and the percentage of chemical fertilizers in the total amount of fertilizers used
(7) Percentage rate of drained land in the cultivated land
(8) System of irrigation or water management system
(9) Potential intensity of agriculture in conventional oral units
(10) Landuse system
(11) System of land or crop rotation
(12) Principal tools and machines applied in tillage and harvesting
(13) Rate of land producing feeds to the number of animals in large animal units
(14) System of livestock breeding
(15) Final production per hectare of agricultural land and per employee
(16) Selected crop or animal products per hectare of agricultural land or per person employed in agriculture
(17) Orientation in agricultural landuse, livestock breeding and gross and commercial production identified in a more accurate way either with one of the techniques that define leading elements in structure or with the following indices showing the percentage rates of:

Vegetative crops *versus* seed crops,
Intensive (row) crops in total arable land,
Structure (leguminous) crops in total arable land,
Cattle in total number of livestock,
Pigs in total number of livestock,
Grain crops (in gross agricultural production),
Fodder crops (in gross agricultural production),
Vegetables in commercial production,
Fruits in commercial production,
Industrial crops in commercial production
Milk in commercial production.

Now it becomes quite obvious that with the increase of variables, the typology will certainly increase but that will also certainly reduce the importance of the other variables. In no case should the investigators be allowed to lessen the number of primary variables irrespective of their non-existence because that would reduce the chance, of comparability for wider dimensions of space.

## MODEL TYPES OF WORLD AGRICULTURE

This has been determined after considering the various agricultural classifications and the case studies collected from the literature and field work done by Kostrowicki and his collaborators.

In the scheme presented below the four groups of characteristics have been grouped in the following manner as the types of agriculture available in different parts of the world:[6]

T=Traditional Small Scale Agriculture

$T_s$=Shifting Cultivation

$T_{st}$=Shifting forest fallow agriculture

$$\text{C,C,3-4,0,1-2, 2-3}\frac{1,1,1,1,1,1,1,1,1,}{1,1\text{-}2,\ 1,1,1}\ 1,1$$

Where T is types of agriculture, s is shifting, f is forest, G.C. are common use of common land, by larger or groups of people (3-4), size of holding acreage cannot be identified (0). Number of animals per group very small to small (1-2), gross output small to medium (2-3), labour and capital inputs very low (1) and share of perennial crops and permanent grass lands (1). Density of livestock population very low (1), land productivity very low to low (1-2), commercialization very low (1), specialization (1) and crop products clearly dominate (1). Central Africa, South West Asia, Amazonia.

$T_{sb}$=Rotational bush fallow agriculture

$$4221221\frac{1111121}{2211111}\ 215111$$

West Africa, South East Asia, Latin America

$T_o$=Extensive subsistence to semi-subsistence crop-growing and migratory herding

$T_{oc}$=Current fallow, subsistence to semi-subsistence agriculture with migratory herding

$$3431221\frac{2211132}{1122111}142331$$

Mediterranean countries, Middle East.

$T_{on}$=Nomadic herding with subsidiary crop growing

$$4211211\frac{1111121}{1122111}142451$$

North Africa, Middle East.

$T_n$=Nomadic herding

$T_{nn}$=Nomadic herding (1)

$$5111000\frac{1100001}{1122111}151551$$

North Africa, Middle East

$T_e$=Small scale, extensive, current-fallow, subsistence to semi-subsistence agriculture

$T_{em}$=Current fallow, semi-subsistence mixed agriculture

$$2341211\frac{3311132}{2222221}123221$$

Some parts of Europe, Middle East

$T_{ec}$=Current fallow, semi-subsistence crop agriculture

$$2441211\frac{2111131}{2222221}215111$$

Parts of Africa, Latin America

$T_d$=Labour intensive, subsistence or semi-subsistence, medium productive agriculture

$T_{ds}$=Labour intensive semi-subsistence, semi-irrigated, medium productive agriculture

$$1151211\frac{4411244}{2211222}114111$$

South Asia.

$T_{dh}$=Labour intensive, subsistence to semi-subsistence,

semi-irrigated, medium productive mixed agriculture

$$1251211\frac{4511344}{2211111}123341$$

South Asian mountains.

$T_i$=Labour intensive, subsistence to semi-subsistence agriculture

$T_{ia}$=Labour intensive, semi-irrigated, semi-subsistence, medium productive agriculture

$$2231212\frac{4211441}{3311222}214111$$

North Africa, Middle East.

$T_{in}$=Labour intensive, semi-subsistence, highly productive non-irrigated agriculture

$$1241212\frac{4311142}{4411112}214111$$

South and East Asia.

$T_{ii}$=Labour intensive, subsistence to semi-subsistence, highly productive irrigated agriculture

$$1251212\frac{5411452}{5511112}115111$$

South and East Asia.

$T_j$=Labour intensive, semi-commercial agriculture

$T_{jn}$=Labour intensive, semi-commercial, non-irrigated agriculture

$$1151212\frac{4223142}{4422332}214221$$

Parts of South East Asia, Japan.

$T_{ji}$=Labour intensive, semi-commercial, irrigated, highly productive agriculture

$$1151212\frac{5323452}{5522333}114111$$

Parts of South East Asia, Japan.

$T_{jm}$=Labour intensive, semi-commercial, irrigated, medium productive agriculture

$$1151112\frac{4422454}{3322332}214112$$

Northern India.

$T_s$=Semi-commercial, specialized crop agriculture

$T_{sv}$=Semi-commercial, specialized crop agriculture

$$2241222\ \frac{3412241}{2222332}\ 313113$$

West and East Africa, South Asia, Latin America, Oceania

$T_{se}$=Semi-commercial, specialized extensive crop agriculture

$$2241131\ \frac{1111131}{1122331}\ 312114$$

Parts of Latin America.

$T_{sc}$=Semi-commercial, specialized arid land crop agriculture

$$1351221\ \frac{3311232}{2211332}\ 324113$$

North West India, Pakistan, Sudan.

$T_m$=Semi-commercial or semi-subsistence, small-scale, mixed agriculture

$T_{mk}$=Semi-subsistence labour intensive mixed agriculture

$$1151212\ \frac{4423144}{4422223}\ 122331$$

Parts of Europe, mostly mountains.

$T_{mo}$=Semi-commercial, mixed agriculture with crops prevalent

$$2241212\ \frac{4422243}{3322333}\ 224221$$

Mediterranean countries, Latin America.

$T_{mm}$=Semi-commercial mixed agriculture

$$1151222\ \frac{3323143}{3322332}\ 123231$$

Some less developed parts of Europe, Latin America.

$T_{ml}$=Semi-commercial mixed agriculture with livestock breeding prevalent

$$1151232\ \frac{2223143}{3333333}\ 131344$$

Parts of Europe.

**L. Traditional Large-scale Agriculture (Latifundia)**

$L_c$=Traditional latifundia with crop growing prevalent

$L_{cc}$=Traditional latifundia with crop growing prevalent

$$1351454\ \frac{2222132}{2222332}\ 124211$$

Southern Europe, Latin America.

$L_l$=Traditional latifundia with livestock breeding prevalent.

$L_{ll}$=Traditional latifundia with livestock breeding prevalent

$$1251354\ \frac{2211132}{2233442}\ 141341$$

$L_p$=Traditional plantation agriculture

$L_{pp}$=Traditional plantations

$$1351444\ \frac{4411142}{3322443}\ 333213$$

Latin America, some other tropical countries

**M. Market-oriented Agriculture**

***Intensive***

$M_i$=Commercial, small-scale non-specialized agriculture

$M_{ii}$=Commercial, small-scale irrigated non-specialized highly productive agriculture oriented toward food production

$$1151211\ \frac{4234551}{5533444}\ 114111$$

Japan, Taiwan.

$M_s$=Commercial small-scale specialized crop agriculture

$M_{ss}$=Commercial small-scale specialized crop agriculture

$$1151212\ \frac{3123341}{3333443}\ 411114$$

Japan, Taiwan, Malaya, some parts of Africa, Latin America.

$M_h$=Commercial small-scale horticulture

$M_{hg}$=Commercial small-scale vegetable growing agriculture

$$1151222\frac{4145351}{5533555}115111$$

Europe, predominantly sub-urban zones.

$M_{ht}$=Commercial small-scale fruit growing agriculture

$$1151222\frac{4134241}{4444554}415111$$

Europe, North America, some parts of Latin America.

$M_{m}$=Commercial small-scale mixed agriculture

$M_{mc}$=Commercial small-scale mixed agriculture with crop growing prevalent

$$1151222\frac{4244242}{4433444}213222$$

Southern Europe, Latin America.

$M_{mm}$=Commercial small-scale mixed agriculture

$$1151222\frac{3255144}{4444444}123231$$

Central and Western Europe.

$M_{mn}$=Commercial small-scale mixed agriculture with industrial crop growing prevalent

$$1151222\frac{3155342}{4444554}112214$$

Central and Western Europe.

$M_{ml}$=Commercial mixed agriculture with livestock breeding prevalent

$$1151232\frac{3454145}{4444443}131351$$

North-Western Europe.

$M_{x}$=Commercial large-scale mixed agriculture

$M_{xm}$=Commercial large-scale mixed agriculture

$$1151344\frac{2154143}{4455444}123231$$

Western Europe.

$M_{xg}$=Commercial mixed agriculture

$$1151232\ \frac{3154143}{3344553}\ 133331$$

General farming in USA and Canada.

$M_{xc}$=Specialized grain crop growing and livestock breeding.

$$1151254\ \frac{1152132}{3355553}\ 132331$$

Australia, South America, USA.

$M_{xl}$=Specialized livestock breeding (both of dairy and beef orientation).

$$1151243\ \frac{2143143}{3355443}\ 131451$$

North America, Australia and New Zealand.

$M_{c}$=Specialized grain crop agriculture

$M_{cc}$=Specialized grain crop agriculture

$$1151254\ \frac{1152131}{2255552}\ 125111$$

USA, Canada, Australia, South Africa.

$M_{v}$=Large-scale intensive crop growing agriculture

$M_{vm}$=Large-scale irrigated mixed crop growing agriculture.

$$1151244\ \frac{2154441}{4455554}\ 113112$$

USA, South Africa, Southern Europe.

$M_{vh}$=Large-scale horticulture

$$1151345\ \frac{2154241}{4444554}\ 315111$$

USA, Australia, South, Africa, Latin America, Western Europe.

$M_{vn}$=Large-scale agriculture specialized in industrial crop growing

$$1151255\ \frac{2154242}{4455554}\ 311114$$

USA, Australia, South and East Africa.

$M_p$=Modern plantation agriculture

$M_{pp}$=Modern plantation agriculture

$$1151455\ \frac{3133241}{4433554}\ 411115$$

Latin America, Africa, Southern Africa, Oceania.

$M_r$=Commercial livestock grazing

$M_{rr}$=Commercial livestock grazing (ranching)

$$1151354\ \frac{1100001}{1144551}\ 151551$$

Western USA, Australia, South Africa.

$M_u$=Highly industrialized livestock breeding

$M_{uu}$=Highly industrialized livestock breeding

$$1151323\ \frac{5100005}{5555555}\ 151551$$

Various industrialized countries.

$M_g$=Highly industrialized crop growing.

$M_{gg}$=Highly industrialized horticulture

$$1151323\ \frac{5155551}{5555555}\ 114111$$

Various industrialized countries.

**S. Socialized Agriculture**

$S_e$=Socialized incipient mixed agriculture.

$S_{ec}$=Socialized incipient mixed agriculture with crop growing prevalent

$$1115443\ \frac{3432231}{2222442}\ 124121$$

South-East Europe, South-Western part of USSR, Mexico.

$S_{em}$=Socialized incipient mixed agriculture

$$1115443\ \frac{3233132}{2233332}\ 133331$$

East-central Europe, USSR.

$S_m$=Socialized mixed agriculture

$S_{mc}$=Socialized mixed agriculture with crop growing prevalent

$$1115554\ \frac{3233242}{3333443}\ 223221$$

South-Eastern Europe.

$S_{mm}$=Socialized mixed agriculture

$$1115555\ \frac{2144143}{3333443}\ 132341$$

East-Central Europe.

$S_c$=Socialized agriculture specialized in grain crop growing.

$S_{cc}$=Socialized agriculture specialized in grain crop growing

$$1115555\ \frac{1143131}{2233552}\ 115111$$

Southern parts of the USSR, South-Eastern Europe.

$S_l$=Extensive livestock breeding with subsidiary crop growing.

$S_{ll}$=Extensive livestock breeding with subsidiary crop growing

$$1115555\ \frac{1142132}{2244442}\ 142441$$

South-Eastern parts of the USSR, Central Asian and Caucasian Soviet Republics.

$S_r$=Socialized agriculture specialized in livestock grazing.

$S_{rr}$=Specialized livestock grazing

$$1115555\ \frac{1100001}{1133551}\ 151551$$

Southern parts of USSR.

$S_t$=Socialized agriculture specialized in industrial/crop growing.

$S_{tc}$=Socialized agriculture specialized in cotton growing

$$1115555\ \frac{2143342}{2233552}\ 321224$$

Central Asian Soviet Republics, Caucasus.

$S_{th}$=Socialized agriculture specialized in sugarcane growing

$$1115555\ \frac{2143341}{3344553}\ 411115$$

Cuba.

$S_h$=Socialized horticulture.

$S_{hv}$=Socialized agriculture specialized in vegetable growing

$$1115545\ \frac{4154342}{4433555}\ 115111$$

USSR and the countries of East-Central Europe, mainly suburban zones.

$S_{hf}$=Socialized agriculture specialized in fruit growing

$$1115545\ \frac{3134241}{4444554}\ 415111$$

USSR and the countries of East-Central Europe.

$S_o$=Socialized irrigated mixed agriculture.

$S_{oc}$=Socialized irrigated mixed agriculture

$$1115555\ \frac{2154443}{4455554}\ 122222$$

Southern parts of USSR, South-Eastern Europe.

$S_i$=Socialized labour intensive agriculture.

$S_{in}$=Socialized labour intensive non-irrigated agriculture

$$1115555\ \frac{4333143}{4422222}\ 224221$$

China.

$S_{ii}$=Socialized labour intensive irrigated agriculture

$$1115555\ \frac{5433454}{5522223}\ 114111$$

China.

$S_g$=Socialized highly industrialized crop growing.

$S_{gg}$=Socialized highly industrialized crop growing.

$$1115545\ \frac{5155551}{5555555}\ 114111$$

Various socialist countries.

$S_u$=Socialized highly industrialized livestock breeding.

$S_{uu}$=Socialized highly industrialized livestock breeding

$$1115555\ \frac{5100005}{5555555}\ 151551$$

Various socialist countries.

The typology made above by Kostrowicki does not claimed to be static but a change is inevitable with the change of variables. The modification of environmental character may change the characteristics of agriculture but with the introduction of one new variable the change in agricultural typology is not possible.

## CONCLUSION

The above study shows that agricultural typology may be devised with the help of the association of numerous variables available in particular environmental conditions or in a given territory. According to Kostrowicki it is a dynamic concept like the scene of a film. A change may take place with the change of environmental conditions in general but the introduction of a single variable may not affect the results or change the agricultural typology.

## REFERENCES

1. J. Kostrowicki, "The Typology of World Agriculture, A Preliminary Scheme", Mimeograph, International Geographical Union, 1972.
2. K.J. Beak and J. Bennema, "Land Evaluation for Agricultural Land-use Planning, an Ecological Methodology" Mimeograph, 1972.
3. J. Kostrowicki, "The Typology of World Agriculture Principles, Methods and Model Types" *Agricultural Typology and Land Utilisation*, Verona, 1975, p 431.
4. J. Kostrowicki, op cit, 1975.
5. J. Kostrowicki, op cit, 1975.
6. J. Kostrowicki, "World Types of Agriculture"—International Geographical Union, *Commission on Agricultural Typology*, Warsaw, 1976, pp 24-35.

# 20

# A Rural Decision-making Model and its Research Implications in Agricultural Geography

—*Joseph T. Uyanga*

THE PROBLEM of rural development engages the attention of countries at all stages of development. In the developed countries special attention is paid to a number of problems arising from the rapidly changing pattern of agricultural production, the falling prices of many rural products, the declining population and the loss of cultivable land to urban-industrial expansion. The developing countries face a different set of problems. In most of them, the base for economic development lies mostly in agriculture. It is agriculture and the industries which supply raw material that particularly in the early stages of development must ensure a nation's economic growth. It is widely recognized that national development cannot progress far without wider utilization of the human and physical resources of the rural areas. By accelerating the rate of agricultural growth and at the same time expanding agricultural processing industries, it should be possible to mitigate the problems of unemployment and under-exploitation of resources which most developing countries face. Rural areas are important suppliers of the food and raw materials required by urban population and at the same time represent a source of labour for industrial and service activities.

In the developed countries, the inter-dependence between the different sectors has grown over the years and each sector has helped in the advancement of the others. The importance of these inter-connections is reflected both in the share of agricultural inputs originating in industry and services and the changing proportions of agricultural output intended directly for private consumption and for industrial use.

The importance of the rural areas and their growth is also being increasingly recognized by international bodies. The Food and Agricultural Organization (FAO) has been engaged for over 25 years now in promoting agricultural development. Currently it is giving special attention to rural development and the integration of agriculture in the total effort to develop rural areas. The International Labour office (ILO) has been active in promoting policies and strategies which relate to rural employment, the setting of minimum wages for rural workers, rural labour welfare, and the promotion of workers, organisations. The United Nations has been involved in rural community development and welfare as well as in the integrated approach to rural development as methods of raising the levels of living of the rural population, particularly the rural poor. The World Health Organisation (WHO) has been giving special attention to rural health services and the United Nations Education, Scientific and Cultural Organisation (UNESCO) to rural education, both formal and informal. Very recently, the World Bank has come out strongly in favour of rural development and is developing policies that will enable it to support and finance rural development projects in African countries.

## CHARACTERISTICS OF RURAL AREAS IN DEVELOPING COUNTRIES

The phenomenon of unemployment and underemployment is rife ; a large proportion of the population growth occurs in the rural areas where almost 76%, of the population lives. The state of the rural economy does not provide sufficient job opportunities for those entering the labour market, and the limited urban economy is unable to provide more than a fraction of what is needed. Consequently, the majority of the people continue to live in the rural areas trying to earn an income through casual employment which often fails to provide them with even a minimum standard of living.

Unemployment and underemployment in rural areas have been seen as results of such factors as population growth and stagnant rural economies. It is also at the same time a contributory cause to low productivity and depressed incomes in rural areas, the parlous state of many of the social facilities and services as well as their increasing poverty.

Low productivity is typical of Nigerian rural areas. The majority of the farmers and craftsmen in the Basin operate at a level of technology that has not changed for centuries. Capital investment is directed in large measure towards the industry rather than agriculture. Consequently, there is a considerable difference in productivity as between the farmer and the industrial worker. This difference is in the order of 1 : 6 in many African countries (Weitz, 1971).

Because of this low level of productivity, the rural producer lacks the purchasing power for the products of industry either for personal consumption or for investment in his enterprise whether farm or workshop. It is only when he produces over and above what is needed for his family's basic consumption that the surplus can be converted into purchasing power for non agricultural commodities. Since subsistence farming still consitutes the major pattern of production in the rural areas of most developing countries, the capacity of the rural market to absorb the products of domestic industry is strictly limited. It is further limited by the fact that many of the small farmers are in debt, and any unforeseen event, such as a flood or crop failure, leaves them destitute.

A further characteristic of the low productivity of our rural sector is the inability of agriculture to meet the growing demand for food by the increasing population. The rate of growth in the demand for food exceed the rate of growth of food production. Consequently, many basic items have to be imported from abroad. Such imports absorb a large part of the limited resources which these countries could have otherwise utilized for capital formation and development activities.

Incomes at the farm level are determined by a host of factors that include the quantity and quality of inputs such as land, labour and water, the technology used, the prices received for outputs, and the prices paid for inputs. While the relative use of inputs varies widely, land remains the most important of the factors of production that determine levels of outputs and income. Studies indicate that most of the small holdings are used for traditional low-yielding subsistence production, and that very few farms of less than two hectares of land, producing traditional crops, generate incomes above the poverty line (Uyanga, 1975). This situation underscores the need for far-reaching reforms in agrarian structures and relationships, if agricultural productivity and rural incomes are to be improved.

Another factor which directly affects rural productivity and incomes in the developing countries is the role of women in agriculture. In many parts of the area women are heavily engaged in agricultural activities. In Nigeria, for instance, it is the women who are responsible for the production most subsistence crops, for instance, vegetables, rice, cassava, maize and cocoyam, which form the staple food of the family. The man assist the woman only in the heavier operations.

The cumulative effect of low productivity resulting in low incomes in agricultural and other rural enterprises is that the majority of the rural population finds it difficult to meet even their basic needs. Add to this, the relative lack of public services in rural areas, especially community facilities such as roads, water, housing, education, health, electricity and sewerage, and the living conditions of this majority cannot be described as anything but poor. Rural dwellers constitute small-scale farmers, tenants, share-croppers, landless workers and a growing number of young persons entering the labour market for whom no prospects of gainful employment are available in the stagnant rural economy (Essang and Ogunfowora, 1975).

In most cases, this poverty-stricken segment of the rural population is found living alongside the prosperous. The poor suffer from limited access to natural resources, particularly land. But more frequently they suffer because they have little access to technology and services and because the institutions which would sustain a higher level of productivity are inaccessible to them. In many cases, vested interests operate to ensure not only that the benefits of productive activity are distributed inequitably, but also that the poor are denied access to inputs, services and organisations which would allow them to increase their productivity.

Thus, the socio-economic system particularly as it operates in the developing countries is often an obstacle to rural development. It reinforces rural poverty and frustrates the efforts of the poor to move up. In the past, the increase in rural population could have been accommodated by expanding the acreage under cultivation. This may continue to be the case in areas which have an ample supply of land that can be brought into production at relatively low cost, but in most of the less developed areas the opportunities of such low cost expansion have diminished substantially. With an increasing man-land ratio, increases in output and farm income must come from a widespread increase in yields per acre and from the cultivation of higher value crops. To raise the output and incomes of the bulk of the rural poor means that they should have access to suitable technology and to the capital required to utilize that technology.

The development of agriculture in the less developed countries sometimes dates back to the colonial period when the policy was to encourage exports crop production. For instance, in Nigeria, the post-war efforts culminated in the establishment of the Bamenda-Cross River Resettlement Scheme. The post-independent attempts christened "back to the land", ranged from the settlement schemes by the former Eastern Nigerian Government, to the plantation projects undertaken by the Agricultural Development Corporations. "The Operation Feed the Nation"is the last in the series of attempts to boast agriculture by various governments in this country.

Unfortunately this paper is not in a position to assess the successes and failures related to the various Agricultural schemes. Instead its aim is the elucidation of the basic considerations which such schemes have been glossing over for one reason or another. After representing a model of on-the-farm decision-making processes the paper discusses the practical aspects which could be emphasised in the policy context.

## DECISION MAKING IN THE RURAL CONTEXT

The rural sector is a decentralised decision-making system made up of numerous small farmers each of whom receives market information and each of whom is constrained or stimulated by direct and indirect government controls. The market and policy sectors thus form an environment not controlled and not well understood by farmers but to which the farmers adapt by responding to various variables such as prices, tax, and price supports which are exogenous from their point of view. The farmer's rules of adaptation also link his current actions to his past behaviour and performance. From the formal point of view these mean that the decentralised decision-making process within the sector is recursive in nature, myopic, and based on limited information. It is one in which the various farmers' actions depend on one another through the feed-back effects of the market and their own past behaviour.

The fundamental concepts here are decision and feedback. The farmer acts on decisions, generally, in a disequilibrium situation based on limited information and on a severely circumscribed understanding of its total environment. Its environment, which consists of all other farms, markets, political, cultural and institutional organisations, and nature, is affected to some degree by its decisions. The farmer's ideas usually are also modified by his own past actions. As a result, he perceives a new state in the next decision period. It is necessary to treat a part of the environment as an exogenous force which affects the farmers' decision but which is unaffected by what the farmer does.

The complex system of farms and markets is illustrated in Figure 20.1 where the feedback loops connecting a given agent in the present period to its experience (loop 3) and its "environment" (loop 1, 2, 4-12) are indicated.[1] Loops 1 and 2 show the indirect feedback that comes through the market, something that affects financial feasibilities, profit, and commercial consumption goals. Loops 6 and 7 indirectly connect each farm to the outside economy. For purposes of reference, we refer to the system illustrated in Figure 20.1 as an adaptive micro-system.

*Farms and Markets System*

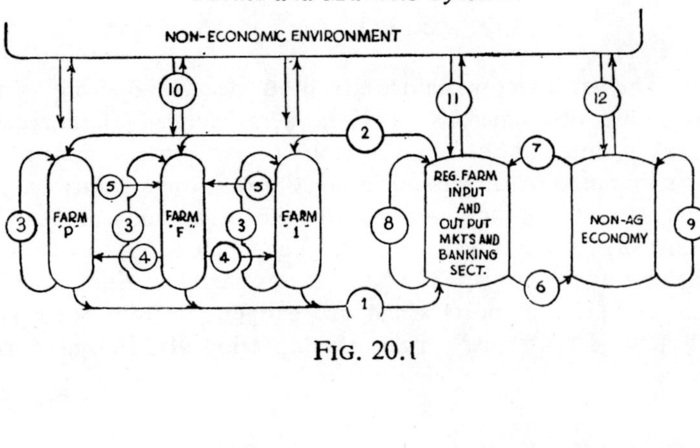


FIG. 20.1

We can construct a general model of how a given representative farmer behaves within this system. It is general because we do not discuss production, investment, household, and other choices in particular, but rather activities in general and how choices between them are made and may be modelled from the adaptive point of view.

Rural activities planned at the beginning of a given production period include preparation, production, sales, preservation, investment, trading and household activities to be undertaken immediately or in future periods. Some activities may be chosen, others not pursued at all. This choice is constrained by technology, that is, by the set of available and perceived opportunities and by resource availabilities and financial considerations. At any given time these constraining factors and the degree of their limitativeness circumscribe the individual's opportunities within a feasible region of activity levels.

Associated with each constraining factor or influence is a constraint function that determines how the various activities are constrained by that factor and a limitation coefficient that describes how limiting the particular constraint is. The latter is inherited from the past and is fixed for the time being. However, their effective limitativeness can be reduced by the choice of certain activities within, of course, the bounds of their own limitations. Thus, labour capacities inherited from the past can be augmented by the hiring of more labour. The supply of money and seed available for investment in the farm can be augmented by the borrowing of funds. However, investment in a given year is limited by reinvestment funds and borrowing by lenders' credit rationing rules so that the choices in any one year are bounded. Adaptation in response to feedback from past decisions and the external environment has the effect of modifying the feasible region from one period to the next. For a given period, it is fixed.

When all of the technical and financial feasibilities are accounted for, one has a set of feasible alternatives on which rational (or arbitrary) choice may operate. It is conventional to assume the existence of a complete preference structure over all alternatives as a fundamental axiom. However, a possibly more realistic concept of rationality is "target planning" (Cyert and March 1953).

Target planning is based on two elements : first, a desired choice or set of choices, that is, a target; second a concept of distance from the target that defines or describes how "close" to the target a given choice is thought to be. This distance becomes disutility function to be minimized or alternatively its negative and becomes a utility function to be maximized. If the target is feasible or contains a feasible subset, then an element from this subset is chosen; if not, the closest is selected. Thus target planning is formally analogous to constrained economic optimizing with utility functions.

Target planning is illustrated in Figure 20.2 In Figure 20.2(a) the target cannot be met. The solution of the planning problem is then the unique feasible alternative closest to the target set. In Figure 20.2(b) the target and feasible region intersect so that there are many "best solutions". Thus, choices equidistant from the target are equally "good" according to the target criterion. Adaptation in response to feedback has the effect of modifying the location and "shape" of target set, but we suppose that for purposes of formulating a given plan the target is "given" by some adaptive rule based on information about the past.

Behavioral theorists suggest that decision-makers choose good enough solutions when they can;and only when they cannot achieve a "satisfactory" level do they go all out to maximize. This important hypothesis of satisfying behaviour can be readily accommodated into the target planning framework by posting a "close enough distance, aspiration level" or "satisfying level" which defines an acceptable choice in relation to the target (Simon, 1972).

Faced with extreme uncertainty about the physical and market environment, real rural decision-makers may be described as behaving cautiously. One approach to modelling cautious behaviour is to follow the expected utility portfolio approach in which decision-makers are assumed to behave like sophisticated gamblers on the basis of well defined "odds" or probabilities. An alternative approach (Day, 1971) uses the concept of target planning just defined. According to this later approach, the adaptive agent forms what he regards, for purposes of the current forming exercise a target of safe-enough decisions, a "zone of flexible response" within which he is

*Target Planning*
(Solution Unique)

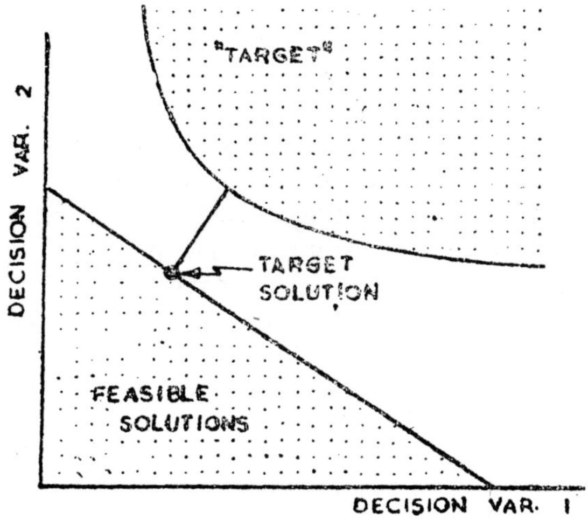


FIG. 20.2(a)

(Multiple Solutions)

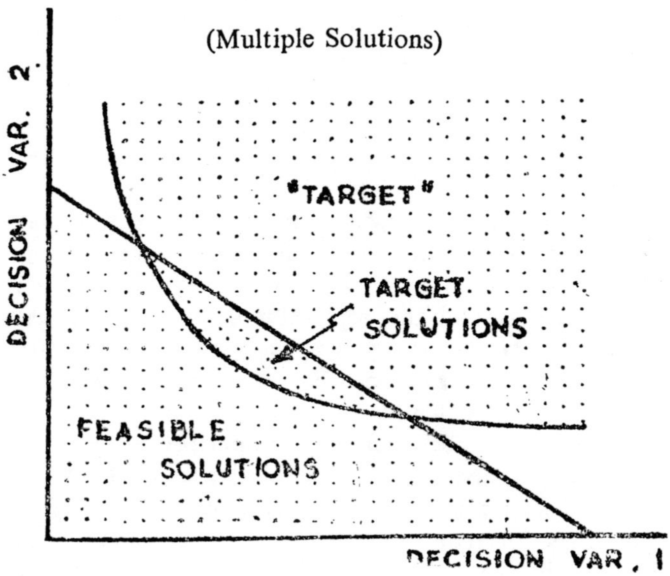


FIG. 20.2(b)

willing select alternatives according to another criterion such as profit or utility maximizing of the usual kind. This zone is determined adaptively by behavioral rules that incorporate previous experience. If the zone contains no alternative that is perceived to be feasible, then the decision-makers chooses a feasible alternative as close as possible to those he regards as safe enough. In this case, caution dominates his behaviour.

Most behavioral scientists regard behaviour as determined by many different goals, utility functions, or targets and hold that these are arranged according to some (perhaps temporary) hierarchy or priority order. A first goal dominates comparison of alternatives until a satisfactory solution is obtained according to it; then a less important goal is maximized over the remaining alternative and so on until a single unique choice is reached and no more "room" for considering further goals is left. This concept is of considerable use in understanding micro-economic behaviour. It explains for example why farmers in one stage of development might behave very much differently from individuals in other stage of development and why many of the considerations that dominate rich farmers' choices often seem trivial to the poor (Georgescu-Roegen, 1954; Ferguson, 1965).

The concepts presented here are meant to describe the behaviour of the adaptive rural agent, the peasant farmer in his choice of activities for a given period of time. At the beginning of each farming period, he must observe environment, reflect on his experience and then with as much information as he can effectively exploit, he must determine his constraints, his goals and his aspiration level for each of them and proceed to make a new choice according to his current, now modified situation. A complete adaptive theory then must describe how the feasible region, the targets, and the aspiration or satisfying levels are all modified adaptively. It is safe to say that relatively little is known about this adaptive process and it is here that we can provide little theoretical guidance for policy modelling.

Some sources of feedback are straightforward. For example, the effect of investment on capital accumulation is well understood in economic theory and is the basis indeed of much of what passes for dynamic economics. Financial constraints introduce considerable complications beyond those involved in physical resource limitations for they depend on current price and profit expectations, which depend on past experienced returns, which in turn depend on total sales and purchases in the region, which finally depends on what each and every ruralite has done in the past.

Obviously, it is impossible in quantitative work to develop an empirical micro-systems model of the kind described so far for an entire economic sector. In agriculture, thousands upon thousands of individual micro-components would be required. To overcome this problem interactions may be suppressed, and a few type or a single model for the sector as a whole may be employed. The greatly simplified structure of decision and feedback, which is referred to as an adaptive macro-system, is shown in (Figure 20.3).

*The Adaptive Macro-system*

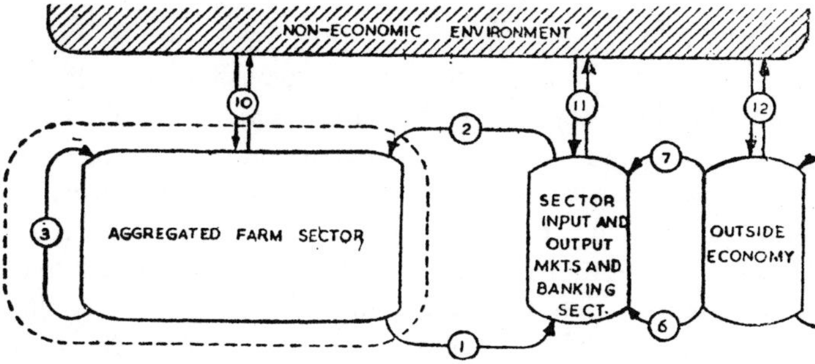


FIG. 20.3

It is often useful to model a given setor somewhat coarsely, the outside economy may be treated as exogenous in which case the analysis focuses entirely on the internal structure of the sector and on its adaptive response to outside influences.

When one treats the non-economic environment and the outside economic linkage sectors and suppresses inter-agent interactions one obtains an open adaptive macro-systems that represents the behaviour of all aggregate of adaptive decision-makers in response to outside influences which are themselves not a part of the model. We thus arrive at the sub-system within the dotted line in Figure 20.3.

The question that must be raised in this connection is, therefore, under what conditions of technology and behaviour will a model, perceived as an open adaptive macro-system truly represent or approximate with reasonable accuracy the aggregate implications of the more complex, adaptive micro-system for which it is the supposed analogue? Progress has been made in answering this question but the analysis is highly technical. Briefly, two approaches have been suggested. The first is to discover how much alike rural activities must be in order that a single model, applied to this aggregate of their resources, will predict the same aggregate allocation as one would obtain by solving a model for each individual firm and then adding up the results to obtain the sector aggregate.

If the technology is linear, it is found that farms need not be identical, but they must be similar in the same sense that we say triangular or other geometric figures which are not of the same size are similar. This approach leads to a representative firm aggregate approach in which all farms estimated to be similar or almost similar are grouped together. A second approach suggests that even though the farmers envisaged may not be technically similar, they behave as if they were through a socio-economic process of managerial leadership and imitation. The theory of such an approach involves explaining how decision-makers not skilled in the optimal selection of complex choices might prefer to imitate the choices of others who they estimate to be more rational than they are themselves. Imitation is then argued to be the optimal course for many even though the managerial leaders whom the many imitate would, in their followers' shoes, do something quite different than they do in their own situation. Either or both of the two lines of thought support the use of an adaptive macro-system whose structure is determined by a theory of behaviour at the micro level but which is empirically estimated and simulated for the sector as a whole.

## PRACTISING THE MODEL

To operationalize the model to reflect the adaptive economic theory presented, the following components must be specified :

(1) Technologically available production alternatives—old and new,

(2) Financially and agriculturally feasible alternatives,
(3) Current limits on adopting new rural activities and farming method,
(4) A set of goals pursued by farmers,
(5) A set of satisfying levels for each goal,
(6) A hierarchy among the several goals,
(7) A behavioral feedback structure that describes adjustment of current goals and constraints to current conditions and past experiences,
(8) Physical feedback structures that relate current resource and financial availability to past investment and financial activities and to market and environmental conditions, and
(9) Indirect feedback to past regional behaviour through the market on current price and budget expectations.

Given the acute data famine in many developing countries, the best that can be done is a fair knowledge of these parameters in their interactive role in the process of rural decision-making; to describe, analyse and assess their effects on rural development—but so far what has been done in this area in policy context is grossly inadequate. For this reason, the rest of this paper will not attempt measuring the unquantifiables, but will be devoted to appraising how some of what has been presented above could be viewed by policy markets.

Most important among these concerns not only more research but also giving priority to uncovering the "myths" of decision-making and adaptive processes among ruralities. These constitute the bottlenecks to rapid agricultural development to the extent that most rural policies are definitely misdirected. For instance, an agricultural development policy could involve :

(i) A sound knowledge of each of the existing farm systems in the area, why peasant farmers behave as they do, and what are the cultural, economic, technical and social constraints on the expansion of output;
(ii) A knowledge of the new inputs-methods tools, systems, services, institutions which can relax those constraints; and

(iii) The means to demonstrate to the farmer that all of this is both possible and rewarding—how do you motivate the farmer to adopt new technology ?

Virtually nowhere in the developing countries have these three conditions been fully met. In most cases, the mass of the farmers have been barely touched by new ideas. The priority research issues can be briefly analysed as they relate to the basic requirements of agricultural development.

**The Economics of Farm Systems and within that the Economics of the Smaller Farmer and Land Tenure System**

Before a peasant farmer is "forced" in the direction of increased productivity, it is necessary to know the exact nature and rationale of the existing farm system. An incredibly high percentage of effort so far expended in has been based on inaccurate or incomplete awareness of these systems. There are relatively very few reliable farm management studies in which appraise the inputs of farmers (financial, technical and labour), outputs and productions, the seasonality of the process and management practices and decision-making. Without this knowledge, extension and technical research are severely emasculated. There are many reasons for the poor policy attention to the economics of peasant farming. Perhaps the most important has been the almost exclusively technical orientation of agricultural developers. The analytical sociology and economics of peasant farming have been virtually ignored.

The typical organisation of agricultural staff has also often precluded a proper appraisal of farms and their constraints. The lowest level extension workers, those with immediate contact with the farmers have been undereducated in these respects, besides each has hundreds of farmers to deal with. They are usually not competent to appraise the problems of even the few farmers they may be in touch with. But even if they were, communications within agricultural departments and rural development agencies are normally downward not upward. The most senior officers who have often had sufficient field experience and established support with farmers, issue directives and the lower echelons try to carry them out—the flow is downward, the feedback is poor.

**The Economics of Motivation and Incentives to the Rural People, and how the Farmer Views his Own Development Problems**

An examination of the growing maize of planning policy, documentation, of schemes, programmes, projects and so on all designed by agencies and department to raise output and boost growth in the developing countries, strikes one with the growing gulf between ambitions and efforts, on the one hand and the meagre results on the other hand. Policy makers politicians, planners and administrators, seem to be attacking the problems or bottlenecks as they see them. But that is rarely how the farmer sees them, almost everywhere the rural people and the official see the same problem differently. Most farm projects fail because the farmers could not relate the effort to their own aspirations, their own resources, or their own sense of economic viability. There is another dimension to the problem. For a variety of reasons, some farmers and their communities are riper for development than others. We require some methodologies of rapid survey and appraisal which will tell us which farmers or villages are the most prepared to innovate and change socio-economically.

The concept of "frustration gap" could be a useful starting point in this respect and can be linked to the concept of "target planning" earlier discussed. For example, a typical rural area has needs for a certain level and composition of income. It obtains these from a variety of sources. Over time the pressure of the village population on its resources will increase; that means an increase in the felt needs. If there are alternatives to meeting these felt needs and pressure regardless of the rate of population increase then there is no frustration gap, otherwise the gap increases. This gap may be a function of the level of expectation (Mabogunje, 1970). Should that gap get too large of course for very many people, many social problems may arise. If it is very small, then no amount of persuasion or pressure will induce the villagers to adopt new production systems. Many livestock societies in Nigeria are in this category so are the terribly cultural bound villagers of the Cross River areas of Nigeria. Development policy for them should emphasise first the raising of their aspirations.

## BASIC CONSIDERATIONS FOR EMPHASIS

In general ideas and opinions differ regarding agricultural models and development policies. Some people believe that existing technology will provide the basis for dramatic increases in agricultural output in many underdeveloped areas. Others feel just as strongly that the necessary prerequisites cannot be put in place in a short period of time even under the most favourable conditions. This school argues that appropriate tehnology is not available and ready for immediate application. These prerequisites can be likened to capital investment, although expenditures for some purposes are not usually defined as investments.

Policy-makers agree that agricultural technology does not transfer readily from one culture to another, and may not always be transferable from one sub-culture to another within the same society. There is rather widespread agreement that modern agricultural technology must be "adapted" before it can be incorporated in agricultural production in many countries. But there is little agreement on the nature or the difficulties of this "adaptive" research, on its qualitative requirements on the time it will require or on the volume of "adaptive" research that will be needed.

Investment that will increase the supply of land available for cultivation are ruled out of their approach. The problems normally considered are how to increase output-input ratios for the most common inputs-land, man-hours of labour, water, fertilizer, productive livestock and so on.

The discussion which follows has in focus some factors in the long-run measures to transform traditional agrarian structures in the typical developing countries. These include:

(a) Human and material investments which will rule out other investments that would also contribute to gross domestic product in agriculture.

(b) Measures to improve the environments of decision-making, especially measures that improve the incentive system bearing upon agricultural production decisions.

All investments have opportunity costs and these aimed at improving the decision-making environment have one thing in common, both involve information.

New knowledge is created by basic research in all fields of knowledge. It is almost always an indispensable input to applied research. Applied research produces what we have chosen to call "new technology", which becomes incorporated in the social production process through at least three paths :

(a) By means of development processes that are carried out largely by private enterprise, new technology is embodied in capital goods and services that will be used in private and public economic activities.
(b) A second path arises because the application of these more productive goods and services often requires the services of more skilled human resources, although the total human effort required per unit of output is likely to be diminished.
(c) Another class of new technology consists of new knowledge that must be adapted to particular production situations before it can be incorporated directly in a production process.

An information dissemination stage may also be required. This function should be performed by the Cooperative Extension Services. Adaptation of new agricultural technology could be combined either with applied research or with the dissemination of information. In many instances, new technology will become fully integrated into a production process through being incorporated concurrently into a land reform policy capital goods, additional human skills, and into widely disseminated information, concerning possibilities and advantages of new techniques.

An underdeveloped country like Nigeria can import some but not all of the needed basic knowledge from developed countries. What can seldom be borrowed is knowledge of the society itself—information on its social, cultural, administrative, political, and economic structures and on its natural and human resources. Particularly important for agricultural development is knowledge about agrarian structures. The potential results from imported technology may not be reached until this information is available and in the right hands. New technology—the output of successful applied research—can seldom be applied directly in an underdeveloped society such as studied. Information adaptation should be the only connector leading away from the stock of new technology. Those involved in development. Educational and dissemination processes would obtain inputs of information from the stock of new technology in its unadapted form.

The important considerations then are:

(a) What volume and quality of human resources must be allocated to these information processes, and to human capital formation ? How will information be properly disseminated ?

(b) How should this investment capital and high-level human capital be distributed among these information processes and human capital formation ?

(c) What length of time and socio/political structure are required for inputs into these information processes to work their way through the structure and produce changes in traditional agriculture and large increases in agricultural products ?

In the developing countries the factor most limiting the expansion of the information processes is the lack of well-trained, dedicated, and creative individuals. These processes must share the short supply of trained people with the unavoidable requirements for highly trained people to train more highly trained people.

The second basic consideration involves public and private investments, which may improve the quality existing inputs, or make available new inputs of a type not in general use before. One example of input improvements is the education and training of the labour force in agriculture, or of the youth that will be the agricultural labour force of the next generation. New inputs might include pesticides, mechanical equipment, marketing facilities, and so on.

These are considered below in general terms.

**Genetic Inputs**

More than any other single programme element, better plant varieties and more productive breeds of livestock can probably pave the way for larger increases in output. Nearly 36 per cent of the 1939-1961 increase in corn yield per acre in the United States was attributable to corn hybridization, 31 per cent to fertilization, 18 per cent to changes in regional specialization, and 15 per cent to other sources, Head et al, (1964). It seems likely that some proportion of the gains due to fertilization, to regional specialization, and to other sources could be attributed to interactions between fertilizer and better varieties, between regional specialization and better varieties, and between other sources and better varieties respectively. Crop yield increases due to genetic factors are potentially greater in the case of crop hybridisation

There is little conflict between the long and the short run in the development of better genetic materials. A long-run research programme is likely to spin off useful results almost from its initiation. Testing available varieties under local conditions will produce useful results quickly and is an essential step in achieving genetic progress in the long-run. The long-run goal is to develop a steady flow of improved genetic materials for all biological processes that have economic significance. Among the activities that have to be established are basic genetic research, applied genetic research, higher education to provide the human capital for the needed research and for educating the educators, enterprises to multiply the genetic material produced by the research, distribution systems for the multiplied material and for information on its efficient use, credit to make optimum use possible, and so on. Each of these activities has its own requirements for inputs flows, especially for human resources. These flows will have to be enlarged and maintained.

**Soil Management**

Problems of soil management including fertilization are difficult if for no other reason than the quantity and quality of the needed information that is available. Gains from fertilization come only after much information on soils had been developed by means of research, and only after improved varieties had not only been developed but also distributed.

TABLE 20.1

*Results of Linear Programming Models Using Different Levels of Technology in the Zaria Area of Northern Nigeria*

| | | Linear Programming Models | | | |
|---|---|---|---|---|---|
| | | Indigenous technology | | Improved technology | |
| | Average farm A | Labour restriction (1) B | Labour restriction (2) C | Labour restriction (1) D | Labour restriction (2) E |
| Land availability (acres) : Upland | 8.1 | 8.1 | 8.1 | 8.1 | 8.1 |
| Lowland | 1.0 | 1.0 | 1.0 | 1.0 | 1.0 |
| Results : | | | | | |
| Total labour used (hrs.) | 1753 | 1543 | 1822 | 1588 | 1996 |
| Months in which no surplus labour is available | — | Apr., June, July, | May, June, July, Nov. | Apr., June, July, Nov., Jan. | May, June, July, Nov. |
| Cultivated acres : | | | | | |
| Indigenous technology Sole crops (lowland) | 0.8 | 0.5 | 1.0 | 0.6 | 1.0 |
| Sole crops (upland) | 1.2 | 0.9 | 1.8 | 0.0 | 0.0 |
| Mixtures (upland) | 5.3 | 4.7 | 4.5 | 4.6 | 5.5 |

| | | | | | |
|---|---|---|---|---|---|
| Improved technology | | | | | |
| Sole crop (upland) : | | | | | |
| Sorghum | — | — | — | 0.8 | 1.8 |
| Total (acres) | 7.3 | 6.1 | 7.3 | 6.0 | 8.3 |
| Net return | 173.36 | 185.61 | 218.17 | 187.60 | 222.67 |
| Per cent increase in net return over B | — | — | 17.83 | 1.01 | 20.00 |
| Are food needs satisfied | Yes | Yes | Yes | Yes | Yes |

(1) and (2). The monthly labour distribution reflected the degree of agricultural activity. With labour restriction (1) the labour availability was the same as that actually used on the average farm. With labour restriction (2) labour availability in each month was the non-family labour actually hired in each month on the typical farm, i.e., a Plus the time actually spent by family members on farm A during June, the peak labour month for family labour. In other words it assumes that in any month family members will be prepared to spend as much time on farm work as they do during June. Sour : (Nor, 1970).

Neither development could be characterized as instantaneous nor can it be achieved by second-line scientists. The recipe for using chemical fertilizer as a means for increasing output parallels the use of genetic inputs, and begins with soil research. Human resources are indispensable and the production period for the human inputs is longer than for other inputs.

Chemical fertilizer plants also require time to establish and to get into efficient operation, but the gestation period for adapting basic soils information for useful application in an underdeveloped country is probably longer and may be much longer. As with genetic inputs, the complementary activities for soil management are many and important. Fertilizer manufacture and distribution, simple soil testing devices, and information dissemination are quite essential if output is to approach its full potential. These should be highly decentralised so as to deal with farmers at the grass root level.

### Water Management

Many agriculture regions in the Cross River Basin must rely on irrigation, where water management takes on great importance. As with other factors, the cutting edge for improvement lies in information. Much more knowledge is needed on the physical relations between varieties and various applications of fertilizer and water. The large gains probably lie in the interactions rather than in the direct effects of a single input or a single changed practice. Water is seldom treated as a limiting factor in designing irrigation systems or the supporting institutions.

### Pesticides

The potential yields from better plant varieties with plant nutrients and moisture in adequate supply can be achieved only if weeds, diseases, insects, and other biological predators are kept in check. Genetics can assist with disease-and insect-resistant varieties, and biological controls may become much more useful in the future. Yet chemical controls may have greater immediate utility in the study area. The pressing needs are for more information that will come from "adaptive" research, and for the materials themselves to be readily available when their usefulness has been clearly indicated. As before, information dissemination, credit, and the availability of necessary inputs assume great importance.

**Product Marketing and Regional Specialization**

Some areas in developing countries have tremendous potential for specialization. If regional specialization in agricultural production is to increase agricultural output and economic welfare substantially marketing of agricultural products must be made more efficient and will require more resources. A high degree of specialization requires much larger exchanges of products among different regions. Two dimensions of efficiency are urgently needed. Product marketing must be efficient in its use of scarce resources, particularly capital resources. Secondly, there needs to be some certainty that the exchanges will take place; breakdown in market exchanges among specialized regions can subject populations to serious risks of food shortages or starvation.

**Mechanical Cultivation**

Complete mechanization of agricultural production in developing countries can hardly be considered seriously. With large masses of unemployed and underemployed labour, there are situations when mechanical equipment makes increases in output possible that can be achieved in no other way. Uncultivated lands may be brought into cultivation. Double- or triple-cropping may be made feasible by more tractors and machines which speed up work. Mechanisation must be related to credit facilities, servicing, information dissemination and land tenure systems.

**Adoption of Improved Technology**

Improved technology improves the efficiency with which conversion takes place. Types of improved technology may be classified into (1) Chemical, e.g. fertilizer, insecticides, herbicides, fungicides etc., (2) Biological e.g. seeds, planting materials, breed of livestock etc., and (3) Mechanical e.g. oxen, tractors etc. (Krishnaswamy, Norman, 1976). Chemical and biological are in general yield increasing in nature. Fertilizer is the most widely used of all these inputs. The level of fertilizer consumption however is still very low. In Nigeria, for example, it is about 1.28 kg per cropped hectare. This is far below the recommended average of 18.18 kg per cropped hectare (3rd National Development Plan 1975-80). The rate of adoption of new practices is very low because most technologies lack relevance to particular situations and farmers are often exposed to a number of problems in the attempt to adopt them.

Some of these which need to be solved are :

(a) Farmers must be convinced of the need to change to new methods. Krishnaswamy and Norman have emphasised that an improved technology is likely to be attractive to farmers if it satisfies their felt needs, it is technically feasible, economically profitable and dependable socially acceptable.
(b) Technological inputs must be available in the right sized units and at the right time.
(c) Most of the inputs of the new technology call for additional costs which often the farmer cannot afford unless has access to a loan. A Guided Change Programme, can be designed to solve these by making inputs available to the farmers and backing this up with a credit and extension programmes.
(d) The resulting increase in ouput may cause a surplus of any given crop. If the demand does not expand, a fall in price would follow. A price stabilization programme or storage facility should be planned or developed for the excess.
(e) If the new techniques are not properly applied, yield reduction may result. Any failure may mean financial ruin or loss of livelihood to the farmer. This calls for thorough testing of new technologies under farmers' conditions before final recommendation.

## INPUT MARKETS AND CREDIT

These are the agricultural elements for which information is important but cannot, however, be generated by adaptive research alone. Grafting imported institutions on the less developed agrarian structures will bring about the unsatisfactory results that have often characterized institution-building efforts. Not only can viable agricultural institutions from advanced countries seldom be transplanted successfully in quite different cultural environments, but the information on which successful adaptations might be based is simply not now available in the western world.

One requirement is basic information on the social, political and economic structure of rural units, including knowledge of motivating forces that might be used to increase individual and group support for the changes needed to increase agricultural output. The Comilla experience in East Pakistan represents at least a limited success in building effect institutions for credit, extension, and marketing in a cultural environment. The crucial element is the considerable effort that went into creating these co-operative institutions. Not least in importance is the patient creation of real self-confidence among the villagers and among the bureaucracies that reach down to the villages. The essential requirements for institution building include information and human capital. These requirements apply to all indispensable rural institutions.

## IMPROVING THE DECISION-MAKING ENVIRONMENT

All the investment discussed above will fall short of their potential increase in productivity unless the incentive system can be reinforced. Many of the development programmes around the world assume that private decision-makers do respond to price incentives to increase output; in no sector of the economy does this assumption prevail more than in agriculture Empirical evidence supports that peasant farmers do respond in a rational manner to their entrepreneurs serving agriculture are less sensitive to economic signals.

If this responsiveness is as pervasive as it seems to be, there is good reason for an analytical re-examination of a great many public policies—transport and storage, processing and marketing, inputs prices and interest rates, product prices, consumer goods prices, taxes and subsidies, import and export regulations, co-operatives, social services, and perhaps many others. Their effects as incentives and disincentives need to be determined rigorously, and this knowledge needs to be incorporated in the machinery for making public policy decisions.

The Government bureaucracy responsible for agricultural development is a second aspect of the incentive system that urgently needs critical and analytical re-examination particularly the bureaucracy connected with rural areas and agriculture. Factors affecting incentives and bureaucratic effectiveness include income, prestige, and rewards attached to different positions; personnel selection and training practices; policies toward higher education; administrative relations among different public agencies; and many other policies. In particular, it seems unreasonable to expect workers in the agricultural bureaucracy to pull off difficult and miraculous transformations if they are at the bottom of the totempole in almost every measure of income and prestige, excepting, perhaps, the frequency with which they are exhorted to do better.

The discussion has been based on the contention that adequate understanding of the aspirations, processes and the basis of decision-making by rural people should be the bedrock of any policies and strategies framed for developing the rural areas in the less developed countries.

## SUGGESTED READINGS

R.M. Cyert and J.G. March, *The Behavioral Theory of the Firm,* Prentice Hall, 1963.

R.H. Day, "Rational Choice and Economic Behaviour", *Theory and Decision*, 1971, Vol. 1, pp 30ff.

R.H. Day and I. Singh, "A Dynamic Model of Regional Agricultural Development", *International Regional Science Review*, 1975, Vol. 1, No. 1, pp 27-48.

S.M. Essang and O. Ogunfowora, *Plantation Agriculture and Labour Use in Southern Nigeria*, University of Ibadan Press Nigeria, 1975.

C.E. Ferguson, "The Theory of Multi-Dimensional Utility Analysis in relation to Multiple Goal Business Behaviour: Synthesis", *Southern Economic Journal*, 1965, Vol. 32, pp 17-24.

Georgescu-N. Roegen. "Choices, Expectati ns and Measurability." *Quarterly Journal of Economics*, 1954, Vol. 68, pp 35-50

E.O. Heady and L. Auver, "The Contribution of Weather and Yield Technology to Changes U.S. Corn Production 1939-61." *Weather and Food Supply*, CAED Report 20, Iowa State University, Ames, 1964.

A.L. Mabogunje, "A Typology of Population Pressure on Resources in West Africa" in W. Zelinsky, L.A. Kosinki, and R.M. Prothero, (eds.) *Geography and a Crowding World,* New York, Oxford University Press, 1970, pp 114-128.

D.W. Norman, *Inter-disciplinary Research on Rural Development. The Experience of the Rural Research Unit in Northern Nigeria* : Overseas Liaison Committee, American Council on Education, 1974.

H. Simon, "Theories of Bounded Rationality" In C.B. McQuire and Rander (eds.) *Decision and Organisation,* North Holland Publishing Company, 1972.

J.T. Uyanga, "Dimension of Unemployment and Superfluous Labour in Rural Cross River of Nigeria". Paper presented at Department of Geography Seminar, University of Ibadan, March 1975.

Raaman Weitz, *From Peasant to Farmers : A Revolutionary Strategy for Development,* New York, Columbia University Press, 1971, p 123.

# Index

# RECENT TENDS AND CONCEPTS IN GEOGRAPHY

VOLUME THREE

## About the Editors

**Dr. R.B. Mandal** (*b.* 1948) teaches geography at R.D. and D.J. College Monghyr and at the Post-graduate Department of Geography Bhagalpur University. He had his education from graduation to post-doctoral work at the Patna University. He has also been associated with research projects sponsored by the University Grants Commission and the Rockefeller Foundation.

**V.N.P. Sinha** is currently on the faculty of the Department of Geography, University of Cape Coast, Ghana. He holds a Ph.D. degree from Patna University.

# Recent Trends and Concepts in Geography

## Volume Three

*Edited by*
Ram Bahadur Mandal
Vishwa Nath Prasad Sinha

**CONCEPT PUBLISHING COMPANY PVT. LTD.**
**NEW DELHI-110059**

**ISBN-13: 978-81-7022-853-0 (Set)**
**ISBN-13: 978-81-7022-854-7 (Vol. I)**
**ISBN-13: 978-81-7022-855-4 (Vol. II)**
**ISBN-13: 978-81-7022-856-1 (Vol. III)**

First Published 2016

*Published and Printed by*

Concept
Since 1974

**Concept Publishing Company Pvt. Ltd.**
A/15-16, Commercial Block, Mohan Garden
New Delhi-110059 (India)
*T*: 25351460, 25351794, *F*: 091-11-25357109
E: publishing@conceptpub.com, W: www.conceptpub.com
**Editorial Office:** H-13, Bali Nagar, New Delhi-110 015, India.

**Cataloging in Publication Data**--Courtesy: D.K. Agencies (P) Ltd. <docinfo@dkagencies.com>

**Recent trends and concepts in geography** / edited by Ram Bahadur Mandal, Vishwa Nath Prasad Sinha.
3 volumes cm
Reprint. Originally published: 1980.
Includes bibliographical references and indexes.
ISBN 9788170228530 (set)
ISBN 9788170228547 (vol. I)
ISBN 9788170228554 (vol. II)
ISBN 9788170228561 (vol. III)

1. Geography. I. Mandal, R. B., 1948- editor. II. Sinha, V. N. P. (Vishwa Nath Prasad), editor.

G62.R43 2016 DDC 910 23

# Preface

GEOGRAPHY is a unique science in the sense that it integrates various disciplines, encompassing as it does the physical, economic, social, political, historical and medical sciences in relation to man, space and environment.

With the widening scope and increasing dimensions of geography, both in respect of its conceptual and systematic branches, today, there is available a huge amount of literature on this subject. Considerable changes have been made in the methods of study adopted by modern geographers who have spared no effort to make theirs as exact a science as possible.

In trying to make choices between old fashioned and modern geography, descriptive and quantitative geography, spatial analysis and inter-disciplinary approaches and so on, geographers today reveal an eagerness to enlarge their horizons and to keep abreast with the latest developments in the subject.

There is also visible a good deal of enthusiasm for deeper research into topics already touched upon with the application of newer techniques to geographical findings such as statistical methods, model building, remote sensing, the use of data matrix, and a host of other sophisticated tools. All this adds up to what can be termed as a conceptual revolution in the field of geography.

What is important today is the dissemination of such knowledge so that both concepts and techniques can be studied and utilized by an ever widening group of geographers and not limited to a coterie of professionals and specialists. It is with this purpose in mind that the essays in these three volumes have been collected together and published. The volumes readings suggested at the end of chapters as well as the references, provide a comprehensive bibliography of research materials.

The three volumes in this series cover the conceptual, physical, resources, agricultural landuse and human geography aspects in their broad outlines. Woven into the main fabric are the social, political, historical and allied problems which interact and provide complexity to the main themes.

All the three volumes contain brief introductions to each part which gives the reader glimpses into the topics dealt with in each of them.

It is hoped that this unique collection of essays will serve the purpose of enlightening the reader regarding the kind of research and the variety of topics which are being taken up by modern geographers today.

The extensive use of maps, figures and diagrams will help illustrate the problems that have exercised the minds of scholars and researchers in this wide-ranging subject.

These volumes have been made possible because of the unstinting cooperation of a host of people too numerous to acknowledge individually. Gratitude must be expressed however, to Prof. P. Dayal, Head, Department of Geography and Dr L.N. Ram, from the same department at Patna University for their valuable suggestions and the latter for sponsoring this project on behalf of the Geographical Research Centre, Patna. The credit goes to the contributors who contributed scholarly papers on a variety of topics with great enthusiasm in spite of other urgent engagements. To friends and critics like M. Jha, G.N. Singh, P.N. Mandal and Suresh Kumar go special thanks for constructive suggestions and generous help at all times. Without the cooperation of all of them these volumes would not have seen the light of day.

MONGHYR
*18 May 1979*

RAM BAHADUR MANDAL
VISHWA NATH PRASAD SINHA

# Contents

# List of Contributors

1. *Adhikari, Sudeepta,*

Lecturer in Geography, Patna Women's College, Bailey Road, Patna, India.

2. *Amoah, F.E.K.,*

Senior Lecturer in Geography, University of Cape Coast, Ghana.

3. *Brahmachari, S.K.,*

Department of Geography, R. D. & D.J. College, Monghyr, India.

4. *Devi, Sunita,*

Lecturer in Political Science, B.R.M. College, Monghyr, India.

5. *Gwebu, Thando D.,*

Department of Geography, The University of Zambia, P.O. Box 2379, Lusaka, Zambia.

6. *Lal, Hira,*

Lecturer in Geography, Gorakhpur University, Gorakhpur, India.

7. *Jain, N.G.,*

Lecturer in Geography, Nagpur College, Nagpur, India.

8. *Kulkarni, K.M.,*

Department of Geography, Gujarat University, Navrangpura, Ahmedabad, India.

9. *Kumar, Alok,*

Research Scholar, Department of Geography, Patna University, Patna.

10. *Kumar, Capt. Ashok,*
Neelam Road, Monghyr, India.

11. *Lal, M.M.,*
Department of Zoology, Bhagalpur University, Bhagalpur, India.

12. *Mandal, R.B.,*
Lecturer in Geography, R.D. & D.J. College, Monghyr, India.

13. *Mishra, B.P.,*
Lecturer in Geography, R.D.S. College, Muzaffarpur, India.

14. *Najma Khan,*
314, Godavari Hostel, Jawaharlal Nehru University, New Delhi, India.

15. *Routray, J.K.,*
Department of Geography, Utkal University, Vani Vihar, Bhubaneshwar, Orissa, India.

16. *Sharma, N.,*
Reader in Geography, Mithila University, Darbhanga, Bihar, India.

17. *Sinde, S.D.,*
Department of Geography, Shivaji University, Kolhapur, India.

18. *Singh, B.P.,*
Department of Geography, B.H.U., Varanasi, India.

19. *Singh, D.K.,*
Reader in Geography, Utkal University, Vani Vihar, Bhubaneshwar, Orissa, India.

20. *Singh, R.D.,*
Lecturer in Geography, T.N.B. College, Bhagalpur, India.

21. *Singh, R.L.,*

U.G.C. Scientist, Department of Geography, B.H.U., Varanasi, India.

22. *Singh, Rana P.B.,*

Lecturer in Geography, B.H.U., Varanasi, India.

23. *Sinha, B.L.,*

Bhagwan Bazar, Chapra, Bihar, India.

24. *Sinha, K.K.,*

Lecturer in Zoology, R.D. & D.J. College, Monghyr, India.

25. *Sinha, S.P.,*

Department of Zoology, Bhagalpur University, Bhagalpur, India.

26. *Uyanga, J.T.,*

Senior Lecturer in Geography, University of Calabar, Calabar, Nigeria.

27. *Violita Sister, A.C.,*

Lecturer in Geography, Patna Women's College, Bailey Road, Patna, India.

# List of Figures

# HUMAN GEOGRAPHY

# Introduction

HUMAN GEOGRAPHY is a scientific analysis of the spatial distribution of man, cultural landscape, cultural grouping, cultural phenomena of the earth surface in relation to the physical, economic and social environment. This branch of study is directed towards the analysis of complex problems of human groupings in rural and urban areas and the hierarchy of settlements as human habitation and service centres. It also includes integrative conceptual appraisal and integrated area development and planning. Further, this study tends to analyse the spatial shift of the centres of population, spatial variation in human genetics of different social groups besides the geography of health, geo-politics in relation to state idea, historical perspectives of a region, the geography of music and the spatial analysis of the problems of warfare.

Rural settlement geography is not a new subject but its investigation is a recent phenomenon, especially in the underdeveloped world, particularly for the sake of the resurgence of rural masses and rural settlement planning. The works of Ritter, Ahlman, Demangeon, Ratzel, Blache, Stone, Jordan and R.L. Singh have changed the dimensions of this branch of geography and a whole lot of approaches, methods, principles, model, types and techniques have been included to study the form and types of rural settlements in different parts of the world. The present study is an outcome of the continuous efforts of field work and data analysis relating to the origin and settling process of rural settlements in rural areas with special reference to India.

The functions performed by the service centres for their surrounding zone of influence and for their own consuming population are known as "Central functions" and the centres which perform such functions are known as 'central places'. The Central Place theory has two key concepts—range and threshold of goods. Threshold defines the condition of entry of a business, measured as the minimum sales volume needed to support the business, whereas the range relates to the area occupied by this threshold sales volume. The theory states that the group of functions set a place above the level of towns because the latter have a more restricted hinterland. The theory of central place is not meant only for patterns of cities, towns and villages within larger economic systems but also for the multiple internal pattern of service centres. The central place theory presented by Christaller and Lösch has no temporal dimensions and is generally applicable as theory of retail trade and dealing with the urban functions.

In a paper on review of central place study in India the author has remarked that most of the studies on central places have been duplicated from the west. The criteria adopted for centrality measurement in the west are of limited utility for macro-regional investigation in India due to differences between the urbanization process in highly industrialised and less developed countries. But the author, in an attempt to study the centrality of Mithila region Bihar, has used Davies location coefficient and location quotient in order to get the hierarchy of central places.

The paper on temporal dimensions of urban geography deals with the relevance of the subject, its evolutionary trend, concepts and approaches, and frontiers of research, etc. It is felt that in urban geography studies are made about towns, their location and site, functions, patterns, layout, origin, evolution and hierarchy of urban places. This study is a modest attempt to highlight the recent trends of research in urban geography.

The urban fringe is a comparatively new concept in urban studies, and several geographers, such as Wehrwein, Golledge, Wissink, Pahl, Pryor and others have attempted to define and delimit the fringe of towns. Though their approaches are a little different, they arrive at a common conclusion, viz., that the landuse influx and pattern and the structure of population are the main determinants. Following these lines of investigation in delineating the urban fringe of Bareilly, it was found that the fringe comprises 77 villages around Bareilly within an area of 134.8 km. The development of the fringe has generally taken place along the main arteries in a star-shaped formation extending more towards the west and the north of the city, avoiding physical barriers such as low-lying areas and rugged topography in the south-west and east respectively.

Further, the concept of the *rural-urban fringe* is introduced as a cognisable entity in the settlement continuum. It is a tentative and empirical zone. Its delimitation demands an assessment of its physical, functional and cultural features best expressed in the value of the land. This temporal land deserves due consideration in contiguous urban-rural planning.

The smaller towns or larger villages are known as rurban centres. Most of the smaller towns are overgrown villages and still retain their rural character. They show a combination of economic, social and educational function. While all towns have a city region around them, the small towns function as the economic and social capitals of their surrounding essentially agricultural areas. By virtue of its intermediate location and the nature of evolution, the small town has certain advantages and offers unique opportunities in the general scheme of rural-urban and agricultural-industrial continuum. Thus, the study of smaller urban settlements is highly important 'as they are urban islands in the vast sea of rurality' and thus represent the basic connection between the dispersed agricultural population and the agglomerated urban population. The primary objective of this paper is to examine the factors responsible for the spatial distribution of smaller urban settlements and their role *vis-a-vis* the rural settlements in the Bihar Plain. Such urban units are marked by the growth of certain significant urban functions. In fact they are neither purely rural nor purely urban but a mixture of both.

Nowadays, cities are expanding in close correspondence with the growth of population and diversification of urban functions which has brought about significant morphological changes. Further, the growth and development of new residential colonies on a sectoral pattern in the outer zones of the city have almost aggravated the situation. In this paper a brief account of the neighbourhood concept and its implication in urban planning in the light of modern residential growth pattern have been interpreted.

Considering the higher order settlement, "A Spatial Model of a medium sized Indian city" has been analysed with special reference to Nasik city. It has been realised that the Smailes model of the Indian city appears to be getting outdated as our cities are changing in space and characteristics. Further, it has been felt that the distinction between the indigenous part and the anglicized westernized part is fast disappearing as maps and cartograms of the area suggest.

In discussing the micro-level planning in the Indian context it has been emphasized that the national and state planning without proper diagnosis of the problem and without considering the requirements of the people aggravated the real situation in space and on the economic front. Here the pattern of development in Baudh-Khondmal district of Orissa has been analysed on a micro-level with respect to existing development potential and coordinate spatial and sectoral planning of smaller areas for integrated development.

The paper on "Integrated Rural Development: A Geographical Approach" examines the economic growth and social change after considering the recent trends of rural development through the establishment of *development blocks*, rise in urbanisation, education and agricultural expansion programmes, etc.

"Integrated Rural Area Development : A Geographical Approach with Nigerian Example" examines the role of economic investments in the spatial structure in Nigeria with special reference to the problems of integrated area development. The differential areas of economic development may be determined by (1) areas of very energetic and flourishing rural economy located near major urban centres and the core areas of the major export and food crops ; and (2) the areas of low and passive economic activities located either far away from the major urban centres or near the most parasitic urban centres.

In "Systematizing Spatial Organization in Tropical Africa : An Integrative Conceptual Appraisal" examines the problems of underdevelopment and spatial reorganization in the Third World. The problems are associated with the relocation of administrative centres, construction of efficient communication and transportation lines, land tenure systems, initiation of regional development schemes, migration and urbanization. This paper analyses these components as substructures that are dynamically interactive.

The paper on dynamics of population centres analyses the centrographic method of Hart (1943). This method helped a lot in explaining the regional, national, economic, and population planning. This shows the centre of population on land in relation to population distribution, where the population of all directions is balanced according to the weight of the number of persons.

The purpose of the paper "Population Potential" is to recognise the distribution of population in eastern Uttar Pradesh with the help of a potential model. The resulting map reveals that the maximum number of people are within the reach of the places around Gorakhpur, situated in the north central part of the region. Major concentration of high potential in this part shows the influence of fertility of soil, cultivation of cash crops, establishment of sugar factories, communication lines and habitability over a long period. The population potential surface shows a steady decline in the values of potential away from the highest potential centre. A steep gradient is marked in the north-west direction. Further population potential of the districts has been compared with general population density and the variations between observed and expected density of the area have also been analysed.

Analysing the geo-medical approaches of human health problems within a time perspective, it has been found that the environmental and ecological synthesis is more influential in the causation of diseases. These approaches can be classified into three groups : ecological, spatial, and eco-spatial, among which the latter is the result of the emergence of the former two. Taking this view as the base, the "Geography of Health" can be defined as the proper subset of four factor groups, i.e., environmental, biotic, behavioral and socio-economic.

The concept of work-load provides a fairly objective and useful device for examining the distributional pattern of medical facilities in developing countries like India. The formula devised by medical geographer McGlashin is employed here to identify spatial variations in the available medical facilities in Maharashtra State.

In delineating the nature of political geography, the discussions have been made about states, their frontiers, the boundaries, inter and global relations, contacts and groupings, the variation of political phenomena from place to place considered in relation to other features of the earth as the mainstay of human society. Further, this paper analyses the meaning, scope, concepts, method of study and approaches of political geography.

In historical geography the history of geographical explorations discoveries, reconstruction of past environment as a series of successive sections in time and a sequential and retrospective appraisal of change through time are generally included under its scope of study. This paper deals with the following aspects : scope and methods of historical geography and the current problems and prospects of historical geography.

Music has been affected by geography and consequently it varies from place to place. Broadly speaking, Eastern music differs from Western music in one important respect and that is the absence of harmony. In India, North Indian music differs from South Indian music. Besides this there also exist state to state variations in music. Many of the countless musical instruments are the result of the geography of the place.

"Internal Migration : Concept, Theories and Methods of Analysis" has been recognised as one of the most important areas of contemporary research in population geography. This paper analyses after considering migration under the following sub-heads : concepts, theories, size, direction, motivation, mechanism, differential processes, consequence, typology and methods of migration analysis.

In the field of military geography, the military evaluation of political areas from the strategic, tactical and logistic points of view have to take into account the physical, economic and human elements and their mutual relationship.

The effectiveness with which a government can fulfil the purposes of a state is the result of a balance between two opposed sets of forces, those of integration, leading to unity and coherence and those of disintegration, leading towards division and the break-up of the state. Out of the forces of integration operating in the state, there emerges and develops a distinct political idea which becomes the *raison d'etre* of the state. The present work is an attempt at analysing the various forces of integration and disintegration as well as the state-idea of the Indian state-area.

The study on human genetics of India reminds us that various groups of people are widely diversified in their general make-up, groups of small to big, modest fertilities, variable degrees of in-breeding and there is a variegated range of eco-systems of the Indian sub-continent where the study of man is possible under varying environmental conditions.

Finally, it can be said that geographical studies are now heading towards the inclusion of the pursuits of natural, social, economic, political and historical sciences which are associated especially with the problems of land development and the cultural features of the earth in relation to man.

# 1

# Nature of Rural Settlement Geography

*—Ram Bahadur Mandal*
*—Sanjay Kumar Brahmachari*

RURAL settlement geography is a branch of human geography where the human occupance is studied in detail for the well-being of society in order to know the cultural advancement of people and the man-land relationship of the environment. Settlement as the primary residence of human society is the nerve centre of all geographical studies because transport geography, military geography, population geography, political geography, economic geography and agricultural geography each has an intimate relationship with patterns of human living.

The concept of 'settlement geography' is derived from the German *Siedlungs geographie*[1] as the cultural landscape created by man in the process of occupancy of land is the product of the ecological balance of different variables obtained in the environment. The science of settlement is now-a-days known as *ekistics*, which was originally devised by C.A. Doxiadis[2] of Greece, an architect of Islamabad city in 1969. In this way it is said that rural settlements are the topographic expression of groupings and arrangements of two fundamental elements of human geography—houses and highways[3]—in the rural environment. In this way rural settlements show the reciprocal relationship of human occupance features and environment.[4]

Aurousseau has expressed rural settlements as the arrangement of houses as geographical entities in relation to highways. A rural settlement is in the main, an agricultural workshop which cannot be separated from the land for certain special characteristic features. The shape and arrangement of rural dwellings are often in strict accord with the kind of work, the agricultural technique and the manner in which the soil is used.[5]

This paper aims to analyse the recent trends and concepts in rural settlement geography.

## SCOPE AND OBJECTIVES OF STUDY

Settlement geography includes permanent buildings or houses grouped around agricultural fields along with the temporary camp of nomadic herders up to the town, city, metropolis, conurbation, tyranopolis and ecomonopolis of total world urbanization. Keeping this view in mind, the totality of the human community in rural areas includes the social, material, organizational, spiritual and cultural elements are necessarily required for sustaining of human living. Under physical requirements as housing, work, energy supply, transport, communication, water availability, education, health, protection and social welfare, system of territorial organization, local self-government, law and economic management and cultural facilities for the development of art, recreation and leisure come under this study. Further, with the growth of population, increasing migration from rural to urban areas and *vice versa*, the demand for housing construction for rehabilitation of displaced persons from floods, etc. are making the scope of rural settlements wider and require an integrated approach for proper solution.

Rural habitations are the living and functional territorial space of man located in the country-side since times immemorial and form an inherent part of the life of man. Hence, the objectives of this study can be of multifarious in nature and the development of the discipline of rural settlement geography has been considerably longer in time and space. Therefore, the consideration of habitations run like a thread through almost the entire life of a man wherever he lives and work besides all branches of geographical thoughts.

Permanent settlement is a distinct product of the cultural advancement of the people in the field of agricultural development and technological advancement. The nomadic and semi-nomadic people of early historic times led an unsettled life in the forests of the plain land. In such a situation they depended for their livelihood on hunting, fishing and gathering of forest products in which the permanent settlement was not necessary. With the beginning of permanent agriculture on a fixed piece of land and the origin of settled life began. At present the types and patterns of rural settlements in the region are the product of agricultural organization, the system of cultivation, historic evolution of settlements after forest cutting and cultural advancement of the inhabitants of the area.

Rural settlements are the product of the cultural traits of an area and time which carry within themselves. Some of the hidden characteristic element. They provide us with the synthesis of dynamic patterns of inter-regional and intra-regional character of the cultural landscape.[6] Hence, settlement geography is an emblem of man-made features of the environment. The attributes of settlement have an enormous range of variation along with the physical and cultural factors of an area as an integrating force. Natural resources and economic development are the integral parts of human establishment but the growth in the patterns of settlements, cultural geography, cultural history, cultural processes, cultural ecology, and cultural area and other factors like typology, distribution, variation, spatial pattern, integration, regional patterns and area characterization along with the field patterns are the prominent aspects of rural settlement geography which naturally come under the scope of this study. Further, the conditions of the rural dwelling, transport routes, ruins of settlement, lines of migration of certain ethnic groups, functional analysis of settlements, cultural regions, religious units like temples, mosques and churches, educational and recreational institutions industrial establishments, store-houses, railway stations and bus depots, fences, and telephones, electric and ropeways all are basically included in the scope of rural settlement geography. Later, as the economist thought about the marginal land of villages with a view to exploring some extra resources of the land, similarly, the rural settlement geographer approached the area of the pioneer fringe to learn about the potentiality of new settlements. "The pioneer fringe is a laboratory of exceptional scientific works because there human intelligence is pitted against extreme or unusual environment conditions."[7] Still further Bowman had remarked that "City geography is at one end of the scale of settlement, pioneering at the other".[8]

## THE CONCEPTS OF RURAL SETTLEMENT GEOGRAPHY

The concepts of rural settlement geography are mainly three in numbers : spatial relationship of man and habitat with the environment; environmental relationship of different factors in relation to human habitat and stages of man-land relationship. Settlement in itself is the wealth of a person without spatial entities, but as soon as the comparisons are made between two houses, space plays a dominant role in order to keep such analyses of settlement in geography. Wherever space is involved between two settlements that becomes a part and parcel of settlement geography. One cannot imagine geography without space. Thus, Watson has rightly said that "Geography is a science of discipline in distance."[9] In this way the concept of spatial relationship includes the relationship of settlement with transportation lines, industrial location, agricultural operation, utilization of land, ranching of cattle, technological advancement through educational institutions, defence from thieves and the knowledge about the hazards of climate in different places for the welfare of human society. All these involve the understanding of the traits of settlements of a place, the causes of their location, consequences towards area development and their interaction with their physical and cultural elements available. Settlement geography is, therefore, multi-disciplinary in nature where geographers study the spatial arrangement of houses, sociologists focus attention on social growth, economists study their optimum location for economic gains, while the planners' and mason's interests centre round their optimum environmental condition for planned growth and architectural style of the time.

Under the concept of environmental relationship, it is generally considered that settlement is the product of the ecological balance of different parameters available in the environment like relief features, soil conditions, climatic conditions, forests, availability of water, historical occupancy of the land, transport development, etc. The concept of man-land relationship involves the variation of man and land in different areas of study and their characteristics of living will naturally reflect in the groupings of human dwellings in which they live. For this fact in the flood affected areas of North Monghyr, the size of settlements is large in relation to the population, while in the flood-free areas of South Monghyr, the size of settlements is small. Basically these are the primary concepts of settlement geography but with the increasing population and technological advancement the concepts of settlement geography are related with the following points :

(1) The use of resources i.e. environment, soil, space, time, energy, goods, money, information and others for the best location and distribution of rural settlements over the earth's surface;

(2) The colonization, spread and competition, change in density and growth of settlement depend upon the economic conditions, history of land occupancy and food productivity of the region;

(3) The concepts of settlement geography include types and patterns, origin, growth and mobility of different clan groups, material of wall and roof, morphology of settlements, facilities of the professional services, and the migration of the working force from rural to urban settlements that play a measurable role in the growth of human habitations. Planned development of rural settlement is felt almost necessary for regional economic development and the welfare of people.

The settlement evolved only with the use of resources available locally according to the sources of information of a person and the society. The concept of the evolution of rural settlement also includes colonization, spread and competition of different clan groups in the process of origin and growth of rural settlements. Generally speaking, settlements originated on higher ground amidst fertile agricultural areas where water supply was plentiful and life was secure. In this way rural settlements originated with the emergence of permanent agriculture. The infighting of plundering hordes, migration of clan and the succession of one political power after the other, during historic times, led to the evolution of settlements of varied nature with respect to monumental designs, architectural styles of the time, place names, territorial organization and the union of clan groups into rural settlements.

The concept of density and distribution of population are co-related and their studies are not possible in isolation. The high density of settlements has been found to depend upon agricultural prosperity and suitability of the region for human settlements since historic times. On the other hand, the distribution of settlements may indicate the agglomeration or dispersion of settlements in a particular ecological condition. This gives an idea of types of rural settlements in which the compact, semi-compact, semi-sprinkled and sprinkled settlement forms are generally observed in a regional analysis of Indian rural settlements.

The concept of size and spacing of rural settlements is again correlated with each other, as the one is dependent upon the other. In areas of large size settlements, their spacing is abnormally high while in areas of small settlements the spacing is very close, except hilly, forested and flood affected areas.

The concept of patterns and processes of village morphology include the measurement of the shape of rural settlements variant from the circle, identification of the external shape of settlements from a bird's eye view and internal layout of streets and lanes.

In the concept of house types, the regional variation of houses in relation to the material of wall, roof, construction method, architectural style, cost involved and placings of courtyard, inner verandah, eavesline, female apartment and male apartment are important aspects for consideration. The density of houses, the occupancy rate, and the demand and shortage of houses are also the subject matter of this study.

The concept of regional patterns of rural settlements includes the regional conditions of their density, as also residential houses, settlement types and other elements of settlements in totality.

The concept of the rural service centre includes the primary, secondary and tertiary service centres. The concept of ruban centres, centres of trade and commerce and central place hierarchy all come under the category of service centres. The centres which provide goods and services to their zone of influence are known as service centres. The functions which are performed by a centre for the hinterland are known as central functions, while the functions centred in urban areas are called nodal functions.

The concept of hierarchy of service centres ranges from roadside settlement to megalopolis, fabroblast and sacroma settlements. A roadside settlement is an isolated house of a motel or petrol pump station or eating place for the passers-by. In India roadside settlement is a common feature especially that of a betel shop. A hamlet is larger than a roadside settlement and this includes some residential houses besides commercial establishments. A village is larger than a hamlet with varied functions of agricultural, residential craftsmanship and commercial.

A town is a settlement with marked tertiary functions, daily retail market, large population size, development of educational institutions, industries, hospital, telephone exchange, wholesale market, etc. When a town has reached the stage of 1,00,000 population, it is known as a city and when several industrial cities coalesced both physically and economically, even if they are politically independent they are known as conurbation.

The term *metropolis* is applied to such a city which has a population of 10,00,000. This is also known as millionaire city. When several conurbations and metropolises coalesced, the term *megalopolis* was applied for such a contiguous urban area. On the eastern sea coast of the U.S.A. such a megalopolis is discernible.

The terms fabroblast and sacroma settlements are generally applied to the highly congested and over crowded urban places. Similarly, the terms "growth centres" and "growth poles" are generally applied to developing service centres like small towns and developed service centres like state capitals, in India respectively.

Recently, locational decisions have been generally taken on the basis of locational relationships of rural settlements which reflect an ordered adjustment to the factor of distance.

The concept of the rural settlement system may be seen as an adjustment to a combination of physical, biological and cultural inputs, so that a change in one or more inputs may be expected to lead to a readjustment within the system. When the chronological relationship of different grades of settlements is applied with a certain logical principle, that is called system analysis. A system generally works with the process of a feed-back mechanism, whether that is a close or open system. In 1976 Prof. R.L. Singh of India has given a call to the social Scientists to chart a 'national settlement system' for Indian settlements.

There are some common concepts in settlement geography which have linked the rural and the urban settlements together in a single unifying bond : The separation of rural from urban studies may be permissible in geographical work, but it should not stand as a barrier between the two. For example, it is not possible to analyse the service centres purely as a rural phenomenon, they should, on the other hand, be closely associated with urban studies. One effect of such studies is the central place analysis as devised by Christaller in 1933[10] which marks an important concept of settlement geography. The geometric figures drawn up by him as well as recently by several other geographers, give expression to the theoretical spacing of central places of different ranks in an area or evenly distributed rural settlements. For such studies, therefore, it is essential to exceed the realm of rural settlement geography in order to appreciate and do justice to the subject.

## REVIEW OF LITERATURE

General interest in the study of rural settlement goes back to 1920 when Blache, Aurosseau, Auman, Demangeon, Bruhnes and others were laying the foundations of settlement geography. But the evidence of real interest was seen at the time of the establishment of the commission on "Types of Rural Habitation" under the auspices of the international Geographical Union which published its first report in 1928.

Rural Settlement Geography was originally developed as a full-fledged branch of Human Geography in Europe as a specialized field of study. According to lexicographers the settlement has manifold meanings but for the geographer's interest, it is defined as "The place, where one person or more dwells regularly, or the act of establishing permanent residence.[11] This meaning was accepted for geography, specially by French and German geographers in the nineteenth century. In

Germany the work of Ritter was prominent in this direction and his contributions are the base of settlement geography even to this day. According to Ratzel, German geographers mainly examined two aspects : house types and urban centre. In 1895, Mietzen[12] considered the form of European settlement and laid emphasis on morphological studies and the same was followed by Beresford.[13] In 1899, Schluter considered the definition of settlement with special reference to location, size and growth and their relationship with the environment. Wegener emphasized the importance of dwellings in settlement geography. According to Miss Schwarz, "Rural settlement geography includes the study of men, use of plants, an animal, to procure food and clothing, an economic area larger than the dwelling, and of direct relationship between economic area and the dwellings".[14]

Blache was a French settlement geographer who studied human geography in its entirety in detail. Demangeon concentrated his attention on the studies of field ownership, cultivation, building material, house types and the distribution of towns, rural settlements and human habitations.

In Belgium, Lefevre contributed the ideas of rural settlement pattern and density of rural settlement in between 1921-1964. Aurosseau[15] was an Australian settlement geographer who studied the arrangement of dwellings. I.G. Grano was a pioneer worker of Finish settlement regions who had analysed the form and location of rural dwellings besides hamlet and bigger settlements. In Japan Matui[16] (1932) was a pioneer geographer who has analysed rural settlement geography quantitatively. J. Yonekura[17] is very noteworthy for his several works in the field of settlement geography. Up to the year 1978, he has written about 80 research papers and more than 9 books on rural settlements of Japan, India and Asia.

In North America much work have been done on the arrangement, spacing and size of rural settlements. In 1955 C.J. Galpin[18] a sociologist published his work on rural communities of Wisconsin and devised a new concept of rurban community. The classical centre place theory of Mark Jefferson was summarized and developed by Christaller of Germany and Ullman,[19] Dickinson,[20] Brush,[21] Berry and Garrison[22] and especially Beavon and Mabin[23] have shown the derivation of the market areas of central places in full for different size of regions and they give mathematical solutions through equation as put forward by Lōsch.[24] K.H. Stone[25] has written number of research papers of rural settlements and emphasized the study of settling process, establishing and advancing of rural settlements and the pioneer fringe of Swedish and Finish rural settlements. Jordan has studied the settlement geography as integral part of cultural landscape. Bowman[26] has studied the pioneer fringe of settlement while Kant[27] studied the density of rural habitation in several ways following the technique of different investigators. Hillman[28] concentrated his work on community organization and planning but he touched the subject from sociological point of view. On the other hand, Jacobsen[29] of Israel shows a systematic way for conducting research in the field of rural settlement geography. Chisholm[30] has paid great attention in analysing the relationship between landuse and rural settlement while R.C. Eidt[31] of U.S.A. has worked on abandoned settlement analysis. The work of F.S. Hudson[32] is directed towards the introductory idea of rural and urban settlements.

Recently, the work of J.C. Hudson[33] has emphasized the location of rural settlements through colonization, spread and competition of people and he concludes that these processes are cyclic in nature. D. Grossman[34] of Europe emphasized the theory of location of rural settlements with the process of the migration of people from one settlement to the other. Bylund[35] has presented his ideas of settlement diffusion in space through deterministic models especially applied to the Lappland area of Sweden. Morrill[36] and Haggerstrand has developed probabilistic models of settlement diffusion through stochastic processes and random number.

In India, the contribution of geographers in the field of rural settlement geography is new and in most cases the adoption of ideas is novel in nature. C.D. Despande[37] has analysed types of settlement in Bombay-Karnatak in relation to population size of rural settlements. S.M. Ali[38] has discussed rural settlement types, patterns, location, siting and types of rural dwellings in Ghaggar plain in relation to the distribution of population. E. Ahmad[39] emphasized different aspects of settlement with special reference to U.P. His latest work on settlement geography is related with the patterns of rural settlement and compact settlement zones and their regional distribution. K.H. Buschman[40] has examined different aspects of rural settlements with special reference to India. R.L. Singh[41] has elaborated different facts of rural settlement geography including meaning and scope and objectives of study with special reference to the evolution of settlements in the Middle Ganga valley. A.B. Mukerjee[42] has discussed the spacing of rural settlements with special reference to the areas of Punjab, Andhra Pradesh, Rajasthan and U.P. R. Lahiri[43] studied the cycle of settlement growth of Ajoy valley and traced the successive stages of youth, maturity and old age. A.N. Bhattacharya[44] has examined the effects of caste factor on types and patterns of rural settlements in U.P.

R.L. Singh[45] has also examined the typical rural dwellings in the umland of Banaras along with the work of clan rural settlements. L.R. Singh[46] has analysed various aspects of rural settlements with special reference to U.P. V.N.P. Sinha[47] has published his Ph.D. dissertation which is related to the settlement geography of Chotanagpur plateau. R.B. Singh[48] has studied rural settlement types and their distribution and suggested future trends with special reference to Varanasi district.

From the Banaras School of settlement geography proceedings of 1971 a symposium was published entitled *Rural Settlements in Monsoon Asia.*[49] Other publications are *Readings in Rural Settlement Geography* and *Spatial Dimensions of Rural Settlement Geography.* From this centre a special issue of the National Geographical Journal[50] was published in 1975 devoted to the study of rural settlement geography. Rana P.B. Singh[51] is a dedicated scholar of rural settlement geography whose works are mainly directed towards clan rural settlements.

R.K. Mukerjee[52] published his celebrated book *Man and His Habitation* in 1968. R.C. Sharma[53] has published *Settlement Geography of the Indian Desert* in 1972 and *The Structure of Settlement Geography* is under publication. A. Prasad[54] has published his D. Lit. thesis on rural settlement geography in 1973. So far as the regional study is concerned R.B. Mandal has written his Ph.D.[55] and D.Lit.[56] theses on Rural Settlement Geography with special reference to Bihar Plain and North Bihar respectively. He has also published papers on rurban centres and the ecology of rural settlement[57] with special reference to the Kosi Basin, Bihar.

## APPROACHES OF RURAL SETTLEMENT STUDIES

Four main approaches of studies are employed in rural settlement geography :

**The Genetic Approach** : In this approach the original growth of settlements are generally considered with special reference to location and siting conditions. The settlement may have a location in upland plain or lowland, the form may be compact or sprinkled and the pattern may be circular, square, rectangular, linear, etc. Location guides the function of a settlement whereas site determines morphological characteristics.

**The Sample Approach** : Under this approach a rural settlement is generally studied individually with special reference to location, site, boundary conditions, location of settlement, settlement morphology, land utilization around the settlement, house plans of typical dwellings, material of wall, material of roof, distance from one house to another, caste segregation, skyline of the village, migration of dilierent families to the village and outside the village including the location of roads, bylanes, footpath, village ditch, tank. temple, mosque, school building, industrial establishments, shops, etc. In this way a sample study serves the varied purposes of the study of settlement in particular as a case study approach.

**The Regional Approach** : This approach relates to the study of the spatial distribution of rural settlements in a region including settlement types, density, pattern, spacing, conditions of rural dwelling, clan migrations, diffusion of settlements, growth of rurban community and hierarchy of central places. Only regional conditions have been given a due weight under this approach of analysis and naturally no case study approach is needed at this stage.

**The Principle Approach** : This approach considers the location of rural settlements with special reference to the conditioning factors available in the region under study and the formulation of theoretical ideas like central place theory of Christaller polar and anti-polar concept of landuse and location of rural settlements by R.B. Mandal, location of rural settlements by clan migration of D. Grossman and the location of rural settlements by colonization, spread and competition by J.C. Hudson, etc. Hence, the principle approach helps in analysing the integrated development of settlements and their functional zones under varying forms of landscape which may be a physical or cultural landscape as space and time permit.

### PRINCIPLES OF RURAL SETTLEMENT FORMATION

There may be a number of principles for the formation of rural settlement but the most important are as follows :

(a) The maximization of potential contact;
(b) The minimization of effort;
(c) The optimization of man's protective space; and
(d) The optimization of the quality of man's relationship with the environment.

(a) In any formation of rural settlement the pre-requisite is the maximization of potential contact with the neighbours, because man is a social animal and it is impossible for him to live in isolation without social contact. This contact may take place in the form of clan migrations, marriage migrations, contact with kinship group, friendly relations with the neighbours or it may lead towards administrative headquarters, health centres, industrial establishments, educational institutions, religious centres, etc.

(b) Settlements are generally established at a point where functioning, travelling, availability of water, recreational facilities, employment opportunities and social-cum-business contact may be possible with less expenditure.

(c) In any settlement location the optimization of man's protective space is essential from the point of view of defence—from theft, protection from climatic hazards, outbreak of fire and several others.

(d) Generally speaking, the location, types and patterns of rural settlements are strictly in accordance with the land whose use they ensure and the economic progress and technological advancement of the inhabitants for whom they have been built. But due to natural hazards and the problem of social security, each and every location is not suited for human living. For this, the provision of the availability of drinking water, transport facilities, medical facilities, and the nearness to markets seem to be the essential ingredients of the optimization of the quality of man's relationship with the environment.

## HYPOTHESIS TESTING

Hypothesis determines the relationship of two variables and is accepted or rejected after empirical verification. Friedman defines hypothesis as "it explains and abstracts the common and crucial elements from the mass of complex and detailed circumstances.[58] In this way hypothesis is a tentative explanation of phenomena and this framework of tests is needed for the acceptance and rejection of the relationship of two variables. This is a measure in which the selection and organisation and analysis of data are made in such a way that the rejection of hypothesis may be avoided. The level of significance must be known before the hypothesis is verified for acceptance or rejection. In case a new hypothesis is selected after verification, the same results cannot be obtained with the same satisfaction with other set of data. Old research hypothesis must be listed before the facts are introduced in verification. The concept of hypothesis as presented by Friedman "seems to transcend separable comments and tests. The difference is minor as a 'big hypothesis' may be tested indirectly through a series of lesser assertions. The same rules of verification apply to multiple regression for example, each of the partial regression coefficients is separately tested so as to see its significance in the context of the entire model, which itself is a big hypothesis of relationship between and among a set of variables".[59]

It can clearly be said by common knowledge that the density of population in the rural areas of Monghyr District should have a close relationship with the number of rural settlements in any temporal dimensions. The problem can be restated with the help of a hypothesis that the density of population (dependent variable) in 1971 is dependent on the number of rural settlements (independent variable). This simple relationship is illustrated in a scatter diagram (Fig. 1.1) having

*Relationship of Settlements with Density of Population (1971)*

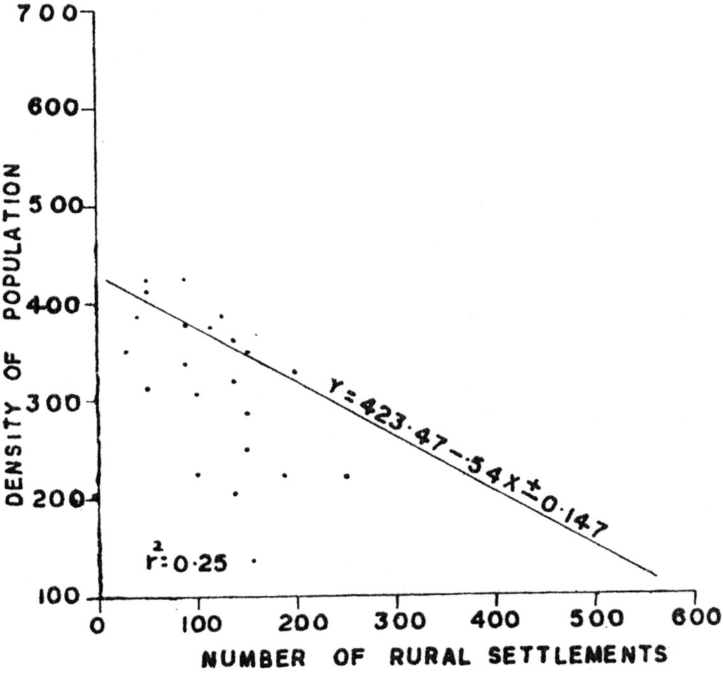


FIG. 1.1

dependent variable on the vertical axis (Y). To know the distribution and magnitude of one variable with the other, the variables may be compared with one another so that their coincidence can be measured. An example, in order to test the hypothesis in rural settlement studies is as follows :

TABLE 1.1
*Number of Rural Settlements and Density of Population in the District of Monghyr (1971)*

| S. No. | Anchals | Number of Rural settlements (X) | Density of population (Y) | X . Y | $X^2$ | $Y^2$ |
|---|---|---|---|---|---|---|
| 1. | Monghyr | 137 | 318 | 43566 | 18769 | 101124 |
| 2. | Jamalpur | 126 | 385 | 48510 | 15876 | 148225 |
| 3. | Tarapur | 149 | 642 | 95658 | 22201 | 412164 |
| 4. | Sangrampur | 119 | 496 | 59024 | 14161 | 246010 |
| 5. | Kharagpur | 146 | 285 | 41610 | 21316 | 81225 |
| 6. | Dharhara | 186 | 222 | 41292 | 34596 | 49284 |
| 7. | Surajgarha | 196 | 327 | 64092 | 38416 | 106929 |
| 8. | Lakhisarai | 153 | 354 | 54162 | 23409 | 125316 |
| 9. | Sheikhpura | 112 | 375 | 42000 | 12544 | 140625 |
| 10. | Ariari | 88 | 334 | 29392 | 7744 | 111556 |
| 11. | Barbigha | 98 | 609 | 59682 | 9604 | 370881 |
| 12. | Barahiya | 48 | 314 | 15072 | 2304 | 98596 |
| 13. | Jamui | 81 | 378 | 30618 | 6561 | 142884 |
| 14. | Khaira | 102 | 224 | 22848 | 10404 | 50176 |
| 15. | Sikandra | 135 | 357 | 48195 | 18225 | 127449 |
| 16. | Halsi | 101 | 306 | 30906 | 10201 | 93636 |
| 17. | Lakshmipur | 138 | 205 | 28290 | 19044 | 42025 |
| 18. | Jhajha | 197 | 222 | 43734 | 38809 | 49284 |
| 19. | Chakai | 600 | 123 | 73800 | 360000 | 15129 |
| 20. | Sono | 250 | 219 | 54750 | 62500 | 47961 |
| 21. | Khagaria | 49 | 449 | 22001 | 2401 | 201601 |
| 22. | Alauli | 45 | 416 | 18720 | 2025 | 173056 |
| 23. | Chautham | 41 | 389 | 15949 | 1681 | 151321 |
| 24. | Beldaur | 29 | 350 | 10150 | 841 | 122500 |
| 25. | Gogri | 49 | 421 | 20629 | 2401 | 177241 |
| 26. | Parbatta | 85 | 422 | 35870 | 7225 | 178084 |
| | Total | 3460 | 9142 | 1050520 | 763258 | 3564288 |

$\bar{X}=133.07$, $\bar{Y}=351.61$, No.=26.

In order to get the correlation value, the Pearsonian correlation coefficient formula has been employed :

$$r=\frac{\Sigma XY-\dfrac{(\Sigma X)\cdot(\Sigma Y)}{N}}{\sqrt{\left[\Sigma X^2-\dfrac{(\Sigma X)^2}{N}\right]\left[\Sigma Y^2-\dfrac{(\Sigma Y)^2}{N}\right]}}$$

where, r is Pearsonian correlation coefficient.

$\Sigma X$ is sum of all values of independent variables.
$\Sigma X^2$ is sum of squares of all independent variables.
$\Sigma Y$ is sum of all values of dependent variables.
$\Sigma Y^2$ is sum of squares of all dependent variables.
$\Sigma XY$ is sum of products of all X and Y values.
N is number of units.

With the help of the above table the product moment correlation coefficient calculation is as follows :

$$r = \frac{1050520 - \frac{3460 \times 9142}{26}}{\sqrt{\left[763258 - \frac{(3460)^2}{26}\right]\left[3564288 - \frac{(9142)^2}{26}\right]}}$$

$$= \frac{-165837}{325468} = -.50\ (r)$$

$r^2 = 0.25$

Calculating from the above, we get the value of $r^2 = .25$. This shows that the relationship is very close. The result has been tested by student "t" test by the formula :

$$t = \sqrt{\frac{r^2\,(n-2)}{1-r^2}}$$

$$= \sqrt{\frac{.25(26-2)}{1-.25}} = \sqrt{8} = 2.83$$

$H_o$ : According to null hypothesis ($H_o$) there is no correlation between number of rural settlements and the density of population in the district of Monghyr in 1971.

$H_i$ : According to research hypothesis ($H_i$) there is correlation between number of rural settlements and the density of population in the district of Monghyr in 1971.

Table value of "t" at 24 degrees of freedom for 5 per cent and 1 per cent level of significance are 2.064 and 2.797 respectively whereas the calculated value of "t" is 2.83. Since the calculated value of "t" is significant at 24 degrees of freedom for 5 per cent level of significance, therefore, $H_o$ (null hypothesis) is rejected in favour of $H_i$ (research hypothesis).

The regression of Y upon X has been calculated by the method of least squares according to the formula :

$$Y = a + bx \pm Sey$$

Where Y is the dependence variable, and X is the independent variable.

The values of a and b are given by the following formula :

$$b = \frac{\Sigma XY - \frac{(\Sigma X) \cdot (\Sigma Y)}{N}}{\Sigma X^2 - \frac{(\Sigma X)^2}{N}}$$

$$\text{and } a = \frac{\Sigma Y - b\ \Sigma X}{N}$$

where X and Y parameters are the same as mentioned earlier.

In the present case the value of $Y = a + bx$ has been estimated as :

$$Y = 423.47 - .54X$$

The accuracy of this estimate of Y can be tested by calculating the standard error of estimate. It is a measure of the error to be expected in estimating the value of Y from a given value of X by means of the computed value of Y. The formula is as follows :

$$\text{Standard Error (Sey)} = \frac{1 - r^2}{\sqrt{n}} = \frac{1 - .25}{\sqrt{26}} = 0.147$$

The standard error of estimate gives the probability of error for two-thirds of all cases on either regression line. Thus the final equation takes the form :

$$Y = 423.47 - .54X \pm 0.147$$

Fortunately, the standard error of estimate is very low.

## CONCLUSION

Rural settlement is a dominant branch of geography which originated from the trunk of human geography. The topographic arrangement of houses and highways in relation to man is expressed as rural settlement whether the same may be a storehouse, residence of human living, railway station, bus depot, electricity generation site, fence, road, etc. This shows that the scope of rural settlement geography ranges from temporary camp of herders to permanent settlements based on permanent agricultural economy on a fixed piece of land. The functions of rural settlements may be agricultural, commercial, mining, transportational, manufacturing, professional services, etc. But up till now the agriculture dominates in almost all parts of the world.

The subject has developed significantly in its multifarious aspects since 1928 when the 'International Geographical Union' established rural settlement geography as one of its important branches of discussion among world geographers. Since then a huge amount of literature, scholarly papers books have been published, methods, principles and techniques have been added to this branch of geography which has either developed indigenously or taken from sister disciplines of sociology, economics, anthropology, zoology, physics, etc. Nowadays, with the introduction of statistical techniques to diagnose properly for the eradication of problems, the balanced development of a region may be attained systematically. In the field of rural settlement geography the frontiers of research are the formation of villages by migration accretion. Secondly, the examination of consumer behaviour pattern on a micro-level so that the alignment of a village can be made properly towards the nearest central place, because in the hinterland of a central place, it has been found that the main thoroughfare is generally aligned towards the neighbouring central place.

## REFERENCES


1. R.L. Singh and K N. Singh, (Ed.) : *Readings in Rural Settlement Geography*, Varanasi, 1975, p vii.
2. C.A. Doxiadis, "The Future of Human Settlement," *The Place of Value in a World of Facts*, Ed., A Tiselius and S. Nilsson, Stockholm, Almquist and Wiksell, 1969, pp 307-338.
3. J. Bruhnes, *Human Geography*, Rand M.C. Nally & Co., New York, 1952, p 152.
4. R.L. Singh, "Meaning, Objectives and Scope of Settlement Geography," *National Geographical Journal of India*, Vol. VII; Part I, March, 1961, pp 12-13.

5. A.V. Perpillon, *Human Geography*, Longmans, Edinburgh, 1966, pp 406-414.
6. T.G. Jordon, "On the Nature of Settlement Geography," *The Professional Geographer*, Vol. XVIII, No. 1, 1966, pp 26-28.
7. I. Bowman, "Settlement by the Modern Pioneer," *Geography in the Twentieth Century*, Second Edition, 1953, Chapter XI, p 266.
8. Ibid.
9. J. W. Watson, "Geography : A Discipline in Distance," *Scottish Geographical Magazine*, Vol. 71, 1955, pp 1-13.
10. W. Christaller, *Central Places in Southern Germany*, translated by C.W. Baskin, Prentice Hall, N.J., 1966.
11. K.H. Stone, "The Development of a Focus for the Geography of Settlement," *Economic Geography*, Vol. 41, 1965, p 348.
12. A. Mietzen, *Siedlung and Agrarwesender Westgermonn and Ostgermanen*, der Kelten, Berlin, 3 Vols, 1895.
13. M.W. Beresford, *The Lost Villages of England*, London, 1954.
14. K.H. Stone, op. cit., 1965.
15. M. Aurosseau, "The Arrangement of Rural Population," *Geographical Review*, Vol. 10, 1920, p 223.
16. Isamn Matui, "Statistical Study of The Distribution of Scattered Villages in Two Regions of the Tonami Plain, Toyama Perfecture," *Japanese Journal of Geology and Geography*, 9, 1932, pp 251-66.
17. R.L. Singh, (Ed.) : *Rural Settlements in Monsoon Asia*, Varanasi, 1975, p 13.
18. C.J. Galpin, "The Social Anatomy of an Agricultural Community," University of Wisconsin, *Agriculture Experiment Station, Bulletin*, 34, 1915.
19. E.L. Ullman, "A Theory of Location for Cities", *American Journal of Sociology* Vol. 46, 1941, pp 853-864.
20. R.E. Dickinson, *City Region and Regionalism*, London, 1956.
21. J.E. Brush, "The Urban Hierarchy in Europe", *Geographical Review*, 1953, Vol. 43, pp 414-416.
22. B.J.L. Berry and W.L. Garrison, "The Functional Bases of the Central Place Hierarchy", *Economic Geography*, 1958, Vol. 34, pp 145-154.
23. K.S.O. Beavon and A.S. Mabin, "The Losch System of Market Areas Derivation and Extension", *Geographical Analysis*, 1975.
24. A.P. Lösch, *Economics of Location*, Yale University Press, 1954.
25. K.H. Stone, op. cit., p 374.
26. I. Bowman, "The Pioneer Fringe", *American Geographical Society*, Special Publication No. 13, New York, 1931.
27. E. Kant, "Some Notes on the Representation of the Density of Rural Habitations", *The Indian Geographical Journal*, Vol. 30, Nos. 1 and 2, 1955, pp 1-8.
28. A. Hillman, *Community Organization and Planning*, The Macmillan Company, New York, 1950.

29. C. Jacobsen, "Modernity in Traditional Villages," *Rural Sociology*, Vol. 38, No. 3 1973, pp 283-295.
30. M. Chisholm, *Rural Settlement and Landuse*, Hutchingon University Library, London, 1962.
31. R. Eidt. and W.I. Woods, *Abandoned Settlement Analysis, Theory and Practice*, Mil Wankee, 1974.
32. F.S. Hudson, *A Geography of Settlement*, Macdonold and Evans, 1970.
33. J.C. Hudson, "A Location Theory for Rural Settlement", *Annals of the Association of American Geographers*, Vol. 59-2, 1969, pp 365-381.
34. D. Grossman, "Do We Have a Theory for Settlement Geography ?—The Case of Iboland", *The Professional Geographer*, Vol. 23-3, 1971, pp 197-203.
35. E. Bylund, "Theoretical Consideration Regarding the Distribution of Settlements in Inner North Sweden", *Geografisca Annalor*, 42, 1960, 225-31.
36. R.L. Morrill, "The Development and Spatial Distribution of Towns in Sweden, an Historical Predictive Approach", *Annals of the Association of American Geographers*, Vol. 53, 1963, pp 1-14.
37. C.D. Deshpande, "Settlement Types of Bombay Karnatak", *The Indian Geographical Journal*, Vol. XVII, No. 2, 1942, pp 115-131.
38. S.M. Ali, "Population and Settlement in the Ghaggar Plain", *The Indian Geographical Journal*, Vol. XVII, No. 3, 1942, pp 157-182.
39. E. Ahmad, "Rural Settlement Types in Uttar Pradesh", *Annals of the Association of American Geographers*, Vol. XLII, No. 3, 1952, p 229.
40. K.H. Buschman, "Settlements and Habitations in India, I, II", *Geographical Review of India*, Vol. 16, 1954, No. 3, pp 19-40, and in No. 4, pp 13-18.
41. R.L. Singh, "Evolution of Settlements in the Middle Ganga Valley", *National Geographical Journal of India*, Vol. 1, 1955, Part-2, pp 69-115.
42. A.B. Mukerjee, "Spacing of Rural Settlements in Andhra Pradesh", *Geographical Outlook*, Vol. VI, 1969, p 1.
43. R. Lahiri, "Settlement Types in Ajoy Barakar Basin", *Calcutta Geographical Review* Vol. 12, 1950, No. 4.
44. A.N. Bhattacharya, "Effect of Tribal and Caste Factor on Rural Settlement in U.P.", *Indian Geographical Journal*, Vol. 31, 1956, No. 3 and 4, pp 94-102,
45. R.L. Singh, "Typical Rural Dwelling in the Umland of Banaras". *The National Geographical Journal of India*, 1957, Vol. III, Part II, p 52.
46. L.R. Singh, *The Terai Region of U.P. : A Study in Human Geography*, Allahabad, 1965, p 37.
47. V.N.P. Sinha, "Houses and House Types of the Chotanagpur Plateau", *The Geographical Knowledge*, Vol. 2, No 2, 1969, p 91.
48. R.B. Singh, "Rural Settlement Types and Their Distributions", *The*

*National Geographical Journal of India*, Vol. XV, Part 2, 1969.

49. R.L. Singh, (Ed.) : Rural Settlements in Monsoon Asia, *The National Geographical Society of India*, Varanasi, 1972.
50. R.L. Singh, (Edt.) *The National Geographical Journal of India, Special Issue and Rural Settlements*, Vol. XXI, Parts 3 and 4, 1975.
51. R.P.B. Singh, *Class Settlements in the Saran Plain*, The National Geographical Society of India Varanasi, 1976.
52. R.K. Mukerjee, *Man and His Habitation : A Study in Social Ecology*, Popular Prakashan, Bombay, 1968.
53. R.C. Sharma, *Settlement Geography of the Indian Desert*, Rajesh Publication, New Delhi, 1972.
54. A. Prasad, *Chotanagpur: Geography of Rural Settlement*, Ranchi University, Ranchi, 1973.
55. R.B. Mandal, "The Geography of Rural Settlements in Bihar Plain", *Unpublished Ph.D. Thesis*, Patna University, 1972.
56. R.B. Mandal, "Planned Development of Rural Settlements in North Bihar", *Unpublished D.Lit. Thesis*, Patna University, 1977.
57. R.B. Mandal, "Ecology of Rural Settlements in the Kosi Basin", *Indian Geographical Studies*, R.B. No. 8, 1977.
58. M. Friedman, *Essays in Positive Economics*, University of Chicago Press, Chicago, 1953, p 14.
59. G. Wunderlich and W.L. Gibson, *Land Economics Research* (Ed. J. Ackerman, and *et al*), Washington, 1962, pp 214-15.

2

# Classical Central Place Theory

*—Ram Bahadur Mandal*

A DYNAMIC radius is drawn by the economy of the present day world affairs where due to the struggle for space among the competing service centres, the idea of central place emerges. Such changing space in relation to the centres which provide services to the surrounding area is the guiding factor of economic development in a particular region whether the same may be a village or city level centre.

This theoretical concept of central place was first envisaged by Mark Jefferson in 1931. According to him : "Cities do not grow of themselves. Countrysides set them up to do tasks that must be performed in central places."[1] Primarily it was observed that cities have their zone of influence in a circular pattern. Similarly villages have their zone of influence from where the "villagers" commute to their fields daily rather than living near their respective fields. The village hinterland of fields and farmers daily trips are in many ways analogous to the conditions of cities and their commuters. Thus the methods applied for measuring the spacing of villages are also helpful for analysing the spatial pattern of even urban settlements.

Later on, Christaller developed the idea on scientific lines in Germany based on telephone connections in the southern part of Germany.[2] One decade later Edward Ullman introduced Christaller's work of central places to the English language. "The essence of central place theory is that a certain amount of productive land supports an urban centre. The centre exists because essential services must be performed for the surrounding land".[3]

Four basic patterns are generally assumed regarding the growth and arrangement of villages into the lattice of central place system. (1) That the land supports a uniform number of people per square kilometre. (2) That the farmers can easily travel to their fields with plough a distance not exceeding 3.20 kilometres which represents the maximum range of a village community. (3) That the land is productive, life is reasonably good and the rate of population growth is slow. (4) That the growth of population favours the hiving off process and new "offspring" villages will be formed possibly closer to the older settlements in order to maintain the social linkages.

Under the classical central place theory, in addition to Christaller, Lösch devised a set of elementary equations and simplified models and developed a set of discrete networks of markets based on the variable K concept in which excess profit is the rule.[4] Where K represents the areas/number of secondary service centres for a larger centre. Isard presented a more realistic picture of central place location through population distribution, agriculture and urban landuse pattern and flow of goods which determine the structure and functioning of central places through location decision of their firms, restrained by the regions total resources, income, gross product, labour productivity and by the living standards and expenditure pattern[5] of their hinterland people.

Thus the dispersed nature of service centres, 'growth poles and growth centres'[6] exist over the countryside because essential services are performed there for their own consuming population, surrounding zone of influence inhabitants and *vice versa* as no central place can survive in isolation. Such places are known as central places and their services as central place functions.[7]

## SPACE PACKING CIRCLES INTO HEXAGON

The present study devises steps in the formulation of concepts and procedures in settlement geography. The consumer's movement and traffic around a central place tend to create circular trade areas, conforming with the hexagon which leave no unserved space in between two central places. This hexagon causes an equal spacing of trade centres with approximately equivalent functions. Within these hexagonal trade areas, smaller trade area with six supporting centres develop around villages and still smaller around the hamlets. Thus, according to this theory hexagonal trade areas of different sizes surround the central places and the roads which serve the hexagonal trade areas coverage on the central place from six directions.

In case A is the original settlement and B and C are secondaries with circular areas of field of influence, than in spatial arrangement of 3.20 kilometres radius D originates from the original settlement, while D′ and F′ colonies form the secondary settlements [Fig. 2.1 (a)]. With the formation of a triangular path in between, the settlements which have originally circular areas of influence will be mitigated as hexagonal arrangement which leaves no unserved and empty land in between the service areas of three settlements. Thus the allocation of unclaimed land is shown in [Fig 2.1 (b)]. This process may give rise to the several generations of settlements with varying population density and village boundaries [Fig. 2.1 (c)].

A number of scientists in Europe and America have criticised the hexagonal concept. The American Sociologists[8] E. des Bruner and J.D. Kolb have observed that the smaller centres do not develop as close to the large centres as larger centres do with one another. The pull exerted by a trade centre varies directly with the size of the centre and indirectly with the distance.

Ultimately it has been realised that the triangular net of points filled by hexagonal areas provides the best geometric solution to our farms, farmers and the location of rural and urban settlements. Only three geometrical shapes can be used to fill space without overlaps i.e. equilateral triangle, square and hexagon. Among all these, hexagons are nearer to the circle. Thus the central place model is devised with a simple set of geographic, social and demographic rules after considering Euclidean geometry.

The above idea produces the central place hierarchy for which Christaller developed three principles K-3 network for marketing principle, K-4 network for transportation principle and K-7 network for administrative principle. In these principles only the marketing principle finds its place in real world application in most of the studies and K-4 and K-7 have not been applied anywhere in the world.

*The Growth of Central Places*

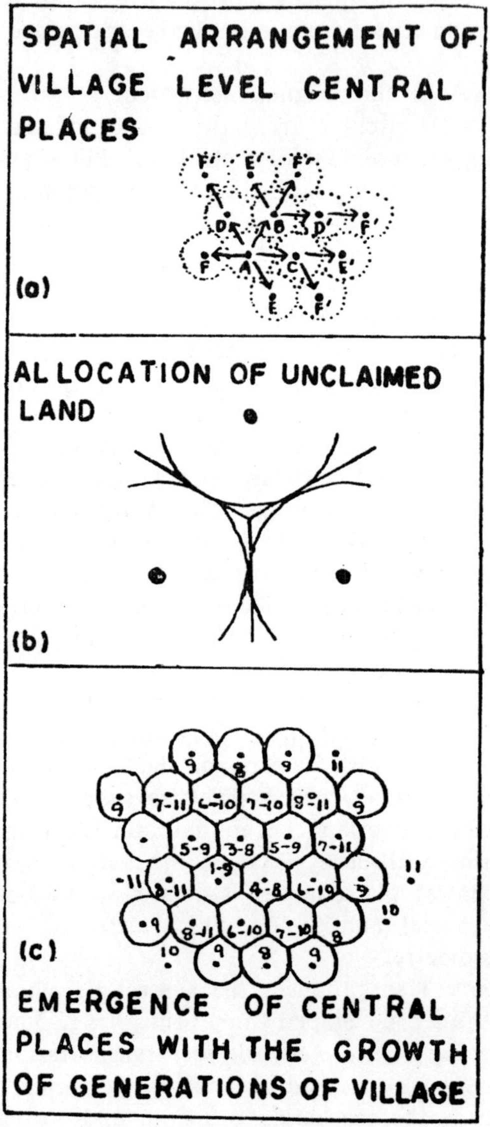


FIG. 2.1

The idea of central place depends much on consumer behaviour, which determines the threshold of goods. The threshold is the lower limit of the range of a central place or the minimum amount of necessary purchasing power for a firm to enter the market. The range of goods represent the distance beyond which a shop is unable to attract customers.

So far as the modern central place theory is concerned the ideas of Berry, Garrison, Marshall and Beavon are pioneering ones. According to Berry, central place theory is the theory of the location, size, nature and spacing of central place clusters and their activity.[9] According to Marshall the fundamental characteristics of the ideal central place hierarchy are as follows :[10]

(1) Spatial inter-dependence of centres.
(2) Functional wholeness of the system.
(3) Discrete stratification of centrality.
(4) Interstitial placement of orders.
(5) Incremental baskets ot goods.
(6) A minimum of three orders.
(7) A numerical pyramid in order of membership.

Beavon and Mabin[11] have shown how it is possible to derive a full set of discrete market areas of Lösch for any sized region and solution of mathematical equations are put forward by Lösch. So far as the K values are concerned, Lösch has suggested that by altering the size and orientation of the hexagons, K values of 9, 12, 13, 16, 19, 21, 27 and so on could also be derived.[12]

A central place may be rural or urban in character, so that hamlets, villages, towns and cities are paralleled by street corner, neighbourhood, community and regional shopping centres and C.B.D. This force is best observed on isotropic-plain surface where topography is uniform so that no site has any advantage in terms of slope, physical influence and transport routes. The economy is such that no site has any advantage in producing such primary products as grain, timber or other raw materials. Based on these assumptions,[13] a theory of central place location has been evolved having evenly spaced centres which will be approachable within two hours time on foot and seven hours on automobile transport from all sides.

After Jefferson,[14] who envisaged the term 'central place', this theory was developed by Christaller (1933),[15] Lösch (1954)[16] Berry and Garrison (1958),[17] Vance (1970),[18] Thakur (1971),[19] and others. After Second World War there was almost a flood of ideas[20] on central place studies.

In India N.R. Kar,[21] A.K. Dutt and Banerjee,[22] S.M. Singh,[23] O.P. Singh,[24] Steen Folke,[25] L.N. Ram,[26] P. Deka,[27] R.B. Mandal[28] and L.S. Bhat *et al*[29] have studied different aspects of central places in keeping with the ideas of the regional framework of Christaller. The work of P. Deka, however, is on Lösch. The present study is a modest attempt to analyse the central places on 'Löschian Landscape' for planned rural settlement growth in a multi-directional and integrated pattern.

For planned rural settlement growth, analysis, description and interpretation, the study of systems of central places is absolutely essential so that one can allocate the resources properly in a particular system of central places according to his reach over a central place. This would also help in eradicating regional imbalances, stagnant and sagging economy, unbalanced service facilities, inefficient transport growth and inconvenience caused to rural consumers.

To overcome these difficulties, the central place study may provide the basic framework for a rational distribution of a cluster of central functions accessible to the maximum profit area in the command at different hierarchical levels of settlements, rational organization of flows of people, goods, services and the location of growth centres for regional development.

Lösch has considered the study of central places from lower-order centres and goods to higher-order. The present study pays more attention to explain the "Löschian Economic landscape" than to Christaller's view.

## THEORETICAL DIFFERENCES BETWEEN CHRISTALLER AND LÖSCH

The theoretical bases for central place studies by Christaller and Lösch[30] are given here in a nutshell Table 2.1.

TABLE 2.1

| View of Christaller | View of Lösch |
|---|---|
| 1. Region—S-W Germany. | Iowa. |
| 2. Isotropic Plain surface. | Isotropic Plain surface. |
| 3. Begins from city level centres. | Begins from village level least specialized centres. |
| 4. Suggested hierarchy from highest-order goods to the lowest. | Suggested hierarchy from village level goods upwards. |
| 5. Considered triangular arrangement of production sites, retail stores and hexagonal market areas, having uniform population density, unbounded plain and equal access in all directions. | Considered triangular arrangement of production sites, retail stores and hexagonal market areas, having uniform population density, unbounded plain and equal access in all directions. |
| 6. Christaller required that all lower-level centres' locations take into account locations of larger centres. | Lösch derived a variety of optima and tried to assemble these into an overall pattern. |
| 7. Fixed K values of K-3, K-4 and K-7 | No fixed K values. |
| 8. Location and functions of central places obtained for one 60° sector radiating from metropolis. | Location and functions of central places obtained for one 60° sector radiating from metropolis. |
| 9. Christaller produces central place hierarchy in which there are levels of centres and groups of goods. | Lösch derived a more diversified economic landscape in which centres specialize. |
| 10. Relevant for analysing retail and wholesale business. | Provides a framework for analysing the spatial distribution of market oriented manufacturing. |
| 11. Christaller agglomerative requirement is for travel cost minimisation by rational consumers on multipurpose trips. | Not satisfied by Lösch. |
| 12. In Germany smaller central places were larger in number relative to higher order centres. | Such a condition is not found in Iowa. |

TABLE 2.1—*contd.*

| View of Christaller | View of Lösch |
|---|---|
| 13. All places in a particular tier are of the same size and have the same function | The same size central place need not have the same function. |
| 14. Number of central places be minimized. | All excess profits be eliminated. |
| 15. All higher order centres contain the functions of the smaller central places. | Larger centres need not have the functions of lower ones. |
| 16. Different hierarchical levels have differential supply of goods. | One place may be a supply centre for all goods. |
| 17. Model developed by the market areas of consumer behaviour pattern. | Model developed by superimposition of fishnets (demand cone) of different goods threshold range. |

*Source* : Compiled from Lösch (1954), Isard (1956), Christaller (1966) and Berry (1967).

## CENTRALITY AND NODALITY

According to classical studies the centrality value incorporates the aggregate amount of services that a central place provides to its hinterland around it. In most world studies this measure has been taken on the basis of functions located in the central place itself. But R.E. Preston has called such central place functions as the measure of 'nodality'[31] not 'centrality'. According to him, the centrality value should be considered only after considering basic functions or city-forming services which bring capital and resources from outside to the central place providing its fundamental means of sustenance and it is for this reason that geographers have paid far more attention to city forming than to city serving population.[32] Thus centrality value should only be considered after measuring the intensity and amount of supply of goods and services directly to the population of a surrounding contiguous area on a regular year round basis. Only this type of analysis should be considered for measuring the centrality value in central place research. But due to lack of data the present study is based on nodality value and not centrality.

### CRITERIA OF CENTRALITY

For central places, the number of persons alone is not a satisfactory answer to determine the centrality. Christaller has considered telephone connections while studying the central places of southern Germany as centrality index.

Christaller takes the telephone and derives the degree of centrality of a place, that is, amount of services performed apart from those relating to the needs of its own inhabitants by means of the following formula :

$$\text{Centrality of a place} = \text{Number of telephone in place} - \text{Number of inhabitants in place} \times \frac{\text{Number of telephone in area served by place}}{\text{Number of inhabitants in area served by place}}$$

Or

$$C = T_2 - E_2 \times \left(\frac{T_g}{E_g}\right)$$

Or

$$C = \left(t - \frac{PT}{P}\right)$$

where, C is centrality, t is number of telephone in a town, P is population of the town, T is number of telephone in thc whole region and P is population of the whole region.

This method has been heavily criticised in recent years as telephones are also used for non-central functions and most of the lower-order places have no connections at all.

The centrality of a centre can be measured after considering the goods and services for which the zone of influence depends on the centre. Bracey has applied this method numerically in southern England. After considering the intensity of bus services Green has measured the centrality of places in England and Wales. Godlund of Sweden has selected retail trade as an index of centrality and has given the formula[33] :

$$\text{Centrality} = \frac{St \,.\, 100}{Pt}$$

where St is number of persons employed in retail trade and Pt is population of the region.

Kashi Nath Singh has put forward a more feasible proposition in the shape of formula as :[34]

$$C=\frac{N.100}{P}$$

where, C is centrality

N is the number of persons dependent upon commerce, and

P is the total regional population in commerce.

M.S. Vishwanath while studying the growth pattern and hierarchy of urban centres in Mysore has used the following formula to measure the centrality index : [35]

$$C=\frac{P\times 100}{\text{Total State population}}$$

where, C is centrality, and

P is the total regional population in commerce.

V.N.P. Sinha while studying the hierarchy of urban settlements in Chotanagpur Plateau devised the following formula to measure the centrality of a place :[36]

$$C=\frac{T\times 100}{P}$$

where, C is centrality

T is the number of persons dependent upon tertiary services of a centre.

P is the total regional tertiary population.

According to Davies before deriving the centrality value for a central place, one should find out the location coefficient of shops of the region with the help of the following formula[37]

$$LC=\frac{t}{T}\times 100$$

where, LC is the location coefficient of shops in a central place,

t is the number of shops in that centre, and

T is the total number of shops in the entire, region under study.

Multiplying the relevant location coefficient by the centrality index (derived from the weight given to the different functions performed by a centre) will give the final degree of centrality in terms of 'centrality value' :

$$CV = CI \times LC$$

where CV is centrality value,

CI is centrality index for services, and

LC is location coefficient for retail trade.

### IDENTIFICATION OF CENTRAL PLACES

The author investigated the central places in North Bihar by means of functional indicators covering the categories of education, medical facilities, recreation, communication, administration, transport, finance, trade, religious nodes, industrial establishments and employment in tertiary services. Each category is represented by one type of central function and most of the types have been subdivided according to their position in the functional hierarchy.

TABLE 2.2

*Weighing and Classification of Central Functions*

| Functional Indicators | Class I 20 points | Class II 15 points | Class III 10 points | Class IV 5 points |
|---|---|---|---|---|
| 1. Education | — | — | High School | — |
| 2. Health | — | — | Hospital | — |
| 3. Recreation | — | Cinema | — | — |
| 4. Communications | — | — | Telephone Exchange | Public call office or post office |
| 5. Administration | — | — | Block development office | — |
| 6. Transport | — | — | Railway Station (within 1 km) | Metalled road |
| 7. Financial institutions | — | Bank | — | — |
| 8. Trade | — | Village Hut | Wholesale Grain Market | — |
| 9. Religious Node | — | — | — | Religious centre of repute |
| 10. Industry | — | — | Large scale industrial establishment | — |
| 11. Tertiary workers percentage | — | — | 50% to 100% of total workers | — |

*Source* : Compiled by the author.

The following criterion has been adopted for selecting central places :

Only those centres are listed where at least one of the above functional indicators is present irrespective of the functional units, but such centre must have a well developed retail market in command. A score has also been allotted to each of the functional units, irrespective of the number of its occurrences.

## CENTRAL PLACE HIERARCHY

In 1933, Walter Christaller, a German geographer had noticed the hierarchy and regularities in the settlement pattern of Bavaria in southern Germany. He explained the regularity of settlement model for even distributed population in different sized villages, towns and cities deductively.

Similarly in 1945, August Lösch, a German economist derived an answer of central place hierarchy which is closely related with the Christaller's view. They came to a conclusion "that the best way to spatially provide a population with all types of services is to place such activities in a nested hierarchy of increasingly larger central settlements located at the nodes of hexagonal network".[38] The models presented by Christaller and Lösch are essentially static having no reference as to how they might change through time. Christaller himself observed that "the State of rest is only fiction, whereas motion is reality".[39] Thus in order to have a dynamic model it seems that their contributions have stimulated geographic investigation so that the absolute explanation of real world situation can be achieved.

### The K-3 Hierarchy

This represents the hierarchy of central places according to the marketing principle of Christaller. In each case the number of village-sized market areas served progression is 1, 3, 9, 27, 81. . . . Here one higher order central place area is surrounded by the two equal sized areas (one-third of each) of the second order centres. (Fig. 2.2). They are called K-3 networks of marketing principle.

*The Hierarchy of Central Places*

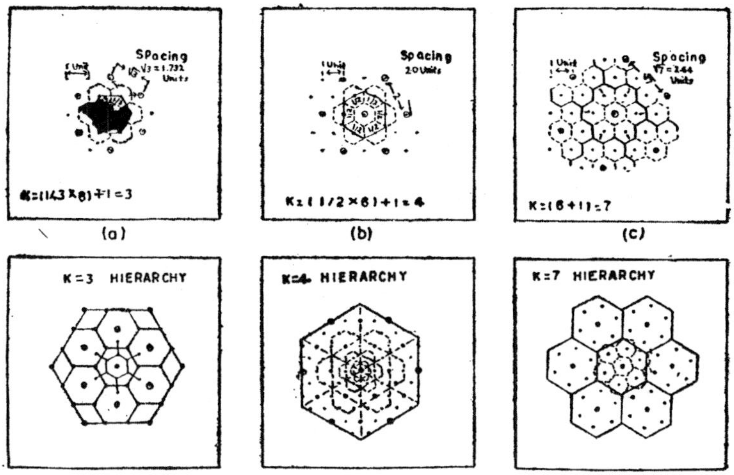

FIG. 2.2

**K-4 Hierarchy**

This represents the hierarchy of central places according to transportation principle of Christaller. The progression of market areas in K-4 system is 1, 4, 16, 64. . . . In K-4 network all settlements are located in the side of road leading directly away from the highest order centres. In K-4 hierarchy central places also provide a ranked order of goods and services. Here the lower order central places are located exactly halfway between places of the next highest order and lie on the boundaries separating the market areas of highest order (Fig. 2.2). Here one higher order centre area is surrounded by the three equal sized areas (one-half area of the six surrounding centres) of the second order centres.

**K-7 Hierarchy**

The progression of market areas in K-7 central places system is 1, 7, 49. . . . Christaller and Lösch both had called it administrative principle. Under this principle the hierarchy of administrative centres are found to be true, where one higher order centre area is equivalent of the areas associated with seven central places of the next smaller order [Fig. 2.2 (c)]. In this system of central places, each central place is located in the centre of the area with which it is associated and is never divided by the boundary lines of higher order places for this fact it is advantageous for administrative purposes. Here the centre lies in between two larger central places, the consumers can go to whichever place they choose as long as their travel costs remain unchanged.

Considering the city as a dimensionless point beyond K-3, K-4 and K-7 networks it may be possible to recognise that the basic unit of hexagonal pattern may contract around larger central places and expand away from the larger places [Fig. 2.3(a)] shows a modular hexagonal pattern of this type which can be repeated in all direction without any break. Such hexagonal network is suitable for areas having sparse and dense population.

*Modular Hexagonal Network in Areas of Varying Density of Central Place Customers*

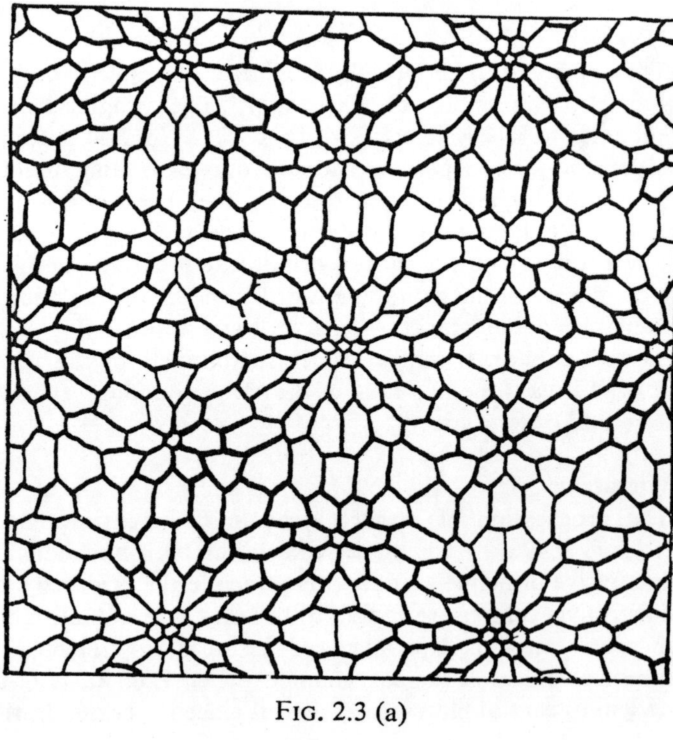

FIG. 2.3 (a)

Population density varies from one place to another besides urbanisation in different parts of the earth surface. The causes are physical, economic and cultural variations. An examination of the ecological variations in relation to population distribution reveals that population becomes more sparse, in very cold, very dry and in rugged terrain conditions, while "thinning out can be shown graphically by the gradual change in size of a hexagonal network with smaller hexagons representing Wetter or Warmer or more productive areas with dense population whereas larger hexagons indicating colder, drier, or hilly areas with very dense population [Fig. 2.3 (b)]

*Logarithmic Expansion of a Hexagonal Lattice from High Population Density to Low Population Density*

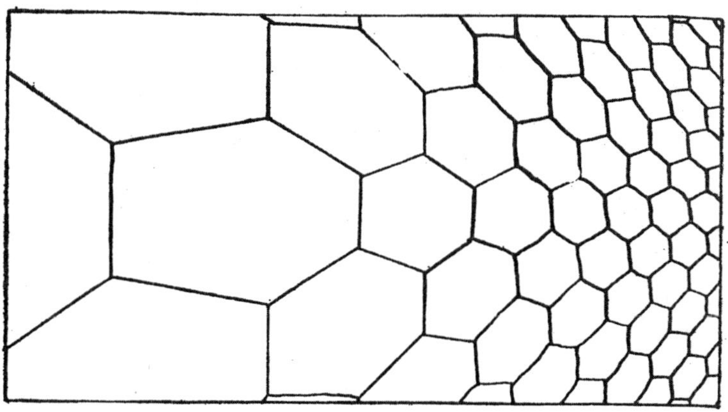

FIG. 2.3 (b)

## CENTRAL PLACE HIERARCHY IN NORTH BIHAR

The aggregate weight of localization of functions at a place is known as nodality value. For central places of North Bihar this measure has been devised after giving proper weight for different central functions (Table 2.2) plus the number of permanent shops which that centre possesses.

The final list of all central places in North Bihar shows a total of 342 centres which include 53 towns and rest are villages. Therefore, the proposition of a hierarchy, should be first made for providing the framework upon which other laws rest.[39] A closer examination of the figures of nodality value which ranges from 7283 for Muzaffarpur to 24 for Rajkhand, reveals that the data has a very wide range. To obtain the range of central places in different hierarchy levels according to K-3 network of Christaller the natural breaks have been found out after dividing the highest nodality value to that of K-3 network progression.

The following breaks in the nodality value emerged from highest to lower level: 7283, 3641, 1214, 405, 135 and 45. The result is given in the table below :

TABLE 2.3

| Range of Nodality value | Order | Actual No. of centres | Theoretical No. According to K-3 |
|---|---|---|---|
| 7283 and over | I | 1 | 1 |
| 7282 to 3641 | II | 2 | 2 |
| 3640 to 1214 | III | 17 | 6 |
| 1213 to 405 | IV | 48 | 18 |
| 404 to 135 | V | 146 | 54 |
| 134 to 45 | VI } | 128 | 162 |
| 44 and Below | VII } | | |

According to K-3 network, the number of centres in each size class should progress as 1, 2, 6, 18, 54. . . . . . and the progression of market areas of each level should be 1, 3, 9, 27, 81. . . . . It will be seen that the actual distribution (Fig. 2.4) closely corresponds with the theoretical distribution for higher level centres and fails in the lower order central places. On the basis of the above groupings of nodality values, (Table 2.3) a map of central place hierarchy in North Bihar has been prepared, which gives a clear idea of K-3 network of central places in the region (Fig. 2.5).

## LOSCHIAN ECONOMIC LANDSCAPE

The Development of settlements is closely related to the growth and dispersal of rural central places. In a densely settled region like North Bihar, the lowest order of central places provides a link between the villages and the towns. Such central places fit with the concept of "Löschian Economic

*North Bihar Central Place Hierarchy K-3 System*

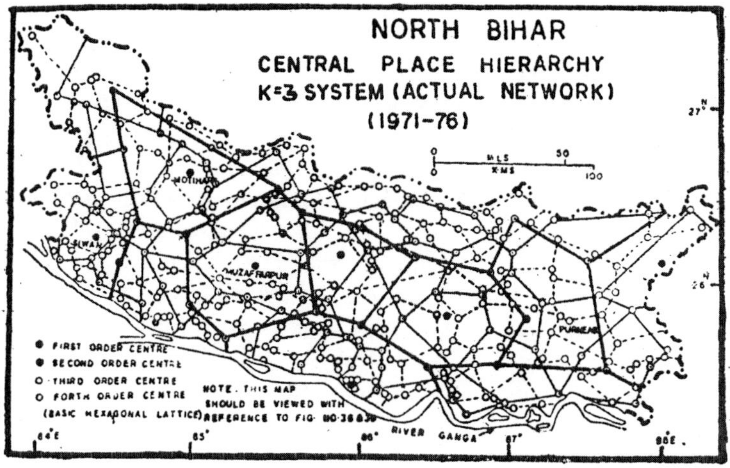


FIG. 2.4

*Central Place Hierarchy (Theoretical Model) K-3 System*

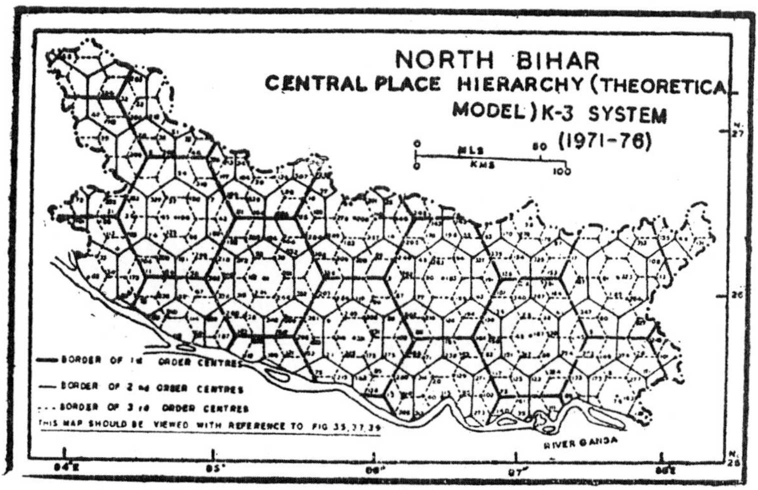


FIG. 2.5

Landscape" which has been formulated in the context of "densely settled areas undergoing economic change".[40]

Lösch (1954) hypothesized an orderly principle of the central place theory to include zones of nucleated settlements, pattern of agricultural villages and their arrangement in a triangular fashion. The growth and multiplication of rural settlements on isotropic plain surface, localization of resources and utility services in all directions, along with other assumptions[41] may lead to a theoretical tendency for rural settlement distribution towards uniform hexagonal pattern. This pattern leads to optimum utilization of resources and adequate distribution of service facilities in evenly spaced honey-combs of central places.

Among the rural settlements of North Bihar there are central places which serve people in their respective zone of influence. As the travel is equal in all directions from the central place, it is possible to draw a demand cone around the central place, according to which quantity consumed drops off with distance, because of increasing price to the consumer due to increased transport costs.

Considering an isotropic plain as discussed above, with a 60° lattice of points, which represents the densest packing of

*North Bihar Frequency of Central Place Coincidence on Löschian Landscape*

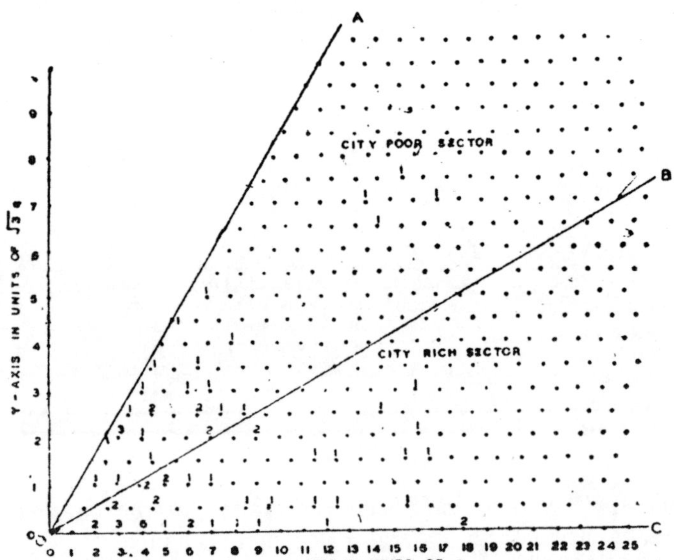


FIG. 2.6

the basic settlement forms[42], separated by distance 'a' for any pair of adjacent central places, Lösch defined small hexagon as $H_o$ with its sides 'S', contiguous with a set of similar size hexagons. Again he constructed 150 (342 in case of North Bihar) successively larger size hexagons in between $H_o$ and 151 ($H_o$ and 341 in case of North Bihar) (Fig. 2.6). Then superimposition was made on one another in such a way that in each size of network at least one centre coincides with the basic hexagonal honey-comb $H_o$. Thus based on $H_o$ (the basic hexagon) the rotation of all other networks or original location of central places gives at least one centre in the 30° sector of the basic lattice so that as many centres as possible with coincide.

The geometric solution of the above facts is as follows: The arrangement of the given lattice of points is on a set of right angled axes so that the Y-axis is measured in units of $a\sqrt{3}$. Then the area of a hexagon with side 's' is given by

$$H_i = \frac{3S^2\sqrt{3}}{2}$$

$$\text{or} \quad \frac{1}{2}S \, . \, \frac{\sqrt{3S}}{2} = \frac{1}{4}\sqrt{3S^2} = \frac{6}{4}\sqrt{3S^2} = \frac{3\sqrt{3S^2}}{2} \qquad \ldots (1)$$

Then the basic hexagon $H_o$ will have sides $S = \frac{a}{\sqrt{3}}$ and

$$\text{area } H_o = \frac{6}{4} \cdot \sqrt{3} \, . \, \frac{a^2}{3} = \frac{1}{2}\sqrt{3}\, a^2$$

$$\text{or} \quad \frac{a^2\sqrt{3}}{2} \qquad \ldots (2)$$

The relative size n of any other hexagon, $H_i$ in the system can be obtained by comparing it with $H_o$. Thus

$$n_i = \frac{H_i}{H_o} \qquad \ldots (3)$$

The side S of any hexagon $H_i$ is dependent on the distance $b_i$ between the centre of such a hexagon at the origin of the system and the centre of a contiguous hexagon of the same size. This is given by:

$$b_i = S_i\sqrt{3} \qquad \ldots (4)$$

and so, $$S_i = \frac{b_i}{\sqrt{3}} \qquad \ldots (5)$$

As 'a' distance between a pair of adjacent points can be set to unity it follows from equations (1) and (2) that

$$n_i = b_i^2 \qquad \ldots (6)$$

For the 30° sector of the basic lattice, it is possible to calculate $b_i$ between origin and each point in turn. It is, therefore, possible to calculate $n_i$ for each point. If $H_0$ is termed area zero, then the ranked values of $n_i$ or ranked nodality values of originally located central place on a map, can be termed as $A_r$(r—1 . . . . 342). It is then possible to assign area number $A_r$ for each point P(x, y) which is the centre for a particular hexagon $H_i$ (Fig. 2.7). It is possible to plot the area numbers associated with each point in the sector AOC and the remaining five sectors will be identical with the 60° sectors so completed. (Fig. 2.7).

The identical area of $H_i$ (each different sized hexagon) for two, three or different values of i yielding identical place of location in 60° lattice of Löschian Landscape is known as area numbers ($A_r$). Based on the rank values of $n_i$ (centrality) the area number represents the variable size of central places ($H_i$) so that

$$A_o, \angle A_1, \angle \ldots, \angle A_r, \angle \ldots, \angle \ldots, \angle A_n.$$

These centres have been intentionally arranged in the 60° sector so as to produce city rich and poor sectors. The initial distinction between these two sectors is based on the third central place, other than on the origin point. In the case of North Bihar Siwan comes as the third centre and it lies in ∠AOB city-rich sector. The location of central places of $A_r$(r—1 to 342) for city rich and city poor sectors is shown in Fig. 2.7.

The merit of the Löschian Landscape is that each point having one coincident centre can only earn a normal profit but the coincidence of other centres shows that if necessary or required service facilities are provided, that centre may support the enhanced business of other centres for the welfare of their surrounding hinterland people. Thus in Löschian Landscape

*North Bihar Rotation and Coincidence of Central Places on Löschian Landscape*

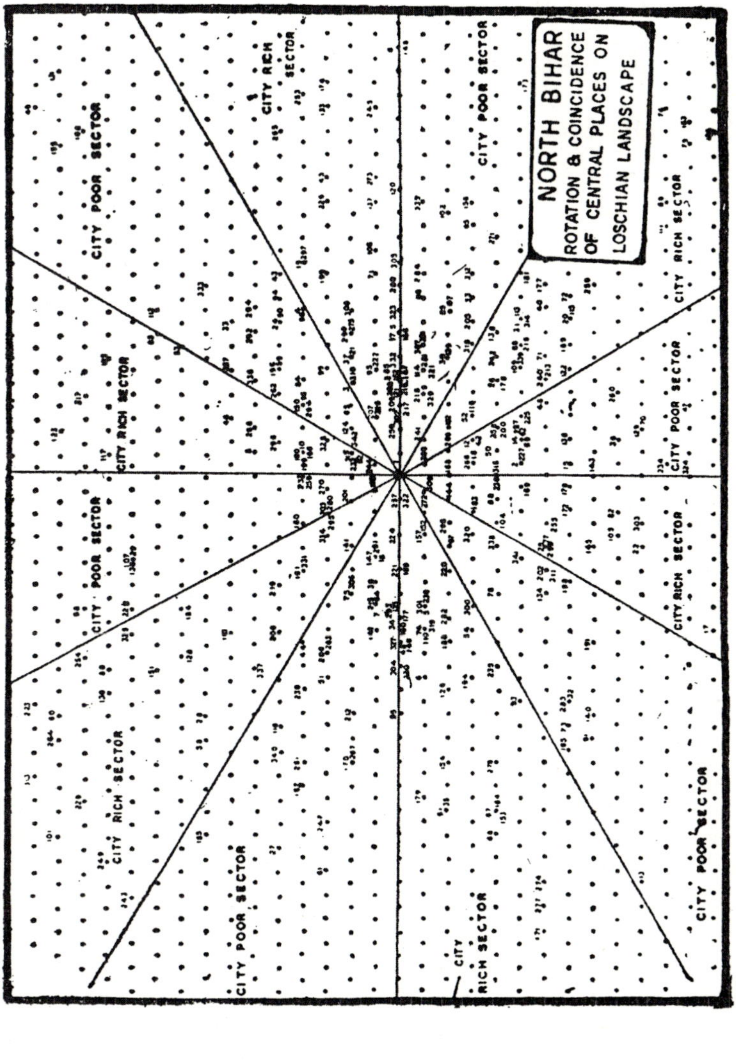


FIG. 2.7

"surplus profits are actually the rule",[43] and this must be accepted for 360° possible values of n markets.

In the Löschian Landscape, the market size of a central place is related to the number of market areas, for it is the centre. Accordingly, the contrast between city-rich and city-poor sectors and the number of coincident centres at each place on the original lattice are summed up in Table 2.4. The comparison of city-rich and city-poor sectors ∠AOB coincident central places is given below in Table 2.4.

TABLE 2.4

*Comparison of the City-rich and City-poor Sectors*

| No. of coincident centres (m) | Coincident centres | | Differences |
|---|---|---|---|
| | City rich | City poor | |
| 7 | 0 | 0 | 0 |
| 6 | 1 | 0 | 1 |
| 5 | 0 | 0 | 0 |
| 4 | 0 | 0 | 0 |
| 3 | 1 | 1 | 0 |
| 2 | 8 | 3 | 5 |
| 1 | 19 | 19 | 0 |

In this way Löschian system seems to be more attractive in comparison with Christaller (1933) with its more rigid spatial order between higher and lower order central places based on restricted trade area (K-3, K-4, K-7). On the other hand, it is difficult on a system of Lösch to equate any two central places as both of them are coincident centres for different sets of areas. For example, the comparison of point in ∠AOB (0-4 and 3-2) in Fig. 2.7, are nodes for central places 0, 8, 76, 121, 162, 187, 211, 285 and for another 0, 96, 150 and 294 number of centres respectively. "The equality of coincidence alone is not ample proof for equating two central places in a hierarchy as the functional types and their trade areas are different. It would appear more accurate to describe the city rich and city poor sectors of the economic landscape as sectors rich in the number of services"[44] and central places offered. This concept appears to be more apposite in the case of Löschian Economic Landscape for lower-order central places in North Bihar.

According to Israd[45] the most serious deficiency of Löschian construction is that it yields different sizes of concentrations of industrial activity and jobs at various production centres and yet it postulates uniform distribution of consuming population.

A second basic limitation of the Löschian argument is that it pertains to situations where raw materials are not required (as in service activities) or are ubiquitous and available everywhere at the same costs. The argument, therefore excludes that production (whether market or material oriented) where material sources exert significant locational pulls.

### CENTRAL PLACE SYSTEMS

In spatial organization the hierarchical ordering of central places gives differential spatial spread of central places in rural areas. Thus, in a central place study, "the system concept emphasizes wholeness-inter-dependence and dynamic relationship"[46] of a number of variables at a place which survives due to infra-structural input and output through feed-back mechanism and its transformation both within and between a system and its environment.

It is over these simple or isolated, multiple, open or closed systems that the dispersed nature of service centres over countryside exists. When only one system, i.e., marketing system gets priority over a larger centre, supported by smaller centres in the surrounding areas, it is called a simple system. But when several systems namely transportation system, marketing system and electric system etc. with multiple nodes coalesce or are mixed together, then it is known as multiple system. Similarly, when a system lies over unbounded plain, it should be recognized as open system and in case it is closed by physical barriers like river, monticle or political boundaries then it will be recognised as a closed system.

North Bihar has several sub-systems of central places, e.g., Muzaffarpur, Darbhanga, Chapra, Purnea, Saharsa, Samastipur, Sitamarhi, Motihari and others. All these larger central places are second order centres of Patna, the larger system. Thus all the larger central places forming sub-systems in North Bihar have several supporting lower-order centres and in turn larger centres provide services to lower-order ones. In this way the process of feed-back mechanism and transformation of input and output according to the regional environmental condition helps in forming several sub-systems of central places in North Bihar.

It is these inter-centre linkages that are the true cement in the structure of the system. Therefore, a central place umland only identifies a complete central place system, if it is delimited in a manner which takes into account the actual spatial interdependences among central places.

## CONCLUSION

A Central Place serves its hinterland people and in turn the people serve the central place. This develops a consumer behaviour pattern in the case of the urban place and daily commuting to the fields in the case of villages and these in turn generally develop a circular zone of influence. Since the circular geometric form fails to cover the entire space and people, Christaller and Lösch have arranged this circular trade areas into a hexagonal principle. At this juncture for the increase of market areas and the number of lower order central places, the K concept has been introduced for filling of space through hexagons. In this way K-3 systems of central places have developed for marketing centres, K-4 for transportation nodes and K-7 for administrative places. Lösch had also introduced variable K concept, but Isard presented a more realistic picture with the analysis of real world situations after considering population density, agricultural and industrial production, etc. In system approach mostly open system has been found to be fit for the central places of North Bihar and not closed or isolated systems.

## REFERENCES


1. M. Jefferson, "Distribution of the Worlds City Folks," *Geographical Review*, Vol. 21, 1931, p 453.
2. W. Christaller, *Central Places in Southern Germany*, translated by C.W. Baskin, Englewood Cliffs, N.J., Prentice Hall, 1966.
3. E.L. Ullman, "A Theory of Location for Cities", *American Journal of Sociology*, Vol. 46, 1941, pp 853-864.

4. K.S O. Beavon, "Interpreting Lösch on an Intra-Urban Scale," *The South African Geographical Journal*, Vol. 56, No. 1, 1974, p 36.
5. W. Isard, *Location and Space Economy* Chapman & Hall Ltd., London, 1956, p 287.
6. R.P. Misra, *Growth Poles and Growth Centres in Urban and Regional Planning in India*, Institute of Development Studies, University of Mysore, 1971.
7. B.J.L. Berry and H.M. Mayer, "Design and Preliminary Findings of the University of Chicago's Studies of the Central Place Hierarchy," *Lund Studies in Geography*, SER. B, No. 24, 1961, pp 247-252.
8. E.D. Brunner and J.D. Colb, *Rural Social Trends*, New York, 1933, p 95.
9. B.J.L. Berry, *Geography of Market Centres and Retail Distribution*, Prentice Hall, N.J. 1967, p 3.
10. J.U. Marshall, *The Location of Service Towns, An Approach to the Analysis of Central Place Systems*, University of Toronto, 1971, p 23.
11. K.S.O. Beavon and A.S. Mabin, "The Lösch System of Market Areas, Derivation and Extension," *Geographical Analysis*, 1975.
12. P.E. Lloyd and P. Dicken, *Location in Space : A Theoretical Approach to Economic Geography*, Harper & Row, New York, 1972, p 23.
13. K.S.O. Beavon, "Generalising the Intra-Urban Model Based on Lösch, "*The South African Geographical Journal*, Vol. 56, No. 2, 1974, pp 1-2.
14. M. Jefferson, 1931, op. cit.
15. W. Christaller, 1966, op cit.
16. A. Lösch, *The Economics of Location*, translated by W.H. Woglom and W.F. Stolper, Yale University Press, New Haven, 1954, pp 109-134.
17. B.J.L. Berry and W.L. Garrison, "The Functional Bases of the Central Place Hierarchy", *Economic Geography*, Vol. 34, 1958, pp 145-154.
18. J.E. Vance, "The Merchants World," *The Geography of Wholesaling Englewood Clifs*, N.J., Prentice Hall, 1970.
19. B. Thakur, "The Changing Functional Make Up of Central Place Hierarchy in South-Western Ontario Between 1871 and 1971," *M.A. Thesis*, Department of Geography, University of Waterloo, 1972.
20. B.J.L Berry and A. Pred, *Central Place Studies : A Bibliography of Theory and Applications*, Philadelphia, 1961-1965.
21. N.R. Kar, "Urban Hierarchy and Central Functions Around Calcutta in Lower West Bengal," *Lund Studies in Geography* No. 24, 1962, p 250.
22. A.K. Dutt and S. Banerjee, "Transportations, Index in West Bengal : A Means to Determine Central Place Hierarchy", *The National Geographical Journal of India*, Vol. XVI; Part 3-4 1970, pp 199-207.

23. S.M. Singh, "The Stability Theory of Rural Central Place Development, *The National Geographical Journal of India*, Vol. II, 1965, pp 13-21.
24. O.P. Singh, "Functions and Functional Classes of Central Places in U.P., "*The National Geographical Journal of India*, Vol. XIV, Part 2-3, 1968, pp 1-2.
25. S. Folke, "Central Place Systems and Spatial Interaction In Nilgiris and Coorg," *Denmark*, Edt. by Jacobson & Jensen; Copenhagen, 1968, pp 55-69.
26. L.N. Ram, "Cartographic Analysis of Central Functions," *Indian Geographical Studies*, R.B. No. 1, 1973, pp 82-91.
27. P. Deka, "On the Cheapest Cost Single Route Location," *The NGSI*, Vol. XX, Part-2, 1974, pp 128-130.
28. R.B. Mandal, "Central Place Hierarchy in Bihar Plain," *The National Geographical Journal of India*, Vol. XXI, Part 2, 1975, pp 120-126.
29. L.S. Bhat, (Ed.), *Micro Level Planning*, Kumar Brothers, Delhi, 1976.
30. B.J.L. Berry, 1967, op. cit., p 68.
31. R.E. Preston, "The Structure of Central Place System," *Economic Geography*, 1971.
32. C.D. Harris and E.L. Ullman, "The Nature of Cities," *Annals of the American Academy of Political and Social Science*, Vol. 242, 1945, pp 7-17.
33. S. Godlund, "The Functions and Growth of Bus Traffic within the Sphere of Urban Influence", Land Studies in Geography, Series B, *Human Geography*, No. 18, 1956, p 12.
34. K.N. Singh, "Rural Markets and Rurban Centres in Uttar Pradesh," *Ph.D. Thesis*, Banaras Hindu University, 1962, p 187.
35. M.S. Vishwanath, "Growth Pattern and Hierarchy of Urban Centres in Mysore", *The Indian Geographical Journal* Vol XLVII; 1972, Vol. 1 and 2, pp 1-11.
36. V.N.P. Sinha, *Chotanagpur Plateau : A Study in Settlement Geography*, K.B. Publications, New Delhi, 1976, p 175.
37. W.K.D. Davies, "Centrality and the Central Place Hierarchy, Urban Studies, Vol. 4, No. 1, 1967, p 63.
38. J.F. Kolar and J.D. Nystuen, *Geography*, McGraw Hill Book Company, New York, 1974, p 83.
39. W.K.D. Davies, op. cit.
40. B.J.L. Berry, op. cit. p 68.
41. W. Isard, 1956, op. cit.
42. K.S.O. Beavon, "A Procedure for Constructing Lösch Regional System of Markets," *South African Journal of Science*, Vol. 69, 1973, pp 377-379.
43. K.S.O. Beavon, 1974, op. cit.
44. Ibid.
45. W. Isard, 1956, op. cit.
46. P.E. Lloyd and P. Dicken 1972, op. cit.

3

# A Critical Review of Central Place Study in India

—*B.L. Sinha*

"The whole problem lies in the quantification of the Central importance of a place."

—*Christaller*

## MEASUREMENT OF CENTRALITY AND HIERARCHY

HIERARCHY—originally a religious term (hierarch—"religious leader") which denotes the organisation of a ruling body of clergy into ranks—is today used to ascertain an order or gradation of phenomena, each grade being subordinate to the one above it and superior to the one below; and can be established on the basis of many criteria and can be applied to many phenomena. The functional organisation of regions and the scientific laws governing them have invoked interest not only in geographers but in political economists and sociologists as well. It was as early as 1809 that the concept of the hierarchical ordering of centres and areas appeared in literature dealing with political economy. A Müller advocated the fundamentals of political science that "the human and economic forces operating on the surface of the earth produce nearly mathematical areas, which are arranged in relation to each other like the cells of a honeycomb. At the centre of each such cell is situated the dwelling of a tiller of the soil. Numerous cells form a larger totality, at the Centre of which is situated a city. The cities further form an economic totality gravitating towards the capital city."[1] But the German scholar Walter Christaller (1893-1969) is acknowledged to be the father of classical central place theory[2] which may be called the most important contribution of geography to theoretical social science. In 1932, he submitted to the University of Erlangen his now-famous doctoral thesis (published in 1933) on Central Places of Southern Germany. His main work on the Centres of Southern Germany and the hierarchical system formed by them represents the first study to result in a truly causal and logical general conception. Christaller's ideas were based on those of locational theorists like J.G. Kohl, Johann Von Thünen, his former teacher Alfred Weber and the German settlement geographer Robert Gradmann. His works had little impact in contemporary Germany and it was not until their introduction into the United States in the 1940s and 1950s that their value was realized. Since then, Christaller's ideas have been verified, improved upon and extended to analyse the pattern of Central Places within cities, such as urban shopping districts, as well as patterns of cities themselves. The term 'Central Place'[3] was, however, first used by Mark Jefferson in 1931 for a settlement which is necessarily a focus of various activities mostly socio-economic in nature for its surrounding hinterland. Nevertheless, the term was standardised only after the classical study of Christaller in 1933.[4]

Central Place theory is concerned with the analysis of spacing, size, distribution and complementary areas of Central Places either urban[5] or rural[6] in character. Central places are not conceived simply as agglomerations of population, as accumulations of shops, as sales totals for trades and services, or as areal expressions of political organisation. The basic element of a Central Place is the role as a source of goods and services for an area larger than itself. Each region portrays a certain system of spatio-functional-organisational-coherence and Centres ranging from small hamlets or village markets to large towns and cities are inter-connectedly diffused over the entire area, creating an economic balance by accomplishing multi-farious functions—primary and secondary, small and large and specialized and diversified—for the area.[7] Central places vary in importance and there exist Central Places of different levels depending upon their *Versorgungsbereich*. Even a village is a Central Place to some extent and to a small area. The term 'Central' is relative and some places are more Central than others. Higher-order Centres stock a wide array of goods and services and are visited for higher types of needs which can hardly be fulfilled by lower-order Centres. Lower-order Centres, on the other hand, stock a smaller range of goods and services, that is, some limited part of the range offered by the higher Centres and thus fulfil local requirements. The higher-order places also possess the lower-order ubiquitous functions and are fewer in number and more widely spaced than the lower-order places and so command a larger trade area. Hence, all the Centres of a region are not equally important in the total areal functional organisation, rather they differ in their service provisioning capabilities or potentialities—whether qualitative or quantitative, as the number, quality and complexity of service or Central Place functions increase much more in higher order Centres. The basic variances regarding the importance of a place has given birth ultimately to a concept of hierarchy and hierarchical class-system of Central Places, as the Centrality or importance of a Centre as a Central Place produces a consecutive arrangement of central places ranging from the lowest or the smallest to the highest or the largest. Many grades of Central Places and of their regions which are areally and functionally interlinked together and form parts of a single large region are, therefore, distinguished, jointly produce an interlocked pattern of spatial organisation.

Such a grading of settlements is based on the centrality of settlements. The Centrality can be described in terms of extent, number, quality and quantity of functions performed by the settlement. All those activities performed for the surrounding area are designated as Central Functions and the Service Centres providing them are "Central Places." They are, thus, essentially distributional and are non-ubiquitous in nature. They are services provided only at a Central Place and made available to the people of the surrounding region, as opposed to those that might be available anywhere without focus and those that are produced for distant, even foreign, markets and are thus of no relevance to the local consumers. The quality of a service or central place function is normally affected by (i) the number of different types of services offered and (ii) by the level at which they are offered. Just as not all the settlements perform all the functions in the same proportion, similarly, every settlement does not perform all the functions at the same level. Hamlets, towns and cities are called 'Central' only in relation to their surrounding areas. The importance of a place is also usually but inaccurately defined by the size of population of the place or by its areal extent. Neither population nor area sufficiently expresses the meaning of the word 'importance' or 'Centrality.' Correlations have, however, been found to exist between size of population and degree of centrality but an estimation of the services of the Centre to the region and of the services of the region to the Centre along with a measurement of the area dependent on the Centre for an array of goods and services will decide the criteria of Centrality. Arrangement of Centres into definite hierarchical classes, however, becomes, in greater proportions, arbitrary and still necessary for practical purposes.

## THE SCALE OF FUNCTIONAL HIERARCHY

The hierarchy of settlements is linked up with the hierarchy of Central functions. The latter could be ascertained by considering individual Central function such as administration, education, health services, cultural institutions and others separately and by distinguishing its component parts. "A Central function is composed of many sub-functions and thus, within a particular Central function it is possible to identify different levels at which it is being performed."[8] For example, schools provide just one good Central Place organisation in any region. The local primary school provides a lower-order Centre (to use Christaller's term) which serves a small part of a large city or a single rural village. There are a large number of such schools in most of the regions and they teach children drawn from only a few sq. km. (i.e. they have a small service area). Above the primary schools come the higher-order services offered by the middle schools, high schools, colleges of varied kinds and at the top of the ladder stands the University. As we move higher up the educational ladder, the number of Centres becomes smaller and their complementary regions become larger. As said, education is but one range of Central goods and services that give character to Central Place organisation and help to distinguish the Central Place functions of one settlement from those of another. On this basis, it is possible to construct a scale of what may be called a functional hierarchy.[9]

In the strict sense the scale of functional hierarchy can be explained in range of relative importance of functions. The higher the level of functional hierarchy, the higher is the Centrality of the settlement with that function. Also, the significance of a Central Place is affected by the number of services or functions offered or performed. A Central Place which offers a number of functions of a distinctive level is more significant than a settlement providing less number of facilities of the same level. If the number of services offered is about the same and if there is a variation in the level of services being offered the place having a set of facilities of a higher order will be more important than the settlement having a similar number of functions or facilities of a lower-order. Thus the level of functional hierarchy becomes a decisive part of the whole question of Centrality. As the level and number of functions are likely to vary from place to place (also from time to time) it is felt that it will be possible to determine Centrality of settlements in any region by measuring the level and number of functions performed.[10]

## DIFFERENT METHODS OR CRITERIA OF MEASURING CENTRALITY AND HIERARCHY IN INDIA

Much of the work relating to Central Place theory is concerned with the identification of hierarchies of settlements as Central Places. And all possible criteria for measuring the Centrality of Centres were broadly suggested by Ullman as early as 1941.[11] The determination of the functional status of a place or Centre has ceased to be a purely academic problem : it has become a technical one. The existence and nature of the Central Place hierarchy have been hotly debated and one criticism has been that methods used to detect the hierarchy are subjective and probably biased. So, a definite need exists to measure Centrality in an objective manner. And it is only within the last two decades that a proposition of a hierarchy of Central Places has been objectively tested,[12] though the problem of measuring Centrality objectively has not been completely solved. The purpose of this paper is to make a review of the various schemes of criteria employed for Centrality measurement within the framework of Christaller's theoretical requirements and to suggest then, a relevant centrality measure. Two key problems which confront researchers in applying the Central Place theory in its classical[13] forms are (i) an inability to ascertain the importance of settlements as Central Places (their Centrality); and (ii) a want of data series providing both for measurement of the importance of settlements as Central Places and for identification of the linkage structure of Central Place system.[14] However, on account of the interest in and the importance of the Central Place, the literature concerned with it has now grown tremendously. There is space here to consider only such works as are related to India.

Most of the Central Place studies have been carried out in the West but the number of analyses from developing and underdeveloped countries is quite large[15] to testify to the utility and applicability of the Central Place theory in these parts of the world. But the number of Central Place studies from India is rather small. Most of the predominantly empirical studies that have been made here are limited only to part-problems—the determination and gradation of Centres in some of the smaller regions. And the need for a macro-regional methodology for measuring the importance and hierarchy of Centres in India, as elsewhere in the world, basically suffers due to specific work done in general. J.B. Parr rightly comments, "There have been relatively few Central Place studies on the scale of larger regions."[16]

As Christaller's single measure—the number of telephone connections—as the most characteristic common denominator of the importance of Central Place—has been turned down, many other parameters have subsequently been suggested at empirical verification of hierarchies following three main approaches : (i) Measuring the importance of a Central Place in terms of the variety and magnitude of functions performed; (ii) identifying the hierarchical breaks on the basis of the magnitude of the market areas served consistent with the threshold and range concepts; and (iii) on the basis of the pattern and intensity of functional interaction among Central Places and their respective hinterlands.[17]

Estimation of functions and institutions existing at the Centre has been one of the simplest, most popular and probably adequate indicators of Centrality which can fully be measured in terms of all the goods and services offered by the Central Places to their surrounding regions, but studies have been based either on single indicator or on totality of functions.

The studies of Kar, Mayfield, Folke, Bhattacharya Krishnan and Chandra, Prakash Rao, Rao and Ramachandran, Mukherjee and others belong to this category.

The investigation by Kar[18] of the urban hierarchy in Calcutta region actually penetrated into the Central Place significance of Calcutta and the towns of lower West Bengal. In his survey of the Central Place System in two districts (Ludhiana and Jullunder) of the Punjab in 1967, R.C. Mayfield[19] used as many as 129 Central functions—not missing even functions like burning ghats, opium sales, astrologers and fortune tellers—to determine the hierarchical orders. He focused attention on the need for research into the great regional variety in India and pointed out that his own method would be of limited help in other, economically less advanced parts of India. Using functional indicators such as administration, banks, post offices, high schools and *shandies* (weekly market)—each category representing one type of Central function and most of the types sub-divided according to assumed position in the functional hierarchy. Steen Folke[20] investigated the Central Place System of Nilgiris and Coorg and determined four grades. Likewise, Bimlendu Bhattacharya[21] selected Central functions such as finance, education, health, transport, recreation and Bata shops and graded the Central Places of North Bengal into six categories. Following the method of H.E. Bracey,[22] Gopal Krishnan and M.M. Chandra[23] identified all the service Centres in Sirmaur district of Himachal Pradesh in the outer Himalayas based on data collected from the 965 villages and three towns in the district by means of questionnaires asking the names of places from which each village was obtaining services such as retail store, ration depot, agricultural market, temple, middle school, secondary school, post office, veterinary dispensary, personal services of a barber, tailor, goldsmith, etc., bus stand, panchayat headquarters, tahsil/sub-tahsil headquarters, etc. They then classified the Services Centres into the following order—(i) highest; (ii) Second; (iii) Third; (iv) Fourth; (v) Fifth; and (vi) Lowest. A simpler and more easily applicable method was used by V.L.S. Prakasa

Rao[24] considering only 8 general indicators for his Mysore study in 1964. Later on, Prakasa Rao and Ramachandra[25] collected data on establishments for 19 selected Central Place functions, irrespective of retail significance for hierarchical planning in 1974. The selected criteria in the West have generally been related to the retailing activity, because published data are available in appropriate proportions, but in India, retailing has not been of particular significance due mainly to paucity of such published data. One more attempt in this regard is that of S.P. Mukherjee[26] who in a bid to establish market hierarchy in Darjeeling district took into consideration retail trade and wholesale business in addition to the Central services like administration, public utilities, etc. and arrived at the following classes of Centres—(i) Tea Estates; (ii) Agricultural village; (iii) Agricultural market place; (iv) Agricultural Collecting Centre; (v) Urbanised village; (vi) Semi-towns; (vii) Ordinary towns; (viii) Sub-divisional towns (ix) District town; and (x) Commercial towns.

Robinson[27] rightly noted that "one is being narrowly definitive in taking the retailing sector alone as an indicator of 'Centrality' ". In a real sense, the classical Central Place theory was "very much a theory relating to service sector location than just retailing location", and hence if Centrality is to be given a "meaning more in line" with Christaller's suppositions, full variety of Central Place functions must be considered[28] as done in the studies of M. Polomaki[29], P. Scott[30] and Bell Leiber and Ruston.[31]

Individually, the studies that have been noted in the above paragraphs concern with the gradation of settlements after a long personal inspection of the study area. Also a certain degree of success in grading centres may have been attained but the difficulty arises when one tries to compare one area directly with another area. This is mainly because of the difference in the grading system. R.H.T. Smith[32] has rightly remarked that a bewildering array of classification techniques have been used so that results seem bound to vary. Some of the above studies use the method of summation of all Central functions. Giving equal status to all kinds of functions invited criticism in all such studies. In doing so, each function is given equal weight and the result is that a small family grocer becomes equivalent to a jeweller. Hence various types of functions must be differentiated. Keeping this in view, the weighting system of various functions was used in some of the studies which has also been gradually refined in other studies in tune with the local conditions. But the basic problem of subjectivity related to the quantification of the aggregate importance of a Central Place remains unsolved. The result, again is that different studies are not directly comparable. The inseparability of the basic, non-basic components of functions is another inherent obstruction in the use of such methods. In fine, all the methods whether identifying several key functions or the grading system of point allocation schemes, or the use of number of active persons in retail trade and service as the basis for Centrality index employ, in some degree, the subjective element in grouping of Centres which severely limits any further use of the hierarchy, if at all identified. It is on such ground that Vining[33] of USA, noted that arbitrary boundaries have been drawn between the categories in many hierarchical classifications and he contended that centres form an evenly diminishing continuous rather than a hierarchical class. It is generally admitted, however, that aggregate analysis inevitably emphasizes the continuous arrangement, whereas elemental enquiry normally identifies a hierarchy.

The functional structure of areas has been studied from many angles and in quite diverse regions. Berry and Garrison[34] brought about a major breakthrough in methodology by applying a highly developed enumerative principle which was followed in India also by Prakasa Rao and others. Prakasa Rao[35] found a five-tier hierarchy in Muzaffarnagar district of India on the basis of threshold population groupings. A good deal of work related to settlement hierarchy has been done at the National Institute of Community Development, Hyderabad by a team of esteemed workers including L.K. Sen, S. Wanmali, S. Bose, G.K. Mishra and K.S. Ramesh,[36] who have considered all possible functions, viz., administration, judiciary, internal security, bus-service, education, health services, communication, finance, trade and shopping. Wanmali[37] ranked 'function' in relation to 'threshold population' for Miryalguda taluk and suggested 'Median population threshold' to rank settlement when the size of the universe is large. Later on,

Narayan and Rao[38] also followed this approach and Sundaram[39] took into consideration all socio-economic facilities in order to ascertain a hierarchy of Centres in Bastar District of Madhya Pradesh. But this view of threshold requirements is also not free from criticism. Several notional insufficiencies of the method have been noted : (i) Thus Bunge[40] criticised the method on the point that it is the population of complementary area that matters for the threshold requirements rather than that of the settlement itself. Every Central Place offers some goods and services to persons living in neither the Central Place nor in its complementary region (and which is designated as irregular consumption[41]). The use of settlement population as threshold also ignored this type of consumption mainly by highway users who also contributed a large part of the true threshold population of settlement. Secondly such study has related itself only to small areas and to small Centres. Therefore, duplication elsewhere is required. Finally, the use of population totals in the analysis is essentially a deficiency because the numbers of people found in any area may be unrelated to the Central Place (tertiary and quarternary[42]) function of a place. They may be dependent upon primary or secondary employment. And it is also worthwhile to note here that investment in public and private infra-structure, though often made in virtually all parts of rural India, the facility-and-programme-centres are not essentially established on threshold grounds. In other words, the commercial establishments and service facilities are not always located on economic viability considerations. Schools are generally located by education departments and the donors of land. Even degree colleges are found at such small place where not even a permanent restaurant or betel shop is traceable. They are not always distributed with respect to population size of centres because their opening depends more on the initiative and zeal of the local elite. Some services like health facilities, family planning clinics and banks are also located on welfare consideration without any consideration of the population of centres. They are opened in case of health facilities by health departments and in case of banks by their regional and national headquarters. Nodality of location is a more important consideration in such cases. Nodal locations are also the main consideration in the opening of shops dealing in cycles and autoparts and repairs in most cases. Cinema houses prefer locations where there is greater assembly of outside population. In some cases, markets are typically developed by private investors and departments of agriculture. The result, in some cases, is that many investments are dispersed that might have been located more efficiently and equitably in hierarchies of service centres. In fine, it can be said that though service activities are located throughout a region to provide convenient access to the population they are designed to serve, yet most facility location patterns for service activities have not, however, explicitly been planned to achieve this goal.

The problem of measuring Centrality has also been tackled by considering the criterion of occupations described in census records. Population Census figures are available for nine 'workers' categories. The first-three being related to primary activities (which are by no means Central functions) and the next three to manufacturing and household industry and construction. Although the latter are important city-building functions, yet their Central importance is very doubtful. The last-three categories are 'Trade and Commerce', Transport, Storage and Communication and 'other services'. All these are, in varying proportions, of Central importance and may be regarded as 'Central occupations'.

R.L. Singh[43] has attempted a tentative classification of towns in the Umland of Banaras taking into consideration both the size and service factors and has determined four grades. Kashi N. Singh[44] has used the following formula which issimilar to that of Godlund :

$$C = \frac{N \cdot 100}{P}$$

where, N=Number of persons depending upon commerce and
P=the total regional commercial population.

A measure similar to that of Davies' relative centrality index was adopted by O.P. Singh[45] making use of commercial population of Centres. He used the following formulae to calculate relative and absolute centrality index :

$$R_{p1} = \frac{P_{c1} - \left( P_{t1} \frac{R_c}{R_t} \right)}{\Sigma P_c - \left( \Sigma P_t \frac{R_c}{P_t} \right)} .100$$

Where $P_{c1}$ = Commercial population of the place or Centre,

$P_{t1}$ = total population of the place,

$R_c$ = Regional Commercial population,

$R_t$ = Total regional population,

$\Sigma P_c$ = Summation of $P_{c1}$, $P_{c2}$ . . . $P_{cn}$ and

$\Sigma P_t$ = Summation of $P_{t1}$, $P_{t2}$ . . . $P_{tn}$

Absolute centrality was calculated thus :

$$A_{c1} = \frac{P_b}{100000}$$

Where, $P_b$=Basic population of the place.

In order to make a conparison between the total size and the Centrality of a Centre the size index was also calculated this way :

$$S.I. = \frac{P_{t1} . 100}{\Sigma P_t}$$

Onkar Singh[46] also used the commercial worker data for his hierarchical study of U.P. towns and his final Centrality index was computed as follows :

$$\frac{E \times 100}{T}$$

where, E=excess commercial worker population of the town and T=total excess commercial worker population of all the urban Centres of the State. Considering that more categories of workers could be of Central significance and modifying the consideration of Godlund and K.N. Singh both, S.N.P. Jayaswal[47] used the following formula :

$$C = \frac{N \times 100}{P}$$

where, C =required Centrality,

N = Number of persons employed in manufacturing, trade and commerce, transport and 'other services' in a Centre and

P = the total number of persons in the region engaged in the above categories.

The relevant Centrality measure proposed by N.P. Saxena and R.C Tyagi[48] is as follows :

$$C = A - \frac{P \times A_r}{P_r}$$

where, C = Centrality measure;

A = Total number of workers engaged in Central functions at a Centre;

P = Population of the Centre;

$A_r$ = Aggregate of workers engaged in Central occupations in the region; and

$P_r$ = Population of a region.

Usha Agrawal[49] and S.S. Bansal[50] also employed the data pertaining to all categories of non-primary workers in their studies.

As we have already seen that the absolute population size of Centres failed to provide a measure of threshold for grouping of function and establishments, it was decided by some workers to use the commercial population of the Centres which is expected to fulfil this end better because it omits the number dependent upon agricultural occupations. But the use of commercial population is basically deficient on so many grounds. One main reason is that enumerators often record the persons belonging to the trading community and not those persons who are actually engaged in trade and commerce. At many places, it is seen, that the number of commercial population is less than the number of establishments. It is an almost impossible proposition. The author, in a study on 'Central Place System in Mithila Plain' has noted this discrepancy on the larger scale. The following Table 3.1 is quite illustrative on this point :

TABLE 3.1

| S. No. | Place | Block | District | Number of workers in Trade and Commerce | Number of shops |
|---|---|---|---|---|---|
| 1 | 2 | 3 | 4 | 5 | 6 |
| 1. | Valmikinagar | Sidhaw | West Champaran | 334 | 350 |
| 2. | Lauria | Laurian | ,, | 144 | 315 |
| 3. | Haripur | Sikta | ,, | 397 | 478 |
| 4. | Majhwalia | Majhwalia | ,, | 41 | 54 |
| 5. | Patilar | Bagaha | ,, | 104 | 240 |
| 6. | Mainatanr | Mainatanr | ,, | 11 | 69 |
| 7. | Chakia | Chakia | East Champaran | 271 | 518 |
| 8. | Keshariya | Keshariya | ,, | 162 | 252 |
| 9. | Areraj | Areraj | ,, | 245 | 404 |
| 10. | Ramgarhwa | Ramgarhwa | ,, | 259 | 568 |
| 11. | Ghorasahan | Ghorasahan | ,, | 317 | 650 |
| 12. | Dhaka | Dhaka | ,, | 97 | 225 |
| 13. | Kasba Mehsi | Mehsi | ,, | 129 | 150 |
| 14. | Turkaulia | Turkaulia | ,, | 68 | 137 |
| 15. | Sangrampur | Areraj | ,, | 13 | 260 |
| 16. | Patahijadu | Patahi | ,, | 17 | 50 |
| 17. | Bagahi Bhelwa | Ghorasahan | ,, | 3 | 30 |
| 18. | Pupri | Pupri | Sitamarhi | 751 | 1134 |
| 19. | Riga | Riga | ,, | 199 | 216 |
| 20. | Sonbarsa | Sonbarsa | ,, | 93 | 115 |
| 21. | Sursand | Sursand | ,, | 450 | 635 |
| 22. | Puriharpur | Puriharpur | ,, | 52 | 66 |
| 23. | Choraut | Pupri | ,, | 40 | 56 |
| 24. | Parsauni | Belsand | ,, | 40 | 392 |
| 25. | Athri | Runisaidpur | ,, | 42 | 50 |
| 26. | Baragaon | Bajpatti | ,, | 110 | 142 |
| 27. | Sakra | Sakra | Muzaffarpur | 140 | 148 |
| 28. | Kanti | Kanti | ,, | 50 | 208 |
| 29. | Deoria | Paroo | ,, | 55 | 95 |
| 30. | Dharphari | Paroo | ,, | 7 | 25 |
| 31. | Sahebganj | Sahebganj | ,, | 14 | 65 |
| 32. | Baruraj | Baruraj | ,, | 13 | 30 |
| 33. | Minapur | Minapur | ,, | 5 | 30 |
| 34. | Bhagwanpur | Bochaha | ,, | 5 | 25 |
| 35. | Desari | Shahdai Buzurg | Vaishali | 115 | 195 |

TABLE 3.1—*Contd.*

| 1 | 2 | 3 | 4 | 5 | 6 |
|---|---|---|---|---|---|
| 36. | Biddupur | Biddupur | Vaishali | 72 | 125 |
| 37. | Prataptan | Lalganj | ,, | 41 | 65 |
| 38. | Baheri | Beheri | Darbhanga | 156 | 239 |
| 39. | Singhwara | Singhwara | ,, | 27 | 90 |
| 40. | Nehara | Manigachhi | ,, | 39 | 55 |
| 41. | Supaul | Biraul | ,, | 199 | 268 |
| 42. | Jale | Jale | ,, | 13 | 25 |
| 43. | Mohan Bahera | Benipur | ,, | 17 | 255 |
| 44. | Dalsinghsarai | Dalsinghsarai | Samastipur | 905 | 1025 |
| 45. | Chaklawaini | Pusa | ,, | 129 | 148 |
| 46. | Mohiuddinagar | Mohiuddi-nagar | ,, | 216 | 300 |
| 47. | Harpur Pusa | Pusa | ,, | 25 | 45 |
| 48. | Nayanagar | Hasanpur | ,, | 14 | 80 |
| 49. | Hasanpur | Hasanpur | ,, | 9 | 200 |
| 50. | Madhwapur | Madhwapur | Madhubani | 164 | 275 |
| 51. | Barhi | Babu Barhi | ,, | 47 | 125 |
| 52. | Jhanjharpur | Jhanjharpur | ,, | 186 | 500 |
| 53. | Ghogaddia | Phulparas | ,, | 43 | 221 |
| 54. | Andhratharhi | Andhratharhi | ,, | 9 | 96 |
| 55. | Sakri | Pandaul | ,, | 66 | 358 |
| 56. | Benipatti | Benipatti | ,, | 50 | 97 |
| 57. | Bakhri | Bakhri | Begusarai | 333 | 350 |
| 58. | Sahebpur Kamal | Sahebpur Kamal | ,, | 127 | 225 |
| 59. | Fateha | Bachhwara | ,, | 29 | 90 |
| 60. | Majhaul | Cheria Bariarpur | ,, | 157 | 250 |

*Source* : District Census Hand Books—Champaran; Muzaffarpur; Darbhanga; Monghyr—Census 1971, Series 4.
Number of Shops—noted from Bihar Govt. Secretariat.

The third group of studies have attempted to measure the attractive power of Centres with reference to some measurement of the actual linkage—structures between Centre and the complementary region. This intensity of functional interaction corresponds to an index of the centrality of Central Place. Some particular services may be selected out of those operating between the Centre and its hinterland (activities to and from the Centre and the region). One of such media to measure centrality can be the number of telephone connections. Thus, the telephone index was used by Jain[51] for the Vidarbha region in 1971 in exactly Christaller's fashion where the index of Centrality is :

$$C = t\left(\frac{PT}{P}\right)$$

where, C = index of Centrality,

t = number of telephones in the town,

P = population of town;

T = number of telephones in the whole region;

P = population of the whole region.

But this method would not seem to be valid for a country where telephones are utilized extensively by non-Central establishments and for non-Central functions as well.[52] Some of the Central institutions on the other hand, do not make use of telephones. Some smaller Centres do not have any telephone connections while in larger Centres the number of telephone connections hardly corresponds to the size and efficiency of Central institutions. Such a type of study becomes particularly meaningless in the Indian context where the telephone connections are usually confined to larger towns.

The other media to measure Centrality can be the number of automobiles entering a place, avoiding the sub-urban movement of traffic.[53] Greater intensity of functional interaction as evidenced by the density and distance of traffic converging on a Centre means greater range of goods and services offered at the Centre resulting in greater magnitude of the market area and hence the higher Centrality value of the place. One great advantage of this method is that it takes into notice mostly the basic component of Central functions because the traffic flow from and to a Centre indicates its functional linkages with the complementary region provided care is taken to separate the bridge traffic. At the same time, it also delimits the intensity of the spheres of influence of Centres, and thus defines in an objective style the system coinciding with sphere of influence of the highest order Centre which gives a meaningful frame for hierarchical ranking of Centres using techniques of location co-efficient type or those based on the analysis of threshold requirements of market areas. But we must also realize that the specific type of linkage has historical and spatial limitations and should always remember that 'such a specific linkage is certainly useful within one country (assuming a relatively uniform socio-economic grouping) and may be directly comparable with another state that has similar patterns (for example, as between Britain and Sweden). However, the linkage may perform a completely different function in other countries, for example, the United States of America and India. In the former case citizens are affluent and use private transport, in the latter poverty often rules out the use of public transport for purely local connection,[54] some of the important studies in India in this regard are those of Onkar Singh, Ashok K. Dutt and Sanjay Banerjee, Lata Chatterjee and Reed. Onkar Singh,[55] in his U.P. study, gave it half weight, Ashok K. Dutt and Sanjay Banerjee[56] also took the transportation index in West Bengal and classified Centres into six grades. This concept of flows between the town and country was, in fact, the major theme of investigation by Lata Chatterjee[57] for West Bengal. All the linkage studies in India pertain to lower order centres only. But Reed by paying attention to the upper end of the spectrum analysed the air passenger traffic of India and on this basis of the air connectivity analysis, identified a four level hierarchy of Indian cities. The volume and direction of flows provide proof of the competency of activities in one city to control or dominate activities throughout an urban system. These are used to class the proportion of dominance of control in the hierarchy of Central Place System.

## THE CONCEPT OF 'CENTRALITY'

We have examined different criteria for measuring Centrality and observed that the various schemes employed in hierarchical orderings invoke either directly or indirectly the functions of a place. But the presence of a function in a settlement, as Tarrant suggested with reference to the use of indicator functions, is not compulsorily the indication of a Central function performed there.[58] All those activities performed for the surrounding area are designated as Central functions and hence are offered at a few necessarily Central points in order to be consumed at many scattered points.[59] They are essentially distributional and 'city building' or 'basic' functions of a place. These activities are of two types: (i) those services which require movement on the part of the users or consumers in order to partake of them and are normally, though not necessarily, found at the lower spectrum of the hierarchy and include such services as retailing entertainment and personal; (ii) those classes of functions which are performed on behalf of consumers, both individual and institutional and do not need direct contact between consumers and suppliers. These are usually specialized functions found at higher spectrum of the hierarchy and include wholesaling, mail order business, certain types of postal and banking services, etc. performed both for the metropolitan market and the regional hinterland.[60] It is also legitimate to note here that there appears no consensus (at least in India) as to which functions should be termed as Central functions.[61]

Christaller also differentiated between the terms 'importance' and 'Centrality'[62] which are quite different but related concepts and identify the aggregate and relative importance of settlements respectively. A settlement possesses a certain importance which is usually expressed inexactly by its size (which, in turn, is measured by the population totals of the settlement). So he advocated that if the importance of a place were to be determined only by its size, part of the importance must be attributed to the settlement itself as an agglomeration of population and another part of the settlement as a Central Place. The sum of these two parts he called absolute importance of the town whereas the part that is attributed to its area-serving capacity—as a source of goods and services for a surrounding region—is its relative importance or Centrality. The relative importance thus, shows the degree to which a town is a Central Place. To remove ambiguity, the term nodality is used by some workers[63] for the aggregate importance of a place as a provider of Central goods and services both for its own residents and for other consumers and the term centrality stands for the importance of a place as a supplier of central goods and services in excess of the needs of its own residents. The terms 'importance' and 'centrality' have been substituted for absolute centrality and 'surplus centrality by L.J. Johnson.[64] Though there is a clear correlation between population and centrality of a settlement yet in the real world the large range of non-central, functions adds to the population.[65] The distinction therefore, between importance and centrality is of paramount significance and failure to note this difference may often lead to different results in hierarchical studies.

### THE RELEVANT CENTRALITY MEASURE NOW PROPOSED AND ITS APPLICATION TO MITHILA PLAIN REGION (BIHAR)

Under the impact of history and economics, Central Places develop with varying circles of influence. Distinctive regional characteristics seem to veto the building of any universally valid model of Central Place development. Nevertheless, it is quite necessary to determine the existing hierarchy of Central places so that the areas far off from the important centres of economic activity can be located and then a hierarchical class system of settlements may be proposed in the light of integrated area development because our present economic status does not permit us the development of all the functions in all settlements. This is why the concept of hierarchy has attracted the attention of a number of geographers and American rural sociologists in recent years. Attempts have been made to grade centres for the most part in two ways. The quantitative characteristics of centres, their centrality, have been measured. On the other side, the factors responsible for the centrality of a centre—its central functions—have been studied qualitatively. Many studies have applied both methods. The measurement of the importance of central places in quantitative terms also depends upon the availability of data for the purpose. But as one has to compare the centralities of many places. arrange them in an ascending order and then to make, if at all possible, hierarchical grades, one has to express the centralities in quantitative terms only and there can be three types of criteria for this purpose :

(i) an assessment of the importance of a central place in terms of the variety and magnitude of functions or central institutions and establishments at or about the centre;

(ii) measurement of the area and population dependent for various services upon the centres; and

(iii) determination of the pattern and intensity of functional interaction among central places and their respective hinterlands. Much has already been discussed about all the three types of methodologies in the preceding paragraphs.

As regards the first type of criterion, the total number of functions and functional units existing at the centre may together furnish the actual importance of the central place, provided their varying sizes and differing efficiency are not neglected. Functional units may also be considered as diagnostic keys for all the services. But one should always take into notice the various qualitative and quantitative differences in weights of the central institutions, thus the weight of a university will always be more than that of a high school, while the latter with 500 students will exercise more central influence than that with only 100 students on the rolls.

In the second type, if population is equally and evenly distributed throughout the region under study, a measurement of the area depending upon the centre may furnish a good index of centrality. But if the case be otherwise, population has to be considered and the question as to which of these two (area and population) is more important still remains unanswered. Jagdish Singh[66] has clearly demonstrated that the absolute population size of centres failed to provide a measure of threshold for grouping of function and establishments due to various reasons. Therefore, grouping centres into hierarchical classes on this basis would be meaningless. He showed that the population of central places in Gorakhpur region is not a function of the number and size of commercial and service establishments. This fact is substantiated when census authorities note that in about 25 per cent of the towns in India, mainly those below 20,000 population, agriculture is the predominant function. Thus the basic assumption of Christaller that through the working of income mechanism the population of a centre is a function of the number of types of central goods and services, the central place provided does not hold good in this case. Furthermore, commercial establishments and service institutions are not always located on economic viability considerations. They are opened sometimes on welfare consideration without any consideration of the population of the centres. He also used commercial population of centres since it eliminates the number dependent upon agricultural occupations. But the result was disappointing again, because the enumeration of commercial population in the census is faulty, enumerators often recording persons belonging to the trading community and not those actually engaged in their occupation. The author noted several discrepancies in this regard in Mithila Plain region. Table 3.2 is

TABLE 3.2

*Commercial Population and Number of Outlets Major Central Places*

| Sl. No. | Central Places | Commercial population as counted in 1978 | No. of outlets as counted in 1978 | Commercial population as in 1971 census | No. of shops in 1971 |
|---|---|---|---|---|---|
| 1. | Muzaffarpur | 24065 | 10705 | 9606 | 7108 |
| 2. | Darbhanga | 12236 | 6657 | 8001 | 4452 |
| 3. | Bettiah | 7437 | 3826 | 3819 | 2195 |
| 4. | Motihari | 6750 | 3416 | 2634 | 2415 |
| 5. | Barauni | 5020 | 2475 | 1643 | 1300 |
| 6. | Sitamarhi | 5203 | 2216 | 2276 | 1700 |
| 7. | Samastipur | 5488 | 2733 | 2338 | 1756 |
| 8. | Hazipur | 3822 | 2028 | 2044 | 1403 |
| 9. | Begusarai | 4349 | 2457 | 2643 | 1222 |
| 10. | Madhubani | 4379 | 2511 | 1947 | 1342 |
| 11. | Raxual | 2441 | 1443 | 1044 | 1010 |
| 12. | Khagaria | 2402 | 1243 | 1390 | 1054 |
| 13. | Rosera | 1980 | 1149 | 1306 | 750 |
| 14. | Dalsinghsarai | 2524 | 1236 | 905 | 1025 |
| 15. | Pupri | 2015 | 1166 | 751 | 1134 |
| 16. | Jainagar | 2390 | 1392 | 1049 | 788 |
| 17. | Bairgania | 1885 | 1031 | 758 | 700 |
| 18. | Narkatiaganj | 1424 | 1016 | 1058 | 1032 |
| 19. | Jhanjharpur | 1085 | 645 | 186 | 500 |
| 20. | Bagaha | 933 | 611 | 573 | 500 |
| 21. | Teghra | 875 | 586 | 558 | 486 |
| 22. | Lalganj | 1100 | 731 | 790 | 608 |
| 23. | Mahnar | 846 | 544 | 548 | 418 |
| 24. | Chanpatia | 1053 | 709 | 707 | 600 |
| 25. | Ram Nagar | 960 | 696 | 679 | 670 |
| 26. | Areraj | 699 | 509 | 245 | 404 |

illustrative wherein at several centres the number of commercial population (1971) is less than the number of outlets (1971), which is an impossible proposition. There is also a wide gap between the number of persons engaged in trade and commerce as counted, in most cases, by the author and that of the figures recorded by the census. It should not be surprising then that the relative centrality index based on commercial population of centres would not yield a meaningful grouping of centres.

Much has also been discussed about the third type of methodology. The criteria followed by Christaller and Green belong to this category. But the telephone index is of little use in India because the service is mostly confined to the urban areas. Similarly, the bus-service linkage is of limited applicability here due to the country' vast rural hinterlands and also the poverty rules out the use of public transport for purely local connection.

Criteria for measuring centrality that satisfy most of the requirements of a centrality model have been followed. The study attempts at presenting a view of the general network and hierarchy of central places in Mithila Plain. Both quantitative and qualitative information has been used in establishing the network. And as far the selection of central places is concerned, the analysis, apart from the town and cities considered as such in the 1971 census, covers all places up to and above 800 population which along with contain a high school, a medical centre or at least a private medical practitioner, a post office with savings deposit facilities, a local market with at least 40 permanent commercial establishments and which are connected by rail or bus. Centres which did not meet these basic requirements have been omitted.

To avoid the difficulties encountered with the use of population figures it was thought to base the analysis on the distribution pattern of number (variety) of central functions and the number of establishments alone. The former is a measure of the functional distance (distinctiveness) and the latter indicates the functional magnitude of centres. There is high degree of correlation between these two indices. The location coefficient type analysis discovered by Davies on the basis of these two indices offered a reasonable technique of analysis. Thus assuming the area to be closed system and that the degree of focality, or centrality, of each function varied with the total number of outlets of that particular type of function available in the region the following procedure was adopted to finally arrive at an ordering of central places in the region :

(a) Following Davies,[67] the location coefficient (a measurement of centrality) of each outlet (establishment) of individual function has been calculated using the following construction :

$$C = \frac{t}{T} 100$$

where, C=the location coefficient of function t,
t=one outlet (establishment) of the function,
T=total number of outlets of the function.

Since the location coefficient reduces all function to a common base, each functional type is evidently comparable in an objective manner.

(b) The location coefficient thus obtained has been multiplied by the total number of workers or employees in each type of establishment of individual centre, since a significant correlation has been detected between the turnover of establishments and the number of people employed[68] within each type of establishment. The location coefficient, for each functional type are based, therefore, on the total number of employees and not on the total number of outlets. An establishment with ten persons employed has a location coefficient ten times that of an outlet with only one worker, provided it is the same type of function. Thus, the discrepancy arising out of the varying size of outlets at different central places, has been removed. The material needed for this analysis was obtained from a personal survey of all the establishments in the study area. The field work was carried out in 1978 and one hundred and twenty-nine different functions types were identified. The number of workers in each outlet were obtained by personal interview in most cases and by means of questionnaires in some small centres.

(c) To identify the Central Places in terms of their functional character, an index known as Functional Index F, for a particular place i, has been developed which shows the total location quotient ($L_q$) of various functions to be denoted as the share of employment in a particular function, $Y^E$, at i in relation to its existence in the region, i.e.

$$L_q = \frac{yE\ i}{\Sigma yE}$$

Thus $F_i$ may now be defined as

$$F_i = \sum_{1}^{n} \frac{yE\ i}{\Sigma yE}$$

This functional index measures the functional magnitude of individual Centre and when plotted on a scatter diagram against the number of Central functions present in each Centre, it reveals a tendency of clear breaks at distinct level to warrant a five-level hierarchy.

The functional index does not signify the importance of a place with regard to such qualitative indices as administration, education, health services, cultural institutions, banks, etc. which have only qualitative dimensions in the degree of services rendered and hence not quantifiable in number of commercial establishments. Such activities are significant in themselves as attributes of central places but also pave the way for attracting clusters of commercial functions around them. It becomes inevitable to assign points commensurate with the relative importance of these functions. In the following scheme of scores, emphasis has been laid on the variety of activity rather than the number of population served.

TABLE 3.3

*Mithila Plain : Central Functions indicators 1977-78*

| Types of Central Functions | Score values |
|---|---|
| 1 | 2 |
| **A. Administration** | |
| Block headquarters or Police Stations | 1 |
| Sub-divisional headquarters | 2 |
| District headquarters | 3 |
| Commissionary headquarters | 4 |

TABLE 3.3—*contd.*

| 1 | 2 |
|---|---|
| **B. (i) Communication** | |
| Post office without telegraph or phone | 1 |
| Post office with telegraph or phone | 2 |
| Post office with telegraph and phone | 3 |
| Post office, Head | 4 |
| Radio Station (AIR) | 5 |
| **(ii) Internal Communications offered by Telephones** | |
| Below 50 lines | 1 |
| 51 to 100 lines | 2 |
| 101 to 250 lines | 3 |
| 251 to 500 lines | 4 |
| 501 to 1000 lines | 5 |
| 1001 to 2000 lines | 6 |
| Above 2001 lines | 7 |
| **C. Transport** | |
| (i) *Railways* | |
| Railway station where passenger and ordinary trains stop | 1 |
| Railway station where Mail/Express trains stop | 2 |
| (ii) *Roadways* | |
| Regular bus stop | 1 |
| Bus stop with Bus stand | 2 |
| Bus depot and bus service station | 3 |
| **D. Education** | |
| Primary School | 1 |
| Middle School | 2 |
| High School | 3 |
| Sanskrit Pathshala, Senior Basic School, Junior Basic School, Teachers' Training School, Madarsa, Makhtab etc. | 3 |
| Degree Colleges—Polytechnics | 4 |
| Post-graduate, Law, Medical and Engineering Colleges | 5 |
| University | 6 |
| **E. Health Service** | |
| Registered Medical Practitioners | 1 |
| Ayurvedic, Homoeopathic, Unani dispensaries | 2 |
| Family Planning Centres, Maternity and Child Welfare Centres | 2 |
| Health Centres/Sub-centres | 3 |

TABLE 3.3—*Concld.*

| 1 | 2 |
|---|---|
| Curative and preventive medicines centre | 4 |
| Veneral diseases centres | 4 |
| Small-Pox eradication units | 4 |
| Leprosy control units | 4 |
| Blood and bone banks | 4 |
| *Hospitals including Dispensaries and T.B. Clinics graded by number of beds* | |
| Up to 10 beds | 3 |
| Up to 11 to 25 beds | 4 |
| Up to 26 to 50 beds | 5 |
| Up to 51 to 100 beds | 6 |
| Up to 101 to 200 beds | 7 |
| Over 201 beds | 8 |
| **F. Finance** | |
| Bank–Commercial/Co-operative and/or Land Mortgage | 1 |
| **G. Insurance** | |
| Insurance LIC (DC) | 1 |
| Insurance LIC (SC) | 2 |
| Insurance LIC (BC) | 3 |
| Insurance LIC (Zonal Office) | 4 |
| Insurance others | 5 |
| **H. Recreational and Cultural Facilities** | |
| Cinema | 1 |
| Stadia, Auditoria, Drama halls | 2 |
| Local Museum | 3 |
| Tourist Office | 4 |

The total points secured by a centre for each kind of facility according to the number of establishments present in each case were added up. While plotting on a linear scale these total score points show definite trends of clustering indicating that the facilities of similar status have a greater probability of occurring together. It was quite possible, then, to group these points into five distinct classes on the ground of its placing in the group of a given order or rank, increasing in status from 1 to 5. Now, the functional index of each centre obtained earlier was multiplied, according to its placing here in the service score graphs, by 1, 2, 3, 4, or 5. And we thus obtained integrated functional index, which, when plotted on a scatter diagram against the number of central functions, shows clear breaks in the clustering of centres thus presenting clear hierarchical ordering of central places into five classes—1 fifth order Centre, 3 fourth order Centres, 6 third order Centres, 19 second order Centres and 136 first order Centres. They may, respectively, be designated as major regional centres, minor regional centres, sub-regional centres, Intermediate centres and local or villagenagar centres.[69]

## CONCLUSION

The review suggests that most of the studies in India on central places have been duplicated from the West. Therefore, whatever be the relevance of the criteria adopted in the West for centrality measurements, they are of limited utility for macro-regional work in India. We should always remember that there are important differences between the urbanization process in highly industrialised and less developed countries.[70] The towns and cities of India still bear the stamp of the colonial experience. The concepts of orthogenetic *versus* heterogenetic cities[71] and generative *versus* parasitic cities[72] are illustrative on this point. All the attempts at hierarchies in India represent different approaches. Though a certain degree of success in grading centres may have been attained but the difficulty arises when one tries to compare one area directly with another area and then it is apparent, how limited the current approaches in India to the study of central places actually are. So, a definite need exists to develop a unified methodology in India to measure centrality objectively. The conditions here are quite different from the Western countries. Many functions which are only of lower order in the west are higher order functions here. For example, laundry, hotel, bakery and wine shops are wide-spread in the West. They are found at small focal points and some of them even in rural areas. But hotels, bakery and wine shops are not to be found at smaller places here. The laundry service is not at all required in rural areas. All these suggest sharp differences in socio-economic structures between these regions. There are also wide differences in the size of the outlets. Here the size of outlets, particularly in rural areas, is very small. Tea stalls and betel shops on every small node and along the roadside are examples on this point. They are not at all established on threshold grounds. Unemployment seems to be one of the main reasons for their opening in abundance. It is ridiculous to treat them as separate outlets. One more point to be considered here is that there appears no consensus in India as to which functions should be regarded as Central functions indeed. Bhatt is of the opinion that all functions where the government has 'any say in their development' should be taken as Central functions. But this does not seem to be valid because government projects, e.g., Durgapur Steel Plant cannot be considered a Central function. Prakasa Rao advocates, "The definition of central functions based not only on the rarity or obiquity of functions but also on the preference of the people, both of producers as well as consumers". He considers that highly specialized function like universities and specialized hospitals should not be taken as Central functions. Deshpande maintains that by selecting more significant functions and services and weeding out the others, the problem or identification of Central functions would be streamlined. Then the question of objectivity arises. Also, the problem of data collection remains. In the absence of public or institutional publications, techniques involving the use of scarce data for measuring the Centrality are of limited applicability in India. The probe has, therefore, to be diverted towards simple, unified methodology for a macro-regional study. In an attempt to measure Centrality and hierarchy of Central Places in Mithila Plain, what is followed here is the method which possesses operational relevance and meets most of the requirements of a Centrality model i.e. the technique is quite objective and capable of being tested, the index is not based on a particular data that possesses limited application and finally, this measure incorporates most of the Central functions and the final index is capable of being sub-divided into its component parts.[73]

## REFERENCES

1. Adam Müller, 1809 : *Die Elemente der Staats Kunst. 3, T.* Berlin, pp 300-301 (Cited according to Palomaki 1964).
2. There have been important forerunners, ibid., 1809; La Lanne, 1863: Galpin, 1915; Kolb, 1923 and others.

3. M. Jefferson, "The Distribution of the World's City's Folks," *The Geographical Review*, Vol. XXI, 1931, p 453.
4. Walter Christaller, *Die Zentralen orte in Süddeutschland* (translated by Carlisk W. Baskin, Englewood Cliffs N.J. Prentice Hall, inc. 1966).
5. W.A.V. Clark, "Consumer Travel Pattern and the Concept of Range," *Annals of the Association of American Geographers*, Vol. 58, No. 2, June, 1968, p 386.
6. O.P. Singh, "Functions and Functional Classes of Central Places in Uttar Pradesh," *The National Geographical Journal of India*, Vol. XIV, Part 2-3, June-Sept., 1968, pp 1-2.
7. A.K. Philbrick, "Areal Functional Organisation in Regional Human Geography," *Economic Geography*, Vol. XXXIII, 1957, pp 299-326.
8. Lalit K. Sen, Sudhir Wanmali, et al, *Planning Rural Growth Centre for Integrated Area Development : A Study in Miryalguda Taluka*, NICD. Hyderabad, Jan. 1971, p. 81.
9. Ibid., p. 81.
10. Ibid., p. 81.
11. E.L. Ullman, "A Theory of Location for Cities," *American Journal of Sociology*, Vol. 46, May 1941, pp 835-864.
12. B.J.L. Berry and W.L. Garrison, "The Functional Bases of the Central Place Hierarchy," *Economic Geography*, Vol. 34, April 1958, pp 145-154.
13. 'Classical Central Place Theory' refers to the Contribution of Christaller. Some feel that recent studies have produced a 'Modern Central Place Theory.' The modern version is claimed to have generalized, and thus broadened the applicability of Christaller's approach. To compare classical and modern Central Place approaches see, for the former, Christaller, op. cit. pp 1-83, and for the latter Berry, *Geography of Market Centres and Retail Distribution*, (Englewood Cliffs, N.J. Prentice Hall, 1967), pp 1-88.
14. On the first point see, E.L. Ullman, op. cit. pp 858-859, L.J. King, "Central Place Theory and the Spacing of Towns in the United States," in M. McKas Kill (ed.), *Land and Livelihood* (Christchurch, New Zealand Geographical Society, 1962), pp 246-248, and W.K.D. Davies "Centrality and the Central Place Hierarchy", *Urban Studies*, 4, 1967, pp 61-63. On the second point see, F. Lukermann, "Empirical Expression of Nodality and Hierarchy in a Circulation Manifold, "*East Lakes Geographer*, 2 (1966) and L.H. Russwurm,"Development of an Urban Corridor System Toronto to Stratford Area, 1941-1966", Development of Treasury and Economics Regional Development Branch, *Research Paper* No. 3, (Waterloo : Deptt. of Geography and Planning and Resources Institute, University of Waterloo, 1970), pp 1, 109-154.
15. B.J.L. Berry and A. Pred., *Central Place Studies—a Bibliography of Theory and Applications*, 2nd rev. ed. Philadelphia, 1965.
16. J.B. Parr, "Review Article on B.J.L. Berry's 'Geography of Market Centres and Retail Distribution, 1967" *Journal of Regional Science*,

Vol. 8, No. 1, 1968, p. 124.

17. Dr. Jagdish Singh, "Central Places and Spatial Organisation in a Backward Economy : Gorakhpur Region". (Research Project Financed by ICSSR, New Delhi—Dr. J. Singh, Director).
18. N.R. Karr,"Urban Hierarchy and Central Functions Around Calcutta in Lower West Bengal, India, and Their Significance, *Proceedings of the I.G.U. Symposium* in Urban Geography, London (Sweden) 1960, pp 253-274.
    ——— "Pattern of Urban Growth in Lower West Bengal", *Geographical Review of India*, 24 (3-4), 1962.
19. R.C. Mayfield, "A Central Place Hierarchy in Northern India", *Quantitative Geography* Part-I, Illinois, 1967, pp 120-166.
20. Steen Folk, "Central Place Systems and Spatial Interaction in Nilgiris and Coorg (South India)", *Saertyk of Geografisk Tidsskrift* 66 bind, 1967, pp 161-178.
21. Bimlendu Bhattacharya, "Factors determining the Central Functions and Urban Hierarchy in North Bengal", *Geographical Review of India*, Vol. XXXIV. No. 4, 1972, pp 327-338.
22. H.E. Bracey, "Towns as Rural Service Centres", *Transaction* Institute of British Geographers, 19, 1953, pp 95-105.
    ——— "Social Provision in Rural Wiltshire", London, 1952.
    ——— "A Rural Component of Centrality Applied to Six Southern Countries in the United Kingdom" *Economic Geography*, Vol. 32, 1956, p 40.
23. Gopal Krishnan and M.M Chandra, "The System of Service Centres in the Outer Himalayas" *Ekistics*, 248, July 1976, pp 62-66.
24. V.L.S. Prakasa Rao, "Towns of Mysore State", *Asia* 1964, pp 36-51.
25. V.L.S. Prakasa Rao and R. Rama Chandran, "Planning for an Agricultural Region, "in R.P. Mishra and Others, *Regional Development Planning in India*, 1974, pp 293-335.
26. S.P. Mukherjee, "Commercial Activity and Market Hierarchy in a Part of Eastern Himalaya, Darjeeling", *National Geographical Journal of India*, Vol. XIV, p 6-23, June-Sept., 1968, pp 186-199.
27. K. Robinson, "Central Place Theory II', Centre for Environmental Studies, *Working Paper* 9, 1968, p 5.
28. J.E. Brush, "Review of Marshall", *Economic Geography*, Vol. 46, No. 4, 1970, p 631.
29. M. Palomaki, "The Functional Centres and Areas of South Bothnia", *Fennia* 83, 1964, pp 1-235.
30. Scot Peter, "The Hierarchy of Central Places in Tasmania", *The Australian Geographer*, Vol. 9, 1964, pp 134-147.
31. T.L. Bell, S.R. Leiber and G. Rushton, "Clustering of Services in Central Places", *Annals* Association of American Geographers, Vol. 64, No. 2, 1974, pp 214-225.
32. R.H.T. Smith, "Method and Purpose in Functional Town Classification", *Annals*, Association of American Geographers 55, 1965, pp 539-48.

33. R. Vining, "A Description of Certain Spatial Aspects of an Economic System" *Economic Development and Cultural Change*, Vol. 3, 1955, p 148.
34. B.J.L. Berry and W.L. Garrison, "A Note on Central Place Theory, and the Range of a Good", *Economic Geography*, 34, 1958, pp 304-311.
———op cit, pp 145-154.
35. V.L.S. Prakasa Rao, 1974.
36. Lalit K. Sen and Sudhir Wanmali, et al, op cit.
37. Sudhir Wanmali, "Ranking of Settlements, A Suggestion", *Behavioral Sciences, and Community Development*, Hyderabad, Vol. V, No. 2, pp 97-111.
38. B.K. Narayan and D.R. Rao, "Regional Planning: Growth Centre Technique", *Ind. Jour. of Regional Science*, Vol. 6, No. 1, 1974.
39. K.V. Sundaram, "Planning for a Tribal Region" in R.P. Mishra and others, *Regional Development Planning in India*, 1974, pp 239-292.
40. W. Bunge, "Theoretical Geography" *Land Studies*, in Geography, Series, C, General and Mathematical Geography I, 1962, especially p 146.
41. R.E. Preston, "Two Centrality Models" *Yearbook of the Association of Pacific Coast Geographers*, Vol. 32, 1970, pp 59-78.
42. It is common in some countries to divide services into tertiary and quarternary categories. The former pertains in general, to business, traffic and trade. The latter indicates public services such as health, education and cultural programmes.
43. R.L. Singh, *Banaras, A Study in Urban Geography*, Banaras 1955.
44. Kashi N. Singh, "Spatial Pattern of Central Places in the Middle Ganga Valley", *N.G.J.I.*, Vol. 12, No. 4, 1966, pp 218-226.
45. O.P. Singh, "Towards Determining Hierarchy of Service Centres : A Methodology for Central Place Studies", *N.G.J.I.* Vol. XVII, Part 4, 1971, pp 165-177.
46. Onkar Singh, "Hierarchy and Spacing of Towns in U.P.", in R.L. Singh (ed.) *Urban Geography in Developing Countries* Varanasi, 1973, pp 318-326.
47. S.N.P. Jayaswal, "Hierarchical Grading of Service Centres of the Eastern Part of Ganga-Yamuna Doab and Their Role in Regional Planning", *Urban Geography in Developing Countries* (Proceedings of I.G.U. Symposium No. 15, Varanasi) pp 327-334.
48. N.P. Saxena and R.C. Tyagi, "Criteria for Determining Centrality in a Macro-region", *The Geographical Observer*, Vol. II, March, 1975, pp 55-64.
49. Usha Agrawal, "Functional Hierarchy of Settlements Situated within the Service Area of Meerut Town Group", *Ph.D. Thesis*, Meerut, 1973.
50. S.C. Bansal, *Town Country Relationship in Saharanpur City-Region*, Saharanpur, 1975.
51. N.G. Jain, "Urban Hierarchy and Telephone Service in Vidarbha (Maharashtra)", *N.G.J.I*,, Vol. XVII, Part 2-3, 1971, pp 134-137.

52. E.L. Ullman, op. cit., p 858.
53. Ibid.
54. W.K.D. Davies, "The Ranking of Service Centres: A Critical Review." *Transactions and Papers*, The Institute of British Geographers, No. 40, 1966, p 57.
55. Onkar Singh, op. cit.
56. Ashok K. Dutt and Sanjay Benerjee, "Transportation Index in West Bengal : A Means to Determine Central Place Hierarchy," *N.G.J.I.* Vol. XVI, Part 3-4, 1970, pp 199-207.
57. Lata Chatterjee, "Urban Hinterlands of West Bengal," *Paper* Presented at 21st I.G.U. Symposium on Urban Geography in Developing Countries, 1968.
58. J.R. Tarrant, "A Note Concerning the Definition of a Group of Settlements for a Central Place Hierarchy, *Economic Geography*, 44, 1968, pp 144-151.
59. Christaller, op. cit., p 19.
60. Ibid. pp 139-147
61. N.I.C.D. *Behavioral Sciences and Community Development*, Vol. 6, March, 1972, No. 1, p 151.
62. W. Christaller, op. cit.,
    B.J.L. Berry and Pred., op cit. and H.F. Andrews, "Working Notes and Bibliography on Central Place Studies, 1965-1969" University of Toronto, Deptt. of Geography, *Discussion Paper*, No. 8, 1970.
63. R.E. Preston, op. cit., p. 62.
64. L.J. Johnson, "The Spatial Uniformity of a Central Place Distribution in New England," *Economic Geography*, Vol. 47, 1971, pp 156-170.
65. H. Carter, *The Study of Urban Geography*, 1972, p 89.
66. Jagdish Singh, op. cit.
67. W.K.D. Davies, op. cit., p 63.
68. L. Jay and P. Hirsch, "The Comparative Analysis of Settlements" *Sociologia Ruralis, Vol. 1, No. 1, 1960*, pp 51-65.
69. Following C. Vasudevan, the nomenclature 'Villagenagar' is used to denote an area which, while having the benefits of an urban area without the ill-effects of urbanisation, retains the rustic and sylvan nature of a village, Yojana, Vol. XXII/17, 16 Sept. 1978, p 29.
70. F.R. Pitts, ed. *Urban Systems and Economic Development*, Eugene, 1962.
71. R. Redfield and M.B. Singer, The Cultural Role of Cities, *Economic Development and Cultural Change*, 3 1954, pp 53-73.
72. B.F. Hoselitz, "Generative and Parasitic Cities", *Economic Development and Cultural Change*, 3, 1955, pp 278-294.
73. W.K.D. Davies, op. cit, pp 61-63.
    Requirements for Centrality measurements have been suggested also by R.J. Johnson, "Central Places and the Settlement Pattern" *Annals of the Association of American Geographers*, Vol. 56; 1966, p 545.

# 4

# Temporal Dimensions of Urban Geography

—*Nandeshwar Sharma*

## INTRODUCTION

THE DISCIPLINE of geography today is fast losing its traditional identity and in the present process of diffusion, specialisation and the separation of the fields of knowledge a welter of sub-fields has come up in which each one tries to find its status among the sister branches. Prior to a discussion on trends and approaches to Urban Geography, it becomes relevant to know if this branch stands anywhere in the welter. Of the six encyclopaedic, educational, colonial, generalisation, political and specialisation trends established by Freeman (1965) on the basis of the development of geography during a century and a decade or two from today, the last one seems to be the most natural and recent and Urban Geography owes much to this trend. The recency of the development of this branch of geography acquiring a variegated field becomes obvious by the fact that Urban Geography as a specialist study has existed for barely half a century or so from today as observed by Stamp (1960) and Mayer and Kohn (1967). Its claim of attaining an established status seeks ample justification through the fact that within this relatively short span of its existence there has been such a proliferation of researches and deliberations of such diversity of subject matter and treatment that research papers and books of readings have become not only available but necessarily available in abundance. A compilation effort of bibliography on urbanisation in India by Bose (1976) which covers more than 2,700 bibliographical references exhibiting the dense inter-disciplinary thicket of the subject shows the impressiveness of the volume of work being done in this field. It becomes even more impressive when one considers that these bibliographical references concern the work done only in India where urban Geography is still in its infancy and the period is only between 1946 and 1976. Suffice it to say that by now the bibliography of relevant works must have added to it many items to what existed both in India and abroad.

## WHAT URBAN GEOGRAPHY STANDS FOR AND ITS RELEVANCE

Urban Geography is to stand for the study of urban concentrations and urban phenomena. By urban concentrations is meant the different forms of human occupance in urban setting; and by urban phenomena all the processes contributing to the development of urban centres and allied consequences. This intuitive definition tends to fall in line with Stamp's (1965) observation on urban Geography as "the intensive study of towns and their development" and with Gosal's (1972) recognition of it as "geographic study of urban places which evolve, grow and exist as service centres, largely for their surrounding areas.'' But the problem of evolving a definition complete in all respects still awaits as the problem of standardizing definitions of cities themselves remains unsolved as has been admitted by Haggett (1965). Terms coined as equivalent for urban centres appear to be *town* and *city* in English, *stadt* in German, *cité* in French and 'staden' in Swedish literature. Scholars like F. Ratzel, H. Myklebast, Fr. Von Ritchthofen, W. Christaller, M. Aurosseau, P. George, J. Bruhnes and P. Deffontaines, H.W. Ahlman and W.W. Olsson, G. Taylor and many others evolved specific definitions of cities from time to time, but each with emphasis on some elements and leaving some other ones. It will, however, be rational to assert that a definition of urban centre should carry some basic elements of : (i) a definite territory of location and extent, (ii) a dense agglomeration of sufficiently large population and structures, (iii) an economy predominating in non-primary and differentiated professional activities, (iv) a degree of readiness for change in temporal and regional dimension. Consequently, therefore, the core concern of Urban Geography should be location and areal extent of urban centres and the interaction among various urban functions and groups of people.

Urban centres mean today many things to many people. They have historical, political, economic, social, architectural, engineering and planning significance through ages. They concern geographers and more so in recent years, in their spatial dimensions. Their pulsating growth tends to make Urban Geography of vast relevance to human society and reciprocal relationships between them, their activities and the ground. The relevance of urban geography makes room also as an applied science because its contributions have made valuable aids to problem solving efforts of city planning, public administration, local transportation, etc.

### WHEN AND HOW URBAN GEOGRAPHY CAME INTO EXISTENCE--ITS EVOLUTIONARY TREND

It is not at all easy to tell the date of the inception of any discipline of study and more so in case of Urban Geography, because it concerns the study of the phenomenon of the concentration of people, that is the rise of cities, which shares, as Smailes (1953) observed, common geographical basis with the rise of civilisation itself. The phenomenon of the rise of cities of first stage, associated with 'Agricultural Revolution' which could bring no alarm as growth of centres seemed to belong to the natural order of things. It was, however, only in the closing half of the nineteenth century that the second urban phenomenon associated with the "Industrial revolution" experienced in new and vigorous dimension the development of urban centres which Garnier and Chabot (1967) termed as the human stamp towards ill-defined agglomeration and which became the most obvious, spectacular and geographical aspect of the evolution. The world witnessed unprecedented swelling in the proportion of urban dwellers to total population. The proportion of those living in 20,000+ urban places was only 1 in 40 in 1800, advanced to 1 in 4 in 1970 and is expected to rise to 1 in 2 by the time the present century rolls itself. Trends, however vary from country to country. Not unnaturally, therefore, it was only towards the end of the nineteenth century that urban geography made its appearance, notably in Germany with F. Ratzel. A good treatise of the temporal development of the subject could be found only in Taylor's (1951) *Geography in the Twentieth Century*.

Ratzel in Germany and Reclus in France included many plans of cities in their voluminous writings, but the aspects like the principles underlying urban agglomerations remained untouched. The first quarter of the twentieth century proved to be of significance when several writings dealing with the urban aspects of interest to sociologist and town planners in European countries appeared. The trend had already shown the shift from generalisation to particularisation. This has been very succinctly made known in Aurousseau's (1924) work which represented a valuable summary of the modern approaches to Urban Geography. Of special mention is also the name of Blanchard of France to whose name several studies date back to 1911 when he published the valuable study of Grenoble. His *Methodology of Urban Geography* published in 1922 deserved further appreciation. He was responsible for the establishment of the school of Urban Geography 'through which wide range of studies on city was undertaken. H.J. Fluere of Great Britain in his works emphasised on consideration of historical and commercial tradition of cities. He went to identify two belts of urban centres—one in Germany and the next in Russia in which many of the cities were supposed to be 'Acropolises'. Jefferson's (1909) work pioneered to lead his papers of 1913, 1916, 1917 and 1920 dealing with distributional aspects of Japanese, British and American urban places. During the period of 1919 and 1923 a good number of publications came in the *Scottish Geographical Journal* dealing with the cities of Scotland forming a good example of historical, sociological and geographical researches.

The second good harvest of works in Urban Geography was reaped during the second quarter of the current century. This period achieved success to a greater extent in filling up the vacuum created by the absence of any comprehensive and lengthy study of the ramifications of Urban Geography. Mumford's (1936) volume is of value for geographers in containing important plans of towns and cities and in being of sociological and historical significance. Dickinson's (1947) work went a step forward in evaluating towns and country relationships and umlands. Taylor's (1949) book appeared as a master piece in Urban Geography.

The third quarter of this century made available the publications of Smailes' (1953) book which endeavours to present the contents of Urban Geography in summary and illustration, forms an anthology by Mayer and Kohn (1967) which is to facilitate a comprehension of the character and scope of the field of this branch of geography, both as a scientific discipline and as a body of applied knowledge. A book of Gibbs (1961) in the form of compiled readings providing a vent to the study of demographic and sociological aspects of urban centres and urban phenomenon. An another book by Garnier and Chabott (1967) fulfils the need for a comprehensive survey of the whole field of Urban Geography. Still another book by Northam (1975) present factual information, hypothesis, concepts and theories pertaining to the subject of cities and the phenomenon of urbanisation. Besides there is hardly a book or journal of geographic and sociological interest other than those exclusively serving the cause of physical geography that misses deliberations on the aspects of Urban Geography The works of urban sociologists are much more prolific in this respect. In order to find the roots of Urban Social Geography, one is attracted to pioneer publication of R.E. Park's paper "The City" in 1976 in which he suggested for the investigation of Human Behavioralism in the Urban Environment. His *Race and Culture* and *Human Communities, The City and the Ecology* appeared in 1950 and 1952 respectively. The involvement of social ecologists and urban sociologists has become so inseparable that geographers have been advised by Paul (1970) to learn much from them.

Though recent, contributions of significance in the field of Urban Geography in India started appearing during last three decades : Campbell (1866), Forrest (1903), Irvine (1907, 1909 and 1910), Hodivola (1920, 1923 and 1929), Bhattasali (1939 and 1941), Piggot (1945) and Law (1947 and 1948), Singh's (1955) book, however, proved a landmark of this trend which infused Indian geographers to carry on researches in the field of Urban Geography in profusion. Since then such prolific additions have been made to the volume of work covering many aspects of : (i) Urban places and individual settlement, (ii) Urban places as a phenomenon; (iii) Urban places as economic phenomenon, and (iv) Applied Urban Geography that any effort to compile bibliographic references in a paper of this nature will certainly prove astonishing. The contributions of Gosal (1972) and Bose (1976) are, however, referred to serve the purpose of providing comprehensive bibliographical references. Among very comprehensive works Taneja's (1971) deserves merit on account of encompassing comparability elements in study and that of Sharma (1977) attains exclusiveness in enhancing the applied value of the field by orienting the study to town oriented resource based on urban regional planning.

Developing portions of the world have emerged during a few recent decades as fertile ground for the growth of urban population and urban areas on a scale probably unseen anywhere at any time. The tendency has led to the mass of social problems and Urban Geography in the fitness of things tended towards the socialisation of the field. Evidences of this trend are available in UNESCO (1951, 1956, 1957 and 1961) reports relating to the process of urbanisation and its implications in Africa, Asia and Latin America. A more useful and worth mentioning anthology by Alam and Pokshishevsky (1976) very appropriately tries to provide a proper perspective and realistic appraisal of the characteristics of urbanisation and of urban settlement in the Third World in the context of their historico-socio political setting.

Urban centres, thus having met certain inescapable human needs for centuries, in the wake of recent resurgence of urban phenomenon and associated consequences pressed for a field of study of their own and found it appropriately in the form of the discipline of Urban Geography. The trends of the development of this discipline have ably placed it at a height where it seems poised to acquire perhaps the most impressive and imposing status among the fellow disciplines of geography.

## THE STUDY OF URBAN GEOGRAPHY—CONCEPTS AND APPROACHES

Urban centres constitute the most fundamental elements of Urban Geography. The essence of the concepts and principles of Urban Geography very rightly, therefore, owes much to the formative aspects of settlement which are themselves very much the outcome of the 3Ps principle. The principle accounts for the three elements of Protection, Privacy and Proximity for the initiation, the punctuation and the causation of differing dimensions of settlement forms and systems. Different forms of urban centres acquire a long and close association with emergent human organisations and economic pursuits. A house, a group of houses of varying magnitude and designation have evolved to meet the needs of protection, privacy and proximity of a man, a group of men (family) and a group of families (communities of varying orders). It is their dominant functional traits that cause settlement forms to be identified into rural and urban categories. The dimensions of approaches to Urban Geography too have association with the elements of 3Ps Principle, settlement forms, social organisations and economic character as has been shown in Fig. 4.1. In its temporal dimensions settlement form embellished itself with the evolution of the concepts of town, city, metropolis, megalopolis and the utopian concept of ecumenopolis in urban context. And so did the dimensions of study approaches to Urban Geography.

*The Development of Urban Centres*

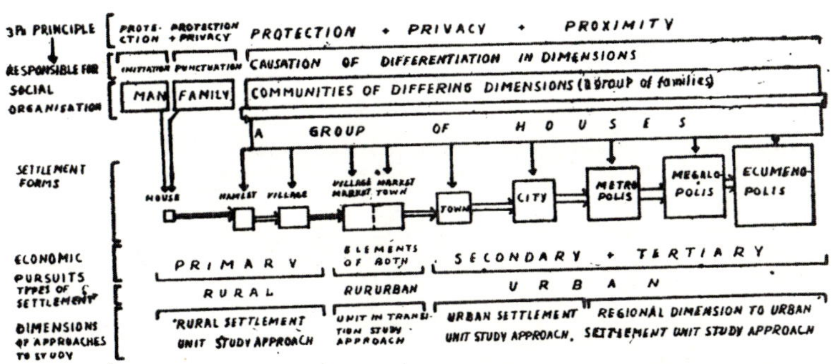


FIG. 4.1

The thrusts of study in Urban Geography have traditionally and dominantly remained tied with the feeding channels of falk to ground, ground to ground, falk to falk, falk within and ground within relationships. Such thrusts of relationship by and large seek place in the temporal dimensions of study approach to Urban Geography, a synoptical diagrammatic presentation of which has been tried to put the whole in one (Fig. 4.2).

*Temporal Dimensions of Study Approaches*

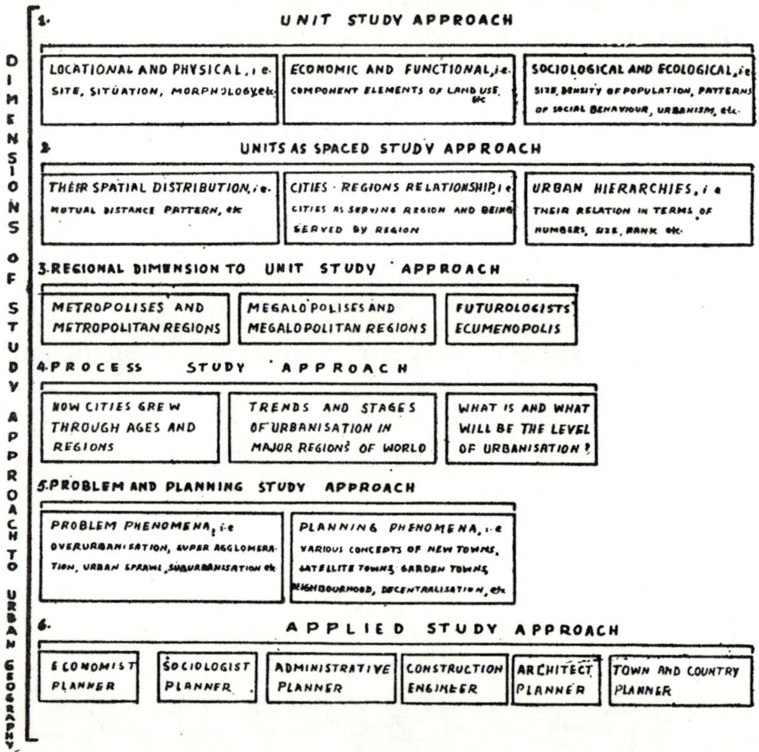


FIG. 4.2

The traditional thrust of study in Urban Geography produced the initial dimension in which specific urban unit study approach predominated. This approach remained involved basically with description and generalisation of physical, economic and ecological aspects of an urban unit as discrete entity. Though traditional, the approach still attaches significance because it is these studies which account for the causation of modern theories of Urban Geography and which exert persistent threat to the authenticity of such theories. The approach fell in line with the classical approach which laid emphasis on the formulation of laws expressing the universal operation of cause and effects has been observed by Wrigley (1970).

With the shift in thrust of study from intra-urban system to inter-urban system the dimensional approach began to study urban centres with respect to their distribution over space, their relation with regions and their hierarchies, etc. Under this dimension the hypotheses and concepts, dealing with the relationships among various centres as spaced over a region represent the line of intellectual descent beginning with Thünen and Kohl and leading down through Weber, Christaller, Zipf, Isard, etc. Central Place theory, Rank-size Rule, Nearest Neighbour analysis and a few other concepts on spatial hierarchies have enabled researchers to handle the issues of the regional ordering and functioning of economy and society. It is, however, to be noted that this approach has proved systematic in nature, but very flexible for application in specific condition.

Regional dimension of approach experienced the emergence of the concepts of metropolis—an extended urban area with a central city, megalopolis—a super city resultant of coalescence of metropolitan boundaries and ecumenopolis—the futurologists' World city. It is under the influence of this dimension of study that Alber, Adams and Gould (1971) venture to declare that in the developed world right now, and in the developing world very soon, the geography of the present and future must stress the urban rather than the rural environment.

Urban Geography includes not only the study of urban settlements, but also the process of urbanisation. Under this dimension of approach is contained the treatise on how cities grew and spread through ages and regions, how and when urbanisation rose to what level and what is and what will be the level of urbanisation. Besides traditional stages of concentration, deconcentration, centralisation, decentralisation, succession and invasion, the role of urban implosion and explosion has recently been enhanced. The process of implosion implies the concept of movement of larger cities closer in relative terms through the effect of differential spatial shrinkage. Rapid growth of larger cities has been attributed by Haggett (1975) to this process.

Modern world urban dimension is featured by ever expanding quantum in which urbanites multiply rapidly and cities appear as dominating force in the organisation of human population. Society faces the ills of hyper-urbanisation, super-agglomeration, urban sprawl, deterioration in habitability. It is naturally to these counts of problems of urban phenomenon that seething swarm of humanity has started repelling and people have began to think of cities as places nice to visit but not to live in. Urban geographers have recently shown their involvement with the study of problem phenomena and to combat them with planning processes. Resultants have been the emergence of planning techniques of urban renewal, slum and blight removal, zoning, and neighbourhood concept which according to Jordan and Rowntree (1976) is to describe small social and ethnic areas where people with shared values and concerns interact on a daily basis as a sequel to intra-urban planning phenomenon and the evolution of concepts of 'new towns', satellite towns, 'green towns', growth foci', etc. as parts of inter-urban planning phenomenon.

The concern regarding urban centres is quite knowledgeable and warranted. The destiny of one-third of the world's population is inescapably linked with these populated places and the other two-thirds of earth's inhabitants rely heavily directly or indirectly on these places of human agglomeration for their economic and socio-cultural uplift. Problems like those associated with the modern unprecedented growth of urban centres, naturally therefore, pose a challenge to sociologists, economists, ecologists, engineers, architects, planners and politicians at one stake. Consequently concern, involved with Urban Geography and which means many things to many people, transcends many disciplines. This transcendency alone speaks of the necessity of the addition of the dimension of study approach which has an applied value. To this approach Urban Geography readily advocates for an interdisciplinary approach to the solution of problems of common concern.

## FRONTIERS OF RESEARCH IN URBAN GEOGRAPHY

Urban Geography as a branch of the mother discipline of Geography has proved to be a very fertile ground for research. It is perhaps one single discipline which can claim the production of the largest number of researches, primarily conducted by urban geographers, urban sociologists, urban demographers and urban economists. The frontiers of research on this account, however, concern the quality of work done. It is not the manufacture of general methodologies rather the production of good research work that should count for the best sign of the health of the discipline. The two should be put into fruitful harmony through the continued devotion to modification, rejection, addition and replacement of past experiences. This alone will make contemporary discussions more productive.

Frontier as associated with the specific city study approach lies in its overemphasis on description and generalisation based on detailed and accurate factual information. The need is to investigate the principles underlying urban agglomerations, social planning of new suburbs and new towns and of industrial and amorphous mining towns. A shift from generalisation to specification is to be effectuated by the researcher in accordance with the individuality inherited by the specific urban centre.

Frontiers of researches in approaches involved in inter-urban system study, process study and regional dimensions are of profound significance and concern. In this context Kashi Nath Singh of Banaras Hindu University very rightly asserted in one of his lectures at the Summer Institute in Geography at Gauhati in 1978 that the modern situations demand the stress to be laid on the 'urban context' than on the 'urban centre' itself. He also challenged researchers to go into the examination of the validity of the terminology e.g. urbanism, urbanisation, pseudo urbanisation etc. and urban classifications as developed in western countries in the perspective of the nature of urbanisation, in developing countries.

Meaningful generalisations and models and paradigms have been attempted to be developed during the few past decades. But due to lack of precision and objectivity in measurement and statement construction, such explanatory models fail to provide accurate comparison and abstraction. Concepts and models have come up in Urban Geography. Their evolution marked the beginning of the methodological foundations of this discipline. The new frontier of research is a decade old or so in which recent studies are preoccupied with testing existing concepts and theories.

The trend is illuminating. Incisive mathematical and statistical models are being introduced. It has been noted by Kumar and Sharma (1974) that in a coming decade or two a very different story may be told. It is, however, necessary to be judicious in not allowing rigorous quantitative models to overshadow thoughts because only then they may be utilised to suit all societies and all places at all periods of time. The value lies in universal applicability alone.

Recent decades have exhibited spatial and regional dimensions in approach. Limitations in these spheres appear in researches being confounded by the need of systematization in spatial structure and relationships of urban networks. Another limitation lies in the inability in providing definitive techniques to classify formally valid regional systems. Various cartographic techniques, measures of spatial associations and cartographic measures are to prove responsive to such needs. The achievement of comparability, maximization of external variation and minimization of internal variation should be the motto while defining a system of regions. Various multivariables systems associated with the name of E. Shevky, R.C. Tryton, C.F. Schmidt, etc. contain most promising techniques of regionalisation in the urban context. They have laid emphasis on identification of social area types considering them of primary significance in the social structure of the urban population.

The most formidable and those of greatest consequence of the limitations confronted by researchers are those associated with planning and applied study approach. In precise terms, planning is a decision-making and resource allocating activity. The greatest frontier remains an attitude which pays scant attention to planning as "a problem-solving method" or to ways of "how to plan and how to plan best". It should be the concern of planner that crucial matter of goal formulation is not only the most important but also the most neglected part of the planning process. Planners are required to refrain from rigidity and utopian thinking of extravagant promises of what can be done for mankind.

Planners today seem to belong to two camps : one to the rational camp consisting of followers of systems approach and another to action planning camp to whom the concern is not of making decisions more rational but of improving the quality of action. 'Structure or Strategy Planning' seems to replace traditional 'comprehensive planning' because problems of today require immediate action. The former has been referred to as 'New' or 'Action' planning and 'Adaptive' planning also. Friedman (1969) advocates for its quality. "It is still the old language, but the perspective is new : planning and action are brought together and fused." Structure planners are required to have 'normative thinking', 'analytical thinking', 'futurist thinking' and 'strategic thinking'.

## CONCLUSION

Urban Geography in its present form has by now made appreciable dent to establish its identity as a separate discipline. Dimensions of approaches to study in Urban Geography have remained contented to urban centre as a singled out unit and have touched the height of applied horizons. Hartshorne (1969) very rightly equates the geography of any particular city to a special form of regional geography and that of individual urban features as found repeated in many cities to systematic urban geography which is tipped to study the cities of the world or any part of it in terms of their fundamental characters in relation to other geographic differences.

In harmony with the expanding dimensions of approaches frontiers of research in Urban Geography too have acquired new dimensions. Many attempts have been done to combat these frontiers, but many more are still to be added to what have been done. A very encouraging trend has been witnessed in the use of techniques based on mathematical and statistical feeders. The role of the computer is further awaited. But the danger of thoughts being overshadowed is to be sensed.

## SUGGESTED READINGS


1. T.W. Freeman, "A Hundred Years of Geography", Gerald Duckwoth & Co., London 1965, pp 11-25.
2. L.D. Stamp, "Applied Geography", Penguin Books, 1960, p 161.
3. H.M. Mayer and C.F. Kohn, "Readings in Urban Geography", Central Book Depot, Allahabad, 1965, p 5.
4. A Bose, "Bibliography on Urbanisation in India", Tata McGraw Hill Publishing Co. Ltd., New Delhi, 1976.
5. L.D. Stamp, op. cit., 1960.
6. G.S. Gosal, "Survey of Research in Geography" ICSSR, Popular Prakashan Bombay, 1972, p 203.
7. P. Haggett, "Models in Geography" (ed) P. Hagget and R.J. Chorley, Methuen & Co., London., 1965, p 241.
8. A.E. Smailes, "Geography of Towns", Hutchinsons University Library, London, 1953, p 7.
9. J.B. Garnier and G. Chabott, "Urban Geography", Longmans, Green & Co. Ltd., London, 1967, p 1.
10. G. Taylor, "Geography in the Twentieth Century", Mathuen & Co., London, 1951, pp 500-527.
11. M. Aurousseau, "Recent Contribution to Urban Geography", *Geographical Review*, 1921, Vol XIV, No. 3.
12. M. Jefferson, "Anthropogeography of Some Great Cities, *American Geographical Society*, Vol. XVI, New York, 1909, pp 537-66.
13. L. Mumford, "The Culture of Cities", Harcourt Brace & Co., New York, 1936.
14. R.E. Dickinson, "City Region and Regionalism"' Routledge and Kegan Paul, London, 1947.
15. G. Taylor, "Urban Geography", Methuen & Co., London, 1949.
16. A.E. Smailes, op. cit., 1953.
17. J.P. Gibbs, "Urban Research Methods", D. Van Nostrand Co., London, 1961.
18. H.M. Mayer and C.F. Kohn, op. cit., 1967.
19. J.B. Garnier and G. Chabott, op. cit., 1967.
20. R.M. Northam, "Urban Geography" John Wiley & Sons, London, 1975.
21. R.E. Paul, Trends in Social Geography, "Frontiers in Geographical Teaching" (ed.) R.J. Chorley and P. Haggett, Methuen & Co. Ltd., London, 1970, p 86.
22. M. Raja and Ahmad, "Survey of Research in Geography", ICSSR, Popular Prakashan, Bombay, 1972, p 157.
23. R.L. Singh, "Banaras, A Study in Urban Geography"; Varanasi, 1955.
24. G.S. Gosal, "A Survey of Research in Geography", ICSSR, 1972, Popular Prakashan, Bombay.
25. A Bose, op. cit., 1976.
26. K.L. Taneja, "Morphology of Indian Cities" NGSI, Varanasi, 1971.

27. N. Sharma, "Strategies for Development of Medium Size Towns in Bihar," Unpublished Ph.D. Thesis, Bhagalpur University, Bhagalpur, 1977.
28. UNESCO, "Social Implications of Industrialisation and Urbanisation in Africa, South of Sahara, Paris, 1956, Urbanisation in Asia and the Far East, Calcutta, 1957. "U.N. Report on World Social Situation", New York, 1957. "Urbanisation in Latin America", Paris, 1961.
29. S.M. Alam and V.V. Pokshishevsky, "Urbanisation in Developing Countries", Osmania University, Hyderabad, 1976.
30. E.A. Wrigley, Changes in the Philosophy of Geography, "Frontiers in Geographical Teaching", Methuen & Co. Ltd., London, 1970, p 5.
31. R. Alber, J.S. Adams and P. Gould, 'Spatial Organisation', Prentice Hall, Inc., New Jersey, 1971, p 379.
32. P. Haggett, "Geography : A Modern Synthesis", Harper & Row Publishers, New York, 1975, p 330.
33. T.G. Jordan and L. Rowntree, "The Human Mosaic", Confield Press, San Francisco, 1976, p 363.
34. A. Kumar and N. Sharma, "Spatial Arrangement and Degree of Specialisation of Urban Centres", *Geographical Review of India*, Vol. 36, No. 2, Calcutta, 1974, p 120.
35. J.R P. Friedman, "Notes on Social Action", *Journal of the American Institute of Planners*, 35, 50, 1969, p 312.
36. R. Hartshorne, "The Nature of Geography", *Ass. of American Geographers*, 1969, p 398.

5

# Urban Fringe—Concepts and Delimitation

—*Hira Lal*

An important development in urban settlement during the past several decades has been the rapid expansion of population and built-up area into the unincorporated suburb and in areas surrounding our large towns and cities. This recent finger-like trend of development at the town margins has been made possible by the mechanised transportation and the extension of the public utility services such as electricity, water and sewage, etc. beyond the city's limit. The process of accretion which takes place outside the city has generated a zone of transition between the rural and urban landscape properly known as the rural-urban fringe. In other words, the centrifugal forces which impel functions to migrate from the central zone of a city towards its periphery, create an amorphous territory around the cities, often called the rural-urban fringe. This is a new phenomenon in the urban landscape of the developing countries of the world in general and of India in particular. The emergence of the fringe zone with its complex problems of adjustment between rural and urban ways of life, has assumed great topical importance these days and has drawn the attention of planners and social scientists (R.L. Singh 1967).[1] Yet the study of the urban fringe has been a neglected area of urban research. Even in the developed countries a few studies have been undertaken in this regard though this developmental phase of urban morphology was experienced there in the very beginning of the present century. Fringe studies have not received much attention among Indians either from geographers or from scholars in any of the other disciplines. Hence, we have one of the major gaps in geographical research in the area of urban studies.[2] The purpose of this paper is to clarify the conceptual framework of the urban fringe with special reference to Bareilly.

## THE URBAN FRINGE : CONCEPTS

A study of the available literature reveals that most of the studies appeared during the period from the mid-1940s to the beginning of 1960s, in which attention were focused mainly to the physical delimitation, identification and the defining features of the rural-urban fringe. The term urban fringe was introduced by Smith (1937)[3] to describe built-up area just outside the corporate limit of the city "though the word 'urban' as synonymous to fringe was used by Gaplin as early as in 1915[4] for 'rural' land in process of conversion to 'urban'. Smith's definition was mainly concerned with the demographic characteristics of the area, therefore, it presented a general rather than a specific conceptualization of the fringe. Further, Salter (1940)[5] defined the fringe as a mixture of landuses that are related to farming and urban interest . . ." In his excellent study "The Rural Urban Fringe" Wehrwein (1942)[6] has called it an "institutional desert" because of the uncontrolled location there of unpleasant and noxious establishments such as slaughter-houses, junkyards and wholesale oil storage, and of utilities such as sewage plants and cemeteries. In his opinion, fringe is the area of mixed landuses lying beyond those sections of the city that are customarily serviced by urban utilities and transport systems. Further, he describes it as an area of transition between well recognised urban landuses and the area devoted to agriculture. Thus, Wehrwein is very clear that the fringe area can better be identified in terms of landuses than in any other way. Andrews (1942)[7] contributed new thought in the fringe studies by attempting to differentiate to the urban fringe from the rural-urban fringe. The urban fringe, as he puts it, is the active expanding sector of the compact economic city and the rural-urban fringe lies adjacent to the periphery of the urban fringe. The year 1942 bore a special significance for such studies as after Wehrwein and Andrews, Klow (1942)[8] and Alpake (1942)[9] were also to emerge on the platform. Alpake has defined the urban fringe as "that cultural development that takes place outside the boundaries of central cities and extends to the areas of predominantly agricultural activities.

After 1942, there has been a gradual but notable widening in the scope of fringe studies. Balk (1945)[10] added a new term, as he found the fringe an area of "rurbanization". Laying emphasis upon the accessibility as a sole factor in the development of the fringe. Walter Fiery (1946)[11] said "the rurban fringe is a marginal landuse area not because of its geographical location, its soil type or its topography, but rather because of its particular degree of accessibility (relative to that of other landuses) to some central transportation point" Rodehaver (1947)[12] in his study "Fringe settlement as a two directional movement" reiterates that new residents converge upon the fringe both from urban places and rural areas. While Dewey (1948)[13] marked that the movement of population into the rural-urban fringe was a movement of urban people. He further adds "... in the fringe area there is mingling of people of some of whom work in and are oriented towards agriculture while at the same time the remainder pursue urban occupations and an urban way of life". The end of the forties saw a well marked deviation from the traditional line of definition and division in rural-urban fringe studies as Aschman (1949)[14] and Bilizzard and Anderson (1952)[15] confined to landuse and Martin (1953)[16] to patterns of population.

Duncan (1956)[17] used the term urban fringe to denote the non-suburban population of the territory in urbanised areas outside the central cities. His urbanised area contained three components—central city, suburb and urban fringe. The urban fringe is limited to the corporate limit of the city and often it is found outside the limit also. Kurtz and Eicher (1958)[18] attempted to differentiate between fringe and suburb. Golledge (1960)[19] used the term "geographical no-man's land" for this contact zone between rural and urban landuses. Similarly, Russwurm (1960)[20] refers to this area as "frontier of discontinuity between the city and country, while Conzen (1960)[21] has recognised the fringe belt as significant determinants of the morphology of urban settlements. Defining its character, he identifies, the fringe belt itself as composed of two sections, the "proximal" or inner margin where development is closer and more continuous and the "distal" or outer margin where growth is more sporadic and the development more dispersed in character. Wissink (1962)[22] in an extensive survey of the fringe areas of American cities identifies "fringe", "suburbs", "pseudo-suburbs", "satellites" and "pseudo-satellites" and in landuse terms has called it an area of "great differentiation." Pastalan (1967)[23] giving an operational definition of the fringe has defined it as an area in transition where rural landuses are giving way to urban landuses. Seeing all these differentiations in the definition, Whitehand (1967)[24] called the fringe a "heterogeneous region". In his view they are examples of regions that derive their unity not from homogeneity of forms but from certain factors that influenced their original location.

Pryor (1968-69)[25] faced another aspect of the problem of definitions when he noted and attempted to resolve the various areal differentiations of the urban fringe which have been proposed. In his view the rural-urban fringe may quantitatively be sub-divided into an "urban fringe" and a "rural fringe" on the basis of its landuse composition. The urban fringe exhibits a density of occupied dwellings higher than the medium density of the total rural-urban fringe and a higher rate of increase in population density, landuse conversion and commuting. By contrast, the rural fringe exhibits a low density and lower rate of increase.

During the last few years different scholars have attempted to analyse some particular character of the rural-urban fringe. Young (1962)[26] studied some geographic features of the urban fringe while Morrill (1965)[27] was more specific as to the mode of its expansion. Harvey (1965)[28] considered the nature and economics of urban sprawl in general. Socio-economic problems pertaining to the fringe and fringe dwellers have attracted the attention of various scholars of this time. The study to throw light on current processes within fringe areas is that undertaken by Pahl (1965).[29] His work focuses largely upon the social character of London's fringe. He recognizes four main processes under way. There is a tendency towards social and spatial segregation, selective immigration, changing commuting character and collapse of geographical and social hierarchies.

In the less developed and developing countries, a slightly different situation has been marked. Some studies of the villages close to the major cities in these countries exhibit two main characteristics : they are closely affected by ideas and innovations flowing from the city, and they provide a permanent labour force by migration or a temporary labour force by commuting to supplement that of the city. None of these commuters are ever former urban residents. (Cornell : 1974[30] and Barelay,[31] Connell,[32] Rao: 1970[33]). Rikkinen (1972)[34] has analysed the way in which ownership boundaries affect urban growth and landuse change, while Schnore and Klaff (1972)[35] have measured the speed of sub-urbanization over the last decade in the United States, compared with earlier decades.

John Hudson (1973)[36] has analysed statistically the demographic characteristics of the suburban fringe of the nine metropolitan centres in the Middle East (U.S.A.). On the other hand, in Martin's (1973)[37] study of spatial distribution of population, cities and suburbs "attention has been focused on entire metropolitan area for measuring the density gradient in which he has found a tendency for population distribution over space to vary between city and suburbs".

The initial problem to emerge from a study of the now vast scholarly literature on the urban fringe is a lack of clarity over what facts of life and landscape are under investigation (Thomas : 1974).[38] Carter (1972)[39] feels that there are two aspects which have attracted most attention. First, there is a notion of the fringe as a distinctive area, primarily designated by characteristic landuse associations. Second, there is the notion that the social characteristics of the population of the fringe are intermediate between those of the town and those of the country.

Thus, it is clear from the above examination that the fringe is a significant development, but due to its vagueness, the areal definition of the term has not yet been standardized. However, the incoherent landuse pattern, which occurs due to outward growth process of the city, may be taken as the representative of the fringe (Lal : 1973)[40]. In other words, the space into which the town extends as the process of dispersion creates the concepts of the rural-urban fringe. Golledge[41] considering all these characteristics has outlined this haphazardly developed zone of twilight as a zone of constantly changing pattern of land occupance, small farm sizes, intensive production of crops, mobile population of low and moderate density, incomplete provision of public utility services, speculative sub-division and building plots, all resulting in a continual movement or dynamism.

## METHODOLOGY

From the foregoing representative definitions it is obvious that the fringe belt is a dynamic area which fluctuates in the span of time. It continues shifting outward from the city with the diffusion of urban innovations in the surrounding localities. As its development is associated with the growth of the city, it becomes more difficult to delineate this overchanging area on the periphery of a growing urban centre. However, among several researchers dealing with the fringe complex, some have discussed the systematic delineation of the fringe. Here some important methods have been examined which were used in the delimitation of the urban fringes of the different cities and towns of the world.

Mayer and Beegle (1947)[42] while distinguishing the fringe townships in the metropolitan region of Detroit used the proportionate size of the NV-RNF (Non-Village, Rural Non-Farm) population with respect to the total population of the township. The use of this index gave them the 'true fringe' consisting of townships (minor civil division unit) with 50 per cent or more NV-RNF population and occurring in an essentially contiguous area outside the city, whereas townships with values of 25 to 50 per cent, normally farther from the city, were called 'partial fringe' areas. Rodehaver (1947)[43] in the study of Medison fringe undertook to ascertain the social, cultural and demographic characteristics of the area. The criteria used for delimiting the fringe were :

— the percentage of non-farm families to the total number of families in a given section,
— density of non-farm population per square mile, and
— assessed valuation of land and buildings per acre.

By using these indices he determined the "greatest intensity zone" in respect of the surrounding area.

Blizzard and Anderson (1952)[44] delineated the fringe area of Williamsport, Pennsylvania by automobile reconnaisance and inspection method. They visualize the inner boundary of the fringe at the points where city services are not in full operation. Normally, it coincides with the corporate boundary of the city. The outer fringe boundary was placed at the approximate point where an urban pattern of living of some concentration yielded to a pattern of widely scattered acreage plots with non-farm type houses among farms, or to pure farming or forest landuses. In the Columbia fringe study, Gist (1952)[45] has selected only open country families which had one or more members employed in the city and in which residential location was at a distance of one and half miles away from the city's corporate limits. Excluded were the families in the immediate suburban zone and in villages or towns within the area. Because of certain difficulties in securing a representative sample by means of accepted sampling technique, an effort was made to locate and interview most of the urban occupied families living in the area.

Instead of using census data, Martin (1953)[46] concentrated on the specific locality, gathering first-hand data from residents of the area. His study of the Eugene Springfield fringe in Oregon was based on probability area sampling in which the location of the single family residences outside the corporate boundary and natural barriers both were considered to delineate the urban fringe. The fringe boundary was drawn at that point "where the pattern of landuse characteristics of the fringe changed to the dispersed pattern of open country farming". But he did not give any indication as to what extent the density of single family dwellings would be considered as an index of fringe characteristics. Russwurm's approach to the London (Ontario) Survey (1961)[47] was more precise and the resultant distribution more detailed. He first classified each 200 acre plots according to their residential owners or tenants engaged in non-farm occupations and then included those plots in the rural-urban fringe where 50 per cent or more families were engaged in non-farm occupations.

The data were obtained from the township assessors Hind-Smith (1962)[48] in the same area identified the fringe by means of four indices : (i) land sub-divided but not yet developed or recorded by the planning board, (ii) farm lands 'for sale' for urban purposes, as indicated by realtors, (iii) land in non-farm ownership, either idle or tenant farmed, and (iv) farm land assessed at higher than normal farm levels as shown by local assessors.

The International Urban Research Unit, University of California, following the U.S. Bureau of Census, has recommended the inclusion in the urban fringe of all those areas which have at least 65 per cent of the labour force working in economic activities other than agriculture. The fringe may be determined on this basis, but certain enclaves, departing from the above standard of the labour force, may die. Such enclaves are the rule, rather than the exception, and, therefore, are treated as part and parcel of the urban fringe. In the study of Orlando, Florida, Mukherjee (1963)[49] has considered undeveloped subdivided land surrounding the city as fringe characteristics. The intensity of urban landuses as indicated by the sub-divisions of land and by the extent to which such sub-divisions are built up was chosen as the major criteria for determining the urban fringe. In a recent study, Pryor Robin J. (1968)[50] has suggested a sampling frame for the rural-urban fringe. In his study, a sampling frame has been developed to meet the complex requirements of a case study of outer sub-urban and urban fringe. He used the built-up area of any city of the previous census as the inner limit of outer suburbs and present as the boundary between outer suburb and urban fringe. Outer limit of the area is fixed with the help of the sampling points taken on the vectors originating in the CBD.

In our country, some geographers have also attempted to delimit the urban fringe. U. Singh (1966)[51] has tried to delimit the urban fringes of Kaval Towns by superimposition of a series of maps covering several geographical factors, i.e., built-up area and house types, types and pattern of streets, occupational structures sited for large institutions and industrial establishments, presence of lime and brick-kilns, and the limits of the essential services, etc. Thus, he finds that the outer limits of the primary fringes generally coincide with the municipal limits while the outer limit of the secondary fringes are more or less coincides with the regulated area of these cities. Alam and Kahn's have studied the (1972)[52] *Metropolitan Hyderabad and its Region* is one of the most outstanding works which deals with the relevant points of urban influences and their metropolitan core and peri-urban zone coincides with the primary and secondary fringes of the metropolis In her Ph.D. thesis, Nangia (1972)[53] demarcated the rural-urban fringe of Delhi on the basis of 19 structural and functional characteristics. The fringe thus demarcated, extends to a distance of 11 to 28 kilometres having 178 villages in total within its bounds. Further, Srivastava and Ramachandran (1974)[54] have also proved those heterogeneous factors that are responsible for the formation of present fringe of Delhi and have worked out a "Stage Model" depicting various stages of which villages around a metropolis may pass through. Occupational structure, landuse pattern, interaction with the main city, availability of urban amenities and locational characteristics are the test variables on the basis of which the villages were ranked. Hira Lal (1972)[55] has analysed the impact of urban influences on the rural settlement concentration with special reference to its variations according to the distance from the centre of the Bareilly city. The density of houses and population, proportion of built-up area to total area and the nature of settlements have been studied in order to trace the gradient in relation to distance from the city. Following U. Singh's indices Raja Ram Singh (1973)[56] has delineated the rural-urban fringe of a small town like Jaunpur. The only difference is that his inner fringe is confined within the municipal limits of the town.

## DETERMINANTS TO DELIMIT THE URBAN FRINGE

It is evident from the studies that there are some common factors which may be taken as the indicators of the urban influence on the surrounding countryside. It would be a rather wise step to decide first those indices or determinants. At the same time it is not advisable to delimit the urban fringes of Indian cities on the basis of the criterion suggested for fast growing urban centres of the west. India is a developing country and its urban centres also show the same trend.

The basic factors to delimit the fringe region are concerned with landuse influx, mixed rural-urban functions, interaction with the main city, socio-economic development, availability of urban amenities and proportion of non-agricultural population. However, besides analysing each and every structural and functional characteristic in delineating the urban fringe, at least preparation of three maps of the same size, showing the present landuse, transportation and occupational pattern of the population are deemed to be helpful. In addition, personal observations in the field would go a long way in fixing the "territorial extent" of the tentative fringe upon which the said factors could be analysed.

## DELIMITATION OF THE URBAN FRINGE OF BAREILLY

The studies by several authors to delimit the fringe have helped in selecting certain determinants to delineate the urban fringe of Bareilly. Here quite a few determinants have been selected which are partly spatial and partly occupational and demographic.

### Spatial Determinants

**Area of Further Extension of the Town** : Based upon the landuse map (Fig. 5.1) and personal investigations into the field, the possibility of further growth of the city towards different directions outside the present limit has been assumed which would be a tentative limit of the urban fringe. There is little possibiiity of the physical extension of the city in the adjoining south and south-west parts of Bareilly because of periodical inundation of the area by the river Ramganga and as a result thereof, the negligible extension of built-up area along Budaun Road which runs in this direction.

The existence of a military cantonment lying in the southeast of Bareilly acts as a strong cultural-barrier for the physical expansion of the town in this direction. Similarly, the river Nakatia flowing in the east of the city restrains the expansion of the city. Due to these factors, the development

*Determinants of the Urban Fringe*

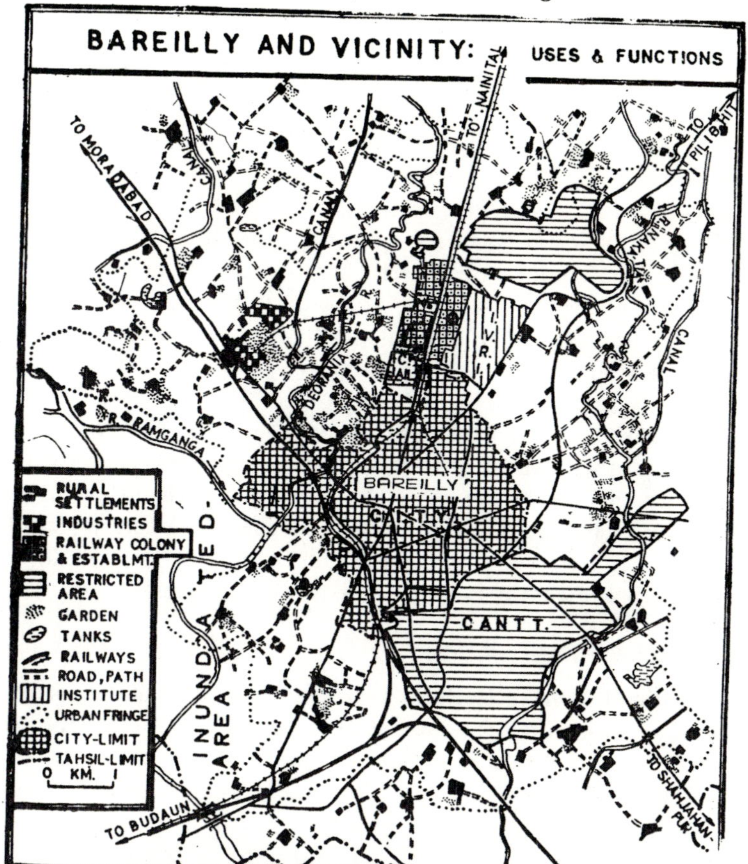


FIG 5.1

of the city along Shahjahanpur and Bisalpur Roads, running in south-east and east direction is little marked. On the other hand, with favourable geographical and economic factors the city is growing very fast towards north-west, north and north-east and most of the future development of Bareilly is expected to take place in these directions. A number of industries are developing fast in a ribbon type manner on all the three highways running in these directions due to availability of better infra-structure for industrial growth in the form of linkage and accessibility, suitability of land, etc. Industrial development is comparatively faster on Rampur Road where a continuous belt of industries has emerged up to Clutterbuck Ganj.

The impact of the city has been hampered on Nainital and Pilibhit Roads because of the defence installations which have been recently constructed near Izatnagar on the land lying in between these roads. However, a number of settlements situated beyond the restricted area like villages Mundia, Ahmadnagar, Chanwar, etc. on Pilibhit Road and Bilwa, Bhura on Nainital Road have strong interaction with the city.

The existing landuse pattern map reveals a mixed use of land in plenty in the west, north-west and the north outside the city limit. Most of the villages lying in between Rampur and Nainital Roads are well connected with minor roads and are easily approachable.

Under the impact of these factors the area having strong interaction with the city is slightly linear with respect to the existing location of the city. The boundary of the urban fringe coincides with these developments. Hence, the western fringe is extended up to 10 kilometres from the city centre, while in the south and east it is hardly up to 7 kilometres.

### Occupational Determinants

**Ratio of Non Agricultural Workers** : One of the important attributes on which the impact of a city may be markedly discernible is the occupational structure of the villages. The settlements in the vicinity of the cities seem to be changing their character more conspicuously than those situated away from it. In fact, a high percentage of non-agricultural workers in the working force of the rural areas, is an approximate measure of the urban zone of influence. This particular group is composed of the persons engaged in household and manufacturing industry, construction, trade and commerce, transport, storage and communications and in other services. These people live in areas within the orbit of cities which are classified as rural yet they are not engaged in farming.

The mean percentage of workers engaged in non-agricultural occupations to total workers for Bareilly tahsil, in which the city exists, is 16.45. Villages having a greater share than the average indicates the urban influence. Hence, they have been included in the fringe region. For further grading the intensity of this factor Nelson's (1955)[57] Standard Deviation (SD) method seems to be helpful. Considering his formula originally used for classification of the cities—that is, mean +1 SD is being taken as the lower mark for fixing the outer limit of the urban fringe. For Bareilly this comes to 24.38 (mean=16.45+1 SD=7.93). Therefore, villages with 24.38 or more percentage of non-farm workers have been included in the urban fringe of Bareilly. However, villages with these activities falling in value in between average and average +1 SD are taken as peripheral rural areas of rural-urban fringe. Further, in the urban fringe, villages having mean +1 SD to 3 SD's values are termed as secondary fringe while village above this mark have been given the name primary fringe, as they show a higher degree of urban character. When plotted on the map, the distribution of non-agricultural activities shows an extension towards the west, north, north-west and south-west. This extension is more along the roads and railway lines which provide the facility to commuters to enter the city daily for various purposes. The eastern part, alongwith least accessibility shows a weaker influence.

So far 31 villages come under the class 40 and above in which 13 villages are showing more than 60 per cent of their working force in non-agricultural activities, while number of villages falling into 24-40 and 16-24 classes are 15 and 8 respectively.

Thus, this index gives satisfactory results and demarcates the urban fringe of the city better. On the basis of those considerations 46 villages come in closest interaction with Bareilly city.

**Demographic Determinants**

The following determinants have been selected from the demographic structure to delimit the urban fringe of Bareilly :

**Density of Population** : The density of population is a fairly good index of the measurement of the zone of influence of the city. The mean rural density for the region is 1.67 persons per acre (4.12 persons per hectare) as per 1971 census. The density of population 2 persons per acre (5 persons per hectare)—it is 3 SD above the mean—has been taken as the lower limit for determining the extent of the fringe and 57 villages come under the fringe region. The isoline of 5 persons per hectare runs approximately along the same distance as the previous one, with elongations towards west, north and south-west from the city.

**Rate of Change in Density** : The rate of change in density during 1961-71 decade has been taken to be the other index of measurement of influence of the city. The mean change in density for the region is 26.3 and the villages showing rate above this have been taken to determine the urban fringe of Bareilly. Though this criterion provides a haphazard picture, the rate of change is more in the area situated comparatively farther from the city limit in the west. In the south also, a few villages fulfil this condition. In all, 40 villages conform to this criterion. There is a greater tendency to increase towards the north of the region as in some villages the rate of growth is more than 100 per cent. These village are either industrial nuclei or situated near industrial suburbs.

**Population Growth** : Like density, population is also an important determinant for delimiting the urban fringe. 25 per cent or more growth during the years 1961-67 has been taken to delineate the fringe boundary of Bareilly. The selection of this value is based on the average growth of the tahsil. 16 out of a total of 43 villages above the average show a 50 per cent or more growth during this period.

**Sex Composition** : Incessant migration of the male population from the rural areas to the cities and their nearest settlements in search of employment increases the male population in the nearby villages. Keeping this view in mind, 800 females per thousand of male population has been selected as an index of the extent of the urban fringe of Bareilly. 42 villages situated near the city show a sex ratio of 800 or less per thousand males. There is a long stretch of low sex ratio in the whole northern part of the region. In 18 villages there are less than 750 females per thousand males.

**Literacy** : With the general observations that the area nearer the city will be more educated and literate than those farther away, the tahsil average 11 per cent has been fixed as the lower mark for determining the urban fringe. About 42 villages are above the average near Bareilly city in which

16 villages enjoy a much higher percentage (20 per cent or above.) The intensity of literates is more in villages lying at the city limit than the remote village.

*Bareilly Boundary—Girdle Identification of Urban Fringe*

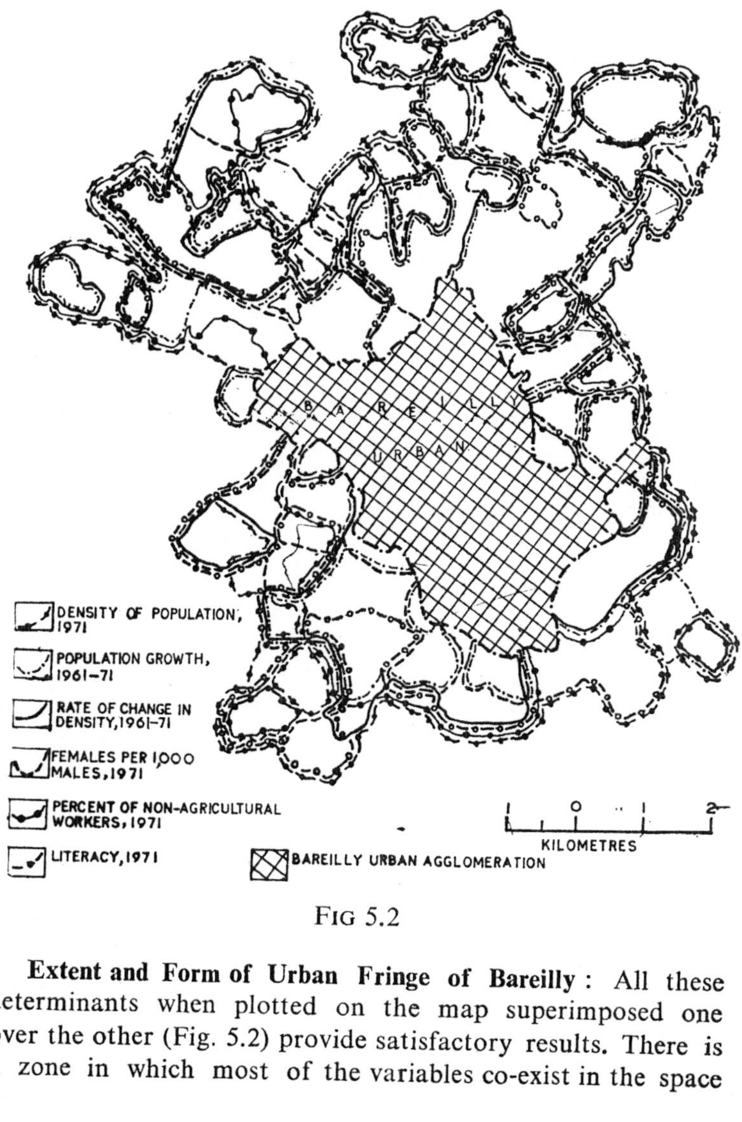


FIG 5.2

**Extent and Form of Urban Fringe of Bareilly :** All these determinants when plotted on the map superimposed one over the other (Fig. 5.2) provide satisfactory results. There is a zone in which most of the variables co-exist in the space and at the same time, there is also a zone where very few are conspicuous by their presence. All those villages which have fulfilled at least any three of the above indices and which are contiguous to the city, have been included in the urban fringe of Bareilly. Towards the inner side (or city), it merges into the city and towards the outerside it grades into the peripheral rural areas. The inner boundary of the fringe coincides with the municipal limits of the town that existed at the time of the previous census viz. in 1961 (Fig. 5.3).

*Intensity of Determinants Around Bareilly*

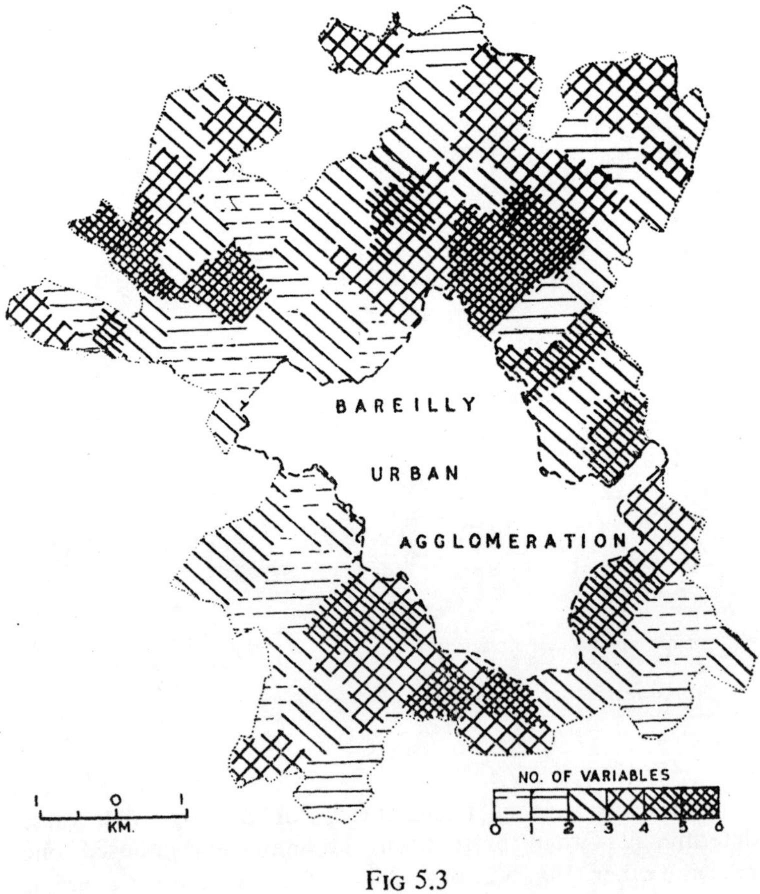


FIG 5.3

Delimited on the basis of attributes discussed above, the urban fringe at Bareilly consists of the 77 villages with an area of 134.5 km². This is not a concentric zone around the city but star-shaped and it extends from 3 to 10 km of the distance from the city. The extension of the fringe is thus approximately along the transport network, especially along the axial routes and in the inter-sectoral spaces, as the fringe tends to shrink.

The urban fringe of Bareilly, thus demarcated may be subdivided into (i) Primary and (ii) Secondary urban fringes,

*Urban Fringe of Bareilly Structural Zones*

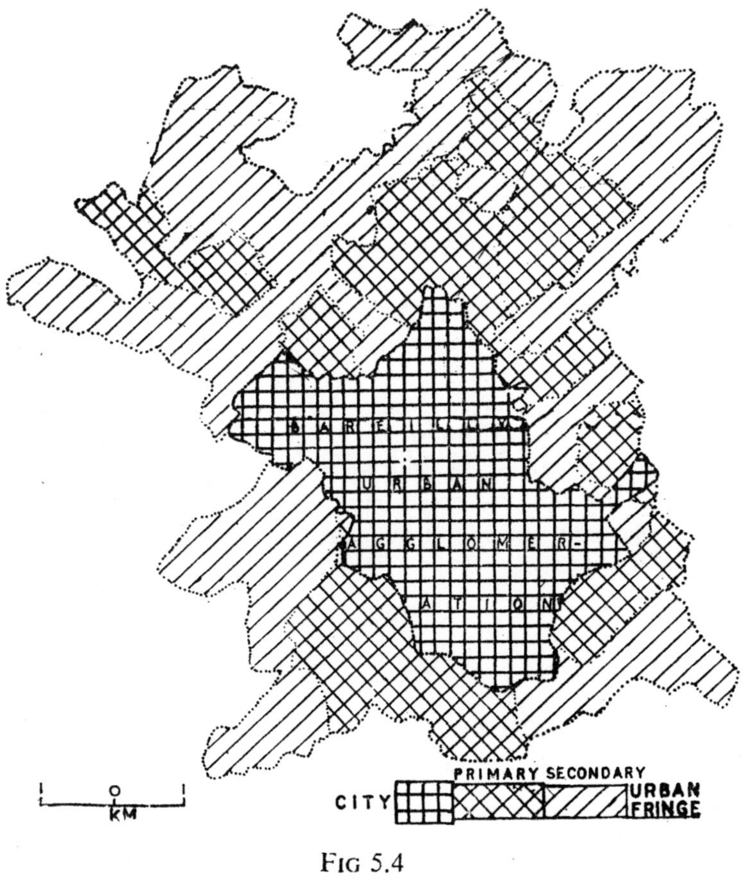


FIG 5.4

according to the intensity of functions. Villages having more than 30% of their working force engaged in non-agricultural pursuits and fulfilling at least three of the discussed determinants have been considered as primary or true fringe. Fig 5.4 reveals the distribution of these villages in a continuous area outside the municipal limits of the city. Actually this zone is intimately connected with the city and is very important for its existence and future growth. There are 24 villages with 5521 ha. (40% of fringe area) of land which fall in this zone. The remaining 43 villages constitute the secondary urban fringe and include 7958 ha. or 60% of the area of the fringe. The urban landuses and functions are, thus more prominent in the case of the former while agriculture becomes an important function in the latter.

Certain villages departing from the norms also exist in the urban fringe of Bareilly. These villages, technically known as village enclaves, which are only exceptions and are treated as part and parcel of the fringe, numbering five, lie just on the border of the municipal limit.

Only 10 out of the 77 villages, may be treated as urban villages which have partly been included in the corporate limit of the city. These villages are undergoing a transformation process in terms of their physical, occupational and demographic structure at a much more rapid rate than the villages beyond. At the same time, they are not only facing a tremendous change in their economic structure, population growth, density and sex composition, but they have also suffered an unplanned growth of the built-up area associated with insufficient sewage and other basic urban amenities which has given rise to various problems at the edge of the city.

## REFERENCES

1. R.L. Singh, "The Rural Urban Fringe Studies," *The Geogr.* Aligarh, Vol. 14, 1967, pp 9-21.
2. B. Srivastava and R. Ramachandran, "The Rural-Urban Fringe" *The Indian Geogl. Jl.* Vol. XLIX, 1974, pp 1-9.
3. T.L. Smith, "The Population of Louisiana : Its Composition and Changes". *Louisiana Bulle.* No. 293, 1937, p 24.
4. C.J. Gaplin, "The Social Anatomy of an Agricultural Community", *Univ. Wis, Res Bull.* No. 34, 1915, p 34.

5. L.A. Salter, "Land Classfication Along the Rural Urban Fringe" *Univ. of Missouri Agl. Expt. St. Bull.* No. 421, 1940, p 24.
6. G.S. Wehrwein, "The Rural Urban Fringe", *Eco. Geo.*, Vol. 18, 1942, pp 217-28.
7. R.B. Andrews, "Elements in the Urban Fringe Pattern". *Jl. of Land and Public Utility Economics*, Vol. 18, 1942, pp 169-83.
8. R.C. Klow, *The Park Ridge—Barrington Area : A Study of Residential Land Pattern and Problems in Suburban* Chicago, 1942.
9. F. Alpake, "Landuse Control in the Urban Fringe of Portland Oregon" *Jl. of Land and Pub. Util. Econ.* Vol. 18, 1942, pp 463-78.
10. H.H. Balk, "Rurbanization of Werester's Economics" *Eco. Geog.* Vol. 21, 1945, pp 104-118.
11. W.I. Fiery, "Ecological Considerations in Planning for Urban Fringes", *Amer, Social Rev.* Vol. 11, 1946, pp 411.
12. M.W. Roherhaver, *Fringe Settlement as a Two Directional Movement, Rural Sociology* Vol. 12, 1947, pp 49-57.
13. R. Dewey, "Peripheral Expansion in Milwaukee Country" *Amer. Jl. Sociology* Vol. 54, pp 119.
14. F. Aschman, "Dead Land" *Land Economics*, Vol. 25, pp 240-45.
15. S.W. Blizzard and W.F. Anderson, "Problems in Rural Urban Fringe Research", Pennsylvania State College *Agl. Expt. St. Progress Report*, No. 89, 1952, p 11.
16. W.T. Martin, "The Rural Urban Fringe : A Study of Adjustment to Residence Location". Univ. Oregon *Studies in Sociology*, Vol. 1, 1953, p 3.
17. O.D. Duncan and A.J. Reiss Jr., *Social Characteristics of Urban and Rural Communities*, 1950 John Wiley, 1956, p 372.
18. R.A. Kurtz and J.B. Eicher, "Fringe and Suburbs : A Confusion of Concept" *Social Forces*, Vol. 37, 1958, pp 32-37.
19. R. Golledge, "Sydneys Metropolitan Fringes : A Case Study in Rural Urban Relations." *Aust. Gwgr*, Vol. 7, 1960, pp 243-55.
20. L.H. Russwurn, "The Rural Urban Fringe with Comparative Reference to London" Referred in Alice Coleman's, "The Planning Challenge of the Ottawa Area", Geogl. Paper No. 42, Deptt. of Energy, Ottawa, 1969, pp 11.
21. M.R.G. Conzen, "Alnwick, Northumberland, A Study in Town Plan Analysis", *Trans Ist Br. Gegr.* 27, 1960, p 58.
22. G.A. Wissink, "American Cities in Perspective, with Special Reference to the Development of their Fringe Areas" *Assen*, 1962, p 201.
23. L.A. Pastalan, "Towards an Operational Definition of the Rural-Urban Fringe" *Indian Social Bulln.* Vol. 15, 1966, p 292.
24. J.W.R. Whitehand, "Fringe Belt : A Neglected Aspect of Urban Geography "*Trans. Inst. Br. Geogr.* Vol, 41, 1967, p 223.
25. R.J. Pryor, "Definition of the Rural-Urban Fringe". *Social Forces*, Vol. 47, 1968, pp 202-15.
"Delineating Outer Suburbs and the Urban Fringe" *Geogragiska Annaler*, Vol. 51-B, 1969, pp 33-38.

26. M. Young, "Some Geographic Features of the Urban Fringe", *The S.En. Geogr*, Vol. 2, 1962, pp 1-6.
27. R.L. Morril, "Expansion of the Urban Fringe : A Simulation Experiment" *Pap. Reg. Sci. Ass.* Vol. 15. 1965, pp 185-199.
28. R.O. Harvey, and W.A.B. Clark, "The Nature and Economics of Urban Sprawl," *Land Eco.* Vol. 41, 1965, p 72.
29. R.E. Pahl, "Urbs in Rue : The Metropolitan Fringe in Hertfordshire", London Scool of Economics and Political Sciences, *Geor. Pap No. 2*, 1965, p 72.
30. John Connell, "The Metropolitan Village : Spatial and Social Processes in Discontinuous Suburbs" in James H. Johnson (ed) *Suburban Growth* John Wiley & Sons, London, 1974, pp 94.
—See for example in three different areas.
31. H.B. Barclay, "*Buuri al Lamaab : A Suburban Village in the Sudan*, Ithaca, 1964.
32. J. Connell, "Mahmood-Abad and Telebabod Villages in Transition", in J. Connell (ed) Semnam : *Persian City and Region*, London, 1970.
33. M.S.A. Rao, "*Urbanization and Social Changes : A Study of a Rural Community or a Metropolitan Fringe*, New Delhi, 1970.
34. K. Rikkinen, "Boundary Changes in a Rural-Urban Fringe Areas", *Acta Geographica*, Vol. 26, 1972, pp 25.
35. L.F. Schnore and V.Z. Klaff "The Suburbanization in the Sixties : A Preliminary Analysis," *Land Eco.*, Vol. 48, 1972, pp 23-33.
36. J. Hudson, "Density and Pattern in Suburban Fringes," *Annals Ass. Mer. Gegr.*, Vol. 63, 1973, pp 28-39.
37. R.C. Martin, "Spatial Distribution of Population : Cities and Suburbs" *Gl. Reg. Sci.* Vol. 13, No. 2, 1973, pp 269-278.
38. David Thomas, "The Urban Fringe : Approaches and Attitudes *Vide in Johnson J.H.* (ed) 1974, op. cit., p 18.
39. H. Carter, *The Study of Urban Geography*; Arnold London, 1972, pp 288-296.
40. H. Lal, "Urban Fringe : An Analysis of Concept," *Uttar Bharat Bhoogol Patrika*, Gorakhpur, Vol. IX, No. 2, 1973, p 71.
41. Golledge, op. cit.
42. R.B. Mayers and J.A. Beegle, "Delineation and Analysis of the Rural Urban Fringe", *Appl. Anthropol*, Vo. 6, 1947, pp 14-22.
43. M.W. Rodehaver, 1947, op. cit.
44. Bizzard and Anderson, (1952) op. cit.
45. N.P. Gist, "Ecological Decentralization and Rural Urban Relationship", *Rural Saciology*, Vol. 17, 1952, pp 328-35.
46. W.T. Martin, (1953) op. cit.
47. Russwurm, (1961) op. cit.
48. J. Hind-Smith, "The Impact of Urban Growth on Agricultural Land : A Pilot Survey" Referred in a *Coleman* 1969, op. cit., pp 11-12.
49. D. Mukherjee, The Concept of Urban Fringe and its Delineation : A Case of Orlando Florida" *Geogl. Rev. India*, Vol. 25, 1963, pp 53-54.

50. R.J. Pryor, "The Sampling Frame for the Rural Urban Fringe" *The Prof. R. Geogr.* Vol. 20, 1968, pp 257-261.
51. U. Singh, "Urban Fringes in KAVAL Town" in R.L. Singh (ed) *Applied Geography*, N.G.S.1. Varanasi, 1966, pp 41-48. See also, K.K. Dube, "Use and Misuse of Land in the KAVAL Towns", NGSI, Varanasi, 1977.
52. S.M. Alam and W. Khan, *Metropolitan Hyderabad and its Region : A Strategy for Development*", Asia Pub. House, London, 1972 pp 2348.
53. S. Nangia, "Pattern of Rural Settlement in Delhi (A Geographical Analysis" *Ph.D. Thesis submitted* to Delhi University, 1972, See her book, *Delhi Metropolitan Region : A Study in Settlement Geography*, New Delhi, 1976 pp 81-102.
54. B. Srivastava and R. Ramachandran, 1974, op. cit., p 1.
55. H. Lal, "Gradients of Urban Influence on the Rural Settlement Concentrations with Special Reference to Bareilly" in R.L. Singh (ed), "Rural Settlement in Monsoon Asia, N.G.S.I. Varanasi, 1972, pp 264-272.
56. R.R. Singh, "Jaunpur and its Rural Urban Fringe : A Study in Settlement Geography", *Unpublished Ph.D. Thesis.* B.H.U., 1973, p 150-160.
57. H.J. Nelson, "A Service Classfication of Cities," *Economics Geog.*, Vol. 31, 1955, pp 189-210.

# 6

# The Concept of Rural-Urban Fringe

*—Bharat Prasad Mishra*

THE FUNCTIONS, forms and patterns of modern cities are undergoing dynamic development and change both laterally and vertically. "The most startling incongruities between fact and administrative fiat are found in the border zones of growing cities. Here urban residences and functions have greatly spilled far beyond the political bounds of metropolis and if urban growth is rapid, no amount of technical alacrity can keep census divisions abreast of the rural-urban frontier."[1] G.S. Wehrwein aptly observes that "the built-up city is not necessarily coterminous with the political city".[2] Indelible imprints of urbanism are found beyond the corporate limits and easily identifiable rural indices are also seen within the urban fence. This transition zone, always in flux, and fickle and full of incongruities of landscape, presents problems of the most acute nature and form. The vertical development in the form of up- and sub-constructions are not as problematic as the lateral expansions are. This is because the vertical expansions are mostly confined to the domain of a single urban authority undertaken with prior approved plans under its direct supervision and control. The peripheral expansions, unlike the above, often take place in the transition zone between urban corporate limits and rural agricultural land under uncoordinated authorities or a weak rural authority never meant for tackling problems of an urban nature. Such problems as unwanted, unwise and premature fragmentations, unguided, uncontrolled and unplanned constructions, haphazard and sporadic establishment of industrial estates along major highways (specially damaging are the brick industry), premature encompassing—rather swallowing—of suburban villages or slums, lack of treatment or piecemeal treatment of problems are crying for planning and control. These mostly fall on deaf ears, but their alarming conditions have prompted authorities to establish Regional Development Agencies. Such a step (meant for development) from municipal or corporation limit to arbitrarily defined regional limit is like the mythological "Hanuman-jump". In this jump, no wonder, if the legs get fractured and plans flop. Anyway, these authorities are supposed to guide, control and plan the areas being eclipsed or likely to be esclipsed in the immediate future by the urban centres. The responsibility of these authorities are great, and naturally, they should note that to plan such areas without the proper assessment of their true pathological conditions, may lead to disastrous results. This necessitates a comprehensive study, proper analysis and spot assessment of the problems and prospects of the rural-urban fringe as evidenced in the form of the impact of the urban centres. Urban centres now-a-days are considered to be partly the indices of the socio-economic and politico-cultural conditions of the community they serve and partly the scale of advancement the nations have achieved economically, technologically and culturally. It is the scales and the indices which constitute the independent personality of the town. The sprawling dissemination of these scales and indices gradually transforms the peripheral belts of the towns and cities. The degree of transformation is the greatest near the town and it diminishes gradually to the distal areas. All this requires the detailed study of the rural-urban fringe.

## DEFINITION

Theterm rural-urban fringe is reserved for the outskirts of the towns and cities. Here the salient rural and urban characteristics are found intermixed. It is a belt possessing neither the qualities of the town nor the village but a smug typifying the features of both. Naturally the fringescape does not resembles either the townscape or the villagescape, but at the same time corresponds to both. To put it in the words of H. Carter, the rural-urban fringe is "an area with distinctive characteristics which is only partly assimilated into the growing urban complex, which is still partly rural".[3] The notion of the fringe is "that where urbanization impinges on rurality . . . where the process . . . indeed should be in operation".[4] In the settlement continuum the rural-urban dichotomy appears conceptually nebulous. It is untenable because "there is no point in the continuum for long agglomeration to small clusters and scattered dwellings where urbanity disappears and rurality begins"[5]. It is gradual and almost imperceptible process operating differentially over time and space and thereby creating a third entity 'the fringe' which may be assumed a by-product of the process. According to H.J. Gans, the simple urban-rural dichotomy cannot be maintained against the existence of the "urban village" and the metropolitan village . . . many studies have shown the existence in city centres of distinct enclaves where the kinship linkages are intensive and where there is high density of social contacts giving a cohesiveness conventionally ascribed to rural settlements".[6] Besides this the dichotomy, an administrative designation, is anachronous and pragmatic. The growing urbanism with shrinking naturality and the interposing of the "fringe", merits a trichotomous division of the continuum. The fringe, although spatially anachronistic, possesses a morphologic permanency and it therefore ought to be considered at par with the two. It is these three together that provide the composite picture of the settlement continuum.

Cities and towns are found almost sharply demarcated by the compact buildings, cultural innovations, relatively higher standard of living, paucity of open spaces, nuisance of slums, sewers and cul-de-sacs, high land values, lack of natural dominants, a characteristic landuse association, a distinctive 'trait complex' of functional institutions, typical demographic structure including distinctive behavioral approach and economic texture, and finally, the tint of the modern civilization.

The universal availability of radio, television, newspaper, facilities of other developed means of transport and communication, banks, dispensaries, educational institutions, flour and rice mills and a host of other developmental agencies and establishments have, no doubt, transformed village life to a great extent; but morphologically villages are quite unlike urban centres. The rural scape is found with compact as well as discrete hamlets with thatched huts and mud walls, low standard of living, abundance of green spots and open spaces, relative lack of congestion, low land values, lack of secondary and tertiary vocation and finally a symbol of our ancient culture. Although both the villages and towns have three dimensions, in the case of the urban centres, the third dimension is very prominent. These rural and urban morphological elements are remarkably distinctive but real and understandable differences exist in their social behavioral patterns and economic structure.

It is the juxtaposition of these two virtually opposed units which gave birth to the third unit known as the suburb or the rural-urban fringe. "Where the urban influences are essentially mingled with rural forces often without striking a proper adjustment as in the case of unplanned growth, designates the nature of the rural-urban fringe"[7]. When the characteristics of the fringe are more akin to or influenced by the urban morphology than by the rural one, the fringe is termed as inner fringe and in the opposite case, i.e., when the fringe is more likened to rural morphology than to the urban one, the fringe is termed as outer fringe. In other words the outer fringe belt exhibits a decreasing tendency of already quasi-urban characteristics towards the periphery. The partly urban and partly rural characteristics starting from the outer periphery of the inner fringe belt begins to vanish gradually until the true rural areas meet at the periphery of the outer fringe belt. Hence this belt is more rural than urban, but the apparent though scanty influences of the urbanity can still be noticed there. The inner fringe may also be called urban fringe, and the outer fringe the rural fringe, and the two together the rural-urban fringe. Now a line or a zone can be assumed where the two morphological units meet the third one and it is this meeting-point or convergence which separates the rural-urban fringe from the urban centres on the one hand and the rural ones on the other. The morphological characteristics of the two may be considered as their intensity indices although they are mixed up rather disproportionately and haphazardly. A line can then be theoretically supposed where rural and urban intensity indices are both equal and opposite increasing and decreasing inversely in reverse direction. The fringe generally advances with irregular tentacles of urbanity and embayments of rurality. The sprawl is clearly haphazard, moving rapidly in one direction, skipping and exploding in another, while in some areas it may be nil in extension.

## THE PROBLEM

The Muzaffarpur Regional Development Authority has recently been established to guide, control and plan the city as well as the eight contiguous surrounding anchals. Till now the Improvement Trust was functioning with all such powers within the watertight Municipal boundary. This sudden, unusual leap that appears like the mythological "Hanuman-jump" has provided the raw material for the present essay. Starting with a brief review of the relevant literature, it looks into the nature and characteristics of the rural-urban fringe. It is also in search of a tool for empirical and tentative delimitation keeping in view the need to divide the rural from the urban. It further suggests that in the event of any urban planning such a study needs to be undertaken. The crux of the problem is to guide and control the successive bands of rapid urban expansion within a limited time and with resources hardly sufficient to cope with the ideal urban regional plans. Conceptual aspects of the rural-urban fringe and its importance in planning alone have been considered here without any empirical study.

## LITERATURE IN CROSS-SECTION

Pioneer studies on the rural-urban fringe date back to the second quarter of the present century specially between the forties and the sixties. The study of Wehrwein (1942) may be termed as a classic attempt on the subject. He has reviewed the entire gamut of the problem in its proper perspective.

Starting from the Von Thunen's isolated state and its central city through Christaller's scheme of urbanisation, he has tried to evaluate the different factors of urban invasion and peripheral dissemination of urbanisation which, according to him, were mainly due to the rail-road as the first decentralizer and the metalled road and automobiles as the source of rapid invasion. He depicted the fringe in USA as "institutional deserts" because it became the location of all sorts of noxious and undesirable activities. He studied the characteristics of the fringe with particular reference to Indianapolis. Chauncy D. Harris (1943) in his article "Suburbs" called it "climax area" the expansion of which is a function of physical dominants of the periphery and the cultural characteristics of the urban centres including their size and functions. He classified the suburbs into six types, i.e. (i) Industrial fringe, (ii) Industrial, (iii) Complex with Industrial more important, (iv) Complex with residential more important, (v) Dormitory, and (v) Mining and industrial. The study of C.F. Kohn (1956) and C.R. Hayes (1957) dealt with differential population growth and suburban residential land values respectively in the different localities of U.S.A. R.A. Kurtz and D.B. Eicher (1958) in their paper "Fringe and Suburbs : A Confusion of Concepts" have attempted on the aspects of definition and the problem of physical delimitation. The fringe being a quasi-urban complex with wide variety of landuses, was termed by Golledge (1960) as "a geographical no man's land". In his study of Sydney, he noted the seven characteristics of the fringe, namely, smallness of the farm, fluid pattern of the land-use, intensive cultivation, mobile population with sparse to moderate density, rapid residential extension, lack of service and facility centres and abundant provision for speculative buildings. Congen (1960) in his paper "Alnwick : A Study in Town Plan Analysis" and Whitehand emphasized the fringe as a potent determinant of urban morphology which is a product of the continuous influence of the former fringes encompassed successively by the advancing urbanity. He visualised fringe as being composed of two sections, "the proximal or inner margin where development is closer and more continuous and the distal or outer margin where growth is more sporadic and development more dispersed in character." He proposed that, depending on the circumstances, the fringe, as it is, can be "preserved" i.e. kept open under parks and playgrounds or can be "translated" i.e. put to different uses. Wissink (1962) in his study called it "an area of great differentiation" while R.E. Pahl (1965) summarised the salient characteristics under the four heads of (1) segregation, (2) selective immigration, (3) commuting, and (4) collapse of geographical and social hierarchies. He contended further that the creation of the fringe was more a process of social transformation that brings about class consciousness through which is woven the dynamism of urban life style. His departure from the usual emphasis on the physical and static features in the form of landuse characteristics is a major one and worth noting. Schnore and Knights (1969) concluded that urban expansion was not a new feature ; it happened earlier too during the nineteenth century. H. Carter (1972) has attempted a good review of the literature on the rural-urban fringe, the residents of which he thinks "live in the country but are not socially and economically of it".[9] Piecemeal literature in the form of passing references is plentiful, but whether it will serve any purpose for detailed academic work is a point and hence it is being deliberately left out.

## CHARACTERISTICS

The distinctive characteristics of the fringe can be divided as (i) physical, (ii) functional or organisational, and (iii) normative or cultural. Physical aspects signify spatial characteristics that include small and fragmenting farms, high priced green belts, haphazardly distributed huts and houses in the different stages of construction and finally friction amongst various spatial uses. Sub-divisions of the farms and fields are the most common feature of the fringe. Realtors, with their agents and brokers start purchasing big plots and layout economic and wildcat parcelizations. With money being the main motive there are petty or plenty gains while amenities are ignored. Buyers with their speculative choice of site and building plan contact and finish at least with fencing or trace-pillars. These sporadic unplanned and uncontrolled constructions are the rule rather than the exception. This often leads to the creation of narrow and one way lanes ending in cul-de-sacs. Industrial establishments and public institutions requiring sufficient cheap space are often located at the distal margin along major route ways which are usually the chief channels of urban invasion. "This invasion would take its start from the central community. Its movement here as elsewhere we may liken to a glacier. It is spreading unthinking, ruthless. Its substance consists of tenements, bungalows, stores, factories, bill-boards, filling stations, eating-stands and other structures whose individual hideousness and collective haphazardness present the unmistakable environment which we call the 'slum'. Not the slum of poverty but the slum of commerce. This substance conceived as a projecting crawling glacial flow, makes its journey along each of the radiating highways."[10] The crawling urbanization demands retail facilities which are usually found near nodal points growing as "satellite, or trailer or shack towns."[11] Temporally growing intensity of the arteries of communication and the increasing transportational facilities are also very important spatial aspects of the fringe. The traffic and street density of the fringe will be of an intermediate nature i.e. more than that of the rural and less than that of the urban areas. "The line at which the fringe is invading the countryside is marked by a basic change in the road pattern."[12]

Functionally fringe is almost a synonym for rapid and random change and transformation. Its most noticeable peculiarity lies in its everchanging pattern of landuse. "In a contemporary context the various land users, older villages, newer residential extensions, commerce, industry, city-services and the underlying farming are not neatly sorted out into homogeneous areas but are intermingled in a random fashion and it is this which gives it distinctive quality to the landuse pattern of the rural-urban fringe."[13] The "ecological complex" of the fringe is defined by the landuse variables which according to Schnore[14] are environment, technology including transport, communication and the friction of space, population, its size, rate of growth, ethnic and social composition and organization including economic base, community and social class composition. These variables are inconsistently complex and these in turn provide greater complexities to the pattern of landuse there. A functional cross-section of the fringe along the green belt to a nodal point begins with the cultivation of mostly marketable products grown with the meticulous care of the intensive cultivator and ends with relatively higher population density, ease of approachability and functional facility. The intermediate section is found replete with a mix-up of well-built as well as half-built houses and factories intermingled with greens and farms. Suburban modality, a point on the circulation and functional intersect is the climax of the fringe. It functions as a diffusing point of urban innovations.

The above mentioned physical and functional aspects combined together result in comparatively higher land values of the fringe. It may be taken as a common but temporally dynamic denominator of the thickening of the entire mesh of economic activities which shadows as the process of urbanization. The economic functioning is imprinted on the pattern of landuse and "the pattern of landuse and land values are mutually determining"[15]. If the pattern of the value of the marginal areas were truly assessed, it will to a great extent conform to the mosaic of the process of urbanization taking place in the fringe zone. The sole function of the fringe is to provide constructional space whose value is summed up in the sole factor of "proximity" to the multiple nuclei and nodalities provided the natural landscape is absolutely uniform and all facilities and services perfectly constant. But the fact is just otherwise and practically "one is forced to conclude that the problems concerning land values are associated with the impossibility of obtaining satisfactory, uniform, objective data. . . rather than with the concept of using values themselves."[16] In the given circumstances the only alternative left is to gather as much reliable information as possible and for the rest subjectivity is the only recourse, while making a study of certain changes in land values in London area. Anstey has noted that "the process by which I arrive at this figure is long and complicated and largely infra-conscious but I am sure it is just about right."[17] Any attempt to draw isovals will really be a gallant effort and this will be a real contribution to the delimitation of the fringe and understanding of the functional pattern and the preparation of plans for it.

The physical and functional aspects as detailed above contribute to the cultural composition of the fringe dwellers. The existing population density is usually low to moderate with a tendency towards concentration. Lampard[18] views urbanization as a process of population concentration which gains momentum with the growing provision of public service facilities which are still incomplete. Most of the fringe dwellers usually travel daily forth and back between the place of employment and their residences. The bulk of the rural unskilled labourers not only cater to the need for a huge mass of cheap labour but also serve in various capacities leaving the household affairs on the shoulder of the older and younger ones left at home. Their inclination and drift to the town is oriented by their folk status, nature, kith and kin already serving in the town, the nature of employment already available and their extra family associations. They are usual as well as occasional and are generally very selective in their vocation. "Naturally the association of this adventitious population in the so-called rural areas are essentially urban."[19] The mobility of this adventitious population is decided by the nature of the work and employment, mode of travel, means of transport and spatial location of the place of work. The changing social and behavioral pattern of the fringe dwellers is one of its remarkable aspects. The growing enrichment of their experience over time and with a cross-section of associations, changing taste and temperament regarding food, clothing and shelter, polished and tactful talk, dextrously developed mental faculty are some of the aspects of the infra-structure of their personality which are thickening and bringing them closer to the concept of urbanity. The urban influence may thus be visualised as a penumbra of cultural and normative integration encircling the urban-umbra. Remarking on the cultural integration Lambert noted that "What should be considered was" not the radiation of influence from cities into rural areas but rather the presumed radial change in life style which confronts the villager when he moves to the city or the city moves out to encompass the village."[20] He further noted that 'urban influences on rural areas appear to depend in India, not so much on the actual position of the village but on the receptiveness of the absorption of new ideas of social groups within the village and the degree of potential flexibility present in the social structure so that curiously many of the villages whose unity and isolation have been emphasized lie very close to towns and cities."[21]

### THE PROBLEM OF DELIMITATION

It is amply clear from the preceding discussion that fringe is neither an umland nor a sphere of influence and is "usually smaller than a commuting area of a given city."[22] R.E. Dickinson while calling fringe "a divide" unhesitatingly admits that the outer limits of the fringe are "indeterminate."[23] The same authority somewhere else voices that "in fact there is an invariability, a fringe of rural-urban uses, a fringe that is wide and irregular, so that it is extremely difficult to run a line around the outer limits."[24] Lack of desirable and reliable data usually stand in the way of objectivity and subconscious subjectivity will be the only alternative left and thereby the opinion and approaches will differ on the practical side of the study. In spite of all these difficulties, we may base our delimitation on such factors as the relative density of population, the density of communicational lines, frequency and volume of transport services, the value of land and the pattern of land utilization, occupational structure of the suburban people and their social behavioral pattern, the topographic conditions favouring and forbidding expansion and finally the degree of dependence of the fringe dwellers on the city or town, etc. All these require an intensive study of the morphological characteristics of the two units and their relative prevalence in the zone of confluence, the outer limit of which should be considered as the distal boundary of the fringe. To speak in the most concrete terms possible, the outer limit of the outer fringe may be defined beyond such sites as educational institutions, industrial estates, store houses, dairy, poultry and such other farms located adjoining to the urban centres, sewerage and sullage disposal sites, milk and vegetable supply limit by pedestrians (and not by vehicles) acquired or proposed residential sites, contiguous urbanized villages, adjoining nodalities providing retail facilities with a tendency to grow as satellites, realtors approach limit, parks, recreation and sports grounds, golf courses missionary and religious institutions, timber stores and saw-mills, aristocratic bungalows, landing-grounds, cemeteries, nearby filling stations, check-posts, normal advertising limit of urban entertainments through mikes and posters, etc. In other words the fringe should be identified and demarcated up to that point where the two intensity indices are contiguous with their respective counterparts i.e. the rural intensity index towards the urban side and the urban towards the rural side can be found in any degree. Actually speaking the determination and demarcation of the fringe on the surface here becomes more a matter of assertion rather than of proof.

## SUGGESTIONS FOR PLANNING

The seemingly annular fringe belt provides a good solution as one of the many similar temporal phases in contiguous rural-urban planning. The device appears to be the single stone killing two birds of temporal urgency and financial bottleneck. This empirically derived belt has the advantages of a close relevance with structural reality in terms of planning and inductive generalization over haphazardly chosen arbitrary areas which may not have the sanction of even deductive approach. The urban proximals require quick as well as deep insight into their problems with particularistic stress which looks possible by this method only. The immediate need is to put a total curb on the independent decision-making of the fringe dwellers to avoid fragmented and incoherent picture prospectively present there. This too is possible only on a small-scale which befits such an approach. In the circumstances as dealt with above, fringe should be recognised as an inevitable phase element in urban and rural planning.

In the preparation and implementation of the plan the viable role of a geographer must be duly recognised. With the inherent composite and holistic perception coupled with a unique capacity to bring about a unified approach out of a diversified vast information, the geographer has the faculty, in the broader sense of the term, unmatched by any other specialist. It is he who can make a simple but cumulative and synoptic presentation of complex factors to be essentially considered in planning. Justice Scott aptly remarked that "Town Planning is the Art of which Geography is the Science." To sum up, employ a geographer who is master of none, and certainly the only master of jack of all trades and he will fulfil a long awaited and much desired vacuum.

## REFERENCES


1. W. Zelinsky, *A Prologue to Population Geography*, Foundations of Economic Geography Series, Prentice Hall, INC Englewood Cliffs, 1966, p 22.
2. G.S. Wehrwein, *The Rural-Urban Fringe, Readings in Urban Geography*, Ed. by H.M. Mayer, and C.F. Kohn, Central Book Depot, Allahabad, 1967, p 533.
3. H. Carter, The Study of Urban Geography, Edward Arnold (Publishers), Ltd., 25, Hill Street, London Wix 8LL, 1972, p 288.
4. Ibid., p 288.
5. Ibid., p 17 (The United Nations Demographic Yearbook for 1955 quoted.)
6. Ibid., p 293 H.J. Gans., 1962, The Urban Villagers (New York) Quoted.
7. R.L. Singh, *Banaras : A Study in Urban Geography*, p. 61, (Banaras, Nandkishore and Bros) 1955.
8. H. Carter, op. cit., p. 140 (Congen, MRG. Quoted).
9. Ibid., p 288.
10. G.S. Wehrwein, op. cit., p 537.
11. Ibid., p 538.
12. J.R. Boochert, *The Twin Cities : Urbanised Area, Past, Present and Future*, The G.R. April, 1961.
13. H. Carter, op. cit., p 289.
14. Ibid., p 181.
15. Ibid., p 182. (Adopted from R.M. Hurd.)
16. Ibid., p 196.
17. Ibid., p 196. (Quoted from B. Anstey, article.)
18. Ibid., p 31.
19. A.E. Smailes, *The Geography of Towns*, Hutchinson University Library, London, Sixth Impression, 1962, pp 35-36.
20. R.E. Pahl, "Trends in Social Geography" *Frontiers in Geographical Teaching* (Ed.) R.J. Chorley and P. Hagget, p 90. (Turner Quoted 1962, p 131).
21. Ibid., p 90.
22. G.S. Wehrwein, op. cit., p 533.

23. R.E. Dickinson, *City and Region : A Geographical Interpretation*, First Published in 1964 by Routledge & Kegan Paul Ltd., Broadway House, 68-74, Carter Lane, London, E.C. 4 (Deptt, Library Cat. No. 911. 72) p 165.
24. Ibid., p 237.

# 7

# The Concept of Rurban Centres

*—Ram Bahadur Mandal*

LARGE rural settlements and small towns are known as rurban centres. Here, the concentration of service facilities for the surrounding hinterland and business proficiency are of overwhelming importance as they are the levers of regional development. The concept of rurban centre is a new ramification of settlement geography, where physical landscape is mostly set in the right direction by social behaviour[1] and the cultural landscape of the society. Economic prosperity plays an important role in the growth and development of rurban centres. Such a small town or small city is essentially equivalent to what Brush has included in his threefold classification of hamlets, villages and towns.[2] The study of rurban centres is of academic interest because it represents the lower end of central place continuum. Any generalizations, theories, or laws developed for central place should hold good not only for large urban places but also for all the hierarchic levels including rurban centres. This essay seeks to identify the factors responsible for the spatial distribution of smaller urban settlements and their role in maintaining the ecological balance with the rural settlements.

## MEANING AND SCOPE OF THE STUDY

Rurban centres[3] have been variously described as "rurban settlements" by R.L. Singh[4], "rurban community" by Galpin[5], "urban villages" by Dickinson[6] and "sub towns" by Smaiks.[7] Recent geographic literature contains some important studies of rurban centres[8]. Among them most praiseworthy are Berry and Garrison's Snohomish county, Washington Studies[9]; King's comparative study of the Canterbury Provincial District, New Zealand[10], and Thomas's study of the functional bases for small Iowa towns.[11]

Monkhouse has used the term "rurban" as being synonymous to that of "rural-urban fringe" and rural-urban continuum," but this is a misuse of the term "rurban". Rurban centre is a small urban centre or larger overgrown village having some importance of secondary and tertiary functions. On the other hand, the rural-urban fringe is "a zone of transition around a town in which urban functions, landuse and other activities are mixed with agricultural ones. This zone separates the area exclusively devoted to urban uses from that exclusively devoted to agricultural uses."[12]

Rurban centres have a distinctive feature of their own. Most of the smaller centres are overgrown villages that still retain their rural character. They show a combination of economic, social and educational functions.[13] Most of the towns have a city region around them; the small towns function as the economic and social capitals of their surrounding essentially agricultural areas. By virtue of its intermediate location and the nature of evolution, the small town has certain advantages and offers unique opportunities in the general scheme of the rural-urban and agricultural-industrial continuum.[14] Hence the study of smaller urban settlements is extremely important as they represent the "basic connection between the dispersed agricultural population and the agglomerated urban population."[15]

There are several characteristic features which separate rurban centres from the list of rural settlements and urban centres as they are urban islands in the vast sea of rurality.[16] The overwhelming growth of population, *laissez-faire* growth of rural service centres and inadequacy of service facilities in rural areas are highly in need of rurban centres study. The problem of social disorder in rurban centres necessitates the need for planning and at present this should be in the forefront of research in geography as this provides the levers of regional development and planning.

Rurban centres may be a commercial, administrative,

transportational, industrial or religious node, but their locational force has severe effects on the surrounding zone of influence for maintaining the "ecological balance".

In general within settlements, the aim of balanced development is equally critical. The aim is the combining of social growth, occupations, enterprises, types of housing and common services that are still to be found in provincial cities and "urban villages." Such intermediate urban centres for marketing, cooperatives, services and industries serving agriculture must be strengthened. Dispersed and desolate rural settlements need to be brought together. Such a policy offers some hope of lessening the pressures of large scale migrations out of agriculture directly into the biggest cities. This may also provide alternative settlement systems designed to achieve a more balanced regional development.

Rurban units are marked by the growth of certain significant urban functions. In fact they are the foci of a number of market villages. Such centres often have significant retail and wholesale marketing facilities. Post and telegraph facilities, a market, a dispensary, a veterinary hospital, rural professional services, a junior high or high school and community block development offices are other characteristic functional units located there. Sometimes these centres are trade centres for the surrounding rural area, or administrative centres due to their being the headquarters of the local administration. On account of their distinct commercial, cultural and administrative elements, such units are characterized by a greater number and variety of services, which are rarely present in a village. Many other functions are attracted to such centres by virtue of their being the seat of local administration.

Many services are found only rarely in rural areas, such as cinemas, hospitals, libraries, etc. These are today available only in towns, as they are too costly to be operated in open country locations or in every small town. Some of these services may be located in the country at the edge of a town but this is basically a town location for all practical purposes. Bus services and metalled roads have lifted considerably the horizons of rural families as far as the accessibility to these services are concerned, yet costs in time and in money are still prohibitive. As a result, most rural families use less of these services than they would had they lived in a town and belonged to the same income group.

Rurban centres are special types of towns based on the services of those involving governmental services of all kinds. Federal and State agencies of various kinds have their offices in rurban centres often in the "country-seat town." These include the Block Development Offices and the circle offices. The farmer will have occasion to visit such offices as part of his business operations, and to a greater extent the rural family will visit them for various aspects of family living. Thus the small towns are quite often seen as the "bridge" between the urban and the rural universe, as the kind of settlement which can offer "the best, and none of the worst of both worlds." Obviously then, a study of rurban centres would form as much a necessary part of rural settlement geography, as the study of the larger villages themselves.

The city level centres have distinct advantages over rurban centres in industrial development. But it is difficult, if not impossible to carry out a systematic review of the role of small towns. Although one might conceive of industrialization of the country mainly through large cities, rurban centres would undoubtedly play an important part in providing levers of development as well as ancillary services, such as educational and health facilities which are essential for economic growth.

The thoroughfares of most of the rural settlements have been oriented towards the hinterland of a rurban centre. This is due to the intimate business and consumer relationship of that market with the surrounding rural settlements. This may be generalized into a law for the location of rural setılements and rurban communities. Thus, the location of a rurban centre guides not only the location of rural settlements but also their transport linkages. This is a new tool for geographers to study the 'law of rural settlement morphology and rurban centres location'.

## RURBAN CENTRES : A REVIEW OF SOCIAL CHANGE

It is useful to make an inventory of the functions of rurban centres of various types in different locations, so that their implications for planned economic development could be reviewed. Two of the functions, actual and potential of the rurban centres are particularly significant. First, rurban centres can act as channels for mobilizing savings and skills which are not available or not easily tapped in large towns. They might be the places, where some portions of income derived from agriculture are invested in industrial establishments such as household industry or small and medium scale units. Secondly, rurban centres might also in certain circumstances, be especially suitable as centres of formal education and specialized medical services.

Rurban centres may be :

(1) religious, transportation and administrative towns;
(2) new industrial or power project towns, such as Barauni and Birpur;
(3) regional or sub-regional centres of services like school, college, hospital, etc.;
(4) agricultural market towns or wholesale and retail trade centres; or
(5) existing small towns in the vicinity of large cities.

These and other types of rurban centres can fulfil specific functions in economic development. Their costs of providing services and maintaining overheads are relatively small. They thus offer several advantages over the larger towns for the country's uplift. They perform useful functions and have obvious potentialities for growth. Community development schemes are at times more effectively carried out in rurban centres than in larger towns. About the social structure of rurban centres we can say that the secondary school age group is the largest component among migrants into rurban centres. On the other hand, the age group of 34 and above show a net migration from rurban centres. Similar trends have been observed in Indonesia and the Philippines,[17] where too, the educational centres draw young people away from the rural areas.

In such centres the occupational structures are certainly varied. Agriculture is invariably the major source of livelihood in a village but it may be negligible in rurban centres. This difference, however, is not necessarily reflected in the social structure. In the rurban centres of the Bihar plain nearly 80% of the people live in areas occupied largely by their own caste or religious groups. The urban character of the settlement has not removed the cultural walls between the groups, instead the urban housing pattern has been modelled upon rural customs and traditions. The study has revealed that rurban centres do not differ very much from villages in terms of the proportion of joint households and of various types of kinship units. Moreover, the migrants always move to the quasi-rural areas of the town, and thus form strong rural pockets in the urban landscape.

## EVOLUTION OF RURBAN CENTRES

There are several parameters which act as 'carriers and barriers' of rurban centre development, but population migration from rural settlements to rurban centres and the orientation of main thoroughfares of rural communities towards the nearest rurban centres are some of the forces which have contributed a great deal to the growth and development of rurban communities.

In any region, the development of rurban centres has been dependent upon minor points of focus that initially were not associated with retail trade. Cross roads, flour mills, post offices, railway stations and police stations provided a meeting place for the peasants. Enterprising businessmen took advantage of these facilities to open retail shops[7] at such centres. This in turn led to the establishment of grain collecting centres, electric lines, telephones and other similar facilities; a combination of these elements led to greater concentration in some of the places, while competition between the growing centres created the hierarchy of rurban communities.

The smallest rurban community is a rural settlement with basic public utility services such as an allopathic doctor, a middle school, a post-office and a metalled road connection. Rurban settlements of this category are closely spaced and may or may not have a retail marketing centre. They do, however, have a weekly or bi-weekly market known as a *hat*.

Rurban centres generally provide services like a high school, telegraph office, dispensary, retail market, block development office and a bus stop or a railway station. Such centres are located on the main transport routes between bigger urban centres and are, therefore, interspersed between towns and tiny rural settlements.

The further development of these facilities and the addition of a telephone exchange, bank, cinema, wholesale market, etc. led to the final growth of rurban centres. Rural settlements which have grown into rurban centres are potential towns. A rise in their population coupled with population segregation in land utilization and increase in employment in tertiary services would make them eligible for classification as towns. It would appear that the evolution and spacing of rurban centres location for a particular period of time and space largely depend on economic prosperity and transport facilities.

## SYSTEM OF RURBAN CENTRES

In the system of rurban centres two sets of variables have deterministic influences on the proper functioning of the centre. The first set of variables includes those having predominance in rural areas and operate through larger villages controlling the satellite hamlets. The nearness of smaller villages to larger villages or rurban centres attracts a large number of visitors. The higher density of population provides a market for local produce, while a lack of population in *sphere of influence* always restricts the development of any rurban centre. The second aspect in rural areas is that of the topographic condition of the region. Highlands with stony wastes prevent the development of rurban centres, riverine productive plains have potentialities for development, while lowlands liable to floods obstruct the movement of transport vehicles. The third group is that of soil characteristics, economic situation and social and cultural factors each of which has an impact upon the development of the centre.

On the urban side three tertiary functions are important for regulating the rurban system. The number of shops, types of shops, utility services and other tertiary services are the main activities of the rurban centre. The other group of factors are modes of transport. The centre may be situated on the road side, railway side or have a riverine location while in some cases three variables may converge at a particular centre and give added impetus to all functions of the rurban system. Still another set of factors is wholesale market control. This regulates the capital investment and is related to the productivity of the hinterland as well as the incentive provided to the businessmen. They together determine the market turnover and in return determine the hierarchy of the rurban centre.

When all these factors are found at one place they give rise to a central place system. Here decisions about the production, distribution and consumption of goods is taken through the system of exchange, the process of distribution, the supplies from producers and the demand of the consumers. Thus a centre is created where social, economic and cultural functions are performed for the surrounding hinterland. When this centre ranks between the purely urban and the rural sectors, it forms part of the rurban group. Further development of this system leads to the development of major trade and urban centres as well as sub-divisional and district administrative headquarters. Again these three help in the development and proper functioning of city level centres reaching ultimately to the regional capital (Fig 7.1). This indicates that rurban centre is part and parcel of central place system as envisaged by Christaller and Lösch.

## METHOD FOR WEIGHTING AND CLASSIFICATION OF RURBAN CENTRES

For this purpose Bihar plain has been considered as a region, where the location of large rural settlements and small towns function as the rurban centres. A rurban centre is

*Model for the System of Rurban Centre*

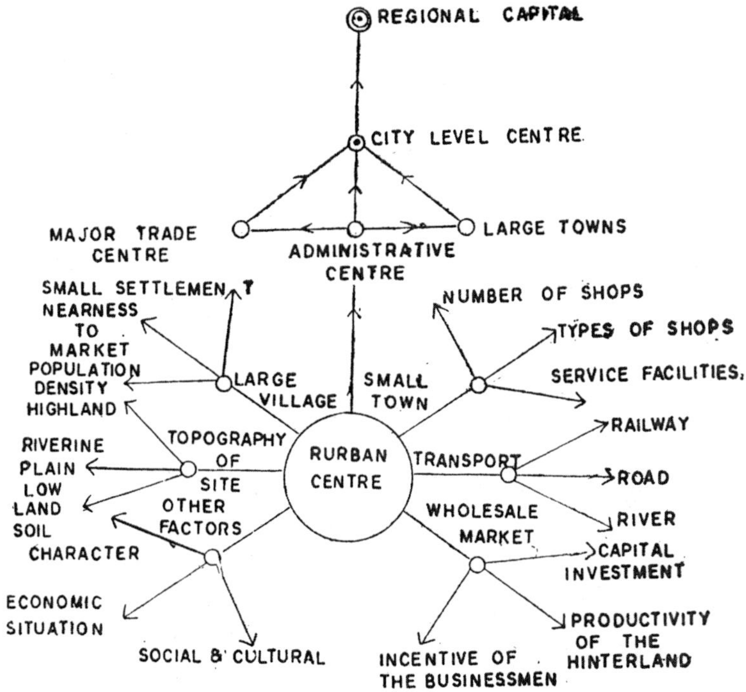


FIG 7.1

essentially a local focus of activities and is marked by commercial, industrial, administrative and cultural institutions. It is, however, difficult to differentiate on a functional basis between the large village and the smaller town, and a combined list of criteria has to be evolved. The measurement of functional indices, particularly in a developing region such as Bihar plain, must be based on the variety and type of nodal functions other than those available in most of the villages. The nodal functions considered in this study have, therefore, been classified by giving adequate weighting as shown in the following table.

Centres having at least three nodal functions given in the Table 7.1 have been considered as rurban centres. Classification of rurban centres on the above basis from the list of villages is simple, the problem arises only in the selec-

TABLE 7.1

*Weighting and Classification of Rurban Centres for the purpose of obtaining Values of Rurban Index*

| Sl. No. | Nodal Functions | Class I 15 points | Class II 10 points | Class III 5 points |
|---|---|---|---|---|
| 1 | 2 | 3 | 4 | 5 |
| 1. | Education | College | High School | × |
| 2. | Medical | × | Hospital | × |
| 3. | Communication | × | Telephone Exchange | × |
| 4. | Finance | Bank | × | × |
| 5. | Transport | × | Railways | Highways |
| 6. | Administration | Sub-Divisional Office | Block Development Office | × |
| 7. | Trade | × | Wholesale Market | × |
| 8. | Recreation | Cinema | × | × |
| 9. | Percentage of total workers engaged in tertiary occupations | × | Over 15% | 25 to 50% |
| 10. | Diversity Index | 9 Nodal functions available | 7 to 8 Nodal functions available | 5 to 6 Nodal functions available |

tion of rurban centres from the list of towns. Both towns and cities have functions of a higher order as compared to the rural, which creates confusion. Thus for the selection of rurban centres from the list of towns, the following method was devised :

$$RI = P_c \times \frac{t_w \times 100}{T_w}$$

where, RI = Rurban index,

$P_c$ = Total population of the centre,

$t_w$ = Workers engaged in tertiary occupations in that centre, and

$T_w$ = Total number of workers in that centre.

After calculating the value of RI by the above method for all towns, the mean was determined, and accepting the mean as the line of demarcation, all towns having an RI value less than mean were classified as rurban centres. Further a detailed list of villages selected as rurban centres on the basis of

Table 7.1 was prepared. The two combined lists of rurban villages and rurban towns were given proper weight as per Table 7.1 to identify 396 rurban centres in Bihar plain. The rurban index values were put to statistical analysis for determining their hierarchy and functional character (Table 7.2).

### FUNCTIONAL HIERARCHY OF RURBAN CENTRES

One great difficulty in determining the hierarchy of rurban centres in the region under study is the lack of relevant data. The services are inadequately distributed, and little published material is available. The author had, therefore, to fall back upon the information contained in the Census of India 1961, as supplemented by fieldwork during 1969-71, during which the lists of 1961 were checked and revised.

The R.I. values were found to range between 20 and 150, with mean at 52.93 and a standard deviation of 32.46. Taking standard deviation as the factor of gradation, 4 groups were recognised. The values were rounded off to the nearest multiple of 5 since the R.I. values increase in steps of 5 only. These groups are as follows :

TABLE 7.2

| Group | R.I. Value | Description | No. of rurban centres |
|---|---|---|---|
| I | 120 to 150 | Above Mean $+2\sigma$ | 29 |
| II | 90 to 115 | Above Mean $+1\sigma$ | 21 |
| III | 55 to 85 | Above Mean | 77 |
| IV | 20 to 50 | Below Mean | 269 |

Thus out of 396 rurban centres, 269 of the centres have a value below Mean, and may be said to be non-specialised, and therefore rural in character. Of the balance, 29 have a high R I. value, 21 have a medium value and 77 have a low value. The detailed list is given below and all these 127 centres have been shown in figure 7.2.

**Group-I Centres (R.I. value 150 to 120)**

Aurangabad T, Jahanabad T, Nawadah T, Buxar T, Siwan T, Gopalganj T, Sitamarhi T, Begusarai T, Khagaria T, Madhepura T, Saharsa T, Jamui T, Madhubani T, Dinapur

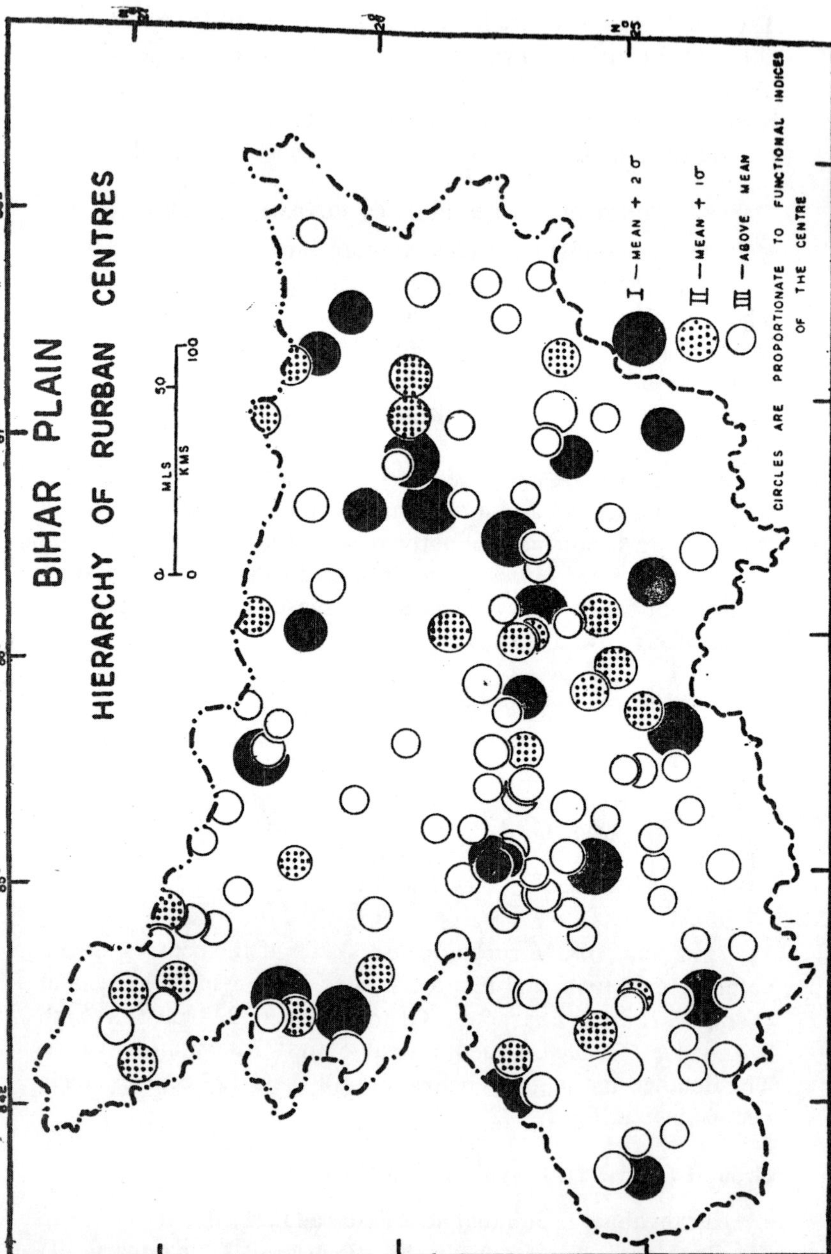
BIHAR PLAIN
HIERARCHY OF RURBAN CENTRES
MLS 0 50 100
KMS
I —MEAN + 2σ
II —MEAN + 1σ
III —ABOVE MEAN
CIRCLES ARE PROPORTIONATE TO FUNCTIONAL INDICES OF THE CENTRE

Fig 7.2

Nizamat T, Bhabhua T, Banka T, Barh T, Araria T, Sultanganj T, Forbesganj T, Supaul T, Khagaul T, Naugachhia T, Rusera T, Jainagar T, Raxaul T, Dumaraon T, Fatwah T, Bikramganj T.

**Group II Centres (R.I. value 115 to 90)**

Sheikhpura T, Lakhisarai T, Murliganj T, Banmankhi T, Bakhtiarpur T, Chakia T, Shikarpur T, Bagha T, Barbigha T, Colgong T, Jogbani T, Teghra T, Maharajganj T, Mirganj T, Chanpatia T, Barauni T, Birpur T, Dalsinghsarai T, Dinapur Cantt. T, Daudnagar T, Warsaliganj T.

**Group III Centres (R.I. value 85 to 55)**

Hilsa, Mohania, Behea, Mairwa, Ramnagar, Jhajha T, Nirmali T, Kasba T, Barahiya T, Bakhtiarpur, Jhanjharpur, Bel, Mahnar T, Nokha, Piro, Ramgarhwa, Sakra, Balia, Maheskhunt, Kharagpur T, Jagdishpur, Rafiganj, Bahadurganj, Bihariganj, Bihpur, Mohiuddinnagar, Dumra T, Bargaon, Kanti, Haripur, Dighwara, Masaurhi, Bihta, Phulwari, Rajgir T, Bodh Gaya T, Bikram, Nabinagar, Sugauli T, Manihari, Lalganj T, Nasriganj T, Itarhi, Jagdishpur T, Koilwar, Sonepur, Mashrakh, Paliganj, Arwal, Hisua, Tilauthu, Kudra, Lauthaha T, Revelganj T, Wazirganj, Tekari T, Sherghatti T, Khusropur, Islampur, Naubatpur, Silao, Barun, Obra, Deo, Belaganj, Khidirsarai, Chenari, Sasamusa, Ghorasahan, Lauriya, Bidupur, Sursand, Bariarpur, Sahebpur Kamal, Singhwara, Barari, Dalan. (Note: T=denotes town).

It is interesting to note that all centres in the first two groups are towns, i.e., urban settlements. They are mostly small towns, but population is no criterion for determining the R.I. values. Thus Barh and Dumaraon with a population over 18,000 have R.I. values of 135 and 120 only, while Madhepura and Bhabhua with a population below 12,000 have R.I. values of 150 and 140 respectively. In the third group of 77 centres, 18 are towns and the rest are villages. It is these villages which have the highest potentiality of growing into towns, particularly if their populations are high. The number of such rurban villages is large in areas where there are fewer towns.

Centres in the first group are mostly sub-divisional headquarters of administration because they have a daily retail market, wholesale market, S.D.O's court for revenue and criminal cases, often a civil court, banks telephone, colleges, cinema and good transportation facilities which combine to put them in the highest rank. All these centres are full-fledged towns with a higher percentage of workers engaged in tertiary functions, as compared to those in primary services. It is interesting to note that this list does not contain a single district headquarters except Saharsa. All the district headquarters and five other sub-divisional headquarters have a marked urban character.

The other towns included in this group are Sultanganj, Forbesganj, Khagaul, Naugachhia, Rusera, Jainagar, Raxaul, Dumraon, Fatwah and Bikramganj. All these towns have excellent transport connections and are leading marketing centres of their region. Forbesganj, Raxaul and Jainagar are on the Indo-Nepal border and therefore, important break-of-bulk towns. Khagaul is a centre of the railways. The others are grain collection and distribution centres. The service area of these rurban centres roughly extends to a radius of 16 kilometres, though in most cases it is strictly a matter of terrain conditions and transport facilities.

The centres of the highest order having R.I. values between 150 and 120 provide such basic amenities as education up to college level, medical facilities, bus or railway transport, telephone exchange and wholesale market, etc., but may not always possess such urban facilities as banking, cinema and sub-divisional courts. Their population usually ranges between 10,000 and 30,000 and the employment in tertiary services is over 50% of the total workers. In such centres, there is usually a marked tendency for a distinct business core to develop, which is conspicuous by its absence in the lower order centres i.e. in rural villages within the same population group.

The second group includes settlements in which the functional index ranges from 90 to 115. These centres generally have a population of 5,000 to 15,000. They have functions and services of the same type as found in the first group, though of a lower degree. The occupational structure of the population also reveals that usually between 35% and 50% of the working force are employed in tertiary services. But the higher order centres of this group are distinguishable by greater nodality and a more pronounced urban character.

The centres of the third group with R.I. values above mean i.e. 55 to 85, are 77 in number. These include 18 towns and 59 villages. Their population generally ranges from 2,000 to 15,000 and the employment in tertiary services is close to 25% of the total workers. These centres have been greatly benefited by development of transport and communication and thus have acquired the status of regional collecting and marketing centres. They are, however, lacking in the services and functions of an urban centre. They are usually found in clusters around the larger cities. This effect is well marked in all cases and is discernible as a node in the overall fabric of rural settlements, though the percentage of the non-agricultural population is not always high.

The centres below the mean number 269, in most of which agriculture directly or indirectly constitutes the occupation of more than 75% of the total workers. The centres of this category are large in size and diversified in nature but are mostly confined to primary activities. They have, therefore, not been considered while preparing the matrix of correlation coefficients. This matrix has been prepared by calculating the Pearsonian correlation coefficient between each of the nodal functions mentioned in Table 7.1.

It is clear from the matrix (Table 7.3) that the Block Development office, wholesale market, hospital and high school are found in close association with each other. These are the primary services available in rurban centres of even the lowest order. Closely associated with hospitals and high schools are road transport facilities, because practically all Block Development Offices today have been connected by metalled roads. The correlation between these services is seen to range from +0.92 to +0.98.

The relationship between the distribution of highways and railways and that between railway stations and telephone exchange are equally high and are close to 0.80. Services of a higher order such as banks, cinemas and colleges, however,

TABLE 7.3

*Matrix of Correlation Coefficients for Rurban Centres in Bihar Plain*

| Rurban functions | $X_1$ | $X_2$ | $X_3$ | $X_4$ | $X_5$ | $X_6$ | $X_7$ | $X_8$ | $X_9$ | $X_{10}$ | $X_{11}$ | $X_{12}$ | $X_{13}$ |
|---|---|---|---|---|---|---|---|---|---|---|---|---|---|
| $X_1$ | 1.000 | | | | | | | | | | | | |
| $X_2$ | 0.138 | 1.000 | | | | | | | | | | | |
| $X_3$ | 0.049 | 0.057 | 1.000 | | | | | | | | | | |
| $X_4$ | 0.632 | 0.909 | 0.532 | 1.000 | | | | | | | | | |
| $X_5$ | 0.228 | 0.248 | 0.268 | 0.192 | 1.000 | | | | | | | | |
| $X_6$ | 0.050 | 0.066 | 0.980 | 0.542 | 0.268 | 1.000 | | | | | | | |
| $X_7$ | 0.898 | 0.303 | 0.553 | 0.640 | 0.175 | 0.549 | 1.000 | | | | | | |
| $X_8$ | 0.581 | 0.682 | 0.961 | 0.527 | 0.171 | 0.952 | 0.563 | 1.000 | | | | | |
| $X_9$ | 0.916 | 0.304 | 0.374 | 0.662 | 0.148 | 0.400 | 0.633 | 0.388 | 1.000 | | | | |
| $X_{10}$ | 0.999 | 0.671 | 0.976 | 0.173 | 0.336 | 0.983 | 0.175 | 0.298 | 0.126 | 1.000 | | | |
| $X_{11}$ | 0.150 | 0.216 | 0.569 | 0.144 | 0.752 | 0.267 | 0.163 | 0.259 | 0.125 | 0.832 | 1.000 | | |
| $X_{12}$ | 0.354 | 0.700 | 0.951 | 0.550 | 0.261 | 0.951 | 0.557 | 0.925 | 0.264 | 0.297 | 0.253 | 1.000 | |
| $X_{13}$ | 0.731 | 0.252 | 0.254 | 0.174 | 0.115 | 0.021 | 0.183 | 0.076 | 0.133 | 0 036 | 0.584 | 0.080 | 1.000 |
| | Tertiary workers Percentage | Bank | Hospital or Dispensary | Cinema | Telephone Exchange | High School | College | B.D.O | S.D.O. | Road | Railway | Whole sale Market | Diversity Index |

show a low relationship with the services of the lower order. There is a high correlation between the distribution of banks and cinemas and that between cinemas and sub-divisional courts. These are located usually in towns rather than in purely rural centres. This is also indicated by the high correlation coefficient of 0.90 between tertiary workers and the distribution of colleges or sub-divisional courts. The tertiary services, however, have a very low correlation with the distribution of hospitals and high schools. Banks which have a high relationship with wholesale markets, show a low correlation with the distribution of high schools, hospitals and roads. The correlation coefficient in each of the above cases is less than 0.10.

It is apparent that there is a clear division between rurban centres of a low order containing Block Development Office, hospital, high school and wholesale markets, and those of the higher order containing sub-divisional court, college, bank, cinema and a high percentage of tertiary workers. The centres of the higher order which contain most of the places mentioned in group I above, contain all services of the lower order and are, therefore, marked by greater diversity. This is apparent from the matrix when we consider that the correlation coefficient between high school, Block Development Office, wholesale market and roads on the one hand and diversity index on the other, is less than 0.10. The diversity index is highest where the percentage of tertiary workers is also high. This fully justifies the hierarchy groups suggested earlier.

## SPACING OF RURBAN CENTRES

The spatial distribution of rurban centres presents several aspects of geographical interest. The dispersed nature of the centre falling in above Mean+2 category are quite evident from Fig. 7.2. The presence of a twin rurban centre in the Dinapur Cantonment and Dinapur Nizamat is noteworthy in the entire plain. Normally the spacing of such settlements in Bihar Plain is irregular mainly due to topographical and other factors affecting their historical evolution. This implies that each centre possesses its own specific sphere of influence within which it dominates over the surrounding villages. The delimitation of the actual configuration of each sphere of influence would entail a detailed study of consumer behaviour in the area which is outside the scope of this paper. Nevertheless, certain generalisations can be made from the pattern obtaining here.

North Bihar Plain shows the highest spacing of rurban centres e.g. 40.40 kilometres in Darbhanga and 37.42 kilometres in Purnea district. Here urbanisation is very slow, because of under-developed resources and annual inundation by the rivers. In contrast, rurban centres are closely spaced in South Bihar Plain, the average spacing being 24 km with a low of 19.26 km in Patna District. This shows that the number and spacing of rurban centres is directly related to the level of regional development.

TABLE 7.4

*Spacing of Rurban Centres* ($D=1.07\sqrt{A/N}$)

| District | Number of centres | Spacing in kilometres | Average population size, 1961 |
|---|---|---|---|
| Patna | 17 | 19.26 | 9,579 |
| Gaya | 18 | 27.88 | 7,683 |
| Shahabad | 15 | 29.32 | 8,066 |
| Saran | 10 | 27.96 | 9,343 |
| Champaran | 12 | 24.97 | 7,853 |
| Muzaffarpur | 10 | 29.72 | 8,585 |
| Darbhanga | 6 | 40.40 | 12,194 |
| Bhagalpur | 6 | 32.50 | 8,475 |
| Saharsa | 8 | 27.40 | 9,225 |
| Purnea | 9 | 37.42 | 8,918 |
| Monghyr North | 6 | 26.88 | 19,770 |
| Monghyr South | 10 | 26.91 | 14,423 |
| Bihar Plain | 127 | 26 08 | 10.343 |

FUNCTIONAL MAGNITUDE OF RURBAN CENTRES

Fig. 7.3 illustrates the functional magnitude of 127 settlements in Bihar Plain. The functional magnitude can be defined as a measure of the importance of a settlement in terms of all the nodal functions considered. The usefulness of this quantitative index in the study of service centres rests on

*Bihar Plain Functional Magnitude of Rurban Centres*
*(Centres : Having R.I. Values above Mean)*

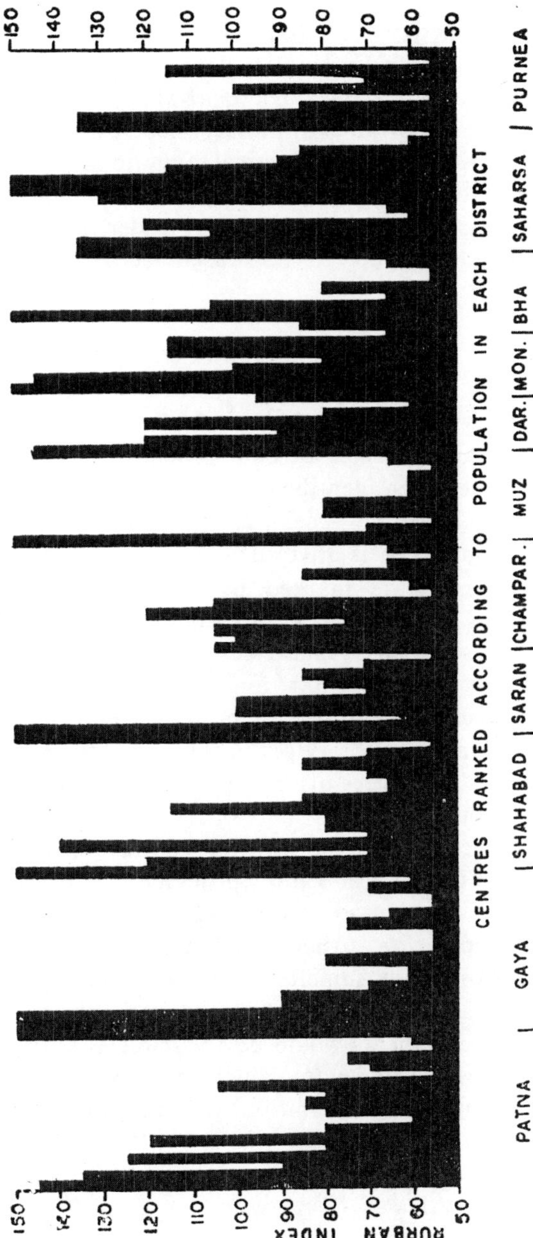


FIG 7.3

its capacity to reveal at a glance the relative importance of the settlements under study in terms of all the nodal functions they perform. In Figure 7.3, eleven major rurban centres in the region with R.I. values of 150 attain prominence. The overwhelming importance of these 11 centres is as expected because centres with almost identical functional magnitude should come within the same order in any hierarchical classification. In contrast to these are the lower order rurban centres with R.I. values of 55 to 65 which occupy the troughs in the graph.

## CONCLUSION

There is thus, a distinct class of settlements in Bihar Plain which may be termed rurban settlements. 127 rurban settlements have been identified of which 68 are towns and 59 are villages. Broadly speaking, they are divisible into three groups on the basis of statistical analysis. The centres of the higher order are marked by specialised services like subdivisional courts, banks and cinemas, while centres of the lower order are marked by Block Development Offices, hospitals, high schools and wholesale markets. The chief distinguishing feature is the high percentage of workers engaged in tertiary services in these centres although broadly speaking the centres of the higher order also have a higher population.

The concept of the rurban centres is useful because it serves to distinguish rural settlements with a bias towards public utility services, commercial services and non-agricultural population structure. Similarly, there are full-fledged towns which lack something in their basic urban character and yet cannot be termed as rural settlements. The geography of rural settlements would be incomplete without identifying such hybrid settlements, which have a high potentiality of urban growth. By definition the rurban centres are partly rural and partly urban and have, therefore, to be viewed as part of the broad settlement geography. In this way rurban centres work as "economic and social capitals" for the neighbouring villages.

## REFERENCES


1. R.B. Mandal, "Rurban Centres in Bihar Plain" *Indian Geographical Studies*, R.B. No. 1, 1973, pp 12-23.
2. J.E. Brush, "The Hierarchy of Central Places in South-Western Wisconsin", *Geographical Review*, Vol. 43, 1953, pp 380-402.
3. R.B. Mandal, "Law of Village Anatomy and Rurban Centres Location on Löschian Landscape", Forthcoming publication in *Proceedings of Small Towns*, Varanasi, 1978.
4. R.L. Singh and S.M. Singh, "Mungra Badshapur : A Rurban Settlement in the Ganga-Ghaghra Doab West", *The National Geographical Journal of India*, Vol. VI, Part IV, 1960, pp 199-206.
5. C.G. Galpin, The Social Anatomy of an Agricultural Community", *Research Bulletin* 34 Agricultural Experiment Station of the University of Wisconsin, 1915.
6. R.E. Dickinson, "The Distribution and Functions of the Smaller Urban Settlements of East Anglia", *Geography*, No. 95, Vol. XVII, Part I, 1932, pp 20-30.
7. A.E. Smailes, "The Urban Hierarchy in England and Wales," *Geography*, Vol. XXIX, Part-2, 1944, pp 41-51.
8. A. Gloward, and Stafford, Jr. "The Functional Bases of Small Towns", *Economic Geography*, pp 165-175.
9. B.J.L. Berry and W.L. Garrison, "The Functional Bases of the Central Place Hierarchy", *Economic Geography*, Vol. 34, 1958, pp 145, 154.
10. L.J. King, "The Functional Role of Small Towns in Canterbury", *Proceedings of Third New Zealand Geographical Conference*, Palmerston North, 1962.
11. E.N. Thomas, "Some Comments on the Functional Bases for Small Iowa Towns", *Iowa Business Digest*, 1960, p 10.
12. F.J. Monkhouse, *A. Dictionary of Geography*, Edward Arnold, London, 1972, p 302.
13. R.L. Singh and S.M. Singh, op. cit., p. 199.
14. K.N. Singh, "Small Towns as Rural Development Centres in Uttar Pradesh", Abstract, *Proceedings of Fifty-Fourth Session*, Indian Science Congress Association, p 11.
15. Thomas N Edwin, "Some Comments on the Functional Basis for Small Iowa Towns," *Iowa Business Digest*, 1960, p 10
16. Kingsley Davis, "Urban Research and Significance," *Urban Research Method*, Ed. J.P. Gibbs, New Delhi, 1966, p XI.
17. "Urban-Rural Differences in Southern Asia," *Report on Regional Seminar*, *Delhi* 1962, London, 1964, p 3.

8

# The Neighbourhood Concept and Its Implication in Urban Planning

—*Ram Dayal Singh*

AMONG the various changes that are taking place in the internal structure of a city as a result of population growth, the emergence of new residential neighbourhoods is a subject matter of great concern. Despite the fact that a large proportion of land in urban areas is given to residential uses, geographers in India have not paid sufficient attention to the recent changes and character of residential locations especially in large metropolitan communities. In modern times, cities are expanding in response to the population growth and diversification of urban functions which subsequently bring about significant morphological changes. Besides, there has been the growth and development of new residential colonies almost on a sectoral pattern in the outer zones of the city. Identification of such modern residential growth patterns is, therefore, essential in the interest of researchers as well as those interested in urban planning. In view of this the paper deals with urban planning in the light of the implication of the concept of neighbourhood in recent residential growth patterns.

Before we begin with the discussion, it seems important to explicate the very concept in the beginning. The term 'neighbourhood' is so frequently used in several contexts of our social life that it is rather difficult to pin point its meaning. Very often the neighbourhood concept is particularly used to refer to a community or a part of the community having distinct socio-economic and cultural traits. Expressing a similar view, Mckenzie[1] (1921) a noted American sociologist, observes that "Probably no other word is used so loosely or with such changing contents as the term neighbourhood and very few concepts are more difficult to define." By neighbourhood he means a patch of common life within the larger community which is sufficiently differentiated from the city as a whole to be thought of as an unit. This is not unnatural because the major concern of a sociologist is on the social rather than spatial dimension of neighbourhood (Riemer, 1949).[2] In fact the concept of neighbourhood is a value-loaded one and is often used synonymously with the 'community', when the term is descriptive of a small, compact and relatively homogeneous place (Chandra, 1977).[3] But the neighbourhood is by no means an exclusively sociological concept. There has always been popular usage of course. In addition, planners have found the neighbourhood a convenient unit for their studies of housing through the neighbourhood, as what they think of it may depart considerably from a sociologist's concept. In any case, it is an areal unit of the city, so it is the direct concern of the urban geographer. Probably every city has some areas around it to be called as neighbourhood at least in the general sense of the term. There are areas that for one reason or the other have come to be known by separate names whether they indicate any considerable degree of social integration or not.

Murdie (1969)[4] defines community or neighbourhood as an economic unit, a polity, a formal social system, a geographic unit of territory or as a society of its own. Gans (1964)[5] employs the expression 'urban village' to denote the intimate and spatially confined neighbourhood ties typical of many such areas. Urban neighbourhoods, whether they are defined as units of territory or of social organisation, differ significantly in their appearance and social character and in their importance in the life style of people who live in them.

Each definition, of course, has been put forward with a different purpose and each has some validity. Anyway, there is no general agreement among the scholars with regard to the definition. The confusion surrounding the neighbourhood concept and neighbourhood planning had led to a considerable debate for a long time about the growing role in the increasing literature on social planning. Nevertheless, it would be rather unwise to emphasize the controversial view points than to offer a definition of the term as a spatial unit to be distinguished from others by some socio-economic correlates. Urban planners, as a matter of fact, take it as a distinct planning area for the purpose of its overall development. In Western countries, there has been a tendency in recent years to use planning districts rather than neighbourhoods. This is because the neighbourhood system is based on residential areas and, thus, leaves out areas of high concentration of factories or other non-residential uses.

Segregation is a dominant process in the origin of residential neighbourhoods. The nature and character of segregation differs remarkably from that in Western countries where residential areas developed entirely on the basis of economic grades and predominance of similar occupational groups. Until recently, residential areas in Indian cities were characterised by highly mixed economic group people (engaged in diverse functions). With the introduction of modern means of transportation facilities, new residential colonies are emerging, based on economic grades and occupations. Such areas are often called as Labour colony, L.I.C. colony, Defence colony, etc. In industrial cities residential colonies are developed by the owners of the industries for their workers.

Another form of residential segregation in Indian cities is found to be developed on the basis of communal groups in contrast with ethnic groups in Western cities (Quinn, 1967).[6] In India some communities have a tendency to reside in the area occupied by the people of their own communities. Concentration of people of the lines of caste, community, religion. occupation and other such considerations have resulted in the segregation of urban people in different socio-cultural groups on the physical space of urban centres. Muslims, Sikhs, Christians, etc., have clustered in compact areas which differ significantly from one another in respect of socio-economic and cultural characteristics. Because of the 'consciousness of kind' and feeling of insecurity among the people, they like to concentrate as closely as possible in a given locality of the town.

## PATTERN OF GROWTH OF RESIDENTIAL NEIGHBOURHOOD

The ever-growing concentration of population in large metropolitan communities is one of the most important features of urbanization process taking place in both developed and developing countries of the world. In India, for example, about 23 per cent of the total urban population in 1901 lived in towns with 1,00,000 population or more. In 1971, this figure rose 52.4 per cent respectively. The census reports reveal that during the period of 20 years (1951-71), the number of big urban centres nearly doubled from 74 to 142 and their population increased from 24 to 57 millions registering a 2.4 fold growth.

The concentration of population in these cities is closely related to the growth and emergence of new residential neighbourhoods and townships often called urban agglomerations. This is, in fact, a single process for the growth of big cities which inevitably leads to an accelerated growth and formation of systems of population centres. A primary role in this process is played by the development and diversification of urban functions, notably industry. It is not only that population is growing at a fast pace, but the towns themselves are spreading over larger territories extending to new areas that were rural in character. The population of Delhi for example increased from 1.4 million in 1951 to 2.4 million in 1961 and 3.6 million in 1971. Between 1951-71 Delhi's population increased more than 2.5 times, while the area expanded from 77.5 sq. kilometres in 1951 to 323.5 sq. kilometres in 1971. Bangalore's population was doubled during these decades, i.e., from 0.78 million in 1951 to 1.65 million in 1971. The metropolitan area during these decades extended nearly six times (69.4 sq. kilometres to 447.34 sq. kilometres). The medium size cities have also started expanding fast with the improvement of transportation facilities in modern times. The population of Allahabad and Patna were 0.30 and 0.28 million in 1951 which shot up to 0.51 and 0.49 million in 1971 respectively while their urban areas witnessed a remarkable increase from 56.5 and 42.74 sq. kilometres in 1951 to 96.7 and 78.17 sq. kilometres in 1971.

Until recently, the residential areas especially of high income group could grow around the central bazar which was attractive to all types of activities because of its accessibility to jobs, transportation node and the retail market. The absence of speedy transport facilities led the wealthy people to build their residence at the central chowk. Urban structure was thus shaped by the assumption of fierce economic competition for central location near the C.B.D. This process resulted in two consequences : first, the central residential land in particular was subdivided so that individuals could afford the rent. It, consequently, led to the general concentration of people around the C.B.D. With further increase of pressure of population on the existing dwellings, the city expanded physically by the growth of peripheral neighbourhoods.

The colonial rule in India was marked by remarkable changes in the character of residential location in Indian cities. British rulers especially in the cities under their administration developed new residential areas at a little distance away from the old congested city centre which latter become the focus of concentration of people and economic activities. As a result of greater security, the latter flourished enormously. Brush (1969)[7] also observed similar features of residential growth pattern in most British built cities like Hyderabad, Bangalore, Patna, etc.

In the post-Independence period, the physical expansion of the city was largely based on the improvement of transport and communication facilities. The introduction of modern public transportation and greater use of the bicycle have brought about tremendous impact on the growth of secondary commercial, industrial and residential neighbourhoods. The traditional tempo of residential crowding at the central city is no longer a dominant pull and the centrifugal tendency to reside or build houses at the fringe has started to operate (Singh, 1974).[8] The massive areal sprawl in the outer zones of the city may be as a consequence of the exhaustion of space for residential development in the central city. This outward shift of residential population in recent years can be viewed as an important modification in the internal organisation of the metropolitan communities. Such a radical change in the spatial dimensions of urban areas apparently represents an adoptive response to the changing conditions of modern urban life. The entire process of structural growth of the city leading to the emergence of planned residential neighbourhoods can be viewed from the schematic diagram (Fig. 8.1).

*Process of Structural Growth of the City*

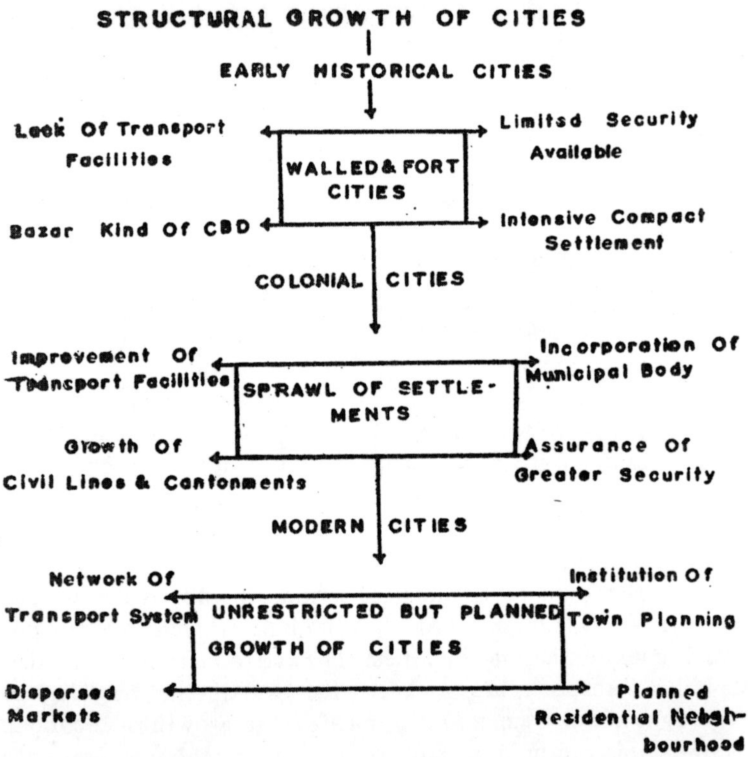


FIG. 8.1

Turning to examine the ecological growth of modern residential areas, it would appear that the overall pattern is on Hoyt's sectoral form, though they may not be entirely on the mechanistic principles adopted by him. Probably the most impressive phenomenon associated with urban landuse in recent decades has been the peripheral growth of residential neighbourhoods, while the central city has been standing still in population or actually declining. The growth of residential areas in the suburbs may be characterized by two contrasting approaches—homes are being built by individuals who buy a plot of land and hire a builder to erect the house that they want. This is likely to be true of the highest priced fringe houses but some low price houses may also be built of the same kind.

In other areas, the whole group of houses are built by the development authority which are most typical in sectoral form, being characterized by a good deal of uniformity of houses. Hoyt (1939)[9] postulates the dominance of economic forces in patterning residential development in American cities while religion and language are most prominent factors influencing residential segregation in Indian cities. The residential neighbourhoods in large urban communities are segregated primarily more on linguistic and religious bases than on economic status. They may, however, consist of sections of predominantly mixed income groups.

The fringe areas of Indian cities also comprise a large number of villages which act as the nucleus of urban residential growth. These rural enclaves drew considerable amount of immigrants particularly of low and medium incomes, which later, coalesced with the cities, and therefore, played a significant role in the evolution of residential neighbourhoods. In view of this concept, it may be generalised that most Indian cities lack the social mobility and economic vitality of American cities and therefore, the processes which are involved in the residential pattern and ecological structure of the cities of U.S.A., are not equally effective in India (Alam, 1976).[10]

## ITS IMPLICATION IN URBAN PLANNING

Planning is a process of all round relevance, leading to the development of a clear urban environment. Though urban planning in India started with the imposition of the First Five Year Plan in 1951, its objectives could not have been achieved as yet. Planned residential colonies may, of course, be the characteristics of large metropolitan communities, but the medium and small size cities are rapidly expanding in a most haphazard way and blighted residential neighbourhoods are emerging at the outer zones of cities.

The villages situated in the fringe areas served as centres for concentration of immigrants because of low residential value and uncongenial surroundings which later turned into slum conditions with further increase in population. These residential neighbourhoods adversely affected the healthy development of cities. The planning agencies, which had to renew, rebuild or rehabilitate these neighbourhoods could not have shown desirable result in this direction. Sectoral planning, the form of planning most commonly used in western and non-western cities may be helpful to develop the areas at the fringe. The success of development planning depends partly on the capacity of the planning agencies at different administrative levels. As explained by Myrdal (1968)[11] "in a successful planning situation, the entire government is involved and begins to operate as a part of a huge planning machinery." The suburban area which is likely to be engulfed into urban environs, the planning agencies should undertake the area into integrated sectors (so-called neighbourhoods). The rural enclaves (*busties*) falling in them must be reorganised and renewed before any immigrants settle there. There is a constant shift of neighbourhoods and as they become older, they tend to be less desirable. Taneza (1971)[12] also emphasized the need for provision of new residential neighbourhoods in suburban areas of the cities in response to rapid growth of population.

As the planned residential area once developed, it tended to retain that character for long distances. The shift of such residential areas is important because it leads to the growth of the city in the same direction. Transportation occupies the most significant place in the evolution of residential and other functional areas of the city.

The sectoral growth strategy thus, has, implications for the development of metropolitan areas. Similar development programmes have been undertaken and suggested by general international agencies. Such growth patterns of residential neighbours can provide opportunities for immigrants and city dwellers to reside in healthy environment. The systematic planned areas also automatically bring education, health and other civic amenities as close to the people as feasible in the present circumstances. In this way, cities in India which are just passing through an intermediate stage of rapid growth may overcome the calamities and glooming environmental hazards that persist in some of India's large metropolitan cities.

## CONCLUSION

Neighbourhood as a planning device was introduced in India for balanced and healthy growth of cities. But due to lack of awareness on the part of planners and poor planning policies, the objectives could not have been achieved. In modern times, medium and small size cities are expanding fast as a consequence of rapid growth of population. This has resulted in the development of blighted residential areas and slums. Incessant immigration into cities and subsequent pressure on the existing dwelling units have not only created insanitation and unhealthy living but have also produced a number of social and economic problems like unemployment, beggary, crime, violence and other such vices, etc. To ameliorate the residential crowding and to provide a congenial environment for city dwellers, development of residential neighbourhood at the fringe of cities is suggested as follows : (i) The existing city limits should be redrawn or extended to a reasonable extent for the development of satellite neighbourhoods; (ii) sites for creation of such townships should be carefully selected for some definite size of city dwellers; (iii) the existing rural enclaves (*busties*) should be rebuilt, remodelled, renovated and renewed so that further overcrowding and congestion may not increase; (iv) satellite neighbourhoods should be made for all low, medium and high income groups people separately; and (v) all neighbourhoods should be interlinked with the parent city and cheap means of communication should be introduced to reach to and from the central city.

## REFERENCES

1. R.D. Mckenzie, *The Neighbourhood* : *A Study of Social Life in the City of Columbus, Ohio*, 1921.

2. S. Riemer, "Frontiers of Housing Research, Land Economics", 1949, Vol. 25, p 69.
3. S. Chandra, *Social Participation in Urban Neighbourhoods,* National Publishing House, New Delhi, 1977, p 105.
4. A.R. Murdie, *The Social Geography of the City : The Critical and Empirical Background,* in L. Bourne (Ed.); Internal Structure of the City, Oxford University Press, Toronto, 1977, p 275.
5. H. Gans, "The Esthetic Slum", Landscape, Vol. 14, 1969, No. 1, pp 16-18.
6. J.A. Quinn, *Urban Sociology,* Eurasia Publishing House Ltd., New Delhi, 1967, p 76.
7. J.E. Brush, "The Spatial Patterns of Population in Indian Cities"; *Geographical Review,* 1968, Vol. 58.
8. R.D. Singh, "Demographic Structure of Gaya, Indian Geographical Studies", *Bulletin-4,* 1974, p 48.
9. H. Hoyt, *The Structure and Growth of Residential Neighbourhoods in American Cities,* Federal Housing Administration, Washington, 1939.
10. S.M. Alam, Indian Cities and Western Theories of Urban Growth' (An Empirical Appraisal), in Urbanisation in Developing Countries; (Eds. S.M. Alam, and V.V. Pokshishevsky), Osmania Univ., Hyderabad, 1976, p 472.
11. G. Mydral, *Asian Drama,* New York, 1968, p 734.
12. K.L. Taneza, *Morphology of Indian Cities,* (Ed. R.L. Singh,) The National Geographical Society of India, 1973, pp 76-90.

# 9

# ASpatial Model of A Medium Sized Indian City

—*K. M. Kulkarni*

THE THEME of the present study was conceived during an empirical study of Nasik city located 181 km N-E of Bombay. The study compares Nasik city with other similar sized Indian cities with regard to demographic and economic characteristics. It, therefore, aims at understanding whether a spatial model of medium sized towns can be attempted by linking size, function and spatial structure. In the present context, medium size refers to a population of one lakh to one and a half lakhs. Several studies have been made on medium and small towns.[1,2] In the present note the spatial dynamics and the underlying common characteristics of medium sized towns have been considered in detail.

## THE MORPHOLOGICAL ASPECTS

In the light of the internal structure of Nasik city, Smaile's model of an Indian city appears to be now getting out-dated as our cities are changing in space and characteristics. The striking contrasts between the indigenous part and the westernized part is fast disappearing. Though Indian cities are still core oriented, the outward expansion is no less significant. The agricultural fields and waste lands are being reclaimed for urban use on a cooperative system. The industrial estates are located farther away from the municipal limits but well within the reach of city dwellers. Many newly constructed annexures are seen side by side with westernized constructions. The annexures indicate our growing needs for an administrative set up. Town planning schemes are implemented outside the indigenous city. Properly laid out sewage lines, drainage lines, electricity lines, wide roads, footpath spacing and construction of buildings etc. are some of the landscape features of city outgrowths. To the extent possible the roads in outer areas are laid straight unlike the innumerable constraints of the indigenous city, such as private property, religious institutions, important historical monuments and above all the traditional and conservative views of people living in such areas. For a town planner the improvement of the indigenous city is a great challenge, as it poses several problems affecting the social and economic life of the city. However, in place of dilapidated and demolished buildings, new concrete structures are coming up both in commercial as well as residential areas. The height of the building is positively more in commercial areas than in the residential areas. The medium sized cities in India are not affected yet by the metropolitan craze for skyscrapers and for that matter not even to the extent of 4-5 storeyed buildings though one can count one or two here and there. City dwellers are proud of such buildings and these soon become the "talk of the town".

Local shopping centres have become a common sight in new residential colonies. Shopping centres of this kind are compact, uniform and more often housed on the ground floor of a residential building. The local shopping centres provide essential commodities and services such as a laundry, grocery, the hair-dresser, the flour mill and fresh vegetables. The traders carrying out subordinate commercial functions to such local centres are *pan*-vendors, cobblers, newspaper-sellers and above all lottery-ticket-sellers ! The city in its stage of growth, acquires varied functions keeping its dwellers occupied. They are left with barely any time to attend to their long list of shopping. As a result wheel barrow-vendors enter the city serving functions. They move from one housing society to another selling all sorts of goods ranging from kerosene to blankets, that too from dawn to dusk not showing the slightest trace of exhaustion from the scorching sun.

The original commercial core has shown considerable decay in the recent past as a result of locational shift in commercial importance. The new commercial lane is closer to banking institutions, post office, bus terminus and more so accessible to newer residential colonies. Probably the newer extensions of the cities influence considerably in the locational shift of commercial importance, and also possibly the old core is not in a position to accommodate the growing complexity of city commercial functions. Hence, the decay of old commercial core and the expansion of new commercial lane or core and many other local shopping complexes.

Slum dwellers are in the main engaged in petty odd jobs but their existence is equally important for city life. They tend to live close to their place of work. They are engaged in the transport of commercial goods as porters at the bus terminal or railway station, in construction works, and quarrying activities where physical strength is required. The slum dwellers engaged in agricultural activity, construction works and quarrying activities stay in the outer areas of the city.

New institutions such as colleges, universities, dairy industry, farm institutions, institutions engaged in poultry-keeping are located in the outer areas as their space requirement is more. The burial ground is another landuse/land misuse which requires more land, and hence it is in the outer areas.

The constraints for city expansion are Physical and Human. The physical constraints are mainly rivers—as many of our Indian cities ars located on the banks of rivers. It does not promote the easy sprawl of a city and effective spatial integration. The other physical barriers are mountains, hills, waterbodies, marshy land and rocky ground. The human factors are the cantonments, the restricted/prohibited areas, the reserved forests and above all the rich agricultural lands, more so the social factors including beliefs and attitudes.

With these changes in the Indian urban landscape of this size and constraints prevailing, an attempt is made to construct a spatial model of a medium sized Indian city like Smailes but more distinctive and limited to this size.

ASPECTS OF A SPATIAL MODEL OF AN INDIAN CITY

| *Social Areas* | *Landscape Features* | *Functional Aspect* |
|---|---|---|
| 1 | 2 | 3 |
| (A) *Old Core* | | |
| 1. Priestly Communities | Temples, Schools (Gurukul/Pathashala) | Includes various professional lanes such as |

| 1 | 2 | 3 |
|---|---|---|
| 2. Other Hindu Castes | Fort, Gates | smithy, broom-stick, |
| 3. Lower Castes | Mosques, Tomb | basket makers, potters |
| 4. Muslim Community | Plinth of houses is quite high as a precautionary measure to floods | goldsmiths, copper and brass-smiths, weavers and any other local profession of importance |
| | Artistic wood carving | |
| | Palatial buildings | |
| (B) *British Town/Civil Town* | | |
| 1. Christian missionary colony | Parade ground, golf link, parks, civil hospital, mission school, library (anglicised architecture), court, independent bungalows | Mainly administrative function and other supporting functions like medical, educational services |
| 2. Socially mixed residential area | | |
| 3. Economically better-off social class | | |
| (C) *Newer Developments* | | |
| 1. Social areas with particular regional affinity (showing a trend) | Storied/Flat type housing societies, planned parks, playgrounds, shopping centres, kindergartens, community hall etc. | Mainly residential function. Outskirts of newer residential areas are covered by specialized functions such as educational institutions, industrial units etc. |
| 2. Social areas with particular profession (bank employees colony, govt. employees colony, teachers colony, industrial employees colony) | | |
| 3. Social areas with cosmopolitan outlook by profession, region, language, etc. | | |

LEADING FUNCTIONAL CHARACTERISTICS OF MEDIUM SIZED CITIES

The medium sized cities in the process of their growth added cultural and educational institutions to their already existing religious functions. By doing so it trained a band of intellectual personnel who also had the craze for administrative services. Along with the cultural and educational activities the city acquired administrative responsibilities bringing it a status of administrative headquarters. Even today the medium sized cities are leading administrative headquarters with a high proportion of works engaged in the 'other services' category. Other functions in order of importance are trade and commerce and manufacturing. Manufacturing is gaining importance by way of industrial decentralization policy and location of industrial estates. But the new industrial establishments are not true to all medium sized cities

Most of these medium sized towns are by status District headquarters having District Commissioner's Office, Zilla Parishad, Civil and Criminal Courts, District Police Headquarters, Block Development Office, Taluka Development Board, and other offices. (Fig. 9.1)

*Medium Sized Cities : Functional Characteristics*

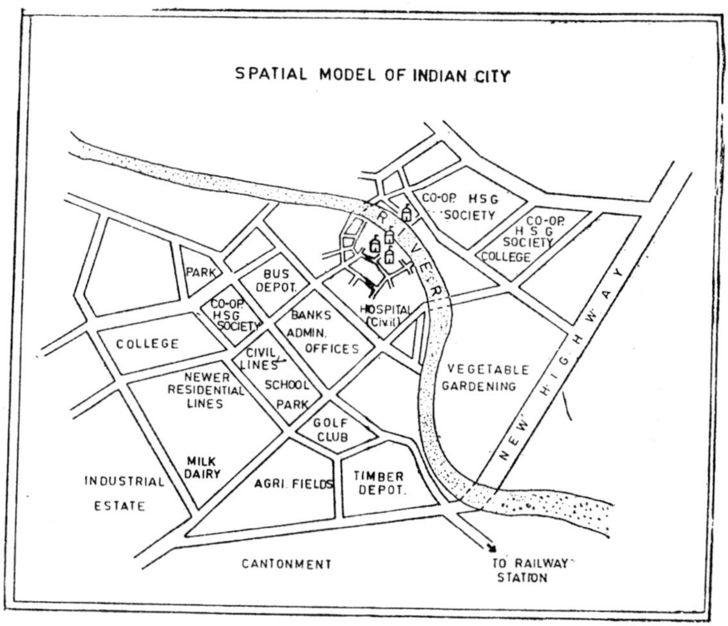


FIG. 9.1

Among educational institutions, Arts, Science, Commerce, Law, Engineering or Medical colleges are the common ones. Apart from the general ones, specialised institutions like police training colleges, staff nursing colleges and industrial training institutes may be found in some of the medium sized cities.

The size class of this kind is neither rural nor highly urban in character. The functional complexity has not reached the climax to realise impersonal social relations and a kind of anonymity. The size is good enough to maintain social harmony by way of cultural activities and cordial relations with the neighbours and also within the neighbourhood. If one analyses the details of the problems of city living one arrives at the functional complexity as the basic factor. The change from agricultural function to non-agricultural (manufacturing, trade and services) activity involves social mobility as the later function requires expertise as well as it makes wage differences. Social mobility brings about a cosmopolitan character to the city social fabric, and along with it social disharmony. William W. Van Groenou[3] brings out the differences between rural and urban life. Whereas urban man experiences stresses and strains, he is somewhat pessimistic, traditional, more strongly self-centred. This kind of social fabric is a result mainly of industrial and service functions which are capable of attracting people from distant places.

Since medium sized cities are dominantly administrative and educational in functional character (services-category), many social gatherings, literary discussions, fairs, religious festivals, take place with gaiety and colour.

## TEMPORAL GROWTH PATTERNS OF MEDIUM SIZED CITIES

Before the advent of public transport on the Indian urban scene cities were compact, huddled together and core oriented. Many medium sized cities were well within walls or fortress. Ancient rulers built the city close to a waterbody covering a small area. During the medieval period under the influence of the Mughal rulers the city grew further leaving imprints on the landscape. During the British period the city grew in spatial dimension as the space requirement of the functions increased. The British introduced a variety of landuse functions, such as parade ground, golf links, police headquarters mission schools, hospitals, library, churches, etc. The anglicised buildings were huge and spacious with compound walls. During the British period these landuse functions, which today we observe well within the city interior, used to be at the outskirts.

The post-independence period, particularly the advent of industrialization with the Five Year Plans in India, reveals considerable expansion of medium sized cities. The newer cooperative housing societies, educational and technical institutions, quite a few new administrative buildings are the landscape features of present city outskirts. An industrial estate or complex is generally away from the municipal limits, but well within the reach of the city workforce.

The direction of city growth is mainly towards that city which exerts an economic pull over the town. The spatial expansion is in the direction of maximum interaction. In the case of Nasik city the spatial growth is towards Bombay, Poona and Trimbak. Similarly, medium sized cities show a growth pattern towards higher order centres. It appears from the trend that medium sized cities showed a distinctive spatial expansion during 1961-1971, in comparison with their earlier growth. The city municipal boundary covers a number of small agricultural villages and hamlets around the urban built-up area. These agricultural 'wadies' or hamlets are connected by city transport but need urban facilities such as electricity, tap water, primary schools, a small library and a few important shops.

The historical imprints on landscape show a distinctive pattern which need the major attention of town planners both in preservation and planned growth. During the ancient period it was a sort of walled city with natural barriers like river, hill during the Moghul period it was a fort town and during the British period it was a planned sprawl with a cantonment town nearby as an 'invigilator'. Now there is no need of such city protection schemes as there is no direct external threat.

The patterns of growth follow a kind of course spreading initially on the major route which connects the higher order settlement. Once the route gets fairly saturated, it starts expanding sideways. This perhaps may be due to the stage-wise construction of roads by the civic authority. Gradually the infilling process takes place in between two built-up linear areas. The most accessible area gets built-up first and then the next best accessible area. The spatial expansion of a city follows this principle other factors being even.

## COMPARISON BETWEEN A MEDIUM SIZED INDIAN CITY, A SOUTH EAST ASIAN CITY AND A WESTERN CITY OF THE SAME SIZE

### Horizontal and Vertical Distribution of Urban Population

India has larger cities than any country in south-east Asia. In comparison with the south-east Asian countries, the urban population is relatively evenly distributed in large, medium and small cities of India. South-east Asian countries have a situation indicative of urban primacy. Though this phenomena is a typical of countries having a low level of economic development, some countries such as Denmark having high level of economic development also demonstrate this primacy of a city.

### Territorization of Economic Base

South East Asian cities as T.C. McGee[4] puts it, grew not because of industrialisation as they did in the western countries. The tertiary sector of the cities occupational structure grows significantly in excess of what is required. This is true of medium sized cities of India where on an average 59.82% of the working population is engaged in the tertiary sector.

### Tertiarization in the Eastern Countries Over and Above Manufacturing Base

The current trends in the economic base of western cities also show a tendency towards tertiarization. For example during 1955-1970, the proportion of persons engaged in commercial finance and services increased from 43 to 50 per cent. This kind of western tertiarization is certainly different from the one we have in India or in South East Asia. Hawkers, pedlars, wheel-barrow vendors, rickshaw pullers, shop assistants and government employees proliferate in India and South East Asian cities. The increasing tertiarization in Western cities includes a provision for more and more social amenities and services.

**Urban Immigration**

The growth of cities in India and South East Asian countries is a consequence of high natural increase and rural-urban migration. Western cities grow mainly by immigrant populations as the natural increase is very low. In France, the smallest towns suffered stagnation, because their rural hinterlands lost their population through migration to the larger cities. The important element in urbanization policy since 1960 has thus been the attempt to give advantage to small and medium sized towns, trying to promote the industrial decentralisation and endowing them with services and centralised facilities comparable with those of larger cities.

**Role of Cities in the National Building Process**

Cities of Western Europe played an important role in the process of national building as Murphy[5] puts it. The city has instigated most of the great changes in western society and has been the centre of its revolution. In western Europe, the city has been the base of an independent entrepreneur group which has successfully challenged and broken the authority of the traditional order. On the other hand Indian and south East Asian cities have inherited their economic structure from an era of colonialism, characterised by an excessive specialisation in the production of materials, and do not show that vigour in their economic basis as is the case with western cities.

**Functional Specialization within the City Fabric**

The residential segregation is another spatial phenomena of western cities. The core and residential suburbs are quite distinct and even within the core area clear cut functional zones are observed. On the other hand Indian and South East Asian counterparts are still core oriented. There is a definite trend towards the concentration of activities inside the central areas, due to clustering of shopping, religious and social and entertainment activities. The general urban poverty should also be included among the causes of concentration. As a result in the Central area, congestion and mixed land-use are growing. Differentiation based on functional use is difficult to make in Indian and South East Asian cities. However, old cores of Indian cities do reveal functional specialisation and spatial segregation. The areas of commercial function had specialised lanes dealing in different commodities such as Jewellery, brass and copperware textiles, cutlery and hardware and so on. The household industries also had specialised lanes such as Weavers' lane, Blacksmiths' lane, Goldsmiths' lane Coppersmith's lane and other lanes manufacturing baskets and mats, earthen pots and tiles and broomsticks. But the present trend in the shopping patterns is one of the 'Mix' small innumerable individual shops are growing up, selling articles ranging from postal stationery to hardware. This probably reflects the general poverty of an urban dweller of South and South East Asian countries.

## MEDIUM SIZED CITY GROWTH AND ITS PLACE IN THE WORLD URBAN PROCESS

The annual population increase of medium sized cities of India during 1951-61 was 3.4%. Small towns are losing population to the medium and large cities and to some extent medium sized towns to bigger metropolitan cities. A moderately vigorous growth of new office buildings, many of which are financed from private sources are reserved in central areas of comparable size, shows stagnation and declining growth in central core areas. Small towns in France could not develop industries or urban facilities as they had lost their population to larger cities. The expansion of population in medium sized towns is less expensive than in the case of large cities and the problems of traffic, housing and commuting are easier to resolve. In France, efforts are underway to make small and medium sized towns, a sort of intermediate stage between large cities and the countryside.

In Brazil, an attempt has been made to reduce the over-dependence on medium sized cities as such cities are not always able to cope with such an inflated demand. In the USSR, the number of medium sized cities continues to grow in the process of the gradual emergence of urban settlements out of rural ones. In Japan, also the small and medium sized urban centres are losing their inhabitants, especially to the fast growing metropolitan regions of Keihin (Tokyo-Yokohama) and Hanshin (Osaka-Akobe). The provincial urban centres are modest creations in Australia. Larger ones are mainly state stimulated. It is also argued out that their is hardly any significant difference between rural and urban life in Australia. In Israel, medium sized cities are not growing significantly as the lower order settlements have centralised marketing and purchasing organisations controlling the whole economy. In Czechoslovakia, at the present time, it is the medium sized and large cities that are growing most rapidly.

Most leading cities of the third world, share the same problems of rapid city growth without a sufficient economic base : obvious ones are problems of unemployment, under-employment, unplanned occupational structure, poverty and overcrowding. The tertiary sector of the cities' occupational structure grows tremendously in excess of what is required.

Because of such acute urban problems do exist and have been recognized to do so, there is a great deal of talk in favour of decentralization of functions. Questions like the shifting of administrative functions or industries from an exploding metropolis do occur to present-day politicians and academicians. Some experts have suggested employment passes for entry into a bigger metropolis. This system eliminates unnecessary and unproductive population from entering the city. The shifting of centralised functions of bigger cities and their subsequent benefits to a smaller and medium sized cities with an efficient transport network brings much desired spatial development in the country. The employment pattern should be heavily rural and household oriented like that of Japan and China. The importance of population control, mass education, effective mass participation, female literacy and work participation should be viewed seriously as they influence the urban growth, structure and processes. These factors though basic have very wide implications.

## CONCLUSIONS

The above discussion leads us to make certain observations regarding the medium sized towns of India.

(1) The landscape features of the past—the Moghul and the British—are fast losing their significance and also their quality. The old core and the civil lines created under British rule is fast disappearing.

(2) By and large, medium sized Indian towns have remained dominantly as administrative and educational centres though an element of industrial function is entering the townscape.

(3) Medium sized towns show an increase in the growth rate pattern at the loss of population in the small towns. Similarly medium sized towns lose to some extent to large metropolis. This phenomena is true of other countries also.

(4) An attempt is being made in most of the countries to strengthen the economic base of the medium sized towns, develop characteristics independent of the metropolis and yet retain the high standards of living.

The medium sized towns are viable in size, neither facing the acute problems of the metropolis nor having the traditional, backward, primitive look. The medium and small sized towns may also prove capable of absorbing the future urban shock.

## REFERENCES

1. Hemlata Acharya, "Urbanizing Role of One Lakh City", *Sociological Bulletin*, Vol. V, 1956.
2. Kusum Lata Taneja, *Morphology of Indian Cities*, B.H.U. Varanasi.
3. W. Van Groenou William, "Crowding : A Report on India" ; *The Journal of Social Work*, Vol. XXXVIII, No. 1, April, 1977.
4. T.G. McGee, *South East Asian City*, A Praegev Fredrick ; New York ; 1967.
5. R.E. Murphy, *The American City*, New York, 1966.

# 10

# An Approach to Micro-Level Planning

—*D. K. Singh*
—*J. K. Routray*

PLANNING and development as it exists today at micro-level in the Indian context, does not seem to have any specific or special consideration; it operates within the main stream of national and state planning principles without a proper diagnosis of the problems, prospects, potentialities and requirements of the people and the economy concern of the sub-region. As a matter of fact the eccentric development, both in the sphere of spatial and sectoral has aggravated the real situation. The phenomenon has in turn given rise to mal-appropriation of the existing infra-structure facilities, inoptimal utilization of resources and diseconomies of disproportionate growth on the one hand, and on the other concentration of activities at certain points causing unbalanced development within the micro units.

In spite of indifferent data base and non-availability of relevant information, this paper attempts to analyse the present pattern of development in a district region with respect to existing development potential ; and finally, it has tried to coordinate spatial and sectoral planning for an integrated development.

## THE STUDY AREA

The micro-level unit under study is Boudh-Khondhmals, one of the backward districts in Orissa. The whole district is

The authors have duly acknowledged Mr. S.K. Mishra for his kind help in computer programming.

an undulating plateau except for the narrow valley plain along the Tel and Mahanadi river system in the north. About 60% of the total population constitutes scheduled tribes and scheduled castes taken together, out of which 40% of the total population belongs to the scheduled tribes. The district is underdeveloped in primary, secondary and tertiary sectors as well as in social and infra-structure development. Major constraints associated with the district development process are drought and large scale shifting (*podu*) cultivation. There is very limited scope for industrial development because of insignificant mineral resources availability like clay and graphite and secondly lack of entrepreneurial skill among the people. The district is potentially rich in forest resources which are yet to be organized and properly managed for a meaningful exploitation and utilisation. The forest area covers about 80% of the total geographical area and it is the primary source of income for a large section of the tribal population. This district comprises fifteen community development blocks with an area of 11,070.00 sq kms and a population of 6,21,675 in 1971.

## METHODOLOGY

For measuring the intensity of inter-sub-regional disparity in the study area, 21 indicators of the levels of development have been selected. The data for indicators were recorded in 1977-78 on a Block level. The selection of more representative indicators are constrained by non-availability of important information.

The indicators have been selected from the sectors like agriculture, industry, professional services, infra-structure facilities and socio-cultural development.

Indicators are as follows :

1. Intensity of cropping.
2. Area sown more than once, as percentage to net area sown.
3. Area under high yielding variety as percentage to net

area sown.

4. Percentage of irrigated area to net area sown.
5. Consumption of chemical fertilisers per hactare of net area sown (in kg.).
6. Sectoral concentration of secondary labour force.
7. Sectoral concentration of tertiary labour force.
8. Percentage of female labour force in tertiary activities.
9. Sectoral concentration of labour force in trade and commerce.
10. Sectoral concentration of labour force in transport and communication.
11. Population density.
12. Literacy.
13. Road density (all types) per 100 sq km of area.
14. Percentage of villages electrified.
15. Primary and middle schools per 100 sq km of area.
16. Primary and middle schools per 1000 population.
17. Secondary schools per 100 sq km of area.
18. Secondary schools per 10,000 population.
19. Number of hospital beds per 10,000 population.
20. Number of hospitals and dispensaries per 100 sq km of area.
21. Number of veterinary hospitals, veterinary dispensary and livestock aid centres per 100 sq km of area.

For constructing composite indices, the technique of principal component analysis has been used. This technique justifies its application as it does explicitly take account of the problem of multicollinearity among original indicators by orthogonalisation of the whole set of variables and also given objective mathematical weight for constructing a composite index by linear aggregation.

The following correlation matrix tables represent the relationship between the conformal groups along with the characteristic vector and corresponding objective weight.

The inter-correlation matrix tables are symmetrical. Multiplying these weighting values with respective indicators and adding up together, we get composite sectoral indices.

The overall development of any spatial unit is a function of its existing development potential along with some deliberate

| | $Y_1$ | $Y_2$ | $Y_4$ |
|---|---|---|---|
| $Y_1$ | 1.00000 | 0.29455 | 0.32553 |
| $Y_2$ | | 1.00000 | 0.43736 |
| $Y_4$ | | | 1.00000 |
| Eigen Vector | 0.86527 | 0.97659 | 1.00000 |
| Weight | 0.40265 | 0.45444 | 0.46534 |

The large root of this matrix is 1.70884..

| | $Y_6$ | $Y_8$ | $Y_{10}$ |
|---|---|---|---|
| $Y_6$ | 1.00000 | 0.10641 | 0.13237 |
| $Y_8$ | | 1.00000 | 0.44261 |
| $Y_{10}$ | | | 1.00000 |
| Eigen Vector | 0.47392 | 0.98688 | 1.00000 |
| Weight | 0.26102 | 0.54355 | 0.55078 |

The large root of this matrix is 1.49938.

| | $Y_7$ | $Y_9$ | $Y_{11}$ |
|---|---|---|---|
| $Y_7$ | 1,00000 | 0.13975 | 0.48632 |
| $Y_9$ | | 1.00000 | 0.25324 |
| $Y_{11}$ | | | 1.000C0 |
| Eigen Vector | 0.93529 | 0.62614 | 1.00000 |
| Weight | 0.48906 | 0.32741 | 0.52290 |

The largest root of this matrix is 1.61338.

| | $Y_{13}$ | $Y_{15}$ | $Y_{20}$ |
|---|---|---|---|
| $Y_{13}$ | 1.00000 | 0.24034 | 0.53514 |
| $Y_{15}$ | | 1.00000 | 0.39816 |
| $Y_{20}$ | | | 1.00000 |
| Eigen vector | 0.90776 | 0.77615 | 1.00000 |
| Weight | 0.43495 | 0.37189 | 0 47915 |

The largest root of this matrix is 1.78505.

| | $Y_{12}$ | $Y_{14}$ | $Y_{16}$ | $Y_{20}$ | $Y_{21}$ |
|---|---|---|---|---|---|
| $Y_{12}$ | 1.00000 | 0.17616 | 0.04277 | 0.66839 | 0.32203 |
| $Y_{14}$ | | 1.00000 | 0.35464 | 0.33241 | 0.86725 |
| $Y_{16}$ | | | 1.00000 | 0.20399 | 0.45258 |
| $Y_{20}$ | | | | 1.00000 | 0-62074 |
| $Y_{21}$ | | | | | 1.00000 |
| Eigen Vector | 0.61818 | 0.84998 | 0.55685 | 0.83749 | 1.00000 |
| Weight | 0.21279 | 0.29258 | 0.19168 | 0.28828 | 0.34422 |

The largest root of this matrix is 2.70833.

and forceful development policy issues. The development potential, is the combined interaction of demographic potential, resource potential and infra-structure potential, out of which the role of resource potential is an important one. The resource potential can be specified as mineral resources, forest resources, soil resources measured in terms of fertility status and water resources. Keeping these views in the background, the following development potential indicators have been framed on the basis of available information to identify potential blocks for planning purposes and to establish a relationship between development potential and levels of actual development.

The indicators for this purpose are (1) population density ; (2) percentage of working force ; (3) availability of cultivable land per agricultural family in hactares ; (4) coverage of land as percentage to total geographical land ; (5) irrigation potential areas can be created by different sources as per cent to total available cultivable land ; (6) coverage of irrigation potential area by groundwater development as percentage to total available land ; (7) coverage of forest land as percentage to total geographical area, (8) coverage of pasture and grazing land as percentage to total geographical areas ; (9) road density (all types) per 100 sq km of area ; (10) spatial concentration of primary and middle schools per 100 sq km of area, and (11) spatial concentration of agricultural cooperatives per 100 sq km of area.

The methodology has been adopted to find out the development potential composite index as similar to levels of economic development. Though the first variable is not conformal, it is fused with the composite index by Pal's method.

The following correlation matrix tables represent the relationship between the conformal groups along with characteristic vector (eigen vector) and corresponding objective weighting.

The pattern of economic development *vis-a-vis* the development potential by development blocks can be compared from the following statement. The index values for levels of development and development potential by blocks are given in

| | $Dp_2$ | $Dp_4$ | $Dp_6$ | $Dp_9$ | $Dp_{11}$ |
|---|---|---|---|---|---|
| $Dp_2$ | 1.00000 | 0.23354 | 0.30130 | 0.72345 | 0.55080 |
| $Dp_4$ | | 1.00000 | 0.18653 | 0.11984 | 0.58155 |
| $Dp_6$ | | | 1.00000 | 0.60357 | 0.47728 |
| $Dp_9$ | | | | 1.00000 | 0.53300 |
| $Dp_{11}$ | | | | | 1.00000 |
| Eigen vector | 0.92947 | 0.60048 | 0.82367 | 0.99447 | 1.00000 |
| Weight | 0.28235 | 0,18241 | 0.25021 | 0.30210 | 0.30378 |

The Eigen value for this matrix is 2.78431.

| | $Dp_3$ | $Dp_5$ | $Dp_7$ | $Dp_8$ | $Dp_{10}$ |
|---|---|---|---|---|---|
| $Dp_3$ | 1.00000 | 0.42592 | 0.07357 | 0.22954 | 0.61046 |
| $Dp_5$ | | 1.00000 | 0.64208 | 0.14206 | 0.66107 |
| $Dp_7$ | | | 1.00000 | 0.14910 | 0.39249 |
| $Dp_8$ | | | | 1.00000 | 0.22005 |
| $Dp_{10}$ | | | | | 1.00000 |
| Eigen Vector | 0.77462 | 1.00000 | 0.75286 | 0.42519 | 0.99463 |
| Weight | 0.26653 | 0.34408 | 0.25904 | 0.14629 | 0.34223 |

The Eigen value for this matrix is 2.53129.

Appendix No. 1. Four classes have been identified by the standard deviation grouping technique and presented in the map.

| Orders of Development | Levels of Economic Development | Levels of Devclopment Potential |
|---|---|---|
| 1st order (High) | G. Udayagiri, Boudh, Phulbani | Chakapad, Boudh |
| 2nd order (Medium) | Kantamal, Chakapad, Naugaon | G. Udayagiri, Naugaon, Phulbani, Kantamal. |
| 3rd order (Low) | Tikabali, Tumudibandh, Daringibadi, Hara-bhanga, Khajuriapada, Raikia | Tikabali, Khajuriapada Raikia, Baliguda, Hara-bhanga Tumudibandh. |
| 4th order (Very Low) | Baliguda, Kotagarh Phiringia | Daringiabadi Kotagarh, Phiringia. |

From the analysis it is revealed that the present pattern of economic development is more or less concomitant with the existing pattern of development potential partly created or to be created by blocks. The blocks like Chakapad and Baliguda have not been developed up to the expectation level as depicted in the development potential ladder (Fig. 10.1). G. Udayagiri,

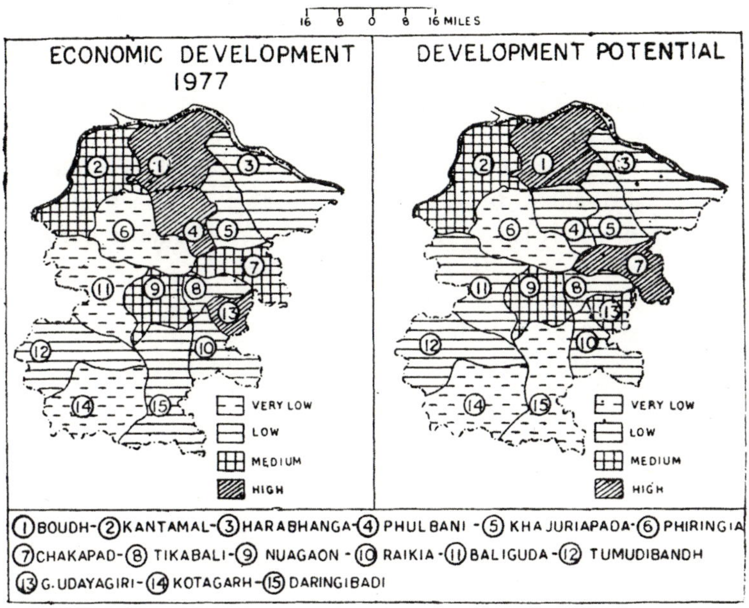


FIG. 10.1

Phulbani, Daringibadi blocks have been developed beyond their scope of development potential. It is very interesting to observe that the development pattern in the case of 10 blocks namely Boudh, Kantamal, Naugaon, Tikabali, Tumudibandh, Harnghanga, Khajuriapada, Raikia, Kotagarh and Phiringia, conform with the development potential factor. That most of the tribal blocks remained backward is due to the fact that whatever opportunity was offered by development potential for upgrading the standard of living was not properly availed of in those areas on the one hand ; and on the other, the exploited tribals are deprived of availing of the facilities which are already at hand because of institutional bottlenecks, political and social factors. In contrast to this G. Udayagiri being a tribal block has developed beyond the expectation limit because of the Christian missionary activities.

Boudh block stands out prominently both in economic development as well as in development potential. Partial concentration of development is seen at G. Udayagiri, Boudh and Phulbani blocks in their backward district. The aim of this study is not to isolate these particular blocks from the development and planning process, but rather to strengthen further with sound infra-structure facilities. Simultaneously, due emphasis would be given for the development of the undeveloped tribal blocks in an integrated way.

The peculiarity observed with Kantamal and Harabhanga blocks is, singular. In spite of favourable physical location and uniform fertile soil track, these two blocks have not been developed much as it is generally expected.

The analysis of this type is not mere identification of problems ; it serves to fix a priority area for planning purposes taking into consideration the problems as well as the availability of the possible development potential factor.

## PROPOSALS FOR DEVELOPMENT AND PLANNING

1. The district is very often prone to drought in spite of a good amount of rainfall during monsoon. Due to topography and internal structure, the ground water potential is not sufficient to fight drought. The most important aspect of planning for development here is the watershed management.

2. At present agriculture is the predominant activity and will continue to be so because of resource endowment structure. The important aspect for agricultural development may be emphasised on the soil structure and fertility status. For this, detailed soil survey and mapping are the first priority, along with water management which will lead to an alternative best suited cropping pattern for the district.

3. The only resource of the district is the forest, which covers 80% of the geographical area. Paradoxically enough the earning and dependency for livelihood is much below the reality. Proper management and a special programme for forest resource development is the basic requirement. In reality major portions of the forest areas are not economically important. Those areas should be identified immediately and replaced by commercial plantation. Day by day the area is being depleted with the practice of shifting cultivation and injudicious exploitation. Whatever may be the case it is certain that next to agriculture the forest is the only alternative economy.

4. Planning for the backward population is equally important along with sectoral and spatial planning. An integrated development approach must be adopted in which sections of people, sectors of the economy and spatial units are to be combined to arrive at desirable solutions to overcome the existing problems and to minimize future problems.

## APPENDIX 1

| Name of the blocks | Percentage of backward population | Percentage of S.T. population | Index of Economic Development | Index of Development Potential |
|---|---|---|---|---|
| 1. Phulbani | 54.39 | 41.36 | 1.2959 | 0.1034 |
| 2. Khajuriapada | 83.16 | 54.61 | 0.8280 | 0.9375 |
| 3. Phiringia | 79.83 | 61.87 | 0.5662 | 0.6620 |
| 4. Boudh | 25.85 | 10.29 | 1.4420 | 1.3553 |
| 5. Harabanga | 32.84 | 15.01 | 0.9242 | 0.8790 |
| 6. Kantamal | 33.04 | 17.07 | 1.2561 | 1.0678 |
| 7. Baliguda | 67.91 | 51.79 | 0.6426 | 0.9256 |
| 8. Naugaon | 67.62 | 51.65 | 1.0597 | 1.1540 |
| 9. Tumudibandh | 81.55 | 50.28 | 0.9656 | 0.7491 |
| 10. Kotagarh | 82.15 | 62.16 | 0.5854 | 0.7010 |
| 11. Tikabali | 77.98 | 56.68 | 0.9944 | 1,0094 |
| 12. Chakapad | 73.64 | 50.08 | 1.2115 | 1.4083 |
| 13. G. Udayagiri | 72.96 | 54.86 | 1.4857 | 1.1664 |
| 14. Raikia | 69.81 | 53.49 | 0.8189 | 0.9288 |
| 15. Daringibadi | 74.96 | 61.50 | 0.9253 | 0.7207 |

# 11

# Integrated Rural Development—A Geographical Approach

—*N.G. Jain*

## GEOGRAPHY AND ECONOMIC GROWTH WITH SOCIAL JUSTICE

THE CONCEPT of economic growth with social justice has recently been introduced in our national planning.[1] This concept is further elaborated by adding the dimension of growth with ecological balance so as to harmonise the short and long term goals of development. This clearly directs us to a symbolic interaction between science and society and the faithful adherence to it is the requirement of modern geography.

There have been significant recent developments in the application of geography to the cause of promoting social welfare. In the nature-man relationship, nature is the passive and man the active component. Whatever be the nature of man's habitat, he is trying to satisfy his basic and secondary needs through the employment of technology and, as a result, this man-made change in the landscape has become a matter of great importance to modern geographers.

The traditional view of geography as landscape analysis and synthesis restricts its scope to a series of natural and cultural elements. According to this view, "Geography is the science of distribution of phenomena on the earth's surface and mutual relationships." It is not surprising that the emphasis throughout is on the understanding process because of man's deep desire to explain what he had observed on our planet. But this traditional view is to be modified in view of the present requirements of human society. Much emphasis is needed on the future social welfare. Out of this need, the subject is becoming increasingly action-oriented, and scientific in its analysis and interpretation. As a result, now we have the system analysis, the "quantitative" geographers, the "regional" planners, "applied" geographers and computer-mapping specialists. These non-traditional professional groups are expected to make their contribution to the planned national development.

In the post-independence period of economic progress much emphasis was given on industrial development in the country. But at the same time it has created many social problems. Migration on a large scale is taking place from rural to urban areas. It is generating environmental pollution and a steep rise in the consumption of non-renewable forms of energy. It has been agreed by one and all that going too far on the road of urbanisation shall prove to be dangerous. Even in highly urbanised countries there is a marked movement back to the countryside. If this process continues there would be integration of some of the basic benefits of urban life with rural living. In India, 80 per cent of the population still lives in rural areas. Because of migration, this percentage may decline but even then the rural population is expected to grow in absolute terms. Of the existing rural population, nearly 50 per cent live on the border-line of poverty. The fast rate at which population is growing in the rural areas indicates that the future solutions to these basic problems of under-development must be found in the countryside. The main battle in the war on poverty has to be fought in the rural areas and not in the urban ones. What is necessary at this time is the rapid rural development based on the scientific utilisation of all our resources, both natural and human. A planned re-grouping of services and amenities throughout the country according to the agro-ecological potential and socio-economic needs of each area is the necessity of the present.[2]

Economic growth with social justice is confronted with many problems. The modern geographer, with his interdisciplinary and integrated approach and training in survey and analysis is quite competent to make an effective contribution in solving these problems.[3] A geographer can identify the problems; he can conduct a survey and collect data; he can analyse the problems; he can determine the objectives; he can identify and recommend the alternatives for action; and lastly, he can programme the steps for effectuation process.[4] Thus, he can be an expert at all the stages of the planning process. Planning always concerns men, space, time and resources. Geography analyses their varying spatial patterns and applies the principles of synecology for the working balance so that the desired goals for optimum development may be attained.

In most of the advanced countries, we find geographers are actively associated with the planning process. In our country also geography is entering a new era of 'involvements' and becoming a problems-and-process-oriented social science. Any science today that wishes to lead has either to answer some of man's inherent question or has to contribute to his welfare. Geography can, no longer, concern itself only to the former.[5] It has to contribute to the welfare of society in its everchanging landscape.

## MACRO-LEVEL TO INTEGRATED MICRO-LEVEL PLANNING

Post-independence development in our country so far was based on macro-level sectoral planning. It has positively taken the nation on the path of overall development; but it has created regional and social imbalances. Certain areas and a significant proportion of the population have remained outside the main stream of economic progress. These imbalances, can be corrected, as the Planning Commission has emphasised, by grass root level or micro-level planning based on available local resources and the village community.[6] The question is how the disadvantageous groups in the rural areas can be brought into the main stream of national development. The answer lies in rural change in an orderly fashion.

## INTEGRATED AREA DEVELOPMENT

The rural development effort in India has so far proceeded on two assumptions: one, that the existing village communities are compact enough units to merit concentration of planned development to bring about their socio-economic transformation; and two, that the administrative and other infra-structure for such development has to be provided for larger territorial area. Thus it is that the local unit for development administration has been conceived as a grouping of some hundred contiguous villages in a compact area for providing the administrative apparatus, the extension services, monetary and material inputs, the people's institutions and other infra-structure for development. As a means to bring a minimum frame of development to the millions in the country-side, therefore, the Community Development Blocks supported by a network of Panchayat institutions have been established all over the country. In addition, special programmes of intensive agricultural development were undertaken. Since these sectoral and temporal programmes have seldom had a strong methodological foundation aimed to promote growth and integrated development, it was but natural that their impact would be uneven. The spatial nature of functional integration as implied in the geographical approach has been found neglected in these programmes.

"Planning in the rural areas must now be increasingly designed so as to serve continued agricultural progress and, at the same time, to preclude the evils which could result from loosely-managed agricultural prosperity and to ensure optimal economic and social returns on the investment made within the sector. For this, it is necessary to relate sectoral and temporal programmes to spatial constraints and physical needs."[7] This can be achieved by Integrated Area Development Planning which refers to functional and spatial integration. Functional integration refers to the integration of all economic and social activities which influence the life of the people.[8] Thus, health, education, agriculture, industry, transport and several other aspects of our day to day living overlap. A change in one function almost invariably brings change in another. This inter-relationship is significant in integrated area development as it implies planning for a given geographical area in all the aspects needed for modernisation and betterment of rural living.

The nature and degree of inter-relationships among various socio-economic activities depend mainly on where they are located. The actual location of a specific function in relationship to other functions, depends on several factors such as the general level of development, demand and supply of specific functions, degree of their mobility in terms of time and distance, level of income of the people and so on. It is, therefore, proposed locations for new functions in the process of development that are significant. An appropriate location of a new function starts a chain reaction of development. An understanding of functional inter-relationships in space, therefore, is integral part of the process of development.[9]

## GROWTH CENTRES

Integrated area development requires a framework for decentralising economic and social activities by locating specific functions in appropriate location, which though selective cannot be discriminatory. Development, viewed as a process of innovation as well as growth, does not appear everywhere at the same time but manifests itself at favourable points from which it tends to propagate outward.[10] The centre along with its dependent territory is involved in this process. Such selected locations are termed as 'Growth Centres', which would provide the occupational diversity, the social services and amenities necessary for a higher standard of modern rural life in India and would also absorb much of the urban and occupational migration. The settlements selected as growth centres are of different orders creating their hierarchy (Fig 11.1). Higher order settlements have their own hinterlands which include these of the lower order settlements. The network thus created provides a meaningful infrastructure which can attract and sustain a diversified but a growing economy.[11] Such settlements are necessarily in the central place. Now since the emphasis is to be given on the development of rural areas, it is not desirable to utilise the existing hierarchy of settlements in a particular region for formulating a development plan. The areas far away from the important developed new centres in such regions should be

*Kalamnuri Tahsil Growth Centres*

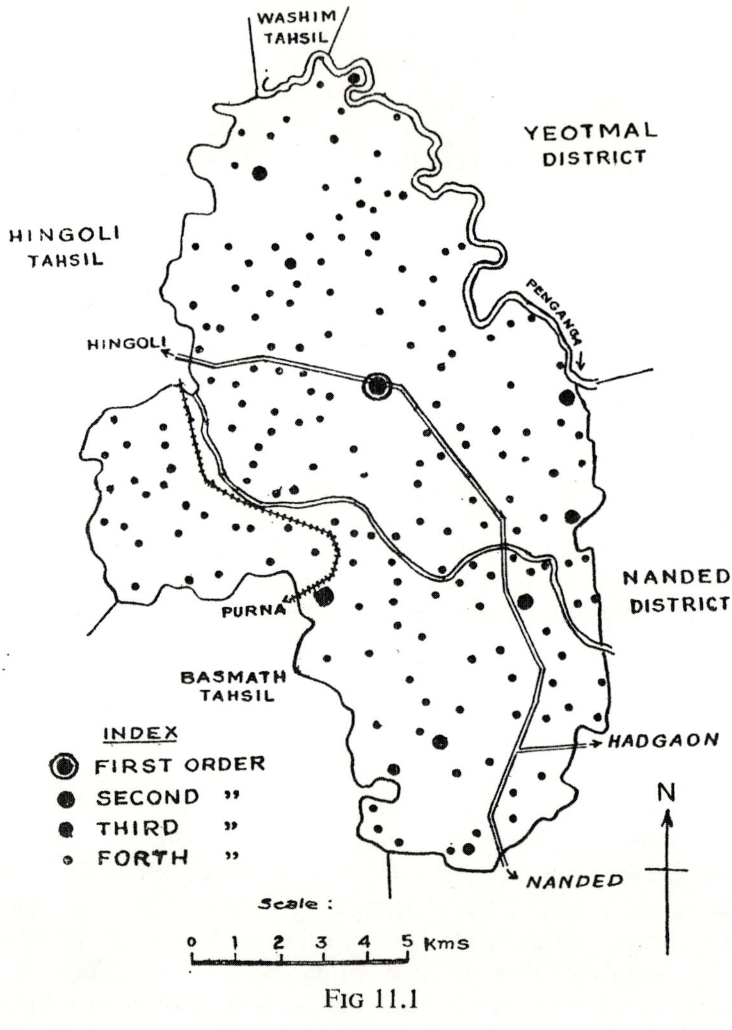


FIG 11.1

identified as having potential growth capacity. When these growth centres are identified, their hinterlands can be a region for planning of economic growth, and, through it, of creating social equality.

It has come to be recognised that the village, in the present stage of development, scarcely provides, in a great majority of cases, sufficient resources and viability to sustain comprehensive development. Even within the given development set-up and without disturbing its pattern, it has been found necessary, therefore, to study the growth factors which enable the rural community to pick up its viability. Thus planning for development must be based on the study of areas with social, economic and spatial inter-relationships which have functional or systematic wholeness, on a given scale. Within India, the minimum scale at which functional and systematic wholeness exists, is an area consisting of a growth or service centre and a surrounding growth of related villages. Such localised rural communities contain a growth centre and a group of villages are the nodal points in larger infra-structural networks which form the activity systems of a macro-region and ultimately, of the nation. It is important to keep in mind that in the larger framework of national integration, the cohesive functional rural community is the essential building block. A number of these functional communities constitute a sub-region which will also provide a set of more complex services. The central service or functional village forms an integral element in the rational allocation and mobilization of resources. Thus one can visualise a hierarchy of progressively higher ones with increasing functional complexity.[12]

Thus integrated area development refers to the appropriate location of social and economic activities over a physical space for the balanced development of a region. Therefore, growth centres provide a basic tool for micro-level development. Normally villages want immediate access to all required goods and services while the facilities require a large number of people to support their operation. The demands of these two components in any spatial system are often contradictory. Central place theory is concerned with the contrast and can form the basis for location of growth centres. The pattern of settlements in india is characteristically a vast number of villages, predominantly devoted to cultivation, dispersed rather uniformly in space, with a very small number of trade and service centres and with poor linkages for the movement of goods and people among these places.

## MICRO-LEVEL PLANNING REGION

Once the areas with identified central place are demarcated, they form the lowest level area for planning. This region forms a base for planning the development and for solving the socio-economic rural problems such as water supply, disposal of human and animal wastes, effect of pesticides and fertilizers, deforestation, soil erosion, improper housing facilities, transportation and communication, rural industrialisation, cooperative and credit institutions, banks, schools, health and unemployment.[13] This region is to be used to assess the existing conditions potential and its prospects by way of survey. As J.N. Jackson remarked.[14] "The role of survey is to study the situation factually and objectively, to present clear and carefully reasoned reports, to provide the necessary understanding before decisions for development are made and to evaluate the effects of development." R.L. Singh remarked[15] that "such a procedure involves both functional and spatial integration which are inter-related to each other in various complex forms" and leads to socio-economic development of micro-level rural areas. Geographers with their inter-disciplinary wholistic approach contribute substantively to integrate the socio-economic development of rural areas.

The geographical approach to integrated rural development has been applied here to Kalamnur tahsil of Parbhani District of Marathwada. The region is most backward and it can provide an example of planning for the most backward and economically deficient rural areas. This tahsil is the north-eastern tahsil of the district stretching along the southern boundary of Vidarbha—a region which has also been classed as backward. The tahsil lies between 19° 30′ N and 20° 00′ N latitudes and 77° 00′ E and 77° 00′ E longitudes, with an area of 1510 sq km.

The terrain of the region is mostly of hills belonging to the Ajanta system. These hills run in NW-SE direction and form a dissected plateau. The average elevation is 450 to 540 metres a.s.l. The plateau slopes gradually towards the south. The region is traversed on the north by R. Penganga forming a boundary between Kalamnur tahsil and Washim tahsil and Yeotmal tahsil of Vidarbha. A tributary R. Kenad flows in west-east direction through the central part of the region.

The average temperature is 21.90°C. Rainfall is not uniform in all parts of the tahsil. The average rainfall gradually increases in a west-east direction. The annual average rainfall is 900 mm. It begins about the first week of June and lasts till the end of September. Rainfall variability is slightly low as the coefficient of variability is found to be 30.89 p.c. The climate on the whole is healthy.

The soil of the tahsil is black cotton soil but it is of relatively light and medium type and being shallow with high gradient is subjected to heavy erosion.

The tahsil is devoid of any important minerals. Black basalt and granite are available everywhere and are used for building and road metalling. The area under forest is scattered. The forests contain valuable trees such as teak, *dhavda, salai, khair, tiwas* etc. Cotton and *jowar* are the main staple crops. There are no irrigation facilities. The construction of dams at Sidheshwar and Yeldari on R. Purna has provided electricity to this region. The classification of landuse of the region is as under: forest 2.72 p.c.; barren and uncultivable 2.63 p.c.; land put to non-agricultural use is 1.44 p.c.; cultivable waste 1.84 p.c.; permanent pastures 6.57 p.c.; tree crops and grooves 0.15 p.c.; current fallow 19.50 p.c.; other fallow 0.24 p.c.; net area sown 64.0 p.c.; gross cropping area 67.22 p.c. This shows that the economy of the region is based on agriculture only. The percentage of workers working on the land to the total workers comes to 88.83 p.c. and the gross area sown per worker comes to 4.09 acres.[16] The labour participation ratio is 54.38 which is greater than that for Maharashtra (47.91) and for Parbhani district (51.75). This shows a high dependency of the people on the land.

The total population of the tahsil is 1,27,459 out of which 1,19,871 is rural and 7,588 is urban. The region has only one town and 199 villages.

The region is traversed by the Hingoli-Nanded road in a north-south direction and by a metre-gauge railway line on the western margin.

The total revenue from this region is approximately Rs. 2,75,556. This shows that the community living in this region is mostly poor and backward. They are to be provided with modern amenities. This will mean, naturally, an adverse budget; yet it has to be done to bring the community into the national stream. But its development should be systematically planned in such a way that maximum services at minimum cost are provided to the most economically backward people. It is necessary sometimes to create conditions to induce growth for the development of the backward areas.

An attempt is made here to identify the potential growth centres in the tahsil. Various procedures and methods have been devised or identified for growth centres. Here I have adopted a two-fold procedure i.e. first identification on the basis of special functions discharged by the particular settlement and then I have used the Relative Centrality Index.

It is found, as shown in the map, that there is only one higher level centre namely Kalamnuri as tahsil headquarters. Second order centres are eight namely Sirsam B.K., Gaul, Yelegaon Tukaram, Akhada Belapur, Jawala Panchal, Nandpur, Borda and Takali Nandpur. Surprisingly all these second order centres are on the periphery of the tahsil but are more or less equidistant from the first order centre. These centres can provide the growth centres at the lowest level commanding approximately 90 km of an area, a group of 18 to 20 villages with 12,000 inhabitants. The purpose of this paper is not to prepare a development plan for this backward region; but the hinterland of each of these growth centres can form the lowest level region for the integrated development of a rural and backward community hitherto neglected and unaware of the conditions of modern living.

## REFERENCES

1. M.S. Swaminathan "Science and Integrated Rural Development", General Presidential Address, Sixtythird Session, ISCA, Waltair, 1976, pp 11-41.
2. N.G. Jain, "Geography and National Development". Presidential Address, Sixth NUTA Conference, Amravati, 1976.
3. R.L. Singh, "Concept, Method and Technique of Applied Geography", *Applied Geography*, NGSI, Varanasi, 1966, p 7.

4. H. Peter, "Planning, Geography and Ekistics". *Studies in Applied and Regional Geography*—Edtd. by M. Shafi and M. Raza, 1971, p 23.
5. Ibid, p 55.
6. Planning Commission, Govt. of India, Fourth Five Year Plan, 1969-74, New Delhi, pp 229-30.
7. "Pilot Research Project in Growth Centres (India)", *Project Design*, pp 2-3.
8. L.K. Sen, "The Need for Micro-level Planning in India". *Readings on Micro-level Planning and Rural Growth Centres*, pp 4-9.
9. Ibid, p 5.
10. Andrade Preston "The Growth Centre Concept". *Readings on Micro-level Planning and Rural Growth Centres*, p 13.
11. S. Wanmali, et. al., "Planning Rural Growth Centres for Integrated Area Development, A Study in Miryalguda Taluka," *National Institute of Community Development*, Hyderabad, 1971, p 3.
12. Andrade Preston, op. cit., p 10-12.
13. S.L. Kayastha, et. al., "Geographical Environmental Approach for Integrated Rural Development." Proceedings of 65th Session of ISCA (Geology and Geography Section), p 3.
14. J.N. Jackson, *Survey for Town and Country Planning*, Hutchinson University Library, London, p 27.
15. R.L. Singh, et. al., "Spatial Approach Towards Rural Development, Case of Jhagadia Taluka, Gujarat" op. cit., p 5.
16. *District Census Hand Book*, Parbhani, 1971.

# 12

# Integrated Rural Area Development—A Geographical Approach

—*Joseph T. Uyanga*

THIS ESSAY examines the role of economic investments in the spatial structure in Nigeria, and how the associated development problems can be solved integratively. The factors of rural development and investment pattern normally create a spatial pattern which is characterised by differential economic fortunes. There may be (i) areas of very energetic and flourishing rural economy located near major urban centres and the core areas of the major export and food crops; and (ii) areas of low and passive economic activities located either far away from the major urban centres or near the most parasitic urban centres. The last point can be illustrated by a recent government statement in Nigeria. "The exodus of young persons from the neighbouring rural areas of Ijebu and Abeokuta to Lagos, has affected the rural economy and the revenue collecting capacity of these areas".* Rural depressing effects can also be seen in the empty houses, broken bridges, roads in disrepair and derelict oil mills in parts of these areas, (ILO[1] 1970).

In South Australia, Smailes (1969) has delimited areas adversely affected by the dominance of Metropolitan Adelaide; other evidence abound in Moseley and Woolmington[2].

Between rural typologies (i) and (ii) in Nigeria, on the one hand and urban and rural areas on the other hand, differences in inter-regional income and development have been amplified

* Press statement by the Military Governor of the former Western State of Nigeria on April 19, 1971.

during the last few years, Uyanga.[3] On this point also Essang's[4] study shows that 9% of all the cocoa farmers in Western Nigeria account for 58% of the total cocoa earnings and receive most financial and other benefits from the existing government programmes. The majority in this group are in the largest urban centres, pay taxes and incur most expenditures there. The Marketing Boards have enhanced and reinforced the tendency for rural-urban transfer of wealth and income. Far from helping to reduce the urban-rural income gap, the "Nigerian Marketing Boards have become the primary vehicle of widening this gap . . . a substantial proportion of the funds accumulated by the boards are invested in projects which largely benefit the non-farm urban sector" (Essang 1971, p 8).

Investments in the process of national development had largely lost sight of the "backwash effects" its sum total could have on the space economy. Helleiner[5] for instance estimates that between 1949 and 1962 about 80 million of the Marketing Board funds (10½%) had been invested initially in agriculture but later on these were diverted to large urban centres thus consolidating the core-periphery spatial system.

It appears that past rural strategies have always lost sight of the side effects of ramifications. They were mostly single-purposed, and lacked comprehensive co-ordination. Rural development requires a multi-faceted approach. Olatunbosun[6] (p 157) points out that, in order to improve the quality of life of the neglected rural majority in Nigeria, programmes of agricultural production, health with a more efficient rural health delivery; investment in rural non-farm activities; nutrition; education and training, both formal and informal; rural electrification; co-operatives, water supply; rural credit, entertainment; and road construction, should be planned and implemented in an integrated manner, with due consideration for projects that mutually support and interlock with one another in an overall rural development plan.

The primary objective is the mobilization of rural resources to cope with the complexity of rural problems; the stimulation, active support and participation of the rural people and the creation of basic institutions to foster integrated rural development programmes. This has always been suggested as the essential approach to Nigerian as well as any other country's rural planning, (Olatunbosun).

It is in this context that the discussion of integrated rural development is relevant because its basic assumptions have been realized elsewhere in many areas of economic growth and modernizing of production processes, as well as in the social improvements for the rural masses. One is tempted to cite examples related to the Tanzanian rural development programmes. Nigeria, therefore, needs a new strategy.

Integrated rural development is not denying the need for other approaches to rural economic growth and modernization, rather the strategy emphasizes the fact that the economic base in rural areas has to be broadened, through efforts to mobilize and better utilize human and natural resources by giving the less privileged people assumed access to means of production and services. It tends to create motivation and purchasing power through better distribution of income and employment opportunities in the rural areas. It could establish closer links between the agricultural, industrial and service sectors in the rural areas, boost efforts at improving the conditions of the living regarding housing, water supply, roads, through assistance to self-help actions.

The geographical aspect of integrated rural development is different from general rural development because as a matter of policy, it places greater emphasis on the mobilization and development of the natural and human resource potentials and on achieving a more equitable access to resources and a fairer geographical distribution of income. With regard to planning, integrated rural development goes beyond the agricultural sector. In order to apply the strategy, the tenets must be properly enunciated and understood. It embraces at the national level an overall policy conducive to achieving social improvements among the rural masses and takes up the complexity of regional/area development when it comes to implementation, including aspects of administrative, decentralization and people organizations. It can incorporate a national philosophy for planning with the local concept of self-help and self-reliance.

However, a geographical discussion of integrated rural development is still confused and it will be some time before the "pragmatists" and the "theoreticians" speak the same language. The pragmatists, whose conceptual arguments are often based on the principles of a market economy and the conclusions arrived at in analyzing the historical processes in developed countries, may not deny the objectives of Integrated Rural Development, but over-emphasize the difficulties encountered in its implementation. The "theoreticians" are challenged because of their unrealistic expectations regarding the political flexibility of the governments and not seeing the administrative, technical and human obstacles occurring if comprehensive programmes are to be implemented. Lele[7] and Ruttan[8] bring this out quite clearly. Ruttan asks the question "Why is it possible to identify a number of relatively successful small-scale or pilot rural development projects, but so difficult to find examples of successful rural development programmes." He analyses three major lines of, thinking :

(a) the urban-industrial impact hypothesis,
(b) theory of induced technical change,
(c) and the new models of institutional change.

The urban-industrial impact hypothesis starts with Van Thunen's work and from deductions from the American situation as studied by Schultz[9] concludes:

> Economic development occurs in a specific locational matrix; these locational matrixes are primarily industrial-urban in composition; the existing economic organization thus works best at or near the centre . . . . The basis for this hypothesis is efficient and well-functioning geographical factors and product markets in areas of rapid urban-industrial development. This hypothesis is also the backbone of the growth pole theories, the urban-industrial trickle down approach.

However, the ability of rural areas to respond to the opportunities for economic growth generated by local urban-industrial development depends on technical innovations which are capable of generating substantial new income flows in response to new economic opportunities, therefore, the capacity of rural areas to successfully respond to opportunities for growth depends critically on the achievement of rapid technical change and as Ruttan concludes, "Significant growth in agricultural productivity can rarely be realized by the reallocation of resources within the traditional agricultural system", (Ruttan 1975, p 11).

With regard to integrated rural development the point is that both the urban-industrial impact hypothesis as well as the hypothesis of induced technical change are relevant as far as production is concerned, but the beneficiaries are those who are part of the urban-rural exchange process and those own or have access to means of production in order to apply new technology and who have the know-how to do so. But already this, even without considering the fact that a large number of rural people are not part of the system, requires the existence of efficient and well-organized institutions which are not yet available, to them. Therefore, institutional change and their bureaucratic characteristics become the most critical element in rural development, particularly if the involvement of the less privileged and issues of employment and income distribution are major objectives.

Lele[10] asserts that because of the extreme constraint of trained manpower . . . mass participation rarely seems feasible in Africa in the shortrun through delivery systems oriented toward individual farmers. It seems therefore that delegation of responsibility to the rural people is necessary. Past programmes indicate that if the process of rural development is to be viable, there is not only the need for expansion of administrative coverage, but also for emphasis on improvements in the performance of the indigenous administration systems.

The overall success of rural development projects depends to a large degree on local action taken by small farmers to complement outside development, management and resources. Local action, on the other hand, depends on the small farmers' involvement in decision-making, in the implementation phase of a development project and the small farmer resource commitment to the project. Projects which take the time and efforts necessary to build in active participation and a cooperating role for the small farmer are likely to be more successful than those projects which follow traditional, externally-dominated development approaches.

The participation of rural people in rural development thus depends not only on project design, but also on the overall administrative and organizational setting, power structure and resource base of the various geographical and social groups in the project area. Usually three distinct strategies to development in general and rural problems in particular have been very popular—the technocratic strategy, the reformist strategy and the radical strategy. These three strategies differ in their objectives and in the priorities they attach to certain objectives, in the ideology which is used to mobilize support and action, in the dominant form of land tenure institution and in the patterns of property rights, as well as in the way the benefits of the economic system and growth process are distributed.

The technocratic approach remains valid in many aspects regarding functional, technical and economic interrelationships, but what has not worked is the distribution of the benefits through the "trickle down" effect because a profit-seeking market economy is not conducive to the involvement of people and, in the absence of appropriate institutional and political structures, a growing portion of the people were left outside the streams of development. One wonders if Nigeria is not an upholder of this strategy !

The radical approach, on the other hand, denies the need for modern technology, which does not fit existing institutions and social structures, but emphasizes the need for self-reliance, that is, mobilizing people on the grounds of a common ideology, using local resources and achieving capital accumulation by extensive utilization of existing human resources. The emphasis is on mobilizing people in a locally interacting system of agricultural and non-agricultural activities aimed at satisfying the needs of a larger community. The system is basically "inward-looking" and because of the low priority attached to technology and the exchange of goods, no strong outside linkages are necessary. However, also in that case, institutional changes are required and collectivization is called for.

Therefore, an inward-looking concept of self-reliance is not very acceptable. Mobilizing people and a combination of self-reliance with modern technology and appropriate institutional, administrative and political changes, including prospects for real participation seems to be a true alternative. This, then, is the category where Integrated Rural Development as a concept for planning has its position as a strategy for rural development.

## THE CONCEPT AND OBJECTIVES

As discussed earlier, the geographical perspectives of Integrated Rural Development is different from general agricultural and rural development because as a matter of policy, it places greater emphasis on the development and mobilization of the human and geographical resource potential and on achieving a more equitable access to resources and a fairer distribution of income. This should be achieved through policies, project design, rules and regulations which pay particular attention to the involvement and gradual absorption of the poorer rural areas in an expanding regional economy.

This requires that new yield increasing and advanced local technology is made available to the agricultural sector at large, which implies the need to develop local institutions capable of research and extension. It implies also a deliberate re-orientation and geographically located set of administrative system, local organization and production-supporting institution toward the small producer and their needs. Experience has shown, however, that this will not materialize unless provisions are made for a greater participation of the less privileged of the rural people at the village level and through appropriate local organization so that their voice will be heard. This can come out of local initiative, but the government must provide a policy which (a) prevents the suppression of these organizations, and (b) motivates through educational activities the rural population to engage themselves in such a more active role. Government will move in this direction only if there is a firm regional policy commitment. The key to any design of a comprehensive development strategy is the establishment of objectives for rural development as a common denominator and a basis for policy design.

Integrated rural development can be defined as a series of mutually supporting, inter-related, agricultural and non-agricultural and geographically specified set of activities oriented towards a stated regional objective which involves the improvements in the rural system as a whole. The special emphasis on broadening the rural economy and aiming at the involvement of the rural poor in the development process through which they should get a larger share of the benefits clearly shows that the objectives of integrated rural development must go beyond those of agricultural development because "such factors as increased social justice, eradication of absolute poverty or changes in the power structure are not simply good things to be pursued after economic growth has been achieved, but they are basic requirements for setting economic development in motion." Therefore, integrated rural development aims at the following objectives :

(1) Increased agricultural production with special emphasis on increasing yields of subsistence farmers, but avoiding environmental degradation;
(2) Improving the distribution of income and non-material benefits including social security;
(3) Improved consumption patterns, particularly in regard to food and nutrition of the most vulnerable groups;
(4) Progress in regional integration and improved basic amenity conditions as a means to increased productivity,

for which:

—the provisions of more equal access to productive resources and employment opportunities in order to contribute to and to benefit from the development process, and

—the mobilization and motivation of people, particularly

at the local level, to achieve a wider participation, are the most essential areas in which concrete action programmes are required.

Integrated rural development must therefore be considered as a concept for planning and implementation, providing a framework and understanding within which the linkages and inter-relationship between the various components can be identified considering :

(i) how they affect each other and each region in view of the objectives,
(ii) what technical, institutional and administrative means are required in order to link the various activities, and
(iii) the role of the people themselves within this process.

Such elements as the package programmes for increasing agricultural output, the development of rural industries and decentralization of industrial growth, popular participation, and the equitable income distribution have generally been recognized as important elements in national development strategies and have been pursued in one form or another. What distinguishes other approaches from Geographical Integrated Rural Development is that these have often been pursued independently of each other, and their inter-relationship have not been clearly grasped. The new dimension in Integrated Rural Development is the emphasis on the mobilization of human and natural resources in its widest sense, the inter-relationship between overall increase in output and more equitable distribution of income and access to resources and the resulting linkages between the agricultural and non-agricultural sectors, and the various geographical regions at large.

Basic to any efforts in implementing Integrated Rural Development is a consistent spatial policy framework in which the rural sector assumes a central role in the economy. This would have implications for a wide range of policies at the national level. Such policies may require national legislation which may take the form of land reform, location, pricing and income policy, cooperative, marketing, labour organization and overall orientation towards the involvement of the rural masses. The formulation and implementation of programmes and projects within the overall strategy will need to take into account the constraints imposed by resource availability and the capability of the administrative system. By pursuing these objectives of integrated rural development, a country will have to accept the need for major structural changes. The speed and extent of this structural change will depend on political and other constraints, and therefore strategies will have to be adapted to achieve results in a wide variety of political and regional settings.

Integrated, rural planning thus has two dimensions—institutional integration and spatial or geographical integration. The integrated geographical planning thinks in terms of spatial interactions for instance rural-rural, rural-urban and urban-urban. The institutional integration focuses on the links between developmental institutions—social, cultural, economic, religious, political and educational. The two levels of integration are not mutually exclusive, rather both should be integrated in the total process of rural development planning.

## PLANNING THE SPACE INTEGRATIVELY

Deducting from several Nigerian local studies Ajaegbu[11] indicates that in the operation of the urban-rural system, various inter-acting processes work in both directions, but the most important of such processes is the way that information on new ideas (i.e., social and economic innovations) is spatially transmitted from the lively or economically active urban centres to their rural hinterlands. Such urban centres constitute the largest single markets and often offer the highest prices for the local products from the rural areas. Thus, the nearer a rural area is geographically located to such an urban centre, the greater is the volume of information and urban influence generally received by the rural inhabitants.

The extent of influence emitted by an urban centre will be related to its size (or hierarchical order) and to its social and economic buoyancy. How far the urban influences go within the hinterland areas will, of course, depend largely on the nature of accessibility, the efficiency of the communication system, and the readiness of the rural hinterland people to accept and adopt innovations. Above all, the influence diminishes with geographical and time distance from the centre of transmission.

Once reached by the influences which make for economic and social change, the Nigerian rural inhabitants are found to be very rational in their economic decisions and behaviour and respond particularly favourably to market opportunities for their products. Some may gear their economies almost entirely to the needs of the urban market centres, and thus have generally achieved greater economic development than the more distant rural areas. Indeed the greatest rural economic development in Nigeria today seems to occur in the immediate hinterlands of the major urban centres, while such development often falls off very markedly with increasing distance away from the centres. Examples of this situation have been identified and described in the rural hinterlands of Lagos (Ajaegbu, 1967), Abeokuta (Camara),[12] Kano (Mortimore,),[13] Zaria (Norman),[14] and several other places in the country. Moreover, the spread of the growth influences of active urban centres, as well as the decreases in the extent of economic development with distance from the centres of transmission, have been reported in several other parts of the world, (Schultz).[15]

It is because of this growth influence, exerted by active urban centres on their hinterlands, that Nigerian regional planning should be integratively urban-rural oriented. Any investment in urban centres should have due regard to this vital function which they can perform for their spheres of influence. The contiguous urban-rural functional region should be recognized and planned as a unit.

In the more distant rural places in Nigeria Ajaegbu (1975) suggests that prohibitive effects of distances are experienced in these rural areas located far away from the major urban market centres. It is also observed that even though some of these rural areas may be more favourably located than others, growth and development is more efficient at the rural areas located near the urban-rural development thrust matrix. This argument could as well be extended to rural areas which could be classified as special project areas such as The Bacita Irrigation Area, the Delta oilfield areas, and the tobacco producing areas of Oyo State. It must be emphasised however that the favourably located rural areas derive increased growth stimulus and orientate their economies to the urban market demands. The distant rural areas suffer from diseconomies of long time-distances as well as of large money and effort-cost-distances from the developmental foci.

The aspect of the people's responses or adjustments to the discouraging effects of distance from the consuming markets has been the growth of a valuable middle-men system in marketing local products. This marketing system flourishes, although many rural inhabitants in Nigeria today still produce on a very small-scale, for eager and active middlemen buy up the products from the rural inhabitants who produce them, and then bulk and take the goods to the markets, particularly the urban centres, for sale. Thus, the beginning of a system suited to a more advanced economy is already to be seen at work. The process may be thought of as an effort by the middlemen, who have the capital to operate on a relatively larger scale, to make a living out of both buying and selling, and by transporting the goods from the production points in the rural areas to the consuming centres.

This is a desirable state of affairs in rural areas where the small-scale of production does not justify the farmer going to market himself, since little or no profit would thus accrue to him. At best this marketing system as it is now being operated ensures that the products are eventually evacuated from even the remote rural areas and sent to the consuming or grading centres. At worst, because the highest market prices are today obtained mainly in the urban centres, it could transfer an undeniably high proportion of the profit on the products to the middlemen, while the actual producers receive relatively little for their goods.

The people may therefore suffer from disincentives in their economic pursuits. Moreover, because of the long distances the amount of information which they receive, either concerning market conditions at any time or in the shape of new ideas and market is often a mere smattering and does not usually lead to acceptance, adoption and eventual rural development. In consequence, the gap in economic and social development between the immediate rural hinterland areas and the more distant rural districts is indeed wide; one of the major problems of regional and/or economic planning in Nigeria today is how to bridge this gap.

The need thus arises for providing several rural centres of information and innovation which will be proximal enough to ensure effective interaction and development. To this end, the existing rural central places could be identified and equipped with the necessary infra-structures to influence their surrounding areas. This suggestion envisages a network of urban-rural centres covering the whole country which will influence the rural areas and lead to effective regional development. These centres and their immediate zones of interaction would constitute the spatial units for integrated development planning. The arguments here premise that the selection of these units should be purposeful and rational, leading down to rural villages at strategic localities. Each will be provided with basic amenities and accessibility. Basic facilities such as primary schools, dispensaries as well as footloose services such as training colleges, and agro-service centres may also be sited there.

## INTEGRATING RURAL AREA DEVELOPMENT PLANNING INSTITUTIONALLY

The geographical integration of planning is incomplete without an institutional framework. The existing institutions for rural development activities consist of the Ministries, parastatals, semi-government and private agencies at the State and Federal levels which have to do with education, co-operative and community development, local government, health and agriculture. Often each of these bodies operate their variously defined objectives as though the others do not exist. By their very nature they function independently within their programmes of development. The result is an haphazard and uncoordinated development planning. Four basic ideas should underlie an integrated rural development which is institutionally oriented.

(i) The state functions which touch intensively and directly on the rural areas should be decentralized;
(ii) The strengthening and re-organisation of the existing rural institutions;
(iii) Integration and dove tailing of plans for rural development; and
(iv) Involving the villages in the conception and execution of development projects.

With all four of them it is possible to maintain an orientation towards the rural poor, at least as far as small farmers are concerned. The more restricted the approach, the less the chances are of involving the lower groups, the under or unemployed and the landless people and making them part of the on-going development process.

A number of other basic principles, however, have to be taken into account. Within the conventional planning process, investment allocation is very much based on considerations of how to improve production geared to the existing farm structure, landuse patterns, land ownership, infra-structure, marketing facilities, and, usually, receive very high priority with the argument that social improvements can come later. While this approach has some merits in many cases, particularly where no serious population pressure and a reasonable distribution of land exists, it leaves out two essential points:

(a) the mobilization of human resources, their motivation, their initiatives and their technical know-how matters most in making full use of the scarce natural local resources. If people are motivated, then the provision of new inputs and perhaps low level technology becomes a minor affair of much larger effect than it would be if the provision of new technology is the main effort.

(b) this, however, has its limitation where the majority of the people have no access to land or other enumerative employment so that their creative and productive potential cannot be utilized. And this is the case in most parts of Nigeria.

It is striving toward self-reliance and local level planning of motivated people with perhaps minor inflow of outside technology or investment capital which matters, but which can be seriously constrained by the existing socio-economic structure and distribution of power and leadership.

While people can best be motivated if they see the benefits actually accruing to them from within their own areas, stipulated by their own local organizations, a distribution of power and resources can hardly be organized at the local level, but needs a government decision. Therefore, local level performance and government decisions have to go together, not only in respect to access to land.

In the final analysis, a re-orientation of the strategic approach giving greater emphasis to the involvement of the rural poor, to the mobilization and development of the human resources, to the areas as an operational unit and keeping the notion of self-reliance in mind, has far-reaching implications which can be dealt with only if a number of supporting policies become an integral part of the entire policy concept.

Integrated rural development aims at a process of socio-economic change in rural areas based on the mobilization of local natural and human resources.[16] This process is characterized by the existing inter-relationships between the objectives for (a) increased agricultural production, (b) improved income distribution and employment opportunities, and (c) improved conditions of living and social integration.

In order to initiate this process the governments must set a policy which brings a more dynamic tune in the rural society by greater involvement of the rural masses and small producers in the production process and improved opportunities for people's participation stimulated by access to resources, services and employment. The aim must be equity and broad based development.

Regardless of whether one is representing the government itself, a national adviser to the government or an international organization, the implementation of the broad-based concept of integrated rural development requires three different sets of activities which are (a) promotion of the concept, (b) identification of basic policy issues, and (c) actual implementation of specific tasks.

Each new idea has to be promoted and made known to the people concerned. If it is a government leader who wants to go along the lines of integrated rural development, he has to seek support for the objectives underlying the concept and the consequences implied, particularly in regard to structural changes concerning distribution of power, resources and income. But he has also to point out the long-run implications if such changes do not take place in due time.

Besides the basic decision towards a poverty-oriented approach and socio-economic integration of the rural masses, consistent policy framework must include at least the following four major issues to make the whole concept operational.

### ACCESS TO MEANS OF PRODUCTION AND EMPLOYMENT

This includes such questions as agrarian reform, rural industries, the role of the rural economy as a whole, as well as investment decisions.

Considering the fact that the majority of the population in Nigeria still lives in rural areas, it is most important to make use of their productive potentials, but also to give them a chance to become creative and self-employed, the latter being most crucial because it includes also motivation and acquisition of appropriate skills. The above statement implies a fundamental aspect : people must have access to means of production, otherwise they cannot make use of their creative potential.

Therefore, some form of land reform combined with state assistance to help people to invest in small-scale non-agricultural activities are essential. Both, however, require a whole programme of supporting activities such as training, delivery of material and inputs. In the case of non-agricultural activities, people will have to be made aware of appropriate activities and the possible employment effect of divisions of labour within a community. However, this is more likely to succeed if it can emerge together with the establishment of rural development centres.

On the other hand, the employment capacity of large-scale private land holdings must remain relatively low. There is not only the intention to make profit, but there is also a must to make profit in order to maintain the production process, to cover losses and risks, to fulfil tax obligations and to reinvest in order to maintain or improve the capital stock. Under the conditions of high population pressure on land, a poverty-oriented approach to rural development is in clear contrast to large private/public agricultural enterprises, since a high level of "granted employment" cannot be generated or maintained over a long period of time. Salaries always have the tendency to increase, which needs to be matched by increasing labour productivity, either by increasing output or reducing labour per unit of land.

Likewise, generating employment directly through investments in industrial jobs will not create the necessary labour-absorption capacity at reasonable costs. Investing in structures or projects in which labour can "move in" and stay will not constitute a success in the long run.

What is required is an employment generating initiative by the individuals themselves which requires that they have access to means of production, land, credit, know-how, incentives and a basic infra-structure supporting the production process. In the earlier phases of development, the employment generating effect will be realized only on small-scale to family size adopted units.

However, for a number of reasons, the labour intensive economic structure composed of agricultural and non-agricultural activities in rural areas will not survive if there is a parallel strategy of an urban-industrial expansion. Therefore, the political decision to be made is to give the rural economy as such a prominent role in the overall strategy.

## POPULATION POLICY AND THE ROLE OF WOMEN

The growth of rural population and the availability of productive resources are two sides of the same coin. A growing population needs to have a growing output. As far as agriculture is concerned, this output is limited by accessibility to land and can be increased only by improved technology. However, gains in output through new technology are usually slower than growth in population. Furthermore, the introduction of new technology happens usually only in certain already more productive areas, while the growth in population is taking place unilaterally, causing increased economic disparities and different levels of living among geographic areas. This then stimulated internal migration with negative effects on the family structures as a base for social stability, depopulates backward areas and worsens the situation in urban slums.

It is sometimes argued that population growth stimulates also economic growth or that over population is not necessarily a cause for food shortage Singh however it is a fact that improving the generally low level of living and shortage in food supply will be more difficult with an increasing population than with zero growth. One also has to take into account that already the present youth-age structure in the rural areas, a long-run rural development strategy, therefore, will have to take into account very seriously the population component. However, past experience has shown that population and birth control by itself does not work simply because it cannot be expected that a family decision is made on grounds of what is good for the nation as a whole is also good for the family. It is not the macro-economic indoctrination to birth control which will work, but rural families must be induced to smaller families by becoming aware of the benefits which they themselves can experience from fewer children. This, however, will not happen if nothing changes in rural life style. At least two major elements have to be realized. First of all, the economic value of children as labour force and as a base for social security have to be reduced which by itself implies a whole range of complementary activities. Secondly, it is particularly the women for whom new and more attractive prospects outside the family need to be created. The association with activities outside agriculture is what matters and which will create the incentive for smaller families apart from the fact of material gains.

## ADMINISTRATIVE DECENTRALIZATION

The actual handling of broad-based rural development is ideally not a matter for the central government, but it can observe and evaluate its implementation by an appropriate monitoring system. Local initiative, local level planning and a responsive institutional and administrative system is required. For most governments, this implies major revisions of the traditional system of planning and execution, new types of local organization, different patterns of decision-making and communication, etc. Administrative decentralization, regional planning and local participation are the crucial elements in implementing an integrated approach to rural development Ajaegbu. Without doubt, these are major political decisions in order to move in that direction, but they are essential prerequisites if a government wants to pursue a policy of developing the local resource potential.

Although any effort need to be part of a larger development plan and must be supported by the overall political objectives, actual implementation of broad-based rural development must be based on the area level because :

It is only at the area level possible to arrive at operationally feasible programmes, derived from a thorough analysis of the area resources potential and the functioning inter-relationships between production structure, social structure and administrative structure, so that workable measures can be found for integrating the various social groups in the development process.

It is also here that the economies of scale and the complex nature of poverty-oriented programmes require a certain coverage of geographic areas and population.

Finally, the mobilization and motivation of people is more likely to be achieved at the area and local level. It is here too that can be created the local linkages to other institutions so that a vertical and horizontal integration can develop.

One of the elements of decentralization will be the promotion of local grass-roots organization and functional groups who could effectively articulate their demands and improve their bargaining position within the rural power structure.

## REFERENCES


1. ILO, *Socio-economic Condition in Ife, Otta and Ilaro Districts of Western States of Nigeria*, Ibadan, 1970.
2. P.J. Smailes, "Metropolitan Trade Shadow", *Tijds voor Ecom. en Soc. Geog.*, 1969, Vol. 60, pp 329-335.
   G. Moseley, The Impact of Growth Centres in Rural Regions *Regional Studies*, 1973. Vol. 7, pp 57-94.
   G. Woolmington, "Centralise or Decentralise" Institute of Australian Geographers, Urban Studies Group, Symposium held at Flinders University, February 1973, p 3.
3. Uyanga, *Towards a National Urban Policy for Nigeria*, Ibadan, Ibadan University Press, 1977.
4. S.M. Essang, "The Impact of Marketing Boards on the Distributions of Cocoa Earnings in Western Nigeria". A paper presented at the NISER/Department of Economics Joint Seminar, University of Ibadan, March 1971.
5. G.K. Heleiner, "A Wide Ranging Development Institution" *Nigerian Journal of Economic and Social Studies*, 1954 Vol. 6, pp 98-123 and 239-259.
6. D. Olatunbosun, *Nigeria's Neglected Rural Majority*, Ibadan Oxford University Press, 1975.
7. U. Lele, "The Design of Rural Development : Lessons from Africa" IBRD., Johns Hopkins University Press, 1975.
8. V.W. Ruttan, "Integrated Rural Development Programmes : A Skeptical Perspective", *International Development Review*, 1975, Vol. XVII, No. 4, pp 9-16.
9. T.W. Schultz, "A Framework for Land Economics—the Long View" *Journal of Farm Economics*, 1951, Vol. 33,. p 204-215.
10. OIL, Office Mission, Report on Field Visits for Selection of Suitable Project Areas on Rural Employment Survey, Ministry of Labour and Social Welfare, Ibadan, 1965.
11. H.I. Ajaegbu, "Regional Planning and the Rural Areas in Nigeria" in K.M. Barbour (ed) *Planning for Nigeria*, Ibadan, Ibadan University Press, 1972, pp 100-109.
12. C. Camar, "L'organisation de la space geographique par les villes dans le pays Yoruba" cited by Ajaegbu (1975) op. cit., 1969, p 37.
13. M.J. Mortimore and J. Wilson, "Land and People in the Kano Close-settled Zone", A.B.U., Zaria, Department of Geography, Occasional Paper, No. 1, 1965.
14. D.W. Norman, "An Economic Study of Three Villages in Zaria Province", A.B.U., Zaria, Samaru Miscellaneous Papers, Institute of Agricultural Research, 1967.

15. N. Singh, "Overpopulation is No Excuse" CERES, FAO. *Review on Development*, 1976, No. 51, Rome.
16. H.R. Kotter, "Some Observations on the Basic Principle and General Strategy Underlying Integrated Rural Development". *Monthly Bulletin of Agriculture and Statistics*, 1974, Vol. 23, No. 4, FAO, Rome.

# 13

# Systematizing Spatial Organisation in Tropical Africa—An Integrative Conceptual Appraisal

—*Thando D. Gwebu*

THE ELIMINATION of contemporary problems of underdevelopment in the Third World is very much contingent on spatial reorganization. Relocation of administrative centres, construction of an efficient communication and transportation infrastructure, revision of obsolescent land tenure systems and initiation of various regional development schemes are but a few visible attestations to this premise.

Concern with these and related problems has been the object of the social sciences in general with geography taking the lead in the resolution of those specifics involving man and space. Hence geographic research and literature on such crucial topics as migration, development and urbanization has grown to unprecedented magnitudes during the last decade. It is unfortunate though that the inter-relatedness among these three processes has been overlooked. Policy-wise, this has often resulted in the fabrication of what others might characterise as a system of apparent wisdom in the folly of hypothetical delusion. A completely sterile and self-defeating policy invariably ensues from a poor understanding of these three sub-systems and their linkages.[1]

The aim of this paper is, therefore, to articulate these components as sub-structures that are dynamically interactive.

Helpful comments were made on the original drafts of this paper by M.E. Harvey, S.M. Bhardwaj and Bari J. Epstein, all three Professors of Geography at Kent State University.

Such a task requires first, a clear delineation of what each sub-system involves, and secondly, an analysis of how these three sub-systems are inter-connected. In order to achieve this task the paper is divided into five sections. The first deals with the migration sub-system. Basically, this is a brief synopsis of migration determinants. The second is an overview of urbanization. The third highlights the development subcomponent according to alleged causes and examines some of the proposed solutions to the problem of underdevelopment. Because of the relatively divergent academic views on this sub-component, its coverage is perforce comparatively more extensive. The fourth section is essentially synthetic. It details the macro and the micro components of space economy. The fifth and final section is a conceptual overview which constitutes a framework for the operationalization of this systematization.

## THE MIGRATION SUB-SYSTEM

Conventionally defined, migration is the permanent relocation by an individual or a group of individuals from one administrative unit to another. This research will however employ the term as a 'convenient catchall categorization' of all forms of intra- and inter-regional moves irrespective of their duration This is appropriate within an African context where rural to urban migrants, as a rule, treat their home as an alternative destination should an urban life fail. This is demonstrated, for example, by occasional visits, monetary remittances or even maintenance of what are regarded as traditional roles in the home area even after an individual has moved to an urban environment. From a policy view-point our pre-occupation should be with the determinants of migration because once the causes are identified we could either stem the migration tide, attempt to reduce it, or redirect it.

### The Causes of Migration

Cross-sectional studies have failed to provide either intelligent or intelligible propositions on causes of migration.

This is basically due to the glaring oversight afforded by a historical perspective which falsely assumes the present to be isolated from the past. Fortunately some researchers have been aware of the inaccuracies contingent upon cross-sectional approaches. Harvey in his analysis of social change and ethnic relocation in Sierra Leone[2], adopted a longitudinal perspective that was sensitive to the traditional, precolonial situation, and the colonial situation. His latest paper dealing with inter-regional migration studies in tropical Africa identified the tribal-based, traditional state-based, and colonial related migration typologies before considering contemporary migration types and processes.[3] Tribal-based migration was attributed to either deteriorating resources to population ratios or fear of enslavement by a militarily superior enemy. The traditional state-based phase is closely associated with the rise of both centralized and non-centralized states of post-eighth century Africa. Migration during this period was basically due to inter-regional trade, resources to population ratios, persistent fission and fusion of states and the Islamization process. Inter-regional migration during the colonial period was due mainly to unconformities in the socio-economic surface. We should add here that initially the human response to these regional disparities was not voluntary. We will now focus on migration causes in present day Africa.

Contemporary Western theories on migration have attributed this process basically to stressors or push factors at the migrant's point of origin and the pull or superior place utility at the point of destination. The typical push factors that are usually cited are environmental deterioration, unemployment, or some from social persecution or ostracization. The main pull factors include, better socio-economic facilities at the destination and all those factors aggregately designated under the "psychological bright light" concept.

Several models have been devised to try and predict the number of migrants from origin to destination areas in tropical Africa. Perhaps the most ambitious is Todaro's because it is supposed to be sensitive to both the purely economic and non-economic determinants of the migration process.[4] As such, it will be reviewed here for illustrative purposes. His basic model of rural to urban migration hinges on the differential in expected incomes between two developmental sub-systems; the rural and the urban sub-components.

At its crudest form, the model is :

$$\frac{M(t)}{L_u} = f[V_u(t)] - V_r(t)\, V_r(t)^{-1}$$

In words this means that the net number of rural to urban migrants scaled by the size of the urban labour force, $M(t)/L_u$, is a function of the difference between $V_u(t)$, the present value of the expected urban real income over an unskilled worker's planning horizon and $V_r(t)$, the discounted present value of the expected rural real income stream over the planning horizon.

The basic assumption of this model is rational maximization of place utility based solely on income. This model does not take into account the barrier effects such as physical and socio-psychological distance which is likely to be encountered should the move indeed eventuate. The size and therefore complexity of the destination in terms of possible job opportunities is also ignored. Moreover, the basic notion of demographic self-selection based on such factors as educational attainment, age and sex characteristics of the migrants are ignored. Finally, the model is incognizant of the importance of either previous moves by the migrant or the catalytic effects attributable to the presence of members of the same ethnic group at the migrant's potential destination. A modified version of his model took some of these considerations into account. The linear model was of the form[5]

$$M^E_{ji} = a_i\,(V^E_i(t) - V^E_j(t)^{-1} + a_2 D_{ji}(t) + a_3\, C^E_{ji}(t) + a_4(A_i(t)$$
$$(N_i\,(t)^{-1}\, A_j(t)\,(N_j(t)^{-1} + a_5\, N_i(t) + a_m\, T_{ij} + e$$

Where $M_{ji}$ is the percentage of the relevant population in area j that migrates to centre i;

$V_i$ is the expected real-income stream in urban centre i;

$V_j$ is the expected income in a typical rural area j;

$D_{ji}$ is the cost of moving from area j to centre i;

$C_{ji}$ is the measure of clan contacts from area j already resident in centre i;

N is the number of people in the relevant population;

A is an index of available amenities;

T indicates ethnic affiliation; and

E is a superscript indicating that functional relationships exist within a group having achieved a particular level of educational attainment.

Despite the exclusion of sex-selectivity, it must be admitted that Todaro's model is very fascinating and plausible. It however raises certain operational problems. The main difficulty is data. Of course, one could, easily resort to aggregate data. Nevertheless this would plunge one into an intellectual quagmire of ecological fallacy. The second problem relates to the *ex post facto* nature of the model. Two fundamental theoretical difficulties stem from analysis of this generic nature. The first is that of *ex post* justifications made after the move whilst the second relates to dissonance reduction which is a function of the time lapse between the move and the survey interview. To circumvent these conceptual problems it seems desirable to adopt an *ex ante* or prospective model which relies on disaggregate data for its analysis. Not only would the later approach eschew the thorny theoretical problems mentioned above; it would also be crucial policy-wise making it feasible either to halt or divert moves as dictated by the systematic regional policies. Briefly, what is really required is a shift in emphasis from *ex post* models which assume optimization to *ex ante* models which focus on satisfying.

## THE URBANIZATION SUB-SYSTEM

The colonial era generated schisms or geographical disparities in employment and infra-structural availability. Developments of plantations and urban centres spelt regional inequalities in the distribution of educational, medical, entertainment and occupational opportunities. The bulk of migration in Africa is responsive to what the urban centres have to offer. It seems appropriate, therefore, to focus on urbanization so as to understand its implications in contemporary spatial dynamics.

There seems to be no concensus as to whether the concept of city-building should be attributed to diffusion from Mesopotamia or to independent origins. However, the definition of a city in terms of size, population density and heterogeneity and preponderance of non-agricultural specialists is accepted by Western Scholars. The process of contemporary urbanization is interpreted as involving a shift from purely agricultural forms of living to those forms mainly associated with towns.[6] Table 13.1 is a 1969 U.N. survey table, extracted from Herbert. It summarizes the general global urbanization trend.

As yet the processes underlying urbanization in non-Western countries are not very well understood although it is generally agreed that they appear to be different from those operating in the Western countries. For example, despite a long history of urbanization in many underdeveloped countries most of their populations still remain rural. This is evident in Table 13.2 which shows a predominance of non-urban population in Third World nations.

TABLE 13.1

*Aggregate Global Urbanization Trends*

| Year | Per cent in cities 20,000+ | Per cent in cities 500,000+ |
|---|---|---|
| 1920 | 14 | 5 |
| 1940 | 19 | — |
| 1960 | 25 | 12 |
| 1980 | 32 | 17 |

*Source* : Herbert, D., *Urban Geography* : *A Social Perspective,* Praeger Publishers, New York, 1973, p 32.

During the colonial era, most of the development infrastructure was concentrated in the alien enclaves of the colonies, particularly the urban areas. In tropical Africa these urban centres have witnessed mercurial population growth due to natural increase and especially, migration (see Table 13.3). Basically, these cities have, by and large, remained as ; rowth poles that are very well-integrated with the colonial

TABLE 13.2

*Global Urbanization by Region Over Years*

| Area | 1920 | 1940 | 1960 | 1980 |
|---|---|---|---|---|
| Europe | 32 | 37 | 41 | 50 |
| North America | 38 | 45 | 57 | 61 |
| Oceania | 34 | 38 | 50 | 50 |
| USSR | 10 | 24 | 36 | 46 |
| East Asia | 7 | 13 | 20 | 28 |
| South Asia | 6 | 8 | 14 | 21 |
| Latin America | 14 | 19 | 32 | 41 |
| Africa | 5 | 7 | 13 | 18 |

*Source*: Herbert, D., *Urban Geography: A Social Perspective*, Praeger Publishers, New York, 1973, p 28.

and former colonial field of economic forces. Socio-economic polarization with their geographical hinterlands and parasitism has been their distinguishing hallmark. Worse still, the profits they have generated have not been ploughed back, as it were, to improve and dynamize the basic/non-basic relationships. As such, concepts like the multiplier effect seem to have no relevance for these cities. In brief the saga of Third World urbanization portends an unhappy ending.[7]

The bulk of the research on urbanization has preoccupied itself primarily with squatter settlements, slums and the entire spectrum of those problems that are associated with over-urbanization. The persistently melancholy conclusions have alerted some governments to take a hard look at the *laissez faire* paradigm which cherishes unequal development including the whole belief-system which views urbanization as an appropriate tool for forging rapid economic development.[8] Perhaps what is really needed is an integrative approach which pivots the problems generated by migration and over-urbanization within the framework of a dynamic developmental formulation. It is for these reasons that this investigation will underscore the importance of the links between migration, urbanization and development. The salient components of the development sub-system have to be outlined before the synthetic structure composed of migration, urbanization and development is articulated.

TABLE 13.3

*Pre-war and Post-war Population of Selected Cities of Tropical Africa*

| City | Country or Territory | Pre-war(a) (b) | Post-war a (b) | Percentage Increase |
|---|---|---|---|---|
| Abidjan | Ivory Coast | 10 (1931) | 251 (1963) | 2410 |
| Accra | Ghana | 70 (1931) | 522 (1966) | 645 |
| Addis Ababa | Ethiopia | 150 (1959) | 560 (1965) | 274 |
| Bamako | Mali | 20 (1931) | 165 (1965) | 725 |
| Bangui | Ubangi Shari | 14 (1937) | 111 (1964) | 693 |
| Brazzaville | Republic of the Congo | 40 (1937) | 136 (1962) | 240 |
| Conakry | Guinea | 7 (1931) | 175 (1964) | 2400 |
| Dakar | Senegal | 93 (1936) | 375 (1961) | 303 |
| Dar-es-Salaam | Tanganyiki | 34 (1937) | 190 (1965) | 458 |
| Douala | Cameroun | 28 (1931) | 200 (1965) | 614 |
| Freetown | Sierra Leone | 55 (1931) | 148 (1966) | 169 |
| Ibadan | Nigeria | 387 (1931) | 627 (1963) | 62 |
| Kano | Nigeria | 89 (1931) | 295 (1963) | 231 |
| Khartoum | Sudan | 45 (1938) | 173 (1964) | 234 |
| Lagos | Nigeria | 126 (1931) | 665 (1963) | 420 |
| Kinshasa | Republic of the Congo | 36 (1938) | 508 (1966) | 1311 |
| Luanda | Angola | 40 (1934) | 224 (1960) | 460 |
| Monrovia | Liberia | 10 (1938) | 81 (1962) | 710 |
| Nairobi | Kenya | 65 (1939) | 266 (1962) | 309 |
| Salisbury | Southern Rhodesia | 33 (1936) | 325 (1965) | 884 |

*Source* : R. Weitz and Y.H. Landau (eds.), *Rural Development in a Changing World*, (Cambridge; Massachusetts, The M.I.I. Press, 1971), p 176.

(a) Figures in parentheses are the year of census or estimate.

(b) Pre-war data from Walter Yust (ed.) *Encyclopaedia Britannica World Atlas*, 1942, pp 182-211, except for Abidjan, Bamako, Conakry and Douala, which are from France, Minister d'Outre-Mer, 1957, p 27, and Accra, which is from Gold Coast (Ghana), Office of the Government Statistician, 1950, *Census of Population*, 1948, Report and Tables.

(c) United Nations, *Demographic Year Book*, 1966, pp 176-178.

## THE DEVELOPMENT SUB-SYSTEM

Two interesting approaches have been adopted in the study of development. The first describes the determinants of development, whereas the second explores ways on how development can be achieved.

Studies of the former type fall into two broad and divergent categories—the cross-sectional and the longitudinal types. Typical of the former are those studies which tend to relegate the underdeveloped countries into a low equilibrium trap characterized by a vicious cycle of poverty such as that depicted in Figure 13.1 Writers of the cross-sectional persua-

*Cycle of Poverty*

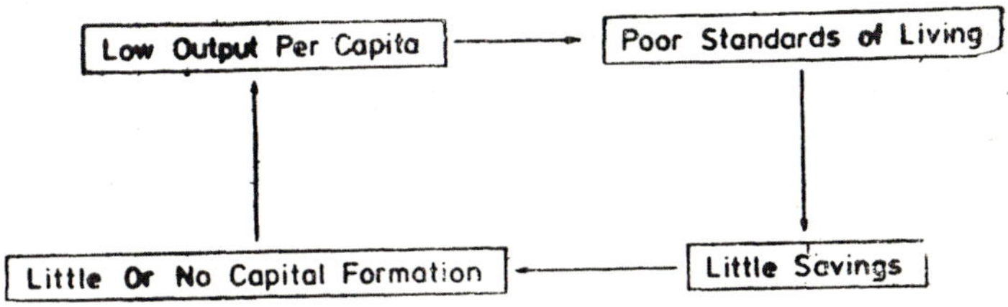


FIG. 13.1

sion tend to emphasize either specific physical or socio-economic-cultural variables and their roles in under-development For example, environmental factors such as enervating climate pathological determinants and the general fragility of the tropical ecosystem have been propounded by structuralist-oriented writers such as Gourou.[9] Cultural determinists such as Talcott Parsons and his pattern variables,[10] and Weber with his ideas on the protestant ethic and its relationship to Capitalism[11], have appealed to cultural idiosyncracies as a basis for the rationalization of differential development. McClelland[12] has attempted to link culture and economic behaviour patterns in a society. His arguments have been restated rather succinctly by Pareek.[13] A comprehensive reappraisal of the structural studies is critical because most of the previous development formulations have embraced their premises. Only Pareek's work will be briefly reviewed here because it typifies the cross-sectional treatment of the development problem. Briefly it regards under-developed societies as closed systems whose untoward fate is endogenously predicated (see Figure 13.2).

*A General Paradigm of Social Behaviour*

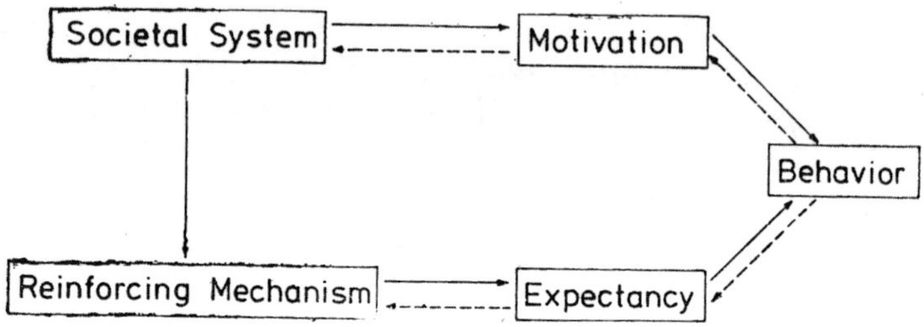


FIG. 13.2

According to Pareek[14] societal behaviour is assumed to be a function of motivation and values which are in turn determined by the given social system. The reinforcing mechanisms which operate at the societal level are, in the main, responsible for the institutionalization and routinization of those processes which produce reward and punishment systems. They include child-rearing practices and the school system. These institutions and mechanisms, inculcate behaviour which is evaluated as either desirable or undesirable. He maintains that motivation and expectancy are both important because they directly determine and cause behaviour and that they influence the social system and its reinforcing mechanisms. Pareek then views poverty as a structure and a product of a given societal system (see Figure 13.3). Supposedly poverty causes low need for achievement, low need for extension, and high need for dependency. Such motivational patterns cause behaviour characteristic of what he calls cultures of poverty.

Academically this is all interesting. However, it raises eertain fundamental questions which beg for substantive answers, particularly with regard to motivation and its cross-cultural operationalization. For example, how does one measure need or concern for excellence, success and failure cross-culturally. Pareek has this to say on this score :

. . . .persons with a high level of achievement motivation

*Paradigm of Culture of Poverty (Pareek)*

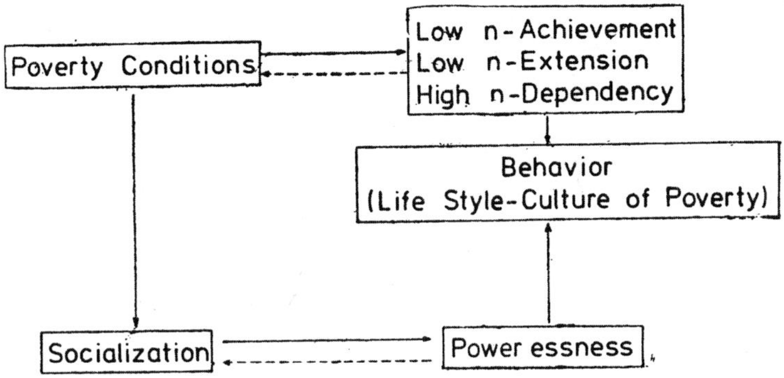


FIG. 13.3

have been studied in detail and have been found to have characteristics for entrepreneurs (p 305).

He is obviously thinking of those economies where competition rather than cooperation is stressed. In those societies where there is imperfect competition the need for achievement would definitely be manifested in a fundamentally different way.

Extension need refers to concern for the common welfare of all and is reflected by genuine concern for other persons, unselfishness, and cooperation for the purpose of achieving a common goal. His argument that extension need is low in poor countries has to be cross-checked against concrete reality. For example, despite the cut-throat competition encouraged by imperialism with its associated emphasis on personal aggrandizement and greed, most of these poor countries still honour communalistic and extended family traditions.

It would be reasonable to admit that apart from the several categorical and usually eurocentric socio-anthropological studies, little has been done to try and understand the value systems of Third World countries. In many instances uncritical trans-cultural operationalization of value indices has had unwarranted, naive, although plausible academic imputations. That is why scholastic refutations[15] of these claims might furnish fresh insight about development.

Structural rationalizations of under-development are still held with esteem in academic circles. However, their inability to provide insightful answers to the basic tenets of development has generated unprecedented intellectual ferment among both the Western and the Third World scholars.[16,17,18,19,20,21,22] First, is it really valid to claim that underdeveloped countries are closed poverty systems ? Secondly, if for some reason they are not closed systems, are the prominent and scholarly explanations of under-development accurate ? Finally, have these countries been always poor, if not then what are the true rationalization of their predicament ? The scholars who are earnestly probing into these issues interpret geographic space of under-developed countries not as a detached static unit but as a dynamic system reflecting "a multiplicity of forces ..... external and internal, past and present, political, economic and social."[23] As such, they argue that an intelligent analysis of the development of these countries lends itself best to conflict analytics rather than to consensus theorizing.

Consequently a whole range of concepts such as development, modernization, growth, dual economies and several others, which are related to socio-economic change, is under scrutiny. McGee[24] has, for instance questioned.

> ... the one-sided emphasis on the investigation of the process of development as a penetration of traditional systems by elements of modernization which are thought to be equated with development...

David Slater[25] has likewise reassessed the theoretical underpinnings of "development geography". His contra-puntal analysis has focused on spatial differentiationist, spatial diffusionist and spatial integrationist perspectives of development geography. It is instructive to reiterate some of his views because they might offer alternative insights to the study of development. Unlike the structural approaches we have discussed above, these reflect a relational and dynamic approach to the problem.

**Spatial Differentiation Approaches**

Slater has observed that the concept of "dual economy" has furnished a theoretical pivot for the spatial differentiation studies. He traces the roots of this concept to sociological research where it was originally employed to designate a cleavage between imported societal norms and those of the indigenous culture. The belief that the developed (imported) and underdeveloped (traditional) sectors had independent economic and social structures was tacitly assumed. Social scientists adopted this "dualistic" idea and imputed to it the dimension of a dynamic modern sector juxtaposed against a stagnant traditional sector. The currency of this belief entailed construing of development as merely involving a transference of the factors of production to the modern sector. Labour from the supposedly morbid sector was unlimited and always eager to enter the robust sector since the latter paid better wages. It was also assumed that the modern sector would generate a multiplier effect in terms of industrial and urban employment.

Drawing his evidence from Griffin[26] and Beckford,[27] Slater demonstrated that the thesis of unlimited and eager labour from a supposedly morbid sector is unfounded. Plantation economies, for example, have always been plagued by labour shortages. Initially imperialist governments had to resort to coercive tactics such as taxation or naked brutal force to make indigenous work in the modern sector. The validity of notion of unlimited labour being siphoned from the traditional sector with no adverse effects on the latter has to be judged from empirical historical evidence. Many underdeveloped countries now realise that the trend of rural to urban migration initiated by imperialists is more than just a nuisance, sociologically and economically. In Sierra Leone, for example the massive exodus of population from the rural areas to the diamond-mining fields has adversely affected the production of staple food, rice. Similar examples of negative effects of labour movement from the traditional to the modern sector have been recently cited in Nigeria.[28] Many other countries such as Zambia are currently trying to reverse population movement trends which were initiated during colonial times, through "go back to the land" programmes.

The modern concept of over-urbanization is an empirical negation of the relevance of economic notions such as multiplier effect in under-developed countries. In traditional economic base theory income generated by the basic sector is supposed to be utilized for the benefit of the given exporting city or region. Table 13.4 shows graphically the effects of colonialist expropriation of basic funds on industrial employment in specific underdeveloped economies.

TABLE 13.4

*Rate of Increase in Non-agricultural Employment for Selected Third World Countries*

| Country | Rate of Increase in Non-agricultural Employment % |
|---|---|
| Cameroons | —1.0 |
| Kenya | —0.5 |
| Malawi | —0.7 |
| Tanzania | —0.1 |
| Uganda | —0.9 |
| Zambia | —0.9 |

*Source*: D. Slater, "Geography and Under-development, *Antipode* 5, 1973, 22.

## The Spatial Diffusion of Modernization

The link between modernization and development in social research is evident from what researchers like McGee[29] have observed:

> . . . there is often a pejorative assumption that traditional systems are, to use a charitable interpretation, often rich culturally but bereft economically. That is to say they are not oriented towards the profit motivated efficiency philosophy of developed capitalistic societies.[30]

Orthodox social research certainly singles our particularism, ascription and functional diffusiveness as causes of underdevelopment. There appears to be a wholesale equation of the traditional practices of these countries with anti-thesis to development. Consequently, modernization becomes defined in terms of involving a total erasure of traditional mores and norms. Modernization defined from that angle becomes a *sine qua non* for development. Towns, railroads, highways, telephones, hospitals, churches, banks and all those other glittering devices from the West are usually employed as an aggregate index for development. In the final analysis provision of the natives with this infra-structure would modernize him and set him in the proper course towards development.

A few problems arise with this type of conceptualization. The theoretically undialectical and unidirectional diffusion of modernization is an empirical impossibility. Cultural interaction and fusion and "syncretic forms of behaviour and organizations" seem to be the rule when two alien cultures confront each other. Another issue which has to be closely examined is whether there was an effort to modernize the native in the colonial era. Now, a study of colonial spatial organization indicates conclusively that the so-called modernizing infra-structure was implemented for the smoother functioning of exploitation rather than the often cited inclination of philanthrophy, such as the advancement of natives. A glance at colonial land and infra-structural allocation in underdeveloped countries will confirm that it was metropolitan rather than native residents who benefited. It is common knowledge that infra-structural diffusion can only occur if the space economy is well-integrated. Colonialism was however notorious for generating resource-to-port-to-mother country infra-structure and thus leaving the rest of the region splendidly isolated.

Few scholars will deny that colonialism induced growth in its favoured enclaves, such as towns. There was however no serious attempt to reach out and upgrade the natives' life. De Souza and Porter[31] have thus truly remarked :

> Through the mechanism of deviation amplification, differences between core areas of the space economy and the periphery increased unremittingly. Growth without development appears to have been the hallmark of the colonial era. Spread effects of development were more than cancelled by backwash effects.

Despite its purported objectivity, the geography of modernization seems to have enhanced the colonial establishment and its concomitant glittering and indeed then novel paraphernalia. Its impoverishing, disruptive, and dehumanizing effect on the indigenous milieux is quietly and conveniently brushed aside and yet rehabilitation of these societies is contingent upon a correct assessment of both the negative and positive socio-economic impacts of what passes as an immacu late conceptualization, modernization. Unless social scientists embark on a critical attitude then their research in this area will not only be sterile but also undoubtedly reactionary. Sister Annette's reflection about presenting a balanced view about some of the impacts of modernizing spatial development should be considered seriously : [32]

> Why have we not described the negative hierarchy, the retrogression of functions which people experience as central places become dominant.

Finally, a wholesale and uncritical acceptance of so-called modernization or westernization by underdeveloped countries might turn out to be an unwise move in the final analysis. Slater's advice[33] on this score should not be dismissed lightly. He warns :

> It would seem, therefore, that rather than adopt modern institutions relevant to the needs of more advance societies, under-developed countries would be better advised to establish institutions and facilities relevant to their own problems, irrespective of whether or not they compare favourably with the existing institutions in the Western nations. Otherwise, adoption will simply lead to increased dependence and dependence is the anti-thesis to development (p 25).

## URBANIZATION DEVELOPMENT AND MIGRATION : A SYNTHESIS

Several important points emerge from the preceding discussion. First the space economy of colonialism was typified by polarization rather than articulation. Secondly, contemporary rural/urban migration owes its genesis primarily from forceful colonial attempts to transfer the factors of production from the traditional (rural) to the colonial (urban, plantation, mining) sector. Such population displacements have continued, and have been consistently reinforced by Westernization, a process which tends to educate the youths out of the traditional psychosocial milieu by inculcating wants and social roles which can only be satisfied or fulfilled in towns.[34] From this perspective any programme designed to solve the problems of under-development would be patently naive if it ignored the dynamic links among urbanization, migration and development sub-systems; they are mutually reinforcing processes.

The recursive relationship among the three subcomponents is sketched out below in Figure 13.4. What are conventionally designated "National Development Plans" have to be sensitive to the mechanics of this recursive system both from a macro- and micro-scale perspective. Basically the macro-level refers to the broad characteristic features of the space economy, whereas the micro-scale relates to the individual decision-making processes. The respective levels are outlined below.

*The Linkages between Urbanization Migration and Development*

FIG 13.4

**The Macro-scale Level**

For a long-settled area it can be rationally argued that the extent of urbanization at any given spatio-temporal point could be expressed as a function of previous immigrations, previous development inputs, previous urbanization impacts and barriers to the urbanization processes. Similarly, the proportion of migrants to that point could be defined in three ways; first, by how urbanized and developed it was in the preceding time periods; second, by the amount of feedback information to the origins; third, by the evaluation of the potential barrier effects such as anticipated physical and psycho-social effects. Finally, the extent of development could be approximated by examining previous development schemes, previous migrations, precedent urbanization and difficulties encountered in trying to achieve development. This implicit inter-relationship among migration, development and urbanization can be symbolically summarized thus :

$$U_t = \sum_{i=1}^{N} (M_{t-i}+D_{t-i}\ U_{t-i}+B_{t-i})$$

$$M_t = \sum_{i=1}^{N} (M_{t-i}+D_{t-i}+U_{t-i}+B_{t-i})$$

$$D_t = \sum_{i=1}^{N} (M_{t-i}+D_{t-i}+U_{t-i}+B_{t-i})$$

where, U is the level of urbanization at time t
M is the level of migration at time t
D is the level of development at time t
and B is the barrier effect at time t

**Spatial Diffusion of Development from Growth Poles**

Few ideas have had as much impact on national development policy formulation as growth pole theory. Perroux,[35] the acclaimed protagonist of pole theory, based it on a propulsive or motor industry associated to a set of smaller but complementary industries by a web of abstract economic backward and forward linkages. Complementary input and output relations among such industries would facilitate their mutual growth. In this case the relationship of the pole with its hinterland was secondary to inter-firm interactions. Perroux's subsequent writings indicate that he cognized dissimilarity between development and growth poles. In the case of development poles the relevance of the forward and backward linkages was geographically circumscribed. That is, the given industrial establishment had to generate socio-economic betterment of the peripheral population. It is critical to appreciate this basic distinction between growth and development pole because these concepts are neither interchangeable not necessarily complementary as is sometimes implied by certain scholars[35],[36],[37]. As a rule they are mutually counteractive. Milton Santos has proposed a two-circuit spatial typology[38] in an attempt to distinguish between those repercussions of growth on one hand and development on the other. Growth, as has already been intimated measures success within the domain of multi-national or corporate space. That is, it occurs within what Santos called the upper circuit. Conversely, development occurs within the perimeters of the lower circuit or communal space. Policies based on growth pole theory *ipso facto* preclude the socio-economic interests of the majority of the population where as those pivoted on development theory are predisposed towards retributive justice. Empirical verification of these propositions is furnished by studies such as those on Nigeria by Mabogunje[39], India by Misra and Sundaram[40], Latin America by Conroy[41] and Zambia by Dietz, van Haastrecht and Scheffer.[42]

The previous expose has dealt with three problem sub-areas requiring immediate attention in underdeveloped countries; migration, urbanization and development. Their separate treatment has hopefully served to highlight their respective salient features. It must however be evident by now that a need exists to integrate these subcomponents particularly if the aim is to achieve balanced and systematic regional development. The following section will, therefore; focus on these sub-areas not individually but rather as a coherent and dynamic unit.

## OVERVIEW OF THE CONCEPTUALIZATION

Figure 13.5 epitomizes the connecting linkages of the three sub-systems discussed above. The numbers in parentheses refer to circled sections of the flow chart.

It is assumed that there exists a factual economic milieu whose lineaments are a function of the interaction of government and foreign decision-making, cultural and environmental determinants (9). The space economy could be segmented into the rural milieu, plantation agricultural areas, mining

*A Scheme of Spatial Decision-making within the Behavioral Space Economy*

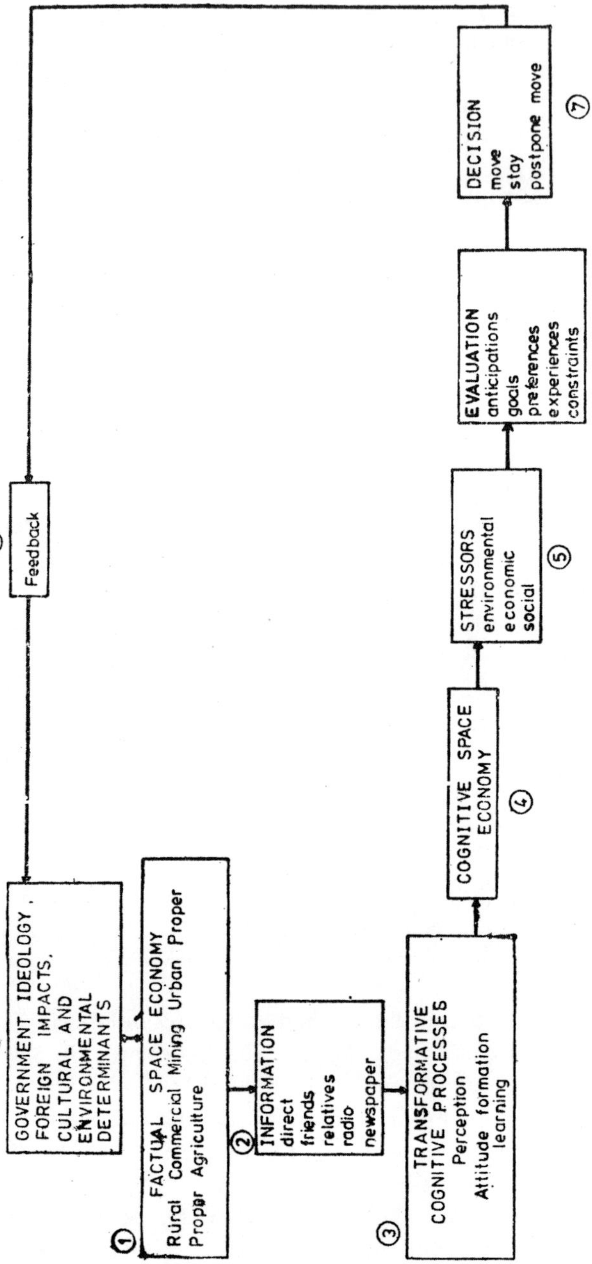


FIG 13.5

sites, and the urban sub-system (1). Each individual becomes exposed to respective components of spatio-economic reality through diverse information of channels. The latter may be either direct or indirect (2). Personal visits to the opportunity areas constitute direct forms whereas vicarious information is obtainable from either friends, relatives, radio, teachers or the newspaper (2).

For all intents and purposes, the factual milieu becomes important, in spatio-behavioral studies, in so far as it furnishes raw stimuli which subsequently undergo transformation by passing through the filter of the individual's cognitive apparatus which is composed of perception, attitude formation, and learning processes (3). The latter are acquired by the individual through the ordinary processes of maturation and socialization.

It is the subjective behavioral space economy (4) which biases individual spatial decision-making. From the subjective behavioral space economy certain stimuli tend to be perceived as being stressful. Such stressors may be of an environmental or socio-economic nature (5). They may be negative or positive. These characterize the perpetual dialectical flux between the already mentioned push and pull factors.

For illustrative purposes it has been assumed that stressors lead an individual to assess his location in space, at a given time. Evaluation (6) for the most part involves comparative place utilities. This may be regarded as being dependent on the individual's anticipations, goals preferences, experiences and sources of information. Each of these are often constrained by financial considerations. This conceptualization assumes that three mutually exclusive decisions are possible. When the cognized opportunities at the potential destination outweigh the inertial forces at the origin, a move may eventuate. However, if the inertial forces predominate the individual may opt to either stay at the origin permanently or temporarily until such a time that the move can be afforded (7).

**The Micro-scale Level**

The micro-scale environment is crucial to the destabilization of the dynamics of any socio-economic system. In the case of the underdeveloped countries it should be realized, for example, that problems pertaining to rural deterioration and over-urbanization derive basically from bright psychological lights generated by the newsmedia and formal education. Below is a detailed discussion of the basic ingredients of the important relationships between the cognitive or micro environment and the absolute environment.

According to Wood, the major division in geography should be between the phenomenal and the behavioral environments :[43]

> The former refers to the absolute or physical environment whilst the latter is a psycho-physical field in which the phenomenal facts are arranged into patterns or structures (*gestalten*) and acquire values in a cultural context.

The importance of the behavioral environment *vis-a-vis* spatial behaviour has been considered by Tuan in the context of topophilia[44] and by Sprout and Sprout in the context of psycho-milieu. According to the latter, this stereotypical or relative environment consists of images and ideas derived from some sort of interaction between what the individual receives selectively from his milieu; *via* his sensory apparatus, and his scheme of values, conscious memories, and subconsciously stored information.[45]

In recent years geographers have been trying to formulate meaningful theories *vis-a-vis* spatial structures, behaviour within those spatial structures and the behaviour of those structures.[46],[47],[48]. Analysis of spatial structures usually answers the question "What is there ?" It is preoccupied with ratiocination of areas, points and lines, relying heavily on mathematical measurements and the formal machinery of logic for explanation. Once the study of spatial structures becomes infused with the time dimension, to consider the change of these structures longitudinally, its subject matter consists of "behaviour of structures." Again, mechanistic processes have been the most frequented sources for the explanation of these structural behaviours. Such rationalizations have tended to minimize any sense of direct human involvement in these changes. However, the very fact that there usually remains a large portion of unexplained variance, when either mechanical or mathematical explanations are involved to explain spatial behaviour, attests to their deficiency in accounting for the behaviour of spatial structures. As such, a need exists to broaden the set of relevant variables to explain the behaviour in question. This involves the analysis of the behaviour within the structures. The argument then is that structural changes over time are a symbolic manifestation of human thinking processes; they are interpreted as a result of individual and group decision-making; an outcome of changing desires for interaction; and they reflect increasing information obtained by the systems members through cognitive processes such as learning, perception and attitude formation.[49] The vital point to emphasize here is the fact that cognitive filtering processes intervene between the phenomenal and the behavioral environment. This can be easily illustrated in Figure 13.6

Koffka's remarks[50] are relevant in the explication of the schematic dynamic relationship depicted by Figure 13.6.

*Diagrammatic Representation of the Cognitive Linkage between the Objective Environment and Overt Spatial Behaviour*

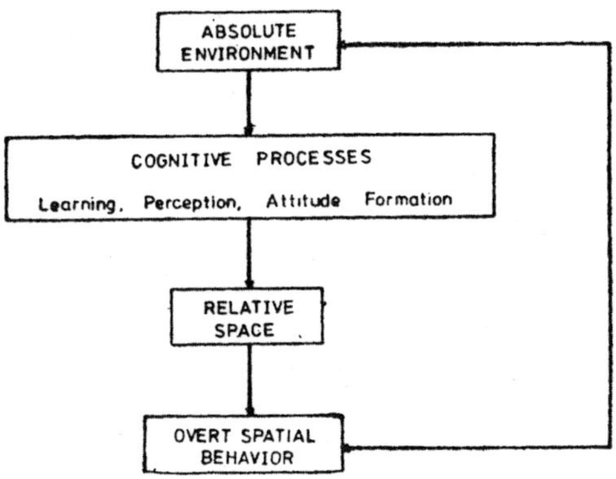


FIG. 13.6

Behaviour takes place in a behavioral environment, by which it is regulated. The behavioral environment depends upon two sets of conditions, one inherent in the geographical environment, one in the organism. . . . The geographical environment, not only the behavioral, is changed through all behaviour.

Finally, Ijomah's observation that "behavioral theories cannot be based only on logic. . . . contingency on the psycho-social world is imperative"[51] also points towards a cognitive rather than a simple stimulus-response approach suggested by behavioralist psychology in spatial theory formulation.

A need for spatio-behavioral studies on Africa cannot be over emphasized. Similar observations have been made by Harvey who recently remarked that there is a dire need to comprehend how individuals cognize and structure geographic space.[52] Mabogunje's excellent paper which focuses on migration and the micro-environment within the rubric of the general systems theory is a move in the right direction.[53] He, for instance, isolates socio-cultural controls which determine observed spatial behaviour of individuals in a hypothetical African Space economy. Also apropos are Gould's findings on "mental maps".[55] Apart from his empirical demonstration of the spatial egocentricism among selected students in the United States he also made interesting and parallel findings about the "mental maps" of some Ghanian and Nigerian students. Rather than review all of these studies, only his findings among-the students from Ghana will be reviewed because they are both instructive and most relevant for this conceptualization. Briefly, Gould observed that the main urban centres in Ghana such as Accra, Kumasi, Ho and Tamale, were the most attractive areas by virtue of their bright psychological lights. Regions of phenomenal economic prospects such as Sunyani were also rated high. In contrast, regions such as the barren Middle Belt were cognized as being unattractive due to environmental reasons such as poor rainfall, pathological reasons such as presence of tsetsefly and for historically related reasons such as the depopulation scars left by European slave-raiding. Now, the question can justifiably be posed as to whether the relatively more attractive areas, to the potential migrants, owe their enviable position to their effective propagandizing or simply to their site-situation idiosyncracies. Policy-wise, the answer is crucial. A footnote from Essang and Mabawonku's paper might throw some light on this issue.[55] They note that in the under-developed countries, urban-related news usually dominates news media such as newspapers. Such news-selectivity must have something to do with exhibited spatial-propensity-behaviour such as in rural to urban migration.

As stated above learning, attitude formation and perception are closely interconnected with the cognitive processes which condition spatial behaviour patterns. It is fundamental to note that all of these processes rely heavily on information. flow. The concept of migration elasticity also becomes important within the information and thus destination—familiarity context. Would it not be valid them to disregard the apparently direct link between the regional ratings and their phenomenal characteristics and concentrate instead on the evaluation of the relationship between cognitive factors such as familiarity of the stimulus-nodes and node-attractiveness ? Evidently these are contentious issues not even amenable to solution by lengthy and ponderous treatises. Canons of science suggest that such arguments be laid to rest empirically. Figure 13.5 constitutes an ideal basis for empirical investigations on this important issue.

The evinced spatial behaviour as already shown in the illustration has repercussions on those mechanisms which are directly responsible for both the macro-scale and the micro-scale components of the above conceptualization (8)(9). Possible impacts on the macro level will be discussed by illustration first. A government, previously disposed to *laissez faire* policies might find itself having to launch deliberate systematic regional planning due to say, overurbanization and rural depopulation. Important population movements within the colonial and neocolonial context have been discussed already. In each of these cases population movements to certain areas are engineered to supply semi-free labour for the production of goods which benefit the country providing the so-called foreign aid nowadays. A graphic illustration of this contention is furnished by Dietz, van Haastrecht and Scheffer in their recent analysis of local effects of two large scale industrial projects in Zambia.[56] Culturally, a given urbanward move might partly deethnicize an individual and serve as a basis for the evolution of a national rather than ethnic identity. particularly if a distinct exploitative urban based foreign out group is homologous to an upper class. In terms of deteriorating population to resources ratio a move from, say a highly pressured area might relieve the pressure somewhat. The impact of the move at the destination will depend on how long such rural to urban, plantation on mine trajectories have been operative. At their earliest phase, such as during the early period of colonialism, all the labour may be absorbed. However with increasing passage of time incoming labour may add to the reserve of unemployed, a symptomatic feature of overurbanization.

At the micro-level spatial behaviour is followed by an appraisal of the outcome of behaving (staying, moving, postponing the move) and is either repeated at future stimulus, modified in compliance with revised spatial cognition, or rejected as a suitable behavioral act.

## CONCLUSION

The above conceptualization has identified the following important aspects of the individual's reaction to the spatial transformation of underdeveloped countries in general and Africa in particular :

(1) The inter-relationships among the three components of space have to be evaluated and critically analysed before the spatial development processes can be understood.

(2) The individual's subjective evaluation and response to the imprints of migration, urbanization and development is the agent that initiates a given type of response. In the aggregate, then, each space economy at t, say, reflects the population's subjective evaluation of the macro-space at $t_1$.

(3) Information flow helps in the crystallization of subjective spatial evaluation.

(4) The decision to move or not to move is based on place utility which is defined by the relative importance of given variables either individually or in combination.

In summary, the tripartite systematic framework linking urbanization, development and migration has to be recognized, by the overall spatial development goals. Thus, the link between spatial decision-making to the macro-space is crucial.

## REFERENCES


1. R.J. Chorley, A.J. Dunn and R.P. Beckinsale, *The Study of Land-forms* Vol. 1, 1964, p 37.
2. M.E. Harvey, "Social Change and Ethnic Relocation : The Sierra Leone Example", *Geografiska Annaler*, Vol. 53, series B, 1971, pp 94-106.
3. M.E. Harvey, "Inter-regional Migration Studies In Tropical Africa," *People on the Move*, edited by Leszek A. Kosinski and R.M. Prothero, (London : Methuen Co; Ltd., 1975), pp 151-163.
4. M.P. Todaro, "A Model of Labour Migration and Urban Unemployment in Less Developed Countries", *American Economic Review*, 1969, pp 183-193.
5. H. Rempel and M.P. Todaro, "Rural-to-Urban Labour Migration in Kenya," in *Population Growth and Economic Development in Africa* ed. by S.H. Ominde and C.N. Ejiogu, (London, Heinemann, 1973), pp 214-216.
6. D. Herbert, *Urban Geography : A Social Perspective*, (New York : Praeger Publishers, 1973), pp 24-28.
7. A.B. Mountjoy, "The Third World : March of The Peasants from Land to City, "*The Geographical Magazine*, Vol. 44, 1974, pp 208-215.
8. R. Nurske, *Problems of Capital Formation in Underdeveloped Countries*, (Oxford : Blackwell, 1953), p 32
9. P. Gourou, *The Tropical World : Its Social and Economic Conditions And Its Future Status*, (New York : 1953).
10. P. Parsons, *Essays In Sociological Theory*, (Glencoe, Illinois, The Free Press, 1953).
11. M. Weber, *The Protestant Ethic and The Spirit of Capitalism*, trans. T. Parsons (New York : Charles Scribner's Sons, 1960).
12. D. McClelland, *The Achieving Society*, (New York : Van Bostrand, 1961).
13. U. Pareek, "Poverty and Motivation : Figure and Ground," in *Psychological Factors in Poverty* ed. by V.L. Allen, Institute for Research on Poverty Monograph Series, (Chicago : Markham Publishing Company, 1970), pp 302-305.
14. Ibid., pp 302-305.
15. B. Magubane, "A Critical Look at Indices Used in The Study of Change in Colonial Africa.', *Current Anthropology*, Vol. 12, 1971, pp 419-446.
16. J.M. Blaut, "Geographic Models of Imperialism," *Antipode*, Vol. 2, (1970), pp 65-85

17. N.E., Hurst, *A Geography of Economic Behaviour* : North Scietuate, (Massachusetts : Duxbury Press, 1972) pp 346-378.
18. I.L. Horowitz, *Three Worlds of Development* : *The Theory and Practice of International Stratification*, (New York : Oxford University Press, 1966).
19. A.G. Frank, *Capitalism and Underdevelopment in Latin America*, (New York : Monthly Review Press, 1969).
20. W., Rodney. *How Europe Underdevelopment Africa*, (Dar-es-Salaam, Tanzania : Publishing House and Bogle—L'Overture Publications, (1972).
21. S. Amin, "Accumulation and Development : A Theoretical Model," *Review of African Political Economy*, Vol. 1, 1974, pp 9-27.
22. K. Ngombale-Mwira, "The Arusha Declaration on Ujaama na Kujitegemea and Perspectures for Building Socialism in Tanzania," in *Socialism in Tanzania* ed. by John S. Saul, (Dar-es-Salaam, Tanzania : East African Publishing House, 1973), p 57.
23. M. Santos, "Geography, Marxism and Underdevelopment," *Antipode*, Vol. 6, 1974, p.s.
24. T. McGee, "In Praise of Tradition : Towards a Geography of Anti-Development," *Antipode*, Vol. 6, 1974, pp 30-49.
25. D. Slater, "Geography and Underdevelopment," Antipode, Vol. 5, 1973, pp 21-32.
26. K. Griffin, *Underdevelopment in Spanish America*, (London, George Allen Union, 1969), p 13.
27. G.L. Beckford, *Persistent Poverty—Underdevelopment in Plantation Economies of the Third World*, (New York : Oxford University Press, 1972), p 50.
28. S.M., Essang and Mobawonku. "Impacts of Urban Migration on Rural Development : Theoretical Considerations and Empirical Evidence from Nigeria," *The Developing Economies*, Vol. 13, 1955, pp 137-149.
29. T. McGee, op. cit.
30. A.R. De Souza and R.W., Porter, "The Underdevelopment and Modernization of the Third World," *Association of American Geographers*, Commission on College Geography, Resource Paper No. 28, (Washington, D.C.) p 46.
31. A. Buttimer, (sister), "Values in Geography", *Association of American Geographers*, Commission on College Geography, Resource Paper No. 24, (Washington, D.C. 1974), p 22.
32. D. Slater, op. cit.
33. F.H. Harbison, "The Generation of Employment in Newly Developing Countries" in James R. Sheffield (ed.), *Education, Employment and Rural Development* (Nairobi, East African Publishing House, 1967), p 7.
34. F. Perroux, "Note on the Concept of Growth Poles," trans. Linda Gates and Anne Marrie McDermott in *Regional Economics* ed. by D.L. McKee, R.D. Dean and W.H. Leahy, (New York : The Free Press, 1970) p 94.

35. J.F. Friedman, *Regional Policy Development. A Case Study of Venezuela,* (Cambridge, Massachusetts : MIT Press, 1966).
36. B.J.L. Berry, "Hierarchical Diffusion : The Basis of Developmental Filtering and Spread in a System of Growth Centers," in Paul Ward English and Robert C. Mayfield (eds.) *Man, Space and Environment : Concepts in Contemporary Human Geography,* (New York : Oxford University Press, 1972), pp 340-359.
37. L. Rodwin, "Metropolitan Policy for Developing Areas," *Daedalus,* (Winter, 1960), pp 132-146.
38. M. Santos, "The Shared Space : The Two Circuits of Urban Economy in Underdeveloped Countries and Their Spatial Implications." Unpublished paper, University of Toronto, 1974.
39. A.L. Mabogunje, *Growth Poles and Growth Centers in Regional Development of Nigeria,* UNRISD, 71.3, Geneva, 1971.
40. R.P. Misra and Sundaram. *Growth Poles and Growth Centers as Instruments of Modernization in India—A Case Study.* University of Mysore, IDA Mysore, 1973.
41. M.E. Conroy, "Rejection of Growth Center Strategy in Latin American Regional Development Planning", *Land Economics,* XLIX, 1973, pp 373-380.
42. A.J. Dietz, J.M. van Hasstrecht, and H.R. Scheffer, "*Local Effects of Two Large Scale Industrial Projects in the Kafue-Mazabuka Area in Zambia,* Geografisch En Planologisch Institute Katholieka Universiteit, Nijmegen, 1977.
43. L.J. Wood, "Perception Studies In Geography", *Trans : Institute of British Geographers,* 50, 1970, pp 130.
44. Tuan Yi-Fu, *Topophilia : A Study of Environment Perception, Attitudes and Values,* Englewood Cliffs : Prentice Hall, Inc. 1974.
45. H. Sprout and M. Sprout, *The Ecological Perspective on Human Affairs with Special Reference to International Politics,* (Princeton, 1965), pp 27-28.
46. R.G. Golledge, "Process Approaches to the Analysis of Human Spatial Behaviour", *O.S.U Discussion Paper* No. 16, 1970.
47. ——, L.A. Brown and F. Williamson, "Behavioral Approaches in Geography : An Overview", *Australian Geographer,* XII, 1, 1972, pp 59-79.
48. ——, R. Briggs, "Decision Processes and Locational Behaviours," *High Speed Ground Transportation Journal,* 7, 1973, pp 81-99.
49. R.G. Golledge, op. cit.
50. K. Koffka, *Principles of Gestalt Psychology,* (New York : Harcourt Brace, 1935), pp 36-37.
51. B.I. Ijomah, Some Problems of Quantitative Research in Africa", in *Survey Research In Africa : Its Applications and Limits,* eds. by William M.O'Barr, David H. Spain and Mark A. Tessler, (Evanston : North-Western University Press, 1973), pp 48-58.
52. M.E. Harvey, "Economic Development and Migration in Sierra Leone", in *Population Growth and Economic Development in Africa*

ed. by S.H. Ominde and C.N. Ejiogu, (London, Heinemann, 1972), p 167.

53. A.L. Mabogunje, "Systems Approach to Theory of Rural-Urban Migration", *Gecgraphical Analysis*, 2, 1970, pp 1-19.
54. P.R. Gould, "On Mental Maps", in *Image and Environment*, edited by Downs and Stea, Chicago : Aldine Publishing Company, 1974, pp 182-220.
55. S.M. Essang and Mobawanku, op. cit.
56. A.J. Dietz, et al, op. cit.

# 14

# Spatial Dynamics of the Centres of Population

—*R. B. Mandal*

ALTHOUGH dynamics of population deals with birth, mortality and migration. But spatial dynamics of the centres of population by moment analysis and centrographic methods for the dispersed location of population in North Bihar has been analysed here. The shift of population is delineated here through spatial change of mean centres of population. In this paper this has been shown for decennial census years beginning from 1951 to 1971.

Johnson (1892)[1] was the pioneer of centrographic method. He had studied the deviation of bullet marks from a target. Later on, this method was presented by Sviatlovasky and Eelle.[2] This methodology is widely utilised in planning the region.

In Russia the centrographic laboratory was established after the famous Russian scientist D.J. Mendelev. The foundation was laid in Leningradn by the Russian Geographical Society in 1925 which was wound up in December 1934.[3] Recently interesting results have been drawn with the help of centrographic analysis in U.S A. and Israel for economic planning where this method has been extensively used by the geographers and regional planners.

## MEANING OF CENTROGRAPHIC METHOD

This method is an ensemble of graphic procedures, a statistical calculus and represents geographic and historic character of human activity.[4] Hence the purpose of this study is to investigate the laws of population distribution through an analysis of the spatial and historical change of the 'centres of gravity' of population distribution.[5] It is a method of 'social physics'. 'Moment analysis' is a subject matter of the principles of mechanics in physics which deals with the mass of the product and distance, which help in the estimation of mean centres and this represents the magnitude of population located at each point. It is an useful method to enquire about spatial distribution of population, settlements, industries, mining centres, agricultural farms and patterns of land utilization.

Nowadays, a good deal of theoretical work has been accomplished on centrographic method. Warntz and Neft (1960) have drawn inferences from statistical mean, median and mode for population distribution and its transformation into spatial mean, median and mode. This "mean centre over space" is a "centre of gravity at a point, at which objects on a weightless surface will balance".[6]

In 1957 Hart[7] explained the computation procedure of minimum aggregate travel and centre of convergence besides spatial mean, median and mode. Porter (1963)[8] had used a speedy procedure which is known as 'least square' method to compute point of minimum aggregate travel. These are quantitative measures of central tendency. Stewart and Warntz (1958)[9] had presented the 'dynamic radius' of population to measure dispersions. They defined the expression as $\sqrt{\{\Sigma(p.\, d^2)\}P}$

where P is the population of a small area; d is the distance from the mean centre, and P is the total population. This is the modified form of Standard Deviation. These studies show that the point of a region from where the number of people is balanced on all sides is known as the mean centre of population of a given region. This study is designed to study the spatial change in the mean centres of population and regional imbalances of population distribution in North Bihar from 1951 to 1971.

According to the 1951 census North Bihar had 51,814 square kilometres of area and 18.17 million people, which comes to a density of 350 persons per square kilometre in districts of the region. In 1971 the region having 51,814 square kilometres of area and 26.05 million people, comes to a density of 503 persons per square kilometre as enumerated in 16 districts of the region.

### ECOLOGICAL CONTROL

Under such a dynamic situation of population change, the mean centre of population is bound to shift in different time and space, due to changing physical, economic, social, political and technical abilities of the region. So far as the physical factors are concerned the terrain conditions like fertile plain land attract people, while the higher rugged plateau, forest cover and marshy land with unfavourable climatic conditions e.g., moist climate of Purnea and swampy land of Saharsa area repel population concentration over a considerable part of the region.

The economic factors like agricultural production, mineral exploitation, industrial development, transport network, development of trade and commerce and literacy have multilateral effects on the centre of population shift. The economic development over a particular side of the region e.g., near Patna and industrialised belts of Chotanagpur plateau have attracted people from North Bihar as labourers in search of employment and this is the cause of the centre of population shift during different census periods.

Among social factors the differences in ethnic composition of varied castes and tribes under characteristic regions, and migration by marriage, growth of urbanisation and advancement of civilization play a dominant role in shifting centres of population in North Bihar. The unbalanced transport development, natural increase of population and pre-determined migration also have considerable impact on mean centres of population shift.

The history of land occupancy also plays a vital role with the chain of events, as it pushes and pulls the people due to a number of causes related with the past events. The old settled areas of any region provide more facilities to new comers than any pioneer fringe. It is a well known fact that historical background of the region has intimate relationship with the centres of population shift in North Bihar. It is a region of long historical background since the prehistoric period which forms the meeting ground of Australoid, Mongoloid, Aryan and non-Aryan tribes. During the Buddhist period it was the playground of Bhikshus and monks. Under the Hindu rule the Bhakti movement of a number of saints like Chaitanya and others had profound effect in attracting people from western Bihar and U.P. to this region. In the Moghul period, the start of the business by Dutch, French and British provides a fresh bid to attract the people around major cities of North Bihar. During the British period, the overwhelming growth of business and sugar mills on the alluvial tract had provided a congenial environment for the poor rural folk to be attracted towards the industrial establishments of the cities. Since historical times there have been two centres of population concentration in North Bihar, one around Muzaffarpur and other around Darbhanga in the district of West Bihar plain. In between these two centres the space may be called interspersed. North Bihar have lower population density in Saharsa and Purnea districts. These two balance the population of North Bihar plain and, for a fact, the mean centre of population always lies in between Darbhanga and Muzaffarpur districts of Bihar from 1951 to 1971.

Due to political decision for compulsory shift of the mean centre of population and the finalisation of industrial location by the politicians also provide a gainful employment to sizeable number of workers to run the industry efficiently. Technical abilities of a particular region due to its advantageous location also help the mean centre of population to shift. The importance of cultural factors is greater in population redistribution during a given period of time than in population distribution at a particular point of time.

## METHOD OF STUDY

The centre of population or mean point of population distributed over an area may be defined as the centre of population gravity for the area. In other words, the point upon which the area would balance, if it were a rigid plane without weight and the population distributed there as, each individual being assumed to have equal weight and to exert influence on the central point proportional to his distance from the point. The pivotal point therefore, would be its centre of gravity.[10] The formula for the co-ordinates of the centre of population may be written as:

$$\bar{X}=\frac{\Sigma P_i X_i}{\Sigma P_i} \text{ and } \bar{Y}=\frac{\Sigma P_i Y_i}{\Sigma P_i}$$

where $p_i$ is the population at point i and $X_i$ and $Y_i$ are its horizontal and vertical co-ordinates respectively.

Thus the mean point unlike the median point is influenced by the extreme items distributed over the total area.

The point of intersection of the horizontal and vertical base line on the map, the equation are:

$$\bar{X}=\frac{\Sigma P_a (X_a-\bar{X}')-\Sigma P_b (\bar{X}'-X_b)}{\Sigma P_i}+\bar{X}'$$

$$\bar{Y}=\frac{\Sigma P_c (Y_c-\bar{Y}')-\Sigma P_d (\bar{Y}'-Y_d)}{\Sigma P_i}+\bar{Y}'$$

where X and Y are the co-ordinates of the assumed mean, $X_a$ is any point east of that mean, $X_b$ is any point west of it, $Y_c$ is any point north of it, $Y_d$ any point south of it and $P_a$, $P_b$, $P_c$, and $P_d$ are the populations in areas east, west, north and south of the assumed mean respectively.

Then with the help of vertical and horizontal tier moment lines of fulera can be obtained after dividing the horizontal and vertical tier moments separately with the total population of the region. On the intersection point of horizontal and vertical lines of fulera values the mean point is located on the map for the successive years of 1951, 1961 and 1971 (Fig. 14.1 ; Table 14.1)

In order to show the shifting centres of population in North Bihar from 1951 to 1971, a map has been prepared (Fig. 14.1) in which centres of gravity or mean points of population for each census year have been determined by the method as

TABLE 14.1

*Summary of Mean Point Location in North Bihar, 1951-1971*

| Years | Total Vertical Tier Moment | Vertical Line of Fulera value | Total horizontal Tier Moment | Horizontal line of fulera value |
|---|---|---|---|---|
| 1951 | 249285000 | 14.32 | 119524000 | 6.86 |
| 1961 | 301058000 | 14.45 | 151704000 | 7.28 |
| 1971 | 359271000 | 13.97 | 187986000 | 7.30 |

*Source* : Calculated by the author.

suggested by Hart (1954). Spatial shifts of population are readily revealed by computing mean points of population at successive intervals of time, even though the mean point may lie in an area completely devoid of population, since mean point is greatly influenced by any movement of population within the distribution. Thus "mean point is based on dynamic analysis of distributions such as the study of area shifts".

*North Bihar : Shifting Means Centres of Rural Population (1971)*

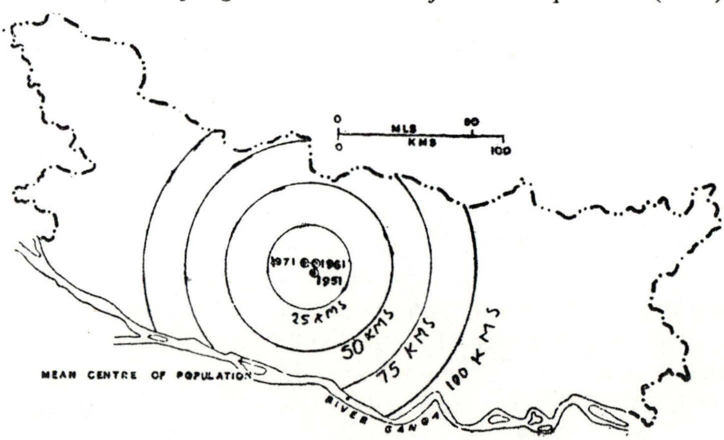


FIG. 14.1

The Map (Fig. 14.1) clearly reveals the shift of mean centre of population in different census years since 1951. In 1951 the mean centre of population was in the southern side of Hayaghat block in Samastipur district. This is the point from where the southern and northern borders are equally distant while the eastern border of North Bihar is further away in comparison with the western border. This is because the unfavourable climate, annual inundation and vast amount of sand deposits repeal population growth in Saharsa and Purnea districts. But now the situation is changing fast with the construction of Kosi dam at Hanuman Nagar near Chatra. On the other hand, the productive, fertile and treble cropped land of Muzaffarpur and Saran plain attract the centre of population towards the west. In 1961, the centre of gravity moved 7 kilometres north in Hayaghat anchal itself which showed northward growth of population during 1951-61 decade. In 1971 the mean centre shifted 7 kilometres west in Gaighatti anchal of Muzaffarpur district which shows the natural growth of population towards higher population density side further west.

The justification for the location of mean centre of population is also provided by the location of Darbhanga and Muzaffarpur cities towards north and west of the mean centre.

## DISTRIBUTION OF POPULATION

North Bihar has a density of 495 persons per square kilometre. The highest rural population density is 897 for Bidupur anchal of Vaishali district and the lowest density of 201 persons is in Sidhaw anchal of west Champaran district. In between these two extremes the density of population has been divided into four categories according to the Median and Quartile values, where $Q_1$ is 389, M is 505 and $Q_3$ comes to 622.5 persons per square kilometre (FIG. 14.2).

The zone of highest density of population (over 622) comprises the Gangetic riparian tracts of Siwan, Saran, Vaishali, Samastipur, parts of Begusarai districts and a continuous belt from Sheohar in the west to Pandaul in the east, in Sitamarhi and Madhubani districts respectively. This high density is the result of easy accessibility and early habitation in the region of fertile soils with three to four seasonal crops in a year. The average density is more than 600 persons per square kilometre and it increases up to 897 in Vaishali, Saran, Siwan and Samastipur districts. As 95 per cent population of the region is rural, there is no significant difference between the total density and rural density of population in North Bihar, where per square kilometre total density is 511 and rural density is 495 persons. This is because, since historic times North Bihar has been "the home of a sedentary agricultural population and the saturation point seems to have been reached long ago." This high density region is interspersed with areas of medium density where the wet and low-lying paddy fields are intersected in parts by sandy uplands and swampy low-lands or *chaurs*. Here the high density co-exists with great agricultural insecurity on account of a single rice crop annually and the periodic floods.

*North Bihar : Density of Population*

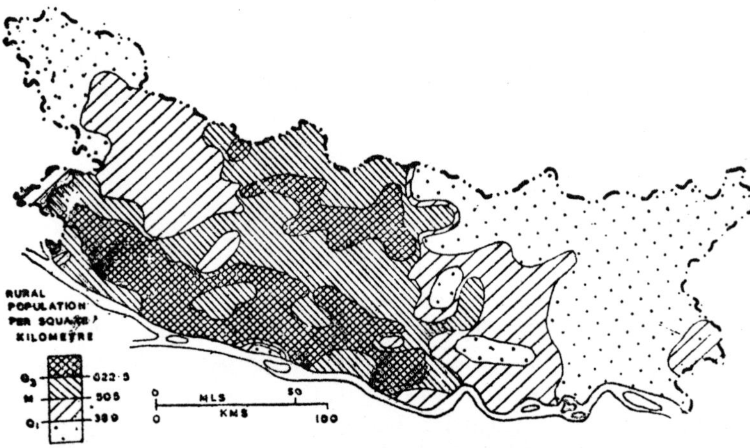


FIG. 14.2

The zone of medium population density has 505 to 622 persons per square kilometre and extends from Sahebganj anchal of Muzaffarpur district in the west to Biraul anchal of Samastipur district in the east (Fig. 14.2). The northern zone of medium population density begins from Ghorasahan anchal of Champaran district in the west of Jainagar anchal of Madhubani district in the east. These two areas are engaged mainly in rice cultivation which naturally has a high potentiality to support dense population. Other areas belonging to this category are Gopalganj district, the Ganges riparian tract of Siwan district, Kahara anchal of Saharsa district and Khagaria anchal of Monghyr district.

The third zone of semi-sparse population density is located in east Champaran and in the eastern half of west Champaran districts. It also covers parts of Khagaria sub-division and Madhubani, Gopalganj and Saharsa districts. Here the population density ranges from 389 to 500 persons per square kilometre. In Saharsa district, the average population density is over 300 persons per square kilometre. Here due to heavy rains, unhealthy malarial climate, marshy land, unstable conditions due to disastrous floods and frequent changes in the course of the Kosi river, the population distribution is sparse and its density is low. Commonly, one can meet with vast patches of land, unfit for cultivation due to the deposition of a huge amount of sand every year, having a depressed population growth. The Kosi has shifted 122 kilometres towards the west in 250 years, and this has converted 7680 square kilometre of fertile rice land into sandy waste and grassy marshes. This has also disrupted communication lines which resulted in large scale depopulation by outmigration. In the Kosi belt, thus, we find the crops and cattle destroyed and people made homeless and on the verge of starving at frequent intervals. Epidemics like malaria, kalazar, cholera, dysentery, typhoid and pneumonia, etc., create havoc. It is rightly said that "a poor man in the Kosi belt is very much like a helpless cat shut up in a room to be killed by the angry owner of the house."

Now the situation has changed and the Kosi project with its protective embankments and canals has at least temporarily checked the menace of floods. It has also increased the population supporting capacity of the region.

The sparse population density comprises parts of west Champaran and the whole of Katihar and Purnea districts. In these areas, the population density is less than 389 persons per square kilometre. The least populated areas are found both in the north-western and north-eastern corners of North Bihar, where the forest covered rugged topography and flood affected sandy soils of the Mahananda, the Panar and their tributaries are to be found. In the whole of North Bihar the higher proportion of the rural population reflects on the one hand the high productive capacity of land, and on the other, the high reproductive capacity of the population as compared to urban areas. In such a situation, it is difficult to check this high rate of population growth and economic change in the region.

## CONCLUSION

The study of dynamics of the centres of population through centrographic method is a subject of interest in U.S.A., U.S.S.R., Israel and India. This method of study has given a great fillip to regional, national, economic and population planning in different parts of the world as the review of literature shows. In north Bihar, this method has been applied due to one sided mobility of population from surrounding rural areas of Bihar. This has created regional imbalances of population distribution and economic growth of the region.

Dynamics of centres of population is thus an objective study. This shows the distribution of population in its right perspective and the point on land where the population of all directions is balanced by its weight. For North Bihar such a study is very useful from the planning point of view as this region has the problems of over population, flood, etc. Hence the dynamic character of population needs the proper allocation of population in a certain region by providing certain employment and job opportunities.

## REFERENCES

1. W.W. Johnson, *The Theory of Errors and Method of Least Squares*; New York, 1892.
2. E.E. Sviatlovasky and W.C. Eells, "The Centrographical Method and Regional Analysis"; *Geographical Review*; Vol. 27, 1937, pp 240-254.
3. N.B.K. Reddy, "Moment Analysis and Centrographic Study of Area, Urban Settlement and Population of the Krishna and Godavari Deltas", *The Indian Geographical Journal*, Vol. XLV; 1970, pp 1-14.
4. P.W. Porter, "What is the Point of Minimum Aggregate Travel", *Annals of the Association of American Geographers*, Vol. 53, 1963, pp 224-232.
5. N B.K. Reddy, op. cit., 1970.

6. W. Warntz and D. Neft, "Contributions to a Statistical Methodology for Areal Distributions," *Journal of Regional Science*, 1960, 2, pp 47-66.
7. J.F. Hart, "Central Tendency in Areal Distributions", *Economic Geography*, Vol. 30, 1954, pp 48-59,
8. P.W. Porter, op. cit., 1963.
9. J.Q. Steward and W. Warntz, "Some Parameters of the Geographical Distribution of Population," *Geographical Review*, Vol. 49, 1958, pp 270-273.
10. U.S. Bureau of Census : Statistical Atlas of United States, 1924, p 7.

# 15

# Potential Model and Spatial Distribution of Population

—*H. Lal*

DISTRIBUTION of population helps to understand the geo-economic prospects and potentiality of the area and hence different methods of investigation into the distributional pattern of population need proper attention. Such investigations use either absolute or relative measures. Simple measure of man-land ratio does not give clear picture of the future potentiality of the area. A better method for investigating the population distribution is the population potential measurement which is based on physical laws known as gravitational-model, frequently used in socio-economic sciences. First introduced by an astrophysicist J.Q. Stewart (1947)[1], this potential of population model gives a statistical explanation of the variations in population density in smaller areas within a large region. It considers both the number of inhabitants of other places and their distances from the given place. The present study aims to analyse the distribution of population in eastern U.P. with the help of this model. The population potential of the districts has been compared with general population density and the variations between them observed and the expected density of the area have also been analysed.

## METHODOLOGY

The potential model is the simplest and the most widely applicable of all interaction models. It postulates that interaction between any two groups of people or locations increases in proportion to the size of the groups (populations) and decreases with some function (unit or exponent) of the distance between the groups. Potential at a point may be thought of as a measure of the proximity of that point to all other places in the system. In other words potential at a point is simply an aggregate measure of the influence of all distant places on that point (Stewart and Warntz 1958)[2]. Mathematically the potential ($V_i$) created at i by each point j is equal to the population (p) at j divided by its distance (d) from i. The symbol is $V_i = \frac{P_j}{d_{ij}}$. Thus, $V_i$ is a summation of the effect of all places on place i including the effect of i itself on itself.

For preparing the potential map for eastern U.P. all the district headquarters were located and their population figures inserted on an outline map of the region. Here the population of each district was considered to be centred at the point of district headquarters. The next step was to find out the distances between various districts [Table 15.1(A)]. For this straight line distances in kilometre were noted from the map [Fig. 15.1(A)]. The distance within the districts was also included which is half of the distance between point i and its nearest neighbour. Thereafter the population potential of each district upon all other districts was calculated by dividing the population of each district by the distance of that district from every other district as well as from itself [Table 15.1(B)]. The population potential of each district is the sum of its relationship, derived as described above, to all districts. Considering the potentials as the control points, equipotential lines were interpolated on the map [Fig. 15.1(B)]. Further, a population potential surface map was prepared which expressed the percentage of potential of each point below the point of highest population potential [Fig. 15.1(C)].

## SPATIAL DISTRIBUTION OF POPULATION POTENTIAL

The population potential map portrays a general picture of the original distribution on which it is based. It gives an idea of the number of people within reach of different localities. The maximum number of people are within reach of the places around Gorakhpur situated in the north central part of the

*Eastern U.P. : Population Potential (1971)*

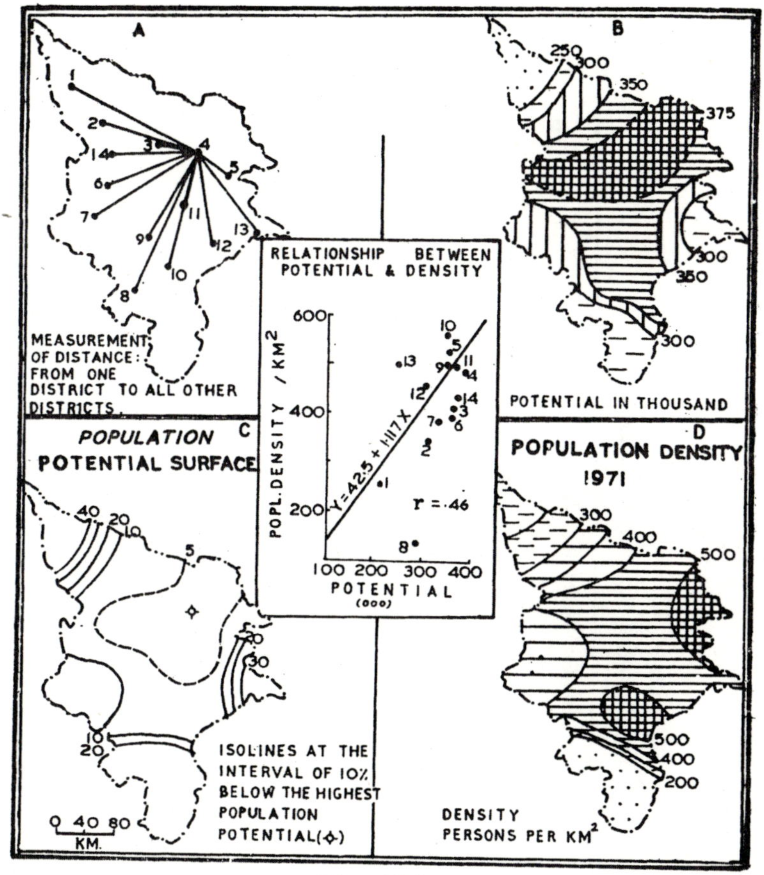


FIG. 15.1

region. The population potential ranges from 218 thousand persons per kilometre in Bahraich district to 384 thousand persons per kilometre in the Gorakhpur district [Fig. 15.1(B)]. The map gives a view of the relationship of each place to the total population of eastern U.P. Barring some districts of the border areas, most of the central districts show comparatively high potential values. A striking feature of the potential map of this region is a major peak existing at Gorakhpur from which runs a dominant axis descending gradually towards the south-west and steep towards east and north-west. The major

TABLE : 15.1 (A)

*Population Potential, Eastern U.P., 1971*

| Districts | | Distance in Kilometre (d) | | | | | | | | | | | | | |
|---|---|---|---|---|---|---|---|---|---|---|---|---|---|---|---|
| | | $D_1$ | $D_2$ | $D_3$ | $D_4$ | $D_5$ | $D_6$ | $D_7$ | $D_8$ | $D_9$ | $D_{10}$ | $D_{11}$ | $D_{12}$ | $D_{13}$ | $D_{14}$ |
| $D_1$ | Bahraich | 33 | 65 | 146 | 203 | 254 | 154 | 186 | 288 | 231 | 290 | 232 | 298 | 330 | 105 |
| $D_2$ | Gonda | 65 | 21 | 85 | 145 | 195 | 95 | 130 | 228 | 168 | 225 | 166 | 234 | 267 | 41 |
| $D_3$ | Basti | 146 | 85 | 31 | 64 | 110 | 90 | 122 | 183 | 117 | 165 | 190 | 158 | 185 | 61 |
| $D_4$ | Gorakhpur | 203 | 145 | 64 | 26 | 51 | 142 | 165 | 193 | 130 | 160 | 75 | 128 | 135 | 125 |
| $D_5$ | Deoria | 254 | 195 | 110 | 51 | 26 | 176 | 194 | 195 | 140 | 154 | 77 | 102 | 90 | 170 |
| $D_6$ | Sultanpur | 154 | 95 | 90 | 142 | 176 | 19 | 37 | 135 | 85 | 143 | 115 | 172 | 220 | 60 |
| $D_7$ | Pratapgarh | 186 | 130 | 122 | 165 | 194 | 37 | 19 | 117 | 75 | 126 | 124 | 168 | 223 | 95 |
| $D_8$ | Mirzapur | 288 | 228 | 183 | 193 | 195 | 135 | 117 | 25 | 66 | 50 | 122 | 115 | 175 | 188 |
| $D_9$ | Jaunpur | 231 | 168 | 117 | 130 | 140 | 85 | 75 | 66 | 29 | 58 | 65 | 95 | 150 | 128 |
| $D_{10}$ | Varanasi | 290 | 225 | 165 | 160 | 154 | 143 | 126 | 50 | 58 | 25 | 88 | 65 | 127 | 185 |
| $D_{11}$ | Azamgarh | 232 | 166 | 90 | 75 | 77 | 115 | 124 | 122 | 65 | 88 | 34 | 68 | 115 | 130 |
| $D_{12}$ | Gazipur | 298 | 234 | 158 | 128 | 102 | 172 | 168 | 115 | 95 | 65 | 68 | 31 | 61 | 197 |
| $D_{13}$ | Ballia | 330 | 267 | 185 | 135 | 90 | 220 | 223 | 175 | 150 | 127 | 115 | 61 | 31 | 235 |
| $D_{14}$ | Faizabad | 105 | 41 | 61 | 125 | 170 | 60 | 95 | 188 | 128 | 185 | 130 | 197 | 235 | 21 |

TABLE 15.1 (B)—*contd.*

| Districts | District Population (000) P | Population distance Ratio (P/d) in Thousand $D_1$ | $D_2$ | $D_3$ | $D_4$ | $D_5$ | $D_6$ | $D_7$ | $D_8$ | $D_9$ | $D_{10}$ | $D_{11}$ | $D_{12}$ | $D_{13}$ | $D_{14}$ |
|---|---|---|---|---|---|---|---|---|---|---|---|---|---|---|---|
| $D_1$ | 1727 | 53 | 27 | 12 | 8 | 7 | 11 | 9 | 6 | 7 | 6 | 7 | 6 | 5 | 17 |
| $D_2$ | 2302 | 36 | 112 | 27 | 16 | 12 | 24 | 18 | 10 | 14 | 10 | 14 | 10 | 9 | 56 |
| $D_3$ | 2984 | 20 | 35 | 98 | 47 | 27 | 33 | 25 | 16 | 26 | 18 | 33 | 19 | 16 | 49 |
| $D_4$ | 3038 | 15 | 21 | 47 | 119 | 60 | 21 | 19 | 16 | 23 | 19 | 40 | 24 | 23 | 24 |
| $D_5$ | 2812 | 11 | 14 | 25 | 55 | 110 | 16 | 14 | 15 | 20 | 18 | 37 | 28 | 31 | 17 |
| $D_6$ | 1643 | 11 | 17 | 18 | 12 | 9 | 89 | 44 | 12 | 19 | 12 | 14 | 9 | 8 | 27 |
| $D_7$ | 1423 | 8 | 11 | 12 | 9 | 7 | 39 | 82 | 12 | 19 | 11 | 11 | 8 | 6 | 15 |
| $D_8$ | 1541 | 5 | 7 | 8 | 8 | 8 | 11 | 13 | 62 | 23 | 31 | 13 | 13 | 9 | 8 |
| $D_9$ | 2005 | 9 | 12 | 17 | 15 | 14 | 24 | 27 | 30 | 69 | 35 | 31 | 21 | 13 | 16 |
| $D_{10}$ | 2852 | 10 | 13 | 17 | 18 | 19 | 20 | 23 | 57 | 49 | 114 | 34 | 24 | 22 | 15 |
| $D_{11}$ | 2857 | 12 | 17 | 32 | 38 | 37 | 25 | 23 | 24 | 44 | 32 | 84 | 42 | 25 | 22 |
| $D_{12}$ | 1532 | 5 | 7 | 10 | 12 | 15 | 9 | 9 | 13 | 16 | 24 | 23 | 50 | 25 | 8 |
| $D_{13}$ | 1589 | 5 | 6 | 9 | 12 | 18 | 7 | 7 | 9 | 11 | 13 | 14 | 26 | 52 | 7 |
| $D_{14}$ | 1927 | 18 | 47 | 32 | 15 | 11 | 32 | 20 | 10 | 15 | 10 | 15 | 10 | 8 | 94 |
| Population Potential of each district (000) | | 218 | 346 | 364 | 385 | 354 | 361 | 333 | 292 | 355 | 353 | 368 | 310 | 252 | 375 |

area of high population potential in this area shows the influence of communication lines, fertility of soil and cultivation of cash crops like sugarcane and the establishment of numerous sugar factories. The peak corresponds with the regional city Gorakhpur, which along with the highest nodality index, is an important industrial and transportational centre of the whole of the Saryupar plain. The growing importance of this peak is also reflected in the increase in population which was 109.4% for this district since 1901.

The area of high potential is surrounded by an equipotential line of 350 extending more towards south and west. In the south, it extends up to Varanasi and in the west it includes Basti, Faizabad and Sultanpur. The existence of a sizeable number of towns and the prevailing high rural density in the whole central region have weighted heavily in causing a higher population potential. Indeed, the potential line is influenced by all the three biggest cities, i.e., Varanasi, Gorakhpur and Faizabad, which fall in this potential area. It contains more than 60% of the region's urban population and 56% of the total population. The region, of course, holds the promise of development of agriculture, irrigation and agro-based industries in the future.

The southern most part of the region is a plateau area whose rugged terrain offers strong deterrents to the rural settlement though some industrial towns have emerged in this rugged portion. The region supports a low density of population and also presents a low population potential. The north-west triangular area of Tarai belt located away from the centres of population concentration and inaccessibility, dense forest together with flood havoc causing the lowest population potential in Bahraich where the potential declines from 350 to 218 thousand persons per kilometre.

The potential surface map prepared on the basis of the potential values, shows the rate of change in population potential in per cent in relation to the highest potential centre [Fig. 15.1(C)]. It shows a steady decline in the values of the potential away from the highest potential centre Gorakhpur to most of the south and south-east part of the region. The reason for such gradient is quite apparent. It reflects the overall isotropic nature of the suitability of homogeneously diffused habitability conditions, high percentage of arable land and generally good conditions for making a living out of the local and regional resources. A steep gradient is marked in the north-west direction of the region, where from Gonda to Bahraich the value drops from 345 to 218 only in 50 kilometre distance which is more than 40% lower than the highest potential value. The region offers meagre opportunities of livelihood even for a small number of population. It also coincides with the lower population density. Another steep gradient is visible in the eastern border areas of the region. It shows a 30 per cent deviation from the highest potential. The trend has been proved statistically. The relationship between distance from the highest potential centre and population potentials indicates a substantial negative correlation (r=.55).

## POPULATION POTENTIAL AND DENSITY OF POPULATION

Population potential may be compared with the general population density. In fact, density (D) is found to vary as the square[3] of the potential (V). In equation Density (D) equals $kV^2$, where k is a constant derived from the formula $k = \frac{\Sigma\left(\frac{D}{V^2}\right)}{N}$. For eastern U.P. this constant k comes to 0.003667. On the basis of this formula the expected density from the potential of population was calculated for each district and homogeneous regions were identified by their systematic deviation from a general regularity (Table 15.2). Four districts in the central part were observed to have a consistently high population density. Figures denote that the actual density values, barring a few, are lower than the expected. With the increase in density, the difference between the actual and expected values decreases. Comparatively low population potential in this region is mainly a function of increasing distance from the peak of potential. Because of the

TABLE 15.2

*Potentials of Population and Population Densities, Eastern U.P., 1971.*

| Districts | Potential of population (in thousands of persons/km) | General Population Density (in persons per sq km) | |
|---|---|---|---|
| | | Actual | Expected |
| Gorakhpur | 384 | 481 | 540 |
| Faizabad | 374 | 435 | 513 |
| Azamgarh | 368 | 497 | 496 |
| Basti | 364 | 408 | 486 |
| Sultanpur | 361 | 371 | 478 |
| Jaunpur | 355 | 496 | 462 |
| Deoria | 354 | 521 | 459 |
| Varanasi | 353 | 560 | 457 |
| Gonda | 345 | 314 | 437 |
| Pratapgarh | 333 | 381 | 407 |
| Gazipur | 310 | 453 | 352 |
| Mirzapur | 292 | 136 | 312 |
| Ballia | 252 | 499 | 233 |
| Bahraich | 218 | 251 | 174 |

favourable physico-economic conditions population density in the central part of the region is considerably higher.

Another homogeneous region identifiably by its systematic deviation from general regularity, is that the actual population density is lower than the expected. With the decrease in population density, the difference between actual and expected values in this region increases. Comparatively low population density corresponding to a high potential is the result of less favourable agricultural conditions and low percentage of urban population which reduces the population density.

Thus, the map of population potential illustrates a certain relation to the map of the density of population. However, this relation is not very close. As in the graph regression line showing relationship between population potential and density for eastern U.P. indicates a positive trend with a lower value (r=.46) which is an indication of substantial relationship. After all the most densely populated part of the region (Varanasi 560 and Deoria 521) [Fig. 15.1(D)] lie in the second category of potential values and Ballia having density of 499 persons per sq km ranking third in order, shows one of the lowest potentials of 252. It is perhaps owing to the fact that the state border weakens the influence of the strong centres to a considerable extent, but not completely.

## APPLICABILITY OF THE POTENTIAL MODEL

The method suffers from all those drawbacks to which the concept of population potential is itself susceptible. The potential of a district upon another was considered simply as a function of its population size and the distance between the two, ignoring several other vital factors, such as pattern of actual accessibility, regional differences in socio-economic development.[4] These changes can be accommodated modifying the mass (population) or distance terms in the equation. The population can be evaluated (weighted) by introducing the criterion of the labour-force or per capita income ; geometrical distribution can be replaced by economic, if we introduce transport factors or the value of investment in the area[5], etc. Thus enriched with further improved parameters, this model will give a comprehensive picture of distribution and along with its importance in applied geography and territorial planning, the augmented potential model will be used in solving practical problems too.

## CONCLUSIONS

In an ideal case, i.e. that of an even distribution of population, the highest potential values would be found in the geometric centre of the region. In and around this centre, the potential for interaction with other places is greatest. At the edges of the region, potential falls to a minimum. It should be admitted that eastern U.P., owing to its shape and topography is not very much suitable for the procedure applied, viz., following the spatial distribution of population by means of population potential. Nevertheless, it appears that the situation in this region is much closer to the ideal case, as the peak of potential is nearing to the geographical centre of the area.

## REFERENCES

1. J.Q. Stewart, "Empirical Mathematical Rules Concerning the Distribution and Equilibrium of Population", *Geographical Review*, 37, 1947, pp 461-85.
2. J.Q. Stewart and W. Warntz "Physics of Population Distribution", *Journal of Regional Sciences*, 1, 1958, pp 99-123.
3. Stewart and Warntz (1958) and Prabha (1974) have used this formula to obtain the 'expected' rural density while in this study the same is being used to compute the 'expected' general population density for each districts.
4. R.B. Mandal, "Urban Potentials in Bihar", *Indian Geographical Studies*, Res. Bult. No. 6, 1976, pp 46-51.
5. S. Shastri Prabha, "Potential Model and the Spatial Distribution of Population in Vidarbha" 1971 ; *Geographical Rev. of India*, 1974, 36, pp 138-147.

# 16

# Approaches Towards A Geography of Health—A Synoptic Review

—*R. L. Singh, Rana P. B. Singh and B.P. Singh*

> "A Physician can't understand the essence of disease unless he is a geographer too."
>
> —*Paracelus*

The mechanism which causes the growth and variation of the diseases within a spatial unit has recently drawn the attention of geographers. In fact, such variations are primarily governed by the socio-economic and behavioral factors—the aspects which are mostly neglected by medical scientists. To emphasize these aspects, a new methodology has been advanced being expressed as the geo-medical approach. The conclusions derived from such analysis will provide a base for planning postulates so that the health of human being can be improved through preventive measures rather than medical treatment to a considerable extent.

## PROGRESS OF STUDIES

The ancient medical literatures of India and Greece have dealt mostly with the causes and variation of diseases within environmental perspective, but they are lacking the spatial treatment. Ancient Indian scholars of Ayurvedic medicine like Susruta (B.C. 1500-1000), Charak (c. B.C. 1000), Kasyap, Bhel (c. A.D. 700), Harit and others[1], have described the regional patterning of the various diseases, for example in such texts it has been indicated that in the eastern and northern parts of India there is a preponderance of thyroid disorders.[2]

Grateful thanks are due to A.T.A Learmonth for going through this article and giving constructive suggestions.

Among the ancient Greek scholars, credit goes to Hippocrates (400 B.C.) for presenting the spatial patterning of diseases within the context of geographical factors like place-identification, topographical characteristics, climatic variations etc. For the first time he has been able to divide diseases into two broad types: (i) *endemic* diseases which are always present in the human body, and (ii) *epidemic*, diseases which due to environmental hazards become more frequent at certain period of time.[3] Comparing the prerequisite conditions prevailing in a city and a village, he has also emphasized the behavioral factors, e.g., perception of space, pollution-resistant power, attitudes and thinking, as one of the most significant factors which control the level of the health disorders.

The works done between the Hippocratic era and the nineteenth century have been reviewed by Howe, who grouped these works into two time scales—the works done before A.D. 1848, mostly emphasizing the physical aspects in a generalized form, and the works done after A.D. 1848, which considered the spatial approach along with their representation over the map.[4] The first map of this kind has been prepared by Shaptner[5] using dot method to show the distribution of death caused by cholera in Exeter during 1832. This approach has been followed and advanced by J.M. May, Rodenwaldt and Jusatz[6], who have taken rather large geographical area for mapping, as such these maps show heterogeneity in environmental conditions as well as the prevalence of the disease.

The issue to consider regions or nations as the most viable units for mapping has been raised by Howe[7], the founder of the school of medical geography in the U.K. While preparing the "National Atlas of Disease Mortality in the U.K.," he has considered many diseases like cancer, diseases of the circulatory system, diseases of the respiratory system, etc. for two periods of time 1954-58 and 1959-63. The concept of mapping as propounded by Howe has been further advanced by Stamp during 1964-65. According to him the analysis of causal factors in the context of the disease is the fundamental concern of medical geography.[8] Learmonth, the pioneer in the field of Indian medical geography, has observed that every Indian village is unique in its socio-economic and landuse patterns, as such one can see a village in itself as the distinctive unit. Through the superimposition of maps showing spatial variation of various diseases and the physico-cultural characteristics, it has been concluded by him that there is no distinct correlation among these attributes.[9]

The German school of medical geography was founded after World War II at Hiedelberg. Using medico-cartographic techniques, the scholars have prepared detailed atlases showing spatial variations of selected diseases. Under their series of 'World Atlas of Epidemic diseases, they have published four volumes edited by H.J. Jusatz,[10] i.e., Libya, Afghanistan, Ethopia and Kenya.

Medical geography in the U.S.S.R. claims a far longer basis of research experience than in other European countries, and it is now an independent branch of science. The Russian School has laid great emphasis upon the studies of multi-factorial disease complexes as well as upon public health as a contribution to the productivity of labour along with promotion of its quality, and hence to the development of a national economy within the socialistic approach. Ignatyev has prepared a series of medico-geographic maps showing the relationship between natural-territorial complexes and health attributes.[11]

By the end of 1949, the subject was recognized by the I.G.U., hence a new Commission was formed at Lisbon; this was given a permanent status in 1968 at the New Delhi venue under the chairmanship of A.T.A. Learmonth. The Commission had given particular emphasis to the environmental and ecological system[12], because they have considered this theme as the fundamental base for the development of geographical pathology, so called as geomedicine or medical geography[13] or most recently, as 'Geography of Health'. This discipline provides a scientific basis for the understanding of the wholeness of life which needs "medical care and insight wedded to a knowledge of environmental circumstances that make for health."[14] During the 22nd IGC (Montreal, Canada), the following four topics were considered for special attention : (i) health and environment, (ii) training of medical geographers, (iii) models and theories in medical geography, and (iv) application of medical geography in positive health measures.[15] The scope of the subject has been further widened during the IGU Regional Conference, New Zealand (Dec. 1974), where special attention was paid to human nutrition and etiological problems. Due to the increasing impact of theoretical-cum-empirical methodology, three main themes were selected for the symposium on Medical Geography during the 23rd IGC (Moscow, U.S.S.R., July 1976), i.e., (i) testing and methods of medico-geographical forecasting, (ii) general results and prospects of the application of mathematical and cartographical methods, and (iii) the role of medical geography in the study of the causes and underlying conditions associated with cardiovascular, oncological and major contagious diseases.[16] During this Congress, the title of the working group—"Medical Geography" was renamed as "Geography of Health".

The contributions and approaches of different schools of medico-geographical research have shown many deviations in methodology and techniques, but the prime goal has always been maintained, i.e., the scientific understanding of the human health problem, so that the policies and practical application can be formulated for preventive measures and medical care to promote the quality of life.

## CONCEPTUAL FRAME

It is clear through the foregoing that the prime concern of medical geography is the body of observations pertaining to the problems of health, defined as "a state of complete physical, mental and social well-being and not merely the absence of disease and infirmity".[17] It has been further stressed that health is directly or indirectly controlled by the ecological factors and environmental setting. Hence, any disturbance in the environmental or ecological balance that takes place causes the growth of disease. The level and intensity of disease depend upon the resistance power of the human body. Where the human readjustment along with resistance power is high, naturally there will be healthy human life.[18] This idea is further supported by studies like mortality due to deseasonality in Japan.[19]

Most of the studies done by bio-scientists and medical scientists are related with the human ecology and its problems, i.e., deviations and levels of readjustment within the environmental setting, but in contrast it has also been observed that health imbalances are not uniformly distributed over the earth's surface. In fact, some diseases are very widely spread, e.g., cholera. With this identification, it is needed to justify as to what extent the relationship can be observed between the dispersal and spread of diseases and health-ecological attributes. To solve this challenging problem, the *Geography of Health* provides an interdisciplinary platform for the study of scientific and spatial reasoning working behind, as Howe has put it that this field concerns with the cause effect relationship of human health problems with physico-bio-socio-cultural environment within spatio-temporal frame.[20] The basic drawback of Howe's conception is that he has given much emphasis upon the problem of disease prevailing in the study area, rather than the attitudes towards prevention or their solutions.[21]

GEOGRAPHY OF HEALTH : THE INTERNAL DIALOGUE

The debates and conceptions as postulated by various scholars through various dimensions can be categorized into three groups :

**Ecological Approach**

Most of the medical scientists first seek the vectors of health imbalances in *pathogens* (i.e., pathological enquiry), but human ecologists are first concerned with the etiology of disease within environmental perspective, so called as *geogens*. According to May, the founder of the American school of medical geography, geogens are responsible for the pathogenesis of disease.[22] He defined disease as "an alteration of living cells or tissues that jeopardizes survival in their environment".[23] His concepts of disease ecology are process oriented and he has paid particular attention to natural environmental risk factors. Elaborating his theme he remarked that three factor-groups converging into spatio-temporal frame and causing the growth of disease, are : (i) stimuli from the environment, (ii) response from a host, and (iii) conglomeration of thoughts and traits (cultural).

It is evident through many empirical studies that environmental factors are very important in the understanding of health problems. For example, a Japanese study has demonstrated strong associations between mortality rates for heart diseases and stroke and seasonal temperature variations.[24] And Howe observed that debilitating effect of polluted air is one of the responsible factors for the growth of cancer.[25] Pyle concludes that the diffusion of cholera had followed the water routes in nineteenth century U.S.A.[26] Similar observations have also been made in the case study of diffusion of cholera in England and Wales during 1957.[27]

Ignatyew, a leader of the Russian school, has advocated that there may be certain natural prerequisites within a territory which cause health imbalances. These prerequisites consist of certain properties of the geo-system that assume the form of pathological states when interacting with the human organism These manifestations take place under certain conditions dictated by the reactivity of the organism and by the character of man's economic activities.[28]

**Spatial Approach**

Although the ecological bases provide a more scientific reasoning for the analysis of problems, yet it lacks the regional identification. This gap has been fulfilled by the Hiedelberg school of medical geography.[29] In England, it has been initiated by Stamp, who stated that with the help of maps "a medical geographer can analyse environmental relationship of an organism either in disease etiology or in health ecology"[30] According to Jusatz "the best way to describe the various correlations of environment with the incidence of disease and epidemics as well as their vectors in a country is the medico-cartographical method. Without the help of geomedical maps we shall not be able to develop a regional geomedical study."[31] Jusatz's views have been further supported by McGlashan and Harington, who stated that "since pattern definities should form the beginning of etiological search, it must be followed by a rational plan of environmental monitoring."[32]

With the growing interest in computer techniques, scholars have become conscious about the computer graphics as well as statistical testing. For trend surface analysis complicated isopleths are being used, notable as examples are the medico-statistical maps of Australia prepared by Learmonth and Nichols.[33] Using standard deviation of the arrays of rates, Armstrong prepared the mortality-maps with the help of computers.[34] Some scholars have used computer graphics for various types of map preparations like droughting, diabetes mellitus, mortality ratios etc.[35] The remark of Howe is notable in this context that "regional variations in the distribution of diseases may reflect the habits, the tempo, the mental tensions and the anxieties of our twentieth century existence, our diet or the increasingly sedentary nature of our work."[36] Such maps will provide the territorial pattern which will show entirely consistent incidence ingredient of the local population, and thus it may throw light upon the spatial aspects of human health problems.[37]

**Eco-Spatial Approach**

The increasing interest of scholars of different disciplines paid special attention to the spatial perspective; in this way has grown the concept of multi-dimensional analysis.[38] The basic assumption of the ecological approach in the environmental imbalance as the causative factor. Some studies reported that technological advances bring more environmental control,[39] while some other studies observed that some technological attainments have caused the growth of different health problems.[40] According to Doll, the findings of geographical pathology have never proved to be a medical truth in the individual case since such findings necessarily represent community generalizations. Irrespective of individual proof, finding on these geographical bases represent the ecological context of disease occurrence and as such, represent one step towards the alleviation of overt human suffering.[41] Through the geographic scale of spatio-temporal frame, the cause and effect relationship of human health problems could also be analysed in a more scientific way.[42] Furthermore, it is notable that some cultural groups have particular behavioral habits which can be directly associated with specific diseases. McGlashan correlated the incidence of esophageal cancer in southern Zambia and south-eastern Malawi with the use of contaminated alcoholic spirits.[43]

At a particular point society reaches a state of socio-cultural complexity where in certain aspects the man-environmental relationship changes to cultural behavioral identification which within space become more apparent,[44] which is also to be studied by geographers through their medico-geographic techniques, thus finally one can find more about the etiology of a particular disease.[45] Within this perspective medico-geographers have shown their interests towards the empirical-inductive approach, formerly developed by Hirsch during the nineteenth century and later advanced by Markovian.[46] This approach has provided a new stimulation for model building and theory construction in the geography of health, through which the epidemiological as well as environmental hazards could be understood very clearly.[47]

At present even the ecologists have drawn attention to the spatial approach. This approach can be depicted through the model showing the anatomy and physiology of environment—its structure, processes and pathways of energy transfer. According to Dansereau the environmental regimes can be classified into three groups—minerotrophy, phytotrophy and zootrophy, in which the circulation system is processed by metabolism, elaboration, excretion, death and weathering (Fig. 16.1). The pathway of energy transfer within each level

*Environment : Anatomy and Physiology*

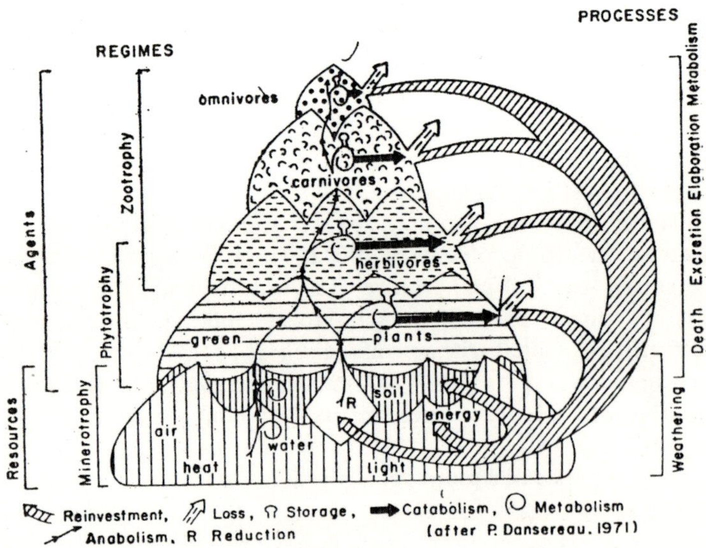


FIG. 16.1

and between levels results in storage, reinvestment and loss; hence at a certain period of time, due to environmental imbalances within a particular space, diseases start growing.[48] This model can be transformed within the geographical scheme of environmental arrangement which concern the range of physical, physico-human and human environments.[49] The latter two may be sub-divided into four groups—(a) environmental, (b) biotic, (c) behavioral, and (d) socio-economic, which are analysed through the four approaches—analytical, theoretical, spatial and applied (Fig. 16. 2). All the three environments are joint together with the human dimension in the core and the temporal dimension around it. If any disturbance takes place at any level in any period of time, that will create a disturbance in the human life cycle. Here the eco-spatial approach is fundamentally necessary because it is the only way which can provide the findings dealing with the association of disease with environmental and ecological variables.

The emergence rate of diversification and multi-dimensional analysis in such a newly advanced field is so high that now it becomes difficult to define and place the subject. To define the core concern of the *Geography of Health*, the help of the Set theory and Venn diagram (Fig. 16.3) can be taken; thus four sets of factor-groups have been considered which also contain different elements in themselves; these are (a) environmental, (b) biotic, (c) behavioral, and (d) socio-economic. The study area of these factor-groups jointly form the core concern of the geography of health. By overlapping allied interdisciplinary disciplines can also be placed, e.g., Patho-physiology, Psycho-somatic Studies, Bio-sociology and Human Ecology, etc. Taking this view in a broader sense, the Geography of Health can be defined as "*the orderly description and interpretation of the causes, processes, patterning, inter-relations, and the spatial organization of human health attributes, with varying distribution over the earth surface in the frame of environmental, ecological, and behavioral perspective*". This definition represents the concept of "holistic" approach. Consequently this discipline is interpreted as the study of total eco-spatial system. Moreover, in its own way it works towards the search for such an ideal environment, where man will get full peace as well as fullness of life.[50] Within this holistic approach 'health' is considered as

Geo-Spatial Environmental System

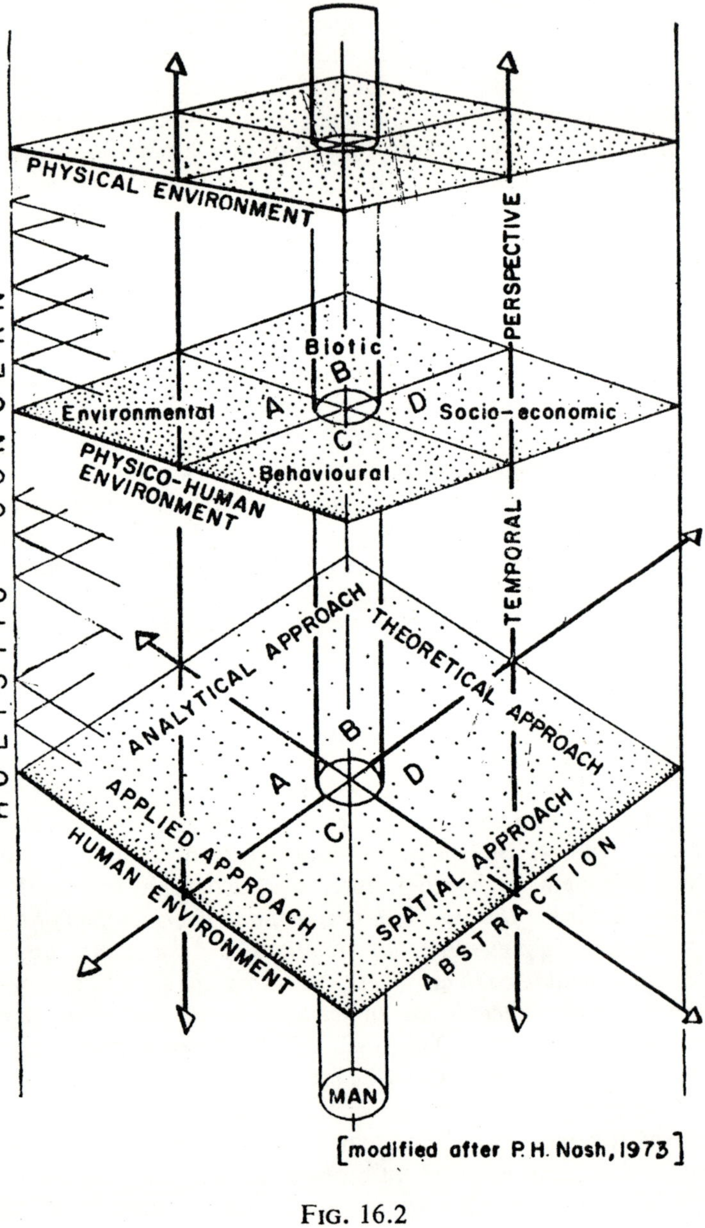


FIG. 16.2

*Place of Geography of Health in terms of Set Theory*

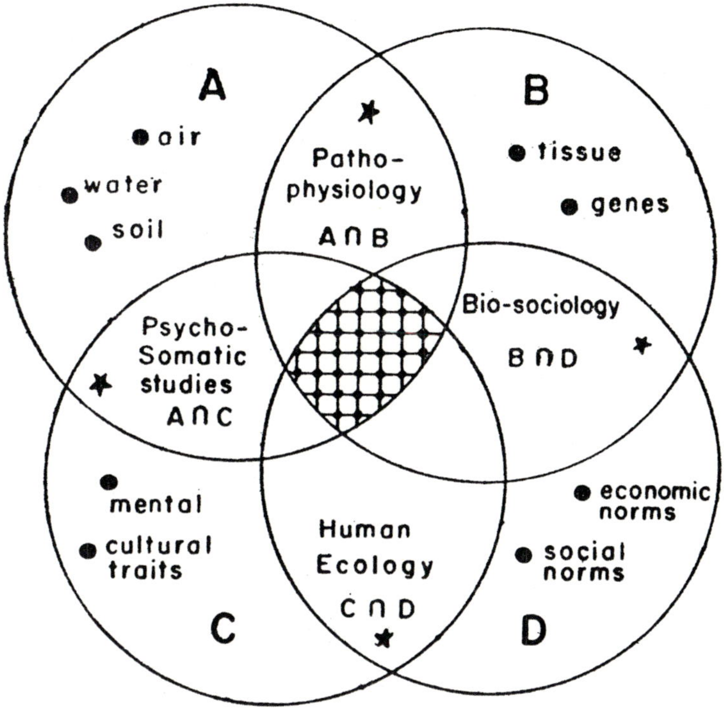


A ∩ B ∩ C ∩ D
GEOGRAPHY OF HEALTH

A Environmental Factors

B Biotic Factors

C Behavioural Factors

D Socio-Economic Factors

FIG. 16.3

"the function of pattern, processes and degrees" parallel to that of geomorphologists' "Landscape is the function of structure, process and stage" and settlement geographers'

"Morphology of settlement is the function of form, function and genesis".[51]

REFERENCES

1. See *Susruta Samhita*, ed., K.K. Bhishagratna, Varanasi: Chaukhamba Sanskrit Series, 1963, 2 Vols; *Kashyapa Samhita*, Varanasi : Chaukhamba Sanskrit Series, 1953; *Bhel Samhita*, Varanasi : Chaukhamba Vidya Bhavan, 1959.
2. *Bhel Samhita*, *Sutra* 13/2.
3. F. Adams, (transl.), *The Genuine Works of Hippocrates* 2 Vols., London-Sydenham Society, Vol. 1, 1849, p 179-222; see also W.H.S. Jones, *Hippocrates Treatise*, London, 1923.
4. G.M. Howe, *National Atlas of Disease Mortality in U.K.*, 2nd ed., London-Nelson 1970, pp 10-12.
5. T. Shapter, *History of the Cholera in Exeter in 1832* (1849), quoted in G.M. Howe, *ibid*, p 12.
6. See J.M. May, "Atlas of Diseases" *Geog. Review*, 40-4, 1950, 17 plates; E. Rodenwaldt and H.J. Jusatz *World Atlas of Epidemic Diseases*, Hamburg-Falk-Verlag, Hiedelberg-Acd. of Sc., 1952-1961.
7. G.M. Howe, op cit, ref. 4, p 1.
8. see L.D. Stamp, *Some Aspects of Medical Geography*, London, Oxford Univ. Press, 1964, *idem. The Geography of Life and Death*, London, Collins, 1964.
9. See A.T.A. Learmonth, "Medical Geography in India and Pakistan", *Geog*, JI, 127, 1961, 10-26; *idem*, "Health in Indian Subcontinent 1955-64" *Occasional Papers* No. 2, Australian National Univ., School of General Studies, Canberra, 1964.
10. H.J. Jusatz, (ed.) *Geomedical Monograph Series, Regional Studies in Geographical Medicine.* Hiedelberg—Acd. of Sc., Berlin, Springer-Verlag, Vol. I, Libya 1967, Vol. 2 Afghanistan, 1968, Vol. 3, Ethiopia, 1969, Vol. 4. Kenya, 1970.
11. See V.P. Byakov, et al., *Soviet Geography : Accomplishments and Tasks*, New York : American Geog. Soc., 1962; E.I. Ignatyev "Medical Geography and Practice in "J. Ducolot (ed.) *Melanges de Geographic Offerts a M. Omar Tulippe Liege*, Gembloux, 1966; V.V. Vorobyev, *et al*, *Principles and Methods of Medical Geography, Mapping*, Irkutsk 1968; E.I. Ignatyev, *Key Problems in Mapping in the Field of Medical Geography* Washington, Deptt. of Army transl. No. J-6384, 1969; A.A. Shoshin, *Fundamentals of Medical Geography*, Washington, Dept. of Army transl. No. J-5264, 1968.
12. See. N.D. McGlashan, Medical Geography : An Introduction in *Medical Geography, Techniques and Field Studies*, ed. N.D. McGlashan, London, Methuen & Co., 1972, pp 13-14.
13. See N.D. McGlashan and J.S. Harington, "Some Techniques for

Mapping Mortality, "*South African Geog., Jl.*, 58-1976, pp 18-24.

14. *Newsletter—Commonwealth Human Ecology Council*, No. 3, May, 1976, p 1.
15. *General Circular*, 22nd I.G.C., Montreal, Canada 1972, p 31.
16. See, *Third Gircular*, 23rd I.G.C. Moscow, U.S.S.R. 1976.
17. *W.H.O. Constitution*, Geneva : W.H.O., 1964, p 3.
18. G.M. Howe, "Plotting the Scale of Diseases" *Geog. Magazine*. 44-5, 1972.
19. S.M. Momiyama, "Signs seen in Japan of Deseasonality in Human Mortality, "*Papers in Meteorology and Geographics*, 26, 1975, 9-34.
20. G.M. Howe, "Some Aspects of Medical Geography of the United Kingdom" Paper read in *I.G.U. Commission on Medical Geography*, 21st. I.G.U., New Delhi, 1968.
21. G.M. Howe, "The Geography of Death in England and Wales", *The Lancet*, 1, 1963, 818; *idem*, *National Atlas of Disease Mortality*, ref. 4.
22. See J.M. May, "Medical Geography—Its Method and Objectives", *Geog. Review*, 40, 1950, 9-41; *idem* "Medical Geography" in *American Geography : Inventory and Prospects*, ed. P.E. James and C.F. Jones, Syracuse University Press, 1954.
23. See J.M. May, "The Ecology of Human Diseases "*Annals, N.Y. Acd. Sc.*, 84, 1960, pp 789-94.
24. K. Katyama and M.M. Sakamoto, "A Biometeorological Study of Mortality from Stroke and Heart Disease, "*Papers in Meteorology and Geophysics*, Tokyo, 21-2, 1970, pp 127-129.
25. G.M. Howe, "The Geography of Long—Bronchus Cancer and Stomach Cancer in the U.K., "*Scottish Geog., Magazine*, 87-3, 1971, pp 202-220.
26. G.F. Pyle, "The Diffusion of Cholera in the United States in the Nineteenth Century", *Geog. Analysis*, 1, 1969, pp 59-75.
27. J.M. Hunter and J.C. Young, "Diffusion of Influenza in England and Wales, "*Annals, Asso. Am. Geog.*, 61-4, 1971, pp 637-653.
28. See T.M. Gelyakov, et al., "The Present State of Medical Geography in U.S.S.R., "*Soviet Geography*, 8-4, 1967, pp 228-234; also E.I. Ignatyev, ref. 11; *idem, Medical Geography in U.S.S.R.*, Paper read in 23rd I.G.C., Moscow (July 1976).
29. See H.J. Jusatz, op. cit., ref. 6 and 10.
30. L.D. Stamp, "Geography in the Modern World", *Geog. Magazine*, 37-3, 1972, 165-167; also G.M. Howe, ref. 4.
31. H.J. Jusatz, "Method for the Completion a Regional Geomedical Study" in *21st I.G.C. 1968, Selected Papers, Vol. I. Physical Geography*, ed. S.P. Chatterjee and S.P. Dasgupta, Calcutta-N.G.G., 1970, p 413.
32. N.D. McGlashan and J. S. Harington, ref. 13; also M.J. Joshi and C.D. Deshpande, "Geographical Distribution of Some Disease Common in Southern Asia", *Geographia Medica*, 3, 1972, pp 5-29.
33. See A.T.A. Learmonth and G.C. Nichols, "*Maps of Some Stan-*

*dardized Mcrtality Ratios for Australia 1959-1963*, Canberra A.N.U., Deptt. of Geog., 1965.

34. R.W. Armstrong, "Standadrized Class Intervals and Rate Computation in Statistical Maps of Mortality", *Annals, Asso, Arn. Geog.* 59, 1969, pp 382-390.
35. See R.A..Armstrong and J. Strock, "Medical Section", *U.S. National Atlas*, Washington, D.C. Govt. Printing Office, 1970, pp 251-254; also N.D. McGlashan and D.H Bond, "A Method for Using Computer Assistance in Mapping Geogrphical Synoptic Data", *Die Erde*, 98, 1967, 292-7; G.M. Howe, "Some Recent Developments in Disease Mapping, "*Royal Scottish Health Jl.*, 90, 1970, pp 16-20.
36. G.M. Howe, op. cit., ref. 4, p 1.
37. E.F. Rose and N.D. McGlashan, "The Spatial Distribution of Oesophageal Carcinoma in the Transkei, South Africa" *British Jl, of Cancer*, 31, 1975, pp 197-206.
38. See J.M. Hunter (ed.), *The Geography of Health and Disease*, Chapel Hill-Univ. of North Carolina, Deptt. of Geog., Studies No. 6, 1974.
39. See R. Doll (ed.), *Methods of Geographical Pathology*, Oxford : Blackwell, 1959.
40. A.L. Banks, "The Study of Geography of Disease" *Geog. Jl.*, 125, 1959, pp 199-210.
41. R. Doll, op. cit., ref 39.
42. J.S. Harington, "A Spatial and Temporal Analysis of Four Cancers in Africa", *British Jl. of Cancer*, 31 1965, 665-678; also K.M. Kwofie, "A Spatio-temporal Analysis of Cholera Diffusion in Western Africa", *Econ. Geog.*, 58-2, 1976, pp 127-135.
43. N.D. McGlashan, "Esophageal Cancer and Contaminated Alcoholic Spirits, An Exercise in Medical Geography" *International Pathology*, 9-3, 4, 1968, pp 50-51.
44. G.F. Pyle, "Introduction, Foundation of Medical Geography", Econ. Geog., 58-2, 1976, 99; *idem Measles as an Urban Health Problem*; The Akron Example, "Paper Presented in 22nd I.G.U. Commission on Medical Geography, 1972, also G.M. Howe and J. Loraine (ed.), *Environmental Medicine*, London—Heinemann Medical Book Ltd., 1974.
45. R.W. Armstrong, "Medical Geography—An Emerging Speciality", *International Pathology*, 6, 1975, pp 61-63.
46. A. Hirsch, *Handbook of Geographical and Historical Pathology*, 3 Vols, London—New Sydenham Society (1833, 1885, 1886); also A.P. Markovian, "Historical Sketch of the Development of Soviet Medical Geography", *Soviet Geog.*, 8, 1962, pp 3-19.
47. See A.A. Brownlea, "Modelling the Geographic Epidemiology of Infectious Hepatities" in *Medical Geography*, op. cit., ref. 12, p 279-300; P. Haggett, "Contagious Processes in a Planer Graph : An Epidemiological Application" in *Medical Geography*, op. cit., ref. 12, pp 307-324; R.W. Armstrong, "The Geography of Specific Environ-

ments of Patients and Non-patient in Cancer Studies with a Malaysian Example" *Econ. Geog.*, 58, 1976, pp 161-170.

48. P. Dansereau, "Dimensions of Environmental Quality" *Sarracenia*, 14, 1971, 9-24, pp 78-79.
49. See P.N. Nash, "Emerging Trends in Environmental Training of Geography in North America : Antipode to Academiasis and Diplomania" in *Applied Geography and the Human Environment*, ed. R.E. Preston, Waterloo, Deptt. of Geog., Univ. of Waterloo, 1973, 326-329.
50. See R.P. Mishra, *Medical Geography of India*, New Delhi : N.B.T. (1970), 9; also Razia Khan, "Purpose, Scope and Progress of Medical Geography", *Indian Geog. Jl.*, 46, 1971, pp 1-9.
51. See R.L. Singh, K.N. Singh and Rana P.B. Singh, eds, *Readings in Rural Settlement Geography*, Varanasi—National Geog. Soc. India, Res. Pub. No. 14, 1975, viii.

# 17

# A Note on Medical Facilities in Maharashtra

—*S.D. Shinde*

## INTRODUCTION

HEALTH care in India like several developing countries of the world currently faces a number of serious crises in cost, quality of care and equitable distribution of modes and standards of service to the population as a whole. Systematic planning to improve existing services and to implement new ones has been gaining momentum only since 1951 under different schemes of governmental assistance. But a notable lack in most health planning is a concern for locational factors, especially regarding resources. Health plans, according to Frieden and Peters are 'map shy'. This is where medical geographers could make valuable contributions.[1] This study attempts to portray regional disparities in the medical facilities available in Maharashtra, the third largest State of India in size as well as population.

## DATA AND METHODOLOGY

Demographic and medical data relevant for the present study have been abstracted from publications[2,3] of the Central and State Government.

Medical data refers to an arbitrary measure, viz., the number of general use in-patient beds in any hospital. Doctors in private practices largely concentrated in the relatively fewer urban centres are obviously left out. Hence, inaccuracies are

Thanks are due to A.T.A. Learmonth of the Open University, U.K. for his suggestions in the preparation of this paper.

inevitable. One such stems from the fact that hospitals whose staff and facilities are not proportional to the number of beds; second would occur if a certain hospital had an out-patient disproportionately larger (or small), when compared to its bed numbers. Moreover, the 'sphere of influence' of any hospital is also difficult to define. For, it often differs with the patient's status—physical as well as monetary. Several potential patients within the sphere of influence (however defined) will not be present at a given hospital for treatment. A majority of these will be those who seek customary remedies or those who, not understanding the nature of their complaint, are put off attending by such practical considerations, as the hardship, inconvenience or cost of the journey.[4] It may be pointed out that lack of research facilities and the large size of the study area left little scope to undertake a geographical pathology survey. Hence the analysis of the distributional pattern of medical facilities is largely based on the documentary data mentioned above.

The dots from Fig. 17.1 within each district give a figure for total population for comparison with the in-patient facilities (see Table 17.1).

TABLE 17.1
*Maharashtra Hospitals*

| Sl. No. | Hospital | (a) General beds | (b) Population served | b/a | 'wl' Factor | Class* |
|---|---|---|---|---|---|---|
| 1 | 2 | 3 | 4 | 5 | 6 | 7 |
| | *Districts* | | | | | |
| 1. | Greater Bombay | 13,540 | 5,970,575 | 441 | 4 | A |
| 2. | Thana | 2,579 | 2,281,664 | 885 | 9 | A |
| 3. | Kolaba | 192 | 1,263,002 | 6,578 | 66 | D |
| 4. | Ratnagiri | 1,060 | 1,990,583 | 1,878 | 19 | B |
| 5. | Nasik | 651 | 2,369,221 | 3,639 | 36 | C |
| 6. | Dhulia | 374 | 1,662,181 | 4,444 | 44 | D |
| 7. | Jalgaon | 1,033 | 2,123,121 | 2,055 | 21 | C |
| 8. | Ahmadnagar | 959 | 2,269,117 | 2,366 | 24 | C |
| 9. | Pune | 5,992 | 3,178,029 | 530 | 5 | A |
| 10. | Satara | 816 | 1,727,376 | 2,117 | 21 | C |
| 11. | Sangli | 1,151 | 1,539,820 | 1,338 | 13 | A |
| 12. | Sholapur | 1,350 | 2,253,840 | 1,669 | 17 | B |

TABLE 17.1—*contd.*

| 1 | 2 | 3 | 4 | 5 | 6 | 7 |
|---|---|---|---|---|---|---|
| 13. | Kolhapur | 668 | 2,048,049 | 3,066 | 31 | C |
| 14. | Aurangabad | 927 | 1,971,006 | 2,126 | 21 | C |
| 15. | Parbhani | 152 | 1,506,771 | 9,913 | 99 | D |
| 16. | Bhir | 506 | 1,286,121 | 2,542 | 25 | C |
| 17. | Nanded | 189 | 1,397,762 | 7,396 | 74 | D |
| 18. | Osmanabad | 197 | 1,896,687 | 9,628 | 96 | D |
| 19. | Buldhana | 481 | 1,262,978 | 2,626 | 26 | C |
| 20. | Akola | 831 | 1,501,478 | 1,807 | 18 | B |
| 21. | Amravati | 936 | 1,541,209 | 1,647 | 16 | B |
| 22. | Yeotmal | 357 | 1,426,677 | 3,996 | 40 | C |
| 23. | Wardha | 288 | 7,79,562 | 2,707 | 27 | C |
| 24. | Nagpur | 3,793 | 1,942,688 | 512 | 5 | A |
| 25. | Bhandara | 303 | 1,585,580 | 5,232 | 52 | D |
| 26. | Chandrapur | 245 | 1,640,137 | 6,694 | 67 | D |
| | Maharashtra State Total | 39,570 | 50,412,235 | 1,274 | 13 | State Average |

*See Fig. 16.1 key for class intervals.

From these two figures in columns a and b a calculation has been made to show the number of people theoretically served by each in-patient bed. The 'work-load factor'[5] defined as:

$$wl = \frac{\text{Population served in hundreds}}{\text{General use in-patient beds}}$$

is tabulated and then classed. The work-loads are classed by scatter diagram and are based on comparison with the State average which is obtained from the fact that some 39,570 beds are available to some fifty million people.

## DISCUSSION

Applying the above yardstick, it is observed that the national average work-load is twenty. It is of interest to note that the State average is above the national average. Maharashtra as a whole thus appears to be a relatively better served State of the Indian Union in respect of medical facilities. However, when district-level work-loads are examined tremendous regional disparities are revealed. Only five of the total twenty-six districts belong to 'A' class, namely, Greater Bombay, Pune, Thana, Sangli and Nagpur districts. Here medical facilities appear to be relatively better mainly bacause of the larger proportion of urban population. Next in rank are the four districts Ratnagrii, Sholapur, Akola and Amravati which belong to 'B' class wherein the work-load averages are slightly above the State average. 'A' and 'B' classes if taken together for discussion purposes thus involve nine districts of the State and may be described as regions of Maharashtra having better medical facilities.

Rest of the seventeen districts covering most of Maharashtra and its population are worse in respect of medical facilities as can clearly be seen on Fig. 17.1 where

*Maharashtra Hospital Areas and Work-Load Factors*

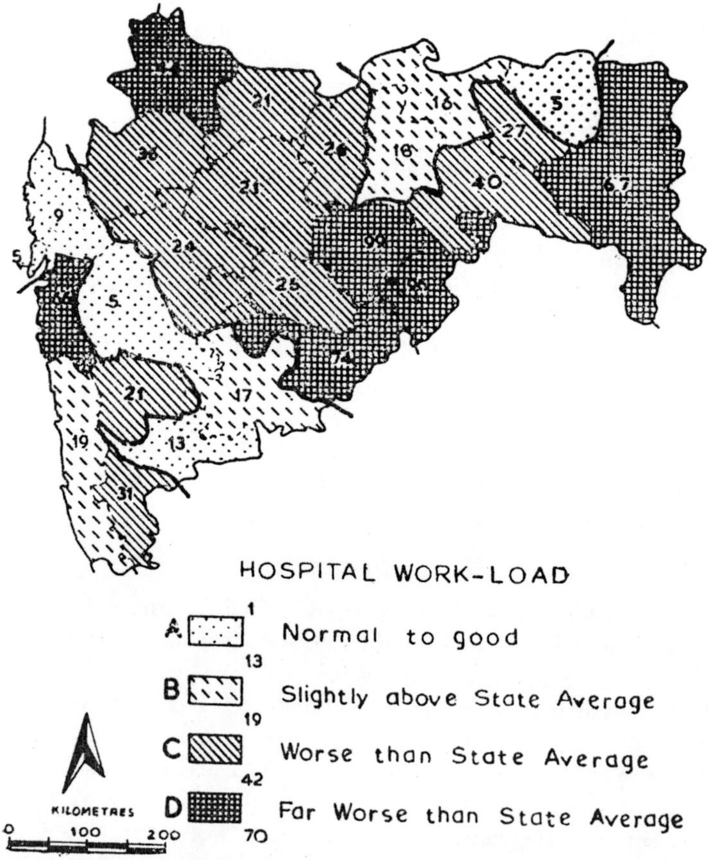


FIG. 17.1

'C' class dominates the distributional patterns. Parbhani and Osmanabad districts of Marathwada are the worst of the lot indicating virtual absence of medical facilities. On the whole large sectors of the State particularly rural areas have poor medical facilities much below the average for the State. There is thus a concentration of medical facilities amongst the urban population leaving majority of the rural population without any semblance of medical facilities.

## CONCLUSION

The analysis of the distributional pattern of medical facilities clearly reveals that only nine per cent of the total population is relatively better served, whereas over ninety per cent remains under-served. Though such glaring disparities are highlighted by the work-load concept, it must be emphasized that it has two weaknesses.[6] Firstly, it fails to allow for the number of out-patient clinics/dispensaries which effectively reduce the actual load on hospital facilities in most developing countries. Secondly, several types of diseases mainly experienced in the study area are such that they are fairly adequately treated by clinic/dispensary rather than hospital attendance. All the same some such device of suggesting areas of population under-served by medical facilities is undoubtedly a useful and necessary study.

## REFERENCES

1. R.W. Armstrong, 'Medical Geography and Health Planning in the United States : Prospects and Concepts', in N.D. McGlashan, (ed.), *Medical Geography : Techniques and Field Studies* (London), 1972, pp 119-132.
2. Census of India, *Population Tables*, Maharashtra State, Government Publication, 1971.
3. Maharashtra State, *Statistical Abstract*, Government Publication, 1972.
4. N.D. McGlashan, (ed.), '*Medical Geography* : *Techniques* and *Field Studies*, (London), 1972.
5. McGlashan, op. cit., 1972, p 94.
6. McGlashan, op. cit., 1972, p 95.

# 18

# Nature of Political Geography

—*Sunita Devi*

POLITICAL Geography is a branch of human geography but this subject is mainly concerned with the areal unit of the State. Geography deals with the earth as the home of man. On the other hand, man is the creator and designer of politics. The word 'political' has multiple meanings within the defined areal limit, the condition of government, sovereignty, territorial organization of the State and law-abiding people of which the politics and geography both are devised by man. In order to know the nature of political geography, one has to learn the nature of geography and political science both as systematic disciplines and present a view to generate some incentive for inter-disciplinary research.

The origin and growth of the political system or sub-system in any country is dependent upon the social systems prevailing there. Political systems are the outcome of the economic and social development of any society. In a developed social condition political unity among different regions is clearly discernible, which is heterogeneous in areas of less developed society with a number of local feudal heads.

Political units are more defined than regional units. Political units will peril a man if he were to ignore them. Thus political units and boundaries are real. The creation of a wall on the boundaries is the work of power politics made by man.

Political geography utilizes a great deal of material provided by political science and other social sciences. Hence political geography deals with the political phenomena in space, and space is the power of geography. Geographical and ecological factors are responsible for political phenomena. Writing the geography of a political region should not be included in political geography. The purpose of this paper is to highlight the nature and concepts of political geography with a view to analyse the political and geographical problems as an inter-disciplinary approach in social science study.

## MEANING

The meaning of political geography is mainly concerned with the spatial relationship of states and cultural relationship of people in a particular territorial organisation as functional, structural and behavioral aspects which determine the internal and external relationship of the inhabitants of one particular area with the other. According to Hartshorne, "It is a subject which studies the areal differentiation and similarities in political character as an inter-related part of the total complex of areal differences and similarities."[1]

Political Geography is also concerned with the spatial interaction between political and geographical phenomena, and one of the fundamental issues concerning the location of the subject in the spectrum between geography and political science.[2]

According to Alexander "Political Geographers are not related with the form and structure of the Government by itself or the division of functions among the executive, legislative, and judicial branches or whether the same may be democratic or totalitarian."[3] In 1971, Cohen and Rosenthal felt that without greater attention to political science, geographical insights are likely to be limited and sterile.[4] Political geography creates a greater link between geography and political science and it has greater importance as a separate discipline of study for both geographers and political scientists as the spatial perspectives of political phenomena are the hallmark of the subject. It is a subject which is overwhelmingly concerned with the practical and theoretical nature of problems with special reference to man made boundaries, States organisation, power development, war and the growth and expansion of the State in both the vertical and horizontal dimensions.

In this way it can easily be said that the development of political geography is largely dependent upon the inter-disciplinary research, a perspective training in history, geography, humanities with greater emphasis of social sciences.

Recently the studies of political geography reveal that it is very close to the international relations, statistics, sociology, international law, economics, psychology, ethnology, historical relations of trade etc. This shows that it is beyond the control of man to have a practical and thorough grounding in all these disciplines. Thus the strength of the subject lies not only in the context of regional geography but all branches of political science, geography and history combined together.

Recently, the subject matter of political geography shifted towards the analysis of the political process in geographical areas. The political system is a technique to analyse the political variation between various countries of inside a State which are the result of conscious ideas, decisions, movements and fields of problems. Thus the environmental consequences of political decisions and actions governed the methods, principles, approaches and concepts of the political geography. Thus it can be said that political geography is intimately related with national problems.

## SCOPE

The scope of political geography includes the political map of the world, distributional patterns of States of various kinds which have no resemblance with the natural regions formulated by the geographers. In the field of political geography, the relationship between community and physical environment is considered as most of the States are constituent elements of the world political structure which are intimately related with the States and in case of integration the test of the ability of policy stands as real carriers or barriers amongst the various countries. The division of internal and external relationship should not be over emphasized but it provides a useful line of approach in the analytical treatment of the subject. With the failure of internal harmony among the constituent States result in disunity and instability which may culminate as international warfare among the States. The States having geographical environment possess three essential elements, i.e., territory, people and organization in general which may be analysed, described and mapped clearly with greater accuracy and precision. But in each political system the inhabitants' activities are generally guided by the available environmental conditions of a given territorial space.

There cannot be a State without people, but there may be people without State, which is termed as Stateless persons. There cannot be an uninhabited State because territory and people are integral parts of the State. It may be regarded as 'geographical determinism', because inhabitants of a State are the product of its environmental conditions, cultural, political and economic organisations which will perhaps continue as long as the world exist. In political affairs it is unwise to impose one's political system in other States, if the same is found unsuccessful in case of the other. Thus each and every State is the outcome of territory, people and their relationship which are generally considered out of the scope of political geography. The physical aspects of the States come under the purview of physical geography. Population distribution and density are generally studies in demography. Similarly, economists, historians, sociologists and psychologists all have their interest in analysing the different aspects of States but political geographers study the above characteristic features of States with a view to analyse the social and ecological conditions as symbolic characters of a State.

In dualism the relationship of political geography with other subjects is considered significant in the internal and external functions of a State for the welfare of the people because any unsatisfactory external relationship may disturb the internal condition up to the point which is not tolerable to the inhabitants. Thus the political geographer's duty is to analyse and record those geographical and political factors of human relationship without which the existence of society is impossible.

In political geography, studies are generally made about the dynamic character of the State during different historical periods, because changes through external relationship are associated with the precision of the adoption of environment by man which is described as the "Conquest of Nature."

With the cultural advancement of one society in material wants a man can be mitigated in the present day scientific and technical world with achievements which would never have been the possibilities in the past. Millions of Chinese and

Indians, every year starve due to the lack of food but it cannot be said in this regard that the material culture has reached a satisfactory level, because the lack of transport facility and the organizational power of the people have decreased the hope of agricultural and industrial development. Only through which the desired level of social welfare is possible, however, the development of political and social consciousness developed the internal and external conditions in relation to States and they cannot be separated from the internal condition as in North America and Europe; the white Australian policy has no repercussion but the same may foster the military in Japan which may lead to a condition of warfare.

The political, moral and economic supports are generally derived with the increase of communication lines, printing press, propagation of ideas, means of communication like the radio and cinema, which have both direct and indirect relationship in changing the conditions of the State. It cannot be ignored because all this forms the part of the environment in which people live and conversely it has a direct influence on the form and character of the State in the external relationship. In India the mode of existence of States of the old order have now completely changed because with the change of administrators the political unity of the country has changed and one cannot say with firm belief that the same condition will remain in future. In this way the political and historical conditions including the geographical framework may alter or it may not, but no modification can be forbidden with the change of environmental conditions because no change in political organization in society can keep pace with the economic change. In any change of political structure the three basic needs of mankind food, cloth and shelter must follow the line because "the main functions of the political structure of the State is to organize and facilitate the economic activities of its inhabitants."[5]

Thus politicians and economists should consider the concepts of changing conditions in a changing world. The political geographer is considered with the observation, recording and analysis of the changes for the present day world affairs. The State is consciously and critically created by the inhabitants as a political unit which is completely different from geographical regions. Uptill now, in practice, it has been completely impossible for a man to devise a State which coincides in all its entirety with the natural regions of a geographer. The political map of India for example illustrates such conflicts clearly. The creation of a State is purely artificial in character. The size, shape and character as created by man on the other hand, are unreal in comparison with the natural boundaries. In this sense a confusion is created in between the physical and artificial boundaries of the State. In this respect the political geographer has the advantage of handling a large number of data whereas the students of physical and human geography are quite often embarrassed with the non-availability of statistical evidence for their geographical regions because in the present day world affairs administrative consciousness is generally measured with the help of political boundaries. In this sense the scope of political geography is never seem to be narrow.

The political societies require to formulate an operative system through the process of feed-back-mechanism which do not imperil the survival of human society. The degree of tolerance must be introduced in a political system as an essential prerequisite which was adhered in the geopolitics of Germany specially in the first and second world war, as in political geography many investigators are searching for evidence in support of the 'Nazi Doctrines'. Ratzel as the father of modern political geography streamlined the aims and objectives of political geography with special reference to State relationship. It is yet not known as how far the geo-political thinking of Houshofer influenced the "German efforts to establish a new world order is not clear, but there can be no doubt that there was a close connection between geo-political thinking and planning in pre-war German."[6]

Each and every State and the political organizations of the world are the result of historical evolution; but in political geography overemphasis should not be placed on history because it "leads to a rigidity in outlook and fosters unwillingly to change shapes and boundaries of States with resultant liabilities to friction and clashes". The development of a political unit must grasp the existing system of approach implied in the past and present conditions, which are helpful in the understanding of political pattern and relationship of

States. The evolution of States comes under the sphere of historical geography but political geography is concerned with the present correlation of external and internal relationship of the State as derived from the past.

Thus there is a close relationship between political and historical geography. Because historical geography deals with the people and environment of various stages in the past whereas political geography is concerned with the present day dynamic evolution of the State. It does not formulate inter-State policies but the result achieved by the politico-geographical studies are considered for field investigation by the politician and statesman. In order to promote the understanding of political systems it is not desirable to give a slogan and myth which has no foundation giving a wider defensive area and a pertinent system of government.

The regularities and hierarchy of different administrative units lead to a system of political boundaries and frontiers under the purview of determinism, possibilism, probabilism, regionalism, rationalism and cognitive behavioralism of any society. Recently, the field of political geography has not wider dimension with the conclusion of several models dynamic descriptive, mathematical and geometrical in nature. The use of cartographic techniques and behaviour analysis of different political systems in the world give rise to a scientific thinking on the subject on a firm statistical footing.

## CONCEPTS

The ideas about political regions is quite pertinent aspect of political geography except uninhabited parts in polar regions, most of the other areas of the world are filled with the sovereign States with political boundaries and people, which are closely related with their geographical and historical milieu. The political regions are also related with the different stages of social and political organization which change the level of development in different politico-geographical circumstances, as for example the Santal of the Rajmahal highlands of Bihar was primarily a hunting and collecting tribe, precariously surviving under hostile environment but now with the cultural advancement they have organized clan territories besides the political boundaries imposed by the State.

The concept of State capital is the place within a State, where the political authorities are concentrated as the place, is the seat of the State Assembly, the residence of the head of the State, Ministers and Government departments including the embassies of the foreign governments.

The concept of core area in political geography indicates the State which forms a nucleus of the State on which the States have grown or which is a focal area binding the other States in geographical thinking.

The concept of heartland denotes the central part or the core area of the old world (the world island of Europe, Asia, Africa) has been introduced by H.J. MacKinder in 1904 and developed by him in 1919. The idea is as follows :[7]

> "Who rules East Europe commands the Heartland,
> Who rules the Heartland commands the world island.
> Who rules the world island commands the world".
> (*Democratic Ideas and Reality*)

The concept of effective State area is a valuable concept through which the Government will control and effective human occupancy on a particular part of land is generally determined. This idea was devised by Whittlesey. He calls it "ecumene". According to James, it is regarded as "effective, national territory", while this concept was more recently employed by Zaidi as "effective State area". All these writers considered it as integrated portion of the State-territory in economic and functional terms while the ecumene is "the portion of the State, that support the high density of population and has the closest transportation line."[8]

The effective national territory was defined, as "only that part of the total territory which actually contributes to the economic support of the citizens of the country."[9]

The concept of enclave is an outlying portion of the State entirely surrounded by the territory of another State. As for example, Belgium territory lying within the Netherlands in Europe. It may be used for other administrative States as outlying part of a State. On the other hand, a small concentration of an ethnic group sorrounded by others, as for example, Italians in Trieste.

The concept of politico-territorial anomalies on the surface of the earth divided into a set pattern of separate sovereignty dividing into a number of their colonies under the conceptual development of political territorial anomalies. The concepts of divided States condominia neutral territories are subject to international political understanding and agreement between any two countries. The division of Vietnam, Korea and Germany into two parts gave the idea of divided States, just like watersheds of the ridge from where two ideological heart lines prevailed instead of the inherent characteristics of the concerned States. In such a situation, the government generally rules over only half of the divided State but the same claimed to represent the State as a whole though in a real sense the temporary division of the State has become entrenched as a boundary between two dissimilar ideologies.

The concept of condominia sovereignty exercised jointly by two States. In Europe new Hebrides was in such a position in 1906, 1914 and 1922. The concept of free cities just like free ports, which generally originated in exceptional circumstances. In Europe the case of Danzig city was such which is part of the Baltic Sea and is free from any custom duty for the ships of any country. Such cities are also known as ports of choice. In a free port generally no duty is charged on imports handled. On such ports it is not possible to prevent the smuggling of goods into the sorrounding areas. Such ports are rarely found and they have very small areal extent with low tariffs e.g. Hong Kong in South East Asia. The area of free-zones are generally known as entrepot ports such as Singapore and London. On such places the entry and exit can be controlled and where goods in transit are loaded and off-loaded without paying tariff to the State in question, although the free zone is the part and parcel of the sovereign part of the State. The concept of extra-territoriality is concerned with rights awarded to one State in the territory of another. This concept found its greatest expression in China following the opium wars, when European powers forced rights of commercial exploitation in the maintenance of their own armed forces by the defeated Chinese forces.

The concept of natural zones and demilitarised areas may be adopted by the State as a deliberate policy in which a compromise of State sovereignty does not seem to be evolved. Neutrality may also be applied only a portion of the State e.g. Norway has agreed not to militarise Spitsbergen.

The concepts of size, shape and location of the State are generally different from one another with some uniqueness from one State to another. As for example one country may be located where it is difficult to penetrate geographically, climatologically, geologically, characteristics of soil, production of crops, vegetation, resource development and approachability and nearness to their neighbours. Under geo-political framework the location of State can be organized into different form which are the outcome of geographical condition, economic personality of the region and international relations with the neighbouring State.

The concept of land-locked State is just like an island which has an uninterrupted area having direct access to all directions of the sea. The micro-States of the oceanic island are examples of land-locked States.

In political geography, the structure, process and stage are defined just like Davisian concept of cycle of erosion in the fluvial landscape. The political structure is determined by the coherency, permanency and efficiency and rigidity of political organization. The process has been defined as the succession of events, actions and operations that human being employed to establish and maintain the political system. The political system is generally composed of a political process in spatial dimensions of space. The political process of landscape changed results from transaction of specific decisions made, but not always through a formal political structure of the government.

In political geography the perception and decision-making process are the result of behavioral analysis of political environment. In States, perception is based on the fact that human decisions are not formed in close relationship with the real world situation but relied upon perception of reality which are always incomplete and inaccurate. The closer the study of the real political decision-makers' perception, the more successful will be the policy adopted. This is the outcome of environment on a spectrum between the opposed extreme environmental determinism, possibilism, racialism and regionalism.

The concept of State in political geography exists when the area has population and territory under its own sovereign government. Thus the State is a form of political organization of people and territory. It is an obvious fact that without an area and population there could not be a State. The concept of nation denotes a group of small provincial States with certain uniform characteristics of administration and government in which case the population of a State need not constitute a nation, which originally coincides with the boundary of the State.

The concept of the State idea was postulated by Preston E. Jomes which described a set of distinctive purposes to which the bulk of the population subscribes, a complex of shared tradition, experience and objectives. The State idea generally varies in potency from time to time and place to place. A State idea of developed nature will continue for a longer period to present new positive goals to the population of the State.

The concept of *raison d'etre* of the State of Hartshorne is like the State idea and political idea of Ratzel which is the part of the perceptional approach to political-geography under the concept of centripetal forces of the States' organizations that are tied together and work for the survival of the State and the concept of centrifugal forces which threaten to tear apart the State into its different constituent units.

The concept of "Nation" implies a common cultural symbol, a particular view of the world which is distinct from other world views. What marks a nation different from other cultural groups, however, is that one of the symbols associated with its values and attitude is a particular piece of territory."[10] Although, "State" refers to an organization, of people and territory whereas nation is a form of human group varies, in objectives of cultural characteristic values, aspirations of political association on particular territorial dimension. The concept of nationalism refers to strong national consciousness which is acquired after the nationalist struggle. Nationalism has developed with the integration of divers cultural strength. The concept of world nationalism refers to the acceptance of a particular programme by people of colonialism who need not constitute nations in the traditionally accepted sense. People of cultural diversities may not generate an on-going nationalism, often the expulsion of the colonial power. The concept of pan-nationalism deals with the larger groups of people having a less integrated nation and their emanity around a limited range of objective known as pan-nationalism, as for example pan African movement. The concept of community conflict nationalism deals with the civil tension which divides one community from another but that can be measured with cultural coherence and the same can be analysed in the social and economic context. There may be interrupted nationalist struggle e.g. black Muslims, Ulster, Catholic, etc. The concept of totalitarian nationalism is exceptionally observed in the history of a nation, when a party may come to power claiming to represent the will of the nation on the basis of myth building, xenophilia and racialism. The concept of national independence is a condition to national sovereignty which produces a far reaching expression in landscape change. The concept of national character is the product of complex fabric of nationhood and the same can be produced with the manifestation of politico-geographical process. The concept of multi-national State refers that not all States are nation States and no nation State expresses the spatial distribution of the nation and ethnic differentiation. Ethnic differentiation is measured with the help of the formula as :

$$S = 100 \sum_{i=1}^{K} \frac{N_i^2}{N^2}$$

Where N is the size of the ethnic group i, ΣN is the total population of the country and K is the number of ethnic group. The lower value indicates the higher degree of ethnic differentiation.

Land frontiers are indicated by a particular State in the expression of effective State area. Frontier refers to the organised portion of boundaries. It is also known as political frontiers or settlement frontiers. The location of borders is the physical structure of the boundary, customs posts, boundary marks and defensive installations. The structure of border is dependent upon the nature of international relations between neighbouring States. Boundary may create a barrier or the limitations of the areal expansion of a country.

In political-geography, the concept of international system establishes a equilibrium between people, territory and the government. The international system is anarchical in nature which backs a supreme authority who impose order upon the sovereign States. The stıuggle for power in an international system has given rise to the concept of State power besides morphological power, demographical power, economic power, organizational power and power from external relationship.

The concept of international organization is very important in political geography which has been divided into the General Assembly, Security Council, the Economic and Social Council and the Trusteeship Council, International Court of Justice and the Secretariat, on the global level. International organization also exists on economic, defence and political levels. When different political regions form associations through joint membership of international organization, an alternative of this form of association of political regions is known as the imperial system.

The political region and State encompasses global, international, regional and local political systems. In political geography, the study of geo-political approaches, geographical approaches, electoral geography, internal administrative division, administrative area reform and the system approaches like open system, closed system, isolated system, sub-system, etc. are generally studied along with the general political system.

## METHODS OF STUDY

Political geography thrived on methods developed in political science, geography, international relations, statistics, sociology and economics. Thus the political geographer must be at once an artist, economists, psychologist, sociologist, geographer and statistician. The work of the political geographer is to examine the land of countries having different political boundaries, frontiers, border disputes, seas and various wealth such as transport linkages, mapping, correlation, land investigation and comparative studies, etc.

In most parts of the present study, a systematic approach has been applied on firm statistical footing to analyse the pertinent aspects of political geography after studying the huge amount of literature. An analytical view has been kept in mind so that the firm description of each element could be examined with a critical eye and its relationship with other elements in space and time could be determined. Finally, quantitative methods have been used and the planning of political systems has suggested integrated development of political frontiers and boundaries on provincial, regional, local and international levels in India.

The purpose of the present study is to correlate a simple and comprehensive model with valid tests so that particular behaviour is explained from these points of view. The discipline of political geography has been examined here after considering the historical perspective analysing the case study, the experimental methods, the statistical analysis, perception and behavioral analysis and improving research technique etc. either to complement or supplement each other.

In preparation of this paper a large number of techniques have been employed in the preparation of maps and diagrams, political maps, scatter diagram, models, choropleth and isopleth techniques are frequently used. Several maps and cartograms have been used to illustrate the result of various statistical techniques to analyse and properly interpret the data available.

The study of political geography through system analysis, comparative study of historical literature and the model building are changing the methods of study fast.

## APPROACHES OF STUDY

In political geography a large number of approaches have been applied in recent years which aim at analysing the generalised conditions of geographical concepts developing in relation to the artificial boundaries and frontiers created by man on the earth surface. Some of the important approaches are as follows :

### Retrogressive Approach

In political geography this approach of study, has been employed after considering the historical chronicles, literature, monumental design, relief features and the picturesque life sketches of countries, States, kingdoms and local boundaries.

In this approach an investigator goes a long way back in reverse direction to know the origin, growth and development of their States boundaries and the systems developed in close connection with the neighbouring States.

**Retrospective Approach**

The investigation is not only entitled to search about the present political system of a nation but he goes a long way back to know the conditions that prevailed there in the past. In this sense it has been rightly said that the present is the key to the past, but it is a reverse thinking because the present landscape features of political system are the development of past political systems on a terrain e.g. the growth of Lothal village in the valley of Indus which reminds us of the glorious past of this valley civilization which came to light with the excavations of Mohan Jodaro and Harrappa in Indus valley itself.

**Prospective Approach**

In prospective approach the future conditions are generally projected after analysing past literature and the present situation for a particular political system developed in any part of the world. In this sense it has been rightly said that the past is not only the key to the present but also for the future. This is the best way of studying political system of any country of the world, because it considers the past, present and future trends at the same time.

**Case Study Approach**

After the selection of representative items from a large number of distributed items in space, an investigator tries to look for the generalised conditions of most of the items, only after studying a number of them. This study is preferred by the investigator because it lacks subjectivity, incurs less expenditure, involves less time in the handling a large number of data to avoid cumbersome studies of each and every individual. In this approach the detailed analysis of political systems of a political territorial space, even a micro level will justify the point in order. In most of the research, this approach is frequently used in almost all branches of social and physical sciences including spatial analysis.

**Regional Approach**

In the regional approach a political geographer tries to cover the physical unit of a landscape having some homogeneous regional coherency. Indeed, it provides a firm base on which the spatial analysis could be judged because heterogeneity will never be studied in a region whenever and wherever the same persists. That should be studied in an exclusive way. This approach is highly suitable in analysing the political systems, i.e., in peninsular India, Lower Ganga Plain, Middle Ganga Plain, Assam Valley, Malabar Coast, Konkan region etc. Under this approach the group of individuals are studied in association as linking thread from one political unit to the other. This idea gave rise to a hierarchy of political systems, beginning from a family level to the tola, village panchayat, thana, anchal sub-division, district, town and national levels.

**Principle Approach**

In political geography the principle approach is generally applied to similar political conditions available as the same had been studied by other scholars in any other parts of the world e.g. the organic and inorganic theory of State development, the concept of heartland, ring land, nationalism, etc. could be examined in detail. Even in Indian conditions the same can be applied with some generalisation after considering the commission and omission of such elements of the rime land, heart land and organic theory of State development.

**Genetic Approach**

This approach implies the origin of political system in relation to a particular environmental condition and landscape feature, cultural elements, social system and economic advancement of certain parts of the earth surface. In order to understand the basic merit and demerit of political systems, this approach is highly used for the enactment and modification of certain regional, inter-regional, national and international political systems of a State. Under this approach the origin and growth of the development of the nation comes to light and that will help a lot in understanding the political enigma of a nation.

**Remedial Approach**

In the remedial approach of the study of Political Geography any previous approach can be applied in order to understand the political systems of a nation in a better way in comparison with the other approaches. Under this approach the enrichment of the subject is continuously in vogue with the inclusion of a large number of techniques and methods of study after discarding the past one. This shows that the growth and development of a subject in relation to cultural advancement of a nation and the people as well.

**Predictive Approach**

In predictive approach the projections and predictions of future about the political system is possible after considering the past trends through arithmetic, geometric and extrapolation methods. Under this approach it is possible to analyse the future conditions of landscape after considering the past trends, struggle for such development and intention to have such development.

**Prescriptive Approach**

In Political Geography there are numerous problems which cannot be solved by analysing the present problem. The investigator has to throw ligh t on the past, present and ensuring problems followed by remedial measures. Such problems can be solved through the detailed planning process. Nowadays, the predictive approach seems to be the best one for planning a particular territorial space in order to have all round development after considering peoples' problem of economic imbalances, social norms, ecological imbalances and haphazard growth of infra-structure facilities and linkages.

REFERENCES

1. R. Hartshorne, *Political Geography in American Geography Inventory and Prospect* 9 (ed. P.E. James and C.F. Jones), Syracuse, New York, 1954, p 178.
2. R. Muir, *Modern Political Geography*, The MacMilan Press Ltd., New York, 1975, p 2.
3. L.M. Alexander, *World Political Pattern*, Chicago, 1957, p 32.
4. S.B. Cohen and L.D. Rosenthal, Geographical Model for Political

Systems Analysis, *Geographical Review*, Vol. 61, 1971, p 6.

5. A.E. Moodie, *Geography Behind Politics*, University, Library, B.I. Publication, Bombay, 1974, p 12.
6. D.S. Whittlesely, *German Strategy of World Conquest*, New York, 1942.
7. F.J. Monkhouse, "A Dictionary of Geography," Edward Arnold, London, 1972, p 173.
8. D.S. Whittlesely, *The Earth and the State*, New York, 1939.
9. B.M. Russett, *International Regions and the International System : A Study in Political Ecology*, Chicago, 1962.
10. G.A. Lanyi and W.C. McWilliams, *Crisis and Continuity in World Politics*, New York, 1966, p 40.

# 19

# Frontier Movements in the Concepts and Methods of Historical Geography

—*F.E.K. Amoah*

THEORETICALLY the problem of location or spatial patterns of earth phenomena may be examined either in a functional or genetic way. A discussion of the distribution of crops or livestock in a country, for instance, may satisfactorily be treated in a functional manner by concentrating on the context of a physical as well as the social, economic and political environment in which little reference need be made to the past. Similarly, a study of migration of population, traffic flows, urban spheres of influence, and input-output analysis of industry may be carried out along these lines. When, however, concern is not so much on the interpretation of *spatial organization* of phenomena but rather on the *evolution* of the features in the present landscape the genetic or historical approach is preferable. The growth patterns of cities and settlements, or changes in agricultural landuse can be better understood or explained in terms of a sustained consideration of the circumstances under which the elements in contemporary landscape came to be what they are.[1]

It is important to note, however, that the functional and genetic approaches are not mutually exclusive but complementary. In practice, therefore, no geographer would insist on a purely genetic approach to, say, the study of cities or fail to recognise the dimensions of time in a functional method. It may also be pointed out how far and in what manner the geographer should trace the present back into the past in a genetic study depends on the nature of the problem and interest of the researcher or teacher. It is also pertinent to note that referring occasionally to the past, while writing about present landscapes does not make one an historical geographer. Indeed it is one thing to recognise the importance of time dimensions in studies of contemporary spatial patterns and it is quite another deliberately to approach all geographical enquiry genetically and label the resultant work as historical geography. The recognition of genetic geography, or for that matter historical geography, as a distinct discipline is capable of being taught in a geography department raises the question as to what precisely is historical geography and how does it fit into the broad field of geography.

## SCOPE AND METHODS OF HISTORICAL GEOGRAPHY

Historical geography, like all other aspects of geography, is an evolving concept. In the past different practitioners of the subject have had their own peculiar views of its content and this in turn has from time to time affected its scope as well as methods of approach.[2] Historical geography today is neither synonymous with the history of explorations and discoveries, and of map making as was the case in the nineteenth century,[3] nor is it about the influences of 'geographical factors' whatever the meaning of the term, on history as was conceived by Ellen Semple[4] about 1903. Moreover, historical geography is not to be confused with the history of geographic thought even though Carl Sauer[5] felt the need to include it within the scope of the subject in 1941. Other works which do not seem to fall in line with current practice of the subject include studies of curious relics of the past which lie more in the domain of archaeology and cultural anthropology. The same may be said of the studies on evolution of landforms which are pertinent aspects of physical geography, especially geomorphology, although studies on the genesis of landscape features such as the clearing of woodland, reclamation of heath, drainage and irrigation have been combined in physical and historical geography in a new and profitable way which seems to demonstrate the absence of any great gulf between the historical and 'scientific' methods of investigation.[6]

Historical geography in the modern sense of the word may be defined variously as the study of changing cultural landscapes, or the reconstruction of past geographies, or the study of geographical change through time.[7] In other words, it is essentially human geography or as an aspect of cultural or regional geography. It is clear then that the term historical geography has come to be increasingly identified with man's role in changing the face of the earth rather than the earth's role in changing man's way of life. This is an area in which geographers and historians may work together in an effective partnership but their respective roles would not necessarily be interchangeable. Indeed writing historical geography characteristically involves approaches in which the data are historical, whereas the methods are geographical. To achieve a balance between successive narratives, descriptions, and analysis, the historical geographer may resort to one of the several methods which are not mutually exclusive but often complementary.[8] In this paper only five principal methods are discussed.

**The Cross-sectional Method**

Whenever it becomes necessary to discuss changing cultural landscapes, the researcher may concentrate on a number of distant or previous date-lines, say, the period before 1850, 1850-1900, 1900-1930, 1930-1950, 1950 and after, in a hypothetical case-study, and then enquire what the region was like at these times. One horizontal cross-section of the different features such as farms, settlements and roads of an area for a specific period would represent a momentary glimpse of the land as taken out of a long film of motion picture. A succession of such cross-sections produces a 'process' that is continuing and seemingly never ending. This is a method often employed by British historical geographers, especially Darby[9] who used it effectively to discuss the historical geography of England before 1800. It was adapted by Dickson[10] for writing the historical geography of Ghana, and may be used for any other country.

The advantage of the cross-sectional method is that it provides treatment in time series and in considerable depth. However, it presents some practical problems because the different elements that make up the total landscape, as indicated above, do not change at the same time or at the same rate. Moreover, some information has to be repeated in cross-section after cross-section and even when no patent change has occurred repetition is unavoidable. A true cross-section may also be said to be unhistorical in being served from the chains of events necessary for its comprehension. In effect the cross-sectional approach fails to adequately represent reality. Despite its shortcomings, the method is quite useful especially if cross-sections are so chosen as to coincide with marked changes in the area as a whole.

**The Vertical or Topical Method**

In a situation where the researcher or teacher has to deal with geographical change through time, it is possible to organize the data, not in horizontal cross-sections but in vertical themes or topics. In this approach the significant factor or set of factors operative in the process of change are examined as part of a whole development. It is an approach which may be employed in discussing, for example, the growth and decline of towns or seaports. Thus in Ghana[11] the major, though not exclusive, determinants of port status in successive periods from about 1800 to 1962 could be examined in relation to (i) early seaborne trade in gold and slaves before 1800, (ii) agricultural and mining activities from 1800-1900, (iii) expansion and improvement of internal transport network from 1900 to 1928, (iv) site, situation and efficiency of ports from 1928-1951, and (v) industrialization and national economic planning from 1951 to 1962. This is an approach which Darby[12] later adopted for his reflections on the changing English landscape about 1951.

The vertical or topical method has the advantage of presenting the material analytically but like the cross-sectional approach it is also open to criticism. In the first place in analysing landscape in relation to selected factors, a picture of the whole developing region is missed. One may wish to know, for instance, not only the effects of local production but also transportation systems on port status before 1900. Secondly, the systematic presentation of successive narratives and descriptions of socio-economic phenomena raises the question as to whether or not historical geography is intrinsically different from economic history. This problem is, of course, an intellectual rather than a practical one for if the researcher or teacher's work is part of a geographical study, the treatment is bound to be largely geographical. On the other hand, if economic change is part of an historical account then the approach would be historical.

**Sequence Method**

In this procedure the practitioner is not primarily concerned with reconstruction of past geographies or analysis of changing elements in the scene, but with perception of the past. This consequently involves an attempt to restore past phases of human occupance that have left vestiges or traces on the land and so continue to exist at present.

Before the ideas and attitudes of past generations can be comprehended the nature of the environment in which people lived and worked must be recreated. This was the basic idea behind Derwent Whittlesey's[13] term, "sequent occupance" in the United States which was brought to its logical conclusion by Preston James[14] in recapturing earlier landscapes of the Blackstone valley in southern New England in 1929. In his study three main phases of change may be recognised. Firstly, the phase of the native Indians with their rudimentary techniques of land occupation which created landscape forms characteristic of their culture. Secondly, the phase of the European settlers and farmers who developed out of the earlier landscape, largely obliterating it, a different and more advanced agricultural economy. Finally, the phase of industrial cities with an entirely new set of cultural forms imposed on a rural landscape. This was an approach which Brown[15] employed in teaching the historical geography of the United States about 1948. It may be adopted for Latin America, Asia or any region with a strong pre-European cultural background.

The sinew of this method lies in the possibility of relying mainly on field work which takes the geographer some way towards reality in explaining the circumstances of change. It also helps one to reflect on the different attitudes of people of various times in the past towards recurrent problems such as soil erosion, water conservation, land ownership, industrial organization and the like. The method is, therefore, of great practical application in giving direction to change. One problem of the method, however, is that it is not always easy to separate the surviving elements of the past phase from associated phenomena. There is consequently the fear that the treatment might lead ultimately to a full-scale reconstruction of past geographies.

**The Retrospective or Retrogressive Method**

In an attempt to provide a historical-geographical description the teacher or researcher can start, not at the beginning, but with the present and describe existing landscape, looking back only when and in so far as these elements cannot be explained in contemporary terms. The geographer, therefore, proceeds from the better known to the less well-known but in trying to restrict historical comment to landscape features he would naturally have to look further back for some origins than for others. However, he must avoid the all too common error of trying to leap at a bound from, say, the nineteenth century to pre-history.

Theoretically there is much to be said for this approach which is often associated with French geographers[16] but its major criticism is that it is not always easy to achieve a completely satisfactory presentation along these lines. To keep looking back is usually unpleasant for the teacher as well as student because of the many interruptions or parentheses and footnotes in case one is writing. Besides the combination of a running geographical account with historical interpolation is unrealistic. It must be remembered also that to refer occasionally or intermittently to the past, while teaching about present landscape does not make one a historical geographer.

**The Model Method**

Traditionally most work in historical geography has been inductive or empirical. The investigators have usually examined the changing character of certain geographical features and have drawn specific conclusions together into general conclusions. More recently, particularly since the beginning of the sixties, interest has shifted to deductive or theoretical approaches by which one proceeds from some *a priori* universal premise to statements about a particular situation or sets of situations. Thus instead of seeking to discover what *there was* in a place at a particular point of time, this relatively new approach theorises, by means of models, on what *ought to have been* the case under a set of assumptions, arrives at theoretical conclusions, and finally compares the result with actual conditions that existed to test the validity of the assumption. In this way, the theory or model can be accepted, rejected or modified according to the rules governing scientific explanation.[17]

Simulation models of the formation of transport network[18] and the growth of towns[19] have been constructed and tested with realistic data from historical records. Specifically, London's importance in changing English society and its economy in the seventeenth century, and the spatial dynamics of urban-industrial growth in the United States from 1800 to 1914 have been analysed on the basis of models of this type by Wrigley[20] and Desai[21] respectively. Peter, Rimmer's[22] study on the changing status of New Zealand ports from 1853 to 1960 and Hargerstrand's[23] works on agricultural innovation and diffusion are also examples of note.

The credit of the deductive model or theoretical approach, which is widely accepted in Modern Science,[24] is that it is bringing historical geography more fully into the mainstream of scientific investigation which geography as a whole is trying to emulate. The shift to the construction and testing of theoretical models in historical geography is also radically transforming the spirit and language of the subject. One fact which appears to be of singular importance is the mathematization of much of its concepts, a change which had earlier affected certain branches of geography notably climatology and geomorphology, population and urban geography.[25] On the other hand, what discredits the theoretical-deductive method is its heavy reliance on quantitative data even though realistic data for the past are often hard to come by. An historical geographer may also be quite literate and imaginative in reconstructing or recapturing past geographies from oral, documentary and other sources but may lack the ability to effectively handle statistics and mathematics. A comparatively new method of selecting key variables in spatial analysis is 'factor analysis' which throws up significant variables out of a large number which account for the variations observed. Factor analysis is a very useful statistical technique of great precision based on a series of correlations, but a high and statistically significant correlation coefficient may not necessarily provide proof for cause-effect relationship. It may only suggest the probability of such a relationship which may have to be carefully verified. Statistical techniques are thus a reflection of something very fundamental, namely the role that they play in testing or verifying ideas already established, however tentatively, by descriptive analysis of several case studies. One may, therefore, question the excessive claim of the theoreticians who maintain that descriptive qualitative studies have no place in the formulation of theories and models.

From the scope and methods just discussed it is clear that historical geography is primarily not a field of topical specialization or a systematic subject for any topic can be examined in terms of its past circumstances. Studies in historical geography, however do require the use of particular procedures, are based on certain kinds of data in considerable part derived from documentary, oral, archaeological and other sources and demand special background in training in geography and history. More recently statistics and mathematics have become new prerequisites for studies in the subject.

## CURRENT PROBLEMS AND PROSPECTS OF HISTORICAL GEOGRAPHY

Recent developments at research level, and to some extent in the structure of University courses in what has come to be known as the 'New Geography', have led to frequent arguments for and against historical geography. The accusations and counter-attacks have usually been centred on the conceptual and organizational questions similar to those hurled against history much earlier but have subsequently been refuted.[26] For want of space, only three main issues related to historical geography and the 'New Geography' are discussed here, namely (1) the logic of genetic explanation (2) development of theory in historical geography and (3) the place of historical geography in the 'New Geography'.

### Logic of Genetic Explanation

Several misgivings have been expressed about the highly fragmentary and incomplete nature of oral, documentary, archaeological and other source materials on which inferences of the past are based and the inevitable differences in interpretation of the evidence. 'Historicism' is synonimous with the approach in which understanding of the past is a matter of knowing past events in their temporal order from which it is hoped some kind of meaning will emerge.[27] Since some facts do not themselves proclaim their meanings, and since the historians judgement of the significance of a particular occurrence cannot be tested, genetic or historical explanations cannot be said to have been responsibly supported.

This view may, however, be challenged. Dickson has correctly pointed out that "history as a discipline is not dependent on the peculiarities and personal biases of any of its practitioners; it has an organised structure of research, widely shared techniques and meanings, and institutional arrangements for its own intelligibility."[28] Individual historians may, nevertheless, acknowledge their ignorance of the sufficient *conditions necessary* for the occurrence of an event by identifying a partial rather than a complete set of determinants and then specify those which they judge as 'important', 'main', 'chief' or 'principal' factors. But even though what historians ultimately propose for a given event may differ, the explanations are not necessarily incompatible. Indeed, the alternative explanations tend to supplement rather than contradict one another. It may be mentioned also that historians sometimes prefer to have as many view points as possible on particular problems simply because they believe that no single account is capable of providing a completely *rational* explanation. What has been said of historians may apply to historical geographers who also require mastery in collecting and interpreting information about the past skills which are similar to, if not the same as, those required by professional historians.

**Historical Geography and the Development of Theory**

Another criticism often levelled against historical geography as a field of specialization in modern learning is that it tends to perpetuate the study of *particular* or *unique* occurrences rather than situations that are repeated in essentially the same manner in large numbers.[29] Studies that make *unique* situations the core of their investigations are said to be idiographic rather than nomothetic.[30] The following statement may elucidate the point with reference to history : "History is about what happened on particular occasions. It is not about what usually happens or what always happens under certain circumstances. For this we go to science."[31]

The implication of the above assumption is that studies of the *particular* or *unique* cannot be of much use in the fulfilment of the ultimate aim of all scientific enquiry, namely the formulation of theories, laws or general truths which increase the ability of scientists to predict an event and, possibly, search for solutions to recurrent problems. Thus, if historical geography is to lay any claims to being a scientific discipline, then it must adopt the 'scientific method' of investigation which involves deduction along with induction in studying *unique* as well as *general* situations.

In his Theoretical Geography Bunch[32] has suggested that whether a study is idiographic or nomothetic can be considered to be a matter of point of view for the disciplines themselves are *priori* not idiographic or nomothetic, but rather it is their practitioners who choose to adopt one attitude or another in their task. Disciplines in principle are thus only primarily idiographic or primarily nomothetic. If this proposition is accepted it would be quite legitimate to assert that historical geography has remained primarily idiographic for a long time because its practitioners have taken keen interest in the study of unique cases at the expense of the search for general laws and principles regarding situations that seem to have occurred again and again in the past.

Recent introduction of theoretical models in historical geography and the verification of their basic principles has opened a new chapter in the quest for scientific techniques in historical investigations. It reflects a shift from the empirical-inductive approach to the theoretical-deductive method which represents the standard model in scientific investigation.

**The Place of Historical Geography in the New Geography**

Recent criticisms of historical geography go beyond the allegation that it perpetuates the idiographic approach which is symptomatic of the early stages of scientific enquiry. There is now also the feeling that historical geography may not even be necessary at all for the advancement of geography as a whole.[33] These misgivings are sometimes justifiable not on the conceptual or methodological grounds but on the grounds of the illustrations given of the usefulness of the subject and of the kind of material presented as historical geography.

It has been shown that the past and the dynamics of spatial change has a major role to play in geography. Consequently historical geography which deals primarily with the past should be constantly reviewed to ensure that it fulfils its objectives. Among the several tasks involved in modernizing the subject are (1) the problems chosen for study, (2) the techniques of analysis used, and (3) the methods of presentation.

In selecting problems for study in historical geography one must be guided primarily by their relevance with respect to the contemporary scene. The question of relevance is debatable, but certainly the historian who selects to study the geography of a place at a certain period or point of time in the past but does not link his study with what happened immediately before and after is not doing much service to the discipline. One must note, however, that the links which connect prehistory with modern geography tend to be long and tedious, debatable and often of little importance in the modern world. It is not being argued, of course, that only those things which one studies in contemporary geography should have their evolution traced. The past may be studied for its own sake if some peculiar geographical relationships in the process of evolving into the contemporary scene are believed to have undergone many changes that should be noteworthy.

Another way of modernizing historical geography is that it should endeavour to make use of all relevant contemporary techniques such as quantitative methods. A regression model or analysis of an element in time series, for example, would confirm the existence of trends and lead to more accurate definition of the problem that has to be explained. Similarly, correlation coefficients are a means to an end; they may give firm support to hypotheses already formulated.

We may emphasize the need for increasing *quantitative* techniques in historical geography without lessening the significance of *qualitative* studies in the subject. This is because the latter form the basis of much *ad hoc* work in historical geography. The introduction of a new and valuable approach in a field of study too often leads some enthusiasts to suppose that all work must hereafter be pursued by a particular method, and that whatever is not amenable to analysis by that method is not worth studying. We are reminded that "in a field in which the phenomena that came under study run the full range of cabbages to kings, from rainfall to religion, it would seem absurd to assert that all that is worth studying can be fully and currently described in quantitative terms."[34] On the other hand, those historical geographers whose studies are less amenable to mathematical-theoretical approaches must appreciate the fact that at best they can only make a contribution by throwing light on problems of the past in a way not gained from other perspectives.

Constant re-evaluation of the methods of presenting studies in historical geography is also necessary if the subject is to make any real impact on modern learning. The study of pictures of past landscapes has long been standard in the repertoire of historical geographers and will undoubtedly continue to be so, for much still remains to be done to fill the gap between what is available in historical records and what is needed by geographers interested in the past. Of particular significance to geographers is the fact that much local history and the larger interest of economic historians has tended to ignore regional differences from one area to another; historians usually are not interested in areal differentiation.

Studies in the genesis of landscape features and geographical change through time have both led to considerably more emphasis in tracing the evolution of agriculture and settlement, trade and industry in what has been called topical or vertical studies. Such studies in the historical branches seem to offer possibilities of fundamental development, particularly in terms of a better understanding of the processes by which geographical change through time takes place. Some of the developments in urban geography, notably those concerned with the central place theory and urban spheres of influence, and statistical techniques in regional analysis would seem to give a new and considerable stimulus to the geography of particular periods. The introduction of the deductive-theoretical approach to the study of urban growth, population migration and diffusion of agriculture will undoubtedly also leave a legacy of considerable importance to future studies in historical geography.

## CONCLUSION

Trends in the literature of historical geography in recent years illustrate diverse themes. The history of geographical explorations, discoveries, map making and other themes which formed the core of the subject in the nineteenth century have now given way to the study of picture periods and geographical change through time as aspects of human geography. Analysis of spatial dynamics in time series is a rigorous exercise that requires special skills and in this lies the justification for specializing in the elucidation of the past.

Without doubt the most significant advance in the subject, particularly since the sixties, has been the formulation of theories and the testing of models with quantitative techniques which is much in line with the scientific method of explanation. One of the ways in which the concept of indeterminancy is being built into a number of locational models is through probability matrices where it is postulated that a number of choices are to be made within a set of locations according to their attractiveness, so that the probability of each area being chosen is weighted. This basic principle of simulation in a probability framework cannot predict what *will* happen but can predict what *could* or *might* happen if conditions remain unchanged.[35] Stochastic models, as they are called, would seem to hold a great future for historical geography which should not only concern itself with the past but also give direction to where changes may lie in the future.

## REFERENCES

1. The importance of genetic explanation in geography has been treated by Carl O. Sauer, "Foreword to Historical Geography" in John Leighly (editor) *Land and Life : A Selection from the Writings of Carl Ortwin Sauer*, University of California Press, 1963, pp 351-79; E.W. Gilbert, "What is Historical Geography?" *Scottish Geographical Magazine*, 48 pp, 129-36; Richard Hartshorne, *Perspective on the Nature of Geography*, Chicago, Rand McNally & Company, 1959, Chapter VIII, pp 81-107.
2. Andrew H. Clerk, "Historical Geography" in Preston James and Clarence F. Jones (editors) *American Geography : Inventory and Prospect*. Syracuse University Press, 1954, Chapter 3, pp 70-105.
3. E.W. Gilbert, op. cit.

4. E.C. Semple; *American History and Its Geographic Conditions*, Boston, 1903.
5. Carl O. Sauer, op. cit., p 355.
6. C.T. Smith, "Historical Geography : Current Trends and Prospects", in R.J. Chorley and P. Hagget· (editors) *Frontiers in Geographical Teaching*. Methuem & Co. Ltd., 1965, p 136.
7. *Ibid.* pp 124-134.
8. F.E.K. Amoah, "Some View points and Methods in the Teaching of Historical Geography", *The Oguaa Educator*, Vol. 5, No. 1, October 1974, pp 42-48; see also H.C. Darby, "The Problem of Geographical Description" *Transactions of the Institute of British Geographers*, Pub. No. 30 pp 1-13.
9. H.C. Darby, *An Historical Geography of England before 1800*. Cambridge, 1936.
10. Kwamena B. Dickson, *A Historical Geography of Ghana*, Cambridge University Press, 1969.
11. F.E.K. Amoah, *The Growth and Decline of Seaports in Ghana 1800-1962.*, Ph.D. Dissertation, Geography Department, University of California, Los Angeles, 1969; (Mimeo) Ann Arbor, Michigan. 1970.
12. H.C. Darby, "Historical Geography, Twenty Years After" *Geographical Journal*, 126, pp 147-59.
13. Derwent Whittlesey, "Sequent Occupance" *Annals of the Association of American Geographers*. Vol. 19, 1929, pp 162-5.
14. P.E. James, "The Blackstone Valley" *Annals of Association of American Geographers*. Vol. 19, 1929.
15. R.H. Brown, *Historical Geography of the United States*, New York, Harcourt Brace, 1948.
16. A.R.H. Baker, "A Note on the Retrogressive and Retrospective Approaches in Historical Geography". *Erdkunde*, 22. pp 243-4.
17. F.E.K. Amoah, "Recent Trends in the Methodology of Geography: An Interpretation, "*The Oguaa Educator*, Vol. 5, No. 2, April 1975. pp 56-67. For greater detail see David Harvey, *Explanation in Geography*, London, Edward Arnold, 1969, Chapter 4, pp 30-43.
18. W.L. Garrison, B.J.L. Berry; D.F. Marble and J.D. Nysteun. *Studies of Highway Development and Geographic Change* Seattle, 1959: see also K.J. Kansky, *Structure of Transportation Networks; relationships between network Geometry and Regional Characteristics*, University of Chicago, Geography Research Paper 84, 1963.
19. R.L. Morrill, "The Development and Spatial Distribution of Towns in Sweden : An Historical, Predictive Approach". *Annals of the Association of American Geographers*, Vol. 53, pp 1-14, 1963; R.L. Morrill, "Simulation and Central Place Patterns over Time" *Lund Studies in Geography. Series B. Human Geography 24*, 1963, pp 109-20.
20. E.A. Wrigley, "A Simple Model of London's Importance in the Changing English Society and Economy", *Past and Present*, 37, pp 44-70, 1967.

21. M. Desai, "Some Issues in Econometric History", *Economic History Review*, Second Series No. 21, pp 1-16.
22. Peter Rimmer, "The Changing Status of New Zealand Seaports 1853-1960", *Annals of the Association of American Geographers*, Vol. 57, 1967, pp 89-97.
23. T. Hagerstrand, "The Propagation of Innovation Waves" in P.L. Wagner and M.V. Mikessel (editors) *Readings in Cultural Geography*, Chicago University Press, 1962, pp 355-68.
24. Ernest Nagel, *The Structure of Science*, London, Routledge and Kegan Paul, 1961, p 21-22.
25. J.W. Campbell and P.A. Wood, "Quantification and the Development of Theory in Human Geography" in Ronald Cooke and James Johnson (editors) *Trends in Geography*, Pergamon Press, New York, chapter 1, pp 81-89.
26. Some of the criticisms have been discussed in F.E.K. Amoah, "Some Methodological Issues in Current Research in Geography : Their Relevance for the Concept of Scientific History" *Ghana Social Science Journal*, Vol. 4, No. 2, November 1977, pp 21-34.
27. Ernest Nagel, op. cit., pp 25-26.
28. K.B. Dickson, "Some Reflections on the Status and Role of Historical Geography" *Ghana Social Science Journal*, Vol. 2, No. 1, 1972, p 6.
29. See David Harvey, op. cit., pp 547-551; and C.B. Joynt and N. Rescher, "The Problem of Uniqueness in History", *History and Theory*, Vol. 1, 1961, pp 40.
30. Richard Hartshorne, op. cit., pp 146-153.
31. Patrick Gardiner, *The Nature of Historical Explanation*, Oxford University Press, 1961, p 40.
32. See W. Bunge, *Theoretical Geography, Lund Studies in Geography*, Series C., No. 7 (Second ed.) 1966.
33. A critical examination of this question is to be found peacemeal in K.B. Dickson, "Some Reflections on the Status and Role of Historical Geography", *Ghana Social Science Journal*, Vol. 2, No. 1, 1972.
34. Richard Hartshorne, op. cit., p 162.
35. P. Haggett, "Changing Concepts in Economic Geography in R. Chorley and P. Haggett, (editors) *Frontiers in Geographical Teaching*, London Methuem, 1965, pp 108-114.

20

# The Geography of Music

*—Sister Violita A.C.*

POETRY is the language of the human heart. It is the rhythmic and verbal music of emotion. That geography can unite and fuse the two. Let John Donne's experience prove :

On a round ball
A workeman that hath copies by, can lay
An Europe, Afrique, and an Asia,
And quickly make that, which was nothing, *All*,
So doth each teare,
Which thee doth weare,
A globe, yea world by that impression grow,
Till thy teares mixt with mine doe overflow,
This world, by waters sent from thee, my heaven dissolved so.[1]
(*A Valediction : of Weeping*)

And when we turn to song and dance, to flute, and pipe, and drum we find this spirit of geography winding in and out among them, stringing them all together interspersed with the glories of nature.

Geography determines the type and character of music of a place. In this sense we would have a typical music of the mountain people as distinguished from that of the desert tribe. The music developed in the plains would definitely be different from that on an island. Indirectly, although the differentiation is apparent to some extent, culture, religion, history of political life have played an important role in the development of the music of any place. This is true of any art. The real effect of geographical factors was more clearly seen in the past when the means of communication were in their infancy and still better before civilization itself began. Now, when the world has shrunk due to jet speed, any clear cut demarcation, fortunately or unfortunately, has disappeared to a great extent. This makes it difficult for us to know the typical music of a particular place and how far environment has played its part in shaping it. With the dawn and spread of civilization and the intermingling of cultures, we have music which has little or no relationship to the geography of that particular place. One such example is that the jazz which has travelled far from the place of its origin, New Orleans.

In order to gauge the influence of physical environment on music one has to go back to the birth of music. Music is a way of expressing man's innermost feelings :

> his attempt to express his relationships with the forces which lie beyond him, with God and the universe and with his fellowmen. It is the chronicle of man's urges to enshrine the feelings thus called forth in his heart by so ordering sounds that they transmit those feelings to others.[2]

When man is too happy for ordinary speech, or too sad he expresses his overflowing heart in song or dance or sounds of flute or drum. In the beginning, man began to discover ways of grouping sounds which we call music. A visit to the tribal areas will help know the themes of primitive music. The difference between tribal music and that of the civilized world is vast. Refinement is typical of a civilized world and can be seen in its music. This refinement today seems to be substituted by the jazzy music that is becoming a fashion and which again is directly related to the hectic life of the people of this jet age. The spirit of primitive music varies greatly from that of tribal music. The dignity and restraint of the Red Indians is far from the unbridled emotionalism of the Africans. The Patagonian Indians and Veddas of Ceylon are restricted to two notes only whilst more advanced tribes employ a full octave or an even larger range.

One of the earliest forms or expression of man's feelings was through folk music. This consisted in a group expressing some happy or sad event of their way of life by means of group singing. Much of this music, which perhaps developed on the spur of the moment, was never written down but transmitted from generation to generation. The language and customs of each race are reflected in its music. English and American folk music generally narrative. They tell the story of a person or of an event. The folk music of a locality remains mostly the same even though its boundaries may change. The songs of Alsace and Lorraine both of which are in France are equally French and German reflecting that they were part of Germany once. Many tunes of Russia, Poland, Czechoslovakia and Hungary have been claimed by Austria and Germany. But the strong racial characteristics of these airs have kept them distinct from Austrian and German folk music. The Wendish Colony of the Spreewald near Berlin has for generations been surrounded by a Teutonic Civilization, but its folk music is as typically Slavic as any to be found in Russia. Gypsies have influenced lands where they made their nomadic homes.

Great political movements too have had their share in shaping the folk music of places. The oppression of the Czars gave rise to the most dramatic of the Russian peasant songs. Slavery in America helped produce the Negro Spirituals. Most American Negro folk songs and spirituals express sadness and nostalgia and began on the plantations of the Old South where the composers were slaves. Folk music has also been influenced by climatic and other environmental conditions. Those who live an outdoor life in a warm climate use music for folk festivals and dances. Their music is louder and more vigorous than that used in colder regions.

The ballad is another way of expressing one's deepest emotions. Ballads were originally meant to be sung. They, therefore, contain a chorus, certain lines meant to be repeated as a refrain. Various changes in the pattern take place as the story progresses. Ballads arise among people who have shared the same habits and ideas for generations. Perhaps one person makes a simple poem in memory of some event such as a marriage or death. Other persons add or change the lines and hand them on to the next generation. The typical ballad develops over a long period which shows that it has been preserved and cherished. Isolated mountain folks in the U.S.A., for instance, have preserved many variations of old English ballads.

Calypso is the type of folk music found in the Caribbean area, South America and West Africa. There are only six to eight basic calypso tunes and the words are often made by the singer as he progresses with the singing. The songs may be about a person, an event, or an opinion.

Sailors in the days of old sailing ships lightened their work by songs. Thus developed the Chantey (Shantih) groups of songs. The crews timed their motions by the rhythm of their song as they hauled up the sails or did other work. These songs unfortunately disappeared when motor driven boats took the place of sailing ships. In New Jersey, being a coastal place, much of the folk music of the colonists is of the sea Chantey type.

Spirituals are religious songs made famous and immortalized by Negroes of the southern United States. They have a strong rhythm and are very emotional. These are considered as the only truly American folk songs and are the direct result of the preaching of the evangelists among the suffering slaves of the southern states. Much of the ship loading and plantation work was accompanied by singing. Spirituals were little known outside the southern states until after the Negro was freed from slavery. The spirituals of the slaves were mostly based upon characters and stories from the Bible. The manner in which these stories are told in Negro spirituals shows a colourful imagination and a simple faith. Many slaves thought of themselves as the modern children of Israel and looked for a black Moses to deliver them from their bondage. Their songs were warmly appealing and very sincere.

The Blues sprang from the combination of field—hollers, work songs and spirituals among the black rural population of the southern states of North America during the latter part of the nineteenth century. This typical vocal folklore fuses musical elements of African and European origin, but in a different way from other Afro-European combinations. The Blues is an individual personal music, a sort lament on life and love, reflecting with singular power the sufferings of an empire, people, the sorrows and the indignation of the freedman after the Civil War of 1861-65. By the end of the century the Blues began to be spread northwards from the "Black Belt" of the deep south by itinerant singers or by Black emigrants moving up the Mississippi River in search of work. By 1920, a Blues craze rapidly extended all over the country. The principal musical characteristics of the Blues is the alteration of two notes, the third and seventh called the "blue notes", which are flattened (lowered a half tone). Although the blues are in a major mode, the repeated inflections of the blue notes seem to move it toward the minor mode, a shift that gives this music its own special colour.

Jazz is the most original contribution of the United States to music. It is an integration of diverse ethnic cultures which makes up America. Jazz, born in the twentieth century is of Negro origin. Here, two great musical traditions have blended to create this unique form—that of Europe with the musical elements of melody, harmony and design, and that of West Africa, with the emphasis on rhythm and the call and response pattern, so characteristic of jazz based on the African tribal music. In this last, a leader shouted a cry and was answered by a group. In jazz the "call" is played by a solo instrument and the "response" by percussion and melody instruments.

Jazz had, as mentioned earlier, a Negro origin. It permitted the musician to express himself, thus offering him some escape from the wretched social condition inherited from slavery and continued by segregation. New Orleans is said to be the cradle of jazz. The music was probably born in more than one American city, but the Crescent city was in a privileged spot as a cultural cross-roads. Some New Orleans musicians, more enterprising than the rest, left that city to travel around the United States. But it is more fundamentally related to the northward migration of southern Negroes in search of jobs. Many of the musicians ended up at the northern end of the Mississippi Valley, and Chicago became the second capital of Jazz during the 1920s. Jazz remains a largely American and especially a black American, invention. Jazz has influenced and changed all contemporary accidental music.

The instruments used in jazz are mostly wind instruments. These instruments were used by the soldiers who used to play while marching past the plantations; and the slaves being musical, were attracted by them. After the Civil War these instruments, remaining unused, were bought by the slaves who created jazz. Jazz in the beginning, was played by Negroes who walked about with their instruments. Hence, the instruments used in jazz are also those which can be carried about and played and can also be played by approximation. Jazz music, therefore, is a tribute to ingenuity and spontaneity.

Music, as we have known it through the centuries, falls into two broad groups—western music and eastern music. Indian music, as well as the music of China, Japan and Java come under eastern music. Much oriental music consists of a single line of melody. There is no harmony as in the west. The nineteenth and twentieth centuries have witnessed a marked change in the musical tradition. It has lost sight of its high position as a religious, political and social force and is gradually being replaced by noisy substitutes, a process greatly hastened by the invasion of western culture with its jazz music. On the other hand, western classical musicians have begun to appreciate the merits of Indian music and the two types are increasingly influencing each other. A singularly satisfying example of fusion of East and West is the partnership of Yehudi Menuhin and Ravi Shanker.

Western music was begun by the early followers of Jesus Christ who were Hebrews and Romans. For their Church services which had to be held at first in secret places, they sang the only music they knew, that is Hebrew and Greek songs. It took a long time to develop music for the early Catholic Church but gradually the first great art music of the Christian era came into being. It was the Plain song. Saint Gregory, Pope of Rome from A.D. 590 had collections made of the music used in the Catholic Church which survives till this day and is known as the Gregorian chant.

Up to the 800s, the music of every nation was of a single melody. Then in the eleventh century polyphony of music with many voices was created. This was the decisive turn in western music. By the time of the Renaissance, harmony had made its appearance in the simple form and then continued to grow till it completely took over the musical scene in the eighteenth and nineteenth centuries.

"Indian music, reflects Indian life having no predetermined beginning or end but flowing without interruption through the fingers of the composer performer"[3]. Indian music differs from western music in that it has no place for harmony, while harmony forms an essential feature in western music. Indian music is melodic in form expressing the natural language of emotion.

Harmonization gives depth to a melody decorating it and adding more charm to it. The reason for harmony in western music and the lack of it in Indian music can be due to the very make-up of the people who live in these parts. The Indian finds it easier to unite himself with the infinite, rather than with his neighbour. The purpose of Indian music, "is to refine one's soul and discipline one's body, to make one sensitive to the infinite within one, to unite one's breath with the breath of space, one's vibrations with the vibrations of the cosmos"[4]. Indians are usually concerned more about their family than about the outside group. For all these reasons the Indian musician is rather a soloist than a player making music with others as in the small groups or large orchestras of western music.

Being predominantly a soloist, skilful improvisation is one of the chief marks of the virtuoso.

By contrast the European's desire has been to form individuals into communities and he is ready in the process to forego his freedom. This is seen in his very life be it in the sphere of worship, work, social life or music where orchestras are characteristic. In orchestras or chorales each instrument or voice has its independence to a certain degree but has to bear in mind the presence of other voices and try to blend his voice or instruments with others in the group. In Indian music the individuals in the group, "remain soloists never coalescing into a harmonic statement"[5] Orchestras, therefore, are contrary to the Indian musical nature.

It is strange that long before harmony was used in western music the hermitages of India echoed with Vedic hymns sung harmonically. The reason could be that *rishis* taught Vedic hymns to a group of students. Also music in those early days was limited to just three notes. Harmony, therefore, came naturally with them. It was only later when music developed into different scales that harmony became difficult.

The music of India differs from place to place but we can distinguish two broad divisions — that of the north, Hindusthani music, and that of the south, Carnatac music. North and south India differ in a multitude of ways. The north had to face many invasions and consequently produced fighting races. The south had a more peaceful existence and people here had time to cultivate and become farmers. They had not many opportunities to mix with foreigners and so remained conservative and isolated. This is reflected in their music, which has remained conservative and classical.

Nature plays a very important part in the music of India. India has different music for different seasons and different parts of the day as also for elements of nature. Day and night music have an impact on the mood of any *raga* (scale). The twenty-four hours are divided into two parts: from twelve midday to twelve midnight forms one part. Certain *ragas* are suited only to day and hence have to be sung or played only in the daytime. Others are suitable only for the evening or night. Certain *ragas* project the image of spring. *Raga Malhar* brings to mind the distant echoes of the sound of thunder and the pitter patter of rain drops. It brings the odour of the earth and of green vegetation, the cry of the peacock and the call of the *koel*. Several *ragas* have evolved in tune with the monsoon season—*Megh*, *Malhar*, *Miya* and others, which give one the feeling of gathering clouds and the downpour of rain. *Raga Bhatiyali* reflects the steady flow of the rivers of Bengal and the fishermen singing in mournful strains as they plod homeward through the twilight. The sustained nature of the notes in this type of music is to let the notes travel a long distance, a means of communication with other boatmen. This type of music originally developed near rivers, where fishermen used to start with a long note, and break off into a falsetto. Each boat of fisherman was a long way off and the only way of communication was through this music.

The strains of ruggedness and the dryness of the rocks and sandy topography of the Marwar region are depicted in *raga Mand*. The echoes of mountains and valleys and the flowing sound of water, as it rushes down the mountain slopes along with boulders and pebbles are amply heard in *raga Pahari*. Here the repetition of the notes makes it melodious and also portrays the echoes of hills and valleys. *Raga Bhairav*, melancholy and sombre, is to be sung in the early morning when everything is quiet.

Early Punjabi folk music may be compared to a fast flowing mountain stream full of life and gladness. Later on this folk music became the medium of telling also tragic tales of legendary lovers and heroes which may be compared to a deep flowing, steady river majestic and slow like the Chenab itself which flows through the land of the Punjab. Punjabi folk music has some similarity with the music of certain west Asian countries. It is closer to these countries and was invaded by its peoples. The people of the Punjab react strongly to joys and sorrows and this is reflected in their songs and music. Vigour and gaiety are the key notes when the mood is a happy one, and deep sadness and mourning when it turns to grief.

*Sravan* for Biharis is a month when married girls visit their parents. Hence the folk songs that are sung have a soulful, wistful air called the *Kajri*. At the same time, this season being the monsoon season, one of hectic activity in the fields for men, they break into the songs known as *Barsati*.

The *Alha* originated in Madhya Pradesh but has gained much popularity among the Magahi speaking people. It is believed that singing the *Alha* brings rain.

When one hears the Bengali folk song one can picture the typical rural life of the Bengali as they perform their ordinary day-to-day work. The diversified nature of this music is due to the changing moods of life, which the music is made to suit. There is also a variation of music from district to district and to suit the variety of physical features of the land. There are songs descriptive of the wide fields and the rivers of Bengal, of seasons in their brightness and in their gloom, songs that express the hopes and dreams and frustrations of its people, their joys and sorrows, their harvest and festivals. It is, in short the music that tells of the customs of a village society.

Speaking of how geography has affected music, we see that the speed of music varies with the environment. In warm climates, music is slower, more expansive and more intricate whereas in colder climates it is vigorous and loud. Culture and civilization affect music. When people had reached a certain standard of civilization they used their intelligence and knowledge to develop music and made it more intricate and expansive. A feeling of satisfaction that comes about in a people who have accumulated wealth is seen in the progress of music which makes it elaborate. Group music developed in plains because people lived in groups, while music on mountains is more individualistic. Music that has developed in rugged terrain, such as deserts and plateaus where people have to fight against nature, reflects this struggle. Most of the music of the Middle East countries is melancholy. The music that the Turks play is sad and monotonous nearly all in a minor key. It seems to be an expression of the Turks' fatalist attitude towards existence. Music brings into his life no gaiety. The peasant may play perhaps upon a little pipe cut by himself out of a reed or upon a very simple form of bagpipes, the bag being of sheep skins and the notes being blown through a reed. Another reason for the monotony of the tunes could be that in a desert or plateau distances being great, it is difficult for notes to travel. Long notes are, therefore, chosen which are able to travel far.

Fertile land, on the other hand, provides a bountiful harvest and where people live in comfort and security, begets a music very different from that of the Turk in the desert. In Ukraine (Russia), a fertile country where nature is bountiful and generous with the people, the music is gay. This land is the home of music and song. Tschaikovski's music, which is delightful, has the spirit of the gracious, smiling landscapes full of colour, charm and surprise.

Language, social, political and religious circumstance affect music, as is seen in the music of Switzerland. Switzerland has sometimes had loose cultural ties and other times close cultural ties with neighbouring regions. As a result of its geographical position, Switzerland has to a large extent been a cross-roads of various cultures. The rollicking music of Ticino, an Italian speaking region differs from the tender sweetness of the Romansch region and the gay melody of Fribourg, a French region differs from that of the German speaking region.

In African tribal areas, music is still in the primitive stage. The Saharan zest for music can be seen in the delicate reed instrument which he blows like a trombone. The African in general is very devoted to his drum. He learns in babyhood to beat it. Egyptians sing while they work. One of them leads the song and the others keep repeating a phrase and clap their hands in rhythm. Egypts love of great size shows itself not only in her pyramids but in her orchestras which consists of six hundred players at a time playing on harps, lyres and rattles.

The Latin people reflect their love of romance, beauty and sparkling gaiety and the Latin speaking countries of South America have this characteristic feature par excellence in their music. This type of music is in direct contrast to the slower more regular music of the Teutons (Germany) or the fiery animated measures of the Slavs (Russia). In Latin America in colonial days, and for some time thereafter, music was largely copied from European sources but later became more and more devoted to native themes, melancholy music of the Indian tribes together with more joyous Spanish strains. In Brazil the music has been influenced by the Portuguese, Indians and Negroes. In the Caribbean countries Negro and Spanish music have combined to produce a characteristic type of which the rumba is one of the best. The traits which are derived more obviously from the Indians are the melancholy and the love of music which is found all over Brazil. The tunes are monotonous, mostly in a minor key produced by native pipes, and a beating of drums which perhaps stirs deep down memories of the summons to war that excited their ancestors. A brass band fills the people with delight.

Spain, enjoying Mediterranean climate with its bright, sunny sky, has given the Spanish a vitality and rhythm which is reflected in their music and is known throughout the world. In the early days music in Spain was influenced by the coming of Christianity. The later type has the imprint of the Moors.

The Hungarians are temperamentally inflammable. They slide from one great passion into another with rapturous delight. This helps to explain their delight in their particular type of music. The instruments known to them are the Tilinko (flute) and the lute. The violin was probably introduced by the gypsies and that is the instrument which has done everything to carry to all lands the Hungarian music which is of gypsy origin—the alternation of wild, reckless delight with slow thoughtful cadences. This music has a positively intoxicating effect.

The Rumanians are musical by nature. The gypsy music of Rumania has a distinct strain of melancholy running through it, very different from the wild and fiery music of the Hungarian wanderers.

The Poles are high spirited. Many of their songs are gay. It is in the nature of Poland that in its music is reflected the soul of the people in all its individuality and manifold lights and shades. It is in its music that the originality and romanticism of Poland take form and expression. Splendidly gifted by nature with a musical temperament, the distinctive taste of the poles in music has long been acknowledged. The early kings of Poland were enthusiastic patrons of music. They were greeted by music as they passed from town to town and hired musicians formed part of their retinue. Frederick Chopin, who won for Poland far reaching fame in the great world of harmony and sound, manifests in his music and originality that knows no bounds and inspired entirely by the mystery and sorrow of his own country. The vein of melancholy that runs through all Chopin's music shows how his country was ever in his thoughts. The inborn sublimity and introspection of a people cries out through Chopin's music, which seems to express the mourning cry of a whole nation following the bier of their dearest hopes. The national music of Poland is of ancient origin and represents the individuality and patriotism of a people which no fetters and no sufferings have ever been able to destroy.

Musical instruments of some sort existed in the world even before there were men to use them. Nature itself gives off musical sounds. It follows the same general rules which man uses to govern the creation, amplification and pitch of sound in musical instruments. Nature's wood-wind instruments include a split reed or piece of loose bark among the branches of a tree. These vibrate or sway when there is a wind. The same principles which affect the reed or bark are used in making clarinets and oboes. Primitive man observed the principle of vibration caused by friction or rubbing among the swaying branches of trees in the forests in early times. The same principle is used in making string instruments.

The basic rule of the rhythm of percussion instruments is closely related to the beating of rain, hail or falling stones against hollow logs or other hollow sound-producing natural objects. Primitive men used these principles to make musical instruments. He began with the drum stage and later he combined these repetitive beats with different pitches which gave rise to the pipe.

Prehistoric man shouted for joy or howled with *raga* which was the beginning of song. He stamped his feet and clapped his hands as he moved about or struck two pieces of woods together in dancing. He hollowed out a tree trunk and stretched the skin of an animal across it to make the first drum. He learned to make whistles out of tree twigs or pieces of bone. These became the first wind instruments. He cut reeds of different lengths, tied them together and played tones that made a scale. Later, he stretched pieces of gut across pieces of wood and invented the stringed instrument.

The bow used in the violin and other such instruments probably originated in the Orient and was introduced into Europe through the Crusades. The plucked guitar may have originated in Spain, where it has become the national instrument. The Bible says that the harp was invented by Jubal. It was used to accompany the Psalms of the ancient Jews, and was played by David. The harp was a favourite of the old Irish poets and it has become one of the emblems of Ireland.

The lute too originated in the East and was brought into Europe and became popular in 1500 and 1600. The lyre came from Asia into Greece. It was a favourite among the Greeks. The mandolin, patterned after the lute, was popular among South Europeans and Latin Americans. It was very much like the Portuguese fiddle. The zither probably developed from an ancient Greek lute-like instrument called cithara.

Greece had two kinds of instruments : the lyre or cithara and the pipe like an oboe.

Early Hawaiian music was a vocal chant accompanied by various simple string wind and percussion instruments the chief of which were the Pahu a wooden kettle drum five feet high, and the Puniu, a small drum made from a cocoanut shell. The Uliuli was a rattle made of gourds or cocoanut shells filled with pebbles. Hollow bamboo tubes produced organ-like tunes when struck on the ground. Smooth stones were used as castanets. Rustling rattles were made of split bamboo. The small stringed instrument called a ukulele was developed from a guitar brought to the island by Portuguese labourers in the late 1800. The modern steel guitar was invented by a Hawaiian musician.

The American banjo seems to have been brought to the United States by Negroes from West Asia, who adapted it from a guitar-like instrument played by the Arabs. The balalaika, a stringed instrument, was invented by the Tartar

Tribes of Russia. The 'celesta' meaning, heavenly, was invented by a Frenchman.

In Java they have weird instruments like the gamelan and the rubab which give forth strange notes. The people love music and have a sort of national piano which gives forth a rather melancholy note. There is a trace of melancholy in the national character which does obtrude itself but is mingled in their docile disposition.

Wherever music may have originated, music as fine art, is the most intimate and fullest expression of human nature. Sound and rhythm are the two important constituents of music, attract everyone. As Coleridge has said, "Music has charms to soothe the savage breast". From the very beginning human beings were attracted to music. Every organ in our body itself is an instrument in an orchestra, the heart beating, the lungs pumping air in and out, the nerves vibrating and every growing cell produces harmonious blending of a variety of notes and rhythm. No wonder then that even a baby is attracted by music.

Music soothes, refreshes and helps man to relax whatever be the music he is familiar with. What appeals to one may not attract another. Music varies from place to place according to culture; and from person to person depending on mood and temperament. Hence we may say that geography affects music and will continue to affect it in the ages to come.

## REFERENCES

1. Grierson H. (ed.), *Metaphysical Lyrics and Poems*, Oxford, 1936, p 11.
2. *Chambers's Encyclopaedia*, Vol. 9, London, 1950, p 616.
3. Yehudi Menuhin, "Understanding Indian Music", Span, Vol. XIX, No. 7, July 1978, pp 11-13.
4. Ibid.
5. Ibid.

21

# Internal Migration—Concept, Theories and Methods of Analysis

—*Najma Khan*

MIGRATION is recognized as one of the most important areas of contemporary research in population. Geographers with their concern for spatially expressed processes are obviously attracted to the study of reallocation of people. However, any attempt to analyse migration inevitably leads to a set of definitions, concepts, theories, typologies and methods of analysis. The knowledge of all these is a prerequisite of migration research. This paper is, however, written with the intention of providing necessary guidance for the research beginners, and other persons who are interested in the field of migration.

## CONCEPTS IN THE STUDY OF MIGRATION

One of the foremost problems with which a researcher on migration will be confronted at the time of data collection and formulation of hypotheses is how to define migration and other related terms. Here, therefore, we will discuss some of the popular terms used in migration researches.

The word *migration* denotes movement of population with a change of residence. The change of residence may occur from one village or town to another village or town, or from town to village, within the district or commune, or between the district or commune, within the country, or outside the country, during the last-mentioned one month, two months, three months, or more. However, the operational definition of migration requires both temporal and locational criteria to be more specifically defined. This operational definition depends both on the type of research and the type of data available.

A *migrant* is a person who moves from one administrative unit to another. In the context of India these administrative units are generally villages in rural areas, and towns in urban. Thus a person who moves out from one village or town to another village or town is termed as a migrant provided his or her movement is not purely temporary in nature on account of casual leave, visit, etc. Sometimes migrants are also called *movers*. Thus at any given time the total population of a place or region may be divided into *migrants* and *non-migrants* or *movers* or *non-movers*.

It sometimes happens that a person willing to migrate is unable to do so because of constraints like lack of money, guidance, or lack of information about job opportunities in the other places. This type of person is called a *potential migrant*. After a certain period away from home, a migrant may go back to his native place; he will then be considered a *returned migrant*. A person returning to his native place with the intention of spending his holidays, or meeting friends and relatives will be called a *visiting migrant*.

In *internal migration* (i.e. the movement of population within the national boundary), to distinguish between the direction of movement the terms *immigrants* and *outmigrants* are used. Immigrants are simply those persons who enter an area and outmigrants are those who leave.

Migration takes place from an area of *origin* (departure) to one of *destination* (arrival) and a group of migrants with the common origin and destination is called *migration stream* (or current). Usually each stream has a *counter-stream* (or counter current). The term *gross-migration* refer to arrival and departure, or total volume of population turnover, and *Net migration* is the balance between immigration and outmigration. If analysis is restricted to two units connected by migration stream, the total size of both stream and counter stream is called *gross interchange* between the two areas and the difference between the two as *net-interchange*. By comparing net migration and gross migration the *efficiency of migration* can be established.

In a study based on data concerning previous residence the term *migration period* or *migration interval* has to be introduced.

If the question asked concerns the residence at specific date this period becomes *migration interval.* If the concern is with *life time migration* than life time becomes synonymous with migration interval. *Duration of absence* from home is counted with the reference of migration interval.

*Migration rates* state the relative frequency with which migration takes place during a specific migration interval. The difference in the rates of migration between various demographic, economic, and social group is termed as *differential migration.*

## THEORIES OF MIGRATION

There is a great deal of theoretical literature on migration. It can be conveniently divided into two parts—Laws, hypothesis, theories and models related with size, direction, motivation and mechanism of migration; secondly, hypothesis and connotations related with differential migration, and consequences of migration. We will discuss them here in sequential order.

### Size, Direction, Motivation and Mechanism of Migration

The first name worth mentioning in this reference is of Ravenstein[1] who in 1880 presented a paper entitled "The laws of migration". His major laws and trends, mainly concerning the value and direction of migration are as follows :

(1) The great body of our migrants only proceed a short distance and there takes place consequently a universal shifting or displacement of population, which produces 'Currents of migration' setting in the direction of the great centres of commerce and industry which absorb the migrants.

(2) If is the natural outcome of this movement of migration, limited in range, but universal throughout the country, that the processes of absorption go on in the following manner :

The inhabitants of a country immediately surrounding a town of rapid growth flock into it, the gap thus left by the rural population is filled by migrants from more remote districts, until the attractive forces of one of our rapidly growing centres makes its influence felt, step by step, to the most remote corner of the kingdom, migrants enumerated in certain centres of absorption will consequently grow less with the distance proportionally to the native population which furnish them.

(3) The process of dispersion is the inverse of absorption and exhibits similar features.

(4) Each main current of migration produces a compensating counter-current.

(5) Migrants proceeding long distances generally go by preference to one of the great centres of commerce and industry.

(6) The natives of towns are less migrating than those of the rural parts of the country.

(7) Female are more migrating than males.

The broad generalization of Ravenstein have been shown to be correct in a number of studies, though his thesis on migration have been severely criticized.

Hill[2] concentrating on the rural districts of Essex, showed that the migration from these districts was mostly short distances into neighbouring countries. In a study of the Glamorganshire Coalfield, Thomas[3] also showed the importance of short distance moves similar to the findings of Greenwood.[4] In the case of migration in India he found that migrants to cities from both rural urban areas tend to come from nearby places. Teauber and Teauber[5] in their study of movement between cities and suburbs, supported the law that each current of migration is accompanied by a counter current in the opposite direction.

Based on the "Distance" as a measure of decreasing probability of contact and increasing cost of movement and population as a measure of attraction, Zipf[6] developed his hypothesis ($P_1P_2/D$) which says that "migration is directly proportional at the products of population of the place of origin and destination and inversely proportional to distance between them."

His model has been the subject of much study. In several empirical studies it is found to be operating.[7] But later Hagerstrand[8] realized that there were many difficulties in its application, because the rate of change in interaction over distance varies in time according to the place and according to social composition.

Though many authors agree that distance is the major determinant of volume and direction of migration. Recent theories looked at migration of population not in mere terms of distance but also in terms of "Opportunities." It is realized that migration is primarily motivated by search for occupational opportunity and that its volume and direction are primarily influenced by job opportunities. In 1940 Stouffer[9] presented his theory of "Intervening Opportunity" in which he proposed :

> ..... the number of persons going to a given distance is directly proportional to the number of opportunities at that distance and inversely proportional to the number of intervening opportunities.

The basic weakness in Stouffer's theory was that he was not able to clarify opportunities well. In his study of Cleveland Metropolitan district he defined it as a number of vacant houses. However, in comparison with Zipf, Grasity model and Stouffer's theory have more predictive power. In several empirical studies where opportunities are defined as occupational theory of Intervening Opportunities was applied with encouraging results.[10] A recent report on population mobility by the Census Bureau has likewise shown that both migration as well as distance moved are related to occupational opportunities in the United States.[11]

In 1960 Stouffer[12] further modified his hypothesis and introduced a variable of "competing migrants" according to which the number of migrants from city A to city B is a direct function of the number of opportunities in city B and an inverse function of the number of opportunities intervening between A and B as well as the number of other migrants competing for opportunities in city B.

Jonson and King[13] tested this modified model in a recent study of internal migration in Belgium. He found that the level of 'explanation' was 'satisfactory' but the term 'competing migrant' is potentially weak.

Rose[14] concentrating on distance and opportunity tested the following hypothesis in Minneapolis. "Higher Studies people seek better jobs or opportunities and have to go over a greater distance to find them on an average than do persons whose skill and aspirations direct them to look for less desirable opportunities". Stub[15] in his study on the occupational characteristics of migrants is a retest to Rose hypothesis and confirms Rose. It shows that professional and managerial workers migrate longer distances than do lower status people.

In a study of the movement of labour from farm Burford[16] considered "psychological distance" to be more important than the geographical distance as a determinant of migration. He constructed the "remoteness index" taking into account the distance between a country seat and the nearest town of 25,000 population, the number of secondary cities (10000-25000) within a radius of 30 miles, and some of the percentage levels of unemployment in the cities involved.

Though the theory of migration which takes into account "opportunity and Distance" as a determinant of size and direction of movement has been successful in describing many migration patterns, it has one major shortcoming. One of the major factors which has not been given due consideration in this theory is that the map of "opportunities actually available" to the migrants is not identical to the corresponding 'mental map.' The migrants do not know about the opportunities, or perceive it in a distorted fashion. Moreover, their "knowledge" is accompanied by varying degrees of uncertainties, a similar observation can be made about, the 'actual costs' associated with attaining each opportunity. Since the transformation from the actual map to the mental map, is very much a function of the information available to and utilized by individual migrants, the 'characteristics of information' are more important in assessing migration.[17]

A number of recent studies[18] show that the directional pattern of migration has been greatly influenced by the "presence of friends and relatives" at the place of destination who have been the major source of information about the potential job opportunities to the migrants.

Based on the above idea Greenwood[19] presented an hypothesis which says: "The previous migrants from one region to another exert a significant influence on the direction of current migration". In a test on the data of India he found that migrants had shown a strong tendency to migrate to localities which had previously attracted natives of their region.

However, the above discussed laws, models, hypothesis are basically concerned with size and direction of migration. They do not say anything about motivation and mechanism which are two important aspects of migration. They also suffer from another shortcoming, in that they tend to describe rather than to explain migration.

The Push and Pull theory was developed in order to explain the cause or motivation of migration. The Push and Pull attributes of the place of origin and destination, are recognised by Bogue.[20] He suggested that there were positive and negative aspects of a migration-provoking situation. Migration may occur as a search for opportunity to improve one's lot in life. In this case destination exerts a Pull on migrants. Migration can also occur as a flight from an undesirable social and economic situation. These situations constitute the explosive Push of the Community". Migration generally takes place when the positive Pull factors at the place of destination are outnumbered by negative Push factors at the place of origin.

There are certain scholars like Myrdal[21] who consider only Push as a potent reason for migration, while there are others like, Trewartha[22] who disagree and assert that migration occurs due to a complex interaction of Push and Pull factors. This has been proved in a recent ILO[23] report on the Census of migration. It concludes : "Though the Push factors of falling income and underdevelopment in agriculture in most of the less developed countries are now very strong, they do not in the absence of strong Pull factors suffice to cause a large shift in manpower between occupations.

In a paper presented at the 1959 population conference, George[24] considers geographical movement of population in two forms—moves caused by necessity or obligation, and moves caused by need. Characteristic of the former type are that they Push certain classes or racial, religious groups who are mostly not suited to the condition in the place of destination. In the second type pressure from the place of origin is accompanied by need (Pull in the place of destination).

Push factors are sometimes labelled as "Centrifugal or impulsive", and Pull factors "Centripetal or attracting". Prothero,[25] recognized both these forces working in the migration of population in South Saharan Africa. He described lack of non-agricultural opportunities in rural areas from a major Push, while many advantages and amenities in urban areas constitute Pull for migrants. In another study of labour migration in Africa south of the Sahara, Mitchell[26] describes the process of migration as the interaction of the "*nexus of centrifugal tendencies*" and the "*nexus of Centripetal tendencies*". The Push and Pull theory is seen from another angle by MacDonald[27] who found that all the Push factors were operating in some rural areas of Italy, still a large proportion of the population did not migrate though free to do so. He found that labour militancy played an important role in keeping the population at home as an alternative to migrate.

There is yet another approach which visualises the Push from home in a purely socio-psychological manner. In a study of migration in Iowa, Zappo[28] presented a theory which recognises the achievement of individual and future aspiration. He says that it is the discrepancy between one's "achievement" (present level of living) and "aspired achievement" (aspired to level of living) which generates migration. The greater the discrepancy the higher are the chances of migration. To determine the discrepancy he constructed a "Pressure index" which was the ratio between expected maximum income divided by scale of income.

However, the Push and Pull theory has long been one of the most important theories in migration. There is no doubt that it has been proved to be a very useful device for listing all the factors affecting the given migrating movement[29] and has produced a lucid and convincing exposition of the underlying factors in migration.[30] But it has a serious limitation which curbs its utility to a great extent. It does not explain "why . . . some migrate and some do not."[31] There is another difficulty with the Push and Pull analysis which emerges when an attempt is made to characterise the combined effect of all the factors at predominantly either Push or Pull.[32] Some scholars have avoided this difficulty by observing that many Push and Pull factors can be matched into pairs, each pair representing two different values of one single variable.[33] Following this reasoning some scholars have concluded that Push and Pull dimensions are subjective characteristics of migration.

Lee's[34] conceptual model on migration which incorporates Push and Pull factors at both origin and destination, tried to overcome this limitation. He proposed that each person is constantly exposed to factors at a place of origin some of which prohibit migration (negative factors) and others which encourage (positive factors) it. In the same way numerous positive and negative impulses come from different potential destinations. Furthermore, between origin and destination stands a set of intervening obstacles in various forms like distance, physical barrier, language and cultural differences, etc. The decision to move or not to move result from the evaluation of all these factors. Such evaluation is not carried out objectively but by exposure to different kinds of conditioning, and basic migration thus takes place when one decides that it is preferable to move rather than stay; then the difficulties of moving seems to be offset by expected reward.

The model proposed by Lee has been very successful in explaining the mechanism of migration, i.e. why one decides to migrate and how migration takes place but it falls short in explaining why one chooses a particular destination. In this respect Wolperts[35] "place utility" or "concept of relative attractiveness" defined as "the net composite utility defined from individuals' integration at some position in place" can be helpful. Perceived utility of a particular place should be greater than the utility (or disutility) of the rest of potential destination and some areas for a migration to take place. If a relatively better area is not available a person will remain in the same place.

Till now we have been discussing various laws, theories, models and hypothesis of migration. A review of all these shows that after Ravenstein Scholars mainly concentrated on distance and opportunity as an explaining factor for migration, which gave birth to the "gravity model," theory of "intervening opportunity", and "competing migrants." Later came the theory of Push and Pull to explain the cause of migration.

Based on this, Lee developed his model which is one of the best so far. However, the overall progress made so far in migration theory is not very impressive. A theoretical framework which can include all the hypothetically relevant factors of migration and can specify the interaction with each other in empirically testable form and thereby serve as a guide to future research work has not yet been produced.

**Differential Migration and the Consequence of Migration**

Differential migration is the selectivity of certain persons or the tendency of certain groups to be more migratory than others. The reason for selectivity has been described by Lee,[36] under the framework of his model. He said that persons respond differently to the set of negative and positive factors at the place of origin and destination. They also have different abilities to overcome the intervening obstacles. It is, therefore, necessary that migration be selective.

In migration studies scholars have repeatedly tried to establish "universal" migration differentials, which would apply to all countries and at all times. But the attempts remained fruitless.[37] Even the generally found age selectivity has certain exceptions, one of the most obvious of which is migration of persons to areas of better climate, which usually is selective of old and retired persons. However, there is no denying the fact that certain differentials like age, sex, marital status, occupation, and social status, are found to be important in a particular place and time.

Trewartha[38] while talking about differential migration said the age of the migrant was the universally accepted differential. The reason in their great adaptability to new conditions and entry into the labour force. Migration is also likely to be sex selectivity but whether those who move out are largely males or females depends on a variety of reasons. There is no universally applicable rule here. In less developed countries migration is predominantly male. Migration selection operates also in terms of marital status. The usual generalisation in that migration is weighted on the side of the single young adult. In less developed countries it is predominantly unmarried males who migrate, but in developed countries married ones seem to be as mobile as single males. Further some occupational groups seem more likely to migrate than do others. Skilled and semi-skilled are those who migrate more than unskilled. Migration selection also operates in respect to intelligence, mental health, educational attainment and race.

There are a number of empirical studies on the selectivity of migrants, which give sometimes similar, sometimes contradicting results.[39] But first of all these studies clearly show that migrants are superior to the non-migrants in various socio-economic and physical attributes, and in the place of origin, while inferior to the non-migrants in the place of destination.

Lee distinguished between negative and positive selection of migration streams and presented the following hypotheses which are worth testing :

(1) Migrants responding primarily to plus factors at destination tend to be positively selected.
(2) Migrants responding primarily to minus factors at origin tend to be negatively selected or, where the minus factors are overwhelming applicable to the entire population group, they may not be selected at all.
(3) Taking all migrants together selection tends to be biomodel.
(4) The degree of the positive selection increases with the difficulty of the intervening obstacles.
(5) The tightened propensity to migrate at certain stages of the life cycle is important in the selection of migrants.
(6) The characteristics of migrants tend to be intermediate between the characteristics of the population of origin and the population at destination.

Another important aspect of migration which will be considered here is the consequences of migration. Because of the great diversity of migration patterns no generalisation about the consequences of migration can be made. One can, however, discuss a few of the possible consequences in relation to sending and receiving areas.

One of the foremost functions of migration is to decrease the population in the community of origin and increase it in the community of destination. By doing so it leaves both positive and negative effects in the places of origin and destination of the migrants. We will first analyse the consequences of migration in the place of origin.

Since migration tends to be both age and sex selection, the migration of younger persons from the places of origin may help to reduce, at least temporarily, the rate of population growth both through the loss of the outmigration and through reduced fertility levels resulting from the movement of persons in the peak productive years. At the same time such age selectivity contributes to an older population at the place of origin, with attendant implications for future social change, health conditions, mortality levels, etc. The selectivity of outmigrants with more educated and skilled persons may also represent a serious loss to the place of origin of trained individuals who could provide talent and leadership essential for social and economic development.

The impact of migration is not restricted to the changes in migrants' places of origin. It also has serious implications for the population in the place of destination. Many problems associated with rapid urbanization are in fact problems which are severely aggravated by rural-urban migration. Overcrowding slums, shanty towns, difficulty of waste disposals, traffic congestion, housing shortage, inadequate educational facilities, poor water supply, and lack of open space for recreation are closely related with heavy movement of population in urban areas, and inabilities of these centres to cope quickly with such multiplicity of social and economic problems. Moreover, the rapid rate of natural population increase augmented by a large inflow of migrants in the reproductive age compound the critical situation. Furthermore, if much of the movement to urban centres is not fully justified by the availability of jobs, underemployment or unemployment results. The income of migrants will be low affecting both the quality of their personal lives and their ability to contribute their proportional share to the costs of providing the basic government facilities needed for a minimum standard of living in an urban environment.

Sovani[40] analysing the demographic consequences of migration in underdeveloped countries said that in these areas migration has a negligible effect, that too for a short time only. The reason is that rates of migration are low, and most of the migrants move temporarily for a short time. In the long run, these migrants return to their place and make up for the loss of migration.

But the reverse was the case in certain regions of the United Kingdom and the United States where migration from rural areas resulted in acute depopulation, shortage of work-force and natural decrease.[41]

Myrdal[42] taken up the topic of migration and economic development. He argued that an advanced region attracts migrants, internal migration tends to widen regional disparity, and this hampers economic development. Hathaway[43], in the case of migration from agriculture in the United States, reached the conclusion that migration has not improved the relative position of low income areas in agriculture. Because most of the people capable of making such improvement were among outmigrants.

A major consequence of migration is the transfer of cash to the areas of origin. Conell, et al[44] in his recent rural study found that in the case of Indian villages proposition of remit-tance to the total rural income is very low ranging from 0 to 8 per cent only. While in the villages of Taiwan it is estimated to be 39 to 76 per cent of the total village income.

Talking about the problems of over-urbanization as a conse-quence of migration in big cities Ganguli[45] said that in many large cities of india the supply of labour exceeds the absorptive capacity of the cities, so that on the whole the rural unemploy-ment and underemployment are converted into large scale urban unemployment. This conversion appears to be affected at a high human cost, because these cities lack the capacity to provide even the elementary social amenities because a mass of unassimilated maladjusted individuals in strange surroun-dings, whose existence is insecure and unpleasant.

**Typology of Migration**

The typologies of migration are devised in order to classify various types of migration into a systematic scheme of presenta-tion. There is no separate typology for internal migration so we will discuss typology for migration as a whole.

There are a number of typologies of migration which have been based on different criteria. An early classification of migration is given by Fairchild[46] in 1925. He classified migration into "invasion", "conquest," "colonization," and "immigration." The criteria underlying this classification were the difference in the level of culture or presence or absence of violence. His typology was followed by many despite certain major criticism raised against it.

Kant [47] presented another typology which was based on circular movement (pastoral, nomadic, transhumance) and areal units between which migration takes place (interlocal or interregional and interlocal or intra-regional). Beltramore[48] persuaded the same line of reasoning and classified migration into temporary and definitive and later on into areal units (extra muros and intra-muros). George[49] gave another classification recognizing only two classes of migration : Migration prompted by economic factors, and that in which need or obligation to move was so strong as to put economic considerations in the background.

Peterson's[50] typology in one of the best known for analysis of migration based on which Kosinski[51] recently developed a more elaborate scheme of migration classification. Peterson utilized a number of criteria. In this typology the most important being the distinction between the migration which is undertaken in order to change the way (innovative) and that which helps to preserve it (conservative). Other criteria include the type of interaction expressed by migration forces which results in classes and type of migration. On the basis of these criteria he recognized five broad classes of migration: primitive, forced, impelled, free and mass.

Kosinski's multi-dimensional scheme of types of migration is, however, one of the recent additions to the existing typologies of migration. As mentioned earlier, he based his classification on Peterson's typology. The major dimensions of his classification are time (temporary, permanent), distance, (long, short), boundaries crossed (internal, external, areal units), decision-making (voluntary, impelled, forced) member involved (individual, mass), social organization of migration (family, clan, individual), political organization of migration (sponsored, free), causes (economic, non-economic), aims (conservative innovative). Kosinski's typology has more utility than Peterson's because of its wider coverage.

## METHODS OF MIGRATION ANALYSIS

Methodology in any study is conditioned by the objectives of the study and the type of data available.

Most of the studies in internal migration have as their main objective the analysis of direction and pattern of migration, characteristics of migrants, causes, and consequences of migration. Among these varied objectives one which is most complicated and difficult to achieve that of the causes of migration. In a study which is based on the census place of birth data, the causes of migration may be analysed by combining the knowledge of the characteristics of the migrants with the data on the characteristics of place of origin and destination. On the other hand, in a study dependent on the field inquiry, where questions are directly asked of the migrants, the causes of migration can be analysed by grouping various reasons given by the migrants into economic, social, and sequential causes and under Push and Pull force of migration.

It should be recognized, however, that both these methods are characterised by certain limitations. In the former method which is most popular among demographers the causes of migration are derived indirectly and are liable to mistake. Moreover, this method is not adequate because it does not reveal anything about the 'personal decision' to move and all that it involves in migration. Similarly, in the second method, the classification of causes is based on the answers given by migrants. These answers are sometimes vague and general like "work", "family reasons", "studies", etc. They do not reveal the exact factors about the social or economic, Push or Pull situation. Many a time each answer may imply both. This vagueness of answers given by migrants is liable to lead to subjectivity in the grouping and analysis of causes. Despite this shortcoming this method is used widely in empirical studies.

Geographers with their special interest in interaction between environmental factors and spatial process usually draw the causal relationship of migration by correlating and regressing various socio-economic and demographic factors with migration. This method is very effective in finding out both intra-unit (household level) and inter-unit (village/town/

district) association of migration with various socio-economic variables.

The characteristics of the migrants is another important aspect of migration analysis. It can be analysed in terms of various socio-economic and demographic attributes of migrants and their related families. The individual characteristics of migrants are generally analysed in terms of age, sex, marital status, religion and caste, language, ethnic structure, work status, occupation and income, while family related or background characteristics of migrants may be looked in terms of the size of family, size of landed property, occupation and income of the family. If the study were to be based on sample data collected in the field, then the characteristics of the migrants can be cross tabulated with each other, and a more deep and meaningful analysis can be made. For instance educational characteristics of migrants can be analysed in relation to age and sex, occupational structure of migrants in relation to age, education and income. These cross tabulations are, however, dependent on the hypothesis developed for analysis. It should also be understood in this connection that place of birth data do not allow any cross tabulation, and limit the analysis to several major demographic characteristics of migrants only.

Migrants differ from non-migrants in many ways. The reason for this selectivity, as we have already discussed in the section on theories and hypothesis, is amply described by Lee who said that persons respond differently to the set of positive and negative factors at the place of origin and migration destination. They also have different abilities to overcome the intervening obstacles. It is because of this reason that migration is generally selective. To analyse the selectivity of migration, the characteristics of migrants can be compared with the non-migrants both at the place of origin and destination. This analysis can be done both on the basis of place of birth and sample survey data, but in the study which is based on population registration data the selectivity of migration cannot be studied. The reason is that data in population registration relates only to those who are moving and not for total population at risk.

In the geographical studies of internal migration keenest interest has tended to centre on the direction of flow and pattern of migration. Earlier, in the second part of this study we have discussed in detail various scales, hypotheses, theories and models explaining the directional bias of migration. These are "linear distance", "intervening opportunities", "competing migrants" and "presence of friends and relatives." By using any of the above mentioned, distance and direction of flow can be analysed. The directional bias of migration can also be looked at in relation to rural-urban places of origin and destination, both within the administrative unit and between the administrative units.

We have also seen earlier that the role of distance in migration differs according to the characteristics of the migrants. This has been pointed out by Rose. However, one can also analyse the distance of migrants' places of destination in relation to characteristics of migrants.

In a migration analysis, beside the spatial pattern of migration which we have considered above there are also temporal patterns. In the temporal pattern of migration, it can be analysed in its historical perspective or in contemporary terms. One of the best examples of historical analysis of migration in India is Zachariah's[52] study based on the period 1901-1931.

It is obvious that an historical analysis of migration can be made only on the basis of place of birth or population registration data. In a study based on a sample survey, migration can be analysed only in contemporary terms, that too for a short period of time i.e., 5 or 10 years. The reason is that in the sample survey, in order to avoid recall lapse in data collection, it becomes necessary to limit the period of study for a short period of time. However, these surveys have one advantage. They can derive the extent of seasonal, periodic, semi-permanent or permanent types of migration on the basis of migrants' intended period of stay out. Such a distinction cannot be made on the basis of the other two sources of data, i.e. place of birth and population registration.

The consequences of migration constitute a very important question from the point of view of policy implication. As discussed earlier migration affects both the community of origin and destination in many ways. It also affects the migrants themselves. The consequences of migration can be observed and measured in demographic, economic and socio-cultural terms.

An overall assessment of the consequences of migration cannot be made on the basis of place of birth data. It allows only the analysis of the impact of migration on the size and structure of the population. On the basis of sample survey data, many other socio-economic consequences can be drawn besides the demographic consequences.

## CONCLUSION

One of the major conclusions, that can be drawn from the foregoing is that migration is too complicated a phenomenon to be conceptualized, defined and analysed.

The theory of migration is still in an underdeveloped stage. Though migrational researchers have put forward a number of hypotheses, theories, and models, there is still a need for more research to test the hypotheses and explain the process. A theoretical framework which can include all the hypothetically relevant factors of migration, and can also specify their interaction in an empirically testable form is the most immediate need.

Because of the scarcity of recorded data the methods of migration analysis are extremely limited. In view of this, more detailed field work needs to be undertaken.

## REFERENCES

1. E. Ravenstein, "The Laws of Migration," *Journal of Royal Statistical Society*, 48, June, 1885, pp 167-235 and June 1885, pp 241-305.
2. A. B. Hill, "International Migration and its Effect upon Death Rates with Special References to the Country of Essex."
3. B. Thomas, "The Migration of Labour in Glamor-ganshire Coal-field", *Economica*, 1930, 10, pp 323-36.
4. M. Greenwood, "Migration in Urban Areas : The Case Study of India," *Journal of Regional Science*, Vol II, No. 2, 1971, pp 261.
5. K. B. Teauber and A. Teauber, "A White Migration and Socio-Economics Difference Between Cities and Suburbs," *American Sociological Review*, 29, Vol 5, Oct. 1964, pp 718-29.

6. G. K. Zipf, "The $P_1P_2/D$ Hypothesis on the Intercity Movement of Persons," *American Sociological Review*, ii, pp 677-86.
7. D. J. Bogue and W. S. Thompson, "Migration and Distance", *American Sociological Review*, 14, 1949, pp 236-44, C. T. Stewart, "Migration and Function of Population and Distance," *American Sociological Review*, 25, 1760, pp 347-56, G. Olsson, "Distance and Human Interaction : A Migration Study", *Geographical Annallar B*, 47, 1765, pp 3-43 and C. F. Claeson, "Distance and Human Interaction" *Geographical Annallar* 50, 1968, pp 142-61.
8. T. Hagerstrand, "Migration and Area: Survey of Sample of Swedish Migration Field and Hypothetical Consideration on Their Genesis" in D. Hauneberg et. al. (ed.), *Migration in Sweden : A Symposium*, Lund Studies in Geography, 13, 1957, pp 27-158.
9. S. Stouffer, "Intervening Opportunities: A Theory Relating Mobility and Distance", *American Sociological Review*, 5, December, 1940, pp 845-67.
10. L.M. Bright and S. D. Thomas, "Interstate Migration and Intervening Opportunities," *American Sociological Reviw*. 1941, pp 773-83, E. C. Is-bell, "Internal Migration in Sweden and Intervening Opprtunities", *American Sociological Review*, 9, 1944, pp 85-91, F. Strodlbeck, "Equal Opportunity Internal", *American Sociological Review*, 1949, 4, pp 490-97 and D. Courgeau, "*Les Champs Migratoires en France*", Paris, 1970.
11. Current Population Reports, Series P-20, No. 14, Internal Migration in United States, April 1940-47.
12. S. Stouffer, "Intervening Opportunities and Competing Migrants," *Journal of Regional* Science, 2, Spring, 1960, pp 85-92.
13. C. J. Jansen and R.C. King, "Migration et Occasions intervenantes" en Belgique', *Recherchcs Eonomiques du lou Vain*, 4, 1968, pp 519-26.
14. A. M. Rose, "Distance of Migration and Socio-Economic Status Migrants", *American Sociological Review*, 23, 1958, 420-23.
15. H. Stub, "The Occupational Characteristics of Migrants to Duluth : A Retest of Rose's Migration Hypothesis", *American Sociological Review.*, Feb. 1962, pp 87-90.
16. R. L. Burford, "An Index of Distance as Related to Internal Migration," *Southern Economic Journal*, 29, 2, Oct. 1962, pp 77-91.
17. P. Gould and R. While, "*Mental Maps*" Baltimore, Penguine, 1974.
18. B. H. Luebke and F. J. Hart, "Migrants from Southern Applachian Community", *Land Economics*, 1958, Feb., p 34, S.J. Schwarzweller, and J. Manglam, "Kentucthy Mountain Migration Stum Family : An American Variation" on the Theme by *Le Play*', in J. C. Janson (ed) *Readings in Sociology of Migration*, London, 1970, pp 93-120 and B. M. Levy and J. W. Wadyckij, "The Influence of Family and Friends on Geographic Labour Mobility : An International Comparison", *Review of Economic Statistics*, Vol. LV, No. 2, May, 1973.
19. M. Greenwood, "The Influence of Family and Friends on Geographi-

cal Mobility in Less Developed Countries : The Case of India", *Review of Regional Studies*, No. 2, 1973, pp 58-68.

20. D. J. Bogue, *Principles of Demography*, New York, 1969, Chapter 19, p 753.
21. G. Myrdal, *Asian Drama*, Vol III, 1969, p 2139.
22. N.V. Sovani, "Internal Migration and Future Trend of Population in India," Paper Presented at World Population Conference, Bel. Yougo, 30 Aug., 10 Sept., 1965, A. Base, "Internal Migration in India, Pakistan and Ceylon," Paper Presented at World Population Conference, Belgrade Yougo, 30 Aug, 10 Sept. 1965 and J. T. Trewartha, "Mobility and Migration", *A Geography of Population : World Patterns*', John Wiley, New York, 1969.
23. ILO, "Why Labour Leaves the Land : A Comparative Study of Movement of Labour Out of Agriculture, *Studics and Reports*, New Series, No. 5, Geneva, 1960, p 209.
24. P. George, "Types of Migration of the Population According to the Proefssional and Social Composition of Migrants", in J. Jansen (ed.), *Readings in Sociology of Migration*, London, 1970, pp 39-47.
25. M. R. Prothero, "Characteristics of Rural Urban Migration and the Effects of Their Movements Upon the Composition in Rural Urban Areas Sub Saharan African", *World Population Conference*, 1965, Vol. II, pp 523-526.
26. J. C. Mitchell, *The Causes of Labour Migration Research Bulletin*, African Labour Institute, 6, 1959, pp 12-46.
27. J. MacDonald, "Migration Versus Non-Migration : Regional Migrational Differentials in Rural Italy", *International Population Conference*, 1961, paper 107.
28. D.F.C. Zappo, "The Relationship of the Pressure Index and Selected Socio-Psychological Factors to the Propensity to Migrate", *Dissertation Abstract International*, A., Vol. 30, March-April, 1970, p 4029-30.
29. G. Mortara, "Factors Affecting Rural Urban Migration in Latin America : Influence of Economic and Social Conditions in these Two Areas", *World Population Conference 1959*, Paper 511.
31. Webber, *The Growth of the City in the Nineteenth Century*, Cornell University Press, 1967, Chapter. 3.
31. W. Peterson, *Population*, New York, 1969, p 607.
32. For Example a Number of Studies have concluded that the Rural Urban Migration in the less developed countries today is activated primarily by Push whereas in the more advanced countries the Pull factors have been strong.
33. Herrick in his work "Urban Migration and Income Differences", London, 1963, p 14. Unifies Push and Pull Hypothesis of lower rural and higher urban income into 'one in which urban migration is the function of expected rural urban income differences', Similarly Kerznets et. al. in his 'Population Redistribution And Economic Growth : United States 1870-1950. Vols. I & III Philadelphia,

American Philosophical Society, 1957-1964, Speaks of Differential Economic Opportunities in Internal Migration and Economic Growth, Later Todro devised a model for an Analysing the determinants of urban labour supply which takes into account differential expected real earning in urban rural Employment after allowance is made for probability of finding an urban job in his 'A Model of Labour Migration and Urban Unemployment in Less Developed Countries,' *American Economic Review*, 1969, p 138-48.

34. S. E. Lee, "A Theory of Migration", *Demography*, Vol. 3, No. 1, 1966.
35. J. Wolpert, "Behavioral Aspects of the Decision to Migrate", J. C. Demko, et al (eds.) *Population Geography : A Reader*, MacGrow Hill Toranto, 1970, pp 298-306.
36. S. E. Lee, 1966, op. cit.
37. D. E. Thomas, "*Research Memorandum on Migration Differential,*" New York, *Social Science Research Council Bulletin*, 43, 1938.
38. T. G. Trewartha, 1968, op. cit., pp 137-138.
39. D. Thomas, "Age and Economic Differentials in Interstate Migration", *Population Index*, 24, 4 Oct. 1958, pp. 313-325, C. H. Hamiltion, "Educational Selectivity of Rural Urban Migration : Preliminary Results of North Carolina Study Proceedings," Annual Milbank Memorial Fund Conference, 1957, Part III New York 1958, E.S. Lee, "Migration Differentials by State of Birth in United States", International Population Conference, 1961, Paper No. 61, K. C. Zachariah, "*Migration in Greater Bombay,*" Demography Research Centre Bombay (Memeo.) 1968, H. Brorming, "Migration Selectivity and the Growth of Large Cities in Developing Societies", in *National Academy of Science, Rapid Population Growth*, Baltimore, 1971 pp 273-314, H. L. Larry, "Migration Differentials by Education and Occupation : Trend and Variation" *Demography*, Vol. 10, No. 2, May 1973 and G. Martin, "Volume Characteristics and Consequences of Internal Migration in Columbia", *Demography*, Vol. 12, 2, May 1975, pp 193-207.
40. N. V. Sovani, *Population Problems in India, A Regional Approach*, Chapter V, Delhi, 1969, p 169.
41. C. Lowenthal, and L. Couitas, "Emigration and Depopulation : Some Neglected Aspects of Population Geography", *Geographical Review*, 52, 2, April, 1962, 195-210 and J. Saville, *Rural Depopulation in England and Wales* 1851-1951, London, 1957.
42. G. Myrdal, *Economic Theory and Underdeveloped Region*, Duck Worth, 1957.
43. D. E. Hathaway, "Migration from Agriculture: The Historical Record and its Meaning", *American Economic Review*, May, 1960, p 380.
44. T. Counell and B. Dasgupta, et al (ed.) *Migration from Rural Areas The Evidence from Village Studies*, Chapter V, Oxford Press, London, 1976, p 96.

45. B. N. Ganguli, *Population and Development*, New Delhi, 1973, p 65.
46. H. P. Fairchaild, *Immigration : A World Movement and Its American Significance*, New York, 1925.
47. K. Kant, "Classification and Problems of Migration", P. L. Wagner and M. W. Mikesell (ed.) *Readings in Cultural Geography*, 1953, Chicago, pp 342-54.
48. A. Bethamove, *La Mobilité geographique d'une Population*, Paris 1966, pp 21-41.
49. P. George, 1970, op. cit., p 39.
50. W. Peterson, 'A General Typology of Migration', *American Sociological Review*, 1958, 23, pp 256-65.
51. L. Kosinski and M. Prothero, (eds.), *Population on Move : Studies on Internal Migration*, M. & Co. Ltd., London 1975, pp 7-9.
52. K. C. Zachariah, '*Historical Study of Internal Migration in Indian Subcontinent*, 1901-1931. Doctoral Dissertation University of Pennsylvania Mimeographed by the Demographic Training and Research Centre, Bombay, 1962.

22

# The Concept of Military Geography

—*Ashok Kumar*

THE GEOGRAPHICAL features of landscape play a major role in modern warfare. Landscape features provide a good camouflage for para military forces, the ocean for submarines and a fair sky with good weather conditions for the air force. Thus military geography is known as the science of the military evaluations of geographical elements of a state which are related to the formulation of its defence policies and war plans and the actual *modus operandi* of war at a time of emergency.

To draw up the defence policy and war plans of a state in support of its political policies, the determination of the following aspects is quite necessary : (1) the size and composition of armed forces for land, sea and air operations; (2) the technological level of war; (3) the geographical extent of war, i.e. global, regional or local; (4) military objectives for the destruction of the enemy war machine (the Indo-Pakistan Conflict, 1965); (5) the political objective to enforce the acceptance of a political demand by the enemy through military action (Anglo-French action against Egypt over the Suez Canal issue); (6) the geographical objective includes pushing out the enemy from a strategically significant territory through military action (the Chinese attack on India to secure the Aksaichin road); (7) physical elements which pose problems of combat are relief features such as mountains, deserts, swampy grounds, climate, hydrology, vegetation, i.e., forest cover as well as cultural landscape; (8) the economic resources structure, technological level, war potentials of the state and its allies and adversaries.

Keeping in view the above ideas military geography regards the geographical features of the earth with a view to military operations, but since whenever man exists he must need modify the original condition of the ground, for the purpose of armed operations, topographic as well as artificial details of a country are as important as those of the physical features of the landscape.[1]

Destruction of the economic core areas, like industrial bases, lines of communication, ports and harbours, can paralyse a nation's war efforts. The lack of emotional integration of the people of a state which can be brought about by the subversive action of the enemy, can also paralyse the war effort.

In the field of military geography the military evaluation of political areas from the strategic, tactical and logistic points of view has to take into account the physical, economic and human elements, in their mutual relationship. The data regarding the spatial environment of the political area and the possible operations are collected for military evaluation in the context of the type of war, the level of military technology, and the military task set by the political policy-makers of the state.

## THE TYPE OF WAR

The war can be either intra-state or inter-state conflicts. Intra-state wars are conflicts between the power-elite of a state and the counter elite, e.g., the conflict in Zaire after its liberation from the Belgian imperial rule is an example of such conflict. Inter-state wars are those where two sovereign states are involved in a conflict. Both these types of war can involve the deployment of guerilla forces, where a factory worker, a farmer, a school boy or an office worker by day turns into a guerrilla fighter at night. Wars can also be conducted by the deployment of nuclear weapons aimed not at the armed forces in uniform; an evaluation of the geographic realities of a state, as well as its potential adversary, should take into acount the type of war that the state is likely to face. The Himalayan mountains can be rightly considered as a barrier to large scale troup movement for launching an effective attack on India. In a conventional type of war the Indian armed forces can draw the enemy forces to a killing ground of its own choice in the plains after allowing the enemy to stretch its logistic lines of support over the difficult mountain terrain. But in a guerrilla type of warfare the enemy can send out small bands of highly trained mobile troops to infiltrate the mountain terrain and neutralise the armed forces defending the area. On the other hand, the enemy can completely ignore the mountain barrier in launching its military offensive and can develop its missile technology and nuclear weapons to achieve its objective. The military evaluation of the Himalayan mountain barrier in the context of the nation's defence has to take into account the type of war that the nation is likely to face.

## THE TECHNOLOGICAL LEVEL OF WAR

Warfare can be reduced to the simple concept of struggle for dominance between *fire* and *movement*.

According to the conventional concept of war, by fire is meant the physical and psychological effect of weapons, from rifles to atom bombs, irrespective of whether projected by gunpowder, dropped from aircraft, or jet propelled like rockets. *Movement* is the change of position of the weapons producing fire, and of the crews serving them, as well as displacement of the material required for their functioning, regardless of whether it is carried out on foot, by means of horse drawn or mechanised vehicles, or by railways or aircraft. Therefore, *movement* is a function subordinate to *fire*. In establishing the ascendancy of movement over fire, military leaders have to study the limitations imposed by geographic factors like terrain and weather, to mention only the most obvious ones. The evolution of military technology from the bow and arrow stage to the missile age was in response to the need to overcome the limitations imposed by geographic factors and to establish the supremacy of either fire or movement. The introduction of cavalry, permitted rapid movement through terrain difficult to negotiate for infantry, man and an offensive weapon. The answer to this mobile force was the development of trench warfare and direct machinegun fire from well defended points. To liquidate the entrenched defensive positions the development of heavy ordnance came up and for some time "artillery captured the terrain, which the infantry was only to occupy". The rising ascendancy of fire over movement led to the evolution of tanks, armoured troop carriers, parachute dropping of troops and supplies, amphibious vehicles, combined operations between land, air and naval forces. This brought about a re-evaluation of geographic elements from the point of view of military technology. The development of air power, missiles and nuclear weapons are merely attempts at establishing the supremacy of fire over movement after the blitzkrieg tactics of highly mobile panzer forces had completely paralysed the 'maginot line' mentality of building up static defences. The military evaluation of the geographic realities of a state takes into account the technological level at which war is to be conducted whether by deploying conventional ground, air and naval forces, or by the concentration of effort in the destruction of enemy war potential by deploying manned bombers or guided missiles or by taking recourse to tactical and strategic nuclear weapons as deterrents or as retaliatory strike or through political, economic and military subversion.

### THE GEOGRAPHICAL EXTENT OF OPERATIONAL THEATRES OF WAR WHETHER VICINAL OR GLOBAL

It will not be far from the truth to mention that the Germans won most of the battles in the Second World War but lost the war owing to their inability to cope with the geographical extension of the war. The successful conduct of military operations depends on the evaluation of the ratio of '*force*' to '*space*'. The word force in this context implies the striking power available to dominate a geographical space.

Hence the basic factor in working out this ratio is the correct military evaluation of the geographical extent of the operations of war and this ratio changes according to the following considerations : the type of war; the level of military technology; the tactical quality of the terrain; the density and quality of the circulatory system; and the quality of the training, leadership and morale.

### The Military Task set by State Policy

The formulation of defence policies are a projection of the political postures and policies of the state. The political power of a nation aims at the maximisation of effort and at the utilisation of the human and other resources of the geographic space in order to ensure that the state can play its assigned role.

Both in the international and domestic spheres the political power of a state depends on its military power. Apart from coming to the aid of the civil power the military commanders are assigned the following military tasks to meet the demands of a war : a military objective implies the destruction of the military potential of the enemy; a geographical objective, implies the neutralisation of enemy control over a territory, which is of strategic significance; and a political objective, implies the creation of a situation through military action which would force the political elite of the enemy country to accept a political decision.

**The Physical Elements of Possible Theatres of Operations of War**

The scientific and technological advances in the recent years have liberated the armed forces to some extent from the limitations imposed by the physical elements of possible operations of war in the strategic, tactical and logistic fields. But the lessons of war in Vietnam have amply demonstrated that in the monsoon-drenched hills, forests, marshy ground and rice lands of the theatres of operations the American superiority in sophisticated weapon system has been neutralised by the effective utilisation of the terrain by highly mobile guerrilla forces. The evaluation of the terrain forms is an essential part of the study of military geography.

## THE ECONOMIC POTENTIAL OF WAR

"Modern war is a total war" on all fronts. In modern warfare the entire energy of a nation has to be devoted to the successful prosecution of war. Today war absorbs the whole of the economic and other resources of the nation. The capacity of a nation to fight a war, depends upon its economic potentials of war. "War now-a-days is an industrial proposition. It is more influenced by the science of economics than by the art of strategy." The term 'economic potential of war' may be defined as the portion of the total production of goods and services which can be devoted to war. In times of emergency, a nation can and does step up its production. The sum total of its production is its economic potential; within this total is the portion that can be used for war; that is the economic potential of war. Military planning is deeply concerned with the economic potential of war. Revolutionary movements with a very weak economic base have to resort to guerrilla type of warfare. The ability of a state to transform the peace-time economy to meet the demands of war operations and its ability to reduce the time lag for such a transformation to the minimum, contributes in a large measure to the success of defence policy and war plans.

## THE HUMAN AND CULTURAL REALITIES

The political stability of a state depends upon a complex set of cultural, historical and demographic factors. The lack of political stability of a state poses problems for its defence and seriously undermines its war potential. The territorial integrity of a state depends upon other factors, such as the stability of its international frontiers. The lack of emotional integration of the citizens of a state in order to support the potential policy of the political-elite of the state may be due to the following factors; linguistic, religious, and other cultural differences; geographic isolation of the ethnic group from the main stream of the nation; isolation of the power-elite from the counter elite on the cultural, political and economic plans; the failure of the ruling-elite to solve the socio-economic problem; and subversion by an external power.

Whatever may be the causes, emotional disintegration may lead to intra-state wars or inter-state limited wars along the nation's borders. Hence a study of the human and cultural realities of geographical environment forms a very important aspect of the study of military geography.

## MILITARY DOCTRINE

Military doctrine is a set of principles that guide the formulation of two basic elements of the defence of a political unit. Firstly, the defence policy, which is operative in the passive phase of international relations of a state, and in the exercise of effective political power within the territorial limits of the state. At this stage it provides the strong arm backing to the state's policies, in both the inter-stage and intra-state spheres. Secondly, the war plan, which is operative in the active phase, when attempts at finding a political solution fail to achieve results in both the inter-state and intra-state spheres.

Politics provides the *raison d'etre* in the formulation of military doctrine. It is not possible to lay down a set of principles for the formulation of military doctrine for universal application. In the first place, military doctrine is national in character as it deals with national, political policies against the background of the domestic, vicinal and global power political scenes. In the second place, a national military doctrine is also relevant to a historical situation.

**Object of Formulating Military Doctrine**

Every state formulates a military doctrine of its own which is in harmony with its political policies and geographic realities. The following are the objectives of formulating military doctrine :

(a) To protect the land frontiers, the air space, and the territorial waters from external aggression;
(b) To provide the strong-arm-backing to the state's foreign policy;
(c) To help the state honour its international commitments; and
(d) To come to the aid of the civil power when specifically called for at the time of emergency.

## MILITARY STRATEGY

Strategy is the art of planning military operations in pursuance of achieving the military task set by the political policy of the state. At this level of planning the need for harnessing the entire war potential of the state requires a correct evaluation of the physical, economic and human realities of the geographic environment. At a lower level strategy is the art of directing larger military movements, operations of campaigns of war potential of the enemy. Strategy deals with the mobilization of all the forces and resources of a country at war. The state draws up a strategic plan, in comformity with military doctrine, to harness the entire war potential of the country in the direction of operations of war. The overall strategic plan has two main facets :

(a) the political strategy of the government in the diplomatic, economic and socio-political spheres to neutralise the enemy efforts and to build up the war potential of the state; and
(b) the military strategy, in support of the state policy, aimed at the most efficient utilisation of the entire war potential of the armed forces in the direction of the war operations.

## TACTICS AND LOGISTICS

Military strategy is closely linked to its tactical and logistic operational plans for the effective handling of defence forces in combat, and these are highly sensitive to geographical factors.

Tactics is the art of deploying and monoeuvring air, land and naval forces when in contact with the enemy. Logistics is the science and practice of moving, lodging and supplying combat troops based on the military evaluation of terrain, weather and the circulatory system. Tactical and logistic problems posed by terrain and weather have activated research in military technology as well as tactics and logistics :

(a) In areas where the terrain prevents the rapid movement of troops to meet a surprise enemy thrust, battlefield mobility can be given by helicopters today;
(b) Where the field of fire and visibility are limited by terrain features, aerial observation removes such a handicap and fire can also be directed on the enemy from the air;
(c) When weather conditions prevent manned bombers to take to the air, guided missiles can be used in support of a combat mission; and
(d) Amphibious vehicles can remove the limitations imposed by water laced terrain on tactical and logistic movements.

Dependence on heavy and sophisticated equipment has increased the significance of logistics in combat operations. Terrain and weather have a great impact on the formulation of tactical and logistic plans. The features of cultural landscape, like roads, railways, ports and harbours, urban centres and industrial bases, canals and inland waterways are of equal significance in this context. The infra-structure of logistic support for such wars becomes highly vulnerable to enemy interference. But in guerrilla type of operations it is the fighting men who are more important than equipment. The dependence of guerrilla fighters on logistic support is very insignificant. The evaluation of terrain and weather features of the cultural landscape in the context of guerrilla operations will be entirely different from that of conventional war operations.

## REFERENCE

1. E.S. May, *Geography in Relation to War*, 1907.

23

# India's State-Idea—An Application of the Functional Approach in Political Geography

—*Sudeepta Adhikari*
—*Alok Kumar*

HARTSHORNE defines[1] Political Geography as "the study of the areal differences and similarities in political character as inter-related parts of the total complex of areal differences and similarities". The main strength of the definition lies in the fact that it sets political geography in the context of all geography, not just regional geography. Certain works[2] reveal that since 1964, the emphasis of political geography has shifted towards an analysis of the interaction of political processes and geographical areas. Thus, 'political systems', seems to refer to technique rather than the subject. Any political variation between different countries, or within a state, results from the conscious decision and action of people working individually or in groups. Political geographers are, therefore, concerned with the geographical consequences of these political decisions and actions, the geographical factors which were considered during the making of any decision, and the role of any geographical factor that influenced the outcome of political actions.[3] One point which must be stressed is, that in addition to analysing the reciprocal relations between geography and politics, political geographers also have a duty to describe the areal patterns which result from them.[4]

The present work is an attempt at analysing the various divisive and cohesive forces of the Indian Union in a geographical perspective, and also to make an analysis of India's state-idea and *raison d'etre.* The state is a system of functioning parts. Many of these functional parts have spatial attributes such as core areas, the frontiers, circulation, etc.

## METHODOLOGY

Political geographers have written very little about methods of research in their field. Wright[5] is one of the few political geographers to have written specifically on this theme, but the others wish to follow the functional method as suggested by R. Hartshorne[6] in one of his well-known articles. In the plethora of books published recently dealing with research methods, and particularly on quantitative analysis, there is rarely any reference to political geography.[7] R. Hartshorne stresses that a central theme in political geography ought to be the functioning of the state. What is it that holds the state together ? How do these binding elements overcome the divisive forces present in any state ? Clearly, this involves several approaches at once : the time element is of importance in that states go through periods of greater and lesser unity and cohesion; the spatial distribution of the various elements of the state (languages, religions, racial groups, etc.) can be of great importance in determining the degree of division that exists in the state. Every state possesses regional differences and similarities, differences that tend to disrupt and similarities that tend to bind together. Hartshorne uses the terms "centrifugal"* and "centripetal,"** which like a number of other terms defined by him, have become standard parts of the vocabulary of political geography. The present paper, therefore, gives more emphasis on the functioning parts of the state-system.[8] Much work[9] has been done on the political problems of the state-areas of the different regions, but none of the works, except those of L.K.D. Kristof and E. Thompson, has tried to find out, and analyse the state-idea of the respective state-areas of the world. The authors have attempted a classical approach to the functional aspect, i.e., through reconstruction of all relevant aspects of a single event or a series of related events. All sources have been used but there has been heaviest reliance on primary sources and documents.

* Centrifugal forces are those which lead toward division and may even threaten the break up of the state-area.

**Centripetal forces are those which lead to unity and coherence and hold the state together.

## INDIA, AS AN INDEPENDENT STATE-AREA

India occupies a curiously symmetrical position in southern Asia, a great rhomboidal land mass hanging like the keystone of the arch formed by the Indian Ocean shores. India is still a convenient geographical expression for all the lands between the Karakoram and Kanya Kumari.[10] Griddled by the young folded mountain on the north-west, north and north-east, and washed by the waters of the Indian Ocean, and its two arms, the Arabian Sea, and the Bay of Bengal, on the south-west and the south-east, India has, since historic times, the privilege of being a well defined geographical and geopolitical unit."[11] India was a part of the British Empire, much tortured, exploited, and oppressed, for more than a hundred and fifty years. Nevertheless, the British administration brought about a cohesiveness, a unitary organisation of the regional administrations in the country. It was during the period of the imperial devolution, after World War II, that the Indian people under the charismatic leadership of Gandhi and Netaji successfully liberated their motherland, on 15 August 1947, and India became a Sovereign Democratic Republic in January 1950. India now occupies a position at the crossroads of the maritime and the continental powers.

**State-idea** : The state-idea is a complex of tradition, experiences and objectives. It is made up of written history, folklore, stories of national heroes, religious beliefs, and the language and art forms in which these things are communicated. It is the body of literature, the painting, the architecture, and the music which are distinctively national. And it is also the characteristic economic, social and political institutions. The state is created to depend on and develop the state-idea. The state must have a *raison d'etre*—a reason for existing. On the other hand, state-area is a section of land and section of humanity, organised on a single unit in terms of a particular, distinctive idea.

In 1947, the people of India did not have any idea of the *raison d'etre* of their country. The Preamble of the Indian Constitution adopted in 1950 resolved to establish a "Sovereign Democratic Republic" with sovereignty vested in the people. The Preamble also referred to justice, liberty, equality and fraternity to which Indians are entitled. Despite everything we owe much to the British, because they laid the foundation of the geopolitical consolidation of the subcontinent, which, Chandragupta Maurya, Ashoka, Akbar and Aurangzeb did during their times. But the tragic part of the story of the geo-political unity of the sub-continent, was the truncation of the political, social and geographical contiguity of the sub-continent, and the creation of Pakistan in 1947. Considerable confusions persisted, soon after Independence, over the nature of the administrative system, because the people had no clear idea of how India could be governed.[12] After years of trial and error, there has emerged a quasi-federal administration in consonance with the regional diversities of the country.

The Indian Constitution has often been described as federal* but with strong unitary features. G. Sawer[13] thinks that it is, "but for one circumstance. . . a federal reaction to a federal situation. . .." This one circumstance is the emergency power of the President which, while it operates, makes the Constitution a unitary one. The Government of the country continued to be carried on during the Emergency in a substantially federal fashion. The central legislature is bi-cameral, consisting of an upper council of states and elected House of the people; the components of the Union, whether ex-British Provinces, old states, or unions of states, are known as States. The capital at New Delhi is in effect, a federal district.

## AN ANALYSIS OF INTERNAL PROBLEMS

The effectiveness with which a Government can fulfil the purposes of a state is a result of the balance between two opposed sets of forces "antifugal" that leading toward division may even threaten the break-up of the state, and "centripetal" that lead to coherence and unity of the state.[14] An analysis of the viability of India being an independent state involves a

*A federal form of government is defined as a form of constitutional government to express imperfect unity and multi-nationalism. Federalism is a particularly complicated form of western democracy (F.G. Carnell, "Political Implication of Federalism in New States" in *Federation and Economic Growth in Underdeveloped Countries* (ed) Hicks, U.K., London, 1961, p 16).

discussion of the balance between these forces. In India, the major population concentrations occur in the effective national territory which is the scene of the interaction of various forces of integration and disintegration, but the unproductive and unoccupied territories do not seem to be potential from the geopolitical point of view. Today, India is passing through a period of great strain, and there are some people who find in it signs of complete political dismemberment of the country in the future. There is one school of thought which emphasizes the view that India had never been a single country in the past. When others talk about the cultural unity of India, this school of thought, ridicules the idea, and holds that such unity lies only in the imagination of nationalistic minded sentimentalists. There is no reason, they say, why we should take sentimental refuge by harping upon Indian's cultural unity, and refuse to face with objectivity, the reality of a growing disunion in our country.[15]

**Centrifugal Forces**

The people of India do not have a common culture and a common language. There are a number of peoples inhabiting compact territories, speaking a common language and having a common culture, a common psychological make-up and economic life. They comprise the various ethnic and cultural groups, many of whom today are organised in linguistic regions of their own within the Indian nation. The topography of the country has played a major role in creating a linguistic and cultural division. The Vindhya range stand as a formidable barrier between the northern plain and the southern plateau. The north and south are themselves divided into several major and smaller linguistic culture sub-divisions by certain topographical lineaments of the landscape. The Indus Valley region is separated from the Ganga-Yamuna plain by the wide expanse of the Rajputana desert, and is correlated with the latter only by a narrow corridor to the north of Delhi. This has resulted in the development of a homogeneous cultural complex from west to east till the northward projecting horn of the plateau in the Chotanagpur region is reached. Bengal, beyond the bottleneck between Chotanagpur and the Himalayas, though culturally one with the rest of the plain, developed its own language and culture. Separated from the Bengali speaking areas by the Shillong plateau is the Brahmaputra valley of Assam with its own linguistic-cultural complex. South of the Indus-Ganga plain area from west to east, Gujarat, Maharashtra, Madhya Pradesh and Orissa—each of them though still Indo-Aryan in their linguistic cultural complex, have developed their own language and script.[16] South of the Deccan lava country of Maharashtra is the Kannada speaking Karnataka. This language belongs to the Dravidian group. Further south of it is the Malayalam speaking Kerala separated from the Kannada and Tamil speaking areas to the east by the ridge of the Western Ghats. The Tamil speaking area is roughly separated by the Mysore Ghats from Karnataka and the Western Ghats from Kerala. North of Tamil Nadu and east of the Kannada and Marathi-speaking areas is the Telugu-speaking Andhra Pradesh which is dominated by the riverine tracts of the Krishna and the Godavari.[17]

Throughout history each of these nuclear areas maintained some kind of separate historical identity, though some of them were divided into a number of rival principalities. These nuclear areas have developed separate iconographies*[18] which are evident from the temple-architecture, pattern of religious beliefs and system, their historical evolution and social history and social hierarchy. Each linguistic-cultural region of the country represents a distinct iconographic landscape which ceases to exist at the regional boundaries of the region. Throughout the historical processes of the geopolitical consolidation of the subcontinent, the regional iconographies had been the major obstacle, and proved to be a resistant and stabilizing factor in their respective regions. Relations between the regions could not develop because there was no effective movement** across the regions in ancient, medieval and even during the British period. The movement factor, according to

*_Iconography_ is the whole system of symbols in which a people believe. These symbols are many and varied. A national iconography in our sense encompasses the national flag, the proud memories of past history as well as the principles of the prevailing religion, the generally accepted rules of economics and the established social hierarchy.

**_Movement_ seems to cover the flow of people, armies, material goods, capital, messages and ideas across the space open to men's activities.

Gottman, makes for change and permits space to be organized. Hartshorne[19] recognizes that movement, i.e., circulation, is a means of overcoming the physical and social barriers to state-integration. It has led to the success of the acceleration process of the Muslim religion and faith in the north, but failed to achieve any success in the south due to various reasons.

Some more problems, in the form of the centrifugal forces, have come out of the linguistic-cultural and religious complexes, which include the following :

(a) **The Communal Cleavage** : This is defined as a religious antagonism caused by the projection of religion into the sphere of politics. Smith defines[20] communalism as the functioning of religious communities or organisations which claim to represent them, in a way which is considered detrimental to the interests of other groups or of the nation as a whole. the term usually implies some kind of political involvement, that there are several communal problems in this country is a known fact.

Prior to British rule, there were wars between the Hindu and Muslim rulers, but these did not take the form of religious antagonism between Hindu and the Muslim masses. Communalism has now assumed a new dimension. Before the partition of the subcontinent, it was Muslim communalism that was the spearhead of reaction.

(b) **The Caste-complex** : The caste system in India is an institution that has no parallel anywhere in the world. Having its origin in a system of division of labour which came into being shortly after the advent of the Aryans into India. Somewhere in the second millennium B.C., it has undergone many changes since then. Originally, there were only four castes—Brahmin, Kshatriya, Vaishya and Sudra which came into being as a result of the vocational structure of society. The conception and practice of caste embodies the aristocratic ideal at present and is obviously opposed to the democratic conception of the country. This may pose a threat to the integration of the country. There are two aspects to this problem. There are conflicts between various upper castes, and between the upper caste and the scheduled and backward castes, particularly prevalent in some states. Caste has become more important factor in the Indian body politic than communalism. It seems to be more a threat to the state than to the secular nature of India. Mahatma Gandhi once said, "one more obstacle which stands in the way of Purna Swaraj is untouchability . . . and Hindus cannot claim Swaraj as long as they do not give independence to the depressed classes."[21]

(c) **Language Problem** : Regional iconographies and the lack of movement across the region during the ancient, medieval and even the British periods have largely impeded the growth and fluidity of a pan-Indian language. There are varieties of scripts, hundreds of dialects and languages, which are completely different from each another. Experiences have shown that the southerners think of northerners as belonging to a foreign country and vice versa. There have been serious language riots in Assam, Tamil Nadu and Maharashtra. Assam has been badly truncated into a number of regions simply on the basis of language. The post-independence reorganization of the states was carved out on the basis of language. "Despite the large number of apparent diversities, after the birth of Pakistan and the linguistic reorganisation of states, language has become the main factor of cleavage between the states of India. As the state boundaries have hardened, language has begun to play mischief, as in Canada, to the detriment of national unity", says Dikshit.[22] The lack of movement has confined the people's intercourse of ideas to their fellow-linguists. This has hindered the inter-state mobility of population.

The Mahatma thought that without a pan-Indian language, the country might disintegrate and the state of affairs that existed before the conquest of India by the British might emerge. He pleaded for a common script for all the Indian languages, which he thought, would bring social and political cohesion in the country. He favoured the Devanagari script instead of the Roman script which lacked the phonetic capacity.[23] The only Indian language which can claim the status of a *lingua franca* of India is Hindi, because out of 72 per cent of the accepted Aryan languages, Hindi is spoken by more than 50 per cent. The trouble arises because the people in the south consider the Devanagri script as a kind of imposition on them.

(d) **Regionalism** : The problem of national integration is not just one of achieving a certain psychological or emotional change in the pattern of thought and behaviour of the different peoples of India. It is primarily a socio-economic problem of removing the obstacles in the path of the development of the people, so that they can advance according to their own genius and culture unfettered and unhampered. In this connection, it has to be realised that one of the most serious problems facing the nation is the question of securing a balanced regional development of India's economy. Regional imbalance in economic development is evident from the lack of initiative towards integration. Major industrial activities are centered around a few metropolitan centres like Calcutta, Bombay, Madras and Delhi, etc., and these industrial nodes appear as islands in the vast ocean of poverty, thus encouraging the centrifugal tendencies in the backward areas. The north-eastern region and the central plateau region of the country are the two major neglected regions. The Central Government have failed to develop the infra-structure of these regions, while steps have already been undertaken to develop the infra-structure in the Yamuna-Sutlej doab, in the Ganga-Yamuna doab, and also in Maharashtra and other southern states. The Green Revolution has been a success in the western U.P., the Punjab and Haryana but a failure in the eastern Gangetic region.

The disunity that we witness today is the result of each region's dominant interest in promoting the welfare of a limited section of the people and a comparative neglect of the larger class of the masses. There is a widespread feeling that in spite of a common political structure, the development of sectional interests has brought a great strain upon India's national unity. But instead of promoting integration at the economic level where the largest amount of corrosion has taken place, we seem to rely more and more on the task of bringing about more unity only at the cultural level. M.P. Shreevastava[24] observes that a wide geographical distribution of the natural resources has caused an unbalanced growth of regional economies. Some regions are rich in mineral resources while others are not; similarly the canal irrigated regions lead the country in agricultural products, thereby causing a wide gap between the surplus regions and the deficient regions which in turn is reflected in the "heterogeneous psychology" of the people of the country towards one another.

(e) **Influx of Foreign Ideas**: Since the campaign of Alexander the Great across the Indus in the trans-Indus region in 326 B.C.,[5] India has been exposed to foreign ideas. The low hills of the Hindukush and the passes in it helped the infiltration of outsiders. These people brought with them their own distinct ideas. The Greek expedition left the imprints of Greek culture albeit over a very restricted region. The Muslims particularly, the Mughals, the wandering nomads of central Asia, brought with them the most static and chauvinistic concept of Islam that they established and it spread across the Yamuna and the Ganga. Mughal India was the geographical expression of the chauvinistic Islamic empire. Similarly, the Europeans, during the eighteenth century, landed on the Indian shores across the Indian Ocean. It was not until the French Revolution of 1789, that the European political and cultural ideas found their expression over the Indian lands. The British conquest of India was not a mere political event. It was also the meeting of three civilizations and of three different value systems, the Hindu, the Muslim and the English. The interaction of such civilizations upset the political, economic and social conditions, setting in motion a chain of unforeseen consequences. "The contact of three distinct cultures brought to the fore the peculiarities of the three systems of philosophy, religious belief and practices and literatures" says Purohit.[26] The cross-profile, formed by the interaction of these three distinct ideas, has confused the geo-political psychology of the country. The post-independence period reveals the impact of foreign ideas on our political system and structure which have resulted in the "geopolitical diversities" of the political structure.

The integration of the tribal people has been a significant political problem in the post-independence decades. Indian tribes had never been exposed to the newer ideas of integration; moreover the British way of political consolidation left them untouched. They had no idea of the kind of British sovereignty being imposed on them. Decades of geographical isolation restricted the psychological integration of the tribes with the enlightened people of the subcontinent. The Naga problem is one of these which once threatened the very existence and the integration of the Indian Union. The Nagas live in a very different country between the Brahmaputra in Assam and the upper Chindwin in Northern Burma; extending as far south-east as Imphal in Manipur and west of Haphlong at the edge of the Khasi-Jaintia plateau, spreading over an area of some 200 by 400 miles. The Nagas are far away from the main Indian ecumene and the movement factor did not play an effective role in binding them with the mainstream of Indian life. Shweinforth[27] observes that the cleavage between the plain people and the hill people here seems to be complete and it has been so aggravated during the past decade of hostilities that any solution other than complete independence may prove unacceptable to the Nagas. The Mizo problem is a similar one.

Being located at a greater distance from the Indian core area with a distinct religious landscape, Kashmir has always been a problem area. Much of its parts are under the illegal occupation of Islamic Pakistan. The infiltration across the cease-fire line often creates and encourages the centrifugal tendencies in the Kashmir valley.[28] The Naga problem and the Kashmir problem once constituted the main "problem areas," for the Indian Union. Much of the Naga tension has been eased but the situation in Kashmir still remains tense. After analysing the centrifugal forces of the country, we come to this conclusion that "through centuries of conquest and infiltration India has become a veritable maze of races, languages, creeds and customs . . . The pattern of language in this country is extremely complex, crossing racial, religious, and political boundaries. Politico-geographically speaking, the diversity in religion and language creates a serious national problem."[29]

### Centripetal Forces

It appears from the above mentioned analysis of the centrifugal forces, that India is an example of unity amidst diversity. What is that inherent force which holds the Indian nation united, creates the feeling of brotherhood and a community of complementary habits when the country has linguistic cultural complexes ?

(a) It is the Hindu religion that acts as a cementing and integrating force, holding together the diverse elements in Indian society and resolving all conflicts into a higher synthesis and integration. What the nation wants is a religion of strength based on the inherent divinity of the human soul, and not a religion of weakness teaching that man is a born sinner. In the words of Swami Vivekananda[30] "..... I am proud to belong to a nation which has sheltered the persecuted and the refugees of all religions and all nations of the earth .... The Hindu religion is tolerant.... The Hindu nation is a spiritual nation. This is the land from where, like the tidal waves, spirituality and philosophy have again and again rushed out and deluged the world, and this is the land from whence once more such tides must proceed in order to bring life and vigour into the decaying races of mankind". A revival of Hinduism which took place in the nineteenth and the early twentieth centuries, was absolutely necessary for the growth of India as a nation.

Griffiths[31] observes that "under favourable geographical circumstances, Hinduism was able to build up throughout India a common culture and a uniform way of life. The foundation thus existed on which, under favourable political conditions, a structure of true unity might be created".

(b) Sanskrit literature brought forth the fluidity and mobility in Hindu religion. It became the uniting bond for the Indian nation, even though there were political divisions.[32] Hardly any language in the world has played as vital a role in the history of a race as Sanskrit has. It is the mother of all the Hindu literatures and languages that are spoken in the country. It has provided a common dialect and a script for the Indian people. In the words of Monier Williams,[33] "India, though it has, as we have seen, more than 500 spoken dialects, has only one sacred language, and only one sacred literature accepted and revered by all adherents of Hinduism alike, however, diverse a race, dialect, rank and creed. That language is Sanskrit, and that literature is Sanskrit literature—the only repository of the Veda or 'knowledge' in its widest sense; the only vehicle of Hindu theology, philosophy, law and mythology; the only mirror in which all the creeds,opinions, customs and usages of the Hindus are faithfully reflected and the only quarry whence the requisite materials may be obtained for improving the vernacular or for expressing important religious and scientific ideals.' One of the major achievements of English literature, on the other hand, is that it exposed the richness of the Hindu religion and civilization, and helped spread the view that the Hindus had a glorious past.

(c) One of the significant post-independent developments in the country has been the growth of an "Indian community" of complementary habits, and facilities of communication. In the words of Karl W. Deutsch[34] "A larger group of persons linked by such complementary habits and facilities of communication we may call a people. The test of complementarity of any set of communications equipment is communicative effectiveness. How fast and how accurately do messages get through." The post-independent development of communicative efficiency and facilities has led to an overall development of a socially standardized system of symbols, acceptable to all communities. The acculturation processes in operation over a long period of time, have, succeeded in creating ethnic complementarity. It must not be forgotten that the methods of working, the utensils used for eating food, the structure and beautification of houses, the uses of spices, the avoidance of eating beef among Hindus of all sections and places, and also among a section of Muslims, marriage customs, religious performances, seniority by age, dresses for home use and methods of cultivation, etc., are somewhat similar if not identical.[35] The communicative efficiency during the 1960s and the early 70s has helped the spread of an Indian culture. The role played by All India Radio and Doordarshan (Television) in this direction, is laudable. The Indian Railways have succeeded in winning over the "distance" (i.e. the intervening areas between the core or ecumene and the peripheries) which once proved to be an important centrifugal force. Moreover, the railways have reduced the communication gap, and removed the human barrier. Indian people are, therefore, held together "from within" by this communicative efficiency which resulted in the community of complementary habits.

(d) The charismatic leadership of Mahatma Gandhi, Pandit Jawaharlal Nehru and Sardar Patel, worked largely as a unifying force in the country removing the geopolitical displacement caused by the lapse of British paramountcy. When India was to become independent, Britain declared that paramountcy was not transferable and that the five hundred and odd princes of India could accede to either of the two dominions. The lapse of paramountcy and its non-transferable character as cunningly interpreted by the British made some of the Indian princes believe that they would be virtually independent after the British left.[36] It was due to the able leadership of Sardar Patel that the geopolitical unification of India was achieved, though in some of the cases police action had to be resorted to. The accession of Kashmir was final on 26 October 1947 when the Maharaja Sri Hari Singh signed the Instrument of Accession, and was accepted by the Governor General on 27 October 1947.[37] Here also Sardar Patel was quick to take the decision of initiating a military operation against the Muslim rajakars, and the Pakistan army. The people's faith in the leadership' was based on the extreme devotion and self-sacrifice that the leadership had shown during the struggle for freedom and after independence. So when the hour of freedom arrived, there was no scramble for power as in some new nations. Further consolidation of India occurred during the regime of Indira Gandhi. The geopolitical accession of Sikkim within the union without any resistance from within or outside was also achieved during her rule.

(e) The role of New Delhi as the federal capital of the Indian nation : "Lying in a narrow lowland passage with a desert on the west, mountains on the north, and where the hills to the south-west give way to the productive valley of the Ganga and to easy routes to the south, Delhi possessed some compelling advantages which no ruler aiming at a tight control over the subcontinent could fail to discern "says Dikshit.[38] New Delhi proves to be a unifying force both psychologically and symbolically. It acts as a balancing force between the different constituents of the country. Its location is neutral as well as non-controversial in the Indian core-land. Historical evidences reveal that the growth of the Indian nation around Delhi, began during the times of Mahabharata, the Pandavas chose it as their capital.". . . . As a conveniently central administrative location it (Delhi) is India's natural capital, and has persisted as the major control point of India since at least the tenth century A.D., and was prominent for a millennium before that."[39] New Delhi lies at the cross-roads of the benevolent Hindu culture and the enlightened Muslim culture, and represents the true "Indian Iconography," incorporating the diverse regional iconographies. It represents the systems, traditions and values of the Indian culture which find their expression in the complementary habits of the Indian people.

New Delhi occupies a neutral site between Calcutta, Madras and Bombay, the three original control points of the British administration. But the decision to shift the capital from Calcutta to the Central core of the country was "an act of administrative expediency on the part of the imperial government" says Dikshit.[40] O.H.K. Spate feels that the national capital in a federal state is almost literally like a key-stone wherever a federation proceeds to bind formerly largely disjointed or discordant units.[41] New Delhi, therefore, balances the emotional attachment of the various provinces and the union territories, and holds the Indian nation united.

## AN ANALYSIS OF INDIA'S STATE-IDEA

Most of the Hindu kingdoms and empires in ancient India and the Indian Muslim empires during the medieval times which were founded over the Indian soils, collapsed and disappeared because none of them had *raison d'etre*, as part of their politico-geographical distinctiveness. It was largely during the mid-nineteenth century that the distinctiveness of Hindu culture found its expression in a series of movements initiated by persons like Swami Dayananda Saraswati, Swami Vivekananda, Bankim Chandra Chatterjee, Aurobindo Ghose and Sri Ramakrishna Paramhansha, who by their writings and preachings, tried to reconstruct the future of a free India in the image of the past. Hindu revivalism was the socio-religious, cultural and political movement in modern India which endeavoured to place before the Hindus, a goal having its roots in their past history and founded on the principles and practices of ancient Hinduism. The tradition was that of Hinduism deeply rooted and the basis of what has been called the fundamental unity of India. The Arab philosopher Albiruni noted in the eleventh century that the Hindus believed 'that there was no country like theirs, no kings like theirs, no religion like theirs, no science like theirs.[42] In the great population cluster of the upper and middle Ganga valley four out of five persons conform to the code of caste. The densely-inhabited south-eastern coast is more than 90 per cent Hindu and on the south-western coast 70 per cent Hindus form more than three-fourths of the population of the peninsular highlands outside the tribal areas. They hold equally large proportions in the Thar cases. In the central Himalayan valleys between Kashmir and Nepal nearly 100 per cent of the inhabitants count themselves Hindu. The two regions of greatest absolute concentration of Hindus—the Ganga plain and the south-eastern coastal plain—have long been the centres of Indian civilization. The northern plain between Delhi and Benaras was the cradle of Hinduism as well as Buddhism and Jainism—the two other native Indian religious creeds which have few followers today in the land of their origin. Whenever a single government gained control over a major portion of the subcontinent its capital was invariably located in the middle of the Ganga valley or upper Ganga plain. Though the political and religious "centre of gravity" was consistently situated in the Indo-Aryan North, civilization flowered in early times in the Dravidian south, which was in communication with the North by routes traversing the peninsular highlands or following the eastern coast.[43]

It is clear from the above analysis that the chief and the predominant state-idea of India is the Hindu religion* though it is not contained in the preamble of the Indian Constitution. The essence of the Hindu religion is a law of ideal perfection for the developing mind and soul of man. The idea is clearly portrayed in the written records of ancient India during two millennia and it is the very life-breath of Hindu ethics. It was the creation of at once ideal and rational mind, spirit-wise and world-wise, deeply religious, nobly ethical, firmly yet flexibly intellectual, scientific and aesthetic, patient and tolerant of

*Any state, to become well established, must present to the population of its as areal parts a distinct *raison d'etre*, its justification for existence, as an areal unit separate from the neighbouring state-areas. This reason must be based upon desires or values of first importance to the populations of the regions included in the state. They include, notably: religion, language and literature, historic memories and the form of government. (Hartshorne, Richard, "The Concept of *Raison d'etre*" and Maturity of States (Abstract) *Annals of the Ass. of American Geographers*, Vol. XXX, 1940, pp 59-60.

life's difficulties and human weakness, but arduous in self-discipline. This was the mind that was at the base of the Indian civilisation and gave its characteristic stamp to all the culture. India is the guardian of the Asian idea and its profound spiritual secrets. Throughout her history India has ever tried to discover this unity in diversity and build a synthesis of the different streams of thought and spiritual discipline that have flowed from her speculative and inward endeavours.[44]

Swaraj was a concept integrally woven with the values of ancient Hindu religion. It was "the national expression of the national mind" says Chittaranjan Das.[45] Thus we come to the conclusion that the Hindu Religion acts as a cementing and integrating force, holding the diverse elements in Indian society and resolving all conflicts into a higher synthesis and integration.

Another distinct political idea of India stems out of its location at the head of the Indian Ocean. Centuries of experience have revealed that the security and the control over the Indian Ocean depend on India. It is believed that one who commands the Indian Ocean, controls the Pacific and the Atlantic Ocean, and gains the mastery over the world Island.[46] Because of its immense strategic location at the crossroad of the "Heartland" and Rimland, it has drawn the attention of the major powers of the world that has resulted in an international rivalry. India is not in a position to counterbalance the growing influences of the foreign powers because she has a small navy. To ensure safe navigation, freedom of movement and to maintain the integrity and the Asian character of the Indian Ocean, India favours the idea that the Indian Ocean should be a zone of peace. It will bring political stability in the littoral regions. India is the main architect of the non-aligned movement in the world. The geopolitical location of the country on the threshold of the major powers leaves no option but to develop the non-aligned movement.[47] There is a causative relationship between India's desire to keep the Indian Ocean a zone of peace, and her attempt to mobilize the non-aligned movement in the region. "The British desire in the late 19th and early 20th century, to control the sea route between the Indian Ocean and the China coast, was the distinct political idea of British India" says Fisher.[48]

The true *raison d'etre* of India is its religious distinctiveness which is evident from the writings on Ashoka's rock edicts :

> The increase of spiritual strength is of many forms.
>
> But the root is the guarding of one's speech so as to avoid the extolling of one's own religion to the decrying of the religion of another, or speaking lightly of it without occasion or relevance.
>
> As proper occasions arise, persons of other religions should also be honoured suitably. Acting in this manner, one certainly exacts one's own religion, and also, helps persons of other religions. Acting in a contrary manner, one injures one's own religion and also does disservice to the religion of the others.
>
> One who reverses one's own religion and disparages that of another from devotion to one's own religion and to glorify it over all other religions, does injure one's own religion most certainly.[49]

Annie Besant[50] identified Hinduism with India. Without Hinduism India has no future. Hinduism is the soul into which India's roots are struck. Very forcefully she has expressed her sentiments : "Make no mistake without Hinduism India has no future . . . Every one might pass away as they come and India will still remain. But let Hinduism vanish, and what is she ? A 'geographical expression of the past, a dim memory of a perished glory." Her history, her literature, her art, her monuments, all have Hinduism written across them."

The degree to which these distinct state ideas are applicable to the various provinces and the union territories of India, is a matter of deep research. The Indian provinces do not have *raison d'etre* of their own. That of the Indian nation is also of the provinces. A majority of the provinces with a few exceptions, are tied up by the common bond of the Hindu religion. It was felt at the time of independence, when the princely states and the British provinces, gladly accepted to accede to India and signed the Instrument of Accession without any hesitation. In some of the princely States, where the rulers were opposed to the idea of accession, the will of the people triumphed over the will of the rulers. In some provinces where the non-Hindus constituted the bulk of the population, India's chief and dominant state-idea did not find its proper expression while, the other political idea of keeping the Indian Ocean a zone of peace, finds its genuine expression. Some of the chronic territorial problems have largely been overcome. The Nagas, the Mizo, the Kashmiris and Sikh have now pledged their faith in India, which they call "indispensable".

India's state-idea is dynamic, forward looking and a future oriented element. It is assertive and claims priority and superiority over the common man's understanding of the *raison d'etre* of India.

## CONCEPT OF INDIAN-NATION

A nation is a soul, a spiritual principle. One is in the past, the other is in the present. One is the possession in common of a rich legacy of remembrances, the other is the actual consent—the desire to live together, the will to continue to value the heritage which all hold in common.[51] A nation is, therefore, an organised section of people occupying a particular and fixed area with a common religion, a common language and culture, a sense of ethnic, kinship, geographic unity and contiguity, a common historical experience and frame of political thought, a common area of economic mutuality and sometimes, the fear of a common enemy.

The primary requisite for the growth of a nation is the certainty, fixity and permanence of places, and when that is assumed, the other formative forces will appear and make themselves felt in due course of time. A common fatherland is preliminary to all national development. Around that living nucleus will gather all those feelings, associations, traditions and other elements which go to make up a people's language and literature, religion and culture, and thereby establish its separate existence and individuality, demanding its preservation and independent development as a valuable cultural unit.[52]

The concept of an Indian nation dates back to antiquity. It seems that the great founder of Indian religion, culture and civilization were fully conscious of the geographical unity of their vast motherland and sought in various ways to impress it on the popular consciousness.

The following River-Hymn of the Rig Veda provides the geographical contiguity of India.

> O ye Ganga, Yamuna, Sarasvati, Sutudria and Purushni receive ye my prayers !
>
> O ye Marudveridha, joined by the Asikni, Vitasta, and Arjithiya joined by the Sushoma hear ye my prayers.

It presents the first national conception of the Indian unity. It was necessarily conditioned by the geographical horizon attained in that age which seemed to have been confined by the snowy mountain in the west, the Indus and the range of the Sulaiman Mountains, and the valleys of the Yamuna and Ganga in the east. These limits practically include the whole of northern India, the geographical unity of which was also recognised and suitably expressed in the designation of the entire territory by the common name of Aryavarta.

The Puranic River-Prayer is but an adaptation of the aforesaid River-Hymn of the Rig Veda to a new environment to an expanded geographical horizon which embraced the whole of India within its limit :

> O ye Ganga, Yamuna, Godavari, Sarasvati,
> Narmada, Sindhu and Kaveri,
> Come ye and enter into this water of my offering !

The Indian nation appears to have been nucleated around the rich Ganga valley which is the modern core-area of India. The process of arealexpansion around the Ganga valley probably began during the Mauryan period. Chandragupta Maurya and Ashoka the Great, expanded their territories across the Vindhyas reaching as far south as the Caveri. Thus for the first time India achieved her nationhood in the true sense of the term : a definite geographical territory with a common Hindu religion and tradition and cultural heritage, and a central authority of the "Chakravarti raja" which reconciled and balanced the centrifugal forces.

The Muslims could not break the homogeneity of the Hindu tradition, though, they made their best efforts to do so. The English being a commercial community, did not bother about it. It was during the British period that Hinduism assumed a new dimension. The "Divine" and "Spiritual" concept of the Indian nation is evident from Bankim's song "Bande Mataram" which he composed for his famous novel "Anand Math". There Bankim saw the golden image of Durga in the form of India.

Mother, I bow to thee !
Rich with thy hurrying streams,
Bright with thy orchard gleams,
Cool with thy winds of delight,
Dark fields, waving, Mother of might,
Mother free.

Glory of moonlight dreams
Over thy branches and lordly streams,—
Clad in thy blossoming trees,
Mother, giver of ease,
Laughing low and sweet !
Mother, I kiss thy feet;
Speaker sweet and low !
Mother, to thee I bow.

It shows how India's history is indissolubly bound up with her geography.

Rabindra Nath Tagore's song *Jana-gana-mana* which he composed in 1912, gives the geographical concept of the Indian nation :

Thou art the ruler of the minds of all people,
Dispenser of India's destiny.
Thy name rouses the hearts of the Punjab, Sind, Gujarat and Maratha,
And of the Dravida, Utkala, Vanga.
It echoes in the hills of Vindhyas and Himalayas,
Mingles in the music of Jamuna and Ganga;
And is chanted by the waves of the Indian Sea;
They pray for thy blessings and sing thy praise.
The Saving of all people waits in thy hands,
Thou Dispenser of India's destiny;
Victory, Victory, Victory to thee !

India is essentially a nation, with a degree of unity underlying the diversity of Indian culture. Centuries of living together and intermingling have given the various races and religious groups a community of interest that is characteristically Indian. A common historical background and common interest are the cement that bind the Indian people together. The underlying of uniformity of life is seen from the Himalayas to Cape Comorin.

In the words of Aurobindo[53] : India is the Bharata Sakti, the living energy of the great spiritual conception, and fidelity to it is the very principle of her existence. By its virtue alone she has been one of the immortal nations ; and this alone has been the secret of her amazing persistence and the perpetual force of her survival and revival. India's social system is built upon this conception, her philosophy formulates it, her religion is an aspiration to the spiritual consciousness and its fruits; her art and literature have the same upward look ; her whole *dharma* or law of being is founded upon it. It is her founding of life upon this existed conception and her urge towards the spiritual and the eternal that constitute the distinct value of her civilization."

## CONCLUSION

India is an organic state. It has grown slowly and over a long period of time from some nuclear, germinal, or core-area. The Indian state has grown and subsequently expanded around the Ganga valley. The process of the territorial accretion around the Ganga valley started during Mauryan times and continued until the advent of modern nationalism. The process of territorial accretion has never been aggressive. The Hindu kings had a conscious understanding over a greater India which they wanted to build regardless of their interests. The accession of the Princely states, and their decision to merge with India marked the final phase of the territorial accretion. The whole phase of the geographical accretion was peaceful except in a few cases, where minimum force had to be used. Each and every state in its early stage has to confront the centrifugal forces, but the organic states resist those forces and maintain their individual personality and geographical existence. States which are created to fill some preconceived geographical framework fail to resist the centrifugal tendencies and they, either collapse or disappear. India is no exception. India is in its youthful stage and there are centrifugal forces operative in the country in the form of communalism, linguistic-cultural complexes and regional iconographies, regionalism, caste system, language problem, problem of tribal integration, distance, etc., but, the centripetal forces, like religion and literature, common complementarity and communication facilities and the charismatic leadership, etc. seem to have overwhelmed the centrifugal forces, and provided two distinct political ideas—(a) "need to protect the basic norms and spirit of benevolent Hinduism"; (b) need to keep the Indian ocean a zone of peace." The spiritual conception of the religious distinctiveness is the *raison d'etre* of India. The Indian nation is an aggregation of men, with a healthy spirit and warmth of heart which creates a moral conscience. It is the end product of a long period of work, sacrifice and devotion. Beneath all the manifold variety there is a fundamental unity, so that the geographic diversity itself far from being a source of neatness is a fertile source of strength and health for India. Its geographic contiguity finds expression in India's spiritual conception as a "Divine Nation" of great people.

## ACKNOWLEDGEMENT

The authors wish to record their grateful thanks to Dr. L.N Ram, Reader in Geography, Patna University, for going through the entire manuscript.

## REFERENCES

1. R. Hartshorne, "Political Geography" in *American Geography: Inventory and Prospect*, (eds.) P.E. James and C.F., Jones Syracuse, 1957, p 178.
2. R.E. Kasperson and J.V. Minghi, *The Structure of Political Geography*, Chicago, 1969.
3. J.R.V. Prescott, *Political Geography*, London, 1972, p 2.
4. (a) Stephen B. Jones, "A Unified Field Theory of Political Geography" in *Systematic Political Geography*, (eds) de Blij, Harm. J., London, 1967, pp 142-154.
   (b) L.K.D. Kristof, "The Origins and Evolution of Geopolitics", *Journal of Conflict Resolutions* IV, 1960, pp 15-51.

5. J.K. Wright, "Training for Research in Political Geography", *Annals of the Association of the American Geographers*, XXXIV, 1944, pp 190-201.
6. R. Hartshorne, "The Functional Approach in Political Geography" in *Systematic Political Geography*, pp 159-194.
7. J.R.V. Prescott, op. cit., p 27.
8. Harm, J. de Blij, op. cit., pp 158-159.
9. Pedeo L. I. Baldovia, "Political Geography of the Philippines", *Philippine Geographical Journal*, I, 1953, pp 15-23.
   Haram. J. de Blij, "Uganda and the Problems of Politicism" in *A Geography of Sub Saharan Africa*, Chicago, 1964, pp 264-278.
   Uladimir, Cervin, "*Problems of Integration of the Afghan.*
   Charles A. Fisher, "The Malaysia Federation, Indonesia and the Philippines : A Study in Political Geography," *Geographical Journal*, 129, Pt. 3, 1963, pp. 311-328.
   Charles A. Fisher, "Malaysia : A Study in the Political Geography of Decolonization" in *Eassays in Political Geography* (ed.) Fisher, Charles, A. London, 1968, pp 75-145.
   L.D. Koristof, "Russian Image of Russia : An Applied Study in Geopolitical Methodology," in *Essays in Political Geography*, pp 345-387.
   M.P. Shreevastava, "Political Problems of India : A Functional Approach," *The Indian Geographical Journal* XXXVII, 2, 1962, pp 35-44.
   Glenn V. Stephenson, "Pakistan : Discontiguity and the Majority Problems," *The Geographical Review*, LVIII, 1968, pp 194-213.
   A. Tayyeb, *Pakistan* : *A Political Geography*, London, 1966.
   Edmund R. Thompson, "Malaya's Stale-idea : An Application of the Unified Theory in Political Geography," Unpublished M.A. Thesis, Deptt. of Geography, Syracuse University, Syracuse.
10. O.H.K. Spate, "India and Pakistan," in *The Changing Map of Asia* : *A Political Geography*, (eds), East, W. Gordon, Spate, O.H.K. and Fisher, Charles, A. Norfolk, 1971, p 127.
11. R.L. Singh and K.N. Singh, et al., (eds) India : A *Regional Geography*, Varanasi, 1971, p 1.
12. M.P. Shreevastava, op. cit,, 36-37.
13. G.S. Sawer, *Modern Federalism*, London, 1969, pp 46-47.
14. Preston E. James, "Some Fundamental Elements in the Analysis of the Viability of the States" in *Essays in Political Geography*, p 34.
15. Nirmal Kumar Bose, *Problems of Indian Nationalism*, London, 1969, p 1.
16. R.D. Dikshit, *The Political Geography of Federalism*, Madras, 1975, pp 120-121.
17. Y.A. Raiker, *Indian History* : *A Study in Dynamics*, Baroda, 1960.
18. Jean Gottman, "The Political Partitioning of Our World : An Attempt or Analysis," in *Politics and Geographic Relationships* : *Readings on the Nature of Political Geography*, (eds) Jackson, W.A.

Douglas; New Jersey, 1964, pp 82-84.

19. R. Hartshorne, "The Concept of *Raison d'etre*" and 'Maturity of States' (Abstract) *Annals of the Association of American Geographers*, XXX (March-1940), p 59.
20. Donald E. Smith, *India As A Secular State*, Princeton, 1963, p 501.
21. Jolly Mohan Kaul, *Problems of National Integration*, New Delhi, 1963, pp 55-56.
22. R.D. Dikshit, op. cit., p 125.
23. V. Venkata Rao, "Language Politics in India," *The Indian Jour. of Political Science*, Vol. XXXI, (July-Sept. 1970) 3, p 203 and 209.
24. M.P. Shreevastava, op. cit., p 39.
25. R.C. Ray Majumdar, H.C. Chaudhuri and K.K. Dutta, *An Advanced History of India*, Part I, London, p 63.
26. B.R. Purohit, *Hindu Revivalism and Indian Nationalism*, Sagar, 1965, p 3.
27. V. Shweinforth, "The Problem of Nagaland" in *Essays in Political Geography*, op. cit, p 162-171.
28. Sudeepta Adhikari and Rajendra Prasad, Kashmir Problems : A Historical Approach in Political Geography," The Current Studies, *A Research Journal of the Patna College*, Vol. XVII, 1977-78.
29. P.P. Karan, Geopolitical Structure of India," *Proceedings of the Eighth General Assembly*, Seventeenth Congress of the International Geographical Union, Washington, 1952, pp 524-527.
30. B.R. Purohit, op. cit., pp 58-60.
31. P. Griffiths, *Modern India*, London, 1962, p. 17.
32. Jawaharlal Nehru, *India Today and Tomorrow*, Calcutta, 1964, p 4.
33. Radha Kumud Mookerji, *Fundamental Unity of India*, Bombay, 1954, pp 28-30.
34. Karl W. Deutsch, "Communication and the Concept of a People" in *Politics and Geographic Relationships* : *Reading in the Nature of Political Geography*, op. cit., p 113.
35. M.P. Shreevastava, op. cit., pp 39-40.
36. V.A. Janaki, *Some Aspects of the Political Geography of India*, Baroda, 1977, p 116.
37. Sudeepta Adhikari and Rajendra Prasad, op. cit., p 68.
38. R.D. Dikshit, *The Political Geography of Federalism* : *An Inquiry into Origins and Stability*, Madras, 1975, p 247.
39. R. Murphey, "New Capitals of Asia" *Economic Development and Cultural Change*, Vol. V, 1956-1957, p 225.
40. R.D. Dikshit, op. cit., p 248.
41. O.H.K. Spate, "Factors in the Development of Capital Cities" in *Systematic Political Geography*, pp 413-422.
42. Percival Spear, *A History of India*, Volume 2, Middle, 1977 pp 158-59.
43. John E. Brush, "The Distribution of Religious Communities in India" in *Systematic Political Geography*, op. cit., pp 267-68.
44. Slsir Kumar, Mitra, *Resurgent India*, New York, 1963, pp 14-15.

45. Rakhari Chatterjee, "Swarajist Revolt : A Forgotten Chapter of Indian Movement," *The Indian Journal of Political Science*, Vol. XXIV, Jan-March, 1968, p 404.
46. Sudeepta Adhikari, et al, "Geopolitics of the Indian Ocean : Struggle for Space and Power," *Indian Geographical Studies*, IX issue.
47. Alok Kumar and Sudeepta Adhikari, "India's External Relations : A Functional Approach," *Geographical Bulletin of India*, No. 2, October, 1977, p 63.
48. Charles A. Fisher, op. cit., p 134
49. Jawaharlal Nehru, op. cit., p. 44.
50. B.R. Purohit, op. cit., p 102.
51. Ernest Renan, "Qu'est-ce Qu'une Nation ?" in *The Dynamics of Nationalism* (ed) Snyder, Louis L., Toronto, 1964, pp 9-10.
52. Radha Kumud Mookerjee, op. cit., p 24.
53. Sri Aurobindo Ghosh, *The Foundation of Indian Culture*, New York, 1953, pp 167-68.

24

# Trends of Human Genetics Research in India

—*K.K. Sinha, Madan M. Lall*
and
*Suresh P. Sinha*

ASSOCIATED with the prodigality of nature, the death and destruction of a part of the progeny is inevitable. In the words of classical Darwinists, those who survive and reproduce are fit ones, and those who get eliminated (die) are unfit ones. And, it is in this manner that more and more individuals, better suited to the environment, are selected. In the language of population geneticists, the gene pool goes on getting richer and richer in respect of such genes which give adaptability to its bearers.

Unfortunately, among human being, owing, to meteoric advance in medicine, surgery, prophylactic and preventive measures etc., sufficiently large number of unfit individuals are retained in the population. To make things worse, many of the better genotypes get eliminated in wars, train or plane accidents etc. A situation, therefore, has now come whereby the human population is abounding in "unfit" genotypes, which share the food and habitat of the fit ones. We cannot and must not get rid of such individuals, as it is against humanity, and may even lead to civil war.

A question therefore arises : Is the human race going to see its doom very soon? Galton in 1905 had said that in order to face up to the challenge of such prospective doom, efforts must be made towards "the studies of the agencies under social control that may improve or impair the racial qualities of future generations either physically or mentally". This branch of genetics has been termed Eugenics.

As far as this science in India is concerned, J.V. Neel, a renowned human geneticist of modern age, has very rightly said that to students of the population genetics of man, India has for some time appeared to offer outstanding opportunities for research. There are numerous well-defined population subgroups, some with a rich historical tradition that provide an unusual documentation of their origins and movements over the centuries, others living deep in the forest under conditions remarkably similar to those encountered by early man. These various groups are widely diversified in their population structure, ranging from groups of relatively small, that is modest fertilities, and variable degrees of inbreeding, to large groups with high fertilities and some of the highest coefficients of inbreeding known in the world. Finally, there is an interesting range of eco-systems of the Indian sub-continent resulting in opportunities to study man in a wide variety of environment.

Indian geneticists have been aware of these opportunities for some time. They have made significant contributions in such theoretical areas as genetic distance and genetic load, and hence, provided a vast amount of original data on diverse topics related to the genetic structure of Indian populations and the extent and effects of inbreeding.

A few of the most popular areas of research in human genetics in India are dealt with below.

## MEDICAL GENETICS

This is the branch used in genetic counselling in the prevention and therapy of genetic disorders, and in the elucidation of environmental mutagens etc. It is heartening to note that recently, a number of centres of medical genetics have been set up in different parts of the country; notably among them are the medical complex situated at Parel of Bombay, Christian Medical College of Vellore, Jawahar Lal Institute of Postgraduate Medical Education and Research, Pondicherry, All India Institute of Medical Sciences, New Delhi, and the Medical Institute of Poona. These institutes are actively participating in the detection of various chromosomal disorders, various haemoglobinopathies, and have suggested remedial measures by treatment of genetic disorders, genetic counselling, and also by making people aware of the hazards of radiation, effects of adverse environment, the causative factors of cancer etc. (Table 24.1)

TABLE 24.1

*Congenital Malformations in India*

(Per 1000 births)

| Locality | Years of study | Neural tube defects | G.I.T. defects | C.V.S. defects |
|---|---|---|---|---|
| Ajmer | 1969 | 7.5 | 3.3 | 1.4 |
| Bombay | 1960-63 | 2.5 | 0.3 | 0.3 |
| Bombay | 1961-64 | 3.6 | 0.6 | 0.2 |
| Calcutta | 1961-64 | 0.6 | 0.2 | * |
| Calicut | 1962-64 | — | 1.1 | 1.1 |
| Chandigarh | 1966 | 13.0 | 4.0 | 4.0 |
| Delhi | 1961-62 | 8.7 | 2.2 | 1.4 |
| Hyderabad | 1970 | 7.6 | 3.7 | 0.0 |
| Hubli | 1976-77 | 6.7 | 1.3 | 2.1 |
| Jaipur | 1965-69 | 11.1 | 1.5 | 0.1 |
| Kanpur | 1967-68 | 6.7 | 1.9 | 1.2 |
| Lucknow | 1963-64 | 4.7 | 1.1 | 1.4 |
| Madras | 1972-73 | 4.2 | 2.4 | 0.9 |
| Madras | 1975-76 | 6.3 | 1.6 | 2.1 |
| Madurai | 1976-77 | 3.7 | 1.0 | 0.9 |
| Mysore | 1967-69 | 1.5 | 1.9 | 0.0 |
| Patna | 1964 | 5.3 | 1.4 | 0.9 |
| Pondicherry | 1963-66 | 2.0 | 0.5 | 0.4 |
| Pondicherry | 1967-76 | 11.2 | 3.8 | 0.8 |
| Bombay | 1966-67 | 2.2 | 1.0 | 0.5 |
| Bombay | 1968-72 | 1.6 | 1.7 | 2.4 |
| Pondicherry | 1971-72 | 5.6 | 0.8 | 0.6 |

— = Not available;

G.I.T.=Gastrointestinal tract includes oesophageal atresia, tracheo-oesophagel fistula, intestinal atresia and exomphalos;

C.V.S.=Cardiovascular system;

* =Frequency less than 0.1 per 1000

(After Verma, I.C. 1978)

## POPULATION GENETICS

Population genetics is the study of the behaviour of genes in population. It involves the investigation of the adaptation of organisms to stable or changing environments. In this discipline inbreeding, population structure, marriage distance and genetic distances etc. are studied.

Consanguineous marriages take place when persons related closely are married. Generally, such marriages, which produce inbreeding, are not considered desirable because it causes the homozygosis of harmful traits, hitherto hidden in the population due to their heterozygosity. Various workers are of the view that such marriages cause loss in fertility, increased mortality, congenital malformations, abnormal sex-ratio (proportion of males and females in the population) while such marriages are not tabooed among the Muslims, they are in vogue among South Indian Hindus. The foetal loss and the after effects of inbreeding are directly proportional to the closeness of the genetic relationship between the spouses. (Table 24.2).

## POPULATION STRUCTURE

This branch deals with the frequency of various alleles of the genetic traits concerning anthropometric parameters, blood groups, behavioral traits etc. Among blood groups, the frequency of various alleles for ABO, MN and Rh types have been obtained for various populations in India. Besides, the occurrence and distribution of genes for various enzymes such as Phosphoglucomutase, G-6-P, Acid phosphatase, 6-phosphogluconate dehydrogenase (6-PGD), Adenylate Kinase, Adenosine deaminase, Diaphorase, Pseudocholinesterase, serum alkaline phosphatase, Malate dehydrogenase, Peptidase A, lactate dehydrogenase, Phosphohexose isomerase, Placental Alkaline phosphatase etc. have been studied. Taking into consideration the frequency of the various alleles of these genes, the genetic distance among various castes and sub-castes are calculated. These distances help in preparing dendrograms. Formerly, morphological traits like skin colour, hair form and metrical characters like stature, head length and breadth etc. were used to distinguish between various populations. Later Mahalanobis developed statistic methods to measure the distance between populations on the basis of variation in a large number of quantitative characters. Now frequently the genetic characters are being used in place of quantitative one. After Mahalanobis, though a measure of angular distance between multinomial populations was developed, later scientists developed a measure of genetic distance analogous to the

TABLE 24.2

*Mortality and Morbidity Rates in Urban and Rural Inbred Populations in India*
(*After Sundar Rao, P.S.S.* 1978)

| Parental relationship | Foetal loss rates per hundred frequency | | Mortality rates per 1000 livebirths | | | | | | Sex ratio (No. of males/100 livebirth | | Congenital Malformations (per 1000 livebirths | |
|---|---|---|---|---|---|---|---|---|---|---|---|---|
| | | | Neonatal | | Post neonatal | | Infant | | | | | |
| | Rural | Urban | Rural | Urban | Rural | Urban | Rural | Urban | Rural | Urban | Rural | Urban |
| Non-consanguineous | 21.1 | 20.8 | 46.8 | 33.7 | 47.7 | 58.8 | 93.6 | 93.3 | 51.2 | 51.5 | 7.7 | 9.7 |
| Consanguineous | 22.9 | 22.1 | 54.6 | 46.4 | 54.1 | 60.3 | 106.7 | 106.9 | 51.1 | 50.2 | 5.9 | 12.2 |
| Beyond FC | 25.6 | 21.7 | 55.3 | 17.9 | 44.3 | 70.2 | 95.1 | 88.5 | 48.6 | 52.5 | 0.0 | 17.9 |
| First Cousins | 21.6 | 22.1 | 53.5 | 55.7 | 61.3 | 46.0 | 112.6 | 102.4 | 50.8 | 50.4 | 7.3 | 10.9 |
| Uncle niece | 23.5 | 22.4 | 56.1 | 44.4 | 48.2 | 90.9 | 102.4 | 133.9 | 52.5 | 47.9 | 6.2 | 11.1 |

chisquare statistic. Most of the findings mentioned above have a theoretical value but a few have practical application in the welfare of human society. As said earlier, this comes within the purview of eugenics, and the application of this branch of genetics to human welfare is enormous. Mention can be made of a few of them as below :

**In Prenatal Forecasting of the Sex of the Child**

All the normal females have 2X chromosomes, and all the males have one X and one Y chromosome in all the cells of their body. Out of the two X chromosomes in a female cell, one remains active and the other undergoes spiralization (Heterochromatinization) and can be easily stained (Acetocarmine or Aceto-Orcein) and observed under the microscope. Similarly, the Y-chromosomes, when viewed under fluorescence microscope, shine. Therefore, by the process of amniocentesis (removal of amniotic fluid from the pregnant mother), cells of the developing child can be identified to have either a condensed X-chromosome or Y-chromosome. In the former case, a female child will be born and in the latter—a male one. This scientific fact has been very successfully used in having the child of required sex, and terminating the pregnancy with the child of non-desired sex.

This very method is also used in forecasting any anomaly concerning sex chromosome number in the developing foetus. Foetus showing spiralized chromosome and also Y-chromosome fluorescence will automatically be of the composition XXY (Klinefelter's Syndrome); or those showing neither additional X-chromosome nor Y-fluorescence will be having XO chromosomal complex (Turner's Syndrome). It is in this manner, the birth of children having abnormal chromosome number and thus destined to doom, can be eliminated.

**Rh Incompatibility**

If a mother is Rh-negative and father is Rh-positive, the child on many occasions will have Rh-positive blood group. The situation now will be a Rh-positive foetus in a Rh-negative mother. As a result of antigen-antibody relationship, the mother's blood get destroyed, consequently she develops anaemia (erythroblastosis foetalis) and finally dies. Such tragedy can be avoided if the blood group of prospective couple is analysed before marriage.

**Haemophilia, Colour Blindness etc.**

These diseases occur due to change (mutation) in specific genes on the femaleness-producing X-chromosome. These mutations are recessive, and hence will not show in the females (whose anothor X-chromosome is having normal gene for blood clotting or colour vision). Marriage with such females (carriers) is expected to produce children having these diseases. Suitable genetic counselling based upon pedigree analysis of such females can help in avoiding the birth of diseased children. J.B.S. Haldane made such an analysis of British Royality.

**Nutritional Deficiencies**

Many children start showing some signs of morbidity right from their birth because they fail to metabolise their protein diet (Phenylketoneuria, Alkeptoneuria). Suitable diet to them can allow them to live a normal healthy life.

All such and many other hereditary and congenital diseases are being predicted and prevented in India.

Thus the recent researches in Human Genetics being carried on in India are playing a vital role shaping our future progeny. However, it is just the beginning of a great philanthropic and social end. The task is very difficult, more so when we look at the huge population size, a very high birth rate, and also the great variability in the population structure, mostly due to the existence of the caste system. Nevertheless, such type of research is very desirable, and it is heartening to note that the challenge has been accepted very sportingly, though very modestly.

SUGGESTED READINGS

J.B. Bresler, *Genetics and Society*, Addison-Wesley Publishing Company, Philippines, 1973.

C.A. Clarke, *Human Genetics and Medicine*. Edward Arnold (Publishers) Ltd., London, 1970.

H. Cummins and C. Midlo, *Finger Prints, Palms and Soles*. Dover Publications, Inc., New York, 1961.

W. Fuhrmann and F. Vogel, *Genetic Counselling*, Longmans. Springer—Verlag, New York, 1969.

V.A. McKusick, *Foundations of Modern Genetics Service.* Prentice Hall, Inc. Engelwood Cliffs, N.J.

L.D. Sanghvi and et al, *Human Population Genetics in India,* Orient Longman Ltd., 1974.

C. Stern, *Principles of Human Genetics,* W.H. Freeman & Co., San Francisco and London., 1960.

I.C. Verma, *Medical Genetics in India,* Vol 1 and 2 Auroma Enterprises, Publications Division, 7, Aurobindo Street, Pondicherry-605001, 1978.

M. Whittinghill, *Human Genetics and Its Foundations.* Oxford and IBH Publishing Co., Calcutta, 1970.

# Index